Sorghum and Millets Diseases

Edited by

John F. Leslie

Iowa State Press
A Blackwell Publishing Company

John F. Leslie is a professor in the Department of Plant Pathology at Kansas State University, Manhattan. Much of the editing of this work was completed during a sabbatical at the University of Sydney (Sydney, Australia) and the Royal Botanic Gardens–Sydney as a Senior Fulbright Scholar under the sponsorship of the Australian-American Fulbright Commission.

Iowa State Press
2121 State Avenue, Ames, Iowa 50014

Orders:	1-800-862-6657
Office:	1-515-292-0140
Fax:	1-515-292-3348
Web site:	www.iowastatepress.com

∞ Printed on acid-free paper in the United States of America. Produced directly from camera-ready copy provided by the volume editor.

First edition, 2002

Library of Congress Cataloging-in-Publication Data

Sorghum and millets diseases/edited by John F. Leslie .–1st ed.
p. cm.
Based on contributions to the Third Global Conference on Sorghum and Millets Diseases in Guanajuato, Mexico September 2000.
Includes bibliographical references and index.
ISBN 0-8138-0389-6 (alk. paper)
1. Sorghum–Diseases and pests. 2. Millets–Diseases and pests. I. Leslie, John F., 1953-

SB608.S6 S67 2002
633.1’7193—dc21

2002192228.

The last digit is the print number: 9 8 7 6 5 4 3 2 1

Table of Contents

Preface

In your hands you hold the global sorghum and millet pathology community's decennial (more or less) summary of diseases and work in progress. It is not a summary of all of the work that has ever been done with sorghum and millet pathology, nor is it a guide to disease identification. Instead, it is a considered summary of the current critical problems and an evaluation of the community's progress since its previous meeting in Harare, Zimbabwe in 1988. It is the third of these efforts, and both the meetings and the books that have resulted from them have been important in defining the research agenda for these crops in the succeeding ten-year interval. The contents of this volume are based primarily on contributions to the Third Global Conference on Sorghum and Millet Diseases in Guanajuato, Mexico in September 2000. Indeed, all but one of the invited talks and all but one of the working group reports are represented here in some form. I also solicited some additional chapters that were not based on any of the presentations to help round out the volume, and I have included as many contributions as possible from African scientists who were unable to attend the meeting due to their inability to obtain visas for travel from the Mexican government.

This volume is somewhat different from its predecessors in that it is not being published by ICRISAT but, instead, by Iowa State Press, with the goal of expanding the readership beyond the immediate sorghum and millet research community. It also is different in that there is only a single editor, and that this editor is not Dick Frederiksen (Fig. P-1), although perhaps it should have been. Dick has been a dominant figure in this field for the last 30-40 years, and a major organizer of what have become nearly institutional meetings. His contacts and efforts in organizing sessions, dispersing participants into effective working groups, and cajoling contributors until they find time to finish a chapter or to strengthen a presentation should be neither underestimated nor undervalued. He has been a good friend and colleague, and as he goes to "fossilize" (his words not mine) in retirement, we can all but wish him the best. The meeting in Guanajuato was his last formal contribution to the sorghum and millet research community, but for those who know him, there undoubtedly remains more to come.

The chapters in this book have all been reviewed by at least one person, and authors have had an opportunity to make revisions in response to these comments. Although I have read each chapter at least four times, the important points raised by other reviewers have made this volume much better than it could have been were one person alone responsible for editing its content. I greatly appreciate the help these colleagues have provided and the advice they have given both to me and to the authors of the various chapters. Persons who reviewed at least one chapter (and some did far more) include: Ranajit Bandyopadhyay, Larry E. Claflin, Thomas Crawford, Jeff Dahlberg, Walter A. J. de Milliano, Richard A. Frederiksen, Liane R. Gale, Laura M. Giorda, Dale E. Hess, Clint W. Magill, Peter G. Mantle, Walter F. O. Marasas, Neal W. McLaren, Gary N. Odvody, Rodomiro Ortiz, Sylvia Pažoutova, John P. Rheeder, Malcolm J. Ryley, Claude P. Seletrinikoff, Ram P. Thakur, Paul W. Tooley, Jeff Wilson, and John Yohe.

FIGURE P-1. Richard A. Frederiksen.

In addition to the reviewers, I owe thanks to the International Sorghum and Millet Collaborative Research Support Program (INTSORMIL) for its funding of my research on fungal diseases of sorghum and millet, to the Australian-American Fulbright Association, the Royal

Botanic Garden-Sydney, and St. Paul's College of the University of Sydney for their funding and for hosting me while on sabbatical when much of the editing of this book was completed, and to my wife, Ingelin, for sharing my time with all of these manuscripts. I also thank the corresponding authors of all of the chapters, who have been (generally) prompt and efficient in their revisions and gracious in helping me as I learn about systems and methodologies that are not those that I use in my own research.

Looking forward to the successor meeting (and volume) to this one, the meeting's global geographic rotation should take us to Asia or Australia around 2010. If this volume wears as well as does its predecessor, which is cited numerous times in this volume, and is still of relevance in 2010, then the work required to assemble it will have been time and effort well spent.

John F. Leslie
Kansas State University
Manhattan, Kansas, USA
November 2002

Addresses of Contributors

A. A. ABBASHER
Abu Haraz College
University of Gezira
Wad Medani, Sudan

IBRAHIM D. K. ATOKPLE
Savanna Agricultural Res. Institute
P. O. Box 52
Tamale, Ghana

A. G. T. BABIKER
Agricultural Research Corporation
P. O. Box 126
Wad Medani, Sudan

DAVID BACKHOUSE
University of New England
Armidale, New South Wales 2350
Australia

CHARLES W. BACON
Toxicology & Mycotoxin Res. Unit
USDA-ARS
Russell Agricultural Res. Center
Athens, Georgia, 30613, USA

RANAJIT BANDYOPADHYAY
International Inst. for Trop. Agric.
% L. W. Lambourn & Co.
Carolyn House, 26 Dingwall Road
Croydon CR9 3EE
United Kingdom

F. P. BEJOSANO
Cereal Quality Laboratory
Dept. of Soil & Crop Sciences
Texas A&M University
College Station, Texas 77843-2474
USA

J. BENADE
Grain Crops Institute
Agricultural Research Council
Private Bag X1251
2520 Potchefstroom, South Africa

SHAMSUL A. BHUIYAN
Australian Quarantine & Inspection Service
Brisbane Airport, Queensland 4007
Australia

MEGAN BLAKE
Fusarium Research Laboratory
Department of Crop Sciences
University of Sydney
Sydney, New South Wales 2006
Australia

AMAURI BOGO
Biochemistry Department
Imperial College of Science, Technology & Medicine
London SW7 2AY
United Kingdom

KHAZAN S. BOORA
Dept. of Biotechnology & Molecular Biology
CCS Haryana Agricultural Univ.
Hisar 125 004, India

P. BOORA
Department of Foods & Nutrition
CCS Haryana Agricultural Univ.
Hisar 125 004, India

ANDREW BORRELL
Queensland Dept. of Primary Industries.
Agency for Food and Fibre Sci.–Farming Systems
Hermitage Res. Station, MS 407
Warwick, Queensland 4370
Australia

LETICIA BRAVO-LUNA
Centro de Desarrollo de Productos Bióticos
Instituto Politénico Nacional
Apartado Postal 24
Yautepec, Morelos 62730, Mexico

WENDY A. BREESE
Centre for Arid Zone Studies
University of Wales
Bangor, Gwynedd LL57 2UW
United Kingdom

LESTER W. BURGESS
Fusarium Research Laboratory
Department of Crop Sciences
University of Sydney
Sydney, New South Wales 2006
Australia

DAVID R. BUTLER
Cocoa Research Unit
University of West Indies
St. Augustine
Trinidad and Tobago

ELIZABETH CÁRDENAS-SORIANO
IFIT
Colegio de Postgraduados
56230 Montecillo, Texcoco
México

CARLOS R. CASELA
EMBRAPA/CNPMS
Caixa Postal 151
35701-970 Sete Lagoas, MG
Brazil

HOWARD. H. CASPER
Dept. of Veterin. & Microbiol. Sci.
North Dakota State University
Fargo, North Dakota, USA

SUKUMAR CHAKRABORTY
CRC for Tropical Plant Protection
University of Queensland
St. Lucia, Queensland 4072
Australia

A. CHANDRASHEKAR
Central Food Technol. Res. Inst.
Mysore 570 013, India

LARRY E. CLAFLIN
Department of Plant Pathology
Throckmorton Plant Sci. Center
Kansas State University
Manhattan, Kansas 66506-5502
USA

S. DELROY COLLINS
Dept. of Soil & Crop Sciences
Texas A&M University
College Station, Texas 77843-2474
USA

MELANIE COLVILLE
Fusarium Research Laboratory
Department of Crop Sciences
University of Sydney
Sydney, New South Wales 2006
Australia

LUIS R. CONCI
INTA IFFIVE
Camino 60
Cuadras, Corboba 5000, Argentina

MARK S. CONDON
American Seed Trade Association
225 Reinekers Lane, Suite 650
Alexandria, Virginia 22314-2875
USA

JEFFERY A. DAHLBERG
National Grain Sorghum Producers
P. O. Box 5309
Lubbock, Texas 79408, USA

J. L. DE LA GARZA
Univ. Autónoma de Nuevo Léon
Facultad de Agronomía
Subdirección de Estudios de
Postgrado
Carreterra Zuazua-Marín km 17
Marín, Nuevo León, México

WALTER A. J. DE MILLIANO
Hoeve Dierkensteen
Bakkersstraat 62
NL-4501 RB Oostburg
The Netherlands

EIGIL DE NEERGAARD
Department of Plant Biology
Royal Veterinary & Agric. Univ.
Thorvaldsensvej 40
DK-1871 Frederiksberg C
Copenhagen, Denmark

KATRIEN M. DEVOS
John Innes Centre
Norwich Research Park
Colney, Norwich NR4 7UH
United Kingdom

C. DIARRA
Institut d'Economie Rurale (IER)
B. P. 262
Bamako, Mali

MAMOUROU DIOURTÉ
Institut d'Economie Rurale (IER)
B. P. 262
Bamako, Mali

GEBISA EJETA
Department of Agronomy
Lilly Hall of Life Sciences
Purdue University
West Lafayette, Indiana 47907-1150
USA

A. E. M. ELZEIN
Inst. for Plant Prod. & Agroecol. in
the Tropics & Subtropics
University of Hohenheim (380)
D-70593 Stuttgart, Germany

KIRSTEN ENGELL
Botanical Institute
University of Copenhagen
Gothersgade 140
DK-1123 Copenhagen K
Denmark

J. PETER ESELE
Serere Agricultural & Animal
Production Res. Inst.
P. O. Soroti
Soroti, Uganda

M. G. FEDORISHCHENKO
Inst. of Agroengin. Problems of
Azov
Black Sea State Agroengin. Acad.
Zernograd, Rostov reg., Russia

ALEXANDRE S. FERREIRA
Embrapa/CNPMS
Caixa Postal 151
35701-970 Sete Lagoas, MG
Brazil

HILDA E. FLORES-MOCTEZUMA
Centro de Desarrollo de Productos
Bióticos
Instituto Politénico Nacional
Apartado Postal 24
Yautepec, Morelos 62730, Mexico

DEBRA E. FREDERICKSON
ICRISAT Matopos Research Station
P. O. Box 776
Bulawayo, Zimbabwe

RICHARD A. FREDERIKSEN
8248 Raintree Dr., NE
Albuquerque, NM 87122, USA

KARINA FRUTERO
CEPROCOR
Cordoba 5000, Argentina

LIANE ROSEWICH GALE
USDA-ARS Cereal Dis. Lab.
University of Minnesota
1551 Lindig Street
St. Paul, Minnesota 55108-6052
USA

NORA E. GARCÍA
Univ. Autónoma de Nuevo Léon
Facultad de Agronomía
Subdirección de Estudios de
Postgrado
Carreterra Zuazua-Marín km 17
Marín, Nuevo León, México

ROGER N. GATES
USDA-ARS Crop Genetics &
Breeding Res. Unit
Coastal Plain Experiment Station
University of Georgia
Tifton, Georgia 31793, USA

GREG GIBLET
Lollybroch, RMB 583
Goonoo Goonoo Road
Tamworth, New South Wales 2340
Australia

LAURA M. GIORDA
Department of Plant Pathology
INTA/EEA Manfredi
Manfredi Ruta Nac. no. 9, Km 636
5988 Manfredi, Cordoba
Argentina

ANTHONY E. GLENN
Toxicology & Mycotoxin Res. Unit
USDA-ARS
Russell Agricultural Res. Center
Athens, Georgia, 30613, USA

IAN D. GODWIN
School of Land & Food Sciences
The University of Queensland
St. Lucia, Queensland 4072
Australia

FERNANDO B. GUIMARÃES
Dow Agrosciences
Caixa Postal 47
14680-000 Jardinópolis, SP, Brazil

ANDY H. HALL
ICRISAT
Patancheru (AP) 502 324, India

C. TOM HASH
ICRISAT
Patancheru (AP) 502 234, India

NICHOLAS J. HAYDEN
Natural Resources Institute
University of Greenwich
Chatham Maritime, Kent ME4 4TB
United Kingdom

ROBERT G. HENZELL
Queensland Dept. of Primary Industries
Agency for Food and Fibre Sci.–Farming Systems
Hermitage Res. Station, MS 508
Warwick, Queensland 4370
Australia

DAMIAN J. HERDE
Queensland Dept. of Primary Industries
Agency for Food and Fibre Sci.–Farming Systems
P. O. Box 102
Toowoomba, Queensland 4350
Australia

MIGUEL HERNÁNDEZ-MARTÍNEZ
INIFAP
Apartado Postal 112
38010 Celaya, Guanajuato, México

DALE E. HESS
Department of Agronomy
Lilly Hall of Life Sciences
Purdue University
West Lafayette, Indiana 47907-1150
USA

C. J. HOWARTH
Inst. for Grassland & Environ. Res.
Plas Gogerddan
Aberystwyth, Dyfed SY23 3EB
United Kingdom

NIPON IAMSUPASIT
Thailand Biodiversity Center
539/2 Gypsum Metropolitan Tower, 15th Floor
Sri-Ayudhya Road
Bangkok 10400, Thailand

SUNKARA INDIRA
National Res. Center for Sorghum
ICAR, Rajendranagar
Hyderabad (AP) 500 030, India

E. V. IONOVA
All Russia Res. Inst. for Sorghum and Other Cereals (ARRISC)
Science District 3
Zernograd, Rostov reg. 347740
Russia

TOM ISAKEIT
Dept. of Plant Pathol. & Microbiol.
Texas A&M University
College Station, Texas 77843-2132
USA

JAYARAJ JAYARAMAN
Department of Biochemistry
Chemistry-Biochemistry Building
Kansas State University
Manhattan, Kansas 66506, USA

STAN G. JENSEN
USDA-ARS
Department of Plant Pathology
University of Nebraska
Lincoln, Nebraska 68583, USA

JEOUNG-MEE JEOUNG
PGEC, USDA-ARS
800 Buchanan Street, no. 2202
Albany, California 94710, USA

MARC A. JOHNSON
Dean of Agriculture Office
Waters Hall
Kansas State University
Manhattan, Kansas 66506, USA

DAVID R. JORDAN
Queensland Dept. of Primary Industries
Agency for Food and Fibre Sci.–Farming Systems
Hermitage Res. Station, MS 508
Warwick, Queensland 4370
Australia

LISE BOLT JØRGENSEN
Botanical Institute
University of Copenhagen
Gothersgade 140
DK-1123 Copenhagen K, Denmark

Z. JURJEVIC
Department of Plant Pathology
Coastal Plain Experiment Station
University of Georgia
Tifton, Georgia 31793, USA

ALIYA S. KASAKOVA
All Russia Res. Inst. for Sorghum and Other Cereals (ARRISC)
Science District 3
Zernograd, Rostov Region 347740
Russia

SERIBA O. KATILÉ
Institut d'Economie Rurale (IER)
Station de Cinzana
B. P. 214
Ségou, Mali

R. R. KLEIN
SPARC, USDA-ARS
College Station, Texas 77845, USA

ISSOUFOU A. KOLLO
INRAN
B. P. 240
Maradi, Niger

BIRTE KOMOLONG
CRC Tropical Plant Protection
University of Queensland
St. Lucia, Queensland 4072
Australia

S. KRISHNAVENI
Centre for Plant Molec. Biol.
Department of Biochemistry
Tamil Nadu Agricultural Univ.
Coimbatore, Tamil Nadu 641 003
India

J. KROSCHEL
Inst. for Plant Product. & Agroecol. in the Tropics & Subtropics
University of Hohenheim (380)
D-70593 Stuttgart, Germany

GRETCHEN A. KULDAU
Department of Plant Pathology
The Pennsylvania State University
University Park, Pennsylvania 16802, USA

HUNTER K. C. LAIDLAW
School of Land and Food Sciences
The University of Queensland
St. Lucia, Queensland 4072
Australia

JILLIAN M. LENNÉ
ICRISAT
Patancheru (AP) 502 234, India

JOHN F. LESLIE
Department of Plant Pathology
Throckmorton Plant Sci. Center
Kansas State University
Manhattan, Kansas 66506-5502
USA

VIBEKE LETH
Danish Govern. Inst. of Seed Pathol. for Developing Countries
Thorvaldsensvej 57
DK-1871 Frederiksberg C,
Denmark

GEORGE H. LIANG
Department of Agronomy
Throckmorton Plant Sci. Center
Kansas State University
Manhattan, Kansas 66506-5501
USA

E. I. LIPKOVICH
All Russia Res. & Develop. Inst. for Farm Mechanization & Electrification (ARRDIFME)
Zernograd, Rostov reg., Russia

CHRISTOPHER R. LITTLE
Dept. of Plant Pathol. & Microbiol.
Texas A & M University
College Station, Texas 77843-2132
USA

CLINT W. MAGILL
Dept. of Plant Pathol. & Microbiol.
Texas A & M University
College Station, Texas 77843-2132
USA

VISWANATHAN MAHALAKSHMI
ICRISAT
Patancheru (AP) 502 324, India

FELISTER W. MAKINI
Kenya Agricultural Res. Institute
Kisii Centre
Kisii, Kenya

GINO MALAGUTI
Natl. Center for Agronomic Res. (CENIAP-FONAIAP)
Faculty of Agronomy, UCV
Maracay, Venezuela

ANACLET S. B. MANSUETUS
Department of Biological Sciences
University of Swaziland
Private Bag
Kwaluseni, Swaziland

PETER G. MANTLE
Biochemistry Department
Imperial College of Science, Technology & Medicine
London SW7 2AY
United Kingdom

WALTER F. O. MARASAS
PROMEC
Medical Research Council
P. O. Box 19070
7505 Tygerberg, South Africa

PAUL S. MARLEY
Institute for Agric. Res. (IAR)
Ahmadu Bello University
P.M.B. 1044
Zaria, Nigeria

MARIA J. MARTINEZ
Department of Plant Pathology
INTA/EEA Manfredi
Manfredi Ruta Nac. no. 9, Km 636
5988 Manfredi, Cordoba, Argentina

KUSUM MATHUR
Department of Plant Pathology
Maharana Pratap Univ. of Agric. & Technology
RCA, Udaipur, Rajasthan 313 001
India

A. BRUCE MAUNDER
National Grain Sorghum Producers
P. O. Box 5309
Lubbock, Texas 79408

JOSÉ LUIS MAYA-DE LEÓN
ITA Roque
Apartado Postal 508
38500 Celaya, Guanajuato, México

DEMBA F. MBAYE
ISRA/CRZ de Kolda
B. P. 53
Kolda, Sénégal

NEAL W. MCLAREN
Grain Crops Institute
Agricultural Research Council
Private Bag X1251
2520 Potchefstroom, South Africa

P. J. MEHTA
Dept. of Soil & Crop Sciences
Texas A&M University
College Station, Texas 77843-2474
USA

IGNACIO MÉNDEZ-RAMÍREZ
Inst. de Investigación en Math. Aplicadas y en Sistemas
Apartado Postal 0-76
100 México D.F., México

LEOPOLDO E. MENDOZA-ONOFRE
IREGEP, Colegio de Postgraduados
56230 Montecillo, Texcoco, México

NOE MONTES
INIFAP
Rio Bravo Experimental Station
Apartado Postal 172
88900 Rio Bravo, Tamaulipas
México

ROBERTO MONTES-BELMONT
Centro de Desarrollo de Productos Bióticos
Instituto Politénico Nacional
Apartado Postal 24
Yautepec, Morelos 62730, Mexico

JORGE L. MORAN
Dept. of Soil & Crop Sciences
Texas A&M University
College Station, Texas 77843-2474
USA

ESTER MTISI
Plant Protection Research Institute
Box CY 550
Causeway, Harare, Zimbabwe

V. MUNIYAPPA
Department of Plant Pathology
College of Agriculture
University of Agricultural Sci.
GKVK Campus
Bangalore, Karnataka 560 065
India

SUBBARATNAM MUTHUKRISHNAN
Department of Biochemistry
Chemistry-Biochemistry Building
Kansas State University
Manhattan, Kansas 66506, USA

V. MUTHUSUBRAMANIAN
11/1424 Chinna Bayalu Lane
West Main Street
Thanjavur, Tamil Nadu 613 009
India

JETTE DAHL MØLLER
Botanical Institute
University of Copenhagen
Gothersgade 140
DK-1123 Copenhagen K
Denmark

GNANAMBAL NAIDOO
Dept. of Plant & Soil Science
368 Ag Hall
Oklahoma State University
Stillwater, Oklahoma 74078, USA

Y. D. NARAYANA
Department of Plant Pathology
Univ. of Agricultural Sciences
Agricultural Research Station
Dharwad, Aland Road
Gulbaraga, Karnataka 585 101
India

JESÚS NARRO-SÁNCHEZ
INIFAP–Campo Exper. Bajío
Apartado Postal 112
38010 Celaya, Guanajuato, México

MIRTHA NASSETTA
CEPROCOR
Cordoba 5000, Argentina

RAÚL NAVA-JUÁREZ
Centro de Desarrollo de Productos Bióticos
Instituto Politénico Nacional
Apartado Postal 24
Yautepec, Morelos 62730, Mexico

SHRISHAIL S. NAVI
ICRISAT
Patancheru (AP) 502 324, India

ADAMA NEYA
Institut de l'Environnement et de Recherches Agricoles (INERA)
Station Farako-Bâ
B. P. 910
Bobo-Dioulasso 01, Burkina Faso

ZACHEE NGOKO
IRAD/NCBE, Bambuy
Box 80
Bamenda, Cameroon

WILLIAM P. NORRED
Toxicology & Mycotoxin Res. Unit
USDA-ARS
Russell Agricultural Res. Center
Athens, Georgia, 30613, USA

S. K. NUTSUGAH
Savanna Agricultural Res. Institute
P. O. Box 52
Tamale, Ghana

GARY N. ODVODY
Department of Plant Pathology
Texas A&M University
Texas Agric. Experiment Station
10345 Agnes Street
Corpus Christi, Texas 78406-9704
USA

ALEXANDER A. OLEYNICK
All Russia Res. Inst. for Sorghum & Other Cereals (ARRISC)
Science District 3
Zernograd, Rostov Region 347740
Russia

NICHOLE R. O'NEILL
USDA-ARS Soybean & Alfalfa Research Laboratory
Beltsville, Maryland 20705, USA

RODOMIRO ORTIZ
International Inst. for Trop. Agric.
% L. W. Lambourn & Co.
Carolyn House, 26 Dingwall Road
Croydon CR9 3EE
United Kingdom

EVELYN ORTÍZ-PÉREZ
Univ. Autónoma de Nuevo Léon
Facultad de Agronomía
Subdirección de Estudios de Postgrado
Carreterra Zuazua-Marín km 17
Marín, Nuevo León, México

SEIGI OSADA-KAWASOE
IFIT, Colegio de Postgraduados
56230 Montecillo, Texcoco, México

TIRSO PACHECO
Hibridos Mejorados C.A. (HIMECA)
Magdaleno, Aragua, Venezuela

V. I. PAKHOMOV
All Russia Res. & Develop. Inst. for Farm Mechanization and Electrification (ARRDIFME)
Zernograd, Rostov reg., Russia

CHARLES K. PALLAGHY
Department of Botany
LaTrobe University
Bundoora, Victoria 3083
Australia

ROBERTO PAREDES-MELESIO
INIFAP–Campo Exper. Bajio
Apartado Postal 112
38500 Celaya, Guanjuato, México

SYLVIE PAŽOUTOVÁ
Institute of Microbiology
Czech Academy of Sciences
Vídeňská 1083
142 20 Prague 4, Czech Republic

BONNIE B. PENDLETON
Division of Agriculture
West Texas A&M University
P. O. Box 60998
Canyon, Texas 79016-0001, USA

DENIS M. PERSLEY
Queensland Horticultural Institute
Queensland Dept. of Primary Industries
Agency for Food and Fibre Sci.–Farming Systems
80 Meiers Road
Indooroopilly, Queensland 4068
Australia

RAMASAMY PERUMAL
Tamil Nadu University
Coimbatore, India

GARY C. PETERSON
Texas A&M Univ. Agric. Res. & Extension Center
Route 3, Box 219
Lubbock, Texas 79403-9803, USA

GARY L. PETERSON
USDA-ARS Foreign Disease-Weed Science Research Unit
Building 1301, Fort Detrick
Frederick, Maryland 21702-5023
USA

JAMES K. PORTER
Toxicology & Mycotoxin Res. Unit
USDA-ARS
Russell Agricultural Res. Center
Athens, Georgia, 30613, USA

JUAN ANGEL QUIJANO-CARRANZA
INIFAP–Campo Exper. Bajio
Apartado Postal 112
38010 Celaya, Guanjuato, México

K. N. RAI
ICRISAT
Patancheru (AP) 502 234, India

PORFIRIO RAMÍREZ-VALLEJO
IREGEP, Colegio de Postgraduados
56230 Montecillo, Texcoco, México

NANDURI K. RAO
ICRISAT
Patancheru (AP) 502 324, India

V. P. RAO
ICRISAT
Patancheru (AP) 502 324, India

BELUM V. S. REDDY
ICRISAT
Patancheru (AP) 502 324, India

JACOB REED
Department of Agronomy
Throckmorton Plant Sci. Center
Kansas State University
Manhattan, Kansas 66506-5501
USA

OSCAR R. RODRIGUEZ-BALLESTEROS
Monsanto Sorghum Breeding
982 Highway 77
Bishop, Texas 78343, USA

R. RODRIGUEZ-HERRERA
Fray Cristobal de Espinosa 208
Saltillo, Coahuila 25051, Mexico

LLOYD W. ROONEY
Cereal Quality Laboratory
Dept. of Soil & Crop Sciences
Texas A&M University
College Station, Texas 77843-2474
USA

WILLIAM L. ROONEY
Dept. of Soil & Crop Sciences
Texas A&M University
College Station, Texas 77843-2474
USA

DARRELL T. ROSENOW
Texas A&M University Agric. Res. & Extension Center
Route 3, Box 219
Lubbock, Texas 79403-9757, USA

MALCOLM J. RYLEY
Queensland Dept. of Primary Industries
Agency for Food and Fibre Sci.–Farming Systems
P. O. Box 102
Toowoomba, Queensland 4350
Australia

JEANETTA SAAYMAN
Grain Crops Institute
Agricultural Research Council
Private Bag X1251
2520 Potchefstroom, South Africa

G. E. SALINAS
Univ. Autónoma de Nuevo Léon
Facultad de Agronomía
Subdirección de Estudios de Postgrado
Carreterra Zuazua-Marín km 17
Marín, Nuevo León, México

FREDOLINO G. SANTOS
Embrapa/CNPMS
Caixa Postal 151
35701-970 Sete Lagoas, MG, Brazil

JOHN F. SCHEURING
Knowledge Management Group
Syngenta Corporation
R1055.128
Basel 4002, Switzerland

N. SEETHARAMA
ICRISAT
Patancheru (AP) 502 234, India

P. SÉRÉMÉ
Institut de l'Environnement et de Recherches Agricoles (INERA)
04 B. P. 8645
Ouagadougou 04, Burkina Faso

HARI C. SHARMA
ICRISAT
Patancheru (AP) 502 234, India

VIRENDER SHEORAIN
Seagram Manufacturing Limited
128/3 Telco Road
Mohan Nagar, Chinchwad
Pune, Maharashtra, India

H. S. SHETTY
Dept. of Appl. Bot. & Seed Pathol.
University of Mysore
Mysore, Karnataka 570 006, India

M. MARIN SILVA
INIFAP
Rio Bravo Experimental Station
Apartado Postal 172
Rio Bravo, Tamaulipas 88900
México

A. SINDHU
Dept. of Biotechnol. & Molec. Biol.
CCS Haryana Agricultural Univ.
Hisar 125004, India

S. D. SINGH
ICRISAT
Patancheru (AP) 502 234, India

S. SIVARAMAKRISHNAN
Dept. of Agric. Biotechnology
ANGRAU, Rajendranagar
Hyderabad (AP) 500 030, India

JILIAN SMITH-WHITE
Royal Botanic Gardens
Mrs. Macquaries Road
Sydney, New South Wales 2000
Australia

JAMES P. STACK
South Central Res. & Extension Center
University of Nebraska
P. O. Box 66
Clay Center, Nebraska 68933, USA

BRETT SUMMERELL
Royal Botanic Gardens
Mrs. Macquaries Road
Sydney, New South Wales 2000
Australia

M. A. TARANOV
Azov Black Sea State Agroengineering Academy
Zernograd, Rostov reg., Russia

GEORGE L. TEETES
Department of Entomology
Texas A&M University
College Station, Texas 77843-2475
USA

TESFAYE T. TEFERRA
Department of Agronomy
Throckmorton Plant Sci. Center
Kansas State University
Manhattan, Kansas 66506-5501
USA

NIABA TEME
Texas A&M University Agric. Res. & Extension Center
Route 3, Box 219
Lubbock, Texas 79403-9757, USA

RAM P. THAKUR
ICRISAT
Patancheru (AP) 502 234, India

PAUL W. TOOLEY
USDA-ARS Foreign Disease-Weed Science Research Unit
Building 1301, Fort Detrick
Frederick, Maryland 21702-5023
USA

J. HERIBERTO TORRES-MONTALVO
Hibridos Pioneer de Mexico S.A. de CV
Centro de Investigación Maíz
Km 1.6 al Poniente San Miguel Cuyutlan
Tlajomulco de Zúñiga, Jalisco 45660, Mexico

DOULAYE TRAORÉ
Institut de l'Environnement et de Recherches Agricoles (INERA)
Programme Coton
01 B.P. 208
Bobo-Dioulasso 01, Burkina Faso

F. K. TSIGBEY
Savanna Agricultural Res. Institute
P. O. Box 52
Tamale, Ghana

MITCHELL R. TUINSTRA
Department of Agronomy
Throckmorton Plant Sci. Center
Kansas State University
Manhattan, Kansas 66506-5501
USA

MARLENE VAN DER WALT
Grain Crops Institute
Agricultural Research Council
Private Bag X1251
2520 Potchefstroom, South Africa

N. S. VASANTHI
Dept. of Applied Bot. & Biotechnol.
University of Mysore
Mysore, Karnataka, India

RODOLFO VELÁSQUEZ-VALLE
INIFAP-Campo Exper. Pabellón
Apartado Postal 20
20660 Pabellon de Art., Ags.
México

R. T. VENKATESHA
Department of Pathology
School of Dental Medicine
University of Pennsylvania
4010 Locust Street
Philadelphia, Pennsylvania 19104-6002, USA

RALPH D. WANISKA
Cereal Quality Laboratory
Dept. of Soil & Crop Sciences
Texas A&M University
College Station, Texas 77843-2474
USA

ELISABETH WESTER
Department of Plant Biology
Royal Veterinary & Agric. Univ.
Thorvaldsensvej 40
DK-1871 Frederiksberg C
Copenhagen, Denmark

HECTOR WILLIAMS-ALANIS
INIFAP
Campo Experimental Río Bravo
Carretera Matamoros-Reynosa, km 61
Apartado Postal 172
88900 Rio Bravo, Tamaulipas
México

DAVID M. WILSON
Department of Plant Pathology
Coastal Plain Experiment Station
University of Georgia
Tifton, Georgia 31793, USA

JEFFREY P. WILSON
USDA-ARS Crop Genetics &
Breeding Research Unit
Coastal Plain Experiment Station
University of Georgia
Tifton, Georgia 31793, USA

K. S. L. WILSON
Natural Resources Institute
University of Greenwich
Chatham Maritime, Kent ME4 4TB
United Kingdom

JOHN R. WITCOMBE
Centre for Arid Zone Studies
University of Wales
Bangor, Gwynedd LL57 2UW
United Kingdom

CHARLES A. WOODFIN
Texas A&M Univ. Agricultural Res.
& Extension Center
Route 3, Box 219
Lubbock, Texas 79403-9757, USA

EMMA M. WRAY
Toxicology & Mycotoxin Res. Unit
USDA-ARS
Russell Agricultural Res. Center
Athens, Georgia, 30613, USA

XIUDE XU
Plant Protection Research Institute
Liaoning Acad. of Agric. Sci.
No. 84 Dongling Road
Shenyang City, Liaoning 110161
China

M. M. YANINA
Russian Academy of Agric. Sci.
Moscow, Russia

DAVID YATES
Department of Botany
University of Queensland
St. Lucia, Queensland 4072
Australia

JOHN M. YOHE
INTSORMIL Management Entity
113 Biochemistry Building
University of Nebraska
Lincoln, Nebraska 68583-0748
USA

FRANCISCO ZAVALA-GARCÍA
Univ. Autónoma de Nuevo Léon
Facultad de Agronomía
Subdirección de Estudios de
Postgrado
Carretera Zuazua-Marín km 17
Marín, Nuevo León, México

JULIO ZYGADLO
FCEFyN-UNC
Cordoba 5000, Argentina

Part I

Transition from the Previous Millennium

1

Transition from the Second to the Third World Review of Sorghum and Millet Diseases

Walter A. J. de Milliano

The present commodity crops, sorghum [*Sorghum bicolor* (L.) Moench] and pearl millet [*Pennisetum glaucum* (L.) R. Br.], are important and sometimes indispensable for the survival of man and domestic animals in austere dry environments. In addition, their genetic diversity and numerous uses (24) are likely to be appealing to citizens of the third millennium searching for food diversification for reasons of enjoyment, health, convenience, eagerness for change, and fashion. This desire for food diversification could extend to other, lesser-known cereals such as *Coix lachryma-jobi* L., *Digitaria exilillis* Stapf, finger millet [*Eleusine coracana* (L.) Gaertn.], *Eragrostis tef* (Zucc.) Trotter, *Panicum* spp., *Paspalum* sp., and *Setaria* spp. Hopefully these markets will develop further, stimulating both higher levels of production and increased farm income.

Sorghum and millets have been affected by many diseases in the past (51) and still are today (17). Potentially, each disease can cause economic losses and jeopardize the food security of the farmer and his/her family. Each previous global review (12, 58, 60) has provided the opportunity to make a unique long-term analysis of different issues over the last quarter century.

Venue, Objectives, and Demonstrations

The first global single crop reviews were in India, where sorghum and pearl millet are both staple foods. "State of the art" discussions were held on diseases and disease resistance breeding (58, 60).

The second review was in Zimbabwe. In this review we were reminded of the widening food gap in the third world, the relative importance of many diseases was established, and strategic plans were developed that were mutually beneficial for both developing and developed nations (12). Technology transfer, seed health, and germplasm utilization were discussed. This meeting also gave many African scientists a rare chance to meet foreign colleagues in both meeting rooms and the field. The practical demonstrations at Zimbabwean research and development stations showed the importance of selective use of ecological zones, trained manpower, and infrastructure development. The unique, international, disease hot spot screening approach of the SADCC (Southern Africa Development Coordination Conference) region also was demonstrated.

The third global review was in Mexico. It focused on integrated management of sorghum and millet diseases for the twenty-first century and the importance of these crops in agro-ecosystems. Disease status was reviewed in national, regional, and global contexts as priorities were altered for the next decade. Also, the importance of proper pathogen identification and the use of population biology approach was emphasized. Panicle diseases, such as ergot and grain mold, and the plant parasite *Striga* received special attention. The Mexican host scientists gave on-farm and on-station demonstrations. In this "home of maize," the area planted to sorghum has been increasing in recent years, and Mexican farmers produce crops with grain yields of up to 15 t/ha. Millets, however, remain of minor importance.

Participation and Publication in the Global Reviews

No national scientist participated in all three global reviews. For sorghum, only three scientists participated in all three global sorghum reviews: Richard A. Frederiksen, Darrel Rosenow, and Nat Zummo. All three of these scientists were from the United States and were participants in INTSORMIL. For millets, S. D. Singh, Ram P. Thakur, and Walter A. J. de Milliano participated in the three pearl millet reviews in association with either ICRISAT or INTSORMIL. Thus, for sorghum, about 5%

TABLE 1-1. Papers by continent, region, and country presented during the previous (12, 58, 60) and current world reviews for sorghum and pearl millet.

	1978	1986	1988		2000	
	Sorghum	Pearl millet	Sorghum	Pearl millet	Sorghum	Pearl millet
Worldwide	n	n	n	y	y	n
Africa	n	y	n	n	n	n
Central Africa	n	n	n	n	y	y
East Africa	y	n	y	n	y	y
Ethiopia	y	n	r	n	r	r
Malawi	y	r	r	r	n	n
Southern Africa	n	y	y	y	y	y
Zimbabwe	n	y	y	y	r	r
Western Africa	n	y	y	y	y	y
Mali	n	y	r	r	r	n
Niger	y	y	r	r	r	r
Nigeria	y	y	r	r	r	r
Senegal	n	y	n	r	n	r
Asia	n	n	n	n	y	y
Bangladesh	y	n	n	n	n	n
India	y	y	y	y	r	r
Japan	n	n	y	n	n	n
Pakistan	y	n	n	n	r	r
Philippines	y	n	y	n	n	n
Thailand	y	n	y	n	r	n
Europe	n	n	y	n	n	n
Eastern Europe	n	n	n	n	n	n
Russia	n	n	n	n	y	n
Australia	n	n	n	n	y	y
Central & South America	n	n	n	n	n	n
Central America and Caribbean Basin	n	n	y	n	n	n
El Salvador	y	n	r	n	n	n
Mexico	y	n	y	n	y	y
South America	n	n	y	n	n	n
Argentina	y	n	r	n	y	n
Brazil	y	n	y	n	y	n
Venezuela	y	n	r	n	y	n
North America	n	r	r	n	r	r

n = no presentation, y = single-country presentation, r = regional presentation.

(first meeting had > 59 participants) of the scientists were still participating 22 years later, and for millet < 3% (first meeting > 100 participants) were still participating 13 years later. Between 30-40% of the participants attended at least two meetings, including both national scientists and those from the international organizations. For both crops, 30-35% of the authors published in at least two of the reviews.

Why did so few national scientists (< 1%) participate actively in more than two of these ten-year reviews? Part of the problem is financial. Simply finding enough money to pay for plane fare, hotel, and *per diem* can be very difficult. A second reason is that active developing-country scientists often are quickly promoted to administrative positions or hired by the private sector. A third problem is inevitable difficulties with visas, schedule conflicts, and travel arrangements. Thus, the fact that over a period of 10 years, 30-40% of the active national and international scientists continue to participate is more impressive than it first appears and, at the least, demonstrates significant continuity. From these statistics, it is clear that the evaluations of regional and continental issues and of long-term trends depend heavily on the published literature. In this sense, the published proceedings of this meeting and its predecessors are of particular importance.

The individual reviews of sorghum in 1978 (58) and pearl millet in 1986 (60) were followed by joint reviews in 1988 and 2000. In the joint reviews, there were distinctly (up to five times) more sorghum than millet articles (Table 1-1). That trend continues in this volume. Beginning in 1988, there also were a few manuscripts on finger millet research (*e.g.*, 1, 28). With the progress in molecular mapping, related cereal crops may benefit through synteny. Thus, progress in one crop may benefit the others. In addition, these "minor" cereals may provide an opportunity to identify disease resistance genes that are not available or do not exist in the major cereals.

Both sorghum and pearl millet have become global crops. Because the crops are grown in many countries, it is difficult to have a complete and balanced report of all the new developments. Initially, there were individual country and regional reviews (Table 1-1), in which, as a rule, the host country received some extra attention. There are now some combined reports, both in terms of having regional summaries instead of those only for individual countries (*e.g.*, 15, 23, 34, 39), and in terms of combined reports for both sorghum and millet (*e.g.*, 23, 34, 50). Stronger efforts to combine reports are recommended.

In this volume, there is a first global sorghum (35) and also a global finger millet disease review (14). For sorghum and pearl millet there is a single continental review for Asia, including, for the first time, reports for China, Iran, Myanmar, and Syria (23). Africa has no continental review, although there is a summary of the working group's meeting (20), but there are regional reviews for southern, western, eastern, and central parts of the continent (15, 34, 39). As a result of political changes since the second world review, South Africa is now actively represented (*e.g.*, 37). Probably as a desire to solve common disease problems, Australia is included in this volume for the first time (50), as is Russia (26). There was no review from Europe primarily because these cereals are not major crops on that continent. Nevertheless, European scientists made significant technical and scientific contributions (*e.g.*, 5, 33, 45).

Publication results from meeting publication standards timely. Thus, some of the presentations from Mexico are not in these proceedings, and some of the invited papers that could not be presented, usually because the scheduled presenter was unable to obtain a visa from the Mexican government, are in this volume.

Disease Control

Method. Duncan and de Milliano (13) suggested that up to modest disease control can be achieved for many sorghum and millet diseases. They compiled assessments by experts in the field for obtainable control and used a scale of 0 to 1, with 0 = no control, 0.1 = limited control, 0.5 = modest control, and 1 = absolute control. Genetic resistance (mean over all diseases = 0.46), seed treatments and tests (mean = 0.42 each), and chemical control in crops and in seed production (mean = 0.39 each) were expected to be the most effective in sorghum. Genetic resistance (mean = 0.5) and chemical control in seed production (mean = 0.35) were expected to be the most effective in pearl millet.

Genetic resistance. In the last two global reviews, genetic resistance received considerable attention through reviews of countries and regions or of specific subjects. Considering the high frequency of this type of review, and the many efforts devoted to obtaining suitable genetic lines, genetic resistance was and is an important control method.

Both for sorghum and pearl millet, there are fewer articles on genetic resistance in this volume than in its predecessor. Major specific subjects for sorghum in the preceding volume were the reviews of the use of genetic resistance (49), the use of the world germplasm collection assembled by ICRISAT and INTSORMIL (41, 49), sweet stem sorghums (61), anthracnose (7), grain mold (40), and downy mildew resistance (9). In this volume, disease resistance breeding and screening for several diseases are reviewed (48, 55), but *Striga* and ergot dominate for sorghum, and for pearl millet, downy mildew (19).

Other forms of control, including "Know your enemy." In this volume, the emphasis has shifted toward more in-depth subject matter reviews. In the preceding volume, there were reviews for specific disease organisms–their biology, taxonomy (*e.g.*, 8), biodiversity (*e.g.*, 7), technology transfer (44), and the toxicology of pearl millet ergot (32). In this volume, there were again reviews of specific disease organisms, but also reviews of interspecific relationships, intraspecific variation (*e.g.*, 45), antifungal proteins (*e.g.* 57), specific disease triangles–host, pathogen and environment (*e.g.*, 37), mycotoxins (*e.g.*, 47), quarantine (43), the use of the Internet (*e.g.*, 11), methods to improve collaboration and funding (*e.g.*, 46), and last, but not least, on the involvement of the private sector (*e.g.*, 10, 25).

In the preceding volume, two significant contributions for the future were reported as short communications. Butler (6) provided the first insight into the host/pathogen relationship that underlies the changes in breeding strategy that have lessened the importance of the *Striga* parasitic weeds. The abstract by Frederickson and Mantle (16) provided the first evidence for the aerial dispersal of conidia of the African sorghum ergot fungus. This research became of global importance when this disease spread to Australia and the Americas in the 1990s.

Crop and disease management, including disease forcasting. In this area, two major new examples are described in this volume: the use of disease prediction models, in particular for the prediction of annual sorghum ergot epidemics (*e.g.*, 37), and a dynamic multiline for pearl millet (59). In the preceding volume (12), the reviews were only for sorghum and were focused on races of anthracnose (*e.g.* 7), and a case study was presented on the epidemiology of sorghum diseases in Central America (56). In this volume, there are reviews of: cultural (including intercropping) and IPM (Integrated Pest Management) strategies for control of *Striga* (2, 21), and of biodiversity as a strategy for both disease and pest management (59). For pearl millet, the management of downy mildew resistance was reviewed (19, 22).

Other methods of control. Overviews of chemical control (38), seed treatments, biological control, and seed tests receive little or no attention in this volume. Chemical control in seed production, which was previously identified as an important disease control method, is barely perceptible in the global reviews. However, seed transmission of sorghum pathogens was a subject of special emphasis in the 1988 conference (36). In the 2000 conference, a determined effort was made to involve the private sector (10, 25, 53), in addition to a presentation on the practical value of a systemic seed treatment against downy mildew in pearl millet (52).

A major change for the control of diseases is the addition of molecular biology-based solutions for all of the crops. Considerable progress has been made since the previous global review and the 1995 strategic meeting in Bellagio (29). Diseases to which such analyses are being applied include sorghum grain molds (3), sorghum anthracnose (18), sorghum head smut (42), sorghum ergot (45), sorghum leaf blight (31), sorghum oval leaf spot (4), "*Fusarium moniliforme*" (30), pearl millet downy mildew (54), and finger millet blast (5). The coming challenge is the practical and efficient use of the basic information now being generated.

Sorghum Diseases

Sorghum has a number of diseases of continuing importance. Many of the important diseases are caused by fungi including grain molds, anthracnose, stalk rots, ergot, smuts, and downy mildew. A novelty at the 2000 meeting was the two-man tag-team review of a single fungal species, *Fusarium moniliforme*, which now is being subdivided into 20+ species and whose present name was retired at this meeting. The importance of various *Fusarium* spp. in relation to human health, food quality and other post-harvest problems, suggests both threats and opportunities. In relation to maize, is sorghum grain a healthier food because of the mycoflora that it does, or does not, support? Therefore, the *Fusaria* deserve continued attention. *Striga* also continued to receive attention. In particular, research at Purdue University has made steady progress, and prospects for the control of these parasitic weeds are greatly improved.

Viruses received attention in both reviews, and their taxonomy is now established. However, the global importance of these viruses is not well demonstrated. Bacterial diseases were reviewed in the preceding volume (8), but not in this one. Nematodes have received little attention, although a review of sorghum nematode problems is included in this volume (27). Insects have been avoided in these global reviews, as have rodents and birds, but these organisms are important, and, in particular, insects can be vectors and facilitators of fungal, bacterial, and viral diseases. With the movement toward IPM, the inclusion of reviews by entomologists and other biologists in the next global review might encourage the development of research and production synergies.

Millet Diseases

Millets also clearly have a number of diseases of continuing importance, of which ergot and downy mildew are receiving the most attention. During the Mexico review, remarkable and somewhat unexpected progress was reported by Breese *et al.* (5) on the development of molecular markers for studying finger millet leaf blast (*Pyricularia* spp.). Blast control in finger millet also may benefit from the recent developments in rice (14). *Striga* also continued to receive attention (2, 21). Though focus in research is desired, and appears to have been attained, the lack of progress caused by the absence or lack of staff is a concern.

Take-Home Messages

In terms of growth measured as planted area or yield, there are few successes with sorghum and millets, as the global popularity of these crops has been decreasing. Yet in Mexico, "the home of maize," growth in both planted area and grain yield is occurring in sorghum. For millets, major successes have not yet occurred, and these crops continue to languish and remain of minor importance.

The supporting technologies provided unexpected tools and unprecedented progress with the development of molecular maps for both the host crops and the causal agents of some of the diseases. Thus, a move from empirical selection for disease resistance to marker-assisted breeding either has or will shortly occur. This shift means that for disease resistance screening, we may no longer be totally dependent on the presence of the pathogen. Dependence on demanding, and often difficult and unpredictable, disease resistance-screening tests also may be reduced. Breeding against quarantine diseases that are not in the country also may be possible if suitably characterized resistant and susceptible germplasm is available. We also can distinguish homozygous and heterozygous resistant plants, without the complications caused by "escapes" in the screening process. DNA analysis is usually a nondestructive test that can be done in the laboratory with material from almost any plant tissue or growth stage, and could accelerate the breeding process. In addition, the molecular technology provides a chance to select systematically for QTL quantitative resistances.

Molecular technology also has provided laboratory tools to identify (quarantine) strains. Through the data generated with these tools we can better understand and evaluate the genetic evolution of pathogens and better understand genetic diversity. In particular, the reviews on ergot (45) and *Fusarium* spp. (30) could become classic examples. Though the development of genetically modified cereals was not a primary topic of the meeting, new opportunities for control of previously uncontrollable diseases and other biological agents, *e.g.*, insects and nematodes, may be possible.

Clearly in the twenty-first century, more than work in the field will be needed for effective disease control. The key for impact will be the management, design, and implementation of multidisciplinary collaborative research. Several issues are of importance to secure this future:

- How can the consumer be involved more effectively?
- Joint planning of regional targets and activities and regular control of the agreed progress.
- Joint planning for actions between INTSORMIL, ICRISAT, and the private sector (*e.g.*, the National Grain Sorghum Producers in the United States) could strengthen sorghum and millet research through complementary efforts and provide a better focus for national and regional targets.
- Food safety, *e.g.*, traceability, methods to measure the quantity of key toxins, food security, and models to predict grain quality all need greater attention.
- Academic training of plant pathologists from developed and developing countries in molecular biology should become essential.

References

1. **Adipala, E.** 1992. Diseases of finger millet in Uganda, p. 289. *In* W. A. J. de Milliano, R. A. Frederiksen, and G. D. Bengston (eds.), *Sorghum and Millets Diseases: A Second World Review*. ICRISAT, Patancheru (A.P.), India.
2. **Babiker, A. G. T.** 2002. *Striga* research in the Sudan: An integrated approach, p. 159-163. *In* J. F. Leslie (ed.), *Sorghum and Millets Diseases*. Iowa State Press, Ames, Iowa.
3. **Bandyopadhyay, R., C. R. Little, R. D. Waniska, and D. R. Butler.** 2002. Sorghum grain mold: Through the 1990s into the new millennium, p. 173-183. *In* J. F. Leslie (ed.), *Sorghum and Millets Diseases*. Iowa State Press, Ames, Iowa.
4. **Boora, K. S., A. Sindhu, P. Boora, and C. W. Magill.** 2002. Identification of molecular markers for an oval leaf spot resistance gene in sorghum, p. 239-241. *In* J. F. Leslie (ed.), *Sorghum and Millets Diseases*. Iowa State Press, Ames, Iowa.
5. **Breese, W. A., C. T. Hash, K. M. Devos, and C. J. Howarth.** 2002. Pearl millet genomics: An overview with respect to breeding for resistance to downy mildew, p. 243-246. *In* J. F. Leslie (ed.), *Sorghum and Millets Diseases*. Iowa State Press, Ames, Iowa.
6. **Butler, L.** 1992. Identification of *Striga* germination stimulants from sorghum-root exudate and their potential significance in control of *Striga*, p. 293. *In* W. A. J. de Milliano, R. A. Frederiksen, and G. D. Bengston (eds.), *Sorghum and Millets Diseases: A Second World Review*. ICRISAT, Patancheru (A.P.), India.
7. **Casela, C. R., A. S. Fereira, and R. E. Schaffert.** 1992. Physiological races of *Colletotrichum graminicola* in Brazil, p. 209-212. *In* W. A. J. de Milliano, R. A. Frederiksen, and G. D. Bengston (eds.), *Sorghum and Millets Diseases: A Second World Review*. ICRISAT, Patancheru (A.P.), India.
8. **Claflin, L. E., B. A. Ramundo, J. E. Leach, and M. Qhobela.** 1992. Bacterial diseases of sorghum, p. 135-151. *In* W. A. J. de Milliano, R. A. Frederiksen, and G. D. Bengston (eds.), *Sorghum and Millets Diseases: A Second World Review*. ICRISAT, Patancheru (A.P.), India.
9. **Craig, J., and G. N. Odvody.** 1992. Current status of sorghum downy mildew control, p. 213-217. *In* W. A. J. de Milliano, R. A. Frederiksen, and G. D. Bengston (eds.), *Sorghum and Millets Diseases: A Second World Review*. ICRISAT, Patancheru (A.P.), India.
10. **Condon, M. S., and J. A. Dahlberg.** 2002. Public-private partnership in international agricultural research: A case for promoting technology transfer and enhancing global trade in sorghum and millet, p. 339-341. *In* J. F. Leslie (ed.), *Sorghum and Millets Diseases*. Iowa State Press, Ames, Iowa.
11. **Dahlberg, J. A.** 2002. Rapid information dissemination on the world wide web, p. 305-308. *In* J. F. Leslie (ed.), *Sorghum and Millets Diseases*. Iowa State Press, Ames, Iowa.

12. **de Milliano, W. A. J., R. A. Frederiksen, and G. D. Bengston, eds.** 1992. *Sorghum and Millets Diseases: A Second World Review.* ICRISAT, Patancheru (A.P.), India.
13. **Duncan, R. R., and W. A. J. de Milliano.** 1995. Plant disease control in sorghum and pearl millet, p. 34-71. *In* J. F. Leslie and R. A. Frederiksen (eds.), *Disease Analysis through Genetics and Biotechnology: Interdisciplinary Bridges to Improved Sorghum and Millet Crops.* Iowa State University Press, Ames, Iowa.
14. **Esele, J. P.** 2002. Diseases of finger millet: A global overview, p. 21-26. *In* J. F. Leslie (ed.), *Sorghum and Millets Diseases.* Iowa State Press, Ames, Iowa.
15. **Esele, J. P.** 2002. Sorghum and pearl millet diseases in the Horn of Africa, p. 383-387. *In* J. F. Leslie (ed.), *Sorghum and Millets Diseases.* Iowa State Press, Ames, Iowa.
16. **Frederickson, D. E., and P. G. Mantle.** 1992. Secondary sporulation of *Sphacelia sorghi* on sorghum, p. 293. *In* W. A. J. de Milliano, R. A. Frederiksen, and G. D. Bengston (eds.), *Sorghum and Millets Diseases: A Second World Review.* ICRISAT, Patancheru (A.P.), India.
17. **Frederiksen, R. A., and G. N. Odvody, eds.** 2000. *Compendium of Sorghum Diseases*, 2nd ed. APS Press, St. Paul, Minnesota.
18. **Gale, L. R.** 2002. A population genetic approach to study variations in *Colletotrichum graminicola*, the causal agent of sorghum anthracnose, p. 191-199. *In* J. F. Leslie (ed.), *Sorghum and Millets Diseases.* Iowa State Press, Ames, Iowa.
19. **Hash, C. T., and J. R. Witcombe.** 2002. Gene management and breeding for downy mildew resistance, p. 27-36. *In* J. F. Leslie (ed.), *Sorghum and Millets Diseases.* Iowa State Press, Ames, Iowa.
20. **Hess, D. E.** 2002. Recommendations from the African working group at the Third Global Conference on Sorghum and Millets Diseases, p. 375-376. *In* J. F. Leslie (ed.), *Sorghum and Millets Diseases.* Iowa State Press, Ames, Iowa.
21. **Hess, D. E., J. Kroschel, D. Traoré, A. E. M. Elzein, P. S. Marley, A. A. Abbasher, and C. Diarra.** 2002. *Striga:* Biological control strategies for a new millennium pathogen, p. 165-170. *In* J. F. Leslie (ed.), *Sorghum and Millets Diseases.* Iowa State Press, Ames, Iowa.
22. **Hess, D. E., R. P. Thakur, C. T. Hash, P. Sérémé, and C. W. Magill.** 2002. Pearl millet downy mildew: Problems and control strategies for a new millennium, p. 37-42. *In* J. F. Leslie (ed.), *Sorghum and Millets Diseases.* Iowa State Press, Ames, Iowa.
23. **Indira, S., X. Xiude, N. Iamsupasit, H. S. Shetty, N. S., Vasanthi, S. D. Singh, and R. Bandyopadhyay.** 2002. Diseases of sorghum and pearl millet in Asia, p. 393-402. *In* J. F. Leslie (ed.), *Sorghum and Millets Diseases.* Iowa State Press, Ames, Iowa.
24. **INTSORMIL.** 2000. INTSORMIL 2000 annual report. Fighting hunger with research. A team effort. Grain Sorghum/Pearl Millet Collaborative Support Program (CRSP). INTSORMIL Publication 00-1. University of Nebraska, Lincoln, Nebraska.
25. **Johnson, M. A.** 2002. Private sector and public institution interactions on sorghum and pearl millet disease management, p. 349-350. *In* J. F. Leslie (ed.), *Sorghum and Millets Diseases.* Iowa State Press, Ames, Iowa.
26. **Kasakova, A. S., and A. A. Oleynick.** 2002. The status of sorghum diseases in Russia, p. 411-413. *In* J. F. Leslie (ed.), *Sorghum and Millets Diseases.* Iowa State Press, Ames, Iowa.
27. **Kollo, I. A.** 2002. Plant parasitic nematodes of sorghum and pearl millet: Emphasis on Africa, p. 259-266. *In* J. F. Leslie (ed.), *Sorghum and Millets Diseases.* Iowa State Press, Ames, Iowa.
28. **Kuwite, C. A.** 1992. *Pyricularia* spp. causing head blight of finger millet (*Eleusine coracana*) and other fungi associated with finger millet in Tanzania, p. 290. *In* W. A. J. de Milliano, R. A. Frederiksen, and G. D. Bengston (eds.), *Sorghum and Millets Diseases: A Second World Review.* ICRISAT, Patancheru (A.P.), India.
29. **Leslie, J. F., and R. A. Frederiksen, eds.** 1995. *Disease Analysis through Genetics and Biotechnology: Interdisciplinary Bridges to Improved Sorghum and Millet Crops.* Iowa State University Press, Ames, Iowa.
30. **Leslie, J. F., and W. F. O. Marasas.** 2002. Will the real *Fusarium moniliforme* please stand up!, p. 201-209. *In* J. F. Leslie (ed.), *Sorghum and Millets Diseases.* Iowa State Press, Ames, Iowa.
31. **Magill, C., R. A. Frederiksen, K. Boora, R. Perumal, and S. Sivaramakrihnan.** 2002. Molecular tags for disease resistance genes in sorghum: Improved prospects for mapping pathogens, p. 247-252. *In* J. F. Leslie (ed.), *Sorghum and Millets Diseases.* Iowa State Press, Ames, Iowa.
32. **Mantle, P. G.** 1992. Ergot disease of pearl millet: Toxicological significance and a mechanism for disease escape, p. 129-132. *In* W. A. J. de Milliano, R. A. Frederiksen, and G. D. Bengston (eds.), *Sorghum and Millets Diseases: A Second World Review.* ICRISAT, Patancheru (A.P.), India.
33. **Mantle, P. G., and A. Bogo.** 2002. Biosynthesis of bioactive honeydew oligosaccharides by sorghum ergot pathogens, p. 91-94. *In* J. F. Leslie (ed.), *Sorghum and Millets Diseases.* Iowa State Press, Ames, Iowa.
34. **Marley, P. S., M. Diourté, A. Neya, S. K. Nutsugah, P. Sérémé, S. O. Katilé, D. H. Hess, D. F. Mbaye, and Z. Ngoko.** 2002. Sorghum and pearl millet diseases in west and central Africa, p. 419-425. *In* J. F. Leslie (ed.), *Sorghum and Millets Diseases.* Iowa State Press, Ames, Iowa.
35. **Maunder, B.** 2002. Sorghum worldwide, p. 11-17. *In* J. F. Leslie (ed.), *Sorghum and Millets Diseases.* Iowa State Press, Ames, Iowa.
36. **McGee, D. C.** 2002. Principles and criteria for seed transmission of sorghum pathogens, p. 297-302. *In* W. A. J. de Milliano, R. A. Frederiksen, and G. D. Bengston (eds.), *Sorghum and Millets Diseases: A Second World Review.* ICRISAT, Patancheru (A. P.), India.
37. **McLaren, N. W.** 2002. Genotype × environment effects on response of sorghum ergot and repercussions for disease screening, p. 95-101. *In* J. F. Leslie (ed.), *Sorghum and Millets Diseases.* Iowa State Press, Ames, Iowa.
38. **Montes, N., G. N. Odvody, and H. Williams.** 2002. Advances in *Claviceps africana* chemical control, p. 105-110. *In* J. F. Leslie (ed.), *Sorghum and Millets Diseases.* Iowa State Press, Ames, Iowa.
39. **Mtisi, E., and N. W. McLaren.** 2002. Diseases of sorghum and pearl millet in some southern African countries, p. 427-430. *In* J. F. Leslie (ed.), *Sorghum and Millets Diseases.* Iowa State Press, Ames, Iowa.

40. **Mukuru, S. Z.** 1992. Breeding for grain mold resistance, p. 273-285. *In* W. A. J. de Milliano, R. A. Frederiksen, and G. D. Bengston (eds.), *Sorghum and Millets Diseases: A Second World Review.* ICRISAT, Patancheru (A.P.), India.
41. **Mukuru, S. Z.** 1992. Using the world germplasm collection in breeding for disease resistance at ICRISAT, p. 325-328. *In* W. A. J. de Milliano, R. A. Frederiksen, and G. D. Bengston (eds.), *Sorghum and Millets Diseases: A Second World Review.* ICRISAT, Patancheru (A.P.), India.
42. **Naidoo, G., and H. Torres-Montalvo.** 2002. Genetic variability among and within host-specialized isolates of *Sporisorium reilianum*, p. 221-225. *In* J. F. Leslie (ed.), *Sorghum and Millets Diseases.* Iowa State Press, Ames, Iowa.
43. **Odvody, G. N., D. E. Frederickson, T. Isakeit, N. Montes, J. A. Dahlberg, and G. L. Peterson.** 2002. Quarantine issues arising from contamination of seed with ergot: An update, p. 123-127. *In* J. F. Leslie (ed.), *Sorghum and Millets Diseases.* Iowa State Press, Ames, Iowa.
44. **Pande, S., and S. D. Singh.** 1992. Successful transfer of ICRISAT downy mildew resistance screening technology: An example of transfer of technology, p. 331-334. *In* W. A. J. de Milliano, R. A. Frederiksen, and G. D. Bengston (eds.), *Sorghum and Millets Diseases: A Second World Review.* ICRISAT, Patancheru (A.P.), India.
45. **Pažoutová, S.** 2002. The genus *Claviceps*: Evolution at work, p. 135-140. *In* J. F. Leslie (ed.), *Sorghum and Millets Diseases.* Iowa State Press, Ames, Iowa.
46. **Peterson, G., B. Pendleton, and G. Teetes.** 2002. PROFIT – *P*roductive *R*otations *O*n *F*arms *i*n *T*exas, p. 365-369. *In* J. F. Leslie (ed.), *Sorghum and Millets Diseases.* Iowa State Press, Ames, Iowa.
47. **Porter, J. K., C. W. Bacon, W. P. Norred, E. M. Wray, G. A. Kuldau, A. E. Glenn, and J. F. Leslie.** 2002. Mycotoxins from fungal-infected sorghum: *Claviceps, Fusarium* and the *Striga* connection, p. 229-235. *In* J. F. Leslie (ed.), *Sorghum and Millets Diseases.* Iowa State Press, Ames, Iowa.
48. **Rooney, W. L., S. D. Collins, R. R. Klein, P. J. Mehta, R. A. Frederiksen, and R. Rodriguez-Herrera**. 2002. Breeding sorghum for resistance to anthracnose, grain mold, downy mildew, and head smuts, p. 273-279. *In* J. F. Leslie (ed.), *Sorghum and Millets Diseases.* Iowa State Press, Ames, Iowa.
49. **Rosenow, D. T.** 1992. Using germplasm from the world collection in breeding for disease resistance, p. 319-324. *In* W. A. J. de Milliano, R. A. Frederiksen, and G. D. Bengston (eds.), *Sorghum and Millets Diseases: A Second World Review.* ICRISAT, Patancheru (A.P.), India.
50. **Ryley, M., D. Persley, D. Jordan, and B. Henzell.** 2002. Status of sorghum and pearl millet diseases in Australia, p. 441-448. *In* J. F. Leslie (ed.), *Sorghum and Millets Diseases.* Iowa State Press, Ames, Iowa.
51. **Saccas, A. M.** 1954. Les champignons parasites des sorghos (*Sorghum vulgare*) et des penicellaires (*Pennisetum typhoideum*) en Afrique Equatorial Francaise. *Agron. Trop. (Paris)* **9:** 135-173, 263-301, 647-686.
52. **Scheuring, J. F., S. O. Katilé, and I. A. Kollo.** 2002. Boosting pearl millet yields with Apron Plus® and Apron Star® seed treatments, p. 47-49. *In* J. F. Leslie (ed.), *Sorghum and Millets Diseases.* Iowa State Press, Ames, Iowa.
53. **Sheorain, V., and A. H. Hall.** 2002. The importance of public-private sector partnership to Indian sorghum farmers, p. 476. *In* J. F. Leslie (ed.), *Sorghum and Millets Diseases.* Iowa State Press, Ames, Iowa.
54. **Thakur, R. P., C. W. Magill, S. Sivaramakrishnan, C. T. Hash, H. S. Shetty, and D. E. Hess.** 2002. Variability of *Sclerospora graminicola*, the pearl millet downy mildew pathogen, p. 51-56. *In* J. F. Leslie (ed.), *Sorghum and Millets Diseases.* Iowa State Press, Ames, Iowa.
55. **Tuinstra, M. R., T. Teferra, L. E. Claflin, R. G. Henzell, A. Borell, N. Seetharama, G. Ejeta, and D. T. Rosenow.** 2002. Breeding for resistance to root and stalk rots in sorghum, p. 281-286. *In* J. F. Leslie (ed.), *Sorghum and Millets Diseases.* Iowa State Press, Ames, Iowa.
56. **Wall, G. C., R. A. Frederiksen, J. Craig, and M. Jeger.** 1992. Epidemiology of sorghum diseases in Central America: A case study, p. 303-317. *In* W. A. J. de Milliano, R. A. Frederiksen, and G. D. Bengston (eds.), *Sorghum and Millets Diseases: A Second World Review.* ICRISAT, Patancheru (A.P.), India.
57. **Waniska, R. D., R. T. Venkatesha, A. Chandrashekar, S. Kreshnaveni, F. P. Bejosano, J.-M. Jeoung, J. Jayaraj, S. Muthukrishnan, and G. H. Liang**. 2002. Antifungal proteins and other mechanisms in the control of sorghum stalk rot and grain mold, p. 287-297. *In* J. F. Leslie (ed.), *Sorghum and Millets Diseases.* Iowa State Press, Ames, Iowa.
58. **Williams, R. J., R. A. Frederiksen, L. K. Mughogho, and G. B. Bengston, eds.** 1980. *Proc. Intl. Workshop on Sorghum Diseases (11-15 December 1978, Hyderabad, India).* ICRISAT, Patancheru (A.P.), India.
59. **Wilson, J. P., and R. N. Gates.** 2002. The dynamic multiline population: An alternative approach to durable resistance, p. 65-69. *In* J. F. Leslie (ed.), *Sorghum and Millets Diseases.* Iowa State Press, Ames, Iowa.
60. **Witcombe, J. R., and S. R. Beckerman, eds.** 1987. *Proc. Intl. Pearl Millet Workshop (7-11 April 1986, Patancheru, India).* ICRISAT, Patancheru (A. P.), India.
61. **Zummo, N.** 1992. Sweet sorghum varieties resistant to anthracnose in USA are highly susceptible to the diseases in Brazil, p. 289. *In* W. A. J. de Milliano, R. A. Frederiksen, and G. D. Bengston (eds.), *Sorghum and Millets Diseases: A Second World Review.* ICRISAT, Patancheru (A.P.), India.

2

Sorghum Worldwide

A. Bruce Maunder

Sorghum of a cultivated nature was known to exist in southern Egypt prior to 6000 BC, and records suggest that this crop existed in China some 5000-6000 years ago. The United States, which currently produces approximately 25% of the world's crop, is a relatively recent sorghum-producing country, with the first importation of note in 1853. Sorghums tolerant to shading and intercropped with maize in Central America, the maicillo criollos, are probably descendants of the earliest introduction of this crop to the Western hemisphere. Sorghums are now found from 40°S to 45°N latitude, primarily in agroecological regions subject to low rainfall and drought. Approximately 90% of the world's sorghum production area occurs in developing countries. Commercialized production, primarily for livestock feed, is characteristic of the developed world and parts of Latin America and the Caribbean.

Sorghum may be the world's fifth most important cereal, but it is the second most important feed grain in the United States and a critical food for the poor in most of the world. Since 1948-1950, the area planted to sorghum worldwide has increased by 71%, and the yield has increased by 160% (3, 6). Recent world sorghum statistics (1989-1999) suggest increases in yield, rather than in area planted, account for the approximate gain in annual production of 5 million metric tons (mmt) (Table 2-1). Of the world's major sorghum producers, the United States, China, Argentina, Mexico, and Australia have yields that are above the world average (Table 2-2). Only Nigeria, Mexico, and Sudan, however, increased total production in 1996-1998 relative to 1979-1981 averages (Table 2-3).

TABLE 2-1. World sorghum statistics from 1989-1999 (after 1, 3).

	1989	1994	1999
Area (million ha)	41.6	38.6	41.1
Yield (mt/ha)	1.33	1.46	1.47
Production (mmt)	55.3	56.3	60.6

TABLE 2-2. Sorghum productivity by country (1989-1999) expressed as the percentage of worldwide production divided by the percentage of worldwide planted area (after 1, 3).

Country	1989	1994	1999
United States	2.7	3.0	3.0
China	2.3	2.6	2.8
Argentina	2.3	2.4	2.7
Mexico	2.3	2.2	2.2
Australia	1.8	1.2	1.5
Brazil	1.5	1.2	1.0
Nigeria	0.6	0.5	0.8
Ethiopia	0.9	0.8	0.7
Sudan	0.5	0.5	0.5
India	0.6	0.6	0.5

Brazil, although not included as a major producer, increased sorghum production during this time by 195% (3, 6). China and Argentina have had the greatest decline in total sorghum production over this twenty-year period. Countries using newer technologies, including hybrids, with the exception of India and Nigeria, have increased yields even when starting from a high base level.

In the early 1980s, an estimated 39% of global sorghum production was used for food, and 54% was used for animal feed (8). The proportion of sorghum used for food has gradually increased because of greater use as a food in Africa and the substitution of sorghum by other grains, primarily maize, as animal feed elsewhere. By 1992-1994, 42% of total sorghum production was used for food and 48% for animal feed. Projections to 2005 suggest increased utilization of sorghum for both animal feed and food (16% and 15%, respectively), but the relative percent utilization probably will remain constant. Growth in demand by 2005 is expected to come from three main sources: (i) increased use as food in Africa resulting from population growth; (ii) increased use as animal feed in Asia as the livestock industry intensifies; and (iii) increased use as animal feed in Latin America

TABLE 2-3. Sorghum production (in million metric tons/year) by the world's major sorghum producing countries (3, 8).

Country	1979-1981 Average	1996-1998 Average
United States	19.2	16.5
India	11.2	9.2
Nigeria	3.8	6.9
Mexico	5.0	6.3
China	7.0	4.8
Sudan	2.7	3.9
Argentina	5.6	3.2

and the Caribbean, as the livestock industry continues to grow.

The current principal animal feed users of sorghum are the United States (11.1 mmt/year), Mexico (7.2 mmt/year), Japan (2.6 mmt/year), China (1.9 mmt/year), and Argentina (1.5 mmt/year) (8). Between 1992 and 2005, food sorghum consumption will decrease in Asia but increase in Africa, reflecting continued dependence on sorghum as a food security crop in Africa.

The principal coarse grains in world trade are maize, barley, and sorghum. The world coarse grain forecast (4) suggests market growth by up to 1.6% annually, and for the market to exceed one billion tons by 2008. A slight growth in harvested area, 0.1% annually, to 307 million hectares by 2008 is anticipated. Yields, however, are expected to increase at a rate of 1.4% annually. Finally, stronger world demand and a declining stocks-to-use ratio could increase world prices modestly. Although sorghum as a crop may benefit from these projections, maize is the more likely benefactor.

Worldwide Sorghum Statistics

Global. Sorghum producers worldwide have increased production from 13.4 mmt in 1948-1950 to an average of 60.5 mmt in 1998-2000. During this 50-year period, yield worldwide increased from an average of 0.57 mt/ha to 1.48 mt/ha, a 160% increase, but the area planted to sorghum has increased by 71%, from 24 to 41 million ha (Table 2-4). Projections through 2007 show that the area planted to sorghum will be relatively unchanged, which is suspicious when considering the increasing production area in Africa that is required to offset yield declines and the opportunities for increased sorghum production in Mexico and the cerrado of Brazil (4). Also, the projected cumulative 6% yield increase for sorghum over the next nine years seems low, especially when compared with maize, where the expectation is for a 13% gain from a much higher base. Together, these area and yield projections suggest only a 4.9% increase in total sorghum production by 2007. The 12% increase in trade (in terms of amount of sorghum grain moving between countries) is the most encouraging long-term prospect for sorghum. In Asia, sorghum production fell during the 1990s, largely because of sharp declines in area planted to sorghum in China. Developed countries, including Argentina, have experienced declines in sorghum production since the mid-1980s, essentially because governmental policy interventions resulted in reduced area planted to sorghum.

United States. The area planted to sorghum prior to hybrids in 1957 did not exceed 3.7 million ha, but by 1958, some 7.2 million ha were planted to sorghum, a level that remained stable until the mid-1980s. In 1985, farm legislation lowered sorghum crop support prices relative to cotton and maize and permitted the interchange of sorghum and maize base acres. Together, these policy changes encouraged farmers to replace sorghum with other crops. Also, more drought-tolerant maize varieties were developed and, in combination with the increased application of no-tillage technology, extended the maize belt further father west into areas that had traditionally been planted to sorghum. As a result, public and private investment in sorghum research has been reduced. A 1994 plant breeding study by Frey (7) found ten times more scientists breeding maize than sorghum, and this number may be an underestimate.

The first hybrids, which essentially replaced varieties within four years of their introduction, resulted in a doubling of yield from 1955 to 1960. Stalk rots and midge had been long term problems with sorghum, but the hybrid sorghums also were affected by downy mildew, maize dwarf mosaic virus, greenbugs, and, most recently, sorghum ergot, a particular concern for seed producers. Current yields average 4.4 mt/ha, compared with 1.3 mt/ha prior to hybrids. In 1999, the U.S. produced some 25% of the world's total sorghum on 8.4% of the world's area, for a production-to-area ratio of 3 (Table 2-2). Area planted, yield, and total production are projected to be flat or increase slightly through 2008. During the 1990s, export prices in the United States have favored maize over sorghum by 5%. The red-seeded hy-

TABLE 2-4. Global area planted (million ha) to sorghum, and average yield (metric tons/ha) (1, 3, 6).

Years	Area Planted	Yield
1948-1950	24	0.57
1958-1960	31	0.97
1968-1970	50	1.10
1978-1980	46	1.35
1988-1990	41	1.33
1998-2000	41	1.48

brids introduced in the 1950s were mostly replaced with hetero-yellow hybrids by 1970 with some acreage of yellow grain sorghum. Current breeding efforts are focused on food-grade white or cream sorghums with tan plant color. Intensive efforts have been made by the sorghum industry to eliminate high-tannin, brown sorghum from the marketplace.

India. Sorghum area, yield, and production have been decreasing, although most of the decrease in the percentage of world production (13.2% in 1999 compared to 20.5% in 1989) is attributed to a 30% reduction in area planted since 1989 (8). At present in India, more than 50% of the sorghum area is planted to hybrids, the first of which became available in 1964. The private sector supplied 60% of the formal seed market by 1996, reflecting a significant research and development effort by ten companies. Sorghum usage patterns also have changed. In 1989, 92% of the crop was used as human food, but in 1999, 83% of the crop was used for animal feed. These changes probably reflect consumer preferences for wheat and rice, and the increase in protein in the diet, *e.g.* with poultry, in many developing countries and especially in Asia. When such dietary shifts occur, sorghum usually becomes a significant coarse grain for animal feed. Two sorghum crops may be grown annually, Kharif in the summer and Rabi in the winter. Maharashtra is the principal sorghum state in India and contains more than 50% of the sorghum area planted in India.

China. Both area planted and total production of sorghum has decreased, although the yield has increased by 50% from 2.72 mt/ha in 1989 to 4.09 mt/ha in 1999. Hybrids, no doubt, account for much of this increase. China's sorghum production has decreased the most of all the major sorghum-producing countries, going from an average of 7.0 mmt in 1979-1981 to an average of 4.8 mmt in 1996-1998. For high yields in China, the recommended inputs include 45 mt/ha of manure, 200 kg of ammonium phosphate, 300 kg of potassium sulfate, and 300 kg of urea as side dressing (13). At present, two-thirds of the crop goes for food and one-third for feed, but the percentage used for food has been drastically reduced (4). Although China has long been a barley importer, projections suggest that imports of maize will be necessary by 2003, providing an opportunity for sorghum depending on price differential and grain availability (4).

Africa. Africa is thought by most to be the center of origin of sorghums, probably in northeast Africa near Ethiopia and Sudan. The dura class of sorghum is found in Ethiopia and to the west, in the driest regions of sorghum culture, in Sahelian Africa, as well as in Arabia, Asia Minor, India, Burma, and the Nile Valley. Guinea sorghum is common in West Africa. Caudatum sorghum

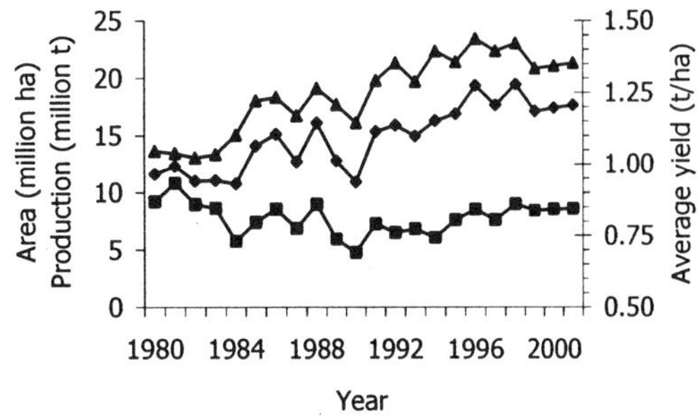

FIGURE 2-1. Annual sorghum growth rates for area (▲, 2.73%), total production (◆, 2.62%), and average yield (■, -1.5%) in Sub-Saharan Africa from 1980-2001 (6).

is most abundant from Nigeria to Sudan and south to Uganda. Kafir sorghum occurs primarily south of the equator in east Africa down to South Africa. Bicolor sorghum is most diverse in Asia, but also is widespread in Africa. Agriculture is the dominant industry in Africa, and sorghum, with its drought and heat tolerance, is an essential part of this industry. As a dryland crop, sorghum is most commonly grown by resource-poor farmers. Commonly, the cultivars planted by farmers are tall (3-4.5 meters) and require 120-180 days to mature. Light-colored grain generally is preferred for food, with red- and brown-colored grains preferred for brewing. High-tannin types frequently are used for food where bird pressure is high.

Some 17.1 mmt of sorghum were produced in Africa from 1992-1994, approximately 27% of the world total, on 21.8 million ha, approximately 48% of the world's production area. The ratio of production to area of approximately 0.5 is much less than most other major sorghum producing areas (Table 2-2). Within Africa, except for Nigeria, yields have been declining slowly in most of the major producing countries, including Ethiopia, Sudan, and Burkina Faso, although the area planted has, of necessity, increased. Even so, production increases are less than the population growth rate in most African countries. From 1980 to 2001, the annual rate of growth of sorghum area in Sub-Saharan Africa was 2.7%, with a production increase of 2.6%, but with a decrease in yield of only 1.2% (Fig. 2-1).

The availability of a dependable supply of good-quality seed of improved cultivars and hybrids is a major constraint in Africa, as is their acceptance. In Ethiopia, for example, only 5% of the sorghum crop cultivated is of improved varieties, 1.1% in Tanzania, 7.5% in Uganda, and 15% in Zambia. Although excellent research has occurred during the last 15 years on varietal improvement and crop management, all too little of the results have left the research stations. Technology transfer, such as that conducted by organizations such as Global 2000, is desperately needed to improve sorghum productivity in Africa. Also, Africa has 12% of the world's population but consumes only 2% of the world's fertilizer. The Republic of South Africa benefits most from advanced seed activity, with lesser benefits in Zimbabwe, Zambia, Malawi, and Kenya, although many of these efforts are focused on maize. Hybrid sorghums were introduced into Sudan in 1983. The hybrid, Hageen Dura-1, in farmers' fields yielded 50% more than the local cultivars. At present, efforts are being made to introduce hybrids in Niger; with NAD-1 being the first introduction. The efforts of Zamseed in Zambia resulted in a steady increase of seed of improved cultivars with the best year, 1993/94, accounting for 1100 metric tons of seed (the equivalent of 49,000 U.S. sorghum units). For hybrids and other improved seeds to be reliably available requires the development of a functioning seed industry.

A major constraint to production of sorghum, or maize, is *Striga*, a parasitic weed that attaches itself to sorghum roots from which it draws its moisture and nutrients, thereby inhibiting host plant growth and reducing yields. *Striga* currently affects an estimated 8 million ha in Africa, some 37% of the total sorghum production area. Annual yield losses are estimated to be worth over 90 million US$ (8). Recently, *Striga*-resistant varieties have been developed and have performed well in Ethiopia, where two cultivars have been released.

Sorghum utilization in southern and eastern Africa has two major uses: food and beverages. In this area, except in Botswana, where sorghum remains the favored cereal staple, the consumption of sorghum porridge has declined and has been superseded by maize. In drier regions of Africa, where wheat availability is limited, sorghum flour can be blended with wheat flour to produce bread, biscuits, pasta, and snack foods. Food use in Africa has increased 42% from 9.0 mmt in 1979-1981 to 12.8 mmt in 1992-1994. In southern Africa, 200,000 mt of sorghum are malted annually for production of opaque beer. Sorghum in beer brewing, both commercial and rural, is easily the largest commercial use of grain sorghum in both southern and eastern Africa.

Argentina/Brazil. In the 1975-1985 period, Argentina had high yields from sorghum planted on 2.5 million ha and was a very significant sorghum exporter. Much of the exported grain, perhaps as much as 75%, was of high tannin, but this grain was readily accepted in the Japanese and Russian markets. Sorghum production in Argentina fell from 8 mmt in 1983 to 3 mmt in 1988 in part because of the drastic decrease in imports by the former Soviet Union. The introduction of soybeans into eastern Argentina also resulted in maize cultivation being moved inland into traditional sorghum-growing regions, and sorghum being pushed even further west into a more hostile environment, where it must compete with sunflowers. The yield potential in Argentina is high, as 4.6% of the world's sorghum is produced on only 1.6% of the world's sorghum production area. Much shuttle breeding has been possible and beneficial to both the United States and Argentina, as the two countries share many objectives and environmental constraints. At present yield, planted area, and total production all are increasing in Argentina, although exports are no longer significant and are projected to decrease even further by 2008.

Brazil differs significantly from Argentina, with only 200,000-300,000 ha, about 10% of the Argentinean area, planted to sorghum annually from 1990-1998, and with yields stable at approximately 1.75 mt/ha. In the *campo cerrado* of central and western Brazil, an area of acid soil and aluminum toxicity, sorghum works well in rotation with soybeans or maize, as sorghum can be used to put biomass back into these fragile soils and to increase performance during the drier May-August period. Agricultural development efforts, especially in Goias, Mato Grosso do Sul, and Mato Grosso states are encouraging this cropping system. Further research is needed to develop sorghums tolerant to soil toxicities and capable of more efficient phosphorous uptake.

In 1999 in Brazil, approximately 402,000 ha were planted to sorghum, boosting production to 620,000 tons, an all-time record. Central-west Brazil (Goias, Mato Grosso do Sul, and Mato Grosso) produced 337,600 mt of sorghum in 1998/99, a 21-fold increase over 1991/92 that accounts for 56% of the country total (5). The south, formerly the principal sorghum production area, has seen sorghum production decrease by 21%, even as the southeast increased sorghum production by 71%. In the northeast, sorghum production declined from 84,500 mt to 19,100 mt over this eight-year period. The south and southeast regions have the highest yields: 2.3 mt/ha, as compared to yields of 1.4 mt/ha for the central west.

Opportunities exist for increased sorghum use, especially in poultry production, but improved cultivars and farmer education are needed. Other constraints to production in Brazil include price, which is set at 80% of maize when nutritionists consider poultry or swine rations. Grain elevators or warehouses in Brazil are reluctant to handle sorghum, as they were designed for maize, and feed processors are better equipped to grind maize than sorghum. Often farmers lack high-quality seed, in part because they use grain from a previous crop to plant their crop for the current year. Much of the sorghum produced has high tannin levels, which reduces protein digestibility. Non-tannin sorghums, such as yellow or cream hybrids, have distinct nutritional advantages.

Mexico. The leading importer of sorghum, Mexico, is becoming the third most important producer of this crop at 6.5 mmt from 2 million ha and a yield of 3.25 mt/ha. This change represents a 47% increase in area from 1989, and, when combined with an 11% yield increase, resulted in a total 67% increase in production. The Bajio area, with its high elevation (2000 meters), ample irrigation, and sufficient fertilizer, consistently produces the world's highest yields (14 mt/ha). The sorghum crop in Mexico is primarily considered animal feed, whereas maize is primarily considered human food, although a food grade sorghum is currently generating considerable interest. High-yielding, early maturing hybrids of above-average drought tolerance grow in much of the Mexican sorghum-producing areas, which often include less than favorable dryland regions. Sorghum outyields maize by 50%, but with a population expected to soon exceed 100 million and inflation largely in check, sorghum imports, now at 3.2 mmt/year, are forecast to increase to 4.2 mmt by 2008.

Australia. Although near Asian and Middle Eastern markets, Australia remains a limited player in international sorghum trade. The area planted to sorghum has increased from 380,000 ha in 1989 to the current level of 600,000 ha, and is expected to remain stable through 2008. Both biotic and abiotic stresses are common in the growing areas. Yields vary considerably, but averaged 2.16 mt/ha from 1989-1999. Research efforts have focused on developing hybrids resistant to the sorghum midge, with the accompanying increases from heterosis receiving less attention. Abiotic stress tolerance for drought, temperature, and soil toxicities also are a high priority. Kramer and Boyer (10) suggest that 93% of the yield difference between average yield and record yield in sorghum in the United States is attributable to physiological or abiotic stress losses. These factors also are important in Australia, but losses caused by midge are a more significant component than in the United States, so the percentage loss attributed to abiotic stress is probably reduced. Unfortunately, market size and quarantine restrictions have limited research input and competition in Australia. Also, a significant infestation in 1997 of sorghum ergot in the late-planted sorghum in central Queensland probably has slowed crop expansion.

Export/Import Trends

World trade in sorghum has decreased from 9 mmt in 1989 to 6.6 mmt in 1999, with a projected increase to 7.6 mmt by 2008. The United States is the largest exporter, with 82% of the world market, with Argentina, Australia, and China accounting for essentially all of the rest. Only 6% of the sorghum in world trade is for use as food. Japan, once the leading sorghum importer with 41% of global imports in 1989, is now second to Mexico, which currently accounts for 49% of global imports, as compared with 36% of global imports for Japan. Increased sorghum trade could result if identity-preserved food grade sorghum can be sold as a barley substitute in Saudi Arabia and north Africa. Also, Mexico is expected to increase its sorghum imports by 32% over the next eight years. The storability of sorghum in tropical climates is being evaluated in Morocco, the Philippines, and Mexico to obtain data to test the hypothesis that fewer microbial problems occur with sorghum than with other feed grains.

World trade in sorghum peaked at 13 mmt in 1985. Current levels are at half this value because of (i) reduced production in the United States and Argentina, (ii) narrowed export price gap with maize, and (iii) the lifting of restrictions on maize imports by some countries, *e.g.* Colombia, Mexico, and Venezuela.

An important trend in sorghum trade is decreasing imports by developed countries and increasing imports by developing countries. Sorghum is almost exclusively traded for feed, so its market price is closely related to that of other feed-quality grains, primarily maize, wheat, and barley. The prices of feed grains are generally influenced by factors such as world production, the size of carryover stocks, and the number of grain-consuming animals.

Sorghum Relative to Maize

Sorghum production is approximately 10% that of maize, wheat, or rice, which are essentially equal in terms of production worldwide (6). Sorghum, because of its stress tolerance and performance stability, can be grown over a larger and more diverse ecological area than maize. The yield advantage for maize normally occurs when available water exceeds 550-600 mm annually (11). Although good evidence exists to rate sorghum's feed value at 95% that of maize (14), cattle nutritionists and feeders are reluctant to consider it much above 92%, less if high tannin, and when steam flaking they suggest that it costs 10% more than maize (12). Test weight, seed size, and uniformity are significant concerns for livestock feeders. The narrower germplasm base of food-grade sorghums may reduce this variability, but environmental influences are likely to remain the most important factor. The price advantage of maize for the farmer is offset by a higher cost of production. Consumer preference for meat color favors maize, with its higher carotene levels for more yellow-colored meat, whereas sorghum usually results in a more white-colored meat.

Frey (7) suggested that some 545 scientist years are devoted annually to breeding maize in the United States compared with 55 for sorghum, with the sorghum figure now probably no more than half that amount due to reductions in both public and private sorghum research. The large investments in maize research have helped increase yields and develop stress-tolerant hybrids that reduce the length of the growing cycles for this primary competitor of sorghum. In the United States, maize has an average annual increase of 1.4%, while the average annual yield increase for sorghum is 0.9% (2). For sorghum to be competitive with maize in the feed grain market, the relative growth of productivity in the two crops should be similar. The prospects for maize achieving the level of drought tolerance inherent in sorghum, however, are limited.

Future Prospects

Sorghum is crucial to the world food economy, as it provides household food security in many of the world's poorest, most food-insecure regions. As geographic regions with high population densities demand a more protein-rich diet, the demand for sorghum as a feed grain is expected to increase, as indicated by the projected 28% increase in coarse grain exports by 2008.

Rising global water constraints and rising water costs should encourage an increase in the proportion of land used to grow feed crops to be planted to sorghum and the production of more grain efficient sources of animal protein, such as poultry. Improvement in nutrient-use efficiency by sorghum plants could accelerate this trend. A greater acceptance of improved cultivars in developing countries, an expansion of the area planted to hybrids, and the potential of increased heterosis suggest that significant opportunities remain for increasing yield and worldwide production. An FAO/ ICRISAT sorghum and millet publication (8) states that, "The single most important technological change in sorghum cultivation since the 1960s has been the development and use of hybrid seed." Establishment of the African Seed Trade Association reflects a much needed effort for technology transfer on this continent.

Often, sorghums are grown with less modern technology related to cultivar selection, plant population, date of planting, soil fertility, water conservation, and pest management than maize. Any of these factors can have a significant impact on both the quality and the amount of biomass produced. These cultural inputs are more significant in developing countries, where yields generally are low. In these countries, inadequate infrastructure for storage and transportation limits export to neighboring countries in need of sorghum grain. At present, some 40% of the world's food comes from the 17% of the cropland that is irrigated. As much of irrigated agriculture currently operates with water deficits, where farmers will find the additional water needed to feed the more than two billion people projected to be added to the global population by 2030 remains a critical question. A water scarcity study (9), implies that by 2025, some 25% of the world's population will experience severe water scarcity with one-third of the population, including India and China. Many countries currently have sufficient water resources but need to produce 25-100% more to meet projected demand.

Finally, an increasing trend in sorghum improvement exists for consideration of the needs of the end-user in addition to those of the producer. Efforts to find new uses for sorghum must be encouraged. In recent years, considerable worldwide interest and demonstrated use of sorghum for ethanol production suggests a potential new demand for increased sorghum production. Food grade sorghums in developed countries need to be marketed as identity-preserved grain, which could influence the use of the crop for animal feed, and as a more desirable export grain for food or animal feed. The consuming public in the developed world is increasingly in search of grains with properties such as those possessed by sorghum, *e.g.* gluten-free, which can be incorporated into antioxidant functional foods. Also, at this time with sorghum the genetically modified organism (GMO) or transgenic issue does not exist. Food security, though, remains the primary goal of efforts to improve the world sorghum economy. For many farmers, increased production will mean higher consumption, better nutrition, and the potential for income growth.

Acknowledgments

I thank Nega Wabeneh for preparing Figure 2-1.

References

1. **Anonymous.** 1995. January 1 World Crops Situation Report. Sparks Commodities, Inc., Memphis, Tennessee.
2. **Anonymous.** 1998. Analysis of average crop yield trends. *Doane's Agric. Rept.* **61(3.6)**.
3. **Anonymous.** 2000. January 1 World Crops Situation Report. Sparks Commodities, Inc., Memphis, Tennessee.
4. **Anonymous.** 2000. *World Feed Grains Forecast*. U. S. Grains Council, Washington, D.C.
5. **CONAB,** as quoted by *Agrianual 2000*. FNP, Brazil.
6. **Food and Agriculture Organization of the United Nations.** 2002. World Agricultural Information Center (WAICENT). FAOSTAT Database. FAO, Rome, Italy.

7. **Frey, K. J.** 1996. *National Plant Breeding Study I*. Iowa State University Experiment Station Special Report no. 98. Iowa State University, Ames, Iowa.
8. **International Crops Research Institute for the Semi-Arid Tropics, and Food and Agricultural Organization of the United Nations.** 1996. *The World Sorghum and Millet Economies*. Patancheru (A.P.), India and Rome, Italy.
9. **International Water Management Institute.** 1998. *Annual Report*. Colombo, Sri Lanka.
10. **Kramer, P. J., and J. S. Boyer.** 1995. *Water Relations of Plants and Soils*. Academic Press, New York City, New York.
11. **Krieg, D. L., R. J. Lascano, F. M. Rhoades, and J. M. Bennett.** 1990. Irrigation of corn and sorghum, p. 569-592 and 719-737. *In* B. A. Stewart and D. R. Nielsen (eds.), *Irrigation of Agricultural Crops*, ASA Monograph 30. Agronomy Society of America, Madison, Wisconsin.
12. **Kuhl, G., K. Dhuyvetter, T. Eck, and D. Simms.** 1995. *Increasing the Use of Grain Sorghum in Kansas Feedlot Cattle Rations: Survey of Current Feeder Attitudes of Grain Sorghum Feed Value, Cultivar Types, and Processing Methods*. Kansas State University Cooperative Extension Service, Manhattan, Kansas.
13. **Liu, H. and W. Bi.** 1993. The study on integrated technology for obtaining high sorghum production. *Agric. Econ.* Suppl. Issue 51-53.
14. **Maunder, A. B.** 2000. History of cultivar development in the United States: From "Memoirs of A. B. Maunder–sorghum breeder," p. 191-223. *In* C. W. Smith and R. A. Frederiksen (eds.), *Sorghum–Origins, History, Technology, and Production*. John Wiley and Sons, New York City, New York.

Part II

Millet Biology and Diseases

3

Diseases of Finger Millet – A Global Overview

J. Peter Esele

In the context of world agriculture, finger millet [*Eleusine coracana* (L.) Gaertn.], known as "ragi" in India, belongs to a group of crops collectively referred to as small millets. The other small millets are foxtail millet (*Setaria italica*), little millet (*Panicum miliare*), proso millet (*Panicum miliaceum*), kodo millet (*Panicum scrobiculatum*), barn yard millet (*Echinochloa colosna*), fonio (*Digitaria exilis*), and teff (*Erogrotis teffi*). Finger millet is agriculturally the most important of the small millets and also is arguably the most nutritionally valuable, but underutilized, of the cereals. It also is sometimes referred to as a traditional crop, and it plays an important role in maintaining stable and sustainable forms of agriculture. Finger millet is grown primarily in the semi-arid portions of eastern and southern Africa (especially Uganda, Kenya, Ethiopia, Tanzania, and Zimbabwe) and south Asia (especially India and Nepal). Where it is grown in Africa, the ability to store the grain for up to 10 years without significant damage was once the basis of household food security, which is controlled by women.

Finger millet is nutritionally superior to rice and wheat. It also has good malting properties, because of high amylase levels, and is rich in vitamins and minerals, especially calcium, iron, potassium, and phosphorus (Table 3-1). Thus, finger millet is a very nutritious food for pregnant women, lactating mothers, and children. Its nutritional superiority, coupled with high yield potential, make this crop a potentially valuable food crop, particularly among the resource-poor communities in the semi-arid areas. The area under production to finger millet has remained more or less stable for the last two decades at about 2.6 million hectares. Its current productivity is about 1200 kg/ha.

Like many other crops, finger millet is subject to a variety of production constraints. Some of these constraints include: lack of improved varieties, traditional production practices, pests, and diseases. Historically undeveloped in terms of commercial processing and national or international markets, finger millet has become an "orphan" crop in terms of infrastructural support and investment in research, development, and markets. The crop has not had high priority in agricultural development, especially at the international level. The International Crops Research Institute for the Semi-Arid Tropics (ICRISAT) center in Hyderabad, India, is the only international agricultural research organization that maintains finger millet germplasm. However, because of the importance of the crop in southern and eastern Africa, the ICRISAT–Southern and Eastern Africa program has begun research work with this crop. National programs, especially those in Uganda, Kenya, and Ethiopia, have strong research programs on finger millet.

Finger millet is susceptible to a wide range of diseases, the most important of which are blast, viral, and bacterial infections, Cercospora leaf diseases, tar spot, and leaf spots. For finger millet, the incorporation of genetic resistance probably is the best choice for disease management. Cultural practices, *e.g.* early planting and appropriate cropping systems, also could reduce disease incidence.

Blast

A wide range of fungal and bacterial diseases have been reported on finger millet (17). The most important of these diseases is blast caused by the fungal pathogen *Pyricularia grisea* (Cooke) Sacc. (teleomorph–*Magnaporthe grisea*). The disease is both economically significant and very destructive, causing over 50% yield losses, especially in wet seasons (5, 11). Field studies have recorded losses of 10-90% in Uganda (1, 9), 64% in Kenya (16), and near total losses in India (11). Environmental conditions of rainfall, temperature (25-30°C) and humidity (90%) are the most important predisposing factors for blast severity.

P. grisea affects finger millet at all stages of plant development, from seedling to grain formation. The

TABLE 3-1. Nutrient composition of finger millet and other cereals (per 100 g edible portion; 12% moisture).

Crop	Protein (g)	Fat (g)	Ash (g)	Crude Fiber (g)	Carbohydrate (g)	Energy (kcal)	Ca (mg)	Fe (mg)	Thiamin (mg)	Riboflavin (mg)
Finger millet	7.7	1.5	2.6	3.6	73	340	350	3.9	0.42	0.19
Rice (brown)	7.9	2.7	1.3	1.0	76	360	33	1.8	0.41	0.04
Wheat	12	2.0	1.6	2.0	71	350	30	3.5	0.41	0.10
Maize	9.2	4.6	1.2	2.8	73	360	26	2.7	0.38	0.20
Sorghum	10	3.1	1.6	2.0	71	330	25	5.4	0.38	0.10

symptoms at the seedling stage appear as small brown spots that eventually coalesce into large elongated or spindle-shaped lesions, the centers of which develop a grayish overgrowth. This overgrowth consists of conidia and conidiophores of the fungus. Severe infections may lead to seedling death. The leaf blast phase of the disease tends to be less severe.

Seedlings are more susceptible to leaf blast than are mature plants (17); however, no relationship is known between the intensity of seedling infection and that of later head infection. Rather, the prevailing weather conditions at a particular stage of crop development determine the intensity of blast infection.

Rath and Mishra (19) reported that neck infection significantly reduces grain number and grain weight and significantly increases spikelet sterility. Crop loss is greatest when the disease appears on the neck and ear during flowering and grain formation. At this stage, the disease may appear on the peduncle and/or on the fingers, causing neck and head blast, respectively (Fig. 3-1A). Depending on the severity of the infection, the disease may result in total inhibition of grain formation or the production of shriveled grain. The pathogen has a wide host range that includes many graminaceous weeds (15). *P. grisea* commonly lives in the glumes and on the straw. The pathogen also is found in the seed, but embryo infection has not been observed (26).

The pathogen is spread primarily by airborne conidia and rain splash. The initial inoculum probably comes from nearby weeds or cereal plants, which act as collateral hosts. The fungus also may persist in plant debris and in seed from infected fingers. There are reports that heavily contaminated seeds germinate poorly and that these infected plants rarely reach maturity. However, there also are reports of suspected seed transmission leading to infection foci within the field. Research in India (22) suggests that seed is a key source of inoculum and that even an extremely low seedborne incidence may initiate an epidemic.

In Uganda and Kenya, the pathogen persists in crop debris and on weeds and wild grasses, *e.g.*, *Eleusine indica*, *Eleusine africana*, *Digitaria* spp., *Setaria* spp., and *Doctylocterium* spp. (16). These hosts could serve as potential inoculum reservoirs. Some workers have suggested that the pathogenicity of a *P. grisea* culture is largely restricted to the host from which it originated, but work in Uganda (7) indicates that some isolates of blast from weeds also will infect finger millet, which implicates weeds/wild grasses as "green bridges" for finger millet blast.

Genetic and pathogenic variation within *P. grisea* from finger millet is a potential problem (16). *P. grisea* is morphologically similar to *P. oryzae* from rice, which is known to be pathogenically diverse and genetically variable. The genetic relatedness and aggressiveness of isolates infecting the different parts of either the finger millet crop or the alternative weed hosts also is unknown. This information is essential for the development of appropriate disease management strategies.

At present, chemical control and the use of host plant resistance are the primary means for control of the disease. Chemical control through spraying with mancozeb or carbendazin once at the seedling stage and twice at ear head stage reduces disease incidence. Ekwamu and Mukiibi (8) in a trial conducted in Uganda found that spraying with Benlate reduced leaf blast by 28% and head blast by 31%, and that spraying with Dithane M45 reduced leaf blast by 16% and head blast by 17%.

Development of resistant varieties is the best means of controlling blast in finger millet. Almost all cultivars are susceptible to blast, although the level of susceptibility varies by cultivar. In general, cultivars with dark seeds and/or compact heads are more resistant than those with white seeds and/or more open heads. Thus, breeding for blast resistance is possible, and a number of varieties have been developed in Uganda with good levels of resistance to blast (10). However, resistance to blast appears to be both polygenic and complex. Seetharam and Ravikumar (21) examined gene action through line × tester analysis and found both additive and nonadditive components to resistance, with the nonadditive components being more common. A five-generation mean analysis carried out in four crosses revealed all types of gene effects–additive, dominant, additive × additive, and dominant × dominant. The plant's reaction to the blast fungus may be a result of the biochemical composition of the plant, *e.g.*, total phenol and tannin contents are significantly higher in resistant cultivars than in susceptible ones (21).

Seedling and Leaf Blight

Leaf and seedling blight is caused by *Helminthosporium nodulosum* (*Dreschlera nodulosum*). This disease has been reported in India, Japan, and Africa. It also attacks other species of *Eleusine*, *e.g. E. indica* and *E. aegyptica*, and *Setaria* spp., *Panicum* spp., *Echinochlea* spp., *Sorghum* spp., and *Zea mays* (14).

The disease may occur sporadically throughout the field or in well-defined patches. It affects all parts of the plant, including the base, culms, leaf sheath, leaf blade, neck, and fingers (Fig. 3-1B). Germinating seeds may be killed before the seedling pierces the soil's surface. On seedlings, symptoms appear as minute, oval, light-brown lesions on the leaf blade. These lesions may coalesce to form large patches on the leaves, which later turn dark brown. Infected leaves wither prematurely, and seedlings may die within two weeks of infection. Lesions are formed on the leaf blade and leaf sheath of other plants. These lesions are well defined and usually are located at the junction of the blade and the sheath. Stem lesions are usually oblong. Once the tissues are infected, necrosis begins as tissues collapse, with the portions of the plant above the lesion usually dying. This process results in breaking and dropping away of leaf tips, spikelets, fingers, or even the entire head. The pathogen also can infect the neck and head, causing chaffiness and discoloration of the seed and consequent reduction in yield.

The seed is the primary source of inoculum, although the fungus also may persist on seed and plant debris. Secondary spread occurs through the air via conidia. High humidity also favors disease development.

As the disease is primarily seedborne, seed treatment with an appropriate fungicide(s) controls pre-emergence damping off and seedling blight. Secondary field infections also can be controlled by fungicidal sprays. In addition to chemical control, studies have been conducted to identify genetic resistance (26). Varieties with green glumes are more susceptible than those with purple glumes (3).

Wilt or Foot Rot

Wilt, or foot rot, is caused by *Sclerotium rolfsii* and is mainly a soilborne disease. The disease occurs in most finger millet growing areas but is more common during the rainy season. The infection occurs at the base of the plants, on the leaf sheaths, and on the culms (Fig. 3-1C), and infected plants appear pale green and stunted. The site of infection often turns brown, and the plants eventually wilt, lodge, and dry up. White mycelial growth may appear between the sheaths and the base of the stem. Sclerotia are formed on the surface of the mycelial growth as small, mustard seed-like, tan-colored bodies.

As a soilborne disease, control by chemicals is both expensive and relatively ineffective. Cultural control practices such as deep plowing before planting or crop rotation with non-graminaceous crops should minimize the disease.

FIGURE 3-1. Common diseases of finger millet. A. Finger millet blast–a single head with neck symptoms. B. Leaf blight on the leaf sheath (caused by *Helminthosporium nodulosum*). C. Foot rot (caused by *Sclerotium rolfsii*). D. Cercospora leaf spot. E. Tar spot (caused by *Phyllachra eleusines*). F. Chlorosis associated with the virus disease complex.

Downy Mildew

Downy mildew, also known as green ear disease, is caused by *Sclerophthora macrospora* (Saccardo) Shaw. The disease has been reported mainly in India, where it first occurred in Mysore in 1930 (25). Symptoms include stunting, shortened internodes, and leaves arising close together, which gives the plants a bushy, bunched appearance. The leaves are pale green, but do not shred as they do in sorghum. The disease is particularly obvious at heading, when the lemma, palea, and sometimes glumes change into leafy structures, beginning with the basal spikelets and continuing until the whole ear develops a brush-like appearance (23).

The fungus has a wide host range that, in addition to finger millet, includes *E. indica*, *Eragrotis* spp., *Digitaria marginate*, maize, rice, oats, and wheat. Physiological specialization is well developed within the species, and some races appear to be confined to a particular host.

Control measures have not been developed, nor have resistant varieties been identified. However, the disease is seedborne, so seed treatment is advisable. Rouging of infected plants also can reduce inoculum.

Smut

Smut in finger millet is caused by *Melano-psichium eleusinis*. This disease appears to be restricted to the Indian states of Maharashtra and Karnataka (17). The disease occurs at the time of grain formation when smut sori develop on the grains, the rachis, and the peduncle. The entire inflorescence commonly is small and shriveled, with the floral organs being converted into sac-like smut sori. Usually only a few sori occur per plant, and these are scattered throughout the spikelets. Floral infection occurs by windborne spores. The pathogen is neither systemic nor seedborne, and control by seed treatment is ineffective.

Cercospora Leaf Spot

This disease was initially described as *Cylindrosporium* leaf spot and has been reported primarily from Uganda (6). In 1992, pathogen samples sent from Serere to CMI (Commonwealth Mycological Institute, Kew, Surrey, U.K.) were identified as belonging to either *Cercospora* or *Gloecercospora*. The disease is common when there is excessive rainfall. Symptoms first appear about 60 days after planting and progress until maturity on the leaves and the leaf sheaths (Fig. 3-1D). Symptoms include dark-colored oval-shaped lesions, the centers of which develop a gray mycelial growth consisting of fungal conidia and conidiophores.

This disease is unknown in India, suggesting that *Cercospora* is not a pathogen of finger millet in that country (20). In Zambia, there has been mention of *Gloeocercospora* on finger millet, but apparently this disease causes little, if any, damage (J. Mulenga, personal communication).

Tarspot

This disease is caused by *Phyllachora eleusines* P. Henn. It too is known primarily from Uganda (6). The symptoms include small jet black and slightly raised spots appearing on the leaves and neck of the plant as it nears maturity (Fig. 3-1E). By itself, this disease is not economically important, but when it occurs with other leaf diseases, *e.g.* blast, *Cercospora*, and leaf blight, then premature defoliation is common.

Bacterial Diseases

Bacterial diseases of finger millet were first reported in India in 1937 (13) and attributed to *Xanthomonas coracana*. At first the foliage is pale green, but later it turns brown. Brownish patches on the stem are slimy to touch, and microscopic examination of the phloem reveals discoloration and bacterial ooze from the intercellular spaces. Desai *et al.* (4) conducted investigations on the control of bacterial blight. They advocated the use of resistant varieties and the application of streptomycin beginning at the seedling stage.

Virus Disease Complex

Finger millet is infected by a wide range of viruses. Several symptoms may appear together in the same fields or plants, giving the impression that the virus diseases are a "complex." Rao *et al.* (18) and Yaraguntaiah and Keshavainurthy (27) reported that finger millet was attacked by a "virus disease complex" comprised of Sugar Cane Mosaic Virus (SCMV), SMV, *Eleusine* Mosaic Virus (EMV), and freckled yellow and "stripe" diseases of sorghum (Fig. 3-1F). Mosaic virus symptoms were reported by McRae (12) in India, where a mottling disease resembling SMV was observed in Madras. Venkatarayan (24) found that *E. coracana* in India suffers extensively from mosaic. Both finger millet and the nearby sorghum were heavily infested by *Aphis maidis* and showed mottling symptoms. The disease caused stunting, severe mottling, chlorotic streaks and general leaf chlorosis. Rao *et al.* (18) found that a sap-transmissible virus caused elongated chlorotic streaks interspersed with light green patches on *E. coracana, E. aegyptica*, and *E. indica*, but was distinct from both

SMV and maize virus and designated as a new virus, EMV. Cherian and Kylasam (2) found symptoms of "freckled yellow" and "stripe" diseases of sorghum on finger millet. In the striped form of this disease, a few parallel, continuous, pale-green streaks are formed, with the veins and intermediate adjacent tissue becoming lighter in color. The streaks are parallel to the veins, and "bunchy top" sometimes results because of the shortened internodes. "Freckled yellow" symptoms include the formation of minute, cream-yellow specks, which are arranged in a linear manner either near or close to the margin(s) of the leaf blade. The symptoms were more pronounced on the top leaves. Both "stripe" and "freckled yellow" types of symptoms can be found on the same plant. The short bug, *Peregrinus maidis*, transmitted "freckling," but not "stripe" on *E. coracana, Sorghum, Pennisetum typhoides*, and maize.

All of these virus diseases have been observed in east Africa (6, 9), but no attempt has been made to identify or characterize them. Losses caused by these viruses can be significant, sometimes resulting in a 100% total loss of grain yield, especially when the crop is planted late in the season. In general, virus diseases are more important in India than they are in Africa.

In summary, blast is the most important disease of finger millet wherever it is grown, with the head infection phase being the most destructive. Understanding this disease will require a significant amount of research, perhaps as much as has been done with the similar disease on rice.

References

1. **Bisht, L. S.** 1987. Blast tolerance and yield losses in finger millet. *Ind. J. Agric. Sci.* **67**: 954-955.
2. **Cherian, M. C., and M. S. Kylasam.** 1937. Preliminary studies on the "freckled yellow" and "stripe" diseases of cholan. *Proc. Assoc. Econ. Biol. (10-15 April 1936, Coimbatore, India).*
3. **Coleman, L. C.** 1920. *The Cultivation of Ragi in Mysore.* Bull. Dept. Agric. Mysore, Gen. Ser. l.
4. **Desai, S. G., M. J. Thirimulachar, and M. K. Patel.** 1965. Bacterial blight disease on *Eleusine coracana* Geertn. *Ind. Phytopathol.* **18**: 139-150.
5. **Dunbar, A. R.** 1969. *The Annual Crops of Uganda.* East African Literature Bureau, Kampala, Uganda.
6. **Ekwamu, A.** 1980. *Diseases of Finger Millet [*Eleusine caracana *(L.) Gaertn.] in Uganda.* M.Sc. Thesis, Makerere Univ., Kampala, Uganda.
7. **Ekwamu, A.** 1989. Host range, morphology, and pathogenicity of the genus *Pyricularia* in Uganda. *East Africa Agric. For. J.* **54**: 101-105.
8. **Ekwamu, A., and J. Mukiibi.** 1985. Fungicidal control of fungal diseases of finger millet. *Proc. 4th EARSAM Regional Workshop (July 22-26, 1985, Soroti, Uganda)*: 100-105.
9. **Esele, J. P.** 1989. Cropping systems, production technology, pests and diseases of finger millet in Uganda, p. 293-299. *In* A. Seethram, K. W. Riley, and G. Harinarayana (eds.), *Small Millets in Global Agriculture.* Oxford and IBH, Delhi, India.
10. **Esele, J. P., and S. E. Odelle.** 1995. Progress in breeding for resistance to finger millet blast at Serere Research Station. *Proc. 8th EARSAM Regional Workshop on Sorghum and Millets (30 October - 5 November 1992, Wad Medani, Sudan)*: 93-99.
11. **McRae, W.** 1922. Report of the Imperial Mycologist 1921-1922, p. 73-76. *Sci. Rept. Agric. Res. Inst. (Pusa, India).*
12. **McRae, W.** 1929. India: New diseases reported during 1928. *Intl. Bull. Pl. Protect.* **3**: 21-22.
13. **Mehta, P., and S. Chakravarty.** 1937. A new disease of *Eleusine oaracana. Mysore Agric. J.* **28**: 783-790.
14. **Mitra, M.** 1931. Report of the Imperial Mycologist 1929-1930, p. 59-71. *Sci. Rept. Imperial Agric. Res. Inst. (Pusa, India).*
15. **Ou, S. H.** 1985. *Rice diseases.* Commonwealth Mycological Institute, Kew, Survey, UK.
16. **Pande, S., S. Z. Mukuru, S. B. King, and R. Karunakar.** 1995. Biology of and resistance to finger millet blast in Kenya and Uganda. *Proc. 8th EARSAM Regional Workshop on Sorghum and Millets (30 October - 5 November 1992, Wad Medani, Sudan)*: 83-92.
17. **Rachie, O. K., and V. L. Peters.** 1977. *Tahe* Eleusines*: A Review of the World Literature.* ICRISAT, Patancheru (A.P.), India.
18. **Rao, D. G., P. M. Varfna, and S. P. Kapoor.** 1965. Studies of mosaic disease *Eleusine* in the Deccan. *Ind. Phytopathol.* **18**: 139-150.
19. **Rath, G. C., and D. Mishra.** 1975. Nature of losses due to neck blast infection in Ragi. *Sci. Cult.* **41**: 322-323.
20. **Reddy, C.** 1975. *Studies on the Phyllosphere Microflora of Ragi [*Eleusine coracana *(L.) Gaertn.].* M.Sc. Thesis, Andhra Pradesh Agric. Univ., Hyderabad (A.P.), India.
21. **Seetharam, A., and R. L. Ravikumar.** 1993. Blast resistance in finger millet. Its inheritance and biochemical nature. *In* K. W. Riley, S. C. Gupta, A. Seetharain, and J. N. Mushonga (eds.), *Advances in Small Millets.* New York International Science Publisher, New York City, New York.
22. **Shetty, H. S., A. Gopinath, and K. Rajashekar.** 1985. Relationship of seedborne inoculum of *Pyricularia grisea* to incidence of blast of finger millet in the field. *Ind. Phytopathol.* **38**: 154-156.
23. **Thirumalachar, M. J., and M. J. Narasimhan.** 1949. Downy mildew of *Eleusine coracana* and *Iseilema lexum* in Mysore. *Ind. Phytopathol.* **2**: 47-51.
24. **Venkatarayan, S. V.** 1946. Mosaic disease of ragi (*Eleusine coracana*). *Curr. Sci.* **15**: 258-259.
25. **Venkatarayan, S. V.** 1947. Diseases of ragi (*Eleusine caracan*). *Mysore Agric. J.* **24**: 50-57.

26. **Viswanath, S., and A. Seetharam.** 1989. Diseases of small millets and their management in India, p. 237-253. *In* A. Seethram, K. W. Riley, and G. Harinarayana (eds.), *Small Millets in Global Agriculture*. Oxford and IBH, Delhi, India.

27. **Yaraguntaiah, R. C., and K. V. Keshavainurthy.** 1969. Transmission of the virus component of the ragi disease complex in Mysore. *Plant Dis. Reptr.* **53**: 361-363.

4

Gene Management and Breeding for Downy Mildew Resistance

C. Tom Hash and John R. Witcombe

Pearl millet [*Pennisetum glaucum* (L.) R. Br.] is now being used in numerous agricultural production systems in Africa, the Americas, Asia, and Oceania (26). This crop usually is cultivated as a subsistence food, feed, fodder, and/or fuel crop in parts of Africa and Asia that are too hot for, too dry for, and/or have soil constraints that prevent economic production of other staple food grain crops (8). Pearl millet breeding programs that target these regions must select for host plant resistance to pearl millet downy mildew, caused by the Oomycete *Sclerospora graminicola* (Sacc.) J. Schröt. (18, 33, 38, 51). Downy mildew resistance is essential to prevent losses that would otherwise negate any productivity gains that might result from improved agronomic practices or genetic enhancement for other traits.

In the agriculturally marginal areas, where most pearl millet is grown, yield potential usually is not the major factor limiting pearl millet production. Downy mildew resistance can relieve some existing constraints to food production at regional and local levels in many pearl millet production areas (8). These constraints are particularly severe in agriculturally marginal areas where human population growth is rapid, *e.g.* western, central, and southern Africa, and southern Asia (5, 7, 46). Thus, breeding to minimize downy mildew losses and reduce the vulnerability of high-yielding, well-adapted pearl millet cultivars should be at least as important a goal as increasing productivity *per se*.

Our objective in this chapter is to review the status of breeding for resistance to pearl millet downy mildew. We discuss conventional resistance sources, screening methods, resistance inheritance, and appropriate breeding schemes, but focus on a range of nonconventional hybrid types–alternatives to composites, synthetics, and single-cross hybrids–that could be more widely exploited to facilitate access to improved seed and extend the use-ful life of productive pearl millet genotypes. At least some of these nonconventional hybrid types should be attractive enough for private sector seed production and distribution, thereby facilitating wider access to improved pearl millet seed with downy mildew resistance made more durable by inherent heterogeneity for disease resistance genes. Although we focus on pearl millet downy mildew, the nonconventional hybrids we describe also should be effective in slowing the evolution of other variable or potentially variable pearl millet pathogens (23), *e.g.* rust caused by *Puccinia substriata* var. *indica*, or leaf spots and leaf blights caused by *Pyri-ularia grisea* and *Bipolaris* spp. The concepts we present also could be used to manage insect and disease resistances, including those conferred by transgenes, in any crop where F_1 hybrids are practical.

Resistance Sources

Sources of host plant resistance to pearl millet downy mildew have been reviewed recently (14, 15), and a catalog assembled of elite inbred pollinators and germplasm accessions, both cultivated and wild, with resistance to pearl millet downy mildew (45). This catalog also contains data on agronomic characteristics of most of the resistance sources listed. Additional breeding materials that are not included in this catalog, but that could be used as downy mildew resistance sources, include elite open-pollinated cultivars, *e.g.*, ICMS 7703, SOSAT-C88, and WC-C75; improved populations; topcross pollinators, *e.g.*, ICMR 312 (49, 61); and elite seed parent maintainer lines, *e.g.*, ICMB 88004 (29) and ICMB 841 (40), and sources of multiple disease resistance (52) that have been bred specifically for downy mildew resistance effective against one or more populations of the pathogen.

Screening Methods

The ability to detect genetic differences in host plant resistance to pearl millet downy mildew has improved tremendously in the last 30 years (9, 19, 33, 42, 45, 54), resulting in much better estimates of the heritability of disease reactions. The first improvements resulted when the role of sporangia in the epidemiology of this disease was recognized (43), which led to improved field screening procedures (55) that were later simplified (45) to reduce labor. Identification of environmental factors that alter the production and infectivity of sporangia and zoospores (19, 36, 37) enabled development (35) and refinement (19, 42, 54) of a greenhouse seedling screen. Data obtained with this screen result in higher heritability estimates of downy mildew resistance and can be used to detect Mendelian segregation of resistance and susceptibility (C. T. Hash and S. D. Singh, *unpublished*) and to map major components of these segregating resistance gene complements (2, 3, 12, 13, 20, 24).

Molecular markers flanking the genomic regions associated with major components of resistance to a particular pathogen population can then be used to select for resistance in a marker-assisted breeding program (3, 11, 16, 17, 31), with or without further greenhouse disease screens to supplement marker-based selection. Until highly polymorphic, well-distributed PCR-based markers are available for pearl millet, alternating generations of RFLP-based marker-assisted selection with greenhouse screening against appropriate pathogen isolates appears to be a time- and cost-effective strategy for screening segregating materials in an applied back-crossing program.

Inheritance of Resistance

Since the 1988 global meeting on sorghum and millet diseases in Zimbabwe, our understanding of the inheritance of host plant resistance to pearl millet downy mildew has improved dramatically. These new findings can be grouped into two complementary groups: results from conventional Mendelian and quantitative genetic studies, and those from quantitative trait locus (QTL) mapping studies.

Conventional Mendelian and quantitative genetics of pearl millet downy mildew resistance have been reviewed recently (14, 15, 33, 38), so we note only work that has not been summarized elsewhere. There remains no evidence for a direct role of cytoplasmic male sterility (CMS) in pearl millet downy mildew reactions (25, 62-64). Clear evidence for both major gene (2, 6, 24, 31, 42, Hash and S. D. Singh, *unpublished*) and quantitative (21, 45) inheritance of pathogen-population-specific host plant resistance is now available. From the limited published information on mapped pearl millet downy mildew resistance QTLs (3, 12-14, 20), it is clear that this technique will greatly facilitate breeding efforts, including the pyramiding of resistance genes in elite inbred lines (14, 15, 60). However, the utility of mapped QTLs in population improvement and the breeding of conventional pearl millet composites is not yet known.

Breeding Methods

The most appropriate breeding scheme(s) for increasing resistance to pearl millet downy mildew depends on many factors (10). These factors include: (i) target trait, *e.g.* available genetic variability, screening methods, and heritability); (ii) target environment, *e.g.* soil characteristics, moisture availability, temperature regime, management intensity, and quarantine restrictions; (iii) the target cultivar type, *e.g.* open-pollinated or hybrid; (iv) the human, physical and financial resources available; (v) alternative sources of improved genotypes; (vi) the urgency with which genetic improvement is to be pursued (32); and/or (vii) the regulatory framework within which new cultivars are made available to farmers (53).

Conventional procedures used to breed improved downy mildew resistance in pearl millet have been reviewed recently in detail (14, 15). Pure line selection (39-41, 44) and mutation breeding have been successful but are no longer the methods of choice. Pedigree selection, recurrent selection, and backcrossing, however, can all be effective when a well-characterized resistance source is used as one of the parents, seedlings are screened against existing target pathogen populations, and the resistance of the survivors of the seedling screens is confirmed in the field.

Marker-assisted selection (65) for specific downy mildew resistances can now be incorporated into an otherwise conventional pedigree breeding program (17). This technique's greatest potential (60) appears to be for pyramiding resistances in the genetic backgrounds of proven elite hybrid parental lines (11, 16, 17, 31).

Conventional Pearl Millet Cultivar Types

A wide array of conventional pearl millet cultivar types has been grown commercially (10). These cultivar types include those that are essentially self-replicating if sown in isolation from other pearl millet genotypes, usually various types of open-pollinated cultivars, and those cultivar types that are not self-replicating, usually various types of hybrids. For hybrids, seed must be purchased for each planting if the genetic potential of the seed sown is not to deteriorate drastically (assuming that the hybrid selected is well-adapted to its production environment). Pearl millet landrace cultivars produced by farmers in Africa and Asia are open-pollinated, as are "improved" crop cultivar classes such as composites, *e.g.*, BN-1, BN-2, and BRS 1501 in Brazil; ICMV 221 in Eritrea and Kenya; WC-C75 and ICTP 8203 in India; Kaufela and Lubasi in Zambia; Okashana 1 in Namibia, Botswana, and Malawi; and PMV 2 in Zimbabwe; and synthetics ICMS 7703 and Mallikarjuna in India, and ICMS 7704 in Pakistan. Pearl millet hybrids initially were based on chance crosses that occur when mixtures of open-pollinated varieties or inbred lines are sown in isolation, *e.g.*, the forage hybrid Gahi 1, but these hybrids are now produced by using CMS systems (22, 30).

Private-sector seed producers prefer CMS-based single-cross hybrids over other conventional pearl millet cultivar types because they:

- Provide a simple method for large-scale multiplication of "genetically pure" products with the highest possible theoretical yield potential.
- Facilitate the control of proprietary intellectual property, *i.e.*, hybrid's parental lines, which is necessary to reduce the potential of price competition among companies having similar products.
- Provide a high rate of return business from satisfied customers.
- Provide a marketing tool–"pure seed quality"–that can be readily monitored by the farmer and his neighbors.

Unfortunately, this package comes with built-in genetic uniformity for resistance gene complements that makes single-cross hybrids especially vulnerable to new strains of variable pathogens, including the causal agents of pearl millet downy mildew and rust.

Since the late 1960s, CMS-based single-cross hybrids have been the most widely commercialized improved pearl millet cultivars in both the United States and India. In nearly all cases in both the United States and in India, the disease resistance in popular single-cross pearl millet hybrids failed before the seed of agronomically superior successor hybrids became widely available (9, 51, 57). These failures occurred when local populations of pearl millet pathogens capable of rapid host-directed evolution, *e.g.*, *S. graminicola* in India and *Puccinia substriata* var. *indica* in the United States, adapted to individual genetically uniform single-cross hybrids, *e.g.*, HB 3, BJ 104, MBH 110, MLBH 104, GK 1004, PG 5822, and ICMH 451 in India, and Tifleaf 2 and HMG 100 in the United States, allowing inoculum levels to build up. When a uniform host genotype and the availability of sufficient inoculum are combined with climatic conditions favorable for an epiphytotic, the resulting pearl millet disease epidemics can be locally devastating (9, 33, 51, 57). Interestingly, widely cultivated genetically heterogeneous improved open-pollinated pearl millet varieties that were bred for increased downy mildew resistance, *e.g.*, WC-C75 and ICTP 8203, have not faced such difficulties (9, 51).

This observation has resulted in considerable effort by public-sector breeding programs in Africa, Asia, and the Americas to develop hybrids that are genetically heterogeneous for their disease resistance gene complements in the hope of extending the useful lifespan of both the hybrids and the disease resistance genes that they carry (48). Genetically heterogeneous hybrids, rather than technically simpler open-pollinated cultivars, were emphasized because private-sector seed producers need to use hybrids to control access to the parentage of improved seeds in the absence of enforceable plant variety protection regulations. Private-sector seed producers are critically important in seed distribution of improved cultivars of highly cross-pollinated crops such as pearl millet (53). Thus, the public-sector emphasis on developing alternative hybrid types, rather than further development of composites and synthetics, that have proven durable resistance, appears justified.

Alternative Pearl Millet Hybrid Types

There are a large number of pearl millet hybrid types that can be considered as alternatives to conventional single crosses. In protogyny-based hybrids, cross-pollination results from the use of a pollen-fertile female inbred line that has exceptionally uniform flowering, and a long protogynous period, *i.e.*, the interval between initial stigma exertion and initial anther extrusion. In CMS-based hybrids, cross fertilization is effected by use of a male-sterile female parent. Protogyny-based pearl millet hybrids were originally conceived as a means to reduce the vulnerability of CMS-based pearl millet hybrids to the grain replacement diseases smut [caused by *Moesziomyces penicillariae* (Bref.) K. Vànky] and ergot (caused by *Claviceps fusiformis* Loveless). They include both line × line hybrids and line × variety hybrids (1). Unfortunately, seed production of protogyny-based hybrids requires very precise agronomic practices to avoid harvesting an

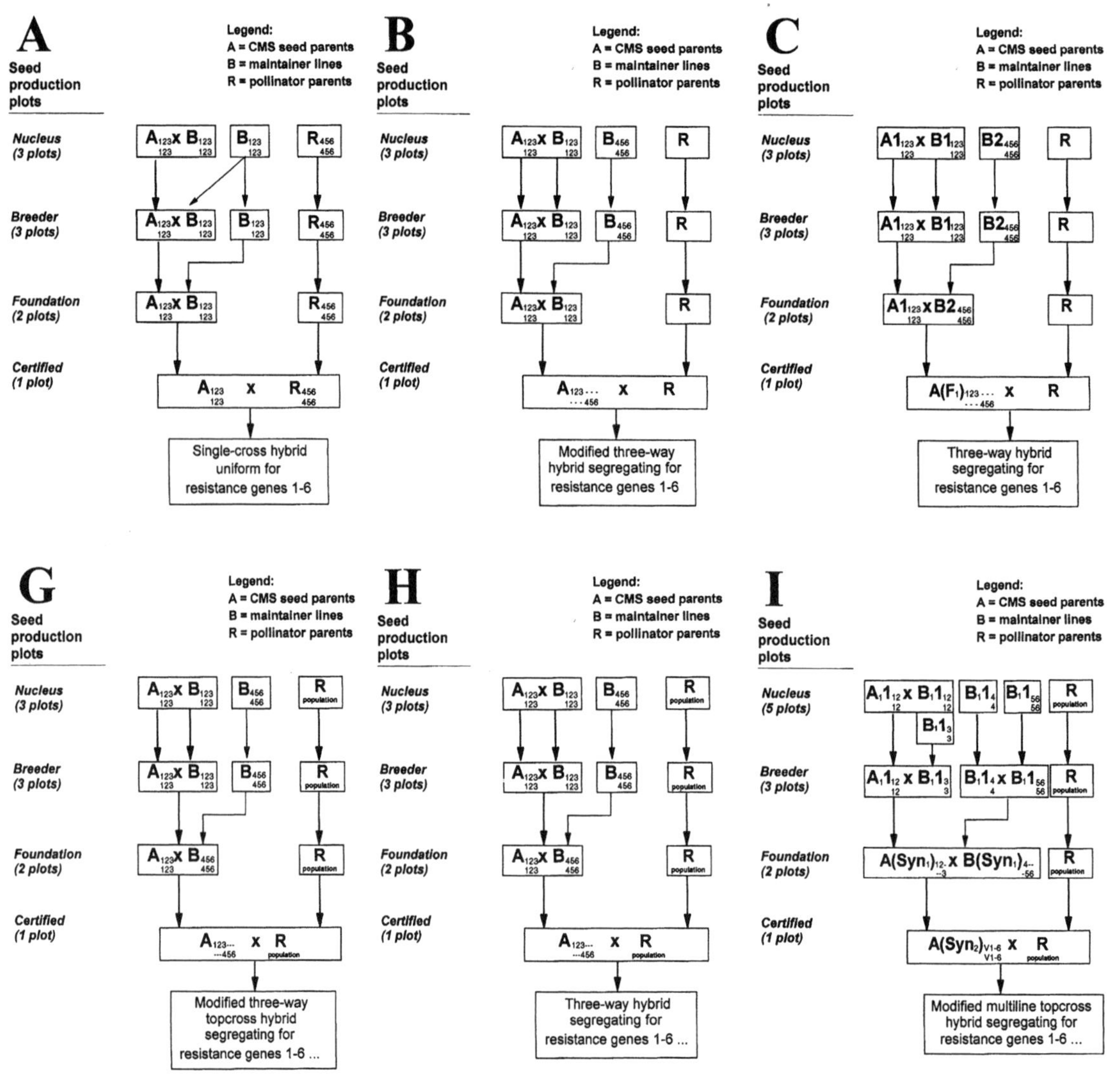
A
Legend:
A = CMS seed parents
B = maintainer lines
R = pollinator parents
Seed production plots
Nucleus (3 plots)
Breeder (3 plots)
Foundation (2 plots)
Certified (1 plot)
Single-cross hybrid uniform for resistance genes 1-6
B
Modified three-way hybrid segregating for resistance genes 1-6
C
Three-way hybrid segregating for resistance genes 1-6
G
Modified three-way topcross hybrid segregating for resistance genes 1-6 ...
H
Three-way hybrid segregating for resistance genes 1-6 ...
I
Nucleus (5 plots)
Modified multiline topcross hybrid segregating for resistance genes 1-6 ...

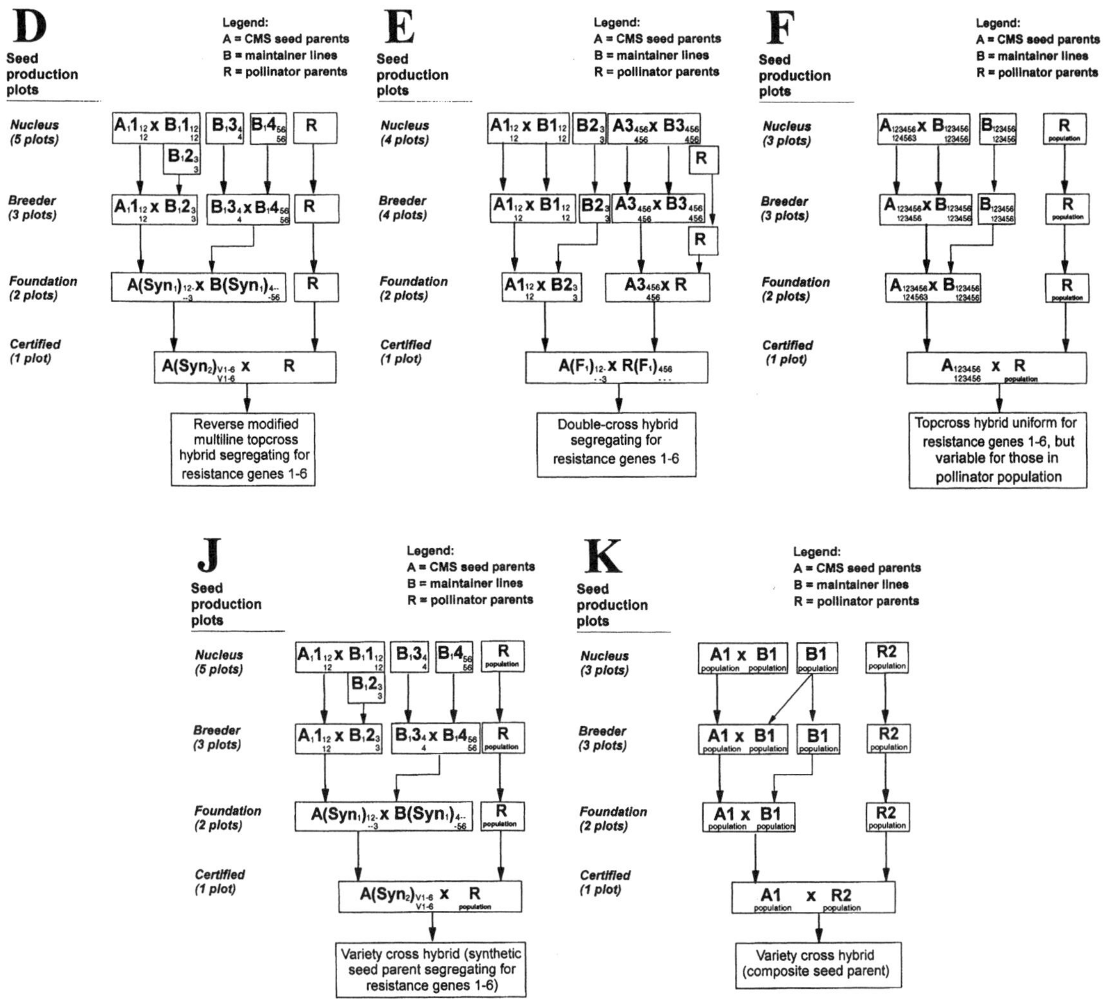

FIGURE 4-1. Cytoplasmic male-sterility based hybrid seed production schemes for genetically homogeneous conventional single-cross hybrids (A), and a range (B-K) of alternative, genetically heterogeneous, hybrid cultivar types for which seed multiplication might be practical. The latter include modified three-way hybrids (B), three-way hybrids (C), reverse modified multiline topcross hybrids (D), double-cross hybrids that require parental A-lines having two different CMS systems (E), topcross hybrids (F), modified three-way top-cross hybrids (G), three-way top-cross hybrids (H), modified multiline topcross hybrids (I), variety cross hybrids with a synthetic seed parent (J), and variety cross hybrids with a composite seed parent (K).

TABLE 4-1. CMS-based pearl millet hybrid types based on the genetic background of their male and female parents; the conventional single-cross hybrid cultivar type is in lower right corner.

Female parent	Male parent				
	Composite population or synthetic population with unrelated parents	Synthetic population with NIL parents	F_1 hybrid		Inbred
			Unrelated parents	NIL parents	
Composite population or synthetic population with unrelated parents	Variety cross hybrid[a]	Reverse modified multiline topcross hybrid	Reverse 3-way topcross hybrid	Reverse modified 3-way topcross hybrid	Reverse topcross hybrid
Synthetic population with NIL (Near Isogenic Line) parents	Modified multiline topcross hybrid[a]	Modified multiline topcross hybrid	Modified multiline 3-way hybrid	Reverse modified 3-way modified multiline hybrid	Reverse modified multiline topcross hybrid[a]
F_1 hybrid with unrelated parents	3-way topcross hybrid[a]	3-way modified multiline topcross hybrid	Double-cross hybrid[a]	3-way modified multiline hybrid	3-way hybrid[a]
F_1 hybrid with NIL parents	Modified 3-way topcross hybrid[a]	Modified 3-way modified multiline topcross hybrid	Modified reverse 3-way hybrid	Modified double-cross hybrid	Modified 3-way hybrid[a]
Inbred line	Topcross hybrid[a]	Reverse modified multiline topcross hybrid	Reverse 3-way hybrid	Reverse modified 3-way hybrid	Single-cross hybrid[a]

[a]Seed production scheme diagrams given in Fig. 4-1.

unacceptably large proportion of self- or sib-pollinated seed from the seed production plot. This problem probably means that commercial multiplication of protogyny-based pearl millet hybrids will not occur unless standards for the acceptable frequency of such seed parent selfs in the seed sold are relaxed considerably.

The range of CMS-based alternative hybrid types for pearl millet is much wider (Table 4-1). The agronomy of their seed production should be much simpler than that of protogyny-based hybrids and comparable to that of conventional CMS-based single-cross hybrids. These hybrid types include those produced on CMS populations having as their maintainers either random-mating populations (27) or synthetics (56, 58, 60), and those hybrids produced on a conventional inbred male-sterile line (A-line) having as its maintainer (B-line) a counterpart B-line with an essentially identical nuclear genotype. Such hybrids also can be produced on CMS F_1 hybrids, which have as their parents an inbred male-sterile line (A-line) and either a genetically distinct inbred maintainer line (28) or a near-isogenic maintainer inbred differing from the A-line only in its cytoplasm and its disease resistance gene complement (60). Similarly, pollinators for these alternative hybrid cultivar types can be populations, F_1 hybrids, or inbred lines (Table 4-1, Fig. 4-1). Variations on the theme of dynamic multiline hybrids (56, 58) correspond to the second column and third row of hybrids in Table 4-1.

Witcombe and Hash (60) established an analytical framework for predicting resistance gene frequency distributions in three-way hybrids, modified three-way hybrids, and their CMS F_1 seed parents produced by crossing an A-line with a B-line having a complementary resistance gene complement. They also made similar calculations for synthetic seed parents and seed parents in which the maintainer is a multiline. All of these hybrid types represent attempts to provide analogs to multiline cultivars in self-pollinated crops (4), while retaining a reasonably practical hybrid seed production scheme. The three-way hybrids, or modified three-way hybrids if the A-lines and B-lines used are near-isogenic except for their resistance gene complements, will segregate for resistance at each of the loci for which their A-lines and B-lines differ. For hybrids based on more genetically complex synthetic seed parents, *e.g.*, dynamic multiline hybrids (56, 58), the route to the heterogeneous hybrid is different, but the resulting resistance gene segregations and frequencies are similar (60). If the nuclear seed stocks of the seed parent maintainer differ at four or more resistance genes, the resistance gene complement of such heterogeneous hybrids should approximate that of an open-pollinated variety with a reasonably high resistance gene frequency.

The seed production schemes (Fig. 4-1) are very similar to those for conventional single-cross hybrids, and differ markedly only in the seed supplied for foundation seed production, which is relatively easy to supervise and control. If the near-isogenic A- and B-lines used to sow the foundation seed production plot for a modified three-way hybrid carry complementary pyramids of resistance genes, then the commercially marketed seed will produce a population of plants that is agronomically similar to a conventional single-cross hybrid, but with heterogeneous disease resistance that is not as easily broken as the uniform resistance complement in a conventional single-cross hybrid. The net effect should be delayed development of epidemics, at least for those caused by pathogens for which host plant resistance gene complements are heterogeneous, and longer useful lifespans for hybrids and for the resistance genes in their parental resistance gene pyramids. The resistance gene complements of the A- and B-lines used also can be changed from season to season, or location to location, which should further reduce widespread selection of pathogen races that could knock an economically important background genotype out of the market.

The USDA-ARS/University of Georgia team working to overcome pearl millet rust has already begun to test and deploy heterogeneous hybrid pearl millets (*e.g.*, 50, 56, 58, 59). The DFID-PSP/ICRISAT/CCSHAU (Department for International Development-Plant Sciences Research Program/International Crops Research Institute for the Semi-Arid Tropics/Chaudhary Charan Singh Haryana Agricultural University) team is just completing demonstrations of RFLP-based marker-assisted selection programs to move additional downy mildew resistance genes into the parental inbred lines of hybrid HHB 67 (= 843A × H 77/833-2), the most widely grown public-bred pearl millet hybrid in India. We have versions of 843A and 843B (47) that are homozygous for two or more resistance genes (Hash and Singh, *unpublished*) from donor ICML 22 (34), as well as donors ICMP 85410 and P 7-3 (Kolesnikova-Allen and Hash, *unpublished*). We also have versions of pollinator line H 77/833-2 into which two additional downy mildew resistance genes have been introgressed from ICMP 451 (31). We are now evaluating hybrid combinations of the new versions of these parental lines for agronomic performance and disease resistance, and have begun larger-scale seed multiplication with the goal of demonstrating the utility of this approach for extending the useful lifespan of the popular HHB 67 hybrid within the next three to five years.

Areas for Future Work

Future work is required to effectively utilize these new pearl millet hybrid types. Areas needing particular attention include mapping additional downy mildew resistance genes (3) and better characterization of virulence gene variation within and between populations of *S. graminicola* (51). This more basic research is required to identify the most effective arrays of host resistance genes for deployment in a target region. The durability of these heterogeneous hybrids must be evaluated. It also would be desirable to compare this strategy with other resistance gene deployment strategies, *e.g.*, deployment as a genetically uniform single-cross hybrid with the full resistance gene pyramid, or serial deployment as conventional single-cross hybrids each having one (or $n + 1$) of the resistance genes of the pyramid. Increasing the number of polymorphic PCR-compatible markers (3) also would facilitate the marker-assisted selection process that is used to develop the parental materials for many of these alternative hybrid types.

Conclusions

We think that a range of nonconventional cultivar types, alternatives to more conventional composites, synthetics and single-cross hybrids, should be more widely exploited to extend the useful life of productive pearl millet genotypes. These nonconventional hybrids probably will be most useful in Africa and Asia, where pearl millet is grown primarily as a staple source of food, feed, fodder, and fuel by risk-averse farmers with limited purchasing power. We think that one, or more, of these alternative pearl millet hybrid types will provide sufficient incentive for private sector seed producers to increase the availability of improved pearl millet seed to farmers in these marginal agricultural production environments. These types of hybrids also could be useful in other crops and/or in more productive agricultural environments, where biotic constraints to crop production are sufficiently important. For example, in sorghum, such heterogeneous hybrids might prove useful in extending the lifespan of host plant resistance to anthracnose (caused by *Colletotrichum graminicola*) and *Striga hermonthica.*

Acknowledgements

This work was supported in part by a research grant from the Department for International Development (DFID) Plant Sciences Programme (work at the University of Wales and at ICRISAT), and by unrestricted core funding to ICRISAT from a number of donors through the Consultative Group on International Agricultural Research (CGIAR). The opinions expressed are not necessarily those of the CGIAR, DFID, or ICRISAT. We thank the pearl millet downy mildew research and development groups in the United Kingdom, at ICRISAT, and in the public and private sectors of national agricultural research systems in pearl millet producing countries of Asia and Africa, without whose contributions the advances reported in this chapter would not have been possible.

References

1. **Andrews, D. J., B. Kiula, and J. F. Rajewski.** 1993. The use of protogyny to make hybrids in pearl millet, p. 208-210. *In* J. Janick and J. Simon (eds.), *New Crops*. John Wiley & Sons, New York City, New York.
2. **Azhaguvel, P.** 2001. *Linkage Map Construction and Identification of QTLs for Downy Mildew (*Sclerospora graminicola*) Resistance in Pearl Millet (*Pennisetum glaucum *(L.) R. Br.)*. Ph.D. Thesis, Tamil Nadu Agric. Univ., Madurai, Tamil Nadu, India.
3. **Breese, W. A., C. T. Hash, K. M. Devos, and C. J. Howarth.** 2002. Pearl millet genomics and breeding for resistance to downy mildew, p. 243-246. *In* J. F. Leslie (ed.), *Sorghum and Millets Diseases 2000*. Iowa State Press, Ames, Iowa.
4. **Browning, J. A., and K. J. Frey.** 1969. Multiline cultivars as a means of disease control. *Annu. Rev. Phytopathol.* **7:** 355-382.
5. **Clarke, A. E.** 2000. Ensuring enough to eat. *Science* **289:** 555.
6. **Deswal, D. P., and O. P. Govila.** 1994. Genetics of resistance to downy mildew (*Sclerospora graminicola*) in pearl millet (*Pennisetum glaucum*). *Ind. J. Agric. Sci.* **64:** 661-663.
7. **Evans, L. T.** 1998. *Feeding the Ten Billion: Plants and Population Growth*. Cambridge University Press, Cambridge, U.K.
8. **FAO and ICRISAT.** 1996. *The World Sorghum and Millet Economies: Facts, Trends and Outlook*. Food and Agricultural Organization of the United Nations, Rome, Italy; ICRISAT, Patancheru (A.P.), India.
9. **Hash, C. T.** 1997. Research on downy mildew of pearl millet, p. 121-128. *In* M. C. S. Bantilan and P. K. Joshi (eds.), *Integrating Res. Evaluation Efforts: Proc. Intl. Workshop (14-16 December 1994, Patancheru, India)*. ICRISAT, Patancheru (A.P.), India.
10. **Hash, C. T.** 1999. Pearl millet breeding, p. 13-30. *In* A. Lopes de Farias-Neto, R. Fernando-Amabile, D. A. Martins-Netto, T. Yamashita, and H. Gocho (eds.), *Proc. Intl. Pearl Millet Workshop (9-10 June 1999, Planaltina, Brazil)*. EMBRAPA Cerrados, Planaltina, D.F., Brazil.
11. **Hash, C. T.** 2000. Concepts for application of marker techniques in Africa, p. 155-168. *In* B. I. G. Haussmann, H. H. Geiger, D. E. Hess, C. T. Hash, and P. Bramel-Cox (eds.), *Application of Molecular Markers in Plant Breeding: Training Manual for a Seminar Held at IITA (16-17 August 1999, Ibadan, Nigeria)*. ICRISAT, Patancheru (A.P.) India. http://www.icrisat.org/text/ research/grep/homepage/mol/S3_1Hash.pdf (last checked March, 2002).
12. **Hash, C. T., and P. J. Bramel-Cox.** 2000. Marker applications in pearl millet, p. 113-127. *In* B. I. G. Haussmann, H. H. Geiger, D. E. Hess, C. T. Hash, and P. Bramel-Cox (eds.), *Application of Molecular Markers in Plant Breeding: Training Manual for a Seminar Held at IITA (16-17 August 1999, Ibadan, Nigeria)*. ICRISAT, Patancheru (A.P.) India. http://www.icrisat.org/text/research/grep/homepage/mol/S2_5Hash.pdf (last checked March, 2002).
13. **Hash, C. T., G. P. Cavan, F. R. Bidinger, C. J. Howarth, and S. D. Singh.** 1995. Downy mildew resistance QTLs from a seedling heat tolerance mapping population. *Intl. Sorg. Millets Newsl.* **36:** 66-67.
14. **Hash, C. T., S. D. Singh, R. P. Thakur, and B. S. Talukdar.** 1999. Breeding for disease resistance, p. 337-379. *In* I. S. Khairwal, K. N. Rai, D. J. Andrews, and G. Harinarayana (eds.), *Pearl Millet Breeding*. Oxford & IBH, New Delhi, India.
15. **Hash, C. T., J. R. Witcombe, R. P. Thakur, S. K. Bhatnagar, S. D., Singh, and J. P. Wilson.** 1997. Breeding for pearl millet disease resistance. *Proc. Intl. Conf. on Genetic Improve. of Sorghum and Pearl Millet (22-27 September 1996, Lubbock, Texas)*: 337-372.
16. **Hash, C. T., R. S. Yadav, G. P. Cavan, C. J. Howarth, H. Liu, X. Qi, A. Sharma, M. A. Kolesnikova-Allen, F. R. Bidinger, and J. R. Witcombe.** 2000. Marker-assisted backcrossing to improve drought tolerance in pearl millet, p. 114-119. *In* J. M. Ribaut and D. Poland (eds.), *Molecular Approaches for the Genetic Improvement of Cereals for Stable Production in Water-Limited Environments. A Strategic Planning Workshop Held at CIMMYT (21-25 June 1999, El Batan, Mexico)*. CIMMYT, El Batan, D.F., Mexico.
17. **Haussmann, B. I. G., H. H. Geiger, D. E. Hess, C. T. Hash, and P. J. Bramel-Cox, eds.** 2000. *Application of Molecular Markers in Plant Breeding: Training Manual for a Seminar Held at IITA (16-17 Aug. 1999, Ibadan, Nigeria)*. ICRISAT, Patancheru (A.P.), India. http://www.icrisat.org/text/research/grep/homepage/mol/molecular.htm (last checked March, 2002).
18. **Hess, D. E., R. P. Thakur, C. T. Hash, P. Sérémé, and C. W. Magill.** 2002. Pearl millet downy mildew: Problems and control strategies for a new millennium, p. 37-42. *In* J. F. Leslie (ed.), *Sorghum and Millets Diseases*. Iowa State Press, Ames, Iowa.

19. **Jones, E. S., W. A. Breese, and D. S. Shaw.** 2001. Inoculation of pearl millet with the downy mildew fungus, *Sclerospora graminicola*: Chilling of sporangial inoculum to delay zoospore release and avoid spray damage to zoospores. *Plant Pathol.* **50:** 310-316.
20. **Jones, E. S., C. J. Liu, M. D. Gale, C. T. Hash, and J. R. Witcombe.** 1995. Mapping quantitative trait loci for downy mildew resistance in pearl millet. *Theor. Appl. Genet.* **91:** 448-456.
21. **Kataria, R. P., H. P. Yadav, C. R. Beniwal, and M. S. Narwal.** 1994. Genetics of incidence of pearl millet downy mildew (*Sclerospora graminicola*) in pearl millet (*Pennisetum glaucum*). *Ind. J. Agric. Sci.* **64:** 664-666.
22. **Khairwal, I. S., C. Ram, and A. K. Chhabra.** 1990. *Pearl Millet Seed Production and Technology.* R. J. Manohar Publishers, New Delhi, India.
23. **King, S. B., and R. P. Thakur.** 1995. Variable pathogens of millets, p. 25-34. *In* J. F. Leslie and R.A. Frederiksen (eds.), *Disease Analysis through Genetics and Biotechnology: Interdisciplinary Bridges to Improved Sorghum and Millet Crops.* Iowa State University Press, Ames, Iowa.
24. **Kolesnikova, M. A.** 2001. *Mapping New Quantitative Trait Loci (QTL) for Downy Mildew Resistance in Pearl Millet.* Ph.D. thesis. Russian Acad. Sci., Moscow, Russia.
25. **Kumar, K. A., R. P. Jain, and S. D. Singh.** 1983. Downy mildew reactions in pearl millet lines with and without cytoplasmic male sterility. *Plant Dis.* **67:** 663-665.
26. **Lopes de Farias-Neto, A., R. Fernando-Amabile, D. A. Martins-Netto, T. Yamashita, and H. Gocho, eds.** 1999. *Proc. of the Intl. Pearl Millet Workshop (9-10 June 1999, Planaltina, Brazil).* EMBRAPA Cerrados, Planaltina, D.F., Brazil.
27. **Rai, K. N., D. J. Andrews, and A. S. Rao.** 2000. Feasibility of breeding male-sterile populations for use in developing inter-population hybrids of pearl millet. *Plant Breed.* **119:** 335-339.
28. **Rai, K. N., S. Chandra, and A. S. Rao.** 2000. Potential advantages of male-sterile F_1 hybrids for use as seed parents of three-way hybrids in pearl millet. *Field Crops Res.* **68:** 173-181.
29. **Rai, K. N., A. S. Rao, and C. T. Hash.** 1995. Registration of pearl millet parental lines ICMA 88004 and ICMB 88004. *Crop Sci.* **35:** 1242.
30. **Schertz, K. F., S. Sivaramakrishnan, W. W. Hanna, J. Mullet, Y. Sun, U. R. Murty, D. R. Pring, K. N. Rai, and B. V. S. Reddy.** 1997. Alternate cytoplasms and apomixis of sorghum and millet. *Proc. Intl. Conf. on Genetic Improvement of Sorghum and Pearl Millet (22-27 September 1996, Lubbock, Texas)*: 213-223.
31. **Sharma, A.** 2001. *Marker-Assisted Improvement of Pearl Millet (*Pennisetum glaucum*) Downy Mildew Resistance in Elite Hybrid Parental Line H 77/833-2.* Ph.D. thesis. Chaudhary Charan Singh Haryana Agric. Univ., Hisar, Haryana, India.
32. **Simmonds, N. W.** 1979. *Principles of Crop Improvement.* Longman, Harlow, U.K.
33. **Singh, S. D.** 1995. Downy mildew of pearl millet. *Plant Dis.* **79:** 545-550.
34. **Singh, S. D., R. Alagarswamy, B. S. Talukdar, and C. T. Hash.** 1994. Registration of ICML 22, photoperiod insensitive, downy mildew resistant pearl millet germplasm. *Crop Sci.* **34:** 1421.
35. **Singh, S. D., and R. Gopinath.** 1985. Seedling inoculation technique for detecting downy mildew resistance in pearl millet. *Plant Dis.* **69:** 582-584.
36. **Singh, S. D., and R. Gopinath.** 1990. Effect of temperature and light on sporangial germination and zoospore infectivity in *Sclerospora graminicola* on pearl millet. *Can. J. Plant Pathol.* **12:** 25-30.
37. **Singh, S. D., R. Gopinath, and M. N. Pawar.** 1987. Effects of environmental factors on asexual sporulation of *Sclerospora graminicola*. *Ind. Phytopathol.* **40:** 186-193.
38. **Singh, S. D., S. B. King, and J. Werder.** 1993. *Downy Mildew Disease of Pearl Millet. Information Bulletin no. 37.* ICRISAT, Patancheru (A.P.), India.
39. **Singh, S. D., P. Singh, D. J. Andrews, B. S. Talukdar, and S. B. King.** 1992. Reselection of a pearl millet cultivar utilizing residual variability for downy mildew reaction. *Plant Breed.* **109:** 54-59.
40. **Singh, S. D., P. Singh, K. N. Rai, and D. J. Andrews.** 1990. Registration of ICMA 841 and ICMB 841 pearl millet parental lines with A1 cytoplasmic male-sterility system. *Crop Sci.* **30:** 1378.
41. **Singh, S. D., and B. S. Talukdar.** 1996. Recovery resistance to downy mildew in pearl millet parental lines ICMA 1 and ICMB 1. *Crop Sci.* **36:** 201-203.
42. **Singh, S. D., and B. S. Talukdar.** 1998. Inheritance of complete resistance to pearl millet downy mildew. *Plant Dis.* **82:** 791-793.
43. **Singh, S. D., and R. J. Williams.** 1980. The role of sporangia in the epidemiology of pearl millet downy mildew. *Phytopathology* **70:** 1187-1190.
44. **Singh, S. D., R. J. Williams, and P. M. Reddy.** 1988. Isolation of downy mildew resistant lines from a highly susceptible cultivar of pearl millet. *Ind. Phytopathol.* **41:** 450-456.
45. **Singh, S. D., J. P. Wilson, S. S. Navi, B. S. Talukdar, D. E. Hess, and K. N. Reddy.** 1997. *Screening Techniques and Sources of Resistance to Downy Mildew and Rust in Pearl Millet. Information Bulletin no. 48.* ICRISAT, Patancheru (A.P.), India.
46. **Smil, V.** 2000. *Feeding the World: A Challenge for the 21st Century.* MIT Press, Cambridge, Massachusetts.
47. **Stegmeier, W. D., D. J. Andrews, K. N. Rai, and C. T. Hash.** 1998. Pearl millet parental lines 843A and 843B. *Intl. Sorg. Millets Newsl.* **39:** 129-130.
48. **Talukdar, B. S., K. N. Rai, P. P. P. Babu, and C. T. Hash.** 1994. Performance of single-cross, three-way, and topcross hybrids in pearl millet, p. 81-83. In *ICRISAT Cereals Program Annual Report 1993.* ICRISAT, Patancheru (A.P.), India.
49. **Talukdar, B. S., and S. D. Singh.** 1994. Stability and heritability for downy mildew resistance in two pearl mil-

let pollinators, p. 70-71. In *ICRISAT Cereals Program Annual Report 1993*. ICRISAT, Patancheru (A.P.), India.

50. **Tapsoba, H., and J. P. Wilson.** 1999. Increasing complexity of resistance in host mixtures through intermating to manage rust of pearl millet. *Phytopathology* **89:** 450-455.
51. **Thakur, R. P., C. W. Magill, S. Sivaramakrisknan, C. T. Hash, H. S. Shetty, and D. E. Hess.** 2002. Variability in *Sclerospora graminicola*, the pearl millet downy mildew pathogen, p. 51-56. *In* J. F. Leslie (ed.), *Sorghum and Millets Diseases*. Iowa State Press, Ames, Iowa.
52. **Thakur, R. P., S. D. Singh, and S. B. King.** 1988. Registration of four populations of pearl millet germplasm with multiple disease resistance. *Crop Sci.* **28:** 381-382. Errata–*Crop Sci.* **29:** 1102.
53. **Tripp, R., ed.** 1997. *New Seed and Old Laws: Regulatory Reform and the Diversification of National Seed Systems*. Overseas Development Institute, London, U.K.
54. **Weltzein, R. E., and S. B. King.** 1995. Recurrent selection for downy mildew resistance in pearl millet. *Plant Breed.* **114:** 308-312.
55. **Williams, R. J., S. D. Singh, and M. N. Pawar.** 1981. An improved field screening technique for downy mildew resistance in pearl millet. *Plant Dis.* **65:** 239-241.
56. **Wilson, J. P.** 1999. The dynamic multiline approach for breeding rust resistant hybrids. *Phytopathology* **89:** S96.
57. **Wilson, J. P.** 2002. Diseases of pearl millet in the Americas, p. 465-469. *In* J. F. Leslie (ed.), *Sorghum and Millets Diseases*. Iowa State Press, Ames, Iowa.
58. **Wilson, J. P., and R. N. Gates.** 2002. The dynamic multiline population: An alternative approach to durable resistance?, p. 65-69. *In* J. F. Leslie (ed.), *Sorghum and Millets Diseases*. Iowa State Press, Ames, Iowa.
59. **Wilson, J. P., W. W. Hanna, and R. N. Gates.** 1993. Stability of forage yield and quality in pearl millet hybrids heterogenous for rust resistance. *Euphytica* **72:** 163-170.
60. **Witcombe, J. R., and C. T. Hash.** 2000. Resistance gene deployment strategies in cereal hybrids using marker-assisted selection: Gene pyramiding, three-way hybrids, and synthetic parent populations. *Euphytica* **112:** 175-186.
61. **Witcombe, J. R., M. N. V. R. Rao, B. S. Talukdar, S. D. Singh, and A. M. Rao.** 1996. Registration of ICMR 312 pearl millet topcross pollinator germplasm. *Crop Sci.* **36:** 471.
62. **Yadav, O. P.** 1994. Influence of A_1 cytoplasm in pearl millet. *Plant Breed. Abstr.* **64:** 1375-1379.
63. **Yadav, O. P.** 1996. Downy mildew incidence of pearl millet hybrids with different male-sterility inducing cytoplasms. *Theor. Appl. Genet.* **92:** 278-280.
64. **Yadav, O. P., M. V. Manga, and G. K. Gupta.** 1993. Influence of A_1 cytoplasmic substitution on downy mildew incidence in pearl millet. *Theor. Appl. Genet.* **87:** 558-560.
65. **Young, N. D.** 1999. A cautiously optimistic vision for marker-assisted breeding. *Mol. Breed.* **5:** 505-510.

5

Pearl Millet Downy Mildew: Problems and Control Strategies for a New Millennium

Dale E. Hess, Ram P. Thakur, C. Tom Hash, P. Sérémé, and Clint W. Magill

Abstract

Downy mildew, caused by *Sclerospora graminicola,* is the most damaging fungal disease of pearl millet in Africa and Asia. In India, new single-cross hybrid cultivars have been withdrawn following epidemics that occurred when pathogens that could overcome resistance genes deployed in these genetically uniform cultivars became common. Resulting yield losses were heavy. *S. graminicola* populations maintain haplotype variation through extensive sexual reproduction, making the breeding of single-cross hybrids with durable resistance difficult. Nonetheless, cultivation of adapted, high-yielding, disease-resistant crop cultivars continues to be the most economical method of disease control. When resistant cultivars are unavailable, the use of metalaxyl as a preplanting seed treatment provides protection for a limited time. Roguing infected plants in the field can reduce production of primary inoculum for the following pearl millet crop. Research priorities for the new millennium should include elucidation of the genetics of pathogen virulence and the development and deployment of cultivars with durable resistance. Marker-assisted selection is a promising new tool that will permit development of agronomically uniform pearl millet cultivars that are heterogeneous for downy mildew resistance. However, if agronomic uniformity is not essential, then topcross hybrids and three-way hybrids should provide more durable resistance while still providing the benefits of hybrid vigor and satisfying the desire of seed producers and distributors for a high seed replacement rate

Introduction

Pearl millet [*Pennisetum glaucum* (L.) R. Br.] is a staple cereal of subsistence farmers in the semi-arid regions of many African countries and India. It is cropped on some of the poorest soils and under harsh climatic conditions where other cereal crops fail (4). Although over 100 diseases have been reported on the crop, only five are economically important. These are downy mildew [caused by *Sclerospora graminicola* (Sacc.) J. Schröt.], rust (caused by *Puccinia substriata* Ellis & Barth. var. *indica* Ramachar & Cummins), smut [caused by *Moesziomyces penicillariae* (Bref.) Vánky (syn. *Tolyposporium penicillariae* Bref.)], ergot [caused by *Claviceps fusiformis* Loveless [syn. *C. microcephala* (Wallr.) Tul.)] (26, 37), and pyricularia leaf spot [caused by *Pyricularia grisea* (Cooke) Sacc.]. Among these, downy mildew is the most widespread and destructive disease of pearl millet (25). Significant progress has been made in understanding the biology, epidemiology, and disease management aspects of all of these diseases over the last three decades (5, 17, 22, 34, 35). Although host plant resistance has contributed to effective control, the disease continues to limit the exploitation of the high yield potential of single-cross F_1 pearl millet hybrids, especially in India (35, 38, 39).

Causal Organism

Sclerospora graminicola is an Oomycete fungal-like organism with sexual reproduction as an integral part of its life cycle (1). Oospores in the soil are the primary inoculum source. Asexual zoosporangia are produced on infected tissues of the maturing host. Epidemics can develop within and between pearl millet crops on tillers, which are produced continuously up to crop maturity (27). Current taxonomic treatment places these organisms in the Kingdom Protoctista (16). This kingdom is defined by exclusion, *i.e.*, its members are not animals, not plants, not fungi, and not prokaryotes. It comprises eukaryotic microbes and their immediate descendants: all nucleated algae, flagellated water molds, slime molds, slime nets, and protozoa. Cells of protoctists have nuclei and most other characteristically eukaryotic properties. Most are aerobic and have respiration in mitochondria, as well as 9 + 2 flagellae at some stage of their life cycle. Within the Protoctistae, Oomycetes appear most closely related to Xanthophytae (yellow algae) and Eustigmatophytae (true eyespot plants), which are yellow-green algae that lack meiosis and, therefore, sexual reproduction. These three groups appear to have branched off from the Phaeophytae (brown algae) some time after the evolution of classical mitosis (16). Thus, the causal organism of pearl millet downy mildew can be thought of, metaphorically, as a brown alga that has "gone astray."

Saccardo first described the downy mildew pathogen as *Protomyces graminicola* on foxtail millet [*Setaria verticillata* (L.) P. Beauv.] in 1876; it was renamed *S. graminicola* by Schröter in 1879 (40). The genus *Sclerospora* has a relatively narrow host range, infecting grasses in tribe Paniceae and, infrequently, maize (*Zea mays* L., tribe Maydeae) (13). *Sclerospora graminicola* was first reported on pearl millet in India by Butler (3). Initially, the disease was sporadic on local landraces under traditional cultivation. It did not become a serious problem on the crop until the cultivation of pearl millet F_1 hybrids became widespread in the late 1960s. The first epidemic appeared in the early 1970s (15, 23), and subsequent epidemics have been reported on a small scale, leading to withdrawal of several hybrids from commercial cultivation (4, 5, 8, 10, 35). Systematic work on biology, epidemiology, and host plant resistance began at ICRISAT in the late 1970s (22, 27). RFLP-based mapping of host plant resistance began in the early 1990s (2, 7, 12), and the first marker-assisted backcrossing programs for downy mildew resistance improvement are just now being completed (14, 21).

Symptoms

Symptoms of the graminaceous downy mildews may be either local or systemic (43). They are generally systemic for *S. graminicola* and other downy mildews of economic importance. The genus is characterized by the white downy growth, frequently seen on both leaf surfaces, with asexual sporangiophores emerging from the stomata (13).

Systemic disease symptoms can appear at any stage of pearl millet development from the seedling through flowering. Systemic infection results from hyphae penetrating the growing point at the base of the shoot (13). Seedlings attacked at a very young age frequently die within several weeks. In the first diseased leaf, the chlorotic diseased portion of the leaf is generally confined to the basal portion of the lamina. Subsequent leaves have larger diseased areas, until the entire leaf is affected. Colonization of the meristem results in the production of characteristic malformed inflorescences known as "green ears," in which the floral tissues are transformed into unproductive leafy structures. *S. graminicola* cannot infect older plants, although tillers and regrowth of cut shoots may be susceptible. Oospore formation occurs only in systemically infected plants (13).

TABLE 5-1. Host range of *Sclerospora graminicola* [after Singh *et al.* (25)].

Tribe	Host
Agrostideae	*Agrostis alba*
Andropogoneae	*Saccharum officinarum* L.[a]
Maydeae	*Euchlaena mexicana* Schrad.
	Zea mays (L.)
Paniceae	*Echinochloa crusgalli* var. *frumentacea* (Roxb.) W.F. Wight
	Panicum miliaceum L.
	Pennisetum glaucum (L.) R. Br.
	Pennisetum leonis Stapf & Hubb.
	Pennisetum spicatum (L.) Roem. & Schult.
	Setaria italica (L.) P. Beauv.
	Setaria lutescens (Weig.) F.T. Hubbard
	Setaria magna Griseb (syn. *Chaetochloa magna)*
	Setaria verticillata (L.) P. Beauv.
	Setaria viridis (L.) P. Beauv.

[a] By inoculation.

Pathogen Variability

Both pathogen and host are allogamous, maintaining a high degree of genetic variability. In *S. graminicola* the concern is not for morphological variation, but, rather, for variation in pathogenicity. The host range of *Sclerospora* is relatively narrow, restricted to panicoid grasses (Table 5-1). In addition to attacking pearl millet, the pathogen is reported to infect maize and teosinte in addition to *Panicum* spp. (13). The host range of pathogenic isolates varies by geographic location, with both pearl millet and/or foxtail millet pathotypes, some of which infect maize (Table 5-2). One report exists of an isolate pathogenic to both millet hosts [Safeulla (1976) cited in 43], but it remains unconfirmed. Pathogenic variability in *S. graminicola* was first reported in Karnataka State, India [Bhat (1973) cited in 25]. Subsequently, variation in the pathogen has been evaluated in the laboratory, *e.g.*, Ball *et al.* (1), and monitored systematically by field surveys and in collaborative nurseries (33). Several pathotypes have been reported to date (33, 36, 38, 39).

TABLE 5-2. Pathogenic specialization in *Sclerospora graminicola* [after Williams (43)].

Origin		Host		
Country	Host	*Pennisetum glaucum*	*Setaria verticillata*	*Zea mays*
United States	*S. verticillata*	-	+	+
Israel	*P. glaucum*	+	-	+
India	*P. glaucum*	+	-	N
India	*S. verticillata*	-	+	N
India	*P. glaucum*	+	-	-
India	*S. verticillata*	-	+	-
India	*P. glaucum*	+	-	N
India	*S. verticillata*	-	+	N
India	*P. glaucum*	+	+	-

- – Non-pathogenic.
[b]+ – Pathogenic.
[c]N – No data.

Economic Importance

Although global grain yield losses to pearl millet downy mildew probably do not exceed 20% in any given year, its destructive potential can be clearly seen in locations where genetically uniform single-cross pearl millet F_1 cultivars are grown in successive years over large areas (5, 34). Under the resulting high selection pressure, virulence allele frequencies change and the pathogen population becomes virulent to the uniform host. As inoculum levels build up, favorable climatic conditions can produce an epidemic with yield losses reaching 30-40% (5, 8, 10, 23). A 1971 epidemic resulted from the widespread cultivation (on 15% of India's pearl millet area) of HB3, and the resulting loss in pearl millet grain production was estimated at ~3 million metric tons (15, 22).

Control Strategies

Host plant resistance. Effective management of downy mildew in pearl millet relies upon host plant resistance. Resistance has been categorized as follows (8, 10):

- *Conventional resistance.* Sources of stable conventional resistance (including both gene bank accessions and elite breeding products) consistently show markedly lower disease incidence than susceptible controls.
- *Reselected resistance.* These resistance sources were developed through pure line selection (for resistance) within susceptible parents at ICRISAT-India. Although resistant in India, these selections are susceptible to downy mildew in West Africa (33).
- *Recovery resistance.* These resistance sources are characterized by their ability to outgrow systemic infection to produce healthy panicles (24). The pathogen completes its life cycle unimpeded, and the infected pearl millet recovers from the disease. West African pearl millet landraces frequently show this type of resistance.

Breeding for resistance contributes to the stability of grain, stover, and forage yields. Research to identify resistance sources and to improve the efficiency of resistance screening has been successful, permitting control of this disease in India (8, 10). Effective field and greenhouse screening techniques have been developed and are being used in conventional and integrated (conventional + marker-assisted) pearl millet breeding programs for both India and Africa. Preliminary screening of seedlings in the greenhouse for reaction to downy mildew allows reduction in the size and cost of field disease nurseries. Refinements of these techniques (11) improve the efficacy not only of conventional selection, but also that of disease resistance Quantitative Trait Loci (QTL) mapping and subsequent marker-assisted selection.

Scientists at Tifton, Georgia, are developing strategies for breeding pearl millet hybrids heterogeneous for rust resistance (31, 44-46). Similar strategies can be used to enhance durability of resistance to pearl millet downy mildew.

Marker-assisted selection employs highly heritable markers linked to regions of the genome controlling the character of interest, in this case downy mildew resistance. At present, molecular markers have been identified in different pearl millet crosses for at least 22 different putative downy mildew QTLs (2). Most of these QTLs appear to be individually effective against only a few pathogen populations. These QTLs can be used to develop pearl millet cultivars with uniform agronomic traits that vary in their downy mildew resistance gene complement. Methodologies available to breeding programs to enhance host plant resistance to downy mildew have been discussed elsewhere (8-10, 18, 28, 42, 47).

Chemical control. Seed treatment with the systemic fungicide metalaxyl is one of the most effective, economical, and technically feasible control methods available to farmers in the tropics for the control of downy mildew on local pearl millet landraces. However, the disease protection provided by the fungicide is limited to approximately the first 35 days of crop growth, and the disease may subsequently appear on nodal tillers and panicles. This promising approach to disease management is described in more detail elsewhere in this volume (20).

Cultural practices. Of the known cultural practices, only roguing of diseased plants is proven to be effective (17, 32). However, in traditional cropping systems where local varieties exhibit recovery resistance or compensate through the development of healthy tillers, farmers do not uproot diseased plants. In fact, in the African Sahel, some farmers interpret the increased incidence of downy mildew in their fields as an indicator of good rainfall and the promise of a good harvest to come.

Research Priorities

Genetics of pathogen virulence. A major constraint to developing resistant pearl millets is the lack of information concerning variability in the pathogen population. Knowledge of existing populations and of the origin of new pathotypes is central to developing the best approach for breeding more durable resistance.

Durable resistance. Recurrent selection has been used successfully to enhance durable resistance to downy mildew in genetically heterogeneous open-pollinated pearl millet cultivars (5, 19, 41). New approaches to using resistance sources, including topcross hybrids and their pollinators (29, 30), three-way hybrids (18), and marker-assisted selection (6, 8, 10), are needed to improve the durability of downy mildew resistance in the future.

Marker-assisted selection. Recent progress has made possible the development of near-isogenic lines and gene pyramids for pest resistance through marker-assisted selection (14, 21). When combined with appropriate seed production technology, this new tool will permit novel, effective deployment of downy mildew resistance genes (9, 47).

References

1. **Ball, S. L., D. J. Pike, and C. Y. Burridge.** 1986. Characterization of populations of *Sclerospora graminicola*. *Ann. Appl. Biol.* **108:** 519-526.
2. **Breese, W. A., C. T. Hash, K. M. Devos, and C. J. Howarth.** 2002. Pearl millet genomics and breeding for resistance to downy mildew, p. 243-246. *In* J. F. Leslie (ed.), *Sorghum and Millets Diseases.* Iowa State Press, Ames, Iowa.
3. **Butler, E. J.** 1907. Some diseases of cereals caused by *Sclerospora graminicola*. *Mem. Dept. Agric. in India. Bot. Ser.* **2:** 1-24.
4. **Dave, H. R.** 1987. Pearl millet hybrids, p. 121-126. *In* J. R. Witcombe and S. R. Beckerman (eds.), *Proc. Intl. Pearl Millet Workshop (7-11 April 1986, Patancheru, India).* ICRISAT, Patancheru (A.P.), India.
5. **Hash, C. T.** 1997. Research on downy mildew of pearl millet, p. 121-128. *In* M. C. S. Bantilan and P. K. Joshi (eds.), *Integrating Research Evaluation Efforts: Proc. Intl Workshop (14-16 December 1994, Patancheru, India).* ICRISAT, Patancheru (A.P.), India.
6. **Hash, C. T.** 2000. Concepts for application of marker techniques in Africa, p. 155-168. *In* B. I. G. Haussmann, H. H. Geiger, D. E. Hess, C. T. Hash, and P. Bramel-Cox (eds.), *Application of Molecular Markers in Plant Breeding: Training Manual for a Seminar Held at IITA (16-17 August 1999, Ibadan, Nigeria).* ICRISAT, Patancheru (A.P.), India. http://www.icrisat.org/text/research/grep/homepage/mol/S3_1Hash.pdf (last checked February 2002).
7. **Hash, C. T., and P. J. Bramel-Cox.** 2000. Marker applications in pearl millet, p. 112-127. *In* B. I. G. Haussmann, H. H. Geiger, D. E. Hess, C. T. Hash, and P. Bramel-Cox (eds.), *Application of Molecular Markers in Plant Breeding: Training Manual for a Seminar Held at IITA (16-17 August 1999, Ibadan, Nigeria).* ICRISAT, Patancheru (A.P.) India. http://www.icrisat.org/text/research/grep/homepage/mol/S2_5Hash.pdf (last checked February 2002).
8. **Hash, C. T., S. D. Singh, R. P. Thakur, and B. S. Talukdar.** 1999. Breeding for disease resistance, p. 337-379. *In* I. S. Khairwal, K. N. Rai, D. J. Andrews, and G. Harinarayana (eds.), *Pearl Millet Breeding.* Oxford & IBH, New Delhi, India.
9. **Hash, C. T., J. R. Witcombe.** 2002. Gene management and breeding for downy mildew resistance, p. 27-36. *In* J. F. Leslie (ed.), *Sorghum and Millets Diseases.* Iowa State Press, Ames, Iowa.
10. **Hash, C. T., J. R. Witcombe, R. P. Thakur, S. K. Bhatnagar, S. D., Singh, and J. P. Wilson.** 1997. Breeding for pearl millet disease resistance. *Proc. Intl. Conf. on Genetic Improvement of Sorghum and Pearl Millet (22-27 September 1996, Lubbock, Texas)*: 337-372.
11. **Jones, E. S., W. A. Breese, and D. S. Shaw.** 2001. Spray-inoculation of pearl millet with the downy mildew fungus, *Sclerospora graminicola*: Chilling of sporangial inoculum prevents premature release of zoospores and their subsequent spray damage. *Plant Pathol.* **50:** 310-316.
12. **Jones, E. S., C. J. Liu, M. D. Gale, C. T. Hash, and J. R. Witcombe.** 1995. Mapping quantitative trait loci for downy mildew resistance in pearl millet. *Theor. Appl. Genet.* **91:** 448-456.
13. **Kenneth, R.** 1981. Downy mildews of graminaceous crops, p. 367-394. *In* D. M. Spencer, (ed.), *The Downy Mildews.* Academic Press, London, U. K.
14. **Kolesnikova, M. A.** 2001. *Mapping New Quantitative Trait Loci (QTL) for Downy Mildew Resistance in Pearl Millet.* Ph.D. thesis. Russian Acad. Sci., Moscow, Russia.
15. **Kumar, K. A., R. P. Jain, and S. D. Singh.** 1983. Downy mildew reaction of pearl millet lines with and without cytoplasmic male sterility. *Plant Dis.* **67:** 663-665.
16. **Margulis, L., and K. L. Schwartz.** 1996. *Five Kingdoms: An Illustrated Guide to the Phyla of Life on Earth,* 2nd ed. W.H. Freeman, New York City, New York.
17. **Mbaye, D. F.** 1992. Les maladies du mil au Sahel: Etat des connaissances et propositions de lutte, p. 42-63. In *La Lutte Intégrée Contre les Ennemis des Cultures Vivrières dans le Sahel.* John Libby Eurotext, Montrouge, France.
18. **Rai, K. N., S. Chandra, and A. S. Rao.** 2000. Potential advantages of male-sterile F_1 hybrids for use as seed parents of three-way hybrids in pearl millet. *Field Crops Res.* **68:** 173-181.
19. **Rattunde, H. F. W., and J. R. Witcombe.** 1993. Recurrent selection for increased grain yield and resistance to downy mildew in pearl millet. *Plant Breed.* **110:** 63-72.
20. **Scheuring, J. F., S. O. Katile, and I. A. Kollo.** 2002. Boosting pearl millet yields with Apron Plus® and Apron

Star® seed treatments, p. 47-49. *In* J. F. Leslie (ed.), *Sorghum and Millets Diseases*. Iowa State Press, Ames, Iowa.
21. **Sharma, A.** 2001. *Marker-Assisted Improvement of Pearl Millet (*Pennisetum glaucum*) Downy Mildew Resistance in Elite Hybrid Parental Line H 77/833-2*. Ph.D. thesis. Chaudhary Charan Singh Haryana Agric. Univ., Hisar, Haryana, India.
22. **Singh, S. D.** 1995. Downy mildew of pearl millet. *Plant Dis.* **79:** 545-550.
23. **Singh, S. D., S. Ball, and R. P. Thakur.** 1987. Problems and strategies in the control of downy mildew, p. 161-172. *In* J. R. Witcombe and S. R. Beckerman (eds.), *Proc. Intl. Pearl Millet Workshop (7-11 April 1986, Patancheru, India)*. ICRISAT, Patancheru (A.P.), India.
24. **Singh, S. D., and S. B. King.** 1988. Recovery resistance to downy mildew in pearl millet. *Plant Dis.* **72:** 425-428.
25. **Singh, S. D., S. B. King, and J. Werder.** 1993. *Downy Mildew Disease of Pearl Millet. Information Bulletin no. 37*. ICRISAT, Patancheru (A.P.), India.
26. **Singh, S. D., and R. P. Thakur.** 1998. Pearl millet diseases and their management, p. 397-417. *In* R. K. Upadhyay, K. G. Mukerjee, and R. L. Rajak (eds.), *IPM Systems in Agriculture. Vol. 3. Cereals*. Aditya Books Pvt. Ltd., New Delhi, India.
27. **Singh, S. D., and R. J. Williams.** 1980. The role of sporangia in the epidemiology of pearl millet downy mildew. *Phytopathology* **70:** 1187-1190.
28. **Singh, S. D., J. P. Wilson, S. S. Navi, B. S. Talukdar, D. E. Hess, and K. N. Reddy.** 1997. *Screening Techniques and Sources of Resistance to Downy Mildew and Rust in Pearl Millet. Information Bulletin no. 48*. ICRISAT, Patancheru (A.P.), India.
29. **Talukdar, B. S., K. N. Rai, P. P. Prakash Babu, and C. T. Hash**. 1994. Performance of single-cross, three-way, and topcross hybrids in pearl millet, p. 81-83. In *ICRISAT Cereals Program Annual Report 1993*. ICRISAT, Patancheru (A.P.), India.
30. **Talukdar, B. S., and S. D. Singh.** 1994. Stability and heritability for downy mildew resistance in two pearl millet pollinators, p. 70-71. In *ICRISAT Cereals Program Annual Report 1993*. ICRISAT, Patancheru (A.P.), India.
31. **Tapsoba, H., and J. P. Wilson.** 1999. Increasing complexity of resistance in host mixtures through inter-mating to manage rust of pearl millet. *Phytopathology* **89:** 450-455.
32. **Thakur, R. P.** 1980. Utilization of downy mildew resistant germplasms and other resources for increased productivity of pearl millet under arid and semi-arid regions of India. *Ann. Arid Zone* **19:** 265-270.
33. **Thakur, R. P.** 1998. *International Pearl Millet Downy Mildew Virulence Nursery (IPMDMVN)–A Summary Report (1992-1998)*. ICRISAT, Patancheru (A.P.), India.
34. **Thakur, R. P.** 1998. Disease management in pearl millet, p. 53-76. *In* T. S. Thind (ed.), *Diseases of Field Crops and Their Management*. National Agricultural Technology Information Center, Ludhiana (Punjab), India.
35. **Thakur, R. P.** 1999. Pathogen diversity and plant disease management. *Ind. Phytopathol.* **52:** 1-9.
36. **Thakur, R. P., C. W. Magill, S. Sivaramakrishnan, C. T. Hash, H. S. Shetty, and D. E. Hess.** 2002. Variability in *Sclerospora graminicola*, the pearl millet downy mildew pathogen, p. 51-56. *In* J. F. Leslie (ed.), *Sorghum and Millets Diseases*. Iowa State Press, Ames, Iowa.
37. **Thakur, R. P., and K. N. Rai.** 2002. Pearl millet ergot research: Advances and implications, p. 57-63. *In* J. F. Leslie (ed.), *Sorghum and Millets Diseases*. Iowa State Press, Ames, Iowa.
38. **Thakur, R. P., V. P. Rao, and C. T. Hash.** 1998. Emergence of a highly virulent pathotype of *Sclerospora graminicola* at Jodhpur, Rajasthan, India. *Intl. Sorg. Millet Newsl.* **39:** 140-142.
39. **Thakur, R. P., V. P. Rao, J. G. Sastry, S. Sivaramakrishnan, K. N. Amruthesh, and L. D. Barbind.** 1999. Evience for a new virulent pathotype of *Sclerospora graminicola* on pearl millet. *J. Mycol. Plant Pathol.* **29:** 61-69.
40. **Ullstrup, A. J.** 1973. An overview of the downy mildews of corn and sorghum. Report of a workshop on the downy mildews of sorghum and corn. *Tech. Rept., Texas Agric. Expt. Sta., Dept. Plant Sci.* **74:** 5-12.
41. **Weltzien, R. E., and S. B. King.** 1995. Recurrent selection for downy mildew resistance in pearl millet. *Plant Breed.* **114:** 308-312.
42. **Werder, J., and S. L. Ball.** 1992. Variability of pearl millet downy mildew (*Sclerospora graminicola*) and its implications for breeding for stable and durable resistance. *Trop. Pest Manag.* **38:** 43-47.
43. **Williams, R. J.** 1984. Downy mildews of tropical cereals. *Adv. Plant Pathol.* **2:** 1-103.
44. **Wilson, J. P., and R. N. Gates.** 2002. The dynamic multi-line population: An alternative approach to durable resistance?, p. 65-69. *In* J. F. Leslie (ed.), *Sorghum and Millet Diseases 2000*. Iowa State Press, Ames, Iowa.
45. **Wilson, J. P., R. N. Gates, and M. S. Panwar.** 2001. Dynamic multiline approach to resistance gene management. *Phytopathology* **91:** 255-260.
46. **Wilson, J. P., W. W. Hanna, and R. N. Gates.** 1993. Stability of forage yield and quality in pearl millet hybrids heterogenous for rust resistance. *Euphytica* **72:** 163-170.
47. **Witcombe, J. R., and C. T. Hash.** 2000. Resistance gene deployment strategies in cereal hybrids using marker-assisted selection: Gene pyramiding, three-way hybrids, and synthetic parent populations. *Euphytica* **112:** 175-186.

6

Identification of Resistance to Downy Mildew and Smut of Pearl Millet in Ghana

Stephen K. Nutsugah, Ibrahim D. K. Atokple and V. P. Rao

Pearl millet [*Pennisetum glaucum* (L.) Leek] is one of the most important staple cereal crops in northern Ghana. The crop is cultivated on 220,000 ha in the savannah agroecological zones of northern Ghana (2) and is susceptible to many diseases (4). Downy mildew [caused by *Sclerospora graminicola* (Sacc.) Schröet] and smut [caused by *Tolyposporium penicillariae* (Bref.)] are the most important diseases of pearl millet in the northern savannah zone of Ghana. The yield losses attributed to these diseases ranges from 10 to 60% (3), with similar losses also occurring in other west and central African countries.

The results of a survey conducted prior to screening for downy mildew and smut resistance revealed that even though farmers could identify downy mildew at both vegetative and reproductive stages of plant growth, they have no effective control measures (3). Rouging of infected plants, as practiced by some farmers in the Upper East Region of Ghana, was ineffective because most infected plants escape identification at the latent stage of infection. Host resistance is an important component of any long-term management strategy for both smuts and downy mildew.

Significant genetic variation in the reaction of millet genotypes to both downy mildew (1, 6) and smut (5) suggests that breeding for resistance or tolerance to both pathogens should be possible. However, host lines respond differently to different sources of inoculum, and pathogen collections of *S. graminicola* from the sub-Sahelian regions of West Africa are generally more aggressive than those from India (1). In particular, pathotypes of *S. graminicola* from Manga (Ghana) and Samaru (Nigeria) were more aggressive than those from Kamboinse (Burkina Faso) and Bengou (Niger) (7). The implication of these findings was that each member country of the West and Central African Millet Research Network needed to screen the millet genotypes at their own downy mildew and smut "hot spots". Our objective in this report is to describe the comparative performances of pearl millet accessions for resistance to downy mildew and smut in Ghana under natural infection and artificial inoculation in the field, respectively.

Materials and Methods

Screening for resistance to downy mildew. Screening for resistance to downy mildew was done at the Manga Agricultural Research Station, Manga (11° 01' North latitude, 00° 16' West longitude), Ghana from 1993 to 1996 with an infector-row system (8). Infector rows were planted every third row and three weeks earlier than the test material. Each accession was planted in two replications; each plot was 5 m spaced 70 cm apart and organized in a complete randomized block design. Each accession had at least 26 plants. Pearl millet cultivar 7042, which is highly susceptible to downy mildew, was planted as a control. Downy mildew incidence (percentage of plants infected) was scored 30 days after planting and at the soft dough stage (60-70 days after planting).

Screening for resistance to smut. We used the technique of Thakur *et al.* (6) for smut-resistance screening. We obtained an aqueous suspension of sporidia by soaking spore balls in water overnight. The suspension was filtered through two layers of cheesecloth before being used as inoculum. To inoculate plants, we inserted a hand-held atomizer into the flag-leaf sheath at the boot stage. Ten heads, in the boot stage, per test entry per replication were covered with parchment paper selfing bags immediately after inoculation. Pearl millet cultivar BK 560, which is susceptible to smut, was inoculated as a control. Smut severity was recorded 20-25 days after inoculation by using a standard scale (5).

TABLE 6-1. Downy mildew (DM) incidence (%)[a] and smut severity (%)[b] of 28 pearl millet accessions at Manga, Ghana from 1993-1996.

Pearl Millet	1993		1994		1995		1996		Mean	
Accession	DM	Smut	DM	Smut	DM	Smut	DM	Smut	DM	Smut
7042	100		100		98		96		99	
BK 560	47		44	5	30	31	27	71	37	27
CMB-32	14	11	24	29					19	20
CMB-40	21	15	11	22	11	0			15	28
CMB-89	7	33	8	8					6	21
CMB-96	32	14	14	9					23	12
CS-1	24	7	11	23	11	4	2		12	11
Ex Borno	34	8	10	15	14	6	8	10	17	10
Gouzouma	19	9	5	7	23	16			16	11
ICMV IS 88217			12	6	10	6	8	13	10	8
ICMV IS 89305			6	15	14	9	6	22	9	15
ICMV IS 90309			11	2	5	7	4	19	7	9
ICMV IS 90311			16	7	7	7	3	23	9	12
ICTP 8203			63	5	5	31	46		47	9
IKMV 8201	25	11	49	5	11	11	21	27	27	14
INMV-77			8	5	10	14	4	13	7	20
Kelague	27	19	15	6					21	13
Local AM-Dam	19	16	7	4					13	10
Manga nara	44	7	21	8	31	9	26	15	31	10
MDN 88			15	1	10	11	1	8	9	7
SE 360			7	3	13	4	2	9	7	5
SE 2124			7	8	10	7	3	16	7	10
SE Composite			13	27	9	12	5		9	20
Sosat			4	< 1	10	9	8	6	7	5
Sosat C-88			11	7	13	6	5	14	10	10
SRR 1	20	8	2		14				12	8
SRR 2	16	14	6		14				12	14
Synth 16CO			12	15	8	6	4	9	8	10
Mean	30	15	18	10	10	18	15	18		
Trial SE ±	6.4	6.3	4.8	5.2	5.2	4.3	3.9	6.0		

[a]Mean of two replications at soft dough stage.
[b]Mean of 20 inoculated panicles from two replications.

Results and Discussion

We evaluated 64 pearl millet accessions in the downy mildew nursery at Manga between 1993 and 1996. Most of the accessions had low levels of downy mildew incidence. Based on 2-4 years of screening, 13 accessions, CMB-89 (Burkina Faso), ICMV IS 88217, ICMV IS 89305, ICMV IS 90309 (Chad), ICMV IS 90311, Sosat C-88 (ICRISAT), INMV-77, MDN 88, SE Composite, SE 360, SE 2124 (Nigeria), Sosat (Mali), and Synth 16 CO (Mali) showed low levels of downy mildew (< 10% mean incidence). Eleven accessions, Ex Borno MDN 88b, SE360, SE 2124 (Nigeria), ICMV IS 88217, ICMV IS 90309, ICTP 8203, Sosat C-88 (ICRISAT), SRR1 (Cote d'Ivoire), Sosat and Synth 16 CO (Mali) exhibited low levels of smut severity (< 10% mean severity). Eight accessions, ICMV IS 88217, ICMV IS 90309, MDN 88, SE 360, SE 2124, Sosat, Sosat C-88, and Synth 16 CO expressed low levels of severity to both downy mildew and smut (Table 6-1). The most susceptible accessions were BK 560, ICTP 8203, and Manga nara in all years with the percentage of plants infected with downy ranging from 21-63% (Table 6-1). The wide variation in disease incidence of these accessions over the years indicates that the seasonal variation in rainfall and humidity affect disease intensity. The reactions of the control for downy mildew, line 7042, provided a standard measure of the reliability of the screening system. This line was consistently susceptible throughout the four years of the study, with a downy mildew severity range of 96-100%. In some cases, the entire plot was destroyed by downy mildew at the vegetative stage and no plants were available for smut assessment. The consistently high level of downy mildew incidence on line 7042 indicates that there is very high selection pressure in the nursery, which should provide an adequate test of the millet accessions screened (Table 6-1).

Our results and conclusions are consistent with those of Werder and Ball (7), *i.e.* that Manga (Ghana) and Samaru (Nigeria) have more aggressive downy mildew isolates than the rest of West Africa. We think that the accessions we have evaluated will not be as severely affected by downy mildew elsewhere in West Africa as they were in these trials at SARI.

Acknowledgements

We thank the West and Central Africa Millet Research Network for financial support, and Ram P. Thakur of ICRISAT (India) for assistance.

References

1. **Ball, S. L., and D. J. Pike.** 1984. Intercontinental variation of *Sclerospora graminicola*. *Ann. Appl. Biol.* **104:** 41-51.
2. **Froelich, W., S. S. J. Buah, and P. B. Tanzubil.** 1993. A review of millet production and improvement in Ghana. *Proc. Regional Pearl Millet Workshop (19-21 September, 1990, Niamey, Niger)*: 110-114.
3. **Nutsugah, S. K., and I. D. K. Atokple.** 1998. Final country report on "Downy mildew control improvement in Ghana". Report submitted to the West and Central Africa Millet Research Network.
4. **Nutsugah, S. K., F. K. Ayensu., and F. K. Tsigbey.** 1999. Survey of seed-borne diseases in Ghana: Initiation of cooperative research programme for seed-borne disease management. *Proc. 1999 World Seed Conf. (Cambridge, UK)*: 50.
5. **Thakur, R. P., S. B. King, K. N. Rai, and V. P. Rao.** 1992. *Identification and Utilization of Smut Resistance in Pearl Millet.* Research Bulletin no. 16, ICRISAT, Patancheru (A.P.), India.
6. **Thakur, R. P., K. V. Suba Rao, and R. J. Williams.** 1983. Evaluation of a new field screening technique for smut resistance in pearl millet. *Phytopathology* **73:** 1255-1258.
7. **Werder, J., and S. L. Ball.** 1992. Variability of pearl millet downy mildew and its implications for breeding for stable and durable resistance. *Trop. Pest Manag.* **38:** 43-47.
8. **William, R. J., S. D. Singh, and M. N. Pawar.** 1981. An improved field screening technique for downy mildew resistance in pearl millet. *Plant Dis.* **65:** 239-241.

7

Boosting Pearl Millet Yields with Apron Plus® and Apron Star® Seed Treatments

John F. Scheuring, Seriba O. Katilé, and Issofou A. Kollo

Abstract

Data from over 900 on-farm tests over 12 years indicate that under both traditional and high-input farming practices in west Africa, pearl millet yields are increased approximately 30% by the application of Apron Plus® or Apron Star® to the seed. The yield increases are due to the additive effects of increased seedling survival, protection against primary and early secondary infection by the causal agent of downy mildew, systemic protection against soil and seedling insects, enhanced tolerance of nematode infections, and more uniform emergence, growth, and pollination.

Introduction

Metalaxyl fungicide was released by Ciba-Geigy Corporation in the 1970s for chemical control of Oomycetes in potatoes, peas, and other crops. The seed treatment forms of metalaxyl were marketed under the trade name Apron®, and by the late 1970s they were being used experimentally by ICRISAT both in India and in Mali. Under heavy primary downy mildew pressure at Koporo-Keniepe, Mali in 1979 (0.25 ha millet plots), Apron® seed treatment resulted in 30-45% yield increases in millet grain relative to control plots treated with the local Thiram seed treatment (1). In the 1980s, Ciba-Geigy identified the more active isomer of metalaxyl, resulting in a product, "mefenoxam," that had almost twice the fungicidal activity of the initial mixed isomer product.

Furathiocarb insecticide was released by Ciba-Geigy in the 1980s as a seed treatment under the trade name Promet®. Metalayxl was combined with carboxin and furathiocarb as a seed treatment and marketed by Ciba-Geigy under the trade name Apron Plus®, a product initially targeted towards maize in Nigeria beginning in 1986. In 1988, Apron-Plus® seed-treated pearl millet was tested on-farm in Mali, where it increased seedling establishment and grain yields approximately 30% even under droughty conditions (2). Apron Plus® became commercially available in Mali in 1992.

In 1996, Ciba-Geigy merged with Sandoz and became Novartis Corporation, and in 2000, the Novartis Crop Protection division was separated from the rest of the company and merged with Zeneca Agrochemicals to become Syngenta Corporation. In 1997, Novartis Crop Protection launched a new insecticidal seed treatment, thiamethoxam (3). In 2000 and 2001, beginning in Nigeria, Syngenta Crop Protection commercially released Apron Star®, which is composed of mefonoxam, difenoconazole, and thiamethoxam.

Materials and Methods

Apron Plus® is a Syngenta Crop Protection seed treatment product targeted for west African markets. The product is primarily targeted towards downy mildew-affected cropping areas of maize and pearl millet; but it also is effective for sorghum, cowpeas, peanuts, and cotton. Apron Plus® has both fungicide and insecticide components: metalaxyl (10%) for protection against Oomycetes, including *Pythium* spp., *Phytophtora* spp., and the causal agents of most downy mildews; carboxin (6%) for protection against seedling blight and seedling smut infections; and furathiocarb (34%) for protection against seedling insects. The recommended application rate is 10 g product per kg of millet seed applied as a dry seed dressing. Apron Plus® has been on the market since 1986.

Apron Star® is a replacement product for Apron Plus®, with ¼ of the application rate and the same or increased efficacy. The components of Apron Star® are mefenoxam (20%) for protection against Oomycetes, including *Pythium* spp., *Phytophtora* spp., and the causal agents of most downy mildews; difenoconazole (2%) for

Table 7-1. Downy mildew incidence and millet grain yields in 1998 and 1999 on-farm trials of Apron Plus® and Apron Star® in Mali.

Year		Downy Mildew Incidence[a] (%)			Plant-Hills/Field (200 m²)			Grain Yield (kg/ha)		
		Apron Star®	Apron Plus®	Control	Apron Star®	Apron Plus®	Control	Apron Star®	Apron Plus®	Control
1998[b]	mean	3.8	4.2	11	530	530	410	1400	1400	930
	median	3.3	3.6	5.0	520	530	430	1200	1100	730
1999[c]	mean	2.3	2.8	8.2	430	440	340	1500	1500	1200
	median	2.2	1.5	6.8	390	410	340	1700	1700	1100
1998/99	median	2.9	3.3	5.5	450	420	360	1400	1300	880
Significance[d]			$p = 0.01$			$p = 0.01$			$p = 0.10$	

[a]Scored 45 days after planting.
[b]Average of 9 farms.
[c]Average of 15 farms.
[d]Kruskal-Wallis One-Way ANOVA on Ranks 1998/1999 combined observations. The data are from on-farm, unreplicated large plots with highly heterogeneous conditions. All sets failed the normality test. This situation is typical for on-farm data in the Sahel. Therefore, we used a nonparametric Kruskal-Wallis One-Way Analysis on Ranks. For all three variables there were no significant differences between the Apron-Plus® and Apron-Star® treatment medians; but the differences between the pooled 1998/99 Apron® treatments and the untreated check were highly significant for mildew incidence and plant-hill counts, with a strong trend in favor of the Apron® treatments for grain yield.

protection against seedling *Fusarium* spp. and seedborne smut infections; and thiamethoxam (20%) for protection against seedling soil and sucking insects; and inert ingredients (58%). The recommended application rate is 2.5 g of product per kg of millet seed applied as a dust application. Apron Star® became available commercially in 2000 and will completely replace Apron Plus® by 2003.

Fungal Control

The most striking and visible effect of Apron Plus® and Apron Star® is the control of primary and early secondary infections by downy mildew fungi. The control of late secondary infection is not as effective as the control of earlier infections. Fortunately, late secondary infections often are "cosmetic" and have little effect on plant growth and grain yield. There are no known reports of Apron® seed treatments becoming ineffective against more virulent strains of the causal agent of pearl millet downy mildew, *Sclerospora graminicola* (Sacc.) Shröet. Disease control (Table 7-1) has been consistent for the life of these products. On average, grain yield from seed treated with Apron Plus® or Apron Star® increased by > 30% relative to the untreated control, and downy mildew incidence was < 5%, with a decrease of > 60% relative to the untreated control.

Insect Control

Soil and seedling insects are controlled by the activity of the systemic insecticides in Apron Plus® (furathiocarb) and Apron Star® (thiamethoxam). Soil pests controlled by Apron Plus® or Apron Star® include millipedes, wireworms, flea beetles, ants, termites, and some coleopteran insect larvae. These products protect millet seedlings against these insects for about four weeks. The persistent effect on sucking insects is more significant with Apron Star® than with Apron Plus®.

Stand Establishment

Uniform stand establishment is a major challenge for millet farmers in the west African Sahel. Germinating seed and young seedlings are threatened by birds, drought, heat, blowing sand and debris, soil and seedling insects, nematodes, and soil- and seed-borne diseases. Apron Plus® and Apron Star® reduce threats pests and environmental factors pose to seedling establishment. In Niger, treatment with Apron Plus® enhanced early growth, which allowed excellent seedling establishment even in the presence of nematodes and blowing sand (Kollo, *unpublished*). The nematode tolerance was unexpected, as none of the active ingredients in Apron Plus® is toxic to nematodes. Protection against primary and secondary infection by *S. graminicola* and other soil-borne pathogens results in healthy seedling establishment and earlier growth relative to infected plants. Finally, the systemic insecticides in Apron Plus® and Apron Star® protect seedlings against the debilitating effects of soil and sucking insects. The combined protective effects and increased early growth vigor result in a net gain in seedling establishment (Table 7-1).

Grain Yield Increase

On-farm millet yields following seed treatment with Apron Plus® or Apron Star® have consistently exceeded the yields in Thiram-treated control plots by 20-40% for over 13 years, in four countries, at 941 on-farm testing locations (Table 7-2). These yield gains have occurred

TABLE 7-2. Effects of seed treatment with Apron Plus® or Apron Star® on millet grain yields in on-farm tests relative to treatment with Thiram[a].

Country	Year	Number of On-Farm Test Sites	Grain Yield Increase (%)	
			Apron Plus®	Apron Star®
Burkina Faso	1987	5	20	-
	1991	16	43	-
Niger	1991	184	26	-
Senegal	1992	2	28	-
Mali	1988	8	37	-
	1990	18	18	-
	1991	22	21	-
	1992	535	31	-
	1993	55	35	-
	1998	9	47	52
	1999	15	24	23

[a]Data are from on-farm trial networks conducted by the Cinzana Research Station in Mali and agricultural extension agencies in Burkina Faso, Mali, and Niger. High yield gains, *e.g.*, 1998, are associated with years of particularly high downy mildew incidence.

under both low (yields ≤ 1 t/ha) and high (yields ≥ 1.5 t/ha) input conditions. The higher yields have resulted from higher seedling establishment, more uniform early growth, and more uniform pollination.

Conclusions

Subsistence low-input millet farms in west Africa average yields of 600 kg/ha. Seed treatment with Apron Plus® or Apron Star® costs the farmer about US$2/ha. With an average yield increase of ≥ 30%, or 180 kg/ha, there will be a minimum net income gain of US$12, for an average return on investment of at least 5:1. Since 1994, over three tons of Apron® products have been sold annually in 10 g sachets to millet farmers in Mali. These products should result in a per annum net increase of ≥ 30,000 t of millet grain in the country with a value ≥ US$2 million distributed amongst the tens of thousands of subsistence millet farmers who purchase the product. The Apron® seed treatments exemplify a win-win partnership between the crop protection industry and the subsistence farmers in the Sahel of Africa.

References

1. **ICRISAT/Mali.** 1980. *1979 Sorghum and Pearl Millet Breeding Annual Report.* IER Document Center, Bamako, Mali.
2. **Institut d'Economie Rurale (IER).** 1985-1999. *Phytopatholigie Mil: Rapports de la Commission Technique.* IER Document Center, Bamako, Mali.
3. **Senn, R., D. Hofer, T. Hoppe, M. Angst, P. Wyss, F. Brandl, P. Maienfisch, L. Zang, and S. White.** 1998. CGA293'343: A novel broad-spectrum insecticide supporting sustainable agriculture worldwide. *Proc. Brighton Crop Protection Conf.–Pests and Dis.* **1:** 27-36.

8

Variability in *Sclerospora graminicola*, the Pearl Millet Downy Mildew Pathogen

Ram P. Thakur, Clint W. Magill, S. Sivaramakrishnan, C. Tom Hash, H. S. Shetty, and Dale E. Hess

Abstract

Sclersopora graminicola, the causal agent of pearl millet downy mildew, is heterothallic and reproduces both sexually and asexually. Pathogenic variability in *S. graminicola* is known in both Africa and India. Host-specific evolution of virulence occurs in the pathogen, especially to genetically homogeneous single-cross F_1 hybrids in India. Virulence monitoring through a collaborative international virulence nursery and field surveys in India have provided useful information on virulence diversity. Genotypic differences in host-specific virulent isolates have been identified by using DNA fingerprinting markers. Quantitative trait locus (QTL) mapping for downy mildew resistance against different pathotypes from Africa and India has confirmed the presence of variable pathotypes. Developing cultivars with resistance to specific pathotypes and resistance gene pyramiding against multiple pathotypes through marker-assisted backcrossing could lead to cultivars with durable resistance.

Introduction

Genetic variability in plant pathogens is important for survival of their species and biotypes. The primary objective of many studies of variation in pathogenicity or virulence is to enable the exploitation of host plant genetic resistance for disease management (3). When selecting breeding material for disease resistance, it is important to know which pathotype to use in the screening process, how the resistance is expressed and inherited, and whether the resistance is likely to be adequate and durable. It also is important to determine whether the pathogen can change in response to the selection pressure resulting from host plant resistance and how rapidly such changes can occur.

The pearl millet downy mildew pathogen, *S. graminicola* (Sacc.) Shröt, is highly variable and reproduces both sexually and asexually (16). Physiological specialization in the pathogen was first reported in 1981 (19). Isolates from Africa and India are known to vary in pathogenicity and to be sexually compatible (1, 2, 10, 27, 29, 34). In India, increased virulence in *S. graminicola* has been attributed to large-scale cultivation of a number of genetically homogeneous single-cross F_1 hybrids (25, 29). Increased virulence diversity in the pathogen population has shortened the useful life of several commercial hybrids, and farmers have suffered losses due to reduced grain and fodder yields. Monitoring virulence changes in the pathogen population, identifying resistance to specific and multiple pathotypes, and directing breeding programs towards strategic utilization and deployment of resistance genes form the basis of long-term downy mildew disease management in pearl millet. In this chapter, we review the results of virulence monitoring, the status of virulence patterns in the pathogen, possible mechanisms of virulence evolution, and strategies for containing the disease through genetic resistance.

Monitoring Virulence Diversity

Pathogenicity is the genetic marker of greatest interest and has been studied intensively since 1981. Pathogenic and genetic variability in *S. graminicola* have been studied through a collaborative International Pearl Millet Downy Mildew Virulence Nursery (IPMDMVN), disease surveys in farmers' fields, inoculations of host differenttial lines in the greenhouse, and DNA fingerprinting markers.

The IPMDMVN. This collaborative virulence nursery was established in 1992 and contains 12 known resistant and susceptible pearl millet lines (inbreds and hybrids) of diverse geographic origin (24). Locations for the nursery were selected in the major pearl millet growing parts of the world where downy mildew is an endemic problem and active pathology research is in progress. These locations cover pearl millet areas from latitudes of 7.6° (Samaru, Nigeria) to 14.8° (Bambey, Senegal) in western and central Africa, and from 11.3° (Coimbatore) to 30.9° (Ludhiana) in India.

The IPMDMVN was planted in a replicated field experiment at four locations in 1992 and at 21 locations in 1998, in both downy mildew sick plots and/or by using an infector-row inoculation system (36). Downy mildew data on disease incidence were recorded at 30 days (preboot stage) and 60 days (soft-dough stage) after seedling emergence. Some pearl millet lines have higher disease incidence at 60 days than at 30 days. Daily maximums and minimums for temperature and relative humidity were recorded for this 60-day period. Collaborators sent infected leaf tissue containing oospores from the heavily infected lines from each location to the coordinating centers in India (ICRISAT, Patancheru, India) and west central Africa (ICRISAT, Bamako, Mali) for studies of variability under controlled environmental conditions.

The results of disease incidence data across 23 locations (3-8 years of data from each location) documents virulence diversity among *S. graminicola* populations (Table 8-1). Seven pearl millet lines (IP 18292, IP 18293, 700651, P310-17, P7-4, MBH 110, and 852B) had differential reactions that could be used to classify 23 populations into one of 15 putative pathotypes. Pathogen populations at several locations in India and Africa behaved similarly and were classified as the same pathotype. For example, the pathogen populations at Bambey (Senegal), and at Jamnagar and Gwalior (India) were in the least virulent category, whereas those at Bagauda (Nigeria) and Cinzana (Mali) were in the most virulent category. These results provide preliminary information on pathogenicity as it relates to spatial and genetic diversity in the population. However, further testing under controlled conditions is needed to understand and appreciate the spectrum of diversity. The field data also are limited by the number of locations used in the various countries, except for India where 14 locations were used. Within India, the 14 populations were classified into nine pathotypes. In Mali, Samanko and Cinzana have different populations, as do Bengou and Sadoré in Niger, and Bambey and Nioro-du-Rip in Senegal. A Western African Downy Mildew Variability Nursery also identified virulence diversity among *S. graminicola* populations at Bambey, Bengou, Kamboinsé, and Samaru (35). This nursery led in 1991 to the West African Downy Mildew and Smut Observation Nursery, in which promising lines from regional breeding programs are evaluated for stability of resistance to downy mildew. Screening locations for this nursery include Bambey, Bengou/Sadoré, Cinzana, Kam-

TABLE 8-1. *Sclerospora graminicola* populations from 23 locations in seven countries divided into 15 putative pathotypes based on seven host differential lines and 3-8 years of field testing from 1992-1999.

Location[a]	Pathogen	Host Differential Line						
	Group	IP 18292	IP 18293	700651	P310-17	P7-4	MBH 110	852B
Bambey, Jamnagar, and Gwalior	I	R[b]	R	R	R	R	R	R
Bengou and Ludhiana	II	R	R	R	R	S[c]	S	S
Kamboinse, Durgapura, and Jodhpur	III	S	S	R	S	S	S	S
Mysore, Pune, and Dhule	IV	R	R	S	S	S	R	S
Bagauda and Cinzana	V	S	S	S	S	S	S	S
Patancheru	VI	R	R	S	R	S	R	R
Hisar	VII	R	R	R	S	R	R	S
Jalna	VIII	R	R	S	R	R	S	R
Coimbatore	IX	R	R	R	R	S	S	R
Aurangabad	X	R	R	S	R	R	R	S
Nioro-du-Rip	XI	S	R	R	R	R	R	S
Samanko	XII	R	S	R	S	R	S	S
Sadore	XIII	S	R	R	S	S	S	S
Mandor	XIV	S	R	S	S	S	S	S
El-Obeid	XV	R	R	R	R	R	S	S

[a]Locations: Bambay and Nioro, Senegal; Bengou and Sadore, Niger; Samanko and Cinzana, Mali; Bagauda, Nigeria; Kamboinse, Burkina Faso; El-Obeid, Sudan; and Jamnagar, Gwalior, Patancheru, Hisar, Jalna, Coimbatore, Aurangabad, Mysore, Ludhiana, Durgapura, Jodhpur, and Mandor, India.

[b]R, < 10% mean disease incidence.

[c]S, > 10% mean disease incidence.

boinsé, Tamale/Bawku, and Samaru/Zaria.

Virulence survey. Farmers' fields in India have been systematically surveyed for downy mildew incidence in pearl millet for the past seven years. Surveys in Maharashtra state documented both the diversity of hybrid cultivars grown by the farmers and the extent of downy mildew incidence (Table 8-2). In 1993 there were three popular commercial hybrid cultivars in Maharashtra. This number increased to 24 in 1999. If a new hybrid is grown on the same piece of land for three consecutive years, then the hybrid generally becomes susceptible if the weather has been favorable for disease development. Hybrids for which very high disease incidence has been reported (up to 80%) in farmers' fields (29), *e.g.*, MLBH 104, MBLBH 267, GK 1004, Proagro 7701, Eknath 201, BK 560, are now slowly being replaced by new resistant hybrids. Similar results have been obtained from limited surveys in the states of Rajasthan and Gujarat (Thakur *et al.*, *unpublished*). In contrast to genetically uniform single-cross hybrids, heterogeneous open-pollinated varieties, *e.g.*, ICTP 8203 and Mallikarjuna, have had durable resistance to downy mildew (29; Table 8-2).

TABLE 8-2. Field survey of pearl millet cultivar diversity and downy mildew incidence (percentage)[a] on selected major hybrids and varieties in Maharashtra from 1993-1999.

Hybrid / Variety	93	94	95	96	97	98	99
BK 560[b]	90	70	-	-	50	-	80
Eknath 201[b]	-[c]	-	15	90	50	-	70
GK 1004[b]	-	-	80	20	75	95	60
MLBH 104[b]	53	20	90	80	70	75	75
MLBH 267[b]	0	40	20	90	60	75	52
Proagro 7701[b]	-	-	40	-	75	55	80
ICTP 8203[d]	0	0	2	0	1	2	0
Mallikarjuna[d]	0	-	0	0	0	-	-
Total entries planted	3	3	6	14	16	21	24

[a]Maximum incidence recorded in any of the observed fields.
[b]Hybrid.
[c]- Entry not planted.
[d]Open-pollinated variety.

Mechanisms of Virulence Evolution

Evolution of virulence in plant pathogens occurs by mutation, genetic drift, recombination, gene flow, and selection. In *S. graminicola*, recombination is important in creating and maintaining haplotype diversity, and considerable pathogenic variation has been reported among single-oospore and single-zoospore isolates from several populations (26, 28, 31). Molecular genetic variation in oospore- and zoospore-derived progeny has been demonstrated using DNA fingerprinting (18, 33) and Random Amplified Polymorphic DNA (RAPD; 15, 22). Host availability serves as a powerful force on pathogen populations and can result in the evolution of new virulence pathotypes (29, 32). For example, a highly virulent pathotype of *S. graminicola* from Jodhpur, Rajasthan appears to have evolved from a *S. graminicola* strain that can attack a local land race, *Nokha* (28). This pathotype caused a downy mildew epidemic during the 1997 rainy season crop at the farm of Central Arid Zone Research Institute, Jodhpur. Another virulent pathotype, specific to commercial hybrid MLBH 104, was identified in the state of Maharshtra (29) and resulted in the withdrawal of this hybrid.

As hypothesized by Leonard (14), selection intensities under greenhouse and field conditions for the same pathogen are not the same. In *S. graminicola*, the differences may reflect more intense selection pressure in the greenhouse due to higher pathogen densities, or that the field estimates include selection during the interval between seasons. The genetic composition of the host genotype has considerable impact on virulence selection, and this process is much faster with inbred lines than with hybrids or composite populations (32). Also, in greenhouse populations of *S. graminicola*, the fungus reproduces only asexually and only on seedlings. This process stabilizes allele frequencies specific to the particular host genotype(s) on which the pathogen is maintained and multiplied. In contrast to the greenhouse situation, sexual recombination can occur in the field, and new virulence gene combinations can result. In a host-parasite system, the amount of genetic variation for resistance and virulence can alter the population dynamics and equilibrium of the interacting species (20).

Use of Host Differential Lines. In host-pathogen systems where near-isogenic lines with known resistance genes are available, the use of host differentials has been very successful in monitoring and identifying new pathotypes or races of the pathogen. There are no such well-defined differential lines available in pearl millet, but we have identified pearl millet lines, *e.g.*, the IPMDMVN, that can discriminate between pathogen populations based on disease incidence. A number of downy mildew resistance QTLs specific to particular pathogen populations have been mapped (7). We are developing near-isogenic lines for several of these QTLs by using molecular marker-assisted backcrossing into a common, agronomically elite background. These inbred lines will be used as differentials for monitoring pathogen virulence.

There has been limited characterization of the pathogen variability present at many locations, and we currently have no good data sets from a number of representative locations in west and central Africa. Temporal and spatial variation in data from field nurseries is commonly reported (24), but these observations need to be confirmed in greenhouse tests.

Use of Molecular Markers. DNA markers are being used to characterize pathogen isolates and to determine the genetic structure of populations and its relationship to virulence diversity. The molecular methods have the advantages of precise scoring, economy when conducted on a large scale, and the simultaneous measurement of variation at several loci including those with co-dominant alleles. At ICRISAT, DNA fingerprinting with Simple Sequence Repeats (SSRs) has been used to detect genetic diversity among pathotypes of *S. graminicola* (18). Of the SSRs tested, $(GATA)_4$ identified the most polymorphism amongst the host genotype-specific pathogen isolates of *S. graminicola*. The same SSRs also were used to confirm the emergence of a new virulent pathotype specific to a recently introduced commercial hybrid cultivar MLBH 104 (29). More recently, Amplified Fragment Length Polymorphism (AFLP) markers also have been used to detect genetic variation among *S. graminicola* isolates from India. We often can score 50-70 bands with each primer-pair combination, and the proportion of polymorphic bands per primer pair ranges from 35-60%.

Studies of *S. graminicola* populations from Africa and India are in progress, but there have been problems due to the need to work with small samples collected from natural lesions in Africa. Oospore samples must be air-dried before being sent to the United States in order to satisfy import regulations. These samples often yield no visible precipitated DNA following standard extraction protocols, which leaves only PCR-based amplification methods for use in detecting variation at the DNA level. Even AFLP analysis requires at least 50 ng of starting material, considerably more than is often available. Consequently, protocols that permit "whole genome amplification," originally developed for use in human diagnostics and forensics, are being tested for use with *S. graminicola*. Recently, our attention has focused on Degenerate Oligonucleotide Primer (DOP)-PCR (23), primarily because this method yields a fairly uniform amplification of sequences throughout the genome. Visible quantities of DNA (> 30 ng/ml) can be amplified from as little as 10 pg to 1 ng of purified *S. graminicola* target DNA. Subsequent amplifications of the DOP-PCR products with RAPD or microsatellite primers yield a high level of polymorphism between the control samples.

Since whole genome amplification primers are not genome specific and field samples are unlikely to contain only fungal material, special precautions must be taken to ensure that DNA polymorphisms that distinguish isolates are not the result of contaminating DNA from other plants or microbes. At present, we hybridize polymorphic bands to *S. graminicola* genomic DNA to confirm the origin of the amplified sequences.

Management of Virulence

Pathogen variation is a major cause of failure for control programs that rely on fungicides (4, 5, 12, 25). For example, studies of late blight of potato, caused by *Phytophthora infestans*, over the last 20 years have shown that resistance can be readily developed to fungicides that target a single active site (6, 12). This resistance can result in decreased or lost activity by such products under field conditions, *e.g.*, the failure of metalaxyl to control late blight of potato (5, 6, 12). Metalaxyl is widely used for seed treatment to control downy mildew in pearl millet, but effective control is obtained for only 30-40 days. Depending on the weather conditions and the susceptibility of the cultivar(s), downy mildew symptoms may appear on late tillers and the ear heads may be malformed. However, neither the failure of metalaxyl nor the emergence of metalaxyl-resistant strains of *S. graminicola* has yet been reported.

Effective disease management can be obtained with cultivars that have durable resistance. The most effective utilization of host plant resistance requires knowledge of the genetics of resistance, the population genetics and evolutionary biology of the pathogen, and the interaction of crop management practices with host plant resistance. The use of multiple resistance genes to produce durable resistance often is effective because the probability that a pathogen will simultaneously mutate to virulence at multiple corresponding loci usually is small (17). Major gene resistance is most efficiently exploited when it is developed either to prevent the pathogen population from easily adapting or to minimize the selection pressure exerted on the pathogen population.

Molecular approaches can be employed either in (i) the pyramiding of two, or more, resistance genes in a host line through marker-assisted selection or (ii) through genetic modification of existing resistance genes. Good sources of resistance to different and/or multiple pathotypes are available (30), and the genetics and inheritance of resistance from some sources are known (21). Genes from several of these sources have been incorporated into elite breeding lines and hybrid parental lines through conventional breeding methods (9). Some of these resistance sources also have been used to develop populations, and the first DNA markers for QTLs against different pathotypes of the pathogen were identified during 1993-1994 (11). At least 25 QTLs have been mapped on seven pearl millet linkage groups. Some of these QTLs account for a significant proportion (20-80%) of the observed genetic variation within a given mapping population and are being transferred into elite hybrid parental lines by using marker-assisted backcrossing. A strategy for gene pyramiding and deployment also has been developed to reduce the damage caused by downy mildew (8, 37). This strategy combines marker-assisted selection, the multiline approach, and hybrid seed production procedures to generate hybrid cultivars that are agronomically uniform but variable in their resistance gene complements (37).

Conclusions

Significant advances have been made in understanding virulence diversity in the pearl millet downy mildew pathogen through field surveys and the IPMDMVN since the last global conference (13). However, the highly heterogeneous nature of the pathogen populations, and their ability to rapidly adapt to new host cultivars, means that new virulence pathotypes can quickly emerge whenever a new host plant resistance gene is deployed uniformly within a popular single-cross hybrid. This diversity in virulence limits the effective life span of high-yielding single-cross F_1 hybrid cultivars in India. Host-directed selection appears to be the key mechanism of evolution of new pathotypes in the pathogen population. Further studies of pathogenic and genetic variation in pathogen populations from diverse pearl millet growing areas will provide better knowledge of virulence diversity. The incorporation of downy-mildew-resistant QTLs specific to pathotypes through marker-assisted backcrossing into elite parental lines of hybrid cultivars should reduce the damage caused by the disease. Virulence monitoring through the IPMDMVN, field surveys, and the characterization of pathogen isolates for their virulence spectrum should continue to ensure optimal utilization and deployment of downy mildew resistance genes in pearl millet.

References

1. **Ball, S. L.** 1983. Pathogenic variability of downy mildew (*Sclerospora graminicola*) on pearl millet host cultivar reactions to infection by different pathogen isolates. *Ann. Appl. Biol.* **102:** 257-264.
2. **Ball, S. L., D. J. Pike, and C. V. Burridge.** 1986. Characterization of populations of *Sclerospora graminicola*. *Ann. Appl. Biol.* **108:** 519-526.
3. **Crute, I. R.** 1981. The host specificity of Peronosporaceous fungi and the genetic relationship between host and parasite, p. 237-253. *In* D. M. Spencer (ed.), *The Downy Mildews*. Academic Press, New York City, New York.
4. **Deahl, K. L., D. A. Inglis, and S. P. DeMuth.** 1993. Testing for resistance to metalaxyl of *Phytophthora infestans* isolates from northwestern Washington. *Am. Potato J.* **70**: 779- 795.
5. **Goodwin, S. P., C. D. Smart, R. W. Sandrock, K. L. Deahl, Z. K. Punja, and W. E. Fry.** 1998. Genetic change within populations of *Phytophthora infestans* in the United States and Canada during 1994 to 1996: Role of migration and recombination. *Phytopathology* **88**: 939-949.
6. **Goodwin, S. B., L. S. Sujokowski, and W. E. Fry**. 1996. Widespread distribution and probable origin of resistance to metalaxyl in clonal genotypes of *Phytophthora infestans* in the United States and western Canada. *Phytopathology* **86**: 793-800.
7. **Hash, C. T., and P. J. Bramel-Cox.** 2000. Marker applications in pearl millet, p. 155-168. *In* B. I. G. Haussmann, H. H. Geiger, D. E. Hess, C. T. Hash, and P. Bramel-Cox (eds.), *Application of Molecular Markers in Plant Breeding: Training Manual for a Seminar Held at IITA (16-17 August 1999, Ibadan, Nigeria)*. ICRISAT, Patancheru (A.P.), India.
8. **Hash, C. T., and J. R. Witcombe.** 1996. Resistance gene deployment strategies for open-pollinated crops using marker-assisted selection. *Downy Mildews Newsl.* **9:** 8.
9. **Hash, C. T., J. R. Witcombe, R. P. Thakur, S. K. Bhatnagar, S. D. Singh, and J. P. Wilson.** 1997. Breeding for pearl millet disease resistance. *Proc. Intl. Conf. on Genetic Improvement of Sorghum and Pearl Millet (22-27 September 1996, Lubbock, Texas)*: 337-372.
10. **Idris, M. O., and S. L. Ball.** 1984. Inter- and intra-continental sexual compatibility in *Sclerospora graminicola. Plant Pathol.* **33:** 219-223.
11. **Jones, E. S., C. J. Liu, M. D. Gale, C. T. Hash, and J. R. Witcombe.** 1995. Mapping quantitative trait loci for downy mildew resistance in pearl millet. *Theor. Appl. Genet.* **91:** 448-456.
12. **Kato, M., E. S. Mizubuti, S. B. Goodwin, and W. E. Fry.** 1997. Sensitivity to protectant fungicides and pathogenic fitness of clonal lineages of *Phytophthora infestans* in the United States. *Phytopathology* **87**: 973-978.
13. **King, S. B.** 1992. World review of pearl millet diseases: knowledge and future research needs, p. 95-108. *In* W. A. J. de Milliano, R. A. Frederiksen, and G. D. Bengston (eds.), *Sorghum and Millets Diseases: A Second World Review.* ICRISAT, Patancheru (A.P.), India.
14. **Leonard, K. J.** 1987. The host population as a selective factor, p. 163-179. *In* M. S. Wolfe and E. C. Caten (eds.), *Populations of Plant Pathogens: Their Dynamics and Genetics*. Blackwell Scientific Publications, Oxford, U.K.
15. **Mahmood, Z., D. S. Shaw, and J. R. Witcombe.** 1996. Molecular variation in *Sclerospora graminicola*. *Downy Mildews Newsl.* **9:** 6.
16. **Michelmore, R. W., M. N. Pawar, and R. J. Williams.** 1982. Heterothallism in *Sclerospora graminicola*. *Phytopathology* **72:** 1368-1372.
17. **Mundt, C. C.** 1991. Probability of mutation to multiple virulence and durability of resistance gene pyramids: further comments. *Phytopathology* **81:** 240-242.
18. **Sastry, J. G., W. Ramakrishna, S. Sivaramakrishnan, R. P. Thakur, V. S. Gupta, and P. K. Ranjekar.** 1995. DNA fingerprinting detects genetic variability in the pearl millet downy mildew pathogen (*Sclerospora graminicola*). *Theor. Appl. Genet.* **91:** 856-861.

19. **Shetty, H. S., and R. Ahmad.** 1981. Physiologic specialization in *Sclerospora graminicola*. *Ind. Phytopathol.* **34:** 307-309.
20. **Simms, E. L.** 1996. The evolutionary genetics of plant pathogen systems. *Bioscience* **46:** 136-145.
21. **Singh, S. D., and B. S. Talukdar.** 1998. Inheritance of a gene conferring complete resistance to pearl millet downy mildew. *Plant Dis.* **82:** 791-793.
22. **Sivaramakrishnan, S., P. Harikrishnan, R. P. Thakur, and V. P. Rao.** 1996. Use of RAPD markers for the detection genetic variability in pearl millet downy mildew pathogen (*Sclerospora graminicola*), p. 51-56. *In* V. L. Chopra, R. P. Sharma, and M. S. Swaminathan (eds.), *Agricultural Biotechnology. Proc. of the 2nd Asia Pacific Conf.* Oxford and IBH Publishing Co., New Delhi, India.
23. **Telenius, H., N. P. Carter, M. Nordenskjold, B. A. J. Ponder, A. Tunacliffe.** 1992. Degenerate oligonucleotide-primed PCR: General amplification of target DNA by a single degenerate primer. *Genomics* **13:** 718-725.
24. **Thakur, R. P.** 1995. Status of international sorghum anthracnose and pearl millet downy mildew virulence nurseries, p. 75-92. *In* J. F. Leslie and R. A. Frederiksen (eds.), *Disease Analysis through Genetics and Biotechnology: Interdisciplinary Bridges to Improved Sorghum and Millet Crops.* Iowa State University Press, Ames, Iowa.
25. **Thakur, R. P.** 1999. Pathogen diversity and plant disease management. *Ind. Phytopathol.* **52:** 1-9.
26. **Thakur, R. P., B. Pushpavathi, and V. P. Rao.** 1998. Virulence characterization of single-zoospore isolates of *Sclerospora graminicola* from pearl millet. *Plant Dis.* **82:** 747-751.
27. **Thakur, R. P., and V. P. Rao.** 1997. Variation in virulence and aggressiveness among pathotypes of *Sclerospora graminicola* on pearl millet. *Ind. Phytopathol.* **50:** 41-47.
28. **Thakur, R. P., V. P. Rao, and C. T. Hash.** 1998. Emergence of a highly virulent pathotype of *Sclerospora graminicola* at Jodhpur, Rajasthan, India. *Intl. Sorg. Millet Newsl.* **39:** 140-142.
29. **Thakur, R. P., V. P. Rao, J. G. Sastry, S. Sivaramakrishnan, K. N. Amruthesh, and L. D. Barbind.** 1999. Evidence for a new virulent pathotypes of *Sclerospora graminicola* on pearl millet. *J. Mycol. Plant Pathol.* **29:** 61-69.
30. **Thakur, R. P., V. P. Rao, S. D. Singh, and S. S. Navi.** 1997. Characterization of downy mildew resistance in pearl millet. *J. Mycol. Plant Pathol.* **27:** 6-16.
31. **Thakur, R. P., and K. G. Shetty.** 1993. Variation in pathogenicity among single-oospore isolates of *Sclerospora graminicola*, the causal organism of downy mildew in pearl millet. *Plant Pathol.* **42:** 715-721.
32. **Thakur, R. P., K. G. Shetty, and S. B. King.** 1992. Selection for host-specific virulence in asexual populations of *Sclerospora graminicola*. *Plant Pathol.* **41:** 626-632.
33. **Thakur, R. P., S. Sivaramakrishnan, J. G. Sastry, and P. K. Ranjekar.** 1996. Pathogenic and genetic variation in *Sclerospora graminicola*. *Downy Mildews Newsl.* **9:** 6.
34. **Werder, J., and S. L. Ball.** 1992. Variability of pearl millet downy mildew (*Sclerospora graminicola*) and its implications for breeding for stable and durable resistance. *Trop. Pest Manag.* **38:** 43-47.
35. **Werder, J., and S. K. Manzo.** 1992. Pearl millet diseases in western Africa, p. 109-114. *In* W. A. J. de Milliano, R. A. Frederiksen, and G. D. Bengston (eds.), *Sorghum and Millets Diseases: A Second World Review.* ICRISAT, Patancheru (A.P.), India.
36. **Williams, R. J., S. D. Singh, and M. N. Pawar.** 1981. An improved field screening technique for downy mildew resistance in pearl millet. *Plant Dis.* **65:** 239-241.
37. **Witcombe, J. R., and C. T. Hash.** 2000. Resistance gene deployment strategies in cereal hybrids using marker-assisted selection: gene pyramiding, three-way hybrids, and synthetic populations. *Euphytica* **112:** 175-186.

9

Pearl Millet Ergot Research: Advances and Implications

Ram P. Thakur and K. N. Rai

Abstract

Ergot, caused by *Claviceps fusiformis* Loveless, is a major floral disease of pearl millet [*Pennisetum glaucum* (L.) R. Br.], affecting both the quantity and the quality of the grain produced because of the neurotoxic alkaloids contained in the fungal sclerotia. The disease has been more severe in single-cross F_1 hybrids than in open-pollinated varieties. Genetic resistance to ergot in pearl millet is mediated through rapid pollination and fertilization. In the last 20 years, important advances were made in understanding the biology and epidemiology of the pathogen and resistance in the host, in developing effective field screening techniques, in identifying sources of genetic resistance, and in developing breeding lines resistant to ergot and other diseases.

Introduction

Ergot, caused by *Claviceps fusiformis*, is an important disease in many pearl millet (*Pennisetum glaucum*) growing areas of Asia and Africa (11, 21). The disease has rarely reached epidemic proportions on a wide scale since its first reported occurrence in 1958 in the Satara district of the Indian state of Maharashtra (17). Prior to the cultivation of hybrids in the late 1960s, research on pearl millet ergot was conducted primarily in India, and more in response to concerns about its toxicological effects on human and cattle (8) than for its effects on grain yield. Like downy mildew [*Sclerospora graminicola* (Sacc.) Schröt], ergot became a major grain yield reducer when commercial cultivation of genetically uniform single-cross F_1 hybrids of pearl millet based on cytoplasmic-nuclear male-sterility (CMS) began in India in the 1970s (33). Research related to disease management, with respect to host plant resistance, was initiated by the All India Coordinated Millets Improvement Project during the late 1960s and at ICRISAT during the late 1970s. About a quarter century ago, a Consultative Group meeting on downy mildew and ergot of pearl millet identified ergot as second in importance only to downy mildew at the global level (9).

At ICRISAT, active research on ergot was conducted for almost 15 years and included studies of pathogen biology, epidemiology, host resistance, and pollen management. An international network of collaborating scientists was established to test the stability of genetic resistance in pearl millet lines through the International Pearl Millet Ergot Nursery (IPMEN). Research on ergot was discontinued at ICRISAT about 10 years ago because of changes in research priorities. King (7), in the Second World Review of Pearl Millet Diseases in 1992, identified significant questions still to be addressed in ergot research, including: (i) determining the relative importance of inocula from two alternative grass hosts (*Panicum antidotale* and *Cenchrus ciliaris*) in disease epidemiology, (ii) determining the relative ability of ascospores and conidial inocula to incite disease, (iii) identifying factors that influence sclerotial germination, (iv) understanding genotype × environment × pathogen interactions and their effects on alkaloid production, (v) evaluating the severity of the ergot toxicity problem, (vi) successfully implementing pollen management methods for the control of ergot in farmers' fields, and (vii) determining the importance and technical feasibility of breeding high-yielding ergot-resistant hybrids. Since 1992, there have been about 40 research publications on pearl millet ergot, dealing primarily with host plant resistance in breeding lines and with biological and cultural control methods. Although not all of the issues King identified have been addressed, some significant advances have been made that have practical implications in the management of ergot, not only of pearl millet but also of sorghum and other cereals.

Floral Biology and Ergot Infection

Flowering in pearl millet is protogynous, which means that the crop is highly cross-pollinated. Large variation exists for the length of time between the emergence of stigmas and the initiation of anthesis from florets both within and between panicles of the same genotype (26). The time course of flowering events and the rapidity of pollination both influence the process of ergot infection and disease development (26, 31, 37). Ergot infection usually is prevented by the difference in time required for pollen germination and pathogen spore germination (26, 31). Pollen takes about 30 min to germinate; whereas spores (both conidia and ascospores) take 12-16 hr. Thus, if pollen grains lodge on a stigma at the same time as spores, or even several hours after the spores, the pollen germinates quickly and the pollen germ tube reaches the ovary much faster than the germ tube of the fungal spore. Once the ovary is fertilized, the ergot pathogen cannot infect (31, 35, 36), since pollination-induced styler constriction and withering of the stigma prevents infection hyphae from reaching the ovary (36, 37).

These results impact our understanding of the role of pollination in disease epidemiology, the mechanism of host resistance, and the development of effective field screening techniques. Under field conditions, ergot becomes severe when pollination is inhibited because airborne pollen is not adequately dispersed due to extensive rain during flowering (21). This result also explains why high-level fertility restoration is important in F_1 hybrid cultivars and why open-pollinated varieties generally incur less ergot infection than F_1 hybrids. Pearl millet lines that have short protogyny and abundant pollen production often escape ergot infection due to rapid pollination.

The ergot screening technique, which is based on our understanding of pollination effects, minimizes or eliminates pollen interference with fungal spore germination on the stigma surface (21). The field screening technique (31, 32) involves: (i) bagging panicles at the boot-leaf stage with parchment selfing bags, which allows stigma emergence in a pollen-protected environment; (ii) inoculating panicles 3-4 days later by briefly opening the bags and spraying the panicles at the full protogyny stage [> 75% fresh (unwithered, unpollinated) stigma] with an aqueous conidial suspension (1×10^6 conidia/ml; 2 ml/panicle) produced from honeydew of infected panicles; (iii) providing high humidity (> 80% relative humidity) with overhead sprinklers twice a day (10:00 and 17:00) for 30 min on rain-free days; and (iv) removing bags two weeks later and scoring ergot severity on a standard scale (31).

Role of Sclerotia and Conidia in Disease Epidemiology

Ergot sclerotia from the infected panicles that fall to the ground and those mixed with seed during threshing serve as the primary inoculum source for the next crop (27). Following rain, these sclerotia germinate and release numerous airborne ascospores that land on flowering pearl millet panicles, where they can germinate and infect the florets through the stigma. Under conditions of high relative humidity (> 80%) and moderate temperature (20-30°C), honeydew symptoms appear within 4-6 days, and sclerotia become visible 15-20 days after inoculation (27). Honeydew contains numerous macroconidia that produce microconidia, which are disseminated by rain splash and wind currents both within and between fields.

At least two grass species, *Cenchrus ciliaris* (18) and *Panicum antidotale* (20), can serve as alternative hosts of the pathogen and help initiate primary infection in pearl millet. Thus, both sclerotia and conidia play important roles in pathogen survival, spread, and epidemiology.

These findings suggest that effective ergot management will include removal of sclerotia from seed, eradication of grasses that serve as alternate hosts, and implementation of cultural practices that reduce sclerotial inoculum in the soil, *e.g.*, crop rotation and deep plowing.

Role of Cytoplasmic-Nuclear Male-Sterility in Ergot Resistance

CMS and ergot susceptibility were correlated in a study involving seven pairs of male-sterile lines (A-lines), their corresponding fertile lines (B-lines), 10 pollinator lines (R-lines), and the resulting 48 F_1 hybrids (25). Significant differences in ergot severity were observed between

TABLE 9-1. Performance of pearl millet lines under field inoculation and artificially induced epidemic conditions during the 1984, 1987, and 1988 rainy seasons at ICRISAT, Patancheru, India [after Thakur *et al.* (23)].

Cultivar	Res.[a]	Ergot Severity (%)		Yield
		Field Screen	Epidemic	(tons/ha)
WC-C75	S	72 ± 6.9	24 ± 11	1.99
ICMPES 8	MR	21 ± 1.4	2 ± 1.2	1.87
ICMPES 9	MR	23 ± 2.9	1 ± 0.2	1.79
ICMPES 32	MR	12 ± 1.5	1 ± 0.2	1.64
ICMPES 5	R	2 ± 0.5	0	1.6
ICMPES 28	R	4 ± 0.6	1 ± 0.2	2.26
ICMPES 34	R	2 ± 1.5	1 ± 0.1	1.93
SE ±				0.19

[a]Resistance: S, susceptible; MR, moderately resistant; and R, resistant.

TABLE 9-2. Ergot severity and selfed seed set on pearl millet cultivars under artificially-induced epidemic conditions in a greenhouse in 1988 at ICRISAT, Patancheru, India [after Thakur *et al.* (23)].

Cultivar	Res.[a]	Ergot Severity (%)	Selfed Seed Set (%)
BK 650	S	41 ± 3.5	6 ± 2.5
841 A	S	58 ± 4.4	0
WC-C75	S	41 ± 5.5	20 ± 6.4
ICMPES 8	MR	14 ± 2.0	36 ± 3.9
ICMPES 9	MR	19 ± 2.4	63 ± 3.6
ICMPES 29	R	1 ± 0.6	38 ± 4.9
ICMPES 34	R	6 ± 1.9	74 ± 3.9

[a]Resistance: S, susceptible; MR, moderately resistant; and R, resistant.

A-lines, B-lines, and their hybrids. The higher susceptibility of A-lines, and the A × R hybrids relative to corresponding B-lines and B × R hybrids, may be attributable to the cytoplasmic, and cytoplasmic × nuclear interacttions that occur during flowering, pollen formation, and fertilization. CMS is associated with increased susceptibility to ergot in wheat, caused by *Claviceps purpurea* (10, 15, 19), and in sorghum, caused by *Claviceps sorghi* (5). These increases could be due to sterile cytoplasm *per se*, or cytoplasm-induced male sterility (12). Rapid production of high levels of pollen by inbred lines and high fertility restoration in hybrids could result in pollination-based escape resistance to ergot in pearl millet (31, 33).

Relevance of Functional Field Resistance in Pearl Millet

For a disease such as ergot, for which high levels of genetic resistance may not be widely available, a search for functional field resistance may be more fruitful. Our artificial inoculation technique for ergot resistance is effective, but it does not simulate natural field epidemics. The technique is quite severe and may eliminate breeding lines with moderate levels of resistance. We used a field screen that simulates natural epidemic conditions (> 90% relative humidity, frequent rain showers, and temperatures of 20-30°C during flowering) to identify pearl millet lines with functional field resistance. Epidemic conditions are created by repeated spray inoculation of protogynous panicles with an aqueous honeydew spore suspension under high relative humidity and moderate temperatures. Highly resistant lines were highly resistant both to artificial inoculation and to the induced epidemic under field and greenhouse conditions. However, lines moderately resistant to artificial inoculation in the field also were highly resistant to the induced epidemic conditions (23). Based on these results, we concluded that an ergot severity threshold level of 20-30% following artificial field inoculation provides an adequate level of functional field resistance to a natural ergot epidemic (Table 9-1).

Grain yields from moderately and highly resistant lines under induced epidemic conditions were comparable to those of a high-yielding, open-pollinated comercial variety, WC-C75. Thus, an ergot-resistant open-pollinated variety might outyield a commercial, open-pollinated variety, such as WC-C75. Under artificial epidemic conditions, there were large reductions in seed set with significant variation among the lines. This reduction generally was smaller in the moderately and highly resistant lines than in the susceptible cultivars (Table 9-2). These results suggest that breeding efforts should aim at moderate levels of resistance in combination with good agronomic background, as combining a high resistance level, which is a complex genetic trait, with good agronomic background may be difficult to achieve in the near term.

Identification and Development of Genetic Resistance

Desirable levels of genetic resistance were not detected in many of the germplasm accessions and breeding lines (24). Resistant lines, however, were developed by crossing less-susceptible plants (< 5% disease severity) from a few lines and selecting for resistant progeny under high disease pressure for several generations (3, 6, 32). At ICRISAT, pedigree selection for improved resistance increased the frequency of resistant plants from 0 in the F_2 to 70% in the F_6 (Table 9-3). A number of lines from several of these resistant plants were developed, and plants from these lineages were then intercrossed at the F_6 stage to enrich for resistance alleles in a second round

TABLE 9-3. Pedigree selection for ergot resistance in pearl millet from crosses involving lines with low susceptibility from 1978-1980 at ICRISAT, Patancheru, India [after Thakur *et al.* (32)].

Ergot Severity (%)	Generation				
	F_2	F_3	F_4	F_5	F_6
< 1	0[a]	0	0	1	14
1-10	0	2	15	14	5
11-20	0	6	21	19	18
21-30	6	11	28	16	9
31-40	9	16	22	20	2
41-50	18	16	9	15	1
> 50	67	49	6	16	0

[a]Percentage of lines in each class.

TABLE 9-4. Ergot severity (%) of pearl millet hybrids formed from ergot-susceptible and ergot-resistant parental lines evaluated during the 1991 and 1992 rainy seasons at ICRISAT, Patancheru, India [after Rai and Thakur (12)].

Seed Parent		Pollen Parent			
Line	Res.[a]	Susceptible		Resistant	
		ICMR 356	H 77/ 833-2	ICMER 92	ICMER 98
81 A	S	90[b]	90	81	67
81 B	S	91	81	37	45
843 A	S	94	94	80	88
843 B	S	87	81	67	38
ICMA 91113	R	96	87	25	31
ICMB 91113	R	82	56	1	11
ICMA 91115	R	84	54	8	19
ICMB 91115	R	72	38	3	3

[a]Resistance reaction: S, susceptible; R, resistant.
[b]LSD ($p < 0.05$) = 14; computed by using the Genstat program following a mixed-model analysis of variance.

of segregating progeny. Resistance stability was tested at multiple locations in India and Africa by including the most promising lines in the collaborative International Pearl Millet Ergot Nursery (IPMEN), which was planted in several countries in western and central Africa. The IPMEN was operated for 10 years, and several lines with high levels of resistance at diverse locations for up to 10 years were identified (21, 24, 28).

About 300 ergot-resistant inbred lines and populations (ICMER numbers) have been evaluated for their agronomic attributes and reactions to diseases (24). Lines with high resistance levels and good agronomic characters have been identified that can be used as donors for developing ergot-resistant varieties (22). A number of these lines and populations have improved agronomic traits and combined resistance to ergot, smut, downy mildew, and rust (22, 24, 29).

Attempts were made to select ergot-resistant lines by using tissue culture techniques, sclerotial extracts, and culture filtrates of *C. fusiformis* as selective agents (2, 16). Regenerated plants were isolated from callus masses of susceptible lines that survived repeated exposure to the culture filtrates. Regenerated plants were more resistant than the original susceptible lines (4).

Ergot resistance is a recessive polygenic trait with significant cytoplasmic × nuclear interactions (30). For a hybrid to be ergot resistant, both parents must carry the same resistance alleles (12, 34). This homozygosity probably will reduce heterosis for grain and stover yield. More detailed genetic studies are needed to better understand the nature and inheritance of ergot resistance.

Breeding Ergot-Resistant Hybrid Parents and Varieties

Management of ergot through genetic resistance is the most effective and economical control method. Utilization of host resistance requires development of a large-scale field screening technique, identification of resistance sources, an understanding of the genetic basis(es) of resistance, and the effective incorporation of the genetic resources into a breeding program. Attempts have been made to incorporate resistance into both the seed parent and pollinator by using backcross and pedigree breeding procedures (1, 24). Ergot-resistant male-sterile lines have been bred through pedigree breeding at ICRISAT-Patancheru (13), and attempts have been made to elucidate the impacts of fertility restoration and disease resistance on hybrid production (12). Resistant hybrids, in which both A-lines and pollen parents were resistant (Table 9-4), have been obtained. The A_1 cytoplasm, with its reduced male fertility, had a large and significant effect on ergot susceptibility of hybrids, but the contribution of the nuclear genome of the A-line was 1.8 times greater that of the cytoplasm (12).

Three promising male-sterile lines (ICMA 91113, ICMA 91114, and ICMA 91115) were evaluated for their hybrid yield potential by using ergot-resistant pollinator parents. Several of these hybrids yielded 11-17% more grain than the commercial hybrid ICMH 423. These lines also had a time to 50% flowering similar to or earlier than that of ICMH 423 (Table 9-5). Two of the 15 hybrids had only 8% and 20% ergot severity, as compared to 45-86% in the others (24).

TABLE 9-5. Performance of F_1 hybrids based on ergot resistant A-lines and resistant (ER) or susceptible (ES) R-lines during the 1990 and 1991 rainy seasons at ICRISAT, Patancheru, India [after Thakur *et al.* (24)].

Pedigree	Line Type	Grain Yield Range (ton/ha)	Time to 50% Flowering (days)	Ergot Severity (%)
A × ER-R	15 hybrids	2.87–3.54	43-50	8-86
A × ES-R	12 hybrids	2.43–3.65	40-47	58-94
ICMH 423	Control	3.11	45	90
ICMH 451	Control	2.94	48	92
ICMV 155	Control	3.17	46	69
SE		0.17	0.03	

TABLE 9-6. Influence of pollen donor (PD) line [SC-2(M)5-4] on ergot incidence and severity in four susceptible hybrids (ICH 118, ICH 220, BJ 104, and ICH 206) cultivated under artificial inoculation in the field from 1978-1981 at ICRISAT, Patancheru, India [after Thakur *et al.* (33)].

	Ergot Incidence (%)	Ergot Severity (%)
Hybrid alone	90	53
Hybrid:PD (2:1)	47	9
Hybrid:PD (4:1)	56	11
Hybrid mix:PD (4:1)	35	2
Hybrid:PD (8:1)	54	8
SE	1.5	1.1

Three synthetics (ICMS 8031, ICMS 8032, and ICMS 8034) constituted from ergot-resistant inbred lines had a grain yield similar to that of a popular open-pollinated commercial variety, WC-C75, and a hybrid, BJ 104, but these synthetics were late maturing. The synthetic varieties had 12-15% ergot severity compared with 24% in WC-C75 and 54% severity in BJ 104 (24). The ICRISAT Ergot Resistant Composite (ERC) constituted from 52 ergot-resistant populations in 1985 is one of the most productive composites in its maturity class and is being used as a source population in crosses to form a new late composite at ICRISAT.

Management Options

Pollen management. Ergot control generally relies on a pollination-induced escape mechanism. Under field conditions, pollination-based ergot control is common for heterogeneous open-pollinated varieties and land races, as flowering continues for a longer time and pollen is available throughout flowering. In contrast, in single-cross F_1 hybrids, where flowering is synchronous and more uniform, ergot control is not effective. Pollen management can effectively control ergot in susceptible hybrids under high artificial disease pressure in the field (33). An early-maturing, ergot-tolerant, high pollen producing line serves as the pollen donor. The pollen donor produces abundant pollen when the hybrid cultivar is flowering, resulting in rapid pollination and fertilization, and thereby protects the hybrid florets from ergot infection. If a hybrid is sown as part of a seed mixture (4 hybrid: 1 pollen donor line) or in alternate rows with the pollen donor line (2:1, 4:1, or 8:1), then under artificial epidemic conditions both ergot incidence and severity are significantly reduced, and grain yield is increased (Table 9-6). These strategies may permit high-yielding, ergot-susceptible hybrid cultivars to be grown in ergot endemic areas. To be successful, the pollen donor line must have the same seed color as that of the hybrid, so that seed color in the grain is consistent. This technology is ready for testing in on-farm trials.

Biological control. Two species of *Fusarium*, *F. sambucinum* and *F. semitectum* var. *majus*, and *Cerebella andropogonis* parasitize honeydew and sclerotia (21). Ergot severity of pearl millet hybrid BJ 104 was significantly reduced when spores of *F. semitectum* var. *majus* were sprayed on panicles at protogyny under field conditions (Table 9-7). This treatment also reduced the number of sclerotia formed per panicle and led to a rapid disintegration of the sclerotia (14). These findings have implications in survival of sclerotia and disease epidemiology. Further research is required to determine the suitability of this biocontrol method at the farm level.

Cultural control. Use of sclerotium-free seed, eradication of collateral hosts, and adjusting planting dates to avoid rain during flowering are among the cultural measures that help to minimize ergot incidence. Use of clean, sclerotia-free seed; eradication of alternative grass hosts, *e.g.*, *C. ciliaris* and *P. antidotales*; and deep summer plowing to bury sclerotia deep enough to prevent germination all help reduce the primary inoculum load. Delaying sowing of pearl millet until August-September often increases ergot incidence, particularly in northern India. This increase occurs because the cooler nights during pearl millet flowering favor ergot development. Crops planted by June generally escape ergot due to prevailing high temperatures and lower levels of airborne inoculum.

Conclusions

Research on ergot of pearl millet at ICRISAT has been successful and has increased our basic understanding of the diseases and led to successful management strategies. Ergot epidemics in pearl millet have been infrequent and geographically limited, so there has been little economic

TABLE 9-7. Effect of *Fusarium semitectum* var. *majus* (FSM) on pearl millet ergot, *Claviceps fusiformis* (CF) during the 1985 rainy season at ICRISAT, Patancheru, India [after Rao and Thakur (14)].

Treatment	Ergot Severity (%)	Sclerotium Formation (%)	Sclerotium Disintegration (%)
CF at protogyny	92	60	52
CF + FSM at protogyny	87	33	77
FSM at protogyny	14	6	–
FSM at honeydew	90	54	76
SE	12	1.9	6.8

pressure for high levels of research. In India, single-cross pearl millet hybrid cultivars occupy about 55% of the total 10 million ha, and ergot occasionally is a major problem in some parts of the country. Large-scale epidemics probably have not occurred because of genetic diversity in pearl millet cultivars, small farm size, and the variation in the cultural practices implemented by farmers. Most pearl millet hybrid seed production occurs during the dry part of the year, when weather is not suitable for ergot development. In Africa, hybrid cultivars are not yet grown by farmers, and ergot has not yet been a major problem. Host plant resistance combined with biological and cultural methods should lead to development of an integrated disease management strategy.

References

1. **Andrews, D. J., S. B. King, J. R. Witcombe, S. D. Singh, K. N. Rai, R. P. Thakur, B. S. Talukdar, S. B. Chavan, and P. Singh.** 1985. Breeding for disease resistance and yield in pearl millet. *Field Crops Res.* **11:** 241-258.
2. **Bajaj, Y. P. S., P. S. Phul, and S. K. Sharma.** 1980. Differential tolerance of tissue culture of pearl millet to ergot extract. *Ind. J. Exp. Biol.* **18:** 429-432.
3. **Chahal, S. S., K. S. Gill, P. S. Phul, and N. B. Singh.** 1981. Effectiveness of recurrent selection for generating ergot resistance in pearl millet. *SABRAO J.* **13:** 184-186.
4. **Chahal, S. S., and S. B. Sharma.** 1988. Development of new ergot resistant pearl millet plants through *in vitro* procedure, p. 260. *Proc. 5th Intl. Cong. Plant Pathol. (22-27 August 1988, Kyoto, Japan).*
5. **Futtrell, M. C., and O. J. Webster.** 1965. Ergot infection and sterility in grain sorghum. *Plant Dis. Rept.* **49:** 680-683.
6. **Gill, K. S., S. S. Chahal, and P. S. Phul.** 1980. Strategy to develop ergot resistance in pearl millet, p. 159-161. *In* V. P. Gupta and J. L. Minocha (eds.), *Trends in Genetical Research in* Pennisetums. Punjab Agricultural University, Ludhiana, India.
7. **King, S. B.** 1992. World review of pearl millet diseases: knowledge and future research needs, p. 95-108. *In* W. A. J. de Milliano, R. A. Frederiksen, and G. D. Bengston (eds.), *Sorghum and Millets Diseases: A Second World Review.* ICRISAT, Patancheru (A.P.), India.
8. **Mantle, P. G.** 1992. Ergot disease of pearl millet: Toxicological significance and a mechanism for disease escape, p. 129-132. *In* W. A. J. de Milliano, R. A. Frederiksen, and G. D. Bengston (eds.), *Sorghum and Millets Diseases: A Second World Review.* ICRISAT, Patancheru (A.P.), India.
9. **Nene, Y. L., and S. D. Singh.** 1976. Downy mildew and ergot of pearl millet. *Pest Article and News Summary* **22:** 366-385.
10. **Puranik, S. B., and D. E. Mathre.** 1971. Biology and control of ergot on male-sterile wheat and barley. *Phytopathology* **61:** 1075-1080.
11. **Rachie, K. O., and J. V. Majmudar.** 1980. *Pearl Millet.* Pennsylvania State University Press, University Park, Pennsylvania.
12. **Rai, K. N., and R. P. Thakur.** 1995. Ergot reaction of pearl millet hybrids affected by fertility restoration and genetic resistance of parental lines. *Euphytica* **83:** 225-231.
13. **Rai, K. N., R. P. Thakur, and A. S. Rao.** 1998. Registration of pearl millet parental lines ICMA 92666 and ICMB 92666 with multiple disease resistance. *Crop Sci.* **38:** 575.
14. **Rao, V. P., and R. P. Thakur.** 1988. *Fusarium semitectum* var. *majus*–a potential biocontrol agent of ergot (*Claviceps fusiformis*) of pearl millet. *Ind. Phytopathol.* **41:** 567-574.
15. **Reitz, L. P., and K. A. Lucken.** 1972. Hybrid wheat–A review of its status and prospects. *Agric. Sci. Rev.* **10:** 7-16.
16. **Sharma S. B., and S. S. Chahal.** 1990. Development of regenerants from pearl millet cell lines resistant to culture filtrate of *Claviceps fusiformis*. *Plant Dis. Res.* **5:** 175-181.
17. **Shinde, P. A., and V. P. Bhide.** 1958. Ergot of *bajra* (*Pennisetum typhoides*) in Bombay state. *Curr. Sci.* **27:** 499-500.
18. **Singh, G. K., L. Vyas, and B. N. Bhat.** 1983. Occurrence of pearl millet ergot on *Cenchrus ciliaris* Pers. in Rajasthan. Ind. J. Agric. Sci. **53:** 481-483.
19. **Stockopf, N. C., and R. K. Rai.** 1972. Cross-pollination in male sterile wheat in Ontario. *Can. J. Plant Sci.* **52:** 387-393.
20. **Thakur, D. P., and Z. S. Kanwar.** 1978. Ability of naturally incident *Claviceps microcephala* from *Panicum antidotale* to produce ergot symptoms in *Pennisetum typhoides*. *Ind. J. Agric. Sci.* **49:** 540-542.
21. **Thakur, R. P., and S. B. King.** 1988. *Ergot Disease of Pearl Millet.* Information Bulletin no. 24. ICRISAT, Patancheru (A.P.), India.
22. **Thakur, R. P., and S. B. King.** 1988. Registration of four ergot resistant germplasm of pearl millet. *Crop Sci.* **28:** 382.
23. **Thakur, R. P., S. B. King, and V. P. Rao.** 1989. Expression of ergot resistance in pearl millet under artificially induced epidemic conditions. *Phytopathology* **79:** 1323-1326.
24. **Thakur, R. P., K. N. Rai, S. B. King, and V. P. Rao.** 1993. *Identification and Utilization of Ergot Resistance in Pearl Millet.* Research Bulletin no. 17. ICRISAT, Patancheru (A.P.), India.
25. **Thakur, R. P., V. P. Rao, and S. B. King.** 1989. Ergot susceptibility in relation to cytoplasmic male sterility in pearl millet. *Plant Dis.* **73:** 676-678.
26. **Thakur, R. P., V. P. Rao, and S. B. King.** 1992. Factors associated with ergot resistance in pearl millet. *Ind. Phytopathol.* **45:** 326-330.
27. **Thakur, R. P., V. P. Rao, and R. J. Williams.** 1984. The morphology and disease cycle of ergot caused by *Claviceps fusiformis* in pearl millet. *Phytopathology* **74:** 201-205.
28. **Thakur, R. P., V. P. Rao, R. J. Williams, S. S. Chahal, S. B. Mathur, N. B. Pawar, S. D. Nefade, H. S. Shetty, G. Singh, and S. G. Bangar.** 1985. Identification of stable resistance to ergot in pearl millet. *Plant Dis.* **69:** 982-985.

29. **Thakur, R. P., S. D. Singh, and S. B. King.** 1988. Registration of four populations of pearl millet germplasm with multiple disease resistance. *Crop Sci.* **28:** 381-382.
30. **Thakur, R. P., B. S. Talukdar, and V. P. Rao.** 1983. Genetics of ergot resistance, p. 737. *Abstr. Contrib. Papers XVth Intl. Cong. Genet., (12-21 Dec. 1983, New Delhi, India), Part 2.* Oxford & IBH, New Delhi, India.
31. **Thakur, R. P., and R. J. Williams.** 1980. Pollination effects on pearl millet ergot. *Phytopathology* **70:** 80-84.
32. **Thakur, R. P., R. J. Williams, and V. P. Rao.** 1982. Development of ergot resistance in pearl millet. *Phytopathology* **72:** 406-408.
33. **Thakur, R. P., R. J. Williams, and V. P. Rao.** 1983. Control of ergot in pearl millet through pollen management. *Ann. Appl. Biol.* **103:** 31-36.
34. **Virk, D. S., S. S. Chahal, and B. K. Mangat.** 1987. Factorial analysis for ergot resistance in pearl millet. *Crop Improve.* **14:** 153-156.
35. **Willingale, J., and P. G. Mantle.** 1985. Stigma constricttion in pearl millet, a factor influencing reproduction and disease. *Ann. Bot.* **56:** 109-115.
36. **Willingale, J., and P. G. Mantle.** 1987. Stigmatic constriction in pearl millet following infection by *Claviceps fusiformis*. *Physiol. Molec. Plant Pathol.* **30:** 247-257.
37. **Willingale, J., P. G. Mantle, and R. P. Thakur.** 1986. Post-pollination stigmatic constriction, the basis of ergot resistance in selected lines of pearl millet. *Phytopathology* **76:** 536-539.

10

The Dynamic Multiline Population: An Alternative Approach to Durable Resistance?

Jeffrey P. Wilson and Roger N. Gates

Abstract

The dynamic multiline population breeding strategy integrates favorable attributes from the gene-stacking and multiline approaches. We evaluated the effectiveness of this approach for resistance breeding in millet. Backcross derivatives of pearl millet Tift 23DB were developed with rust resistance from 18 local varieties from Burkina Faso, from three *Pennisetum glaucum* subsp. *monodii* accessions, and from two elite inbreds. Four cycles of open pollination were made, the last two in the field during rust epidemics. Cytoplasmic male-sterile (CMS) counterparts of the populations in the A_4 cytoplasm were included and advanced simultaneously with the B population. Hybrids with Tift 383 were produced on CMS cycles 1 (C_1) through C_4. The frequency of hybrid seedlings with resistance to single-uredinium isolates of *Puccinia substriata* var. *indica* averaged 18-38% in C_1 and C_4 hybrids, respectively. The experimental dynamic multiline population hybrids and the commercial footage hybrids Tifleaf 1 (TL 1) and Tifleaf 2 (TL 2) were evaluated in three yield trials in 1998-1999. Disease-free biomass yields did not differ among hybrids. Across trials, final rust severity averaged 67% on TL1 and TL 2, 47% on C_1 and C_2, and 30% on C_3 and C_4. Rust severity was reduced by 13% per cycle, and the digestible biomass of rust-infected forage was increased 4% per cycle.

Introduction

Many useful sources for tolerance to abiotic stresses or resistance to pathogenic microorganisms are preserved in germplasm collections. Disease resistance genes are a unique genetic resource in that their use frequently results in an erosion of available resources. As genes for resistance are incorporated into popular cultivars, virulent pathogen biotypes may be selected for and render the resistance gene commercially ineffective. Resistance genes are an important factor in stabilizing crop yields, but their effectiveness can be ephemeral. The "boom and bust" cycles associated with release of new resistant varieties and the accompanying selection for virulent biotypes has resulted in a search for more durable resistance. Developing durable resistance strategies reduces the rate of erosion of resistance gene resources by limiting the ability of pathogen populations to adapt to host resistance.

Generally considered the "holy grail" of resistance breeding and the subject of considerable discussion and research, durable resistance can be identified only in retrospect, after the widespread distribution and extensive cultivation of varieties with resistance for a significant period of time. Durable disease resistance is sometimes obtained by a fortuitous combination of resistance genes, but, given our present understanding of the mechanisms of resistance, these gene combinations cannot be identified before they are deployed. Both reductionist and systems approaches have been used to understand or develop durable resistance. The reductionist approach evaluates the potential utility of various resistance phenotypes, broadly categorized as hypersensitive, partial, adult-plant, or recovery resistances. Systems approaches focus on resistance deployment strategies such as gene stacking and multiline approaches. The gene stacking approach incorporates different resistance genes into a plant selection, which is then grown as a homogeneous population, whereas the multiline strategy disperses resistance genes among different plants, which are grown as a heterogeneous population.

When deliberately attempting to develop durable disease resistance, it can be useful to examine the relative stability of resistance in model systems. Disease resistance, where it exists, of land races and cross-pollinating forages usually is more durable than that found in hybrid cultivars. Resistance is the least durable in the inbred cultivars. Several explanations for this phenomenon exist, one of which is the level of diversity for resistance genes in the host population.

Pearl millet, *Pennisetum glaucum* (L.) R. Br, is an excellent model for evaluating alternative methods for breeding for durable disease resistance. The crop has been or is grown as land races, as a cross-pollinating forage, and as hybrids. As variety development progresses from land races to hybrids, the potential for downy mildew and rust epidemics increases, as does the difficulty of breeding for stable resistance, particularly in regions where research resources are limited.

The dynamic multiline population approach can be used in cross-pollinated hybrid crops (4, 9). This approach integrates the best attributes of both the gene stacking and multiline strategies, in that a large proportion of the plants in the heterogeneous host population have multigenic resistance. The method is likely to provide greater longevity for resistance genes, since the genetic composition of the host varies from year to year.

The general procedure of producing a dynamic multiline population begins with the selection of backcross derivatives in an elite genetic background, followed by random mating in the presence of disease pressure for several generations. A CMS counterpart to the fertile B population can be simultaneously advanced by including rows of a CMS derivative of the elite genetic background in isolation crossing blocks. Dynamic multiline population hybrids can be made by using the CMS-dynamic multiline populations as the seed production parent in a cross with an appropriate pollinator.

In theory, the dynamic multiline population is a potentially powerful approach to managing resistance genes, particularly as the number of resistance genes increases. The number of resistance gene associations in a randomly mating population in Hardy-Weinberg equilibrium is an exponential function of the number of heterozygous resistance loci in the population. For example, with dominant, unlinked genes for resistance, the number of phenotypes is $P = 2^x$ and the number of genotypes is $G = 3^x$, where x is the number of resistance genes in the population. The number of phenotypes, or genotypes, in a standard multiline population is the number of resistance genes ($P = G = x$), if each component of the multiline has a single resistance gene. The Shannon diversity index measures the number of phenotypes, or genotypes, and the evenness of the distribution of their frequencies (2). The index [$I = -p_i\ln(p_i)$, where p_i is the frequency of ith individual] is a linear function of the number of resistance genes for both phenotypes and genotypes if the resistance genes are dominant and segregate independently (Fig. 10-1). The Shannon index of a standard multiline population is a curvilinear function of

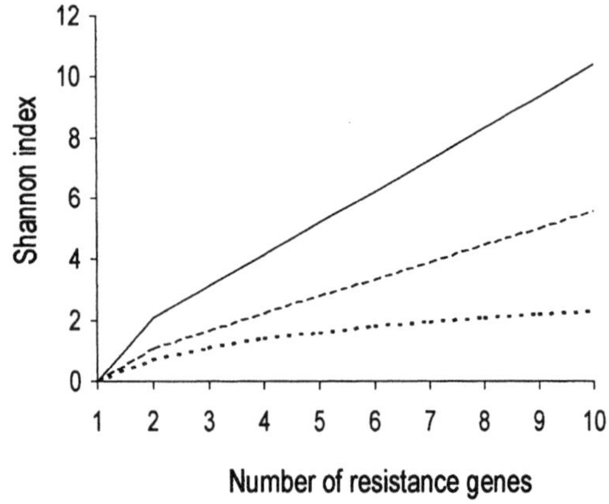

FIGURE 10-1. Shannon diversity index as a function of the number of resistance genes in a standard multiline population (···), and for the phenotypic (- - - -) and genotypic (—) diversity expected in a dynamic multiline population in Hardy-Weinberg equilibrium with unlinked, dominant genes for resistance.

the number of resistance genes, and approaches an asymptote.

Development of dynamic multiline population hybrids is most practical in systems that: (i) have diverse sources of resistance for a genetically diverse pest population, (ii) contain a host in which open-pollinated population development and hybrid production can occur, and (iii) has elite inbreds suitable for producing high-yielding hybrids. A dynamic multiline population with resistance to rust, caused by *Puccinia substriata* Ell. & Barth. var. *indica* Ramachar & Cumm., has been developed in forage pearl millet. The pathogen population is highly diverse for virulence, and the available resistance genes have only a limited effectiveness against the spectrum of races (3). The objective of this study was to determine if the dynamic multiline population strategy was a practical and effective approach to breeding disease-resistant hybrids in pearl millet.

Materials and Methods

Population development. Twenty-three different sources of rust resistance: Burkina Faso landraces–BF 17, BF 19, BF 34, BF 41, BF 49, BF 53, BF 122, BF 137, BF 156, BF 178, BF 191, BF 201, BF 202, BF 221, BF 296, BF 299, BF 334, and BF 337; *P. glaucum* subsp. *monodii*–PS 727, PS 748, and PS 756; and improved inbreds–ICMP 83506 and Tift 8677, were backcrossed into a Tift 23DB background. Inoculations of the isolines with single-uredinium isolates resulted in different race-specific reactions (3), suggesting that the isolines possessed different resistance genes or alleles. Dwarf, rust-resistant progeny from the fourth backcross generation with desirable agronomic traits were selected from field trials in 1994. Seed from these selections were planted in the greenhouse. Plants of each resistance source and of Tift 23DA$_4$ were randomized within a crossing tent in the greenhouse and allowed to cross-pollinate in isolation. Cross-pollination was facilitated by circulating air with a fan and manually shaking the plants daily during pollination.

Equal quantities of seed from each open-pollinated resistance source were bulked to form the fertile dynamic multiline population-B cycle 1. This seed was used to establish approximately 1000 plants. The seed formed on Tift 23DA$_4$ were bulked to create the male-sterile dynamic multiline population-A$_4$ cycle 1, which was planted in pots interspersed among the fertile dynamic multiline population B, and cross-pollinated in the greenhouse. All seed produced on the dynamic multiline population-B plants were bulked to form the fertile dynamic multiline population-B cycle 2. Seed produced on the dynamic multiline population-A$_4$ plants were bulked to form the sterile dynamic multiline population-A$_4$ cycle 2 population.

Seed of cycle 2 of dynamic multiline population-B and dynamic multiline population-A$_4$ were planted in an isolated crossing block in the field in 1995. Four, 2-row plots (42 m long) of the dynamic multiline population-A$_4$ population were separated and surrounded by 4-row plots of dynamic multiline population-B. Seed was combine-harvested and bulked within the fertile and sterile rows to form dynamic multiline population-B and dynamic multiline population-A$_4$ cycle 3. Cycle 3 seed was used to plant a crossing block in 1996 consisting of three 2-row plots (60 m long) of dynamic multiline population A$_4$ separated and surrounded by 4-row plots of the dynamic multiline population-B. Combine-harvested seed were bulked within the fertile and sterile rows to form dynamic multiline population-B and dynamic multiline population-A$_4$ cycle 4. Plantings were made late in 1995 and 1996 to permit selection pressure from natural rust infection to alter the populations.

Hybrid populations for yield tests were developed in 1997 and 1998 by planting five replicates of 6-m long, two-row plots of sterile dynamic multiline population-A$_4$ seed from cycles 1-4 in the field. Each replicate, and the entire test, was surrounded by the moderately rust-susceptible pollinator Tift 383. Seed produced on sterile dynamic multiline population-A$_4$ plots were bulked within cycle across replications.

Greenhouse evaluations. The sterile A$_4$, fertile B, and hybrid cycles were inoculated as seedlings at the 3-4 leaf stage with bulk rust inoculum collected in the field in 1995. Five replicates (pots) of each population were arranged in a randomized complete block design. The experiment was conducted three times, and the proportion of resistant plants in each replication determined. As plants in the host populations were heterogeneous for resistance, and as pathogen inoculum was likely to be heterogeneous for virulence, it was not unusual to observe both resistant and susceptible infection sites on a plant. We scored a plant with mixed resistant infection types as positive for the presence of resistance genes. Data were analyzed within populations. Sums of squares were partitioned into experiment, replication, and cycle components.

Field evaluations. Experiments were planted in 1998 and 1999 to evaluate hybrids for disease resistance and the production of digestible biomass. In 1998, hybrids Tifleaf 1 (Tift 23DA × Tift 383), Tifleaf 2 (Tift 85DA × Tift 383), and the experimental hybrids dynamic multiline population-A$_4$ cycle C$_n$ × Tift 383 (where n = cycle number and ranges from 1 to 4) produced in 1997 were planted in two yield trials on 19 June 1998 and 11 August 1998. Plots were 6 m long, with 2 m spacing between rows, arranged in a randomized complete block design with six replications. Plots were harvested with a forage chopper. The first harvest of the first planting was free of rust and was taken on 12 August 1998. The regrowth and the second trial were naturally infected by rust. Disease severity was recorded weekly from 5-28 October 1998. Rust-infected forage from both experiments was harvested 30 October 1998. Plot green weights were measured. Forage samples were weighed green and after drying at 60°C for 48 h to determine the proportion of dry matter. Dry matter yield was calculated as the product of green yield and proportion of dry matter. *In vitro* dry matter digestibility of samples was estimated by using a filter bag system following the procedures outlined by Vogel *et al.* (6). Three subsamples were analyzed per plot. Digestible dry matter yield was calculated as the product of dry matter yield and *in vitro* dry matter digestibility.

In 1999, one experiment was planted on 8 June 1999. The hybrids planted were identical to those planted in 1998, but the test also included hybrids produced in 1998 in addition to those planted in 1997. Experimental design and plot dimensions were similar to the yield test in 1998. The first harvest, taken on 13 August 1999, was rust free. Rust severity was determined at weekly intervals from 16 September to 8 October 1999. The second harvest was taken on 12 October 1999. Digestible dry matter yield of plots was determined as described above. Data for rust-free and rusted forages were analyzed separately. Sums of squares within the experiments were partitioned into replication and hybrid sources of variation. The statistical analysis found no significant effect of year of hybrid production in the 1999 trial, so the year of production was treated as a subsample during final statistical analyses. Data from an overall analysis of variance were partitioned into trial, replication within trial, hybrid, and subsample within replicate sources of variation.

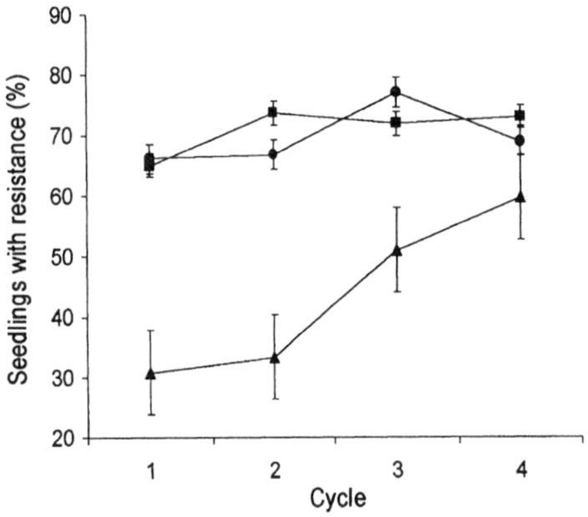

FIGURE 10-2. Frequencies of rust resistance in pearl millet fertile (dynamic multiline population-B, ▪) and sterile (dynamic multiline population-A_4, •) dynamic multiline populations, and the hybrids of dynamic multiline population-A_4 × Tift 383 (▲), when inoculated with a bulk urediniospore rust inoculum.

Results

Greenhouse evaluations. When inoculated with a bulk rust inoculum, the frequency of resistant plants in the fertile dynamic multiline population-B and sterile dynamic multiline population-A_4 populations did not change ($p < 0.05$) (Fig. 10-2). The frequency of rust resistance in seedlings increased in advanced cycles of hybrid populations (dynamic multiline population-A_4 cycles × Tift 383). Cycle 3 and cycle 4 hybrids did not differ from each other, but their means were larger than those for cycle 1 and cycle 2 hybrids. This increase probably reflects the increased frequency of multiple gene resistance (gene stacking) in the dynamic multiline population, since the potential for multiple resistance genes in a hybrid is less than that of the heterozygous parental plant.

Field evaluations. In the harvest of rust-free forage (1998 trial 1, harvest 1 and 1999 harvest 1), there were no significant differences ($p < 0.05$) among hybrids in dry matter yield, digestibility, or digestible dry matter yield. Minor differences in height existed, with cycle 3 and cycle 4 hybrids tending to be slightly taller.

Data trends were similar for diseased plots from both planting dates in 1998; however, planting date was a significant source of variation ($p < 0.05$) for both yield and disease, so the data from the different planting dates were analyzed separately. Results were similar ($p > 0.05$) for diseased plots from both seed lots for the test evaluated in 1999. The year of seed production was not a significant source of variation ($p > 0.05$) for yield and disease; therefore, hybrids from each year of production were considered to be subsamples within treatments for statistical analyses. Hybrids did not differ for dry matter yield in the 1998 trials, but did in 1999. Digestibility, digestible dry matter yield, and the height of the latter cycles of experimental hybrids usually were greater than those of Tifleaf 1 and Tifleaf 2, and increased with advancing cycle. Differences were more pronounced in 1999 (Table 10-1). Linear regression of average digestible dry matter yield against cycle indicated a 4% increase in digestible biomass yield per cycle. The final rust severity in the experimental hybrids generally was less than that observed on Tifleaf 1 and Tifleaf 2, and was reduced as the number of cycles increased (Table 10-1). Linear regression analyses indicated a 13% decrease in rust severity with each cycle.

Discussion

The experiments described here suggest that the dynamic multiline breeding strategy may be effective in developing rust-resistant hybrids. Resistance increased with advancing cycles of the dynamic multiline populations under both field and greenhouse conditions, and the resistance of later-cycle hybrids resulted in increased digestible biomass. Populations of pearl millet with rust resistance that are randomly mated have greater resistance levels (4, 5) and higher forage yields (4). The experiments we describe here demonstrate that developing hybrids from dynamic multiline populations is feasible.

Given the relationship between rust infection and digestible dry matter yield (7, 8), the digestible biomass yield could be improved beyond that obtained by the cycle 4 dynamic multiline population hybrid. At the level of rust severity observed on the cycle 3 and cycle 4 hybrids, significant improvements in digestible biomass yield result from small improvements in rust resistance. Levels of resistance in advanced cycles were stable or improving in inoculations both in the greenhouse and in field trials. Improved resistance of hybrids may be obtained after additional cycles of disease-associated selection on the resistant parental population, or by producing hybrids with a resistant pollinator.

Several principles and attributes of dynamic multiline populations are clear. First, a dynamic multiline population can be developed through the application of the multiline breeding strategy to cross-pollinated hybrid crops, in a manner that integrates the most favorable attributes of the gene stacking and multiline approaches to resistance breeding. Genetic heterogeneity and overlapping gene sets are inevitable in randomly mating populations. If no selection pressure is applied to the population, the theoretical levels of phenotypic and genotypic diversity within the host population are exponential functions of the numbers of hetero-

TABLE 10-1. Digestible dry matter yield of forage and rust severity of dynamic multiline population hybrids at Tifton, Georgia.

Hybrid	Digestible Dry matter Yield (kg/ha)				Rust Severity (% foliage)			
	1998-1[a]	1998-2[a]	1999	Pool	1998-1[a]	1998-2[a]	1999	Pool
Tifleaf 1	3700	3500	2900	3300	75	61	68	68
Tifleaf 2	4100	4000	2900	3500	76	63	62	66
Cycle 1 × Tift 383	3700	4500	3200	3600	60	51	43	49
Cycle 2 × Tift 383	4300	4000	3600	3800	53	52	36	44
Cycle 3 × Tift 383	4100	4700	3700	4000	35	29	28	30
Cycle 4 × Tift 383	4100	4500	3800	4100	39	24	27	29
LSD (0.05)	1100	1100	590	480	17	7.5	7.4	5.7

[a]Two experiments were evaluated in 1998: the 1999 trial was planted with seed produced in two different years.

zygous resistance genes. Disease-associated selection pressure should select for the most effective resistance alleles and resistance gene combinations, presumably correspondding to the most prevalent races of the pathogen, and thereby reduce the effective genetic diversity. However, since pearl millet plants are rarely killed by rust before pollination (Wilson, *unpublished*), it is unlikely that the less effective or recessive resistance alleles will be completely lost from the population.

There are no empirical data on the rate of pathogen coevolution when the pathogen is confronted with a dynamic multiline population. In theoretical models of a complex mixture of genotypes with overlapping gene sets, the number of component lines in the mixture and the number of heterozygous resistance genes are the primary determinants of whether the pathogen population can attain genetic equilibrium (1). The theoretical level of resistance diversity with a dynamic multiline population is unmatched by other conventional breeding approaches. Diverse resistance genes and gene combinations ensure a high level of heterogeneity within the host population and should serve as a buffer against rapid changes in virulence gene frequencies in the pathogen population. As virulence allele frequencies change, the resulting change in disease pressure by prevailing and predominant pathogen races should select for the most effective resistance allele(s) and gene combinations during foundation seed production. Theoretical models of the evolution of complex races generally have focused on static host populations. In contrast to the traditional approach of maintaining resistance in a static host population, the dynamic multiline breeding method can, at least potentially, confront a variable and dynamic pathogen population with a variable and dynamic host population. These advantages should encourage the application of the dynamic multiline population concept in other commercially important pathosystems.

Acknowledgements

This research was funded in part by the U.S. Department of Energy grant DE-FG05-93ER20099, and was a cooperative investigation of the USDA-ARS and the University of Georgia College of Agricultural and Environmental Sciences.

References

1. **Marshall, D. R., and B. S. Weir.** 1985. Multiline varieties and disease control. 5. The "dirty crop" approach with complex mixtures of genotypes based on overlapping gene sets. *Theor. Appl. Genet.* **69:** 463-474.
2. **Pielou, E. C.** 1975. *Ecological Diversity.* John Wiley & Sons, New York City, New York.
3. **Tapsoba, H., and J. P. Wilson.** 1996. Pathogenic variation in *Puccinia substriata* var. *indica* in the southeastern United States and screening for resistance in pearl millet germplasm. *Plant Dis.* **80:** 395-397.
4. **Tapsoba, H., and J. P. Wilson.** 1999. Increasing genetic complexity in host mixtures through intermating to manage rust of pearl millet. *Phytopathology* **89:** 450-455.
5. **Tapsoba, H., J. P. Wilson, and W. W. Hanna.** 1997. Improvement of resistance to rust through recurrent selection in pearl millet. *Crop Sci.* **37:** 365-369.
6. **Vogel, K. P., J. F. Pedersen, S. D. Nasterson, and J. J. Toy.** 1999. Evaluation of a filter bag system for NDF, ADF and IVDMD forage analysis. *Crop Sci.* **39:** 276-279.
7. **Wilson, J. P., and R. N. Gates.** 1999. Disease resistance and biomass stability of forage pearl millet hybrids with partial rust resistance. *Plant Dis.* **83:** 733-738.
8. **Wilson, J. P., R. N. Gates, and W. W. Hanna.** 1991. Effect of rust on yield and digestibility of pearl millet forage. *Phytopathology* **81:** 233-236.
9. **Wilson, J. P., W. W. Hanna, and K. Bondari.** 1993. Directed use of germplasm resources in breeding rust resistant pearl millet. *Trends in Agric. Sci. (Plant Pathol.)* **1:** 67-74.

Part III

Sorghum Ergot

11

Recommendations from the Ergot Working Group at the Third Global Conference on Sorghum and Millets Diseases

Reported by *Gary N. Odvody*

This group met during the Third Global Conference on Sorghum and Millet Diseases at Guanjuato, Mexico, 23-29 September 2000, to update the research recommendations made at the Global Conference on Ergot of Sorghum held in Sete Lagoas, Brazil, 1-8 June 1997. The meeting was open to all interested conference participants, and no list of participants peculiar to this meeting is available.

Genetic Resources

- The screening of sorghum germplasm for sources of resistance should continue. Public researchers are encouraged to submit seed of putative resistant selections to Neal McLaren and Malcolm Ryley for screening under controlled environments and to determine reactions to potentially different pathotypes.
- The following sites should also be considered for screening of accessions against *Claviceps africana*. The best choice may depend on the question being asked because these and other sites have not yet been prioritized by climate and genetic diversity of the pathogen. Countries included are: Argentina, Australia, Brazil, Ethiopia, Honduras, India, Mexico, South Africa, and the United States (both mainland and Puerto Rico).
- Current and future pathogen variability identified either by molecular genetic techniques or by host-pathogen interaction should be evaluated for changes in pathogenicity with a standard set of cultivars and protocols. Where feasible, or in quarantine facilities, these studies should include standard isolates representing pathogen diversity from other global locations.
- Genetic resources should be evaluated using standardized techniques that utilize ergot potential as a measure of resistance.
- Develop standard screening techniques for evaluating disease escape including measures of pollen fertility, pollen shed, fertilization, and seed set.

Control and Management Strategies

Phytosanitary practices

- In areas where *C. africana* is well established:
    - The presence of sclerotia/sphacelia and honeydew-encrusted seed should not be used as a phytosanitary restriction measure, because they pose negligible risk for initiating disease compared to other sources of inoculum.
    - Seed intended for planting should be treated with an appropriate fungicide, *e.g.*, captan or thiram, that eliminates inoculum of the ergot pathogen on seed surfaces and sclerotia/sphacelia.
- Suspected sclerotia/sphacelia should be examined at 10× magnification for confirmation.
- The saline flotation test for identification is not reliable and should not be used as it may yield both false positive and false negative results.
- The demonstrated differences between Australia/Asia and Africa/Americas populations of *C. africana* should be considered in phytosanitary decisions.

Host Range of C. africana

- *Sorghum bicolor*, *S. bicolor* ssp. *bicolor*, *Sorghum halepense*, and other sorghum species are recognized as hosts for *C. africana*. Additional collateral hosts should be recognized only after Koch's postulates have been demonstrated under natural conditions.

Chemical Control

- Improved chemical control methods are needed, especially for hybrid seed production. Alternate fungicides with modes of action that differ from those of triazoles need to be identified to minimize selection pressure for fungicide resistance in the pathogen population. Indiscriminate use of any fungicide should be avoided.

Epidemiology

- Continue development, validation, and implementtation of disease predictive models including regional calibration. Models should be integrated with other disease management components to aid in ergot management decisions including fungicide application, planting dates, and the deployment of cold-tolerant hybrids.
- The epidemiological significance of collateral hosts of *C. africana* should be evaluated.

Monitoring of Sorghum Ergot

- Establish a mechanism to provide for rapid exchange of information on the global spread, distribution, and severity of ergot. An ergot page on the World Wide Web may be the best means to achieve this goal.

Pollen Management

- Pollination/fertilization events are the primary ergot escape mechanism for sorghum. Continue studies of host-pathogen-environment interactions affecting sorghum pollination/fertilization.

Comprehensive and Alternate Control Strategies

- Develop a comprehensive disease management plan or control model from which appropriate components can be selected for application by geographical region. Pursue alternate control methods and evaluate integrated control approaches to complement and conserve other disease control strategies.

Biology of *Claviceps* spp.

- Continue research on the function of sclerotial tissues in the disease and life cycle of sorghum ergot, particularly perennation.
- Determine spatial and temporal dynamics of the production, dispersal, survival, and function of asexual inoculum.
- Toxicological studies should address potential differences in toxicity associated with different hosts, different environments, and variable pathogen populations.
- Determine the role of saprophytic/parasitic fungi and other microorganisms in toxin production, toxicological synergisms, and potential biocontrol.
- Develop multidisciplinary collaborations for toxicological studies and feeding trials of ergot-contaminated sorghum.
- Develop germplasm collections of *Claviceps* spp. from centers of genetic diversity including the highlands of East Africa.
- Evaluate the effects of alternate/collateral hosts on the genetic composition of sorghum-infecting *Claviceps* spp.
- Monitor pathogen populations for variability in virulence and host range

Molecular Studies of *Claviceps* spp.

- Evaluate fertility/incompatibility genetic mechanisms for genes that could be used to transform sorghum for resistance to ergot.
- Continue molecular level approaches for the evaluation of pathogen population biology.
- Refine fingerprinting techniques to enable more rapid determination of the relationships between ergot collections at the species and population levels.
- Confirm the presence of putative new *C. sorghi* strains in India with classical and molecular characters.

12

Distribution and Diversity of the Sorghum Sugary Disease Pathogens in India

Ranajit Bandyopadhyay, V. Muthusubramanian, Paul W. Tooley, Sukumar Chakraborty, Sylvie Pažoutová, and Shrishail S. Navi

Abstract

Until 1991, *Claviceps sorghi* was considered the only perfect stage of *Sphacelia sorghi*, the causal agent of the sugary disease of sorghum in India. The presence of *Claviceps africana*, a distinct pathogen and another perfect stage of *S. sorghi*, was reported during the 1990s in Africa, Australia, and North and South America. In India, the occurrence of *C. africana* was confirmed at a few locations in 1999 and 2000. In this chapter, we describe the geographical distribution and diversity of the sugary disease pathogens in India based on an extensive survey of nine states and examination of infected panicles containing honeydew and sclerotia/sphacelia from seven states. The pathogen was isolated from these specimens and characterized using sphacelial, conidial, and sclerotial morphology; secondary conidia production; colony morphology and growth rate; and molecular markers. Based on these studies, *C. africana* is widespread in seven sorghum-growing states of India and is currently the dominant sugary disease pathogen of sorghum in India. *Claviceps sorghi* is now restricted to a few locations in the states of Andhra Pradesh and Maharashtra.

Introduction

Ergot, or sugary disease, of sorghum is a serious disease that affects the production of F_1 hybrid seed. Sugary disease is particularly severe in male sterile lines (A-lines), either when nonsynchronous flowering of A-line and restorer lines (R-lines) occurs, or when adverse environmental conditions result in a lack of viable pollen and delayed seed set. In India, losses of 10-80% have been reported in hybrid seed production fields (1). Three species have been reported to cause sugary disease on sorghum in different parts of the world. These species are *Claviceps sorghi* Kulkarni, Seshadri and Hegde in India (5), *Claviceps africana* Frederickson, Mantle and de Milliano in Zimbabwe (4), and *Claviceps sorghicola* Tsukiboshi, Shimanuki and Uematsu in Japan (15). Recently, the occurrence of *Claviceps africana* was independently confirmed in India following molecular (10) and biochemical analyses (3). The imperfect stages of both *C. sorghi* and *C. africana* are named *Sphacelia sorghi* McRae (6).

Although both *C. sorghi* and *C. africana* have been reported in India, the distribution and relative abundance of these two pathogens in different sorghum growing areas is unknown. In this chapter, we briefly review the Indian literature on sugary disease of sorghum and summarize the current distribution and diversity of sugary disease pathogens in India. Our current understanding is based on studies carried out at ICRISAT during 1999-2000 to evaluate the diversity of sugary disease patho-

TABLE 12-1. Distribution of *Claviceps sorghi* and *Claviceps africana* cultures isolated from sugary disease affected sorghum panicles collected from different states in India.

State	Number of Locations	Number of Collections	Number of Cultures	
			C. africana	*C. sorghi*
Andhra Pradesh	18	29	25	4
Karnataka	21	29	23	0
Maharashtra	11	13	10	1
Rajasthan	1	3	3	0
Tamil Nadu	4	5	5	0
Uttar Pradesh	7	10	10	0
Gujarat	1	8	8	0
Total	63	97	84	5

gens in different sorghum-growing areas. We carried out disease surveys in eight states in India, recorded incidence and severity of the disease in different areas, collected infected plant materials, cultured the pathogens from field collections, established their pathogenicity, and characterized pathogen isolates using morphological, cultural, pathogenicity, and molecular features.

Disease Distribution

Sugary disease was first recorded on sorghum in 1917 after McRae (6) observed the sphacelial stage in the erstwhile Mysore state in India in 1915. Since then, the disease has been reported in other Indian states such as Maharashtra, Karnataka, Tamil Nadu, Andhra Pradesh and Delhi, and Haryana. The sclerotial stage of the pathogen was first recorded in Koilkuntla, Andhra Pradesh (12). The disease is endemic in Maharashtra (Vidarbha and Ahmednagar regions), Karnataka, Andhra Pradesh, and Tamil Nadu. During 1999-2000, we conducted disease surveys in Delhi, Haryana, Maharashtra, Karnataka, Andhra Pradesh, Uttar Pradesh, Tamil Nadu, and Gujarat. In these surveys, we found no sugary disease in Haryana and Delhi in 1999, and the disease severity was low in the Bundelkhand area of Uttar Pradesh due to dry conditions that prevailed during post-flowering period of sorghum in these areas. However, severe sugary disease was recorded in major sorghum growing areas of Maharashtra, Karnataka, Andhra Pradesh, Uttar Pradesh, Tamil Nadu, Rajasthan, and Gujarat. In September 1999, a severe epidemic occurred in the Rampur and Moradabad districts of Uttar Pradesh, in which almost all of the spikelets of late-flowering plants were infected. In October 1999, towards the end of rainy season, an epidemic occurred in the Machinenapally village of Mahboobnagar district of Andhra Pradesh. In September 2000, several other *mandals* (particularly Kalwakurthy *mandal*) adjoining Machinenapally village also were severely affected. These three epidemics affected local land races of sorghum and led to a total failure of grain harvest.

TABLE 12-2. Cultural characteristics of *Claviceps sorghi* and *Claviceps africana* on T_2 agar medium.[a]

Characters	*C. sorghi*	*C. africana*
Growth rate (mm/day)	0.12 ± 0.005 (0.11-0.12)[b]	0.06 ± 0.014 (0.03-0.09)
Maximum radial growth (cm)	3.35 ± 0.11 (3.25-3.52)	1.71 ± 0.42 (1.12-2.93)
Colony type	Puffy or cottony, submerged with diffused margin	Compact, raised, margin distinct
Puckering nature	None	Low to high
Sporulation	Present	Absent

[a]Average of data from five isolates of *C. sorghi* and 84 isolates of *C. africana*.
[b]Values are mean ± standard deviation, with range in parentheses.

Pathogen Distribution

Until recently, *C. sorghi* was considered the only sugary disease pathogen in India, and *C. africana* was thought to prevail in all other sorghum-growing areas. However, *C. africana* was identified recently from samples collected from a few locations in India (3, 10). From the infected materials collected during the disease surveys in 1999-2000, we cultured 89 isolates and determined their cultural characters. A comparison of cultural and molecular characters (see below) led us to group the isolates putatively as either *C. sorghi* or *C. africana*. We identified *C. africana* in all of the sorghum-growing regions of Maharashtra, Karnataka, Andhra Pradesh, Tamil Nadu, Uttar Pradesh, Rajasthan, and Gujarat (Table 12-1). *Claviceps sorghi* was restricted to northern Andhra Pradesh (Adilabad district) and Maharashtra (Yavatmal district), showing that *C. africana* is now more widely distributed in India than *C. sorghi*.

Fungal Morphology

Cultural characteristics. There are several reports of cultural characters of *C. sorghi* from India (13). On T_2 medium (8), *C. sorghi* produces a white mycelium with abundant aerial hyphae. The colony growth rate is slow compared to that of the rye ergot pathogen *C. purpurea* at 26-28°C (11). On T_2 medium, the culture sporulated, in the form of honeydew secretions, 15 days after inoculation and the color of the honeydew changed from colorless to pale brown or pinkish at later stages (7).

We found that *C. sorghi* and *C. africana* vary widely with respect to the rate of colony growth, colony type, puckering of the colony surface, and sporulation on medium. The culture of *C. sorghi* was white, puffy, cottony, and submerged, with diffused margins, while *C. africana* produced a white, raised, compact colony, with distinct margin (Table 12-2). *Claviceps sorghi* sporulated with honeydew secretions at the center of colonies, but *C. africana* did not sporulate on this medium, confirming the earlier report of Bogo and Mantle (3). *Claviceps sorghi* grew faster than *C. africana* in culture (Table 12-2). A comparison of our results to those obtained previously (7, 11, 13) from India suggests that *C. sorghi* might have been the only causal agent for sugary disease of sorghum recorded in India prior to the 1990s.

Sphacelia. Sphacelia are white, rounded to egg-shaped structures present in infected sorghum florets protruding between the glumes (2). The difference in sphacelial morphology between *C. sorghi* and *C. africana* was described by Frederickson *et al.* (4). The sphacelia of *C. sorghi* are cylindrical, curved or straight, bilaterally grooved, and visible 1-2 days after honeydew exudation begins (8-10 days after inoculation). In contrast, *C. africana* has bulky, soft, white, highly convoluted, oval to spherical sphacelia that were visible 6-7 days after inoculation (4). Variations in size and shape of sphacelia also were observed in our collections. *Claviceps sorghi* had more oblong to elliptical, and larger (4.3-4.7 × 2.9-3.0 mm), sphacelia than did *C. africana*. Sphacelia of *C. africana* were elliptical to conical or sometimes spherecal, measuring 2.9-3.6 × 1.9-2.2 mm (Table 12-3).

Conidia. The fungus produces three types of single-celled, hyaline spores: oblong to oval macroconidia, spherical microconidia, and pear-shaped secondary conidia. Macroconidia of *C. sorghi* are hyaline, elliptical or ovate with smooth ends, and with distinct vacuoles measuring 8-19 × 4-6 μm (5). Frederickson *et al.* (4) describe macroconidia of *C. africana* as oblong to oval measuring 9-17 × 5-8 μm, and slightly constricted at the center, with two polar vacuoles. In our studies, the macroconidia of *C. africana* were more oblong to elliptic and measured 10.2-18 × 6.4-9 μm, whereas *C. sorghi* produced more cylindrical, slender macroconidia measuring 9-18 × 5.1-7.7 μm (Table 12-3). *C. sorghi* and *C. africana* cannot be distinguished based on the size or shape of the microconidia.

At high relative humidity, macroconidia germinate iteratively, giving rise to secondary conidia. Secondary conidiation is a common feature of *C. africana* and is occasionally observed in *C. sorghi* (1). We compared the degree of secondary conidiation in isolates of *C. sorghi* and *C. africana* at different temperature (10, 15, 20, 25, 30, and 35°C) and relative humidity (80, 85, 90, 95, and 100% RH). Isolates of *C. africana* produced 8 to 18 times more secondary conidia at 25°C than isolates of *C. sorghi*. For both pathogens, maximum secondary conidiation occurred at 25°C followed by 20°C, and at 100% RH followed by 95% RH. *C. africana* produced more secondary conidia than *C. sorghi* at all temperature and RH levels.

Sclerotia. Bandyopadhyay *et al.* (1) summarized morphological and biochemical features that have been used to differentiate sclerotia of *C. africana* from *C. sorghi*. The sclerotia of *C. africana* are hard, spherical to oval, and largely confined within the host glumes only rarely protruding for more than a few mm beyond the glumes.

TABLE 12-3. Morphological characteristics of sphacelia, conidia, and sclerotia of *Claviceps sorghi* and *C. africana*.

Structure	*C. sorghi*	*C. africana*
Sphacelia[a]		
Size (mm)	4.3-4.7 × 2.9-3.1[b]	2.9-3.6 × 1.9-2.3
Shape and color	Oblong, cylindrical to elliptical, white	Conical, elliptical to spherical, white
Macroconidia[a]		
Size (μm)	9.0-18.0 × 5.1–7.7	10.2–18.0 × 6.4-9.0
Shape and color	Cylindrical to elliptical, hyaline	Oblong to elliptical, hyaline
Microconidia[a]	Spherical, hyaline, 2.6-3.9 μm	Spherical, hyaline, 2.6-3.9 μm
Sclerotia (with sphacelial cap)[c]		
Size (mm)	8.2-9.6 × 2.0-2.3	4.1-4.4 × 1.4-1.9
Shape and color	Cylindrical to conical, light yellowish brown	Conical to spherical, light yellowish brown
Sclerotia (without sphacelial cap)[c]		
Size (mm)	3.7-3.9 × 1.8-2.0	2.2-2.4 × 1.2-1.3
Shape and color	Oblong to cylindrical, light reddish brown	Conical to spherical, dark reddish brown

[a]Summary of data obtained from two isolates of *C. sorghi* and eight isolates of *C. africana*.
[b]Values are ranges.
[c]Summary of data obtained from four isolates of *C. sorghi* and three isolates of *C. africana*.

Sclerotia of *C. sorghi* are thin, elongate, soft or hard, and protrude as much as 15 mm beyond the glumes. The alkaloid dihydroergosine is synthesized in the sclerotia of *C. africana*, but not in *C. sorghi*. Based on their size and shape, sclerotia of both *C. africana* and *C. sorghi* were found in India during the 1999-2000 field survey (Table 12-3). We did not test the sclerotia of the Indian *C. africana* strains for the presence of dihydroergosine. Collections from varieties CSH 9, JK 22, Yellow Jowar, and MSH 51 all had well-formed sclerotia, suggesting that the production of sclerotia is favored on these genotypes.

Molecular Characterization

Sclerotia of *C. africana* do not germinate easily, and characters associated with its perfect stage have been studied only once (4). Molecular markers have been used to elucidate the phylogenetic and taxonomic relationships between *Claviceps* spp. (8, 9). Differences in nucleotide sequence of the ribosomal internal transcribed spacer region 1 (ITS1) and 5.8S rDNA regions can differentiate *C. sorghi* and *C. africana* (10). Occurrence of *C. africana* at Dharwad (Karnataka state), Patancheru (Andhra Pradesh state), and Akola (Maharashtra state) has been confirmed by the analyses of amplified fragment length polymorphism (AFLP), random amplified microsatellite (RAM) (14), and random amplified polymorphic DNA (RAPD) banding patterns (10). Research is currently in progress in the United States, Australia, and the Czech Republic to characterize the isolates collected in India during 1999-2000. Preliminary data from these molecular studies indicate that *C. sorghi* is present in Adilabad district in Andhra Pradesh and Yavatmal district in Maharashtra. Isolates collected from other locations either belong to *C. africana* or have intermediate genotypes between *C. sorghi* and *C. africana*. Isolates identified as *C. sorghi* and *C. africana* based on morphological and cultural characteristics were consistent with the identifications made by using molecular markers. The cultural characteristics of the intermediate types, as identified by molecular markers, were similar to *C. africana*. These intermediate types might be progeny of interspecific crosses between *C. africana* and *C. sorghi*.

Conclusions

Our survey confirms the presence of both *C. africana* and *C. sorghi* in India. *Claviceps africana* now dominates in all of the sorghum-growing regions in India, and the native *C. sorghi* is restricted to a few locations in Andhra Pradesh and Maharashtra. Molecular characterization of all the isolates collected will provide further information on the diversity of the sugary disease pathogens, particularly with respect to the presence of isolates that are intermediate between *C. africana* and *C. sorghi*. The widespread occurrence of *C. africana*, and its apparent displacement of *C. sorghi* in India, may be explained by the differences in the production of secondary conidia, which largely determines the reproductive and dissemination potential of the two *Claviceps* species. Consequently, *C. africana* can potentially spread rapidly over a relatively large area in a relatively short time. It is not clear when *C. africana* arrived in India and began replacing *C. sorghi* as the dominant pathogen. Based on a re-view of literature, Pažoutová *et al.* (10) concluded that these changes might have begun during the late 1970s. However, the large-scale replacement of *C. sorghi* by *C. africana* as the causal organism of sugary disease in India, makes it necessary to re-evaluate sorghum ergot management options developed prior to the 1980s.

Acknowledgments

This publication is based upon work supported in part by the U.S. Department of Agriculture, under special agreement No. 58-1920-9-F110. Any opinions, findings, conclusions, or recommendations are those of the authors and do not necessarily reflect the view of the U.S. Department of Agriculture.

References

1. **Bandyopadhyay, R., D. E. Frederickson, N. W., McLaren, G. N. Odvody, and M. J. Ryley.** 1998. Ergot: a serious threat to sorghum in Americas and Australia. *Plant Dis.* **82:** 356-367.
2. **Bandyopadhyay, R., L. K. Mughogho, S. K. Manohar, and M. V. Satyanarayana.** 1990. Stroma development, honeydew formation, and conidial production in *Claviceps sorghi*. *Phytopathology* **80:** 812-818.
3. **Bogo, A., and P. G. Mantle.** 1999. *Claviceps africana* discovered in India. *Plant Dis.* **83:** 79.
4. **Frederickson, D. E., P. G. Mantle, and W. A. J. de Milliano.** 1991. *Claviceps africana sp.* nov., the distinctive ergot pathogen of sorghum in Africa. *Mycol. Res.* **95:** 1101-1107.
5. **Kulkarni, B. G. P., V. S. Seshadri, and R. K. de Heg.** 1976. The perfect stage of *Sphacelia sorghi* McRae. *Mysore J. Agric. Sci.* **10:** 286-289.
6. **McRae, W.** 1917. Notes on some south Indian fungi, p. 108-111. In *Madras Agriculture Yearbook.*
7. **Nagarajan, K., and V. Saraswathi.** 1975. Production of honeydew line secretions in the culture of *Sphacelia sorghi*. *Indian Phytopathol.* **28:** 110.
8. **Pažoutová, S.** 2001. The phylogeny and evolution of the genus *Claviceps*. *Mycol. Res.* **105:** 275-283.
9. **Pažoutová, S.** 2002. The genus *Claviceps*: Evolution at work, p. 135-140. *In* J. F. Leslie (ed.), *Sorghum and Millets Diseases*. Iowa State Press, Ames, Iowa.
10. **Pažoutová, S., R. Bandyopadhyay, D. E. Frederickson, P. G. Mantle, and R. A. Frederiksen.** 2000. Relations among sorghum ergot isolates from the Americas, Africa, India, and Australia. *Plant Dis.* **84:** 437-442.
11. **Ramakrishnan, T. S.** 1948. Ergot sclerotia on *Sorghum vulgare* Pers. *Current Sci.* **17:** 218.
12. **Ramakrishnan, T. S.** 1963. *Diseases of Millet.* Indian Council of Agricultural Research, New Delhi, India.
13. **Sangitrao, C. S., S. Indira, and R. Bandyopadhyay.** 1999. Sorghum ergot in India. *Proc. Global Conf. on Ergot of Sorghum (1-8 June 1997, Sete Lagoas, Brazil)*: 41-54.
14. **Tooley, P. W., N. R. O'Neill, E. D. Goley, and M. M. Carras.** 2000. Assessment of diversity in *Claviceps africana* and other *Claviceps* species by RAM and AFLP analyses. *Phytopathology* **90:** 1126-1130.
15. **Tsukiboshi, T., T. Shimanuki, and T. Uematsu.** 1999. *Claviceps sorghicola* sp. nov., a destructive ergot pathogen of sorghum in Japan. *Mycol. Res.* **103:** 1403-1408.

13

Survival of Inoculum of *Claviceps africana* in Zimbabwe: Potential Sources of Initial Inoculum

Debra E. Frederickson and Gary N. Odvody

Abstract

We examined various sources of *Claviceps africana* inoculum as potential initial inocula under natural conditions in Zimbabwe. Inoculum must survive for six months between the cessation of honeydew exudation in June and the flowering of cultivated sorghum (stigma exertion) the following January. Potential inoculum sources include honeydew deposits, sphacelia or sclerotia (all sources of macroconidia), or sclerotial tissue (a source of ascospores following germination). Our data indicate that macroconidia do not survive for more than a few weeks of the off-season in thin deposits of honeydew. At the beginning of the rainy season, macroconidia still retain good viability within sphacelia. However, the moisture and high temperatures of the rainy season do not favor survival of macroconidia, and viability is very low as the rainy season begins. We monitored sources with viable inoculum closely to determine how and when infective inoculum was generated under natural conditions. Wild sorghums and other alternate hosts may play a role in perennation of *C. africana*.

Introduction

The secondary conidia of *C. africana* are primarily responsible for the dissemination of the ergot pathogen and the spread of ergot disease during the growing season. The macroconidia, effectively trapped in the sticky honeydew droplets, can be dispersed only with honeydew, by rain, or by mechanical means (1). Secondary conidia are produced when macroconidia germinate under conditions of high relative humidity above the surface of the honeydew and may become airborne with the potential to spread the disease relatively large distances (5, 7). Windborne dissemination of secondary conidia is responsible for the often explosive pattern of spread exhibited by *C. africana* (7).

In contrast, the source and form of the inoculum at the beginning of each new sorghum-growing season is not known. These potential inocula may stem from ascospores or conidia. Honeydew exuded from parasitic sphacelia is the sole source of macroconidia. Any honeydew-coated surface, *e.g.*, soil, leaves, and seed, also may be an inoculum source, although seed dressings such as captan or thiram, which prevent germination of thin deposits of macroconidia, can eliminate this inoculum source (3). Another potential source of inoculum is sphacelia, which may be incorporated into seed at harvest or carryover as a residual from the glumes or the panicles. Sclerotia (we prefer the term sclerotia/sphacelia because sclerotial tissue is only rarely found without sphacelial tissue attached) also carry a large population of macroconidia in the sphacelial portion. Sclerotial tissue also should not be ignored, although Zimbabwe is the only country in which germinated sclerotia have developed as far as perithecia with immature ascospores (6). Further studies of sclerotial survival and germination are necessary before dismissing sclerotia as a significant method of perennation.

The importance of an inoculum source to future disease transmission depends upon how the pathogen survives the adversities of the "off-season" away from its host. In Zimbabwe, cultivated sorghum is harvested in April but the pathogen may continue to produce honeydew on tillers, regrowth, or feral sorghum through the middle of June. The planting season begins again in November, but sorghum does not reach a susceptible stage until January, so off-season survival must span at least six months. Survival alone is insufficient to guarantee transmission of disease, as both infective propagules and access to the infection court are necessary to cause disease. We hypothesize that any source of macroconidia is a potential source of windborne secondary conidia and that these secondary conidia, and not surviving macroconidia, could move from soil level to panicle height to constitute the initial inoculum. Finally, the timing of production of infective inoculum is critical. Cultivated sorghum reaches the susceptible flowering stage approximately two months into the new season, and infective inoculum must be present at that time. Identification of key stages in the life cycle of *C. africana* may allow us to more effectively target new control measures. The source of initial inoculum is one of these key stages. This chapter outlines the scant research into initial inoculum sources of *C. africana*.

Macroconidia

Zimbabwe–natural conditions. Frederickson *et al.* (6, 7) used macroconidia from sphacelia, stored dry in the "off-season," to initiate infection on sorghum the following season. Initial ergot severities, presumably resulting from the low survival rate of conidia, were consistently very low.

Kansas–natural conditions. In experiments investigating the effect of storage in an unheated, outdoor facility on the viability of dry deposits of conidia, Claflin (2) showed that macroconidia did not survive storage on paper for more than five months. Viability declined from 52% in January to zero (undetectable) in April, coinciding with the rise in temperatures.

These experiments suggest that only low levels of inoculum survive the "off season" under natural conditions; however, Frederickson *et al.* (5, 7) have shown that only a very small amount of conidial inoculum is required to start an epiphytotic because of the rapid amplification of inoculum from short secondary cycles and widespread dispersal of secondary conidia.

Effect of temperature and relative humidity on *in vitro* survival. Surface honeydew was rinsed from sorghum panicles with new sphacelial infections of *C. africana*, and the panicles allowed to dry overnight. These panicles were divided into seven replicate samples for testing after storage either at one of three constant temperatures (6, 20, and 32°C), with relative humidity (RH) ≤ 30%; in a greenhouse where temperature was not controlled (range 18-35°C) and RH ≥ 70%; or at 20°C with a RH of 5.5, 48.5 or 90%. Viability was measured at regular intervals by plating a conidial suspension on nutrient agar and incubating at 20°C for 24 hours.

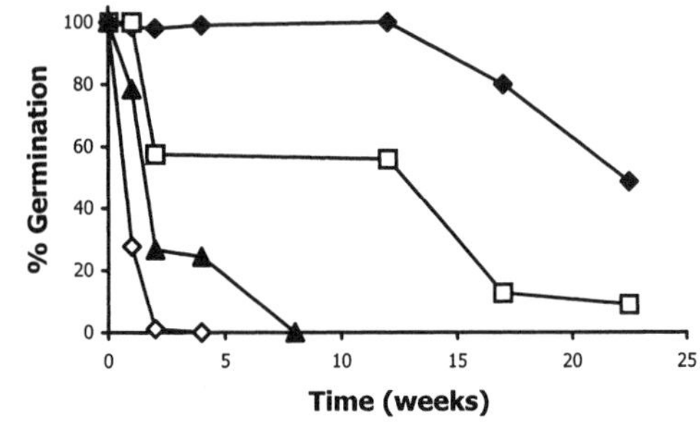

FIGURE 13-1. Effect of storage temperature of sphacelia on survival of conidia of *C. africana.* Except for the greenhouse sample (18-35°C, RH ≥ 70%), RH was ≤ 30%. Viability of conidia was assessed as percentage germination on agar at 20°C (--◆-- 6°C, --□-- 20°C, --◇-- 32°C, and --▲-- greenhouse).

There was a marked effect of temperature on survival of macroconidia (Fig. 13-1). Storage of sphacelia at 32°C or in the greenhouse resulted in a rapid loss of viability. No spores survived after two weeks at 32°C or seven weeks in the greenhouse. However, conidia showed good germination even after 17 weeks at 20°C. Viability did not even begin to decline at 6°C until after 12 weeks and was still at least 50% at 22 weeks.

Low RH, 5.5%, desiccated the sphacelia, but maintained high levels of conidial viability. Viability of conidia from sphacelia, which became moist at the other two RH levels tested, was reduced to zero after four weeks (Fig. 13-2). We think that an interaction between temperature and RH is likely. Viability of conidia from sphacelia stored in the greenhouse declined rapidly, even though temperatures were not constantly high, probably because of the high RH. Viability following storage at 20°C was the same as at 6°C if sphacelia were maintained at 5.5% RH. The effect of high RH in reducing macroconidial viability is apparently manifested through the hydration of sphacelia. Increasing RH increases the moisture content of the sphacelia. Once the RH reaches a threshold, approximately 40%, the hygroscopic sphacelia become fully hydrated and macroconidial viability declines rapidly. Cooler, drier conditions promote survival.

At Matopos, Zimbabwe, where ergot is endemic, mean, maximum, off-season temperatures from May to August do not exceed 24°C and often are much lower for most of the day. Mean minima are 5°C or less. Temperatures rise sharply in September and peak in October/November. The general absence of rain, a RH of 20-30% from May to September, and the generally reduced temperatures during this period should increase conidial longevity.

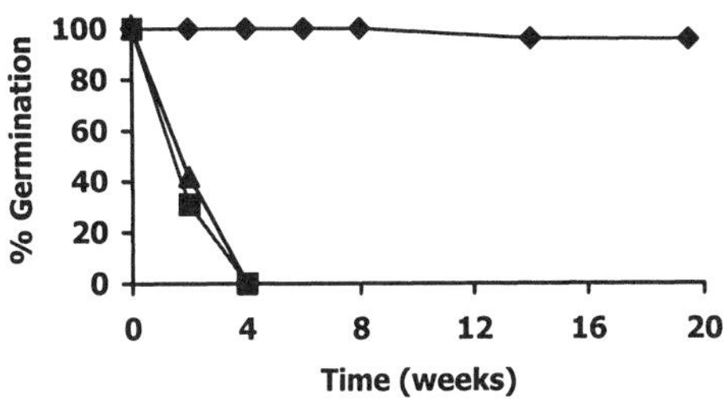

FIGURE 13-2. Effect of storage relative humidity on survival of conidia within sphacelia of *C. africana* at 20°C. Viability of conidia was assessed as percentage germination on agar at 20°C (--◆-- 5.5, --▲-- 48.5, or --■-- 90% RH).

"Off-season" conditions in Zimbabwe. Viability studies of macroconidia complement studies of (i) how viable residual conidia gain inoculum potential, and (ii) the timing of inoculum production under natural conditions. We are monitoring viability changes in (i) conidia in sphacelia in panicles protected from winter rain; (ii) conidia in sphacelia in panicles exposed to (scarce) winter rain; (iii) conidia in sphacelia in panicles in or on soil; (iv) conidia (honeydew) on soil; (v) conidia (honeydew) on paper bags; and (vi) conidia (honeydew droplets) on panicles. Viability is assessed approximately every six weeks as percentage germination of macroconidia on agar at 20°C.

At three months, the viability of conidia in sphacelia protected from the rain remained high, with a germination rate of nearly 50% compared to 75% from refrigerated controls. In exposed sphacelia, germination dropped to 50% at six weeks but was still above 40% at three months. The germination of conidia in honeydew droplets on panicles declined from 80% to 30% in three months and to zero after an additional seven weeks. Viability of conidia in thin honeydew smears on pollination bags fell from 80% to 3.5% in three months. In a second experiment, the viability of conidia in honeydew smears on paper bags fell from 70% to zero in only six weeks. A thin smear of conidia probably is vulnerable to the effects of the sun (data not shown), and even transient humidity increases, *e.g.*, from morning mist or light drizzle, promoted secondary conidiation in the thin deposits, thus reducing the population of nongerminated conidia.

A large proportion of the conidia in the sphacelia remained viable at the beginning of the rainy season. By weeks 19-21, conidial viability from most categories of sphacelia was zero. Only in one category, fully exposed to the elements, were any conidia (3.5%) viable, and the viability of these conidia was reduced to 1% after four more weeks and then to zero after another two weeks. These data support our hypothesis that the most dramatic decreases in conidial viability occur as the "off-season" (dry season) ends and high temperature and rains become common. Moreover, rains help leach macroconidia from the sphacelia with concomitant dilution of honeydew and secondary conidiation. Monitoring changes in conidial viability as the rains begin will provide another important test of our hypothesis.

Sclerotia

Sclerotia designated as "fully-formed" and "incompletely formed" (rind not closed over) are being stored outside, some in soil, under off-season conditions at Matopos, Zimbabwe and in the lab. Their survival (numbers and condition) and germination (numbers, dynamics of germination, and timing) will be investigated beginning with the onset of the rains.

Bridging the Gap between Seasons

If inoculum infects cultivated sorghum directly, then it must remain viable until January, when the sorghums planted in November flower. At least two scenarios are possible. Infective inoculum could be generated (i) as a slow response, or delay, to environmental changes, *e.g.*, if ascospores serve as the inoculum, or (ii) continually, from the start of the rainy season, to provide inoculum as conidia or ascospores for an extended period of time. If inoculum viability declines rapidly at the beginning of the rainy season, an alternative hypothesis is that the fungus first infects another host. The presence of a suitable alternate host could thus be critical for perennation of *C. africana*.

Alternative Hosts

Sorghum species. In Texas, where sorghums flower from late April to August from the coastal bend part of the state to the panhandle, johnsongrass (*Sorghum halepense*) and feral and ratooned sorghums are important for the maintenance of viable inoculum of *C. africana* from September onwards. In February 1998, ergot honeydew was observed on feral sorghum in southern Texas, ahead of the March planting of the sorghum crop. In late 1998 (October-December), ergot was observed on feral and ratooned sorghum and johnsongrass. This pattern was repeated in late 1999, with ergot on johnsongrass in south Texas, especially in areas near fields previously cultivated with sorghum, and in early 2000, with ergot on feral and ratooned sorghum (8).

In Zimbabwe and Zambia, *C. africana* also has been found on wild sorghums (4). Some Zimbabwean species, *e.g.*, *S. halepense* and *Sorghum verticilliflorum*, are perennial and may become established earlier in the season than cultivated sorghum. These species could have a role in the continuity of the life cycle of *C. africana*, especially if inoculum cannot survive long enough to infect cultivated sorghums directly. The distribution of these wild sorghum species, their flowering patterns, and the temporal appearance of inoculum on them have not been studied. Continuous and systematic surveys are needed in each country where ergot has been reported on wild sorghums to gauge the significance these alternate hosts might play in perennation of *C. africana*, and the build-up of inoculum that could infect cultivated sorghum.

Other alternative hosts. At Matopos, Zimbabwe, another host, *Hyparrhenia rufa* (thatching grass), that may be important in the perennation of *C. africana* was identified in 2000. Comparisons of *C. africana* populations from *H. rufa* and sorghum are needed to confirm the importance of this alternative host for the perennation of *C. africana*.

References

1. **Bandyopadhyay, R., L. K. Mughogho, and H. C. Sharma.** 1991. Source of primary inoculum and spread of ergot, p. 24-25. In *ICRISAT Annual Report*. ICRISAT, Patancheru (A.P.), India.
2. **Claflin, L. E.** 1998. Agroecology and biotechnology of fungal pathogens of sorghum and millet from the greater horn of Africa, p. 13-18. In *INTSORMIL Annual Report*. INTSORMIL, Lincoln, Nebraska.
3. **Dahlberg, J. A., G. L. Peterson, G. N. Odvody, and M. Bonde.** 1999. Inhibition of germination and sporulation of *Claviceps africana* from honeydew encrusted sorghum with seed treatment fungicides. *Crop Protect.* **18:** 235-238.
4. **Frederickson, D. E.** 1990. *Ergot Disease of Sorghum*. Ph.D. dissertation, Univ. of London, London, UK.
5. **Frederickson, D. E., P. G. Mantle, and W. A. J. de Milliano.** 1989. Secondary conidiation of *Sphacelia sorghi* on sorghum, a novel factor in the epidemiology of ergot disease. *Mycol. Res.* **93:** 497-502.
6. **Frederickson, D. E., P. G. Mantle and W. A. J. de Milliano.** 1991. *Claviceps africana* sp. nov.; the distinctive ergot pathogen of sorghum in Africa. *Mycol. Res.* **95:** 1101-1107.
7. **Frederickson, D. E., P. G. Mantle, and W. A. J. de Milliano.** 1993. Windborne spread of ergot disease (*Claviceps africana*) in sorghum A-lines in Zimbabwe. *Plant Pathol.* **42:** 368-377.
8. **Odvody, G. N., D. E. Frederickson, T. Isakeit, J. A. Dahlberg, and G. L. Peterson.** 1999. The role of seed-borne inoculum in sorghum ergot. *Proc. 3rd Intl. Seed Test. Assoc.-PDC (16-19 August 1999, Ames, Iowa)*: 136-140.

14

Response of Sorghum B and R Lines to Ergot (*Claviceps africana*) at Celaya, Guanajuato, Mexico

Miguel Hernández-Martínez, Leopoldo E. Mendoza-Onofre, Porfirio Ramírez-Vallejo, Seigi Osada-Kawasoe, Elizabeth Cárdenas-Soriano, and Francisco Zavala-García

Abstract

We compared the response of ten sorghum (*Sorghum bicolor* L. Moench) genotypes to inoculation with *Claviceps africana* at Celaya, Guanajuato, Mexico during the 1998 growing season. We evaluated B and R lines that differed in ergot tolerance, cold tolerance, ergot susceptibility, and good pollen production. We used a split plot design with four replications. The main plots were with and without inoculation with *C. africana*. The subplots were the genotypes. An experimental unit was 10 heads of the same genotype. The inoculated heads were sprayed with 5 ml of a 1×10^4 conidia/ml suspension at the beginning, at 40%, and at 80% of flowering. Heads were covered with bags to prevent inoculation and/or pollination. The variables we evaluated were grain yield, grain number, grain size, fertile pollen percentage, days-to-flowering, head length, flowering period, and number of tillers per plant. The pathogen reduced grain yield 27%, grain number 19%, and grain size 9%. No other variables were significantly affected. Cold-tolerant genotypes had higher grain yield, grain number, and fertile pollen percentage, and fewer days to flowering, shorter flowering periods, and fewer tillers per plant than the ergot-tolerant genotypes.

Introduction

Genetic tolerance to ergot (*Claviceps africana*) in sorghum is scarce or is absent since, under favorable weather conditions, all genotypes are susceptible to the disease. McLaren (1), evaluating germplasm in Africa, found that all male-sterile lines were susceptible, although maintainer lines had varying degrees of tolerance. He identified 30 lines that differed in ergot tolerance, of which RTAM-428, SA-1699, BTx-399, QL-3, BTx-622, BTx-630, and BTx-607 were the most tolerant. In some cases, "escapes" could have occurred due to differences in pollination vigor or in tolerance to cold temperatures (2), resulting in greater viable pollen percentage, without damage to the pollen germ tube (3). In Mexico, more than 95% of the sorghum crop is sown with hybrids, and little or nothing is known of the ergot tolerance of the parental lines. The objective of our study was to evaluate

TABLE 14-1. Sorghum B and R genotypes used in this study.

B-Lines		R-Lines	
Genotype	Character[a]	Genotype	Character[a]
BTx 399	Ergot tolerant	R-111	Good pollen producer
BTx 630	Ergot tolerant	R6956	Good pollen producer
BTx 631	Ergot susceptible	SA1699	Ergot tolerant
B-LIVA X	Cold tolerant	R6-LIVAIT	Cold tolerant
B-LIVA Y	Cold tolerant	R7-LIVAIT	Cold tolerant

[a]Source: ergot tolerant/susceptible, McLaren (1); cold tolerant, Osuna-Ortega *et al.* (4); pollen production, Hernández-Martínez *et al.* (*unpublished*).

TABLE 14-2. Analysis of variance results for grain yield, grain number, grain size, fertile pollen percentage, days to flowering, head length, flowering period, and tillers per plant.

	Degrees of Freedom	Grain Yield	Grain Number	Grain Size	Fertile Pollen %	Days to Flowering	Head Length	Flowering Period	Tillers per Plant
Inoculation	1	93**	29**	21**	0.90	0.52	0.95	0.89	0.07
Genotype	9	69**	44**	34**	300**	1200**	86**	56**	34**
Inoculation × genotype	9	1.14	0.22	0.73	0.07	0.02	0.25	0.31	0.02
CV		13	15	10	1.3	1.0	3.0	6.0	39
r^2		0.93	0.89	0.86	0.98	0.99	0.93	0.90	0.85

** $p < 0.01$.

the response of some commonly used B and R lines, whose ergot- and cold-tolerance and pollen production characteristics had been previously determined.

Materials and Methods

We used 10 sorghum lines (Table 14-1). Plants were grown with irrigation during the 1998 growing season at Campo Experimental Bajío located in Celaya, Guanajuato, México (1560 m altitude). We used a split plot design with four replications. The main plots were with and without ergot inoculation. The subplots were the genotypes. Plot size was a single row 2.5 m long, and rows were 66 cm apart (1.65 m^2). The experiment was fertilized with a 220-40-0 formula (110-40-0 at planting and 110-0-0 at first turning). Ten healthy plants were identified in each plot. Three days before flowering, the heads of the main shoots were bagged. At the start of flowering, those heads were uncovered and sprayed with 5 ml of a 1×10^4 conidia/ml spore suspension (1). These heads were marked and covered again with the paper bag. This procedure was repeated on the same heads at 40% and 80% flowering. On the 10 uninoculated heads, which served as controls, the heads were covered and the bags stapled shut. We scored grain yield (per head), grain number (per head), grain size (grain weight/number), fertile pollen percentage, days-to-50% flowering, length of the flowering period, head length, and tillers per plant. Statistical analysis was carried out using SAS version 6.11 (5). Data were evaluated with Tukey's test at 5% significance for mean comparisons.

Results and Discussion

There were highly significant differences for both inoculation and genotypes for grain yield, grain number, and grain size (Table 14-2). The interaction main plot × subplot was not significant. For the other variables (days to flowering, head length, and tillers per plant), the ergot inoculation effect was not significant, but the genotype effect was. Inoculation with ergot reduced grain yield 27%, grain number 19%, and grain size 9% (Table 14-3). The cold-tolerant genotypes had fewer changes in grain yield and grain number (21% and 12% decreases, respectively) than did the ergot tolerant genotypes (28% and 20% decreases, respectively). There also were significant differences between the genotypic classes for the other variables that we measured (Table 14-4).

The cold-tolerant lines are preferable to the ergot-tolerant lines because of their greater grain yield, grain

TABLE 14-3. Effect of inoculation of sorghum heads with *Claviceps africana* on grain yield (g/head), grain number (per head), grain size (g), and % loss.

Genotypic Class	Grain Yield			Grain Number			Grain Size		
	Uninoc.[a]	Inoc.[b]	% Loss	Uninoc.[a]	Inoc.[b]	% Loss	Uninoc.[a]	Inoc.[b]	% Loss
Ergot Tolerant	25	18	30	1030	830	19	0.0247	0.0226	8
Ergot Susceptible	37	26	30	1210	900	26	0.0307	0.0283	8
Cold Tolerant	42	33	21	1950	1720	12	0.0207	0.0185	11
Good Pollen Producer	30	20	33	1370	1070	22	0.0218	0.0193	11
Mean	33 A	24 B	27	1390 A	1130 B	19	0.0245 A	0.0222 B	9
$LSD_{0.05}$	0.87			27			0.001		

[a]Uninoculated.
[b]Inoculated.
Means followed by the same letter are not significantly different (Tukey, $p = 0.05$).

TABLE 14-4. Effect of genotype on grain yield (g/head), grain number (per head), grain size, fertile pollen percentage, days to flowering, head length (cm), flowering period (days), and tillers per plant.

Genotype	Grain Yield	Grain Number	Grain Size	Fertile Pollen %	Days to Flowering	Head Length	Flowering Period	Tillers/ Plant
Ergot tolerant	22 c	930 c	0.0236 b	81 b	91 c	26 b	11 a	2.7 a
Ergot susceptible	31 b	1060 b	0.0295 a	75 c	98 a	27 a	11 a	1.4 b
Cold tolerant	38 a	1840 a	0.0196 c	92 a	76 d	25 b	8 b	0.4 c
Good pollen producer	25 c	1260 b	0.0205 bc	81 b	95 b	26 b	11 a	1.3 b
$LSD_{0.05}$	4.8	245	0.0023	3.5	0.9	0.8	0.7	0.7

Means with the same letter are not significantly different (Tukey, $p = 0.05$).

number, and percentage fertile pollen, and their reduced number of days to flower, flowering period, and number of tillers per plant. The increase in fertile pollen and reduced flowering period both should reduce ergot incidence. On the other hand, the ergot-tolerant lines had more days to flower, greater head length, and more tillers per plant in addition to being late; all characters that increase exposure to the pathogen's attack. Our results suggest that characters associated with cold tolerance also are important characters in determining the amount of damage ergot can inflict on a plant.

References

1. **McLaren, N. W.** 1992. Quantifying resistance of sorghum genotypes to the sugary disease pathogen (*Claviceps africana*). *Plant Dis.* **76:** 986-988.
2. **McLaren, N. W.** 1999. Host × pathogen × environment interaction in sorghum ergot disease. *Proc. Global Conf. on Ergot of Sorghum (1-8 June 1997, Sete Lagoas, Brazil)*: 122-134.
3. **McLaren, N. W., and F. C. Wehner.** 1992. Pre-flowering low temperature predisposition of sorghum to sugary disease (*Claviceps africana*). *J. Phytopathol.* **135:** 328-334.
4. **Osuna-Ortega J., L. E. Mendoza-Onofre, V. A. González-Hernández, F. Castillo-González, M. Del C. Mendoza-Castillo, and H. Williams-Alanís.** 2000. Potential of cold tolerant germplasm in the adaptation and adaptability of grain sorghum in Mexico. I: High Valleys. *Agrociencia* **34:** 561-572.
5. **SAS Institute.** 1998. *Getting Started with the SAS System. Version 6.11 for Windows.* SAS Institute, Cary, North Carolina.

15

Four Genotypes of *Claviceps* spp. Cause Sorghum Ergot in Australia

Birte Komolong, Sukumar Chakraborty, Malcolm J. Ryley, and David Yates

Abstract

Ergot of sorghum affects seed production and grain usage in animal feed due to concerns about potential animal toxicity. We conducted a radio-labeled amplified fragment analysis of 65 isolates of *Claviceps* spp. and found four genotypic lineages in Australia. Two of these lineages contain only Australian isolates, and the other two contain isolates recovered in Australia, India, the United States, Japan, and Puerto Rico. The origin of this diversity is unclear, but appears unlikely to be due to sexual reproduction within the local population.

Introduction

Ergot of sorghum caused by *Claviceps* spp. increases harvesting and processing costs and affects seed production and grain usage in animal feed due to concerns about potential animal toxicity (8). Sorghum ergot was first reported from India in 1915, and the disease was restricted to Asia and Africa for over 75 years before rapidly spreading to the Americas and Australia between 1995 and 1997 (1). Since its detection in 1996 in Gatton, Australia (7), ergot has become established in all sorghum-growing areas in Australia. Specific weather and host requirements for unfertilized gynoecial infection (1) restricts the window of opportunity for establishment of *Claviceps* spp., and it is unclear how it has spread rapidly between geographically isolated regions.

Based on molecular marker analysis of 28 isolates from Australia, India, the United States, Bolivia, Mexico and South Africa, Pažoutová *et al.* (5) suggested that ergot in Australia came from Asia, but that American isolates originated from Africa. Seventeen of 25 described *Sorghum* spp. are native to Australia. *Claviceps pusilla* was recorded on *Sorghum leiocladum* as early as 1942 (3), and another *Claviceps* sp. has been recorded on other *Sorghum* spp. Therefore, adaptation of *Claviceps* sp. from an endemic host to grain sorghum in Australia cannot be ruled out. The host range of *Claviceps africana* is not clearly defined (1). However, several species of *Sorghum* can serve as alternate hosts, and it is possible that *C. africana* has been in Australia for a number of years but was undetected due to very low infection levels. Strains collected from wild Australian *Sorghum* spp. were not included in the previous study (5). Our objective in this study was to establish an inventory of *Claviceps* genotypes in Australia, to identify sources of the initial introduction, and to help prevent additional incursions of exotic strains.

Materials and Methods

We used an Australian *C. pusilla* from *Dicanthium* sp.; a *Claviceps sorghicola* isolate from Japan; 52 putative *C. africana* isolates from *Sorghum bicolor*, *Sorghum halepense*, *Sorghum almum*, and unidentified wild *Sorghum* spp. from Australia; and 11 *C. africana* isolates from India, Puerto Rico, Japan, and the United States. Most Australian isolates were inoculated on to a susceptible male-sterile line of grain sorghum in a greenhouse to obtain fresh honeydew. Cultures originating from single conidia or sphaecilia were maintained at 27°C on T2 medium (4) or Potato Dextrose agar (Sigma-Aldrich, Sydney, Australia), and stored in 15% glycerol in liquid nitrogen.

Isolates were grown in clarified V8 juice or Potato Dextrose Broth (Sigma-Aldrich) and incubated at 23°C on orbital shakers at 160 rpm for seven days. Mycelium was filtered through two layers of cheesecloth, freeze-dried, and DNA extracted by the method of Rogers and Bendich (6) with minor modifications. Mycelium was ground in liquid nitrogen, suspended in CTAB buffer (50 mM Tris-HCl pH 8.0; 10 mM EDTA; 0.7 M NaCl; 1%

CTAB) and 2× CTAB buffer and incubated at 65°C for 5 min. The mixture was extracted twice with chloroform-isoamyl alcohol (24:1, v:v). DNA was precipitated with two volumes of 100% ethanol at -20°C for 30 min., the pellet was washed with 70% ethanol, resuspended in 0.1× TE buffer (1 mM Tris-HCl pH 8.0; 0.1mM EDTA), treated with RNAse A at 37°C for 1 hr, and cleaned with a DNA clean-up spin kit (Genomed, Stuttgart, Germany) following the manufacturer's instructions.

Ten arbitrary primers of 10 base pairs each were used in a radio-labeled amplified fragment analysis (2). Each reaction mixture contained 1 μl 10× DAF-buffer (1× DAF buffer: 10 mM Tris-HCl pH 8.0, 10 mM KCl, 5 mM $MgCl_2$), 20 μM dNTPs, 1.5 U AmpliTaq DNA Polymerase Stoffel Fragment, 0.125 μM α-^{33}P-dATP, 5 μM primer, and 50 ng fungal DNA. DNA was amplified by PCR in a MJC Research PT100 (Watertown, Massachusetts) thermocycler programmed for 94°C for the initial 5 min, then 30 cycles of 94°C for 30 sec, 57°C for 60 sec, 55°C for 60 sec, 54°C for 60 sec, 53°C for 60 sec, and 57°C for 60 sec, following the final cycle there was a one-time 72°C incubation for 5 min. An equal volume of 98% formamide, 10 mM EDTA pH 8.0, 0.05% bromophenol blue, 0.05% xylene cyanol was added, the reaction products were denatured at 95°C for 3 min, and then immediately put on ice. Reaction products were analyzed by electrophoresis on 4% denaturing polyacrylamide gels, dried on a gel dryer, and exposed to X-ray film (Kodak, Biomax MR-2) at room temperature overnight.

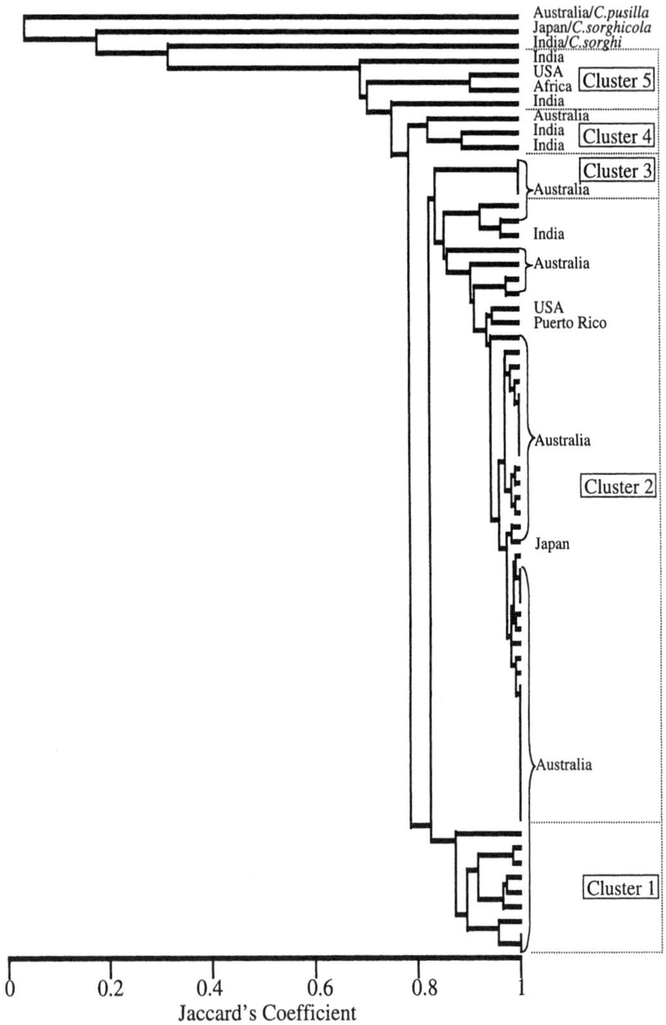

FIGURE 15-1. Genetic relationships between 65 isolates of *Claviceps* spp. from Australia, India, Puerto Rico, Japan, South Africa, and the United States.

Bands consistently present in duplicate samples of each isolate were scored as present or absent. Data were pooled to calculate similarity values from pairwise comparison of isolates using $F = 2m_{xy}\sqrt{(m_x + m_y)}$, where m_{xy} = the number of shared bands, and m_x and m_y are the number of bands present in each isolate. A dendrogram was generated from cluster analysis of the similarity values by using the Unweighted Pair Group method in SAS version 7.2 for the PC (SAS Institute, Cary, NC).

Results and Discussion

The DNA banding patterns from of a *C. pusilla* isolate from *Dichanthium* sp., a *C. sorghicola* isolate from Japan, and a *C. africana* isolate from India were very different from the other *C. africana* isolates we examined (Fig. 15-1). These results suggest that the Indian isolate may belong to a species other than *C. africana*. The remaining 62 putative *C. africana* isolates can be grouped into five clusters, with four of these clusters containing Australian isolates. Clusters 1 and 3 contain only Australian isolates. Nine isolates from grain and forage sorghum and *S. almum* constitute cluster 1. Cluster 2, with 43 isolates from Australia, India, the United States, Japan, and Puerto Rico, constitutes the largest group. Australian isolates in this cluster originated from grain and forage sorghum, *S. almum*, *S. halepense*, and unidentified wild *Sorghum* spp. The overall diversity among isolates in this cluster is greater than that in cluster 1. The two clonal populations in cluster 2, represented by 5 and 10 Australian isolates, respectively, were obtained from fields in southeast and north Queensland several hundred km apart. Cluster 3 is a third clonal population of three Australian grain sorghum isolates from central and southeast Queensland. Cluster 4 contains two Indian and an Australian isolate from grain sorghum, and cluster 5 contains two Indian isolates and one each from the United States and South Africa.

We found similarity among isolates from the United States, Puerto Rico, India, Japan, and Australia in cluster 2, and among isolates from India, the United States, and South Africa in cluster 5, even though only a limited number of isolates from countries other than Australia were used. These results suggest that *C. africana* has been introduced more than once to both the United States and Australia and do not support the hypothesis that ergot in the Americas originated in Africa, while Australian isolates originated in Asia (5). Other markers are needed for diagnostic purposes and to confirm hypotheses concerning pathogen movement between countries.

Pažoutová *et al.* (5) did not detect genetic variation among five Australian isolates of *C. africana*. Those results were consistent with the hypothesis that ergot in Australia was initiated by the introduction of a single isolate. The *C. africana* clone represented by the ten cluster 2 isolates may be evidence for such an introduction. However, we have regularly detected strains with different DNA banding patterns in samples collected from the same field and often from the same inflorescence. Given the lack of demonstrated sexual reproduction by *C. africana* and its relatively short reported history in Australia, since 1996 (7), we think it unlikely that the observed level of genetic variation resulted from only one or two introductions. The host range of *C. africana* is not clearly defined, and often records of a new host are not supported by pathogenicity tests (1). Many *Claviceps* spp. isolates from wild sorghum readily infect grain sorghum in the greenhouse, and the two most common clones of *C. africana* occur on both grain and wild sorghum in the field. More targeted studies of *Claviceps* spp. on native sorghum and other grasses in Australia are needed to test the hypothesis that the diversity observed can be explained by adaptation from endemic local hosts. Such studies should include surveys, host range determination, and genetic diversity studies of *Claviceps* spp. infecting native sorghum and other grasses in Australia.

References

1. **Bandyopadhyay, R., D. E. Frederickson, N. W. McLaren, G. N. Odvody, and M. J. Ryley.** 1998. Ergot: a new disease threat to sorghum in the Americas and Australia. *Plant Dis.* **82:** 356-367.
2. **Harrison, D. K., J. Waldron, C. Ramage, C. Peace, M. Cox, R. G. Birch, and B. J. Carroll.** 2001. New DNA technologies provide insights into the molecular basis of somaclonal variation in sugarcane, p. 580-581. *In* D.M. Hogarth (ed.), *Proc. XXIVth Intl. Soc. of Sugarcane Technologists Cong. (17-21 September 2001, Brisbane, Australia).* International Society of Sugarcane Technologists, Brisbane, Australia
3. **Langdon, R. F.** 1942. Ergot of native grasses in Queensland. *Proc. Roy. Soc. Qld.* **54:** 23-32.
4. **Spalla, C.** 1973. Genetic problems of production of ergot alkaloids in saprophytic and parasitic conditions, p. 393-403. *In* Z. Vanek, Z. Hošálek, and J. Cudlín (eds.), *Genetics of Industrial Microorganisms.* Elsevier, Amsterdam, The Netherlands.
5. **Pažoutová, S., R. Bandyopadhyay, D. E. Frederickson, P. G. Mantle, and R. A. Frederiksen.** 2000. Relations among sorghum ergot isolates from the Americas, Africa, India and Australia. *Plant Dis.* **84:** 437-442.
6. **Rogers, S. D., and A. J. Bendich.** 1985. Extraction of DNA from milligram amounts of fresh, herbarium and mummified plant tissues. *Plant Mol. Biol.* **5:** 69-76.
7. **Ryley, M. J., J. L. Alcorn, J. K. Kochman, G. A. Kong, and S. M. Thompson.** 1996. Ergot of sorghum in Australia. *Aust. Plant Pathol.* **25:** 214.
8. **Wang, E., H. Meinke, M. Ryley, D. Herde, and R. Henzel.** 2000. On the relation between weather variables and sorghum ergot infection. *Aust. J. Agric. Res.* **51:** 313-324.

16

Biosynthesis of Bioactive Honeydew Oligosaccharides by Sorghum Ergot Pathogens

Peter G. Mantle and Amauri Bogo

Abstract

Unique amongst ergot pathogens, the mannitol-containing oligosaccharides in the honeydew of *Claviceps africana* are formed first by cell-associated reduction of free fructose to mannitol, and then by both intracellular and extracellular transferase activity, substituting the glucose of sucrose with mannitol, which becomes linked to a 2-position of fructose. The resulting disaccharide is further linked through the mannitol moiety to the 2-position of another fructose molecule to yield the trisaccharide difructosyl-mannitol, which becomes the principal oligosaccharide in the honeydew. This unusual processing of sucrose accounts for the high concentration of bioactive oligosaccharides in the honeydew exuded from infected florets of sweet sorghums. These oligosaccharides inhibit secondary sporulation, which occurs freely at the lower oligosaccharide concentration commonly found in honeydew on grain sorghum and probably is responsible for disease epidemics.

Introduction

Secondary conidiation of *Claviceps africana* (6) is an important factor in the spread of the ergot disease of sorghum (8) caused by this fungus. Secondary conidiation was first described with respect to *in vitro* germination of macrospores (14) of the related Indian pathogen *Claviceps sorghi,* but *C. africana* is the only ergot fungus to utilize this dissemination mechanism extensively in disease epidemic development on an agricultural crop. The production of F_1 hybrid sorghum seed is now threatened worldwide due to the secondary sporulation and resulting epidemics of *C. africana*.

A strategy to limit disease development in A lines involved in F_1 hybrid production was first proposed at the 13th International Plant Protection Congress in 1995 (11). At this time, the disease was not known to be present in either the Americas or Australia. The strategy involved the use of a sweet character in the A line to elevate sugar concentration in *C. africana* honeydew, thereby suppressing secondary sporulation. The strategy was reiterated (12) at the Global Conference on Sorghum Ergot in Brazil in response to the remarkable simultaneous globalization of the disease through epidemics in Queensland, Australia and in South America. In South America, the disease spread rapidly northwards, aided by the airborne secondary conidia. The pathogen also has become endemic in wild sorghums, most notably johnsongrass (*Sorghum halepense*), which are widespread across these continents. The subsequent rapid northward spread of *C. africana* through Central America and into Texas, where much hybrid sorghum seed is produced, serves as a modern example of the ability of a well-adapted plant pathogen to colonize new geographic areas rich in suitable host plants.

When the "sweet character" strategy was discussed in Brazil in 1997, there already was experimental evidence that the principal honeydew oligosaccharide, a trisaccharide (difructosyl-mannitol), inhibited secondary sporulation at a concentration of 1% or greater. The first field evidence that some sweet sorghums would not support secondary conidiation on the honeydew exuding from ergotized florets also was available (12). This strategy requires several years of research and development, but would result in a purely biological control that requires neither laboratory-based genetic modification nor fungicides (even if these changes are economically feasible). The preliminary step of incorporating a sweet character into a grain sorghum A line, which is needed to test the efficacy of this strategy in limiting epidemic development, however, has not yet been taken. Meanwhile, we confirmed (3) that some accessions in the Brazil sweet sorghum germplasm bank do not support secondary conidiation in any floral infection by *C. africana*.

The principal disaccharide (fructosyl-mannitol) in *C. africana* honeydew also inhibits secondary conidiation (3), suggesting that there are several biologically active oligosaccharides that arise from the sucrose supplied to the fungal sphacelium by the host plant. The *C. africana* oligosaccharides appear to be unique amongst ergot fungal metabolites, with each oligosaccharide containing a single alditol (mannitol or arabitol) component (15). Understanding the biosynthetic origin of these oligosaccharides is important when designing multifaceted approaches to ergot disease control in sorghum F_1 hybrid seed production. To our knowledge, there is no evidence of natural resistance within sorghums to ergot other than the cleistogamous floral biology, which promotes escape in some grain sorghums (9). However, such characters are useless in an A line, which must maximize receptivity to pollen and, consequently, to *C. africana* spores, which behave like pollen.

Ergot fungi have two general metabolic weaknesses as parasites of cereals and grasses. The first is that they cannot directly utilize sucrose, which is the principal saccharide translocated to florets whether they are fertilized and developing seed or are infected by the ergot parasite, which colonizes the ovary and proliferates in its stead. The presence of the parasite does not significantly alter the host's carbohydrate metabolism. The fungus, therefore, must "invert" the sucrose to release the glucose moiety. Once the sucrose is split, the glucose is used preferentially, and the unmetabolized fructose accumulates. Free fructose is generally inhibitory to the growth of ergot fungi, and these fungi have developed biochemical mechanisms to incorporate the unwanted fructose into a series of oligosaccharides.

The best-studied ergot fungus in this regard is *Claviceps purpurea*, which elaborates oligosaccharides by using a fructofuranosidase to successively add one or more fructose units to a molecule of sucrose (5). This process sacrifices some potential glucose, although in practice this reduction in efficiency does not seem to have reduced the success of *C. purpurea* as a plant pathogen. The oligosaccharides produced are lost to the fungus as a potential energy source when they are released into the honeydew exudate as a part of the mechanism for conidial dissemination. Conidia of *C. purpurea* do not germinate in honeydew until it is greatly diluted, and even then germination is limited to a germ tube, used *in vivo* to penetrate the naked unfertilized ovary and establish a new infection. The mechanism by which *C. purpurea* oligosaccharides inhibit spore germination has never been elucidated, although this result usually is attributed to an osmotic effect. The principal oligosaccharides in *C. africana* honeydew, however, do not contain glucose (15), and the mannitol component of the two most abundant oligosaccharides, difructosyl-mannitol and fructosyl-mannitol, could not be derived directly from glucose. Thus, novel mechanisms for their biosynthesis must exist.

Globally, *C. africana* is the most important sorghum ergot pathogen. *C. sorghi* was of minor significance through most of the twentieth century in the Indian subcontinent but, apparently, was not seen in the 1990s and no live cultures of it were thought to exist. It was feared that *C. sorghi* had become extinct, being out-competed by the more aggressive *C. africana* as this pathogen became established in India (4). Recently, *C. sorghi* was rediscovered and characterized by DNA analysis (16) and pathogenesis on sorghum (13). At approximately the same time, however, a different pathogen was recognized in Japan and described as *Claviceps sorghicola* (18). *Claviceps africana* and *C. sorghi* appear to be closely related because their anamorphs (both are *Sphacelia sorghi*) are so similar. However, there is a much closer relationship between *C. sorghi* and *C. sorghicola* in the saccharide composition of their honeydews, as both are rich in glucose and fructose and contain only minor amounts of sucrose-based oligosaccharides, much like *C. purpurea*. Thus, *C. africana* is unusual in its sugar metabolism, which prompted this study of the steps involved. The enzymology of this process remains to be elucidated.

Experimental

We used ^{14}C-pulse-labeled sucrose, fructose, and mannitol added to follow these compounds in fresh cell-free honeydew and in aqueous preparations of the parasitic sphacelial tissues from which the honeydew is produced on sorghum. The plants were male-sterile and grown in a horticultural tunnel at the Chelsea Physic Garden in London. Reactions contained 1 ml of cell-free honeydew or 1 ml of water in which five sphacelia were macerated together with 1 μCi of a test sugar. Reaction mixtures were shaken to mix well at room temperature. Aliquots (50 μl, four replicates) were taken from the reactions over a four-hour period and chromatographed on Whatman 3MM paper (46 × 57 cm) descending, for two days in propan-1-ol:ethyl acetate:water (7:1:2), to resolve the saccharides. (Sugars were located destructively in test strips of chromatograms by heating to 120°C, after dipping the paper in aniline hydrogen phthalate reagent.) Individual sugars were eluted from excised regions of the paper chromatogram with warm distilled water. The radioactivity was measured by scintillation counting, giving qualitative data as to whether or not a precursor was involved in the saccharide's synthesis, and quantitative data on the flow of radioactivity from the precursor to the product(s) during the test period. Prior to this test, we showed that ^{14}C-sucrose (4 μCi) injected into the stem internode of an infected sorghum plant labeled all the honeydew sugars analyzed 20 hrs later. At 40 hrs post-injection half of the ^{14}C detected was in difructosyl-mannitol. This experiment confirmed the validity of comparing the incorporation of fructose and mannitol against sucrose as the standard *in vivo* biosynthetic source.

Results

^{14}C-sucrose and ^{14}C-mannitol in cell-free honeydew were both biotransformed at significant levels into the two major mannitol-containing oligosaccharides. After a 4-hr incubation, 15% of the ^{14}C in the oligosaccharides derived from sucrose was in the trisaccharide difructosyl-mannitol. The corresponding proportion from mannitol was 35%. The disaccharide fructosyl-mannitol became radiolabeled first from each precursor, and this disaccharide was then transformed into the trisaccharide in the honeydew. ^{14}C-fructose was not incorporated into the oligosaccharides in the cell-free honeydew, a result that was somewhat unexpected. We thought that the failure of fructose to serve as a direct precursor of the oligofructosides might be because mannitol was absent from the honeydew we used. We repeated the test of ^{14}C-fructose with the addition of unlabelled mannitol (1 mg), but again the fructose was not incorporated into the oligosaccharides. In whole sphacelia, ^{14}C-sucrose was more extensively incorporated into the oligosaccharides than in honeydew. ^{14}C-mannitol was less extensively transformed, yet more than 25% of the ^{14}C from the ^{14}C-fructose was recovered from the oligosaccharides. We do not think that the labeled fructose was incorporated directly into the oligosaccharides but, rather, that the label was incorporated indirectly after the fructose was reduced to mannitol. Maceration of sphacelial tissue further increased the overall amount and number of the transformations of the labeled precursor into previously characterized *C. africana* oligosaccharides.

Discussion

We found that *C. africana* reduces fructose to mannitol in the sphacelium. This transformation is common in fungi (2, 17). All the steps in the formation of the honeydew oligosaccharides can result from cell-associated metabolism. However, *C. africana* honeydew supernatant also contains an enzyme(s) that can link one molecule of mannitol, presumably arising from the reduction of fructose, to the 2-position of the fructose moiety of sucrose. This linkage mimics a similar linkage of the glucose moiety in the sucrose substrate, occurs without fructose ever being free, and results in the formation of the bioactive disaccharide fructosyl-mannitol. The glucose freed by this process presumably is available for direct utilization by the pathogen if the transformation occurs in the cell. However, if the transformation occurs in the exuded honeydew, then the free glucose is potentially available to the conidia and could serve as an energy source for secondary sporulation. Thus, *C. africana* has a potential extracellular mechanism to enhance secondary sporulation that plays such an important role in ergot epidemics. Based on our NMR spectroscopy analysis of the fructosyl-mannitol disaccharide, the mannitol may be linked either 1-2 or 6-2 to the fructose. Further addition of fructose, also linked by its 2-position to the remaining 1 or 6 positions of mannitol in the disaccharide, gives the trisaccharide difructosyl-mannitol. As in the formation of the disaccharide, mannitol is linked to a 2-position of a fructose in another sucrose molecule with the concomitant release of glucose, but in this case the mannitol is already linked to a fructose moiety.

In *C. sorghi* and *C. sorghicola*, the honeydew saccharides are primarily fructose and glucose, implying that the principal mode of utilization of the host's sucrose is via an invertase that hydrolyzes sucrose to its constituent monosaccharides. *C. africana* probably also has invertase activity, but the saccharide-transferring activity results primarily in the production of mannitol-containing oligosaccharides. Secondary conidiation by *C. sorghi in vivo* is less obvious (13), but its anamorph, *Sphacelia sorghi*, is morphologically indistinguishable from the anamorph of *C. africana*, which means that it is important not to confuse the sphacelial stage of these two species (1). *C. sorghi* may not exhibit significant secondary sporulation in recently exuded honeydew because of the less-bulky early growth habit of its sphacelium (7). This growth pattern requires less of the host's sucrose and could be responsible for the much higher saccharide concentration in the honeydew. In the case of *C. sorghi*, the high fructose level alone probably suffices to explain the fungus's general inability to produce secondary spores on honeydew on inflorescences (3, 13). These results suggest that the ratio between the pathogen's demand and the host's supply may determine whether conidia in the honeydew will germinate to release airborne secondary spores. Even *C. sorghicola*, which has an anamorph that differs from the other two sorghum pathogens, can produce secondary spores *in vitro* (10). Secondary sporulation by *C. sorghicola* has never been reported *in vivo*, however, probably because of the amount and effects of the fructose in the honeydew.

Normally, the developing sphacelium of *C. africana*, which is so large that it forces the glumes apart, utilizes virtually all of the sucrose supplied to a grain sorghum floret. Thus, the concentration of saccharides in the honeydew of *C. africana* usually is low. The glucose moiety of the sucrose is metabolized by the pathogen as a carbon and energy source, and the fructose moiety, partially reduced to mannitol, is incorporated into oligosaccharides. The cellular- and honeydew-associated enzymatic activeity is sufficiently in excess of substrate, such that our experiments were essentially complete within four hr, and that significant transforming activity was evident within a few minutes. The level and efficiency of the biosynthetic mechanism that links mannitol to fructose molecules is consistent with our observations of increased concentrations of bioactive oligosaccharides in *C. africana* honeydew produced if the sorghum plant has a "sweet" character. The present study, therefore, strengthens the hypothesis that the incorporation of a sweet character into A lines for sorghum hybrid production could limit the adverse impact of ergot disease by restricting secondary sporulation on honeydew.

Acknowledgements

We thank CAPES, Brazil, for postgraduate studentship funding for A.B., the Biology Department at Birkbeck College (University of London) for the use of laboratory facilities, and the Chelsea Physic Garden (London) for the use of horticultural facilities.

References

1. **Bandyopadhyay, R., L. K. Mughogho, S. K. Manohar, and M. V. Satyanarayana.** 1990. Stroma development, honeydew formation and conidial production in *Claviceps sorghi*. *Phytopathology* **80:** 812-818.
2. **Blumental, H. J.** 1976. Reserve carbohydrates in fungi, p. 292-307. *In* J. E. Smith and D. R. Berry (eds.), *The Filamentous Fungi. Vol. 2. Biosynthesis and Metabolism.* Edward Arnold, London, U.K.
3. **Bogo, A.** 2000. *Biochemical Physiopathology of Some Ergot Fungi and Other Honeydew-Producing Plant Parasites*. Ph.D. thesis. Univ. of London, London, UK.
4. **Bogo, A., and P. G. Mantle.** 1999. *Claviceps africana* discovered in India. *Plant Dis.* **83:** 79.
5. **Dickerson, A. G.** 1972. A β-D-fructofuranosidase from *Claviceps purpurea*. *Biochem. J.* **129:** 263-272.
6. **Frederickson, D. E., P. G. Mantle, and W. A. J. de Milliano.** 1989. Secondary conidiation of *Sphacelia sorghi* on sorghum, a novel factor in the epidemiology of ergot disease. *Mycol. Res.* **93:** 497-502.
7. **Frederickson, D. E., P. G. Mantle, and W. A. J. de Milliano.** 1991. *Claviceps africana* sp. nov.: The distinctive ergot pathogen of sorghum in Africa. *Mycol. Res.* **95:** 1101-1107.
8. **Frederickson, D. E., P. G. Mantle, and W. A. J. de Milliano.** 1993. Windborne spread of ergot disease (*Claviceps africana*) in sorghum A-lines in Zimbabwe. *Plant Pathol.* **42:** 368-377.
9. **Frederickson, D. E., P. G. Mantle, and W. A. J. de Milliano.** 1994. Susceptibility to ergot in Zimbabwe of sorghums that remained uninfected in their native climates in Ethiopia and Rwanda. *Plant Pathol.* **43:** 27-32.
10. **Hassan, H. A.-G.** 1996. *Control of Ergot Disease in Sorghum with Particular Reference to* Claviceps africana. Ph.D. thesis, Univ. of London, London, UK.
11. **Hassan, H. A.-G., P. G. Mantle, and N. W. McLaren.** 1995. A strategy for constraining ergot disease epidemics in F_1 hybrid sorghum production. *Proc. XIIIth Intl. Plant Protect. Cong. (The Hague, The Netherlands, 2-7 July 1995)*: Abstract 677.
12. **Hassan, H. A.-G., P. G. Mantle, and N. W. McLaren.** 1997. Putative control of ergot disease epidemics in hybrid sorghum production through the inhibition of secondary sporulation by *Claviceps africana*. *Proc. Global Conf. on Ergot of Sorghum (Sete Lagoas, Brazil, 1-7 June 1997)*: 141-146.
13. **Mantle, P. G.** 2002. Experimental parasitism on sorghum of *Claviceps sorghi* recently "rediscovered" in India. *Intl. Sorg. Millets Newsl.* **43:** in press.
14. **Manzarpour, A.** 1985. Aspects of metabolism and physiology in *Claviceps*. Diploma thesis, Imperial College, London, UK.
15. **Mower, R. L., G. R. Gray, and C. E. Ballou.** 1973. Sugars from *Sphacelia sorghi* honeydew. *Carbohydrate Res.* **27:** 119-134.
16. **Pažoutová, S., and A. Bogo.** 2001. Rediscovery of *Claviceps sorghi* (Ascomycotina: Clavicepitaceae) in India. *Mycopathologia* **153:** 99-101.
17. **Smith, D., L. Muscatine, and D. Lewis.** 1969. Carbohydrate movement from autotrophs to heterotrophs in parasitic and mutualistic symbiosis. *Biol. Rev.* **44:** 17-90.
18. **Tsukiboshi, T., T. Shimanuki, and T. Uematsu.** 1999. *Claviceps sorghicola* sp. nov., a destructive ergot pathogen of sorghum in Japan. *Mycol. Res.* **103:** 1403-1408.

17

Genotype × Environment Effects on the Response of Sorghum to Ergot and Repercussions for Disease Screening

Neal W. McLaren

Abstract

Ergot, caused by *Claviceps africana*, is a disease of unfertilized sorghum ovaries. Weather during two critical phases of flowering usually determines ergot incidence and severity. Low night temperatures during the 3-4 weeks preflowering reduce pollen viability and predispose florets to infection. Sorghum genotypes differ in terms of critical threshold temperatures. Similarly, weather during early anthesis affects ergot severity. The optimum temperature for infection and disease development is 19°C, with a limited risk of infection above 28°C. Humidity also has a significant effect on the disease × temperature relationship. The relationship between weather and ergot severity can be used to model the ergot potential of a specific flowering date. Significant differences in the ergot potential required to induce infection, termed the Ergot Breakdown Point (EBP) and determined by nonlinear regression analysis, are known and can be used as a measure of escape resistance.

Introduction

Ergot of sorghum [*Sorghum bicolor* (L.) Moench], caused by the fungus *Claviceps africana* Frederickson, Mantle & de Milliano, is a disease of unfertilized sorghum ovaries. The diagnostic ergot symptom is the development of a sticky, pinkish-brown, translucent exudate at the tips of infected florets. The pathogen emulates pollen tubes and colonizes the tissues of the ovary (10). Infected florets, therefore produce no grain. Additional losses result from harvesting difficulties due to honeydew coating of grain and the resultant stickiness. Grain molds also are increased by the high sugar concentrations in the honeydew, with concomitant reductions in seed viability and malt production and milling quality.

The susceptible period for infection of sorghum by *C. africana* is from flower opening to fertilization (11, 21). Once fertilization occurs, susceptibility declines rapidly. The ability of the plant to produce viable pollen and to fertilize the ovaries before they are colonized by the pathogen is termed "escape resistance." Weather is the primary factor that affects ergot escape resistance and, thus, the incidence and severity of ergot. Predisposition of male normal sorghum to infection by *C. africana* results from reduced pollen viability caused by preflowering weather conditions (14, 15, 19). Manipulation of flowering dates to avoid favorable weather conditions and identification of hybrids and lines with tolerance to these conditions have formed the basis of ergot research and control in South Africa (16). This principle also has been applied to weather that favors pathogen growth, infection, and disease development during the early post-anthesis stage of flowering. In this chapter, I analyze the relationship between weather and ergot incidence and severity, with particular emphasis on the implications for resistance breeding.

Weather Favoring Infection and Disease Development

Effective pollination and fertilization allows pistils to escape or resist further infection (11, 21, 24). Infection of sorghum ovaries occurs primarily through the stigma, and infection hyphae essentially follow the same path as the pollen tube (10). Infection or fertilization depends on environmental conditions that determine whether infection hyphae or pollen tubes obtain the competitive advan-

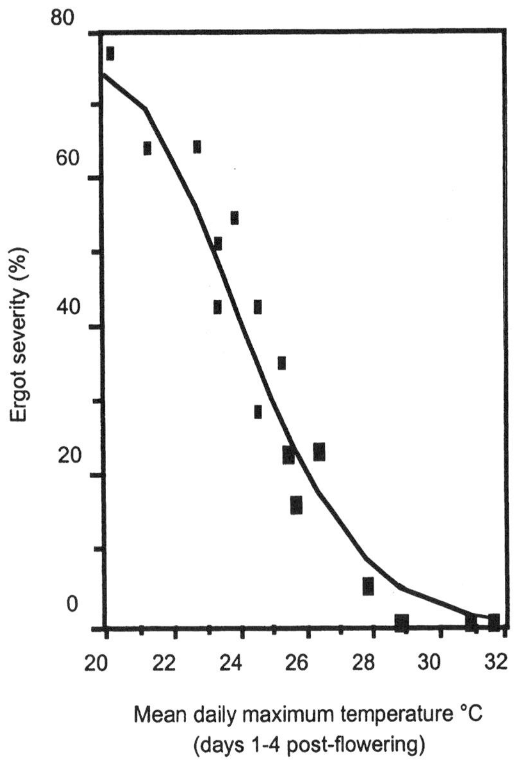

FIGURE 17-1. Relationship between mean daily maximum temperature 1-4 days post-anthesis and ergot severity for NK283; $y = 0.342x^2 + 1.334x - 9.24$ ($r^2 = 0.84$).

tage for the ovary. Under optimum daily temperatures for host development (> 28°C) and night temperatures >17°C (depending on host accession), pollen germinates within 30 min and fertilization occurs within 2-12 hr (2, 26). In contrast, at similar temperatures, conidia require 8-12 hours for germination on the stigma and 36-48 hr to reach the ovary (8, 10, 24). At cooler temperatures, pollen release and growth are retarded and pathogen activity increases. *In vitro* studies indicate that the optimum temperature for germination and growth of the pathogen is 19-20°C (3, 4, 28), with little or no growth occurring above 30°C. Wet, cloudy weather during early anthesis favors rapid development and spread of the disease (27). Kulkarni (13) reported the most severe infections within the range of 21-31°C when humidity was high. Most damage occurred in fields that flowered during the cooler periods. Frederickson (9) found that ergot severities were significantly higher at 20 and 25°C than at 30°C. The latent period for infection also was significantly shorter at lower temperatures.

Cloudiness during anthesis promotes disease development (1), probably due to delayed anther dehiscence and pollen deposition under these conditions (23). Futrell and Webster (12) reported that near 100% relative humidity (RH) for 24 hours during early flowering was optimal for infection of a male-sterile line. Sangitrao and Bade (25) found that when humidity is high, rainfall is not essential for ergot development. Anahosur and Patil (1) concluded that temperatures of 19-21°C with 67-84% RH following panicle emergence were sufficient for infection of male-sterile sorghum. Wang *et al.* (28) found a correlation between mean relative humidity at 9:00 A.M. and ergot severity in Australia. There was a distinct increase in ergot severity with RH > 80%.

The above conditions are similar to those defined by McLaren and Wehner (18), who determined that 19.5°C was the optimum mean daily maximum temperature for disease development during the 1-4 day period post-anthesis and that the upper maximum limit was 28°C, above which disease severity was negligible (Fig. 17-1). They also found an inverse relationship with RH during this period, with maximum disease usually occurring at high humidity. Finally, they identified a strong negative correlation between daily maximum temperature and rainfall (r = -0.71) and sunshine hours (r = -0.86), and suggested that maximum temperature was a suitable indicator of disease-favorable conditions in South Africa.

Weather Favoring Infection

Futrell and Webster (11) suggested that any factor that prolongs the period from flower opening until fertilization could increase the risk of ergot infection. In controlled pollination frequency trials, they found a strong correlation (r = 0.87) between the percentage of unpollinated florets and ergot infection. This extended period from flower opening until fertilization could account for the high risk of ergot in seed production fields where male-sterile A lines are used.

Downes and Marshall (7) found that night temperatures ≤ 13°C during meiosis could induce male sterility in sorghum. Brooking (5) estimated that the critical stage for cold-induced sterility was 2-3 weeks prior to anthesis. The greatest temperature sensitivity was from the late archesporial cell-pollen mother cell development period to the leptotene stage of meiosis. Once meiosis progressed beyond leptotene, sterility was not induced. Development from microspore release to fertilization was particularly insensitive to prolonged night temperature treatments. Inflorescences with low temperature-induced sterility developed anthers that were exerted normally at anthesis but were only partially dehiscent. Pollen grains in these anthers at anthesis were mainly vacuolate, two-celled grains, equivalent to control pollen just prior to the onset of the maturation phase. In contrast, pollen from plants not subjected to cold stress was densely cytoplasmic and packed with starch grains.

Cold-induced sterility and ergot susceptibility are related (19). In field trials with male-normal sorghum hybrids, night temperatures < 13°C 3-4 weeks prior to flowering increased susceptibility to ergot to that of a male-sterile line. Seed set in noninoculated heads under pollination bags also was reduced, suggesting that increased susceptibility was the result of low temperature-induced sterility. Ergot incidence and seed set were inversely correlated (r = -0.92). Low temperature reduced pollen viability and increases ergot severity (Fig. 17-2). We do not know whether induced sterility is attributable solely to reductions in pollen viability, or whether reduced receptivity of stigmas also may be involved. Nonetheless, accessions differ in the ability to tolerate pre-flowering cold stress. For example (Fig. 17-2), PAN8564 maintains viable pollen production and ergot escape at temperatures as low as 13°C. In contrast, a progressive increase in sterility and concomitant increase in ergot was recorded in NK283. This result is similar to the cold-induced sterility described by Brooking (6), who suggested that the linear response of some genotypes to temperature reduction indicates that sterility is a quantitative response and not a qualitative one occurring below some critical temperature.

Molefe (20) reported ergot in Botswana in panicles that flowered when humidity was high, the minimum temperature was 14°C, and the maximum temperature was 27°C. The low minimum temperature may have been a predisposing factor that favors disease development. Similarly, Sangitrao and Bade (25) reported that minimum daily temperatures of 13-18°C, and 76-84% RH were the most favorable for ergot development.

Pepper and Prine (22) shaded sorghum with black plastic fabric (25% light) and identified two critical periods that adversely affected grain yield components. One of these periods was from the pre-boot vegetative stage, when the last leaf is just visible above the whorl, to full panicle expansion or 50% anthesis (approximately 2 weeks before anthesis). Shading during these times resulted in a reduced number of seeds per panicle. Low radiation-induced sterility may, therefore, be an additional factor to be considered in ergot epidemiology and may explain significant increases in ergot severity during prolonged cloudy weather.

Ergot Potential

Relatively small changes in either mean daily maximum temperature during days 1-4 post-anthesis or mean daily minimum temperature days 23-27 preflowering can have a significant effect on ergot severity. This result implies that ergot severity in a single genotype can change on a daily basis, depending on small changes in weather, and that ratings, even those made within a few days, need not be comparable (*e.g.*, Fig. 17-3). McLaren (14) found that the ranking of sorghum lines according to mean ergot severity in multilocation trials differed, which led him to

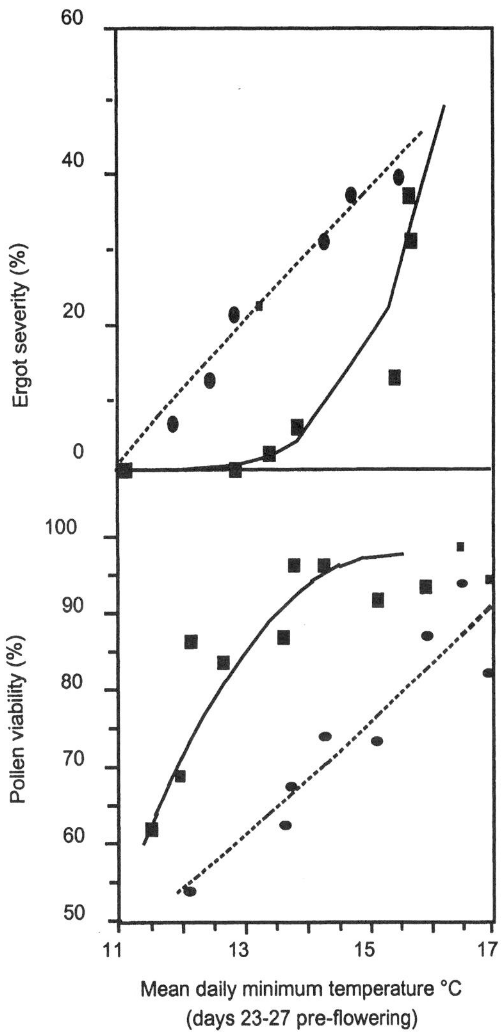

FIGURE 17-2. Relationship between mean minimum temperature days 23-27 preflowering, and (top) ergot severity [NK283 (●) → $y = 1.31x^{0.944}$ (r^2 = 0.98) and PAN8564 (■) → $y = 0.213\times10^{-5}x^{4.156}$ (r^2 = 0.85)], and (bottom) pollen viability [NK283 → $y = 1.33x^{1.494}$ (r^2 = 0.83) and PAN8564 → $y = 69.7x - 2.33x^2 - 424.78$ (r^2 = 0.62)].

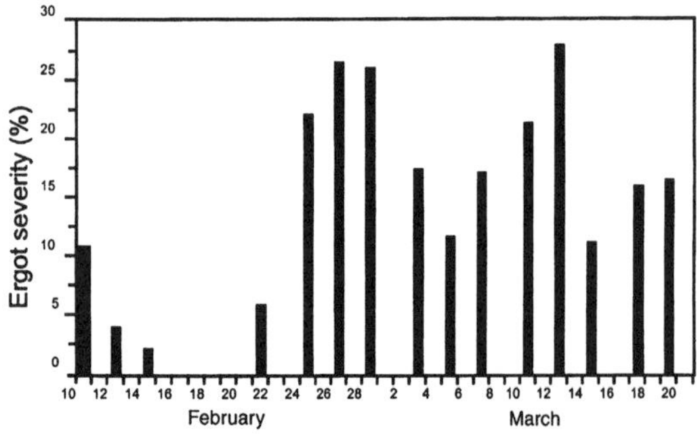

FIGURE 17-3. Ergot severity and flowering date in Potchefstroom, South Africa in 1977.

question whether the differences in disease severity reflected differences in host genotype or differences in weather conditions associated with differences in flowering date.

McLaren also proposed the concept of "ergot potential." Ergot potential quantifies weather effects on observed ergot severity and serves as a reference against which escape resistance in new genotypes can be evaluated, as the deviation from the ergot potential. The ergot potential of a flowering date is the mean ergot severity in a genetically broad-based sorghum population that flowered on that date. In initial screening for ergot resistance, this value was determined as the mean ergot severity over all genotypes that flowered on a specific day. However, as entries were selected for ergot escape resistance, this method of calculating ergot potential failed due to the narrowing of the genetic base. McLaren and Flett (17) developed a multiple regression model that includes weather variables associated with the two critical periods of host development; *i.e.*, (i) preflowering minimum temperature (mean of days 23-27 preflowering), and (ii) mean daily maximum temperature of days 1-5 post-flowering. The relationship was further improved by the inclusion of mean daily maximum humidity (mean of days 1-5 post-flowering). The model equation is:

$$EP = (-3.229\ PF_T) + \mathrm{EXP}(-0.0029\ PA_T + 0.0936\ PA_T + 3.061) + (0.379\ PA_H)$$

where EP = ergot potential, PF_T = mean minimum temperature days 23-27 preflowering, PA_T = mean maximum temperature days 1-5 post-anthesis, and PA_H = mean maximum humidity days 1-5 post-anthesis. Evaluation of the model with independent data sets yielded indexes of agreement (29) between predicted and observed ergot potentials of d = 0.94. The model was developed for use in artificially inoculated screening trials and carries the assumption of high, uniform inoculum levels during the early post-anthesis period. Additional or alternate variables may be needed if ergot potential is to be quantified by using a similar procedure elsewhere (28), since the interaction and correlation of weather variables may differ by location.

Application of Ergot Potential in Quantifying Escape Resistance

Variation in ergot incidence and severity is related to tolerance of sorghum lines to ergot-favorable conditions, *e.g.*, preflowering cold stress, and the ability to escape infection by ensuring effective and rapid pollination and fertilization (3, 4). McLaren (14) incorporated these host characters into a regression screening procedure to quantify ergot escape resistance. He found that the relationship between ergot potential and observed ergot severity was not a constant and that germplasm could be classified into three categories: (i) those in which ergot potential and disease severity were linearly related, (ii) those that are highly susceptible even at low ergot potentials, and (iii) those with varying degrees of resistance despite increasing ergot potentials (Fig. 17-4). These relationships were quantified using the regression model $ES = aEP^b$, where ES = observed ergot severity, EP = ergot potential, and a and b are arbitrary constants. The advantage of this function is that it accommodates all relationships between ergot potential and observed ergot severity. The "b" constant can be interpreted as an indicator of the degree of resistance. When $b > 1$, resistance is indicated, $b = 1$ indicates a linear relationship between observed severity and ergot potential (equivalent to the population mean), and $b < 1$ indicates susceptibility. McLaren (14) also introduced the concept of an "Ergot Breakdown Point" (*EBP*) to indicate the disease potential required to induce ergot in a particular genotype and used this statistic as an indication of ergot resistance. Observed ergot severity of 5% was arbitrarily selected as the point of induced disease and is referred to as the $EBP_{0.05}$. This value can be determined directly from the regression model where $EBP_{0.05} = [ln(5) - ln(a)]/ln(b)$.

In addition, genotypes differ in the rate of ergot breakdown (Fig. 17-4, lower). This rate too can be determined from the regression model for any disease potential as the rate at potential $EP = abEP^{(b-1)}$. A reduced rate of breakdown is of epidemiological significance, particularly at high ergot potentials.

The primary advantage of the regression procedure is that sorghum genotypes with different season lengths, and which consequently flower on different dates, can be compared even though they are exposed to different weather conditions during disease development. This model also permits quantitation of (i) deviation from the population mean, (ii) *EBP* or limits of resistance, and (iii) rate

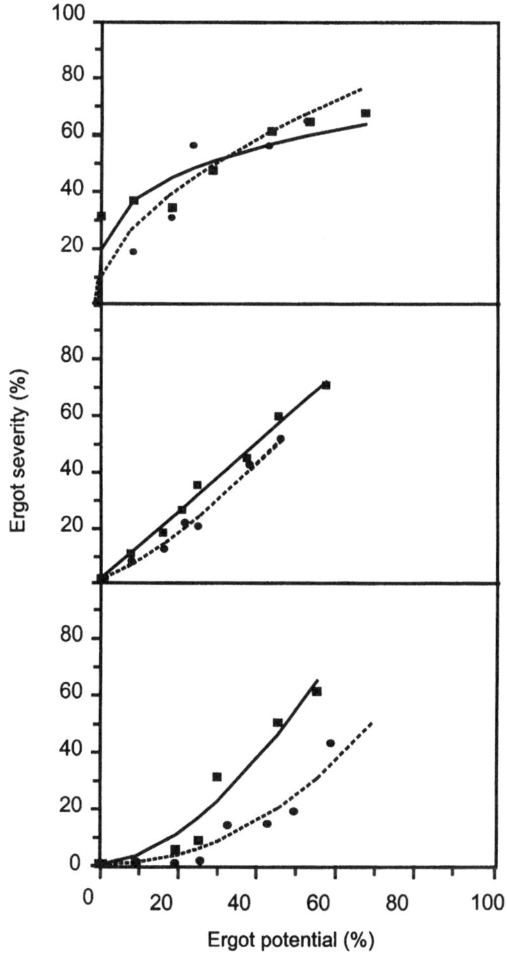

FIGURE 17-4. Host responses to changing ergot potentials: Top, susceptible [20P97 (■) → $y = 21.326x^{0.322}$ ($r^2 = 0.90$) and 12P97 (●) → $y = 23.449x^{0.274}$ ($r^2 = 0.89$)]; Middle, intermediate (population mean) [NK283 (■) → $y = 0.762x^{1.088}$, ($r^2 = 0.91$) and 28P97 (●) → $y = 0.21x^{1.43}$ ($r^2 = 0.92$)], and Bottom – resistant [26P97 (■) → $y = 0.015x^{2.026}$ ($r^2 = 0.91$) and PAN8446 (●) → $y = 0.029x^{1.762}$ ($r^2 = 0.94$)].

of resistance breakdown, all important aspects of ergot escape resistance, from a single data set.

Application of the EBP Methodology to Germplasm Evaluation

We planted ergot evaluation blocks such that there was sufficient variation in flowering date and range in ergot potential for regression analysis. We conducted these evaluations at two locations: Bethlehem (Free State Province) and Potchefstroom (Northwest Province), South Africa. These locations have long-term mean maximum temperatures of 25.8 and 24.5°C (Bethlehem) and 28.5 and 27.1°C (Potchefstroom) for February and March, respectively, when sorghum is flowering.

Single-row evaluation plots consisting of 60-80 plants were replicated three times at each location. Replicates are planted on three different days between mid-November and early January to maximize the range of

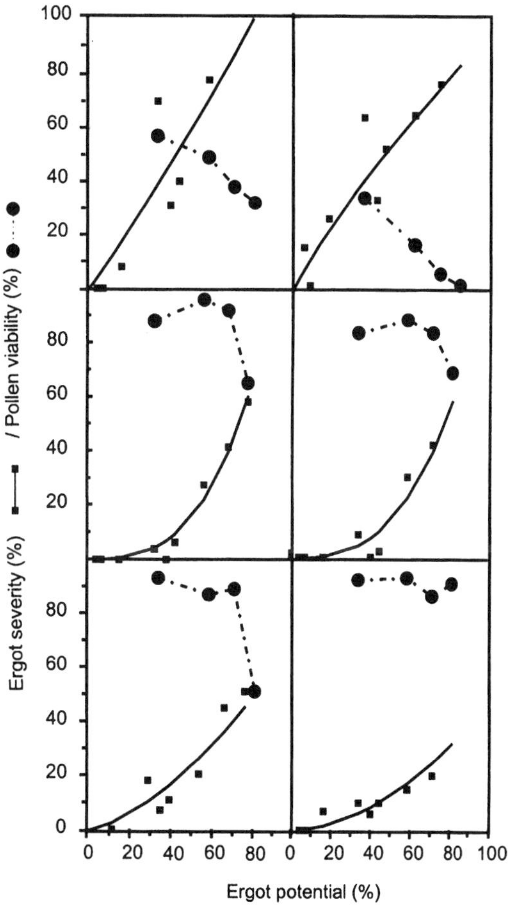

FIGURE 17-5. Ergot severity (*ES*) and pollen viability in experimental hybrids in response to changing ergot potential (*EP*).
Top left, PAN8494 [$ES = 0.742\ EP^{1.123}$ ($r^2 = 0.81$)]; top right, P22-96 [$ES = 3.118\ EP^{0.708}$ ($r^2 = 0.87$)]; middle left, PAN8564 [$ES = 7.01\times10^{-5}\ EP^{3.105}$ ($r^2 = 0.98$)]; middle right, P39-96 [$ES = 1.36\times10^{-5}\ EP^{2.95}$ ($r^2 = 0.95$)]; bottom left, P5-96 [$ES = 0.032EP^{1.653}$ ($r^2 = 0.85$)]; and bottom right, P9-96 [$ES = 7.61\times10^{-5}\ EP^{1.825}$ ($r^2 = 0.92$)].

flowering dates and weather conditions during flowering.

At flowering (±10% pollen shed in male-normal entries or stigma emergence in male-sterile entries), sorghum heads are sprayed until runoff with an aqueous spore suspension (*ca.* 10^4-10^6 spores/ml) obtained from fresh exudate from ergot-infectld plants grown in the greenhouse. Ten heads per entry are inoculated on four to six different dates (spaced 7 days apart) at each location. At the soft dough stage of grain development, ergot severity is quantified as the estimated percentage of florets exuding honeydew. The mean ergot severity per entry per inoculation date is determined, and *EBP* and the rate of resistance breakdown are calculated by using the regression methodology. In the screening of experimental hybrids, *EBP* was stable across seasons (McLaren, *unpublished*).

In advanced screening trials, pollen is collected from ten random heads per entry at the same time as inoculation. This pollen is examined microscopically for starch content, using 2% iodine solution as stain to determine pollen viability, and its relationship to ergot potential.

Results of Screening Trials

Regression models of selected experimental hybrids identified $EBP_{0.05}$s that range from 4.2% in P22.96 to 39% in P9-96 (Fig. 17-5). This range suggests that significant levels of escape resistance are available for exploitation. Similar variation has been identified in lines from both the B and R heterotic groups. The rate of breakdown when ergot potential exceeds $EBP_{0.05}$ differed significantly in genotypes even in those with similar $EBP_{0.05}$s. The rate of breakdown in P39-96 was 0.42 per potential unit compared with 0.37 per potential unit in P9-96 despite similar $EBP_{0.05}$s of 35% and 34%. This difference implies that after the $EBP_{0.05}$ is exceeded, differences in disease severity may be expected with concomitant yield losses. Thus, the rate of breakdown is an additional criterion that should be considered in escape resistance evaluations. Hybrids that continue to produce viable pollen despite increasing ergot potential tend to escape infection more readily, *e.g.* P9-96. Conversely, more susceptible hybrids, *e.g.* P5-96, produce an increasingly higher proportion of nonviable pollen with increasing ergot potential and are more easily infected. Some hybrids, *e.g.*, PAN 8494 and P22-96, appear to be inherently poor pollen producers and are highly susceptible to ergot infection. Thus, this methodology provides a number of criteria that can be used to characterize ergot escape resistance in sorghum genotypes.

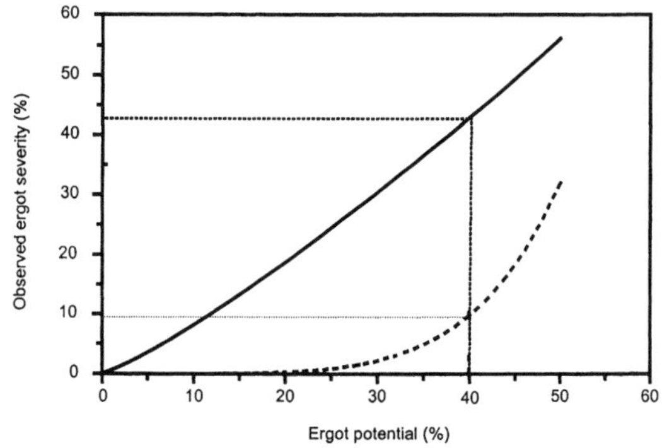

FIGURE 17-6. Comparison of ergot severity in intermediate (NK283, —) and resistant (PAN8564, - - -) sorghum hybrids. At an ergot potential of 40%, the reduction in risk is 32%.

Impact of EBP Methodology

The relationship between ergot potential and observed ergot severity in NK283, currently the most widely used sorghum hybrid in South Africa, is approximately linear (Fig. 17-6). Thus, at 40% ergot potential, approximately 40% (42% observed) ergot severity is expected. This correlation means that in a 500 ha field with a 2 ton yield potential at South African Rand 600/ton (US$1 ≈ 10 Rand), a monetary loss of R252,000 is expected. In contrast, an ergot severity of 10% is expected with the deployment of PAN 8564 under the same conditions, with a concomitant loss of R60,000.

Studies in cooperation with INTSORMIL are in progress to quantify the effect of increased pollen viability at high ergot potentials on seed set and reduced ergot severity in seed production plots. Presumably the deployment of ergot resistant pollinator R-lines will reduce ergot incidence and severity as well as increasing yield and seed germination.

References

1. **Anahosur, K. H., and H. S. Patil.** 1982. Effect of date of sowing on the incidence of ergot of sorghum. *Ind. Phytopathol.* **35:** 507-509.
2. **Artschwager, E., and R. C. McGuire.** 1949. Cytology of reproduction in *Sorghum vulgare*. *J. Agric. Res.* **78:** 659-673.
3. **Bandyopadhyay, R., D. E. Frederickson, N. W. McLaren, and G. N. Odvody.** 1996. Ergot–A global threat to sorghum. *Intl. Sorg. Millets Newsl.* **37:** 1-32.
4. **Bandyopadhyay, R., D. E. Federickson, N. W. McLaren, G. N. Odvody, and M. R. Ryley.** 1998. Ergot–A new disease threat to sorghum in the Americas and Australia. *Plant Dis.* **82:** 356-367.

5. **Brooking, I. R.** 1976. Male sterility in *Sorghum bicolor* (L.) Moench induced by low night temperature. I. Timing of the stage of sensitivity. *Austral. J. Plant Physiol.* **3:** 589-596.
6. **Brooking, I. R.** 1979. Male sterility in *Sorghum bicolor* (L.) Moench induced by low night temperature. II. Genotypic differences in sensitivity. *Austral. J. Plant Physiol.* **6:** 143-147.
7. **Downes, R. W., and D. R. Marshall.** 1971. Low temperature induced male sterility in *Sorghum bicolor. Austral. J. Exper. Agric. Animal Husband.* **11:** 352-356.
8. **Frederickson, D. E.** 1990. *Ergot Disease of Sorghum.* Ph.D. thesis, Univ. of London, London, UK.
9. **Frederickson, D. E.** 1994. Effect of temperature on ergot infection and pollination. *ICRISAT Southern and Eastern Africa Regional Program Annual Report 1993*: 2.
10. **Frederickson, D. E., and P. G. Mantle.** 1988. The path of infection of sorghum by *Claviceps sorghi. Physiol. Molec. Plant Pathol.* **33:** 221-234.
11. **Futrell, M. C., and O. J. Webster.** 1965. Ergot infection and sterility in grain sorghum. *Plant Dis. Rept.* **49:** 680-683.
12. **Futrell, M. C., and O. J. Webster.** 1966. Host range and epidemiology of the sorghum ergot organism. *Plant Dis. Rept.* **50:** 828-831.
13. **Kulkarni, G. S.** 1942. Ergot in India. *Curr. Sci.* 11: 246.
14. **McLaren, N. W.** 1992. Quantifying resistance of sorghum genotypes to the sugary disease pathogen (*Claviceps africana*). *Plant Dis.* **76:** 986-988.
15. **McLaren, N. W.** 1997. Changes in pollen viability and concomitant increase in the incidence of sorghum ergot with flowering date and implications in selection for escape resistance. *J. Phytopathol.* **145:** 261-265
16. **McLaren, N. W.** 1998. Ergot risk analysis. *Proc. 1998 Conf. on the Status of Sorghum Ergot in North America (Corpus Christi, Texas, 24-26 June 1998).* http://www.sorghumgrowers.com/research/confernc/mclaren.htm (last checked November 2001).
17. **McLaren, N. W., and B. C. Flett.** 1998. Use of weather variables to quantify sorghum ergot potential in South Africa. *Plant Dis.* **82:** 26-29.
18. **McLaren, N. W., and F. C. Wehner.** 1990. Relationship between climatic variables during early flowering of sorghum and the incidence of sugary disease caused by *Sphacelia sorghi.* J. Phytopathol. **130:** 82-88.
19. **McLaren, N. W., and F. C. Wehner.** 1992. Preflowering low temperature predisposition of sorghum to sugary disease (*Claviceps africana*). *J. Phytopathol.* **135:** 328-334.
20. **Molefe, T. L.** 1975. Occurrence of ergot on sorghum in Botswana. *Plant Dis. Rept.* **59:** 751-753.
21. **Musabyimana, T., C. Sehene, and R. Bandyopadhyay.** 1995. Ergot resistance in sorghum in relation to flowering, inoculation technique and disease development. *Plant Pathol.* **44:** 109-115.
22. **Pepper, G. E., and G. M. Prine.** 1972. Low light intensity effects on grain sorghum at different stages of growth. *Crop Sci.* **12:** 590-593.
23. **Quinby, J. R.** 1958. *Grain Sorghum Production in Texas.* Texas Agricultural Experiment Station Bulletin no. 912. Texas A & M Univ., College Station, Texas.
24. **Sangitrao, C. S.** 1982. *Studies on Ergot Disease of Sorghum in Vidarbha.* Ph.D. thesis, Punjabrao Krishi Vidyapeeth, Akola, India.
25. **Sangitrao, C. S., and G. H. Bade.** 1979. Meteorological factors associated with honeydew development and sclerotial stage in sorghum ergot. *Sorg. Newsl.* **22:** 107-108.
26. **Stephens, J. C., and J. R. Quinby.** 1934. Anthesis, pollination, and fertilization of sorghum. *J. Agric. Res.* **49:** 123-136.
27. **Sundaram, N. V.** 1971. Possible resistance to sugary disease in sorghum. *Ind. J. Genet. Plant Breed.* **31:** 383-387.
28. **Wang, E., H. Meinke, M. Ryley, D. Herde, and R. Henzell.** 2000. On the relation between weather variables and sorghum ergot infection. *Austral. J. Agric. Res.* **51:** 313-324.
29. **Wilcox, C. J.** 1982. Some comments on the evaluation of model performance. *Bull. Am. Meterol. Soc.* **63:** 1039-1313.

18

Effect of Cold Degree Units on Incidence of *Claviceps africana* in Sorghum Hybrids

Noe Montes, Gary N. Odvody and M. Marin Silva

Abstract

A group of sorghum hybrids were planted periodically for several months to evaluate the effect of cold degree units on natural development of sorghum ergot in northern Mexico. Daily minimum temperatures below 13°C significantly increased ergot incidence, especially if they occurred 9-11 days before half bloom. Cooler temperatures near bloom are important in increasing the potential for sorghum ergot because they probably cause some male-sterility in the developing sorghum florets enclosed by the flag leaf sheath.

Introduction

Sorghum ergot, caused by *Claviceps africana* Frederickson, Mantle & de Milliano (6), was first reported on sorghum, *Sorghum bicolor* (L.) Moench, in northern Mexico in early 1997 (1). This fungus infects only nonfertilized sorghum ovaries (1, 2, 6). In commercial sorghum hybrids, ergot can be a problem if pollen is absent or sterile at the bloom stage (2). The ovary is more vulnerable to ergot due to reduced pollen viability associated with preflowering cold stress (3, 4, 5, 7) and weather conditions during flowering (7). Cold stress, *i.e.*, temperatures below 13°C, prior to flowering increases pollen sterility (3, 4, 5, 7), especially if it occurs during the microsporogenesis stage. The subsequent reduction in self-pollination ability (4, 5) increases the potential for ergot, because it increases the average time that the ovary remains unfertilized. Farmers in northern Mexico commonly use the same sorghum hybrids during the spring and the fall. Sorghum ergot associated with pollen sterility is more likely in the fall growing season due to cool-temperature stress while the crop is flowering.

Materials and Methods

An experimental trial of 33 hybrids with four replications was planted periodically (every 3-4 weeks) at Rio Bravo, Tamaulipas, Mexico from early April to mid-September 1997. Flowering (half bloom) dates of the trials ranged from May to mid-December. Weather data (daily minimum temperatures) were obtained from the INIFAP Rio Bravo Experimental Weather Station. At the soft dough stage, sorghum ergot was recorded as the total number of sorghum heads within each plot with any visible ergot. Triad values (average of 3 consecutive days) of the minimum temperature were recorded up to 30 days before half bloom and six days after half bloom. "Cold degree units" are the number of degrees that the daily minimum temperature was below 13°C for each triad day period.

Results

The triad values of cold stress degree units and ergot incidence were significantly correlated with daily minimum

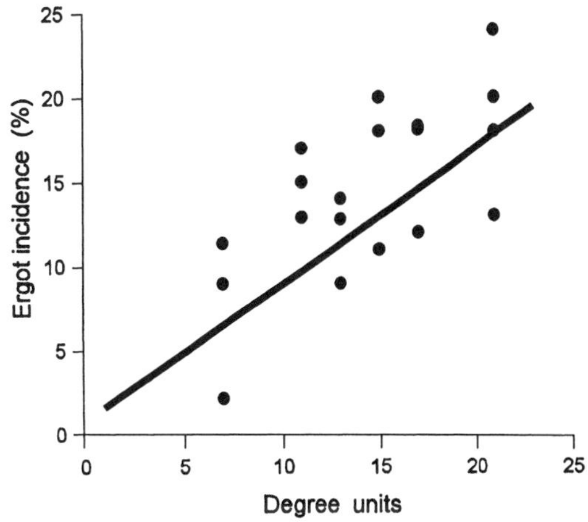

FIGURE 18-1. Relationship ($r^2 = 0.73$, $p < 0.01$) between the incidence of sorghum ergot and degree units less than 13°C recorded 9-11 days before bloom; $y = 0.8336 + 0.8071x$.

temperatures below 13°C, especially if these temperatures occurred 9-11 days before half bloom. The regression model between ergot and cold degree units (Fig. 18-1) accounted for more than 70% of the variation in ergot incidence.

Discussion

Male or pollen sterility can occur in sorghum heads when low night temperatures are present either prior to flowering (4, 7), during pollen mother cell meiosis (5), or between flag leaf emergence and flag leaf sheath elongation (3). Reduced pollen viability caused by low (cool)-temperature stress was a significant factor in the epidemiology of the sorghum ergot in our trials. The highest correlation (r = 0.86, $p < 0.01$) between ergot incidence and cold stress units was at 9-11 days before half bloom. Low temperatures near the eleventh day prior to half bloom appear to have the greatest effect on ergot incidence in self-fertile sorghums, probably by inducing sterility in the sorghum head enclosed inside the flag leaf sheath (3). Our results are very similar to those of McLaren and Wehner (7), who found that ergot and cold degree units at the eleventh day prior to bloom were strongly correlated. The difference between their data and ours is that they observed ergot infection 11 days before bloom initiation, and our window was somewhat larger, 9-11 days before half bloom. The linear response between ergot incidence and cold degree units shows that a continued reduction in daily minimum temperature increases the incidence of ergot due to progression from partial to complete head sterility in the hybrids (5).

References

1. **Aguirre, R. J., A. H. Williams, G. N. Montes, and H. M. Cortinas.** 1997. First report of sorghum ergot caused by sphacelia sorghi in Mexico. *Plant Dis.* **81:** 31.
2. **Bandyopadhyay, R., D. E. Frederickson, N. W. McLaren, G. N. Odvody, and M. J. Ryley.** 1998. Ergot: A new disease threat to sorghum in the Americas and Australia. *Plant Dis.* **82:** 356-367.
3. **Brooking, I. R.** 1976. Male sterility in sorghum *Sorghum bicolor* (L.) Moench induced by low night temperature. I. Timing of the stage of sensibility. *Austr. J. Plant Physiol.* **3:** 589-596.
4. **Brooking, I. R.** 1979. Male sterility in sorghum *Sorghum bicolor* (L.) Moench induced by low night temperature. II. Genotypic differences in susceptibility. *Austr. J. Plant Physiol.* **6:** 143-147.
5. **Downes, R. W., and D. R. Marshall.** 1971. Low temperature induced male sterility in *Sorghum bicolor*. *Austr. J. Exper. Agric.* **11:** 352-356.
6. **Frederickson, D. E., P. G. Mantle, and W. A. J. de Milliano.** 1991. *Claviceps africana* sp. nov., the distinctive ergot pathogen of sorghum in Africa. *Mycol. Res.* **95:** 1101-1107.
7. **McLaren, N. W., and F. C. Wehner.** 1990. Relationship between climatic variables during early flowering of sorghum and the incidence of sugary disease caused by *Sphacelia sorghi*. *J. Phytopathol.* **130:** 82-88.

19

Advances in *Claviceps africana* Chemical Control

Noe Montes, Gary N. Odvody, and Hector Williams-Alanis

Abstract

Infection by *Claviceps africana* can be reduced by triazole fungi-cides that make good contact with the sorghum florets. The relationship of ergot severity to chemical application rate was best represented by exponential decay regression models for the triazole fungicides triadimefon (Bayleton®), propiconazole (Tilt®), and tebuconazole (Folicur®), and for the systemic acquired resistance chemical acibenzolar-s-methyl (Actigard®). Actigard® is no longer used as a chemical control for ergot because it reduces or prevents seed set. Among the triazoles, hand-spray application of triadimefon at the highest rate (250 ppm) could occur up to 36 hours after inoculation and still provide near complete control of ergot. Triadimefon provided the best control and protection with hand spray and head-directed ground applications, but control with ground application equipment was lost after 8 days. Aerial application, even at the highest rates of fungicide, gave poor initial control of ergot, probably because the application method and low spray volume provided insufficient contact and coverage of heads with the fungicide spray mixture. Spray application at or before bloom initiation is necessary to ensure that male-sterile sorghum flowers are protected from *C. africana* by the time they begin bloom. To ensure fungicide contact with all emerging panicles, applications should be made every 3-5 days, with three applications probably needed under warm summer environments and three or four applications needed for cool-temperature environments.

Introduction

Since its appearance in the Americas, sorghum ergot, caused by *Claviceps africana*, has been studied to determine its potential recurring severity and the best method for control. In Brazil in 1995, ergot was found throughout a previously uninfected 800,000 km^2 region within a week, demonstrating the ability of the pathogen to develop into an epidemic rapidly in terms of both time and space (2). By 1997, ergot was observed throughout South America, Mexico (1), and the United States (4), and in some countries in Central America. The triazole fungicide propiconazole (Tilt®, Syngenta, Greensboro, North Carolina) provided good control of sorghum ergot in Brazil and was the only product approved for use in the United States, where aerial application is the primary method utilized. In 1997-1998, fields receiving aerial application of propiconazole developed levels of ergot ranging from none to high incidence and severity (Montes, *unpublished*).

One of our goals was to identify the most effective fungicide and chemical control methodology for use in hybrid seed production fields. Since 1997, ergot chemical control strategies generally have been based on methods developed in Brazil (3), where triazole fungicides are applied to sorghum heads by ground application equipment during the blooming period of the heads. The research in Brazil demonstrated that nearly 100% of the emerged sorghum heads were contacted by fungicide spray in these ground applications. There is concern that the efficacy of fungicides may be reduced with other application systems, *e.g.*, aerial application, which use relatively low volumes of water. Aerial application may provide insufficient contact and coverage of the inflorescence, and the environment at the time of application might determine whether fungicide droplets land on sorghum heads.

Our research focused on the interaction over time between the host (nonfertilized ovaries on male-sterile sorghum heads), the pathogen, and the fungicide. The standard inoculum for experiments was a suspension of 1.6×10^6 macroconidia/ml of *C. africana*, with a typical viability of $\geq 40\%$. Inoculation was accomplished by atomizing the inoculum onto heads with a hand sprayer (Spraymaster, Delta Industries, King of Prussia, Pennsylvania) to provide total coverage of the head but typically not until runoff. Sometimes heads were inoculated immediately after fungicide application to identify the application rates most effecttive for the control of sorghum ergot. This experiment simulated fungicide application in fields where inoculum of *C. africana* was already present and sorghum heads were at various stages of bloom. Other experiments utilized fully bloomed male-sterile sorghum inoculated with *C. africana* and then sprayed with triazole fungicides at different rates beginning at time zero or delayed for specific lengths of time. These latter experiments helped determine the length of time following spore deposition onto the stigma that fungicide application would still prevent ergot development in nonfertilized ovaries.

We also tested different compounds to identify the most effective ones, and the most appropriate rates and application methods for ergot control in the field. Most of this research was done at the Texas A&M University Research and Extension Center in Corpus Christi, Texas, except for a few experiments at the INIFAP-Rio Bravo, Experimental Station (Mexico).

Evaluation of Fungicides

Three triazole fungicides, tebuconazole (Folicur® 3.6E, Bayer AG, Kansas City, Missouri), propiconazole (Tilt® 3.6E, Syngenta), and triadimefon (Bayleton® 50DF, Bayer AG), and the systemic acquired resistance compound acibenzolar-s-methyl (Actigard® 50WG-A, Syngenta) were applied at rates of 0 (control), 31, 63, 125, and 250 ppm to evaluate efficacy against *C. africana* on nonfertilized male-sterile sorghum. Heads were tagged at bloom initiation, and each treatment sequence was applied to 10 replicate heads. Fungicides were applied with a hand sprayer until runoff. After treated heads dried, they were inoculated with a fungal suspension of *C. africana* by using a hand sprayer. Male sterile lines AVAR and ATx623 were used at Rio Bravo and Corpus Christi, respectively.

Ergot severity expressed both as honeydew percentage (percentage of florets on each head with ergot sphacelia and honeydew) and number of sphacelia per head were recorded after the control heads showed infection throughout the heads. Similar results were obtained at both locations, and differences between fungicides and rates were highly significant ($p < 0.01$). At all rates, triadimefon provided the best ergot control (lowest severity), followed closely by acibenzolar-s-methyl (Table 19-1). Propiconazole and tebuconazole gave similar, poor, control of ergot at the lower rates, but at the highest rate (250 ppm), propiconazole provided control similar to lower rates (62 ppm) of triadimefon.

The relationship of ergot severity to individual fungicide or chemical rate was represented by exponential decay regression models. The models for acibenzolar-s-methyl ($y = 69.42e^{-0.06221x}$) and triadimefon ($y = 69.50e^{-0.08075x}$) were very similar, as were the models for propiconazole ($y = 69.29e^{-0.01904x}$) and tebuconazole ($y = 68.88e^{-0.02165x}$). The models for propiconazole and tebuconazole showed that at least 250 ppm fungicide was necessary to have maximum ergot control, while *C. africana* infection and severity were reduced to near zero at rates of less than 100 ppm for acibenzolar-s-methyl and about 63 ppm for triadimefon. For every unit of triadimefon applied to the sorghum heads, ergot severity was about four times less than that with the same level of propiconazole or tebuconazole. Acibenzolar-s-methyl was eliminated from further tests because it reduced or prevented seed set when applied prior to pollination/fertilization.

Inoculum Timing Test

This test was conducted to determine the length of time (hours) needed after application of propiconazole (Tilt® 3.6E) before it was effective against *C. africana* in plant

TABLE 19-1. Efficacy of triazole fungicides against *Claviceps africana* when applied at different rates by hand sprayer to male-sterile sorghum heads (ATx623 at Corpus Christi, Texas, and AVar at Rio Bravo, Mexico) at bloom initiation in 1999.

Conc. (ppm)	Tebuconazole[a]		Propiconazole[b]		Triademifon[c]		Actigard®[d]	
	Honeydew[e]	Sphacelia[f]	Honeydew	Sphacelia	Honeydew	Sphacelia	Honeydew	Sphacelia
31	36[g] a	590 a	38 a	600 a	7 b	176 b	9 b	184 b
62	23 a	390 a	25 a	430 a	0.4 c	33 b	5 b	100 b
125	4 b	160 b	11 a	260 a	0.05 c	12 c	2.8 b	69 c
250	2.5 a	100 a	0.1 c	38 b	0 c	0.8 c	1 b	27 bc

[a]Folicur® 3.6E (Syngenta).
[b]Tilt® 3.6E (Syngenta).
[c]Bayleton® 50DF (Bayer).
[d]Actigard 50WG-A (Syngenta) = acibenzolar-s-methyl.
[e]Average percentage of florets per head with ergot sphacelia and honeydew.
[f]Average number of infected sphacelia per head.
[g]Values with different letters for the same treatment in the same row are significantly different ($p < 0.01$, Tukey's least significant difference test).

TABLE 19-2. Effect of application delay and rate of triazole fungicide on average severity[a] of sorghum ergot on fully bloomed heads of a male sterile ATx623 sorghum at Corpus Christi, Texas in 2000.

Fungicide (ppm)	Hours after Inoculation Until Fungicide Application[b]									
	0		12		24		36		48	
	5[c]	9[c]	5	9	5	9	5	9	5	9
Propiconazole[d] (0)	94 a	99 a	90 a	100 a	95 a	100 a	92 a	100 a	93 a	100 a
Propiconazole (62)	0.6 b	55 b	7 b	73 b	30 b	82 b	60 b	97 a	64 b	91 ab
Propiconazole (125)	0.2 b	20 c	0.6 bc	46 c	14 c	72 bc	38 c	91 b	44 bc	89 b
Propiconazole (250)	0.1 b	6 d	0.1 c	22 d	3 d	60 c	15 d	77 c	28 c	84 b
Triadimefon[e] (0)	95 a	99 a	95 a	100 a	94 a	100 a	94 a	100 a	96 a	100 a
Triadimefon (62)	0.1 b	6 b	0.5 b	52 b	26 b	81 b	31 b	92 b	53 b	92 ab
Triadimefon (125)	0.02 b	2 bc	0.2 b	34 c	5 c	80 b	14 c	84 b	19 c	88 b
Triadimefon (250)	0 b	0.8 c	0.01 b	13 d	0.4 c	24 c	0.4 d	13 c	4 d	41 c

[a]Average ergot severity (percentage infected florets/head) determined from ten sorghum heads/plot.
[b]Fully bloomed heads were inoculated with *Claviceps africana* at time 0.
[c]Number of days after inoculation with *C. africana*. Values in the same column for each fungicide followed by the same letter are not statistically different ($p \leq 0.05$, Tukey's least significant difference test).
[d]Tilt® 3.6E (Syngenta).
[e]Bayleton® 50DF (Bayer).

tissues. After the fungicide contacts the inflorescence, systemic uptake may require some time before levels within host tissues (probably at the base of the ovary) are high enough to prevent infection. This process also may be influenced or delayed by factors such as relative humidity, temperature, and solar radiation. Sorghum heads of male-sterile ATx623 were bagged at bloom initiation until they had completed bloom. At this time, all ovaries on all heads were susceptible to *C. africana* if they retained receptivity and were not yet fertilized. Heads were hand sprayed with propiconazole at one of two rates (125 and 250 ppm), and all sorghum heads were re-bagged. At 6-hour intervals for 36 hours, ten sorghum heads of each treatment rate and the nontreated control were inoculated and then left unbagged. Ergot severity, as percentage honeydew, was evaluated 5 days after inoculation. Compared to the control heads (60-80% ergot severity), propiconazole provided significant protection at both rates beginning at time 0 and throughout the 36-hour duration of this test. Ergot severity in the 125-ppm propiconazole treatment never exceeded 12%, and ergot control was nearly complete (< 2%) in the 250 ppm treatment.

Fungicide Timing Test

Initiation of fungicide application is a major decision because it establishes the onset of control and determines when subsequent applications are to be made. Commonly, the first fungicide application is made following head emergence, either just prior to bloom or at bloom initiation. Two or three additional applications are made at 5-7 day intervals. This strategy was developed because (i) any ovary in blooming male-sterile flowers is vulnerable to infection by *C. africana* if inoculum is present, (ii) 36-48 hours is required for infection to be established, and (iii) approximately 24 hr is required following the application of triazole fungicides for uptake and accumulation into tissues at levels sufficient to prevent infection (5). In seed production fields, the timing of fungicide application can be problematic. If fungicide application is delayed, or if the sorghum plants begin blooming earlier than expected, the sorghum heads may be vulnerable to ergot.

In this test, we determined the risk of ergot infection on blooming heads of male-sterile sorghum when inoculum was present and fungicide application delayed. We used propiconazole (Tilt® 3.6E) and triadimefon (Bayleton® 50DF) at rates of 63, 125, and 250 ppm. Sorghum heads of male-sterile ATx623 were bagged at bloom initiation until they had completed bloom, then inoculated with a conidial suspension of *C. africana* (1.6×10^6 spores/ml) and the heads re-bagged. At time 0 and every 12 hours thereafter for 48 hours, 10 heads for each rate of each fungicide and a control were hand sprayed and re-bagged.

The first ergot severity rating, 5 days after inoculation, indicated some highly significant differences between fungicides, rates, and application delay time (Table 19-2). Both propiconazole and triadimefon at all rates and at all application delay times significantly reduced infection by *C. africana* (90-96%). Propiconazole provided near complete control of ergot at all three rates at time 0, but at only 125 and 250 ppm with a 12-hour application delay. Triadimefon provided near complete control of ergot at all rates at time 0 and with a 12-hour application delay, but only at 250 ppm with a 24-hour application delay. At application delays of 24, 36, and 48 hours, ergot severity progressively increased for each fungicide and rate (Table 19-2). Ergot severity also was rate dependent at each application delay time, with this relationship becoming obvious at the 12-hour and 24-hour application delay times for propiconazole and triadimefon, respectively.

A second ergot severity rating nine days after inoculation showed that only triadimefon at 250 ppm still provided near complete control of ergot (Table 19-2). Other rates of both fungicides still provided significant control, but propiconazole was significantly less efficacious against *C. africana* at all rates and application delay times. Triadimefon had high efficacy at all rates at time 0 and at the 250-ppm application rate through the 36-hour application delay time. Thus, ergot control is maximized by applying the most effective fungicide to the crop at an appropriate rate at the proper development stage.

Evaluation of Fungicide Application Method

To maximize ergot control, good fungicide contact and coverage of the sorghum head is essential. We tested three application methods differing in contact and coverage: hand spray, ground equipment, and aerial application.

In one field test, male-sterile ATx623 sorghum heads that had completed bloom were treated with triadimefon (Bayleton® 50DF) and propiconazole (Tilt® 3.6E) applied with a head-directed spray, ground application equipment, or with a hand sprayer. Fungicide treatment rates were 31, 63, and 125 g a.i./ha in 187 L of water/ha, which is equivalent to 0.028, 0.057, and 0.113 lb a.i./ac or 1, 2, and 4 oz/ac for Tilt® 3.6E and 0.057, 0.113, and 0.226 lb of Bayleton® 50DF, respectively, in 20 gal of water/ac. Each of four replications for each ground application treatment consisted of four rows of 50 m length. The fungicides were applied across all plants, but at least eighty male-sterile sorghum heads of ATx623 that had completed bloom in each replicate were tagged. Prior to ground application of fungicide, forty of the tagged heads within each replicate were bagged for later hand-sprayer treatments.

Hand-sprayer fungicide solutions were adjusted to the nozzle delivery concentrations of each fungicide and rate, *i.e.*, 170, 340, or 670 ppm. After ground application was complete, heads were unbagged and hand-spray applications were made with the appropriate fungicide and at a rate for each replicate that provided 100% contact of the fungicide with the sorghum heads at the same rates as the ground applications. After fungicide applications, all heads in all plots were inoculated with *C. africana*.

Eight days after treatment and inoculation, ergot severity was 42% in the control, 2.3-11% for propiconazole, and 1.3-5.6% for triadimefon in fully bloomed heads of ATx623 (Table 19-3). When equivalent levels were applied by hand sprayer, the ergot percentages were < 0.1% for all rates of both fungicides. At 14 days after treatment, these panicles had 90-91% ergot severity in control plants and greater than 30% ergot severity in the panicles of the most effective fungicide treatments using machine application (Table 19-3). At this time, the hand-sprayed heads at the lowest treatment rate (170 ppm) had less than 0.3% ergot severity for triadimefon and only 2% for propiconazole (Table 19-3).

TABLE 19-3. Comparison of application method and rate of triazole fungicide on average severity[a] of sorghum ergot on inoculated, fully bloomed heads of ATx623 male-sterile sorghum at Corpus Christi, Texas in 2000.

Conc.[b] (ppm)	Propiconazole[c]				Triadimefon[d]			
	Hand		Machine		Hand		Machine	
	8[e]	14[e]	8	14	8	14	8	14
0	42 a	91 a	42 a	91 a	42 a	91 a	42 a	91 a
170	0.1 b	2.1 b	11 b	72 b	0.01 b	0.26 b	6 b	72 b
340	0.03 b	0.31 c	7 c	63 c	0 b	0.02 b	2 c	51 c
670	0 b	0.04 c	2 d	59 c	0 b	0 b	1.3 c	38 d

[a]Average percent infected florets per head based on 40 heads/plot in each of four replicate plots.
[b]Fungicides were applied in 187 L of water/ha. Parts per million (ppm) values are the concentrations used in hand-spray (hand) applications, which are equivalent to the nozzle delivery concentrations in the head-directed spray of the ground-application (machine) treatments.
[c]Tilt® 3.6E (Syngenta).
[d]Bayleton® 50DF (Bayer).
[e]Number of days after treatment/inoculation. Values followed by the same letter in each column are not statistically different ($p \leq 0.05$, Tukey's least significant difference test).

Ergot Control by Aerial Application of Fungicide

We evaluated the efficacy of aerial application of propiconazole (Tilt® 3.6E), triadimefon (Bayleton® 50DF), and tebuconazole (Folicur® 3.6E) plus a control in a uniform field planting of the male-sterile line ATx623. Single, one-pass, aerial application paths were designated within the field at Corpus Christi for each of the fungicides to be applied at 125 g a.i./ha in 47 L of water (0.113 lb a.i./ac in 5.0 gal of water). Per acre, this is equivalent to 4 oz of product for Tilt® 3.6E and Folicur® 3.6E, and 0.226 lb of Bayleton® 50DF. When the majority of the heads reached bloom initiation, three replicate areas were selected within each spray path. In the center of the spray path of each replicate area, 100 heads at the bloom initiation stage were tagged. Of these heads, 50 were left unbagged for aerial fungicide application and the remaining 50 heads were bagged (paper) as a control treatment to prevent fungicide contact during aerial application of fungicide.

TABLE 19-4. Effect of bloom stage, fungicide rate, and aerial versus hand-spray application on average percentage severity[a] of sorghum ergot on heads of male-sterile ATx623 sorghum at Corpus Christi, Texas in 2000.

Bloom Stage Inoculated	Days After Inoculation	Propiconazole[b] 670 ppm			Triadimefon[c] 670 ppm			Triadimefon 340 ppm		
		Application Method			Application Method			Application Method		
		Control	Hand	Aerial[d]	Control	Hand	Aerial	Control	Hand	Aerial
Initiation	8	81[e] a	1.96 a	53 b	90 a	0.05 a	44 b	89 a	0.21 a	49 a
Completed	11	83 a	0.80 a	66 a	86 a	0.02 a	61 a	86 a	0.04 a	59 a
Initiation	14	99 a	2.42 a	96 a	98 a	0.09 a	96 a	98 a	0.30 a	96 a
Completed	14	98 a	1.57 a	69 b	98 a	0.10 a	71 b	98 a	0.14 a	94 a

[a]Average percentage of infected florets/head determined from 40 heads in each of four replicate plots.
[b]Tilt® 3.6E (Syngenta).
[c]Bayleton® 50DF (Bayer).
[d]Aerial applications were made in a water volume of 47 l/ha.
[e]Treatments followed by the same letter in the same column for either the initial (8 or 11 day) or final (14 day) evaluation date are not significantly different ($p \leq 0.05$, Tukey's least significant difference test).

After fungicide application by airplane, bags were removed from the heads and all tagged plants were inoculated with *C. africana*. Ergot severity (honeydew percenttage) was recorded when the control heads showed infection at the base. The first ergot evaluation was made 12 days after fungicide application and inoculation to allow six days for the sorghum heads to complete bloom and another six days to observe the infection at the base of the heads.

At 12 days after treatment, all fungicides reduced ergot below the 25% ergot severity in the control. The best control was provided by propiconazole and triadimefon (reductions of 70 and 77%, respectively), while tebuconazole gave nearly 49% reduction. After one additional week, *i.e.*, 19 days after treatment, heads of all treatments had statistically similar high levels of ergot (84-89%). These data show that the fungicides remain effective for only a short period of time under these conditions.

The aerial application test in 2000 was conducted at the same field location in Corpus Christi. In this study, we evaluated the relative efficacies of aerial and hand spray application for propiconazole (125 g a.i./ha) and triadimefon (62 or 125 g a.i./ha) in a uniform field planting of ATx623. When many sorghum heads reached bloom initiation, four replicate areas were selected within each spray path. In the center rows of the spray path of each replicate area, 120 heads were tagged. When these heads had all completed bloom (3-4 days), an additional 120 heads at bloom initiation were tagged. Forty of the heads in each group (completed bloom and bloom initiation) in each replicate were left unbagged for aerial fungicide application, and the remaining 80 heads at each bloom stage were bagged to prevent fungicide deposition during aerial application. Of the bagged heads, 40 provided a control treatment for each bloom stage and the other 40 were hand sprayed with a fungicide suspension of 340 ppm (triadimefon only) or 670 ppm (both fungicides). These concentrations are equivalent to the rate per acre used in aerial application but in a spray volume (20 gal/ac) for ground-application equipment.

After fungicide application by airplane, bags were removed from control and hand-sprayed heads, and all tagged plants were inoculated with *C. africana*. The ergot (honeydew) severity was recorded at 8 and 11 days after inoculation for the completed bloom and bloom initiation heads, respectively. A second rating of ergot severity was made 14 days after inoculation.

Aerial application of fungicide was ineffective. Initial ergot severity in the heads was 81-90% in the control, and 53-66% and 44-61% across the propiconazole and triadimefon treatments, respectively (Table 19-4). For both fungicides, there was significantly higher ergot severity in the completed bloom heads than the initiation bloom heads. In contrast, all hand-sprayed heads had ergot severities of $\leq$ 2% for propiconazole (670 ppm), $\leq$ 0.2% for the low triadimefon rate (340 ppm), and $\leq$ 0.05% for the high triadimefon rate (670 ppm) (Table 19-4).

Discussion

Triazole fungicides, primarily propiconazole and tebuconazole, have been used in several countries in head-directed sprays from ground equipment to successfully control sorghum ergot on male-sterile sorghums in hybrid seed production fields (3). We found that triadimefon, another triazole, was up to three times more effective than these two triazoles in controlling sorghum ergot when applied at lower rates by using methods that maximized contact and coverage of sorghum heads (Table 19-1). Similar applications on bloomed male-sterile sorghum heads inoculated with *C. africana* showed that fungicide application could be delayed and still provide good control of sorghum ergot (Table 19-2). The efficacy and longevity of ergot control was rate dependent, with triadimefon the most effective triazole.

The most important factor in the initial control of sorghum ergot and the longevity of the control was contact and coverage of the heads with sufficient fungicide. Head-directed spray applications with ground equipment were less efficacious and provided a shorter period of ergot control than did hand-spray application (Table 19-3). More importantly, aerial application of any triazole at rates approved for propiconazole did not provide good initial control of ergot, and even that limited control diminished quickly (Table 19-4). Poor ergot control with aerial application was probably due to insufficient contact and coverage of heads with the fungicide spray mixture. In the United States, propiconazole is the only fungicide registered (Section 18) for use against sorghum ergot in hybrid seed production fields, and airplanes are the primary fungicide application system. We found that, under conditions favorable for sorghum ergot, aerial application of propiconazole or another triazole at approved rates is likely to provide no more than minimal and short-lived control. Several factors complicate measureing the efficacy of aerially applied triazoles in commercial hybrid seed production fields, but the most important is that fertilized sorghum ovaries are resistant to *C. africana*. The timely and consistent presence of viable pollen provides the best protection against sorghum ergot in hybrid seed production fields. In a sorghum ergot epidemic when pollen is absent or ineffective, aerial applications of triazole fungicides as a last resort to control sorghum ergot on the female seed parent could be an expensive false hope. Ground-application equipment should be utilized wherever possible. Alternate application methods for triazoles need to be identified that provide adequate contact and coverage of head tissues. Other fungicides, including other triazoles, should be evaluated for ergot control, but for commercial use they should be more effective than the triazoles that have already been tested when applied either by air or other means. Hybrid seed production fields at risk to sorghum ergot should always include the best pollen management practices.

References

1. **Aguirre, R. J., A. H. Williams, N. Montes, and H. M. Cortinas.** 1997. First report of sorghum ergot caused by *Sphacelia sorghi* in Mexico. *Plant Dis.* **81:** 831.
2. **Bandyopadhyay, R., D. E. Frederickson, N. W. McLaren, G. N. Odvody, and M. J. Ryley.** 1998. Ergot: A new disease threat to sorghum in the Americas and Australia. *Plant Dis.* **82:** 356-367.
3. **de Almeida Pinto, N. F. J., A. da Silva Ferreira, and C. R. Casela.** 1997. Chemical control of sugary disease of sorghum (*Claviceps africana*). *Proc. Global Conf. on Ergot of Sorghum (1-8 June 1997, Sete Lagoas, Brazil)*: 158-160.
4. **Isakeit, T., G. N. Odvody, and R. A. Shelby.** 1998. First report of sorghum ergot caused by *Claviceps africana* in the United States. *Plant Dis.* **82:** 592.
5. **Odvody, G. N.** 1997. Chemical control of ergot in sorghum hybrid seed production fields. *Proc. U.S. Conf. on Sorghum Ergot (June 11, 1997, Amarillo, Texas)*: 28-30.

20

Relationship Between Climatic Variables and *Claviceps africana* Incidence on Sorghum Hybrids in Northern Mexico

Noe Montes, Gary N. Odvody, and Hector Williams-Alanis

Abstract

Sorghum hybrids were planted to evaluate the relationship between weather and ergot incidence in northern Mexico. Ergot was not present before September 1997. Three-day average (triad) values of maximum and minimum temperatures 10-12 days before half bloom (boot stage) had the highest significant negative correlation with the disease. Minimum relative humidity gave a significant and positive relationship with the ergot disease when this factor was high immediately after half bloom.

Introduction

During the first months of 1997 in Mexico, sorghum, *Sorghum bicolor* (L.) Moench, was affected by a new disease for Mexico called ergot (1). Ergot is caused by *Claviceps africana* Frederickson, Mantle & de Milliano (5). This fungus infects unfertilized sorghum ovaries (1, 2, 4). In commercial hybrids, ergot can be a problem if pollen is absent or sterile at bloom stage (2), especially if the plants were subjected to cool, moist conditions prior to bloom. Our objective was to determine the effect of weather on ergot development in sorghum hybrids cultivated under normal conditions that had been planted on customary planting dates.

Materials and Methods

An experimental trial of 33 hybrids with four replications was planted at Rio Bravo, Tamaulipas, Mexico. Replicates of the experiment were made over planting dates spaced from early April to mid-September 1997. The flowering dates ranged from May to mid-December. We used weather data from the INIFAP Rio Bravo Experiment Station Weather Station (daily maximum and minimum temperatures and relative humidity). At soft dough stage, sugary disease incidence was recorded as the number of infected sorghum heads within each plot. Triad values (average of three consecutive days) of maximum and minimum temperature were recorded up to 30 days before half bloom and for six days after half bloom. Triad values of maximum and minimum relative humidity were recorded up to ten days before and 10 days after half bloom. Correlation analyses were conducted to determine the relationship between weather triad values and ergot incidence. Also, regression analyses were used to determine which variables contributed significantly ($p < 0.01$) to variation in ergot potential.

Results

There were significant differences among hybrid type and triad values for each weather factor evaluated. Only early- and medium-maturing hybrids had statistically significant ($p < 0.01$) regression values when maximum temperature was related to ergot incidence. Early-maturing hybrids had the highest significant effect 13-15 days before half bloom, and medium-maturing hybrids the greatest effect 10-12 days prior to bloom (Fig. 20-1, top). The maximum temperature value at which the hybrids did not show any infection was 38-39°C. For daily minimum temperature, the period 10-12 days before half bloom was the most strongly correlated with ergot incidence ($r^2 = 0.75, p < 0.01$) for the early- and medium-maturing hybrids. Ergot incidence in late-maturing hybrids was related to minimum temperatures 1-3 days before bloom (Fig. 20-1, middle). The minimum relative humidity had a highly significant correlation with ergot disease incidence. Early-maturing materials had a period of high risk for ergot infection ($r^2 > 0.70$) from 3 days before half bloom to 6 days after bloom. Medium-maturing hybrids were at high risk for infection 2-7 days prior to the half-bloom stage. Late-maturing hybrids were at greatest

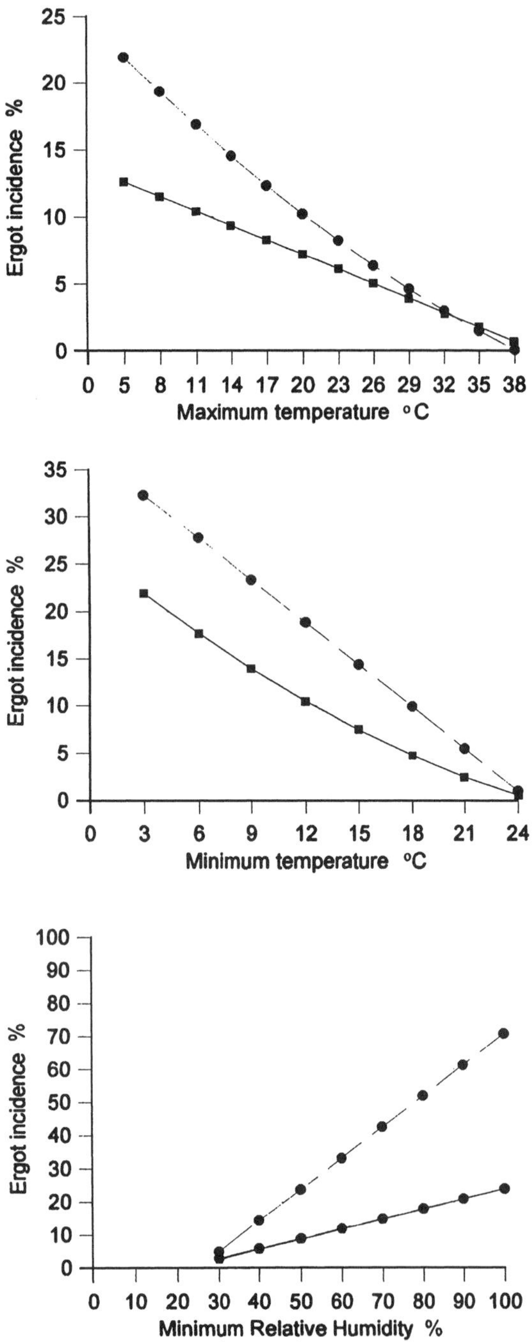

FIGURE 20-1. Relationship between maximum temperature (top, ● – early, ■ – medium), minimum temperature (middle, ● – late, ■ – early and medium), and minimum relative humidity (bottom, ● – late, ■ – early and medium), and ergot incidence in sorghum.

risk at the half-bloom stage. The minimum relative humidity necessary for ergot infection initiation was 23-24% (Figure 20-1, bottom).

Discussion

In 1997 in northern Mexico, sorghum ergot, caused by *Claviceps africana*, was not observed from April to August, but was observed beginning in early September. We interpret this result to mean that the pathogen is present and can become active whenever proper weather conditions occur. We found that susceptibility to infection is very sensitive to temperature and relative humidity. The correlation between ergot incidence, temperature, and relative humidity we observed, and the results of Gupta *et al.* (5), McLaren and Wehner (7), and McLaren and Flett (6) all support the hypothesis that weather is an important factor in ergot incidence in sorghum. Some of the differences in the response to weather could be attributed to the genetic background of the hybrids and to their flowering pattern. Low maximum or minimum temperature values near the boot stage were the most strongly correlated with ergot infection, as reported previously (3). Our present results with crops in farmers' fields suggest that avoiding these weather conditions reduces the risk of ergot infection.

References

1. **Aguirre, R. J., A. H. Williams, G. N. Montes, and H. M. Cortinas.** 1997. First report of sorghum ergot caused by *Sphacelia sorghi* in Mexico. *Plant Dis.* **81:** 31.
2. **Bandyopadhyay, R., D. E. Frederickson, N. W. McLaren, G. N. Odvody, and M. J. Ryley.** 1998. Ergot: A new disease threat to sorghum in the Americas and Australia. *Plant Dis.* **82:** 356-367.
3. **Eastin, J. D., and K. W. Lee.** 1985. *Sorghum bicolor*, p. 367-375. *In* A. H. Halevy (ed.), *CRC Handbook of Flowering*, Vol. IV. CRC Press, Boca Raton, Florida.
4. **Frederickson, D. E., P. G. Mantle, and W. A. J. de Milliano.** 1991. *Claviceps africana* sp. nov., the distinctive ergot pathogen of sorghum in Africa. *Mycol Res.* **95:** 1101-1107.
5. **Gupta, G. K., G. V. Subbarao, and M. B. L. Saxena.** 1983. Relationship between meteorological factors and the occurrence of ergot disease (*Claviceps microcephala*) of pearl millet. *Trop. Pest Manag.* **29:** 321-324.
6. **McLaren, N. W., and B. C. Flett.** 1998. Use of weather variables to quantify sorghum ergot potential in South Africa. *Plant Dis.* **82:** 26-29.
7. **McLaren, N. W., and F. C. Wehner.** 1990. Relationship between climatic variables during early flowering of sorghum and the incidence of sugary disease caused by *Sphacelia sorghi*. *J. Phytopathol.* **130:** 82-88.

21

Differences in Ergot Vulnerability Among Sorghum Genotypes and the Relationship Between Stigma Receptivity and Ergot Vulnerability

Jorge L. Moran, W. L. Rooney, Richard A. Frederiksen, and Gary N. Odvody

Abstract

Claviceps africana was reported in Brazil in 1995 and moved to North America in less than two years. By determining if there are differences in ergot vulnerability between and among types of sorghum germplasm, and by identifying the reasons for these differences, researchers may be able to develop more effective ergot control strategies, *e.g.*, ergot resistant germplasm. The objectives of this research were (i) to characterize the relative rate of ergot vulnerability of a set of publicly available and commercially used A-, B-, and R-lines, and hybrids; and (ii) to determine the relationship between the duration of stigma receptivity and ergot vulnerability. Ergot vulnerability ratings were determined for 12 pairs of sorghum A/B-lines, 12 R-lines, and 12 hybrids at seven environments in four locations during 1998. A-lines were the most susceptible to the disease, followed by hybrids, R-lines, and B-lines. Variation in ergot vulnerability among types of sorghum germplasm also was detected; however, all genotypes evaluated were vulnerable to the disease. We evaluated sixteen sorghum genotypes that resulted from crossing four sorghum B-lines in all possible combinations, an n^2 diallel, to determine the relationship between stigma receptivity and ergot vulnerability. Stigma recaptivity was not the only factor that influenced ergot infection, as some ovules no longer receptive to pollen could still be infected by ergot. Also, inconsistency across environments of the genetic effects related to the "resistance" of the genotypes evaluated, indicates that "pseudoresistance," which is dependent on the environment, may allow some germplasms to escape infection.

Introduction

Since its discovery in the 1950s, the cytoplasmic male-sterility system has provided the only practical method for producing F_1-hybrid sorghum seed on a large scale (14). Essentially all private and public sorghum seed companies worldwide rely on this methodology for the production of hybrid seed. Until a few years ago, the major problem with the utilization of male-sterile lines for hybridization was the lack of good seed set, due to low pollen viability, low pollen production, or nonsynchronization between the R-line and A-line. The sorghum seed industry in the United States was threatened only by the effect of abiotic stresses on the physiology of the sorghum lines involved in the process.

Sorghum ergot poses a new threat to sorghum seed production in the United States. A fungus that causes sorghum ergot, *Claviceps sorghi*, was first described in India in 1915 (4). A similar disease was observed in Africa in 1924 (2), but became economically important only in the 1960s, coinciding with the development of hybrid seed production. Until recently, it had been assumed that the pathogen in Africa was the same as that in India, but Frederickson *et al.* (6) showed that sorghum ergot in Africa was caused by a different species, identified as *Claviceps africana*. In 1995, the African pathogen was observed in Brazil (13), and within two years had spread throughout South America, Central America, and the Caribbean, and had reached the United States (3).

When weather conditions favor the development of ergot or when pollination is delayed or inhibited, *C. africana* infects unfertilized sorghum ovaries. Due to the lack of fertile pollen, the male-sterile lines in hybrid seed production are uniquely susceptible. As soon as the ovary is infected, a fungal mass, a sphacelium, forms and replaces the seed. Soon after, honeydew begins to exude from the sphacelia of infected florets. Current ergot control or management strategies in hybrid seed production systems are expensive. Seed producers have utilized fungicides to control the disease, which is effective but very expensive, if it is practical at all (3). As fertilization confers resistance to infection, effective pollen production and deployment is the best way to manage ergot. Seed producers have increased the number of pollinator rows to ensure adequate pollen at flowering, but this practice reduces the amount of seed produced in a given area. When ergot occurs, honeydew drips onto healthy seeds that must be cleaned, resulting in additional costs.

A more long-term and sustainable approach to ergot control utilizes host plant resistance. Since *C. africana* was described in Africa in 1991 (6), many sorghum breeders in Africa have characterized their sorghum germplasm to identify the most ergot-resistant cultivars. Musabyimana *et al.* (12) and Mukuru (11) identified six and twelve lines with reduced ergot infection, respectively. McLaren and Wehner (10) classified 70 sorghum lines as resistant when evaluated at one location, but when these lines were evaluated in a milder climate, high disease incidence was observed. Since no information is available regarding the levels of resistance in the lines currently used for hybrid production in United States, it is important to screen and quantify the level of ergot resistance in existing inbred lines and hybrids and to determine whether there are differences in vulnerability. Understanding the mechanism by which these differences are expressed and affected by the environment, and relating flowering behavior to the vulnerability of the sorghum germplasm, might allow the development of more ergot-resistant sorghum germplasm. The objectives of this research were (i) to characterize the relative level of ergot vulnerability in a set of publicly available and commercially used A-, B-, and R-lines, and hybrids, and (ii) to determine the relationship between the duration of stigma receptivity and ergot.

TABLE 21-1. Entries, pedigrees, and date of release for the A-/B-line pairs and R-lines evaluated for ergot vulnerability.

Entry	Release Date	Pedigree
A/BTx378[a]	1961	Redlan
A/BTx399	1959	Wheatland-288
A/BTx3197	1957	Combine Kafir –60
A/BTx3042	1964	Redbine Selection
A/BOK11	1966	DwarfHydro × Rice
A/BTx2752[a]	1976	BTx399 × (BTx378 × (4dwTx378 × KS30))
A/BTx623[a]	1977	BTx3197 × SC170-6-4-4
A/BTx626[a]	1985	BTx378 × SC110-6
A/BTx631	1985	(BTx378 × SC110-9) × BTx615
A/BTx635	1991	RS/R(C2)S1-102-1-1-2-1-5-1-Sel
A/B35	Not released	BC1F2 selection from SC35
A/B1	Not released	BTx625 × B35
RTx7000	1957	Caprock
RTx2536	1964	Feterita Derivative
RTAM428	1974	SC110-9
RTx430	1976	Tx2536 × SC170-6-5-1-E2
R.Tx433	1983	RTx414 × SC108-6-6-2-E2
RTx2737	1976	TAM2554 × (Tx2536 × (Tx7000 × SA7536-1))
RTx2783	1979	(((Tx424 × Capbam) × ((Tx412 × SC173-9) × SC326-6)) × SC110-9) × (((RY8 × Tx2536) × (SC110-9 × SC559-6)) × SC110-14)
RTx2862	1989	Tx2783 × GR2-39-1
RTx2880	1989	MR63-12-8-2-4 × Tx2766
RTx436	1992	(SC120-6-sel × Tx7000) × Tx7000
RTx2908	1994	(SC599-6 × Tx430)-2-6-2-BK-1-4-CBK
REON361	Not released	(Tx432 × CS3541) × SC326-6

[a]Genotypes used to generate the $n^2 = 16$ diallel for the stigma receptivity and ergot vulnerability study.

Materials and Methods

Ergot vulnerability ratings. Twelve pairs of sorghum A- and B-lines, 12 R-lines, and 12 hybrids (Table 21-1) composed of selected A- and R-lines were evaluated. These accessions represent a wide range of maturity and pedigrees and are commercially important, publicly available parental lines. The A/B and R/H experiments were conducted in a randomized complete block design with three replications in a total of seven environments at four locations. These environments included Zamorano, Honduras (ZA1 and ZA2); Vallarta, Mexico (VA1 and VA2); and College Station (CS1 and CS2) and Lubbock, Texas (LB). Environments at the same location differed in plating date, with the second experiment planted 2-3 weeks after the first planting. The first trial at Zamorano was planted in October 1997; at Vallarta, in February 1998; at College Station, in April 1998; and at Lubbock, in July 1998. Due to low disease pressure at College Station and Lubbock, the plants in these trials were inoculated with a conidial suspension at flowering by spraying each panicle with 15-20 ml of a conidial suspension (approximately 1×10^6 conidia/ml) at flowering. Five panicles from each plot were evaluated four weeks after anthesis. We used two visual rating scales: (i) ergot incidence, *i.e.*, the percentage of tagged panicles that had at least one floret infected with ergot, and (ii) ergot severity, *i.e.*, the percentage of florets on a panicle that were infected with ergot, where $0 \Rightarrow 0\%$ of florets infected, $1 \Rightarrow$ 1-5% of the florets infected, $2 \Rightarrow$ 6-10% of the florets infected, $3 \Rightarrow$ 11-25% of the florets infected, $4 \Rightarrow$ 26-50% of florets infected, and $5 \Rightarrow > 50\%$ of the florets infected. Severity and incidence ratings were transformed into a single variable termed ergot vulnerability through multivariate analyses (9) utilizing the factor reduction procedure in SPSS® 8.0. Data were statistically analyzed by location and combined as appropriate using the GLM procedure included in SAS® 6.12. Mean comparisons were calculated following the Student-Newman-Keuls procedure with a probability level of 0.05.

Relationship between stigma receptivity and ergot vulnerability. We analyzed 16 sorghum genotypes that resulted from crossing four sorghum B-lines (Table 21-1) in all possible combinations, following a complete diallel (n^2) mating design that included parents, F_1 crosses, and their reciprocals (1). Isogenic male-sterile lines were used to make the F_1 and the reciprocal crosses. The B-lines represent a wide range of maturity and ergot vulnerability. The 16-entry diallel was sown following a randomized complete block design with three replications and two row plots at College Station in April 1998 and at Lubbock in July 1998. Since all entries were male-sterile, we included between plots a mixture of hybrids that varied in maturity to serve as a pollen source and to insure high levels of pollen during anthesis of the diallel genotypes.

Prior to anthesis, 15 panicles per row were bagged to prevent pollination and/or ergot infection; three additional panicles per row were tagged as unbagged controls. Panicles in one row were used to evaluate ergot severity, while plants in the second row were used to evaluate seed set. Prior to bagging, panicles for the seed-set treatments were treated with propiconazole (Tilt®, Syngenta Corp.) at 51.3 g/acre. To insure adequate disease pressure, we inoculated plants at both locations with a conidial suspension. Bags on three plants on the first row (measures of ergot severity) were removed 0, 3, 6, 9, and 12 days a-ter mid-anthesis. Plants were inoculated with a *C. africana* conidial suspension by spraying the panicles until run-off occurred. Unbagged controls were inoculated with *C. africana* when anthesis began. All panicles were evaluated for ergot severity, on the 1-5 scale described above, four weeks after the bags were removed. Severity ratings were taken as a direct measure of the receptivity/vulnerability of the stigma to infection by *C. africana*. Three bags on plants in the second row also were removed 0, 3, 6, 9, and 12 days after mid-anthesis, at which time the panicles were treated with Tilt®, as described above. Unbagged controls were treated prior to anthesis. The duration of stigma receptivity to pollen was measured by visually rating the panicles for percentage seed set at physiological maturity, as there is a linear relationship between stigma receptivity and potential seed set (5).

To include the unbagged control treatment in the analyses, we created a new variable, termed "interval," in which the unbagged control is represented by the interval "0," 0 days after anthesis by the interval "3," 3 days after anthesis by the interval "6," 6 days after anthesis by the interval "9," 9 days after anthesis by the interval "12," and 12 days after anthesis by the interval "15." Thus, each interval represents the number of days from onset of anthesis to bag removal. Individual analyses were performed for each interval for ergot severity and for seed set for all of the genotypes in the diallel. Adjusted least square means were used for diallel analysis of the parents and the F_1s (including reciprocals) following Model I, Method I of Griffing (7), with genotypes treated as fixed effects. This analysis allowed for the estimation of general combining ability (GCA) effects, specific combining ability (SCA) effects, and reciprocal effects for each of the parental lines, by intervals and by location. Standard errors were calculated for each type of effect and used to determine if the estimated effects were significantly different from 0 with a Student's *t* test.

Ergot Vulnerability Ratings

Ergot was observed on all genotypes in all environments, and significant differences ($p < 0.05$) in ergot vulnerability were detected (Table 21-2). Across environments, the highest vulnerability ratings were at Lubbock and in

TABLE 21-2. Ergot vulnerability[a] ratings for A-, B-, and R-lines and hybrids at seven environments in 1998.

Type of Line	Location/Environment[b]						
	College Station		Lub-bock	Vallarta		Zamorano	
	1	2		1	2	1	2
A-Lines	2.8 a	2.7 a	2.9 a	2.4 a	2.3 a	2.3 a	2.0 a
Hybrids	2.3 b	2.2 a	2.1 b	0.6 b	0.7 b	1.0 b	0.6 b
R-Lines	1.3 c	1.6 b	2.0 b	0.5 b	0.5 bc	1.0 b	0.8 b
B-Lines	0.9 c	1.3 b	1.3 c	0.4 b	0.3 c	0.8 b	0.6 b
Combined[c]	1.8 B	2.0 A	2.1 A	1.0 D	1.0 D	1.3 C	1.0 D

[a]Ergot vulnerability rating scaled by factor analysis of ergot severity and incidence ratings measured in each environment.
[b]Values from the same location/environment followed by the same lower-case letter are not significantly different from each other at $p = 0.05$ (Student-Newman-Keuls test).
[c]Environment/location means followed by the same upper-case letter are not significantly different from each other at $p = 0.05$ (Student-Newman-Keuls test).

the second planting at College Station (Table 21-2). Levels of vulnerability between environments were more representative of the source of inoculum utilized (artificial vs. natural) than of the average daily temperatures during the evaluation. Locations at which the plants were inoculated had the highest levels of vulnerability (Table 21-2), but higher vulnerability ratings were not necessarily observed for the environments with cooler temperatures. This result was surprising, given the previous reports that low temperatures affect pollen production and viability, and thus, at least indirectly, the vulnerability of the germplasm to the disease (10).

Vulnerability levels in the second planting at College Station were among the highest observed, even though it was one of the warmest environments evaluated. We conclude that infection can occur readily in warmer temperatures if genotypes are highly vulnerable to the disease. There was no statistically significant relationship between maturity of the genotype and disease vulnerability ($r = 0.28$). This result indicates that the later-maturing genotypes were under disease pressure similar to that for the earlier maturing genotypes, and may reflect that the spreader rows provided inoculum efficiently for the entire duration of the experiment.

Significant differences in ergot vulnerability were detected between the lines evaluated across all environments (Table 21-2). Line × environment interaction was significant ($p < 0.05$), but no major shift in ranks was observed across environments. A-lines were the most vulnerable of the four types of lines evaluated, with an average vulnerability rating of 2.5, followed by hybrids (1.4), R-lines (1.1), and B-lines (0.8) (Table 21-2). Since A-lines are male-sterile, we expected them to be the most vulnerable. We expected the R-lines, not the B-lines, to be the least vulnerable, as R-lines are male-fertile and must shed large quantities of pollen. However, the flowering behavior of some R-lines apparently renders them more vulnerable to ergot. For example, RTx2737 has protogynous flowering behavior, which results in a window in which the pathogen can infect the floret prior to pollen shed. Further research is needed to determine if there are self-incompatibility or low pollen-viability factors that might delay fertilization in the florets of vulnerable R-lines.

The degree of vulnerability observed in hybrids is probably caused by fertility restoration problems attributable to environmental conditions or genetic factors. Pollen viability studies are needed to determine the degree of vigor in the pollen of the hybrids evaluated and the effect of the environment on this trait. Although hybrids in the combined analysis were statistically ($p < 0.05$) more vulnerable than the R- and B-lines, these differences were not large under natural inoculum conditions (Table 21-2). Thus, the combined analysis is significantly affected by the larger differences observed between hybrids and B-lines at Lubbock and between hybrids and B-/R-lines at College Station. Under high disease pressure, hybrids appear to deviate from the "less vulnerable" behavior of fertile germplasm. Under normal conditions, florets in hybrids with fertility restoration problems might complete fertilization even with low levels of viable pollen, since the conidial levels are low enough that there is relatively little competition between the pollen germ tube and the fungal conidia for the ovary.

Variation within A-lines. All A-lines were infected by the disease in all environments, and significant differences in the levels of ergot vulnerability were detected. Disease pressure across all environments was variable. As expected, male-sterile lines in environments with inoculated plants were the most vulnerable. AOK11 had the lowest ergot vulnerability rating, averaging 1.7, which was significantly less ($p < 0.05$) than the other A-lines in the combined analysis (Table 21-3). The most vulnerable male-sterile line was ATx631, but it was not statistically different from A1, ATx623, or ATx626 (Table 21-3). Also, in the combined analysis there was a significant genotype × environment interaction. Most of the changes in the vulnerability of a genotype were minor, and most entries were in a group of genotypes in which no significant differences in ergot vulnerability were detected. Vulnerability in male-sterile germplasm seems to be affected to a greater extent by ergot severity

TABLE 21-3. Combined results of ergot vulnerability[a] ratings for A-lines, B-lines, R-lines, and hybrids across seven environments.

A-Line	Rating[b]	B-Line	Rating	R-Line	Rating	Hybrid	Rating
A1	3.0 a	B1	1.0 abc	REON361	0.8 bc	ATx378 × RTx430	1.1 cd
A35	2.3 bc	B35	1.0 abcd	RTAM428	1.4 ab	ATx378 × RTx2536	1.3 bcd
AOK11	1.7 d	BOK11	0.3 d	RTx430	0.9 bc	ATx378 × RTx2783	1.5 bcd
ATx378	2.0 cd	BTx378	0.6 cd	RTx433	0.8 bc	ATx399 × RTx2536	1.6 bc
ATx399	2.3 bc	BTx399	0.7 bcd	RTx436	1.5 ab	ATx623 × RTx430	1.4 bcd
ATx623	3.0 a	BTx623	0.8 abcd	RTx2536	1.3 abc	ATx623 × RTx2908	1.0 d
ATx626	3.0 a	BTx626	1.4 a	RTx2737	1.8 a	ATx623 × RTX7000	1.3 bcd
ATx631	3.0 a	BTx 631	1.2 abc	RTx2783	0.9 bc	ATx626 × RTx2908	1.0 d
ATx635	2.6 ab	BTx635	1.3 ab	RTx2862	0.9 bc	ATx631 × RTx436	2.0 a
ATx2752	2.2 bc	BTx2752	0.3 d	RTx2880	0.7 c	ATx635 × RTx436	1.7 ab
ATx3042	2.4 bc	BTx 3042	0.6 cd	RTx2908	1.1 bc	ATX2752 × RTx2783	1.3 bcd
ATx3197	2.3 bc	BTx3197	0.3 d	RTx7000	0.8 c	ATx3197 × RTx7000	1.2 bcd

[a]Ergot vulnerability rating scaled by factor analysis of ergot severity and incidence ratings measured in each environment.
[b]Genotype means in the same column followed by the same letter are not significantly different from each other at $p = 0.05$ (Student-Newman-Keuls test).

than by incidence, since all the panicles that were evaluated were infected.

Variation within B-lines. As expected, vulnerability levels were lower in the B-lines than in the A-lines. Vulnerability averages between locations appeared to be affected by incidence more than by severity, since similar trends in cultivar rankings based on environments were observed for both ergot vulnerability and incidence. Ergot vulnerability ratings were highly variable for the B-lines, ranging from 0.3 to 1.4 (Table 21-3). BTx623 and BTx635 were the most vulnerable genotypes, with average ergot vulnerability ratings of 1.4 and 1.3, respectively (Table 21-3). BOK11, BTx3197, and BTx2752 were the least vulnerable, averaging 0.3 on the ergot vulnerability scale. There was a significant ($p < 0.05$) genotype × environment interaction among the B-lines, but the shifts were minor and usually resulted in only small changes in a genotype's rank.

Since A/B lines are isocytoplasmic, the only difference between A- and B-lines is the sterility-inducing cytoplasm in the A-line. The A/B pairs flowered similarly, as shown by the high correlation ($r = 0.88$) between days to anthesis of A- and B-lines. Therefore, each A-line had abundant pollen available from its B-line at anthesis to ensure proper fertilization, and reduce the probability of ergot infection of the floret. If nuclear genetic factors are the sole factors determining ergot vulnerability, then a correlation between A- and B-line vulnerability is expected. We found that the correlation between A- and B-lines vulnerability was low ($r = 0.35$), and think that cytoplasmic or environmental variation, or their interaction with each other or nuclear factors, is responsible for the differential reaction between the A- and B-lines. Interestingly, more recently released A-lines are more vulnerable to ergot than are older A-lines. This difference may be attributable to a longer period of stigma recaptivity in the newer A-lines, which probably also extends the time for which the plant is susceptible to infection by *C. africana*.

Variation within R-lines. Environments with inoculated plants were statistically ($p < 0.05$) more vulnerable than the rest of the environments. Differences between environments were due primarily to ergot incidence rather than ergot severity, and a significant genotype × environment interaction was observed. The specific reaction of the R-lines across environments varied, but shifts in ergot vulnerability between environments were not large, and the combined data generally reflect each genotype's level of ergot vulnerability (Table 21-3).

RTx2737 was the most vulnerable restorer line in five of the environments, with an average vulnerability rating of 1.8. In addition to the higher disease incidence, this line also had an average severity rating of 2.3, which was greater than the severity ratings of AOK11 and ATx378 (Table 21-3). We think that RTx2737 had the highest ergot rating because its stigmas emerged and were receptive to pollen 1-2 days before the anthers shed pollen. This protogynous development might explain the vulnerability of this genotype.

The least-vulnerable R-lines were RTx7000 and RTx2880, with average vulnerability ratings of 0.8 and 0.7, respectively. While none of the genotypes considered as the least vulnerable to the disease were uniformly tolerant across environments, the relative ranking of the genotypes across environments remained similar. This result indicates that the magnitude of infection may vary, but that the genotype's rank relative to the other genotypes does not change. We attempted to correlate pollen shed and ergot vulnerability, and calculated correlation coefficients between a subjective pollen shed rating and ergot vulnerability, but no significant correlations were identified. The correlation between days to anthesis and ergot vulnerability (r = 0.27) also was poor, indicating that flowering date *per se* did not predispose any of the genotypes to ergot infection. This result also is consistent with the conclusion that relative rankings of different lines will remain approximately the same even if the conditions vary.

Variation within Hybrids. In the hybrid combined analysis, there were significant differences in ergot vulnerability across environments and between genotypes. Environments with inoculated plants had higher disease levels than did environments with only natural inoculum. The significant genotype × environment interaction may be more related to the source of inoculum than to weather conditions at the time of the evaluation. Ranks are relatively constant across environments with natural inoculum, and the minor changes observed do not result in significant differences in the statistical analysis. The genotype × environment interaction in the hybrids may be attributable to differences in fertility restoration. This type of variation is expressed within a panicle, which is consistent with our conclusion that severity ratings are more important than disease incidence. This result differs from that obtained with the R- and B-lines, in which differences in ergot vulnerability were attributable to variation between panicles. More research on variation in fertility restoration in hybrids is needed to rigorously test this hypothesis.

In the combined analysis, variation for ergot vulnerability among hybrids was not high, ranging from 1.0, in the RTx2908 hybrids, to 2.0, in the RTx436 hybrids (Table 21-3). In the combined analysis of ergot vulnerability among R-lines, RTx436 and RTx2908 were not statistically different from each other (Table 21-3), yet vulnerability ratings of their hybrids are different. The differential performance of the hybrids could be due to the female parent of the crosses; however, female parents in the least-vulnerable hybrids were ATx623 and ATx626, and the female parents in the most vulnerable hybrids were ATx631 and ATx635. None of these A-lines were statistically distinguishable in the A-line combined analysis (Table 21-3). These results indicate that combining effects play an important role in determining ergot vulnerability, and that there is no simple relationship between the ergot vulnerability of the parental lines and that of the resulting hybrid. Thus, each potential parental line must be evaluated in several combinations before determining its combining ability with respect to ergot vulnerability. In addition, hybrids should be evaluated in an array of environments to identify both fertility and ergot vulnerability problems.

TABLE 21-4. Combined least squares means across intervals of ergot severity ratings and seed set percentages.

Female Parent	Male Parent	College Station Severity[a]	College Station Seed Set	Lubbock Severity	Lubbock Seed Set
Tx378	Tx378	1.2[b] gh	22 fg	2.3 bcd	45 c
	Tx623	4.1 abc	45 a	3.1 a	67 a
	Tx626	3.9 bcd	28 cdefg		
	Tx2752	0.5 h	17 g	2.1 bcd	53 cde
Tx623	Tx378	4.1 bc	41 ab	2.3 abcd	63 ab
	Tx623	4.0 c	22 fg	2.4 abc	57 bcd
	Tx626	4.4 ab	34 bcde		
	Tx2752	3.2 de	39 abc	1.8 d	62 abc
Tx626	Tx378	3.4 de	37 abc		
	Tx623	4.7 a	25 efg		
	Tx626	3.7 bcde	17 g		
	Tx2752	2.5 f	39 abc		
Tx2752	Tx378	2.4 f	29 cdef	2.6 ab	58 abcd
	Tx623	2.9 ef	37 abc	2.4 abc	63 ab
	Tx626	4.1 abc	26 defg		
	Tx2752	1.3 g	20 fg	1.9 cd	52 de

[a]Ergot vulnerability rating scaled by factor analysis of ergot severity and incidence ratings measured in each environment.
[b]Genotype means in the same column followed by the same letter are not significantly different from each other at $p = 0.05$.

Inheritance of Ergot Severity and Seed Set

Ergot severity. Combined ergot vulnerability ratings for the parental lines at College Station ranged from 1.2 to 4.0, with Tx626 and Tx623 being the most vulnerable and Tx2752 and Tx378 the least vulnerable (Table 21-4). Ergot severity means for the crosses ranged from a high of 4.4 (ATx623 × BTx626) to a low of 0.5 (ATx378 × BTx2752), but these ratings were neither significantly higher nor lower than the ratings of the other genotypes evaluated (Table 21-4). ATx623 × BTx626 was statistically more susceptible to ergot than either ATx623 × BTx623 or ATx626 × BTx626, and ATx378 × BTx2752 had a statistically significantly lower ergot severity rating than either ATx378 × BTx378 or ATx2752 × BTx2752 (Table 21-4). Thus, in these crosses there is a heterotic effect for ergot vulnerability, and when progeny from crosses of highly vulnerable lines with least vulnerable lines, *e.g.* Tx623 × Tx378, are compared to the midparent value, the progeny generally were more vulnerable than expected, suggesting that decreased vulnerability is a recessive trait.

In crosses between A- and B-lines in the same heterotic group at College Station, the GCA component of the analysis, usually attributed to additive genes (1), could be both large and statistically significant. Tx2752 and Tx378 had significantly different negative GCA effects, which implies that they can transmit their "resistance" to the progeny, while parental lines Tx623 and Tx626 had similar GCA effect values of 0.77 and 0.66, respectively. At College Station, only cross Tx378 × Tx378 had a significant SCA effect (-0.86), indicating that these progeny were more resistant to the disease than the average. These results suggest that ergot vulnerability in the parents should be an indicator of ergot vulnerability in the progeny, and that both additive and non-additive components play an important role in determining ergot vulnerability.

Reciprocal effects were significant for two of the crosses at College Station, suggesting that ergot "resistance" might involve extranuclear factors. The crosses Tx378 × Tx2752 and Tx626 × Tx2752 had significant GCA values of -0.95 and -0.81, respectively, which indicates that the phenotype of the progeny is influenced by the female parent. Additional studies are needed to determine what factors are responsible for these reciprocal effects.

Variability in ergot severity among sorghum genotypes was not as great at Lubbock, ranging from 1.8 to 2.6, as it was at College Station. Peak infection was reached at interval "6," remaining statistically steady until interval "15," averaging 2.9 on the ergot severity rating scale. Cooler temperature and lower humidity at Lubbock resulted in stigmas that were receptive for a longer time than at College Station (8).

The relationship between ergot severity and the midparent value is not as strong in the Lubbock evaluations as in those made at College Station. For example, at Lubbock, the Tx623 × Tx378 crosses had a significant reciprocal effect value of -0.27, which was not observed at College Station. Such results suggest an important role for environmental factors in determining ergot vulnerability. At Lubbock, SCA accounted for 73% of the variation in a genotype's ergot severity, suggesting that the phenotype expressed there depends upon the environment, genotype, and the genotype × environment interaction. The inconsistency of the SCA estimated effects across locations is consistent with the hypothesis that the observed "resistance" in some of the genotypes is a "pseudoresistance" that results from specific interactions with the environment.

Seed set. Significant differences were detected among genotypes and intervals evaluated at both locations. At College Station GCA, SCA, and R effects were significant, while at Lubbock only GCA and R effects were significant. While a statistically different group of high yielders can be identified at College Station, there is no clear pattern regarding the male or female parent involved in such crosses. Calculated midparent heterosis for the top seed yielder (ATx378 × BTx623) was 104%, indicating that heterosis and nonadditive effects influence seed set in this set of parents. Seed set was significantly lower across intervals at College Station. Tx623 and Tx626 averaged significant GCA effects, 3.32 and -2.11, respectively, at College Station, (Table 21-5). Tx623 significantly increases seed set and stigma receptivity across environments and these effects are heritable. Based on the SCA effects associated with Tx623 and the inconsistency of its performance as a line and in combinations, nonadditive effects or specific interactions are important in determining the phenotypes of this line's progeny. Significant R effects were detected in Tx378 × Tx2752 at College Station and Lubbock, indicating that cytoplasmic-nuclear interactions significantly affect seed in this cross.

Relationship between stigma receptivity and ergot vulnerability. Due to poor germination of one genotype at Lubbock, data from the locations could not be combined for analysis. Therefore, the analyses are reported by individual location.

Since the mechanism by which *C. africana* colonizes the ovary in the sorghum floret is through the infection of the stigma, we expected that longer stigma receptivity for pollen also would predispose the ovule to a higher level of ergot. At College Station there was no consistent relationship between GCA effects for ergot severity and seed set (Table 21-5). At Lubbock, however, all of the parental lines had consistent GCA effects for severity and seed set, and higher stigma receptivity to pollen was consistently associated with higher ergot severity.

At College Station, a sudden drop in seed set occurred at the onset of anthesis, presumably due to a reduction in stigma receptivity that is attributed to high temperatures and humidity. Ergot severity peaked between anthesis and three days later for all parental lines. This result suggests that ovules are vulnerable to ergot infection even after they are no longer receptive to pollen. In this environment, the difference between ergot severity and seed set in Tx2752 and Tx378 was not as great as for

TABLE 21-5. Combined results of six different intervals for estimates of general combining ability (g_i) effects for ergot severity and seed set percentage.

Parent	College Station		Lubbock	
	Severity	Seed Set	Severity	Seed Set
Tx378	-0.56 **	0.18	0.01	-1.46 *
Tx623	0.77 **	3.32 **	-0.07	4.67 **
Tx626	0.66 **	-2.11 *		
Tx2752	-0.87 **	-1.39	0.05	-3.21 **
Standard error ($g_i - g_j$)	0.14	1.49	0.12	1.13

*Significantly different from 0 at $p = 0.05$.
**Significantly different from 0 at $p = 0.01$.

Tx623 and Tx626, which remained highly vulnerable to ergot 12 days after the onset of anthesis.

At Lubbock, a similar pattern was observed with respect to the vulnerability of the ovule after the stigma is no longer receptive to pollen. However, at this location, stigmas were receptive to pollen for a longer time, with Tx623 averaging 60% seed set nine days after the onset of anthesis. Ergot severity at Lubbock peaked 9-12 days after anthesis, nine days after it peaked at College Station. The higher temperatures during anthesis at College Station may have reduced pollen viability and thereby increased ergot severity, while at Lubbock there was greater competition between pollen and the pathogen for the stigma, which delayed peak disease severity. Nevertheless, at both locations, ergot infection can occur even after effective pollination can not. Although we calculated no statistics to determine if there were statistically significant differences in receptivity to pollen and ergot infection, some differences could be detected visually. Additional research is needed to confirm these findings.

References

1. **Baker, R. J.** 1978. Issues in diallel analysis. *Crop Sci.* **18:** 533-536.
2. **Bandyopadhyay, R.** 1999. Geographic distribution and spread of sorghum ergot: causes and implications. *Proc. Global Conf. on Ergot of Sorghum (1-8 June 1999, Sete Lagoas, Brazil)*: 10-20.
3. **Bandyopadhyay, R., D. E. Frederickson, N. W. McLaren, G. N. Odvody, and M. J. Ryley.** 1998. Ergot: A new disease threat to sorghum in the Americas and Australia. *Plant Dis.* **82:** 356-367.
4. **Bandyopadhyay, R., V. Muthusubramanian, P. W. Tooley, S. Chakraborty, S. Pažoutová, and S. S. Navi.** 2002. Distribution and diversity of the sorghum sugary disease pathogens in India, p. 75-78. *In* J. F. Leslie (ed.), *Sorghum and Millets Diseases.* Iowa State Press, Ames, Iowa.
5. **de Vries, A.** 1971. Flowering biology of wheat, particularly in view of hybrid seed production–A review. *Euphytica* **20:** 152-170
6. **Frederickson, D. E., P. G. Mantle, and W. A. J. de Milliano.** 1991. *Claviceps africana* sp. nov.: The distinctive ergot pathogen of sorghum in Africa. *Mycol. Res.* **95:** 1101-1107.
7. **Griffing, B.** 1956. Concept of general and specific combining ability in relation to diallel crossing system. *Austr. J. Biol. Sci.* **9:** 463-493.
8. **Jaster, J. R.** 1985. *Analysis of Seed and Stigma Receptivity Among Cytoplasmic Male Sterile Sorghum* Sorghum bicolor *[(L.) Moench] Cultivars*. M.Sc. Thesis, Texas A&M Univ., College Station, Texas.
9. **Johnson, R. A., and D. A. Wichern.** 1998. *Applied Multivariate Statistical Analysis*, 4th ed. Prentice-Hall, Saddle River, New Jersey.
10. **McLaren, N. W., and F. C. Wehner.** 1992. Preflowering low temperature predisposition of sorghum to sugary disease (*Claviceps africana*). *J. Phytopathol.* **135:** 328-334.
11. **Mukuru, S. Z.** 1999. The ergot disease of sorghum in Africa. *Proc. Global Conf. on Ergot of Sorghum (1-8 June 1999, Sete Lagoas, Brazil)*: 55-58.
12. **Musabyimana, T., C. Sehene, and R. Bandyopadhyay.** 1995. Ergot resistance in sorghum in relation to flowering, inoculation technique and disease development. *Plant Pathol.* **44:** 109-115.
13. **Reis, E. M., P. G. Mantle, and H. A. G. Hassan.** 1996. First report in the Americas of sorghum ergot disease, caused by the pathogen diagnosed as *Claviceps africana*. *Plant Dis.* **80:** 463.
14. **Stephens, J. C., and R. F. Holland.** 1954. Cytoplasmic male-sterility for hybrid sorghum seed production. *Agron. J.* **46:** 20-23.

22

Sorghum Ergot (*Claviceps africana*) in México

Jesús Nárro-Sanchez, Roberto Paredes-Melesio, Juan Ángel Quijano-Carranza, Rodolfo Velásquez-Valle, and José Luis Maya-de León

Sorghum ergot is caused by the fungus *Claviceps africana* (1). In the "Bajío" region of México, which includes portions of the states of Guanajuato, Jalisco, Michoacán and Querétaro, this disease was observed for the first time in commercial grain and hybrid seed production fields during the 1997 cropping season (4). In commercial grain fields, ergot incidence, *i.e.*, the percentage of infected plants, ranged from 1 to 40%, and ergot severity, *i.e.*, the proportion of the head infected, ranged from 0.1 to 30%, probably depending on the hybrid planted. In seed production fields, ergot incidence was 10-70%, and the severity 5-40%. Losses attributed to ergot ranged from 1-30% in commercial grain fields and from 5-40% in seed production fields in this area (3). Ergot is a disease that is well-adapted to the Bajío region, which is located at 1400-1900 meters above sea level (masl), and frequently experiences night temperatures < 13°C, which results in reduced or delayed pollen production or release.

In this study, we focus on the state of Guanajuato because it has the largest area planted to sorghum. In Guanajuato, 239,000 hectares of sorghum were planted for commercial grain production in 1998. Of these, 80,000 hectares were irrigated and 159,000 were grown under rain-fed conditions (2). As ergot is potentially an economically important disease in Guanajuato if measures to reduce its impact are not implemented, we identified the areas with the highest probability for ergot damage based on historic weather data. These risk areas were then monitored closely in case remedial practices were needed to control or limit disease damage.

Weather Database Development

We obtained weather information from the stations managed by the Comision Nacional del Agua (National Committee for Water). Thirty-eight stations were selected from the sorghum-producing area of Guanajuato that differed in climatic type and had ten or more years of historic weather data. Relevant areas were then identified for every weather station included in the study, with the exception of hilly areas.

Irrigated sorghum is planted in Guanajuato from April to May, while rain-fed sorghum planting begins with the onset of the rainy season, usually by mid-June. Thus, sorghum may be flowering in Guanajuato from June to October. We evaluated temperatures for the months in the spring-summer period (June to October) and gave special attention to the critical periods for ergot infection, *i.e.*, four weeks before sorghum flowering and during the flowering period.

We used the minimum temperature recorded at the weather stations to determine the number of days per

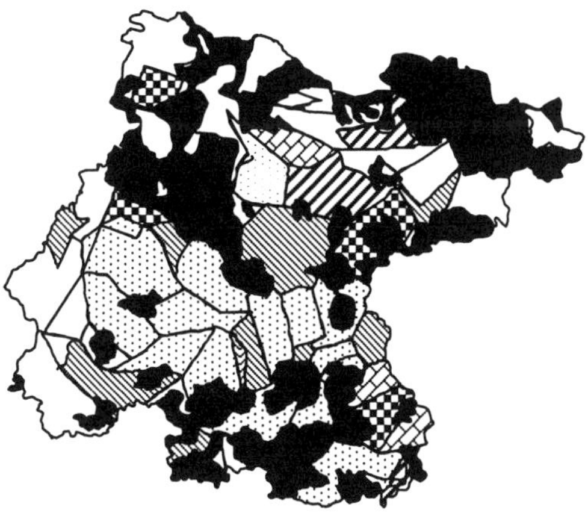

FIGURE 22-1. Number of days in June with a probability of ≥ 80% for a minimum temperature of ≤ 12° C in different portions of the Bajío region of Mexico.

No data; Hills; ≤ 5 days; 6-10 days; 11-15 days; 16-20 days; 21-29 days

FIGURE 22-2. Number of days in October with a probability of ≥ 80% for a minimum temperature of ≤ 12°C in different portions of the Bajío region of Mexico.

No data
Hills
≤ 5 days
6-10 days
11-15 days
16-20 days
21-29 days

month with conditions that induce pollen sterility, *i.e.*, temperatures ≤ 12°C. This information was used to calculate the probability of conditions conducive for pollen sterility occurring for each month from June to October at each of the 38 weather stations. When we analyzed the historical data, we accumulated the relative frequencies for the number of days with conditions that favor ecological sterility. A probability of 80% corresponds to the maximum number of days with conditions for ecological sterility that occurs at least eight of every ten years.

Results and Discussion

We set the minimum temperature for ergot risk at 12°C. Stations with the greatest number of days per month with that temperature or lower were considered to have the highest risk for ergot infection. In June, weather stations 12 (Coroneo), 22 (Apaseo el Alto), and 75 (Dolores Hidalgo) each had 21-29 days with minimum temperatures ≤ 12°C and flowering or preflowering sorghum fields in their vicinity were considered to be at greatest risk for ergot infection (Fig. 22-1). During July, stations 12 (Coroneo), 22 (Apaseo el Alto), 66 (San Jóse Iturbide), 75 (Dolores Hidalgo), and 90, 116, and 132 (Jerecuaro) fell into this category. In August, only six stations, 12 (Coroneo), 22 (Apaseo el Alto), 66 (San Jóse Iturbide), 75 (Dolores Hidalgo), and 90 and 116 (Jerecuaro), were in this category. The number of stations experiencing 21-29 days of sterility conditions is usually less in August than in June or July because the rain generally stops in August, leading to higher ambient temperatures. In September, the six stations at risk in August and station 45 (León) were in the high-risk category. In October (Fig. 22-2), only stations 1, 13, 84, and 95 had fewer than 21 days with minimum temperatures ≤ 12°C. Thus, most of the sorghum flowering in Guanajuato in October is at risk for ergot.

The weather station at Abasolo, Guanajuato had relatively few days with minimum temperatures ≤ 12° during August. For example, in 1972 and 1974 there were single days with a minimum temperature ≤ 12°C, in 1971 there were two days, and in 1973 three days, but in none of the other seven years for which data from this station were available were there any August days with a minimum temperature ≤ 12°C. Thus, for this weather station, the probability of no more than one day with a minimum temperature ≤ 12°C is 80%.

Bandyopadhyay *et al.* (1) suggested that sorghum hybrids lacking cold tolerance commonly show male-sterility, and consequently, are ergot infected because this pathogen infects only unfertilized ovaries. In Guanajuato, cool conditions are more common in August, September, and October, when sorghum planted relatively late, *i.e.*, after May 1, is likely to be in the preflowering or flowering stage. Thus, the incidence and severity of ergot infestations are expected to increase as planting dates are delayed beyond May 1, and to decrease if the planting date is prior to May 1. Sorghum cultivated under rain-fed conditions is not usually planted until after June 1, and incurs a relatively high risk for ergot since the flowering period coincides with a time in which minimum temperatures ≤ 12°C are likely to occur.

Acknowledgments

We thank M. C. Alejandro Rodríguez Guillén and M. C. Santa Ana Ríos Ruiz for critically reading this manuscript before submission.

References

1. **Bandyopadhyay, R., D. E. Frederikson, N. W. McLaren, G. N. Odvody, and M. J. Ryley.** 1998. Ergot: a new disease threat to sorghum in the Americas and Australia. *Plant Dis.* **82:** 356-367.
2. **Centro de Estadística Agropecuaria, SAGAR.** 2000. Situación actual y perspectivas de la producción de sorgo en México 1990-1999. Publicación especial, 40 pp.
3. **Narro-Sánchez, J., R. Paredes-Melesio, J. Á. Quijano-Carranza, and R. Velásquez-Valle.** 1998. Situación actual del ergot *Sphacelia sorghi* McRae en sorgo en el estado de Guanajuato. *Rev. Mex. de Fitopatol.* **16(Suppl.):** 83.
4. **Velásquez-Valle, R., J. Narro-Sánchez, R. Mora-Nolasco, and G. N. Odvody.** 1998. Spread of ergot of sorghum (*Claviceps africana*) in central México. *Plant Dis.* **82:** 447.

23

Quarantine Issues Arising from Contamination of Seed with Ergot: An Update

Gary N. Odvody, Debra E. Frederickson, Tom Isakeit, Noe Montes, Jeff A. Dahlberg, and Gary L. Peterson

Abstract

The recent rapid global spread of *Claviceps africana* raised concerns about the potential for reintroduction and spread of ergot in seedborne inoculum. Quarantines have been ineffective in preventing an eventual initial entry of the pathogen into new geographic regions, and their efficacy has been questioned in regions where the pathogen is already established. Macroconidia in honeydew or sphacelia of *C. africana* are the primary seedborne inoculum because sclerotial germination is rare. Ergot bodies (sclerotia/sphacelia) of *C. africana* can be almost completely removed from sorghum seed by standard seed cleaning and processing operations. Survival of macroconidia is poor at high temperature and high relative humidity, and most macroconidia lose viability within 2-3 months in seed storage environments. Contact fungicides, *e.g.*, captan and thiram, effectively prevent germination of macroconidia in honeydew on seed surfaces. Thus, not only does typical fungicide-treated commercial sorghum seed from ergot-affected fields have very low inoculum potential, but other local sources of inoculum are likely to provide more inoculum at a time when susceptible host tissue is present. The use of quarantines as other than a seed quality issue for sorghum ergot appears unwarranted in regions where *C. africana* is a recurrent pathogen and the existing populations are closely related.

Introduction

Prior to 1995, sorghum ergot, caused by *Claviceps africana*, was restricted to Africa and Asia (5). After the first observance of *C. africana* in the western hemisphere in Brazil in 1995 (22), the pathogen spread rapidly across sorghum-growing regions in South and Central America and the Caribbean, and into Mexico and the United States by early 1997 (1, 5, 17). The global epidemic became complete with the first report of *C. africana* in Australia in early 1996 (24), and its eventual identification as the primary sorghum ergot pathogen in India (7, 21). Windborne secondary conidia from surviving ergot structures of infected hosts are the principal inoculum source for local and regional reoccurrence and spread of ergot (12).

Quarantine issues involving *C. africana* should focus on factors related to the potential for spread and reestablishment of sorghum ergot through seedborne inoculum only in planting seed. These factors include presence, survival, and inoculum potential of *C. africana* propagules in seed; efficacy of control methods, *e.g.*, sanitation through the removal of ergot bodies or seed treatment fungicides; and the likelihood of successful infection of hosts. The additive and synergistic effect of these seedborne factors also must be considered and weighed against the potential for the development of sorghum ergot from other natural sources of inoculum already present in the region. Insufficient epidemiological knowledge of sorghum ergot may produce unwarranted quarantine restrictions that unduly constrain and complicate international seed exchange with no effect on the occurrence of the disease.

Survival of *Claviceps africana*

Macroconidia and secondary conidia are responsible for the dissemination of *C. africana* and the spread of ergot during the sorghum growing season. Macroconidia, spread by rainsplash, carry the pathogen over relatively small distances (6), but secondary conidia, produced by macroconidia, have the potential for spread over much greater distances because they are windborne (10, 12, 14, 20). Thus, any source of macroconidia also must be considered a potential source of windborne secondary conidia. The effect of the environment on the longevity of these different spore types is critical in understanding each spore type's role in disease propagation following extended periods away from the host.

The obvious honeydew exuded from parasitic sphacelia is only one source of conidia. Any surface coated with honeydew, *e.g.*, leaves, seeds, or soil, also may be a conidial source. Similarly, sphacelia incorporated into seed at harvest or that remain in the field after harvest may be a source of conidia. Sclerotia/sphacelia are commonly viewed as a means to produce the teleomorph, but also carry a large population of macroconidia in the sphacelial portion (13, 20). These sclerotia/sphacelia should be considered a more important potential source of secondary conidia than of ascospores.

A concern of the sorghum seed trade industry is the possibility of seedborne transmission of *C. africana* through inoculum associated with sclerotia/sphacelia mixed with seed and honeydew coating seed. The potential of these propagules as sources of inoculum depends on (i) survival of the associated macroconidia in seed storage environments (especially as affected by temperature and relative humidity), (ii) the efficacy of seed conditioning procedures in the removal of sclerotia/sphacelia, and (iii) the efficacy of fungicide seed treatments in eliminating or reducing propagule viability.

Cool (6°C), dry (6% relative humidity) conditions favor the survival of macroconidia in sphacelia (14, 20). Viability of macroconidia inside sphacelia declined to almost zero after two weeks at 32°C and after seven weeks at oscillating temperatures in a greenhouse (14, 20). At 20°C, macroconidial viability declined to near zero after four weeks at $\geq$ 50% or greater relative humidity, but had almost no decline for 20+ weeks at $\leq$ 6% relative humidity (14, 20). Natural field environments have wide diurnal and seasonal fluctuations in temperature and relative humidity that do not promote survival. In stored sphacelia, the viability of conidia did not exceed five months in Kansas, and the sharpest decline was coincident with the increase in temperature in March and April (8). There were almost no detectable viable conidia in the sphacelia after 2-3 months of storage on the laboratory bench (ambient conditions with sphacelia either in covered plastic buckets, burlap and plastic mesh bags, or glass petri dishes), and viability of conidia in honeydew on seed surfaces declined even more rapidly (Odvody, *unpublished*).

Thus, high numbers of viable conidia are not expected to persist in sphacelia stored (with seed) from harvest until the beginning of the next planting season. Several workers in Africa have reported similar results (11, 12, 16, 18). These researchers successfully used conidia from dry- or low-temperature-stored sphacelia to inoculate sorghum the following year, but got very low incidence (< 1%) of ergot. Survival of conidia in sphacelia in the field has been confirmed by several researchers, but its significance as a source of initial inoculum remains to be determined. The airborne nature of secondary conidia and their potential to travel long (but undetermined) distances to infect sorghum hosts also has made it difficult to definitively identify the origin of local sources of initial inoculum.

Survival of *C. africana* in Mexico and Texas

In many areas of Mexico and south Texas, the winter climate favors sorghum growth and flowering outside the normal growing season. This extended growing period also extends the infection cycle of ergot on flowering panicles of johnsongrass, ratooned or volunteer sorghum, and other alternative hosts, easing the transition for the pathogen from one crop season to the next. Sorghum ergot caused by *C. africana* persisted outside of the normal growing season in an active phase predominantly on feral grain sorghum as far north as Corpus Christi, Texas through February 1998 (19, 20) and February 2000 (Odvody, *unpublished*). However, regional weather conditions in 1998 dramatically influenced the incidence of sorghum ergot in commercial sorghum fields. Drought stress throughout the 1998 growing season prevented observation of any naturally-occurring sorghum ergot across northern Mexico and south Texas until the return of wet weather in September. Conversely, cool wet weather in the Pacific coast states of Mexico favored ergot from February–May 1998, and cool weather during September promoted ergot on sorghum in several states in the Bajio and high plains regions of Mexico. A similar pattern of ergot occurrence was observed in the Pacific coast states in February–May 1999 and 2000.

In many areas of Mexico and South Texas, the cool, wet winter months promote widespread development of sorghum ergot over large areas on feral and ratooned grain, forage sorghum, and johnsongrass within fields, along roadsides, and in winter sorghum nurseries (19, 20). The cooler temperatures that prevail at this time extend the bloom period of sorghum and reduce pollen viability on later-blooming sorghum heads, thus increasing the incidence and severity of sorghum ergot. These climatic and host factors have allowed *C. africana* to become a well-established, recurrent pathogen in both Mexico and Texas that can survive extended periods of unfavorable conditions only to quickly attain epidemic proportions over large geographic areas as soon as favorable conditions return (20). Similar survival patterns are likely wherever sorghums are cultivated in climates that are cool but have a reduced risk of frost. These areas of surviving, active reservoirs of sorghum ergot may be the foci for rapid development and spread of ergot over large regions during the growing season and are probably much more important epidemiologically than seedborne inoculum.

Effects of Captan Seed Treatment on Germination of Conidia from Sphacelia

Contact fungicides, *e.g.*, captan (55-140 g a.i./100 kg of seed) or thiram (110 g a.i./100 kg of seed), completely prevent germination of macroconidia of *C. africana* in fresh honeydew encrusting healthy sorghum seed (4, 9). In this same study, Maxim (Syngenta, Greensboro, North Carolina) was nearly 100% effective at the higher rate (5.0 g a.i./100 kg of seed), although a few conidiophores developed at a lower treatment rate (2.5 g a.i./kg of seed). Commercial hybrid sorghum seed is routinely treated with captan, but its efficacy on sclerotia/sphacelia was questioned because penetration of the fungicide into these tissues may be low. Fungicide activity against the innermost conidia might be delayed until after planting, when increased moisture allows greater penetration and activity of the fungicide, especially against conidia that diffuse to the surface of sphacelia.

The efficacy of captan seed treatments in preventing germination of macroconidia on the exterior and interior of ergot sphacelia was determined by using a mixture of fresh sphacelia and healthy seed treated with captan (Captan 400, Gustafson, Plano, TX) at a standard commercial rate (3.0 fl oz rate/100 cwt or 94 g a.i./100 kg of seed) (20). Inner cores of sphacelia were produced from nontreated or captan-treated sphacelia by trimming off the exterior sphacelial tissue. Compared to nontreated controls, captan significantly reduced the number of intact sphacelia with germinating macroconidia (visible production of secondary conidia). The number of inner cores from captan-treated sphacelia with visible germination of macroconidia was slightly reduced, indicating that there was at least some penetration of the captan into the sphacelial tissues. The relative level of sporulation indicated no significant differences between intact control sphacelia, cores from control sphacelia, and cores from captan-treated sphacelia. For captan-treated sphacelia, the number of samples with germinating conidia was so low that mean conidial abundance or mean germination of conidia was just below the detection threshold and was significantly less than that from sphacelia of all the other treatments. Thus, less than half (39%) of the intact, captan-treated sphacelia showed any visible germination, and the average number of germinating conidia on those sphacelia was quite low.

Recognizing Sclerotia/Sphacelia in Seed

Although field symptoms of sorghum ergot are generally well-recognized, recognition of the pathogen in seed is more difficult. Some of the identification problems are due to a poor understanding of the structure and function of sphacelia and sclerotia. However, the process also is complicated by the presence of cracked and moldy seed and other foreign objects that can be mistaken for fungal ergot bodies in seed. If seed is a potential source of introduction of ergot inoculum, then recognition of both sphacelia and sclerotia is critical for the removal of ergot bodies in the seed-processing plant and for minimizing erroneous rejection by the buyer. Sorghum ergot workshops, a training manual (2), and a five-page photo bulletin describing the prominent features of sphacelia and sclerotia (13) have helped educate seed inspection personnel and others interested in correct identification of ergot bodies in seed. Both publications are available in English and Spanish (3, 15).

Seed import restrictions often target sclerotia, but most ergot bodies in seed are probably sphacelia. Even when sclerotial tissue is present, remnant sphacelial tissue usually is associated with it (11, 13). The role of sclerotia and sphacelial remnants in the epidemiology of *C. africana* has not been documented. There are portions of the western hemisphere in which sclerotia can be routinely found, but even at high levels of ergot, little or no sclerotial tissue has been found, *e.g.*, at Corpus Christi, Texas, where high humidity promotes saprophytic degradation of sphacelia before the sclerotial tissue can differentiate. Even when sclerotia/sphacelia are present at high levels, commercial companies in Brazil (23) and Texas have reported near complete removal by using standard seed cleaning and processing equipment (screen-cutting or sieving, air-blowing stages, and gravity table) (20). Variable production and rare germination of sclerotial tissue make sclerotia a poor candidate for the spread of *C. africana* through seed, but these structures may have a local function in terms of long-term survival of the pathogen and the reassortment of genetic variability in the pathogen population.

Conclusions

Seed from commercial sorghum hybrid seed production fields is subjected to a rigorous cleaning process, and nearly all sphacelia/sclerotia of *C. africana* are removed. Seed is then treated with captan (although thiram and Maxim might also be effective) by using equipment that accurately delivers the fungicide and provides good coating of the seed and any contaminating ergot. Most macroconidia of *C. africana* lose viability within 2-3 months in seed storage environments, so the potential germination percentage of conidia in the few remaining sclerotia/sphacelia probably is negligible relative to the number fresh sphacelia. This already low percentage is reduced even further by contact fungicides, *e.g.*, captan, to a level that is likely below the threshold of detection. The planting process buries seed and sphacelia below the soil surface where, in the unlikely event that spores are produced, they would have no inoculum potential. Thus, typical fungicide-treated commercial sorghum seed from ergot-affected fields probably has very low inoculum potential, and other local sources of inoculum probably provide more inoculum at a time when a susceptible host, *e.g.*, nonfertilized ovaries in florets of flowering sorghum, are present.

Seedborne inoclum is one of several ways in which *C. africana* may have been introduced into the western hemisphere and Australia, but it probably is of little, if any, importance in these and other locations where the pathogen is already established. Quarantine restrictions failed to prevent *C. africana* from rapidly becoming a global pathogen because of the pathogen's ability to rapidly increase and spread aerially over large geographic areas. Except as a seed quality issue, quarantine import restrictions for sorghum seed with ergot are unwarranted across regions such as the western hemisphere, where widely established populations of the pathogen are genetically closely related. However, some caution is needed in international seed exchange between Australia-Asia, and Africa or the western hemisphere because of the known genetic differences between these populations of *C. africana* (7, 21).

References

1. **Aguirre, R. J., A. H. Williams, G. N. Montes, and H. M. Cortinas.** 1997. First report of sorghum ergot caused by *Sphacelia sorghi* in Mexico. *Plant Dis.* **81:** 31.
2. **Alderman, S., D. E. Frederickson, G. Milbrath, N. Montes, J. Narro-Sanchez, and G. N. Odvody.** 1999. *A Laboratory Guide to the Identification of* Claviceps purpurea *and* Claviceps africana *in Grass and Sorghum Seed Samples.* Oregon Department of Agriculture, Salem, Oregon. Also available at www.oda.state.or.us/Plant/ppd/Ergotmanual2.pdf (last checked March 2002).
3. **Alderman, S., D. E. Frederickson, G. Milbrath, N. Montes, J. Narro-Sanchez, and G. N. Odvody.** 1999. *Guía de laboratorio para la identificación de* Claviceps purpurea *y* Claviceps africana *en muestras de semilla de sorgo y de otros cereals.* Oregon Department of Agriculture, Salem, Oregon.
4. **Arthur, K.** 1997. Seed treatment. *Proc. U.S. Conf. on Sorghum Ergot (11 June 1997, Amarillo, TX)*: 48-50.
5. **Bandyopadhyay, R., D. E. Frederickson, N. W. McLaren, G. N. Odvody, and M. J. Ryley.** 1998. Ergot: A new disease threat to sorghum in the Americas and Australia. *Plant Dis.* **82:** 356-367.
6. **Bandyopadhyay, R., L. K. Mughogho, and H. C. Sharma.** 1990. Source of primary inoculum and spread of ergot. *ICRISAT Annu. Rept.*: 24-25.
7. **Bogo, A., and P. G. Mantle.** 1999. *Claviceps africana* discovered in India. *Plant Dis.* **83:** 79.
8. **Claflin, L. E.** 1998. Agroecology and biotechnology of fungal pathogens of sorghum and millet from the greater Horn of Africa. *INTSORMIL Annu. Rept.*: 13-18.
9. **Dahlberg, J. A., G. L. Peterson, G. N. Odvody, and M. Bonde.** 1999. Inhibition of germination and sporulation of *Claviceps africana* from honeydew encrusted sorghum with seed treatment fungicides. *Crop Protect.* **18:** 235-238.
10. **Frederickson, D. E., P. G. Mantle, and W. A. J. de Milliano.** 1989. Secondary conidiation of *Sphacelia sorghi* on sorghum, a novel factor in the epidemiology of ergot disease. *Mycol. Res.* **93:** 497-502.
11. **Frederickson, D. E., P. G. Mantle, and W. A. J. de Milliano.** 1991. *Claviceps africana* sp. nov.; the distinctive ergot pathogen of sorghum in Africa. *Mycol. Res.* **95:** 1101-1107.
12. **Frederickson, D. E., P. G. Mantle, and W. A. J. de Milliano.** 1993. Windborne spread of ergot disease (*Claviceps africana*) in sorghum A-lines in Zimbabwe. *Plant Pathol.* **42:** 368-377.
13. **Frederickson, D. E., and G. N. Odvody.** 1999. *Sorghum Ergot, Distinguishing Sphacelia and Sclerotia of* Claviceps africana *in Seed.* Texas Agricultural Extension Service Bulletin L-5315 (6-99).
14. **Frederickson, D. E., and G. N. Odvody.** 2002. Survival of inoculum of *Claviceps africana* in Zimbabwe: Potential sources of initial inoculum, p. 79-82. *In* J. F. Leslie (ed.), *Sorghum and Millet Diseases 2000.* Iowa State Press, Ames, Iowa.
15. **Frederickson, D. E., G. N. Odvody, and N. Montes.** 1999. *El ergot del sorgo, Diferenciación de los Esfacelios y los Esclerocios de* Claviceps africana *en la Semilla.* Texas Agricultural Extension Service Bulletin L-5315S (7-99).
16. **Futrell, M. C., and O. J. Webster.** 1966. Host range and epidemiology of the sorghum ergot organism. *Plant Dis. Rept.* **50:** 828-831.
17. **Isakeit, T., G. N. Odvody, and R. A. Shelby.** 1998. First report of sorghum ergot caused by *Claviceps africana* in the United States. *Plant Dis.* **82:** 592.
18. **Mower, R. L., G. R. Gray, and C. E. Ballou.** 1973. Sugars from *Sphacelia sorghi. Carbohydrate Res.* **27:** 119-134.
19. **Odvody, G. N., T. Isakeit, N. Montes, J. Narro-Sanchez, and H. Kaufman.** 1999. Occurrence of sorghum ergot in Texas and Mexico in 1998. *Proc. 21st Biennial Sorg. Res. and Utiliz. Conf. (21-24 February 1999, Tucson, Arizona)*: 62.
20. **Odvody, G. N., D. E. Frederickson, T. Isakeit, J. A. Dahlberg, and G. L. Peterson.** 1999. The role of seedborne inoculum in sorghum ergot. *Proc. 3rd Internatl. Seed Testing Assoc.-PDC (16-19 August 1999, Ames, Iowa)*: 136-140.
21. **Pazoutová, S., R. Bandyopadhyay, D. E. Frederickson, P. G. Mantle, and R. A. Frederiksen.** 2000. Relations among sorghum ergot isolates from the Americas, Africa, India, and Australia. *Plant Dis.* **84:** 437-442.
22. **Reis, E. M., P. G. Mantle, and H. A.-G. Hassan.** 1996. First report in the Americas of sorghum ergot disease, caused by a pathogen diagnosed as *Claviceps africana*. *Plant Dis.* **80:** 463.
23. **Ribas, P. M.** 1999. Strategy for the control of ergot in sorghum hybrid seed production in Brazil. *Proc. Global*

Conf. on Ergot of Sorghum (1-8 June 1997, Sete Lagoas, Brazil): 26-32.

24. **Ryley, M. J., J. L. Alcorn, J. L. Kochman, G. A. Kong, and S. M. Thompson.** 1996. Ergot on *Sorghum* spp. in Australia. *Australas. Plant Pathol.* **25:** 214.

24

Detection of Sclerotia of *Claviceps africana* in the Western Hemisphere

Gary N. Odvody, Noe Montes, Debra E. Frederickson, and Jesus Narro-Sánchez

Abstract

Collections of sorghum ergot on sorghum and johnsongrass and ergot on grasses were made from different locations in Mexico and the United States beginning in 1997, after initial observations of *Claviceps africana*. Ergot bodies from these samples were analyzed to determine if sclerotia of *C. africana* were present. Sclerotia have been identified at Celaya, Mexico and in northern Texas since 1997 on sorghum and in early 2000 on sorghum in the lower Rio Grande Valley of Texas and Rio Bravo, Mexico. The sclerotial structures were mostly round and thin, with an orange-brown to red-brown rugose surface layer and remnant sphacelial tissue on top. At Celaya, sphacelia-sclerotia structures were 4-6 × 2.5-3 mm in diameter and 2.5-4 mm in length, while at Weslaco, Texas and Rio Bravo the diameters were 3-6 × 2-3 mm and the lengths were 2-4 mm. Sclerotia have never been observed in Corpus Christi and some locations in Mexico, presumably because the high relative humidity favors saprophytic degradation of sphacelia before sclerotial formation can occur.

Introduction

Sorghum ergot was first observed in the western hemisphere in 1995 and spread rapidly through South and Central America and, by early 1997, into northern Mexico (2) and south Texas (5). At that time, the primary symptom was copious exudation of honeydew containing macroconidia and microconidia. This symptom was characteristic of the asexual phase of *Claviceps africana* Frederickson, Mantle, and de Milliano (3). Sorghum ergot has persisted in an active form in 16 Mexican states and several states in the United States since 1997, primarily on feral grain sorghum, both ratoon and volunteer, and on johnsongrass.

Sclerotia are resting or dormancy structures that withstand degradation by the environment (4), and often are used by fungal pathogens to survive the interval from harvest to the next crop season. Sclerotia of *C. africana* develop in, and from the base of the sphacelia. Even when the sclerotia are fully developed, sphacelial tissues remain on top. Thus, most sclerotia contain both sphacelial and sclerotial tissue. Sclerotia are oval to spherical, tapering at the distal end (4-6 × 2-3 mm) with a furrowed, orange-brown surface. The white to cream tissues at the top are sphacelial, while the sclerotial tissue towards the base has a orange-brown to red-brown rind that covers and protects the internal white fungal cells (1). Since sclerotial tissue is typically associated with a thin upper layer of sphacelial tissue, macroconidia and microconidia within the sphacelial remnant also may be found on the surface of sclerotial tissue, but the sphacelial conidia are much more sensitive to heat and desiccation than are their sclerotial counterparts.

Materials and Methods

Since 1997, survey activities to identify sclerotial development of *C. africana* have been implemented in Mexico and Texas. Collections of ergot from sorghum, johnsongrass, and other grasses were made from different locations, and possible sclerotia stored at 8°C in vials (approximately 4 cm in length) containing approximately 1 g of silica gel to absorb moisture. This material was analyzed to determine if *C. africana* sclerotia were present. None of the fungal samples from other grasses with ergot was *C. africana*.

Results and Discussion

Sclerotia have been identified from sorghum at Celaya, Mexico and in the northern (high plains) of Texas since 1997, and, in early 2000, from sorghum in the lower Rio Grande valley at Weslaco, Texas and at Rio Bravo, Tamaulipas, Mexico. Sclerotia were mostly rounded with a thin, orange-brown to red-brown rugose surface layer and a remnant of sphacelial tissue on top (4). Sphacelia/sclerotia structures were 4-6 × 2.5-3 mm in diameter at Celaya, and 3-6 × 2-3 mm in diameter at Rio Bravo and Weslaco (Fig. 24-1), with sclerotial lengths of 2.5-4 and 2-4 mm, respectively. The top portion of structures from all three locations was dry sphacelial tissue, which became very soft when soaked in water for a few minutes, while sclerotia remained firm. Dissected sclerotial tissue was composed of compact white tissue internally, occasionally completely enclosed within an orange rind as described by Frederickson *et al.* (4). These structures were fully developed sclerotia of *C. africana* as defined by Frederickson *et al.* (3). Other ergot structures had various amounts of sclerotial tissue associated with the sphacelial tissue.

This report is the first of the development of *C. africana* sclerotia in the western hemisphere. Formation of sclerotia appears to be favored by low relative humidity and cool temperatures after infection. These conditions also favor survival of intact sphacelial structures of *C. africana* and allow more time for the formation of scleraotial tissue, probably because low relative humidity reduces colonization of *C. africana* tissue by saprophytic fungi, *e.g.*, *Cerebella* spp. that can grow under dead organic matter. There are locations in both Texas, *e.g.*, Corpus Christi, and southern Tamaulipas, Mexico, *e.g.*, Tampico, in which sclerotia have never been observed, presumably because the high relative humidity favors saprophytic degradation of sphacelia before sclerotia formation can occur.

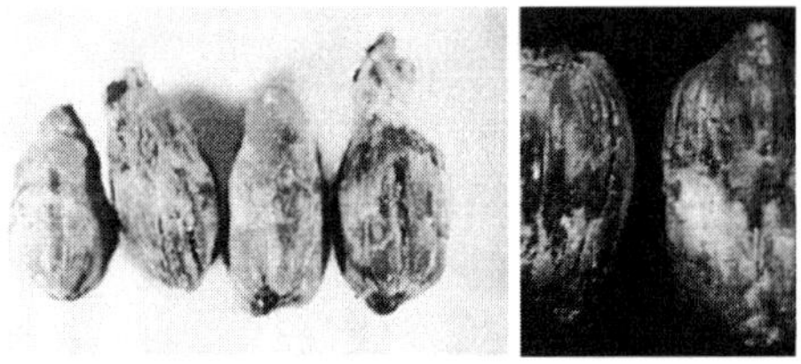

FIGURE 24-1. *Claviceps africana* sclerotia from Celaya, Guanajuato, Mexico (left), and Rio Bravo, Tamaulipas, Mexico (right).

References

1. **Alderman, S., D. E. Frederickson, G. Milbrath, N. Montes, J. Narro, and G. N. Odvody.** 1999. *A Laboratory Guide to the Identification of* Claviceps purpurea *and* Claviceps africana *in Grass and Sorghum Seed Samples.* Oregon Department of Agriculture, Salem, Oregon.
2. **Aguirre, R. J., A. H. Williams, N. Montes, and H. M. Cortinas.** 1997. First report of sorghum ergot caused by *Sphacelia sorghi* in Mexico. *Plant Dis.* **81:** 831.
3. **Frederickson, D. E., P. G. Mantle, and W. A. J. de Milliano.** 1989. Secondary conidiation of *Sphacelia sorghi* in sorghum, a novel factor in the epidemiology of ergot disease. *Mycol. Res.* **93:** 497-502.
4. **Frederickson, D. E., and G. N. Odvody.** 1999. *Sorghum Ergot, Distinguishing Sphacelia and Sclerotia of* Claviceps africana *in Seed.* Texas Agricultural Extension Service Bulletin L-5315 (6-99).
5. **Isakeit, T., G. N. Odvody, and R. A. Shelby.** 1998. First report of sorghum ergot caused by *Claviceps africana* in the United States. *Plant Dis.* **82:** 592.
6. **Odvody, G. N., T. Isakeit, N. Montes, J. Narro-Sanchez, and H. Kaufman.** 1999. Occurrence of sorghum ergot in Texas and Mexico in 1998. *Proc. 21st Biennial Sorg. Res. and Utiliz. Conf. (21-24 February 1999, Tucson, Arizona)*: 62.

25

Factors Associated with Ergot Resistance in Sorghum

Evelyn Ortíz-Pérez, Francisco Zavala-García, J. L. de la Garza, G. E. Salinas, and Nora E. García

Abstract

We evaluated ergot incidence in relation to cytoplasm type and pollen grain production by using nine male-fertile lines and four cytoplasm types. Male-sterile line FA163 produced the most pollen, and line 10251 the least. With respect to cytoplasm type, the male sterile line A_4Tx398 had the least ergot infection, and line A_1Tx398 was the most susceptible. We also found that culturing infected ovaries on potato dextrose agar gave acceptable fungal growth.

Introduction

Sorghum (*Sorghum bicolor* L. Moench) is the third most important crop in Mexico, after maize and dry beans (3). Among the most important factors limiting grain sorghum production are insects and diseases. Ergot is a fungal disease, caused by *Claviceps africana*, that interferes primarily with hybrid seed production. Female parents are particularly susceptible if pollen is not abundant. A male parent that produces a lot of pollen may be the most important factor in reducing ergot infection in commercial fields (1). Thakur *et al.* (4) also suggested that cytoplasmic effects may reduce ergot infection. Ergot is a relatively new disease on sorghum in Mexico, and relatively little local research has been conducted on this problem. The main objectives of this study were: (i) to evaluate different types of cytoplasm for ergot resistance, (ii) to rank a group of male parents by using pollen production, and (iii) to develop a technique to culture *C. africana* from infected plant tissue *in vitro*.

Materials and Methods

Cytoplasm type. We tested four isogenic female lines (Tx398) with different types of cytoplasm: A_1, A_2, A_3, or A_4. Plants were cultivated in 19-L plastic pots and arranged in the greenhouse in a randomized complete block design with three replicates. When the stigmas on the female plants first become receptive, the flowers are inoculated with the fungus, 5 ml of a spore suspension (1×10^6 conidia/ml) per head. The solution was applied directly to the panicles as a spray with a manual sprayer. A second application was made when two-thirds or more of the florets in the panicle were flowering. After the second application, the number of infected florets was recorded every third day.

Pollen production. The male parents we used were FA90, FA163, FA170, 10106, 10249, 10251, 10330, 10331, and 10333. A commercial hybrid, Pioneer 8282, was used as a control. There were three replicates of the experiment, with plants in the field arranged in a randomized complete block design. We measured the number of pollen grains in each flower. Five spikelets were taken from the upper, middle, and lower portions of each panicle (15 total per panicle) 3-4 days before anthesis. The anthers were removed from each flower and placed in a microcentrifuge tube containing 30 μL of 35 N H_2SO_4, to dissolve the anthers and to release the pollen grains. A 20 μl sample was taken and the pollen grains in it were counted at × 100 in a Neubauer chamber.

Fungal development *in vitro*. Ten days after flowering, infected flowers were dissected and the ovaries extracted. Infected ovaries were placed in a sodium hypochlorite (10%) solution for ten minutes. After that, the ovaries were placed in petri dishes containing potato dextrose agar (PDA, Bioxon) plus streptomycin and ampicillin, each at 100 μg/ml. Plates were incubated at 25°C for three days.

TABLE 25-1. Mean number of days to flowering and of infected flowers per panicle of isogenic line ATx398 with cytoplasm A_1, A_2, A_3, or A_4 (winter 1999, Marín, N.L. México).

Cytoplasm	Days to Flowering[a]	Infected Flowers per Panicle[a]
A_1	104 a	75 a
A_3	103 a	17 b
A_2	98 b	14 b
A_4	97 b	9 c

[a]Values followed by a common letter are not significantly different ($p < 0.05$). Based on an analysis of covariance with a complete randomized block design and the number of flowers per panicle as the covariant.

TABLE 25-2. Mean number of pollen grains per flower (spring 1999, Marín, N.L. México).

Sorghum Line	Pollen Grains per Flower[a]
FA163	13,000 a
Pioneer 8282	11,000 b
FA90	11,000 b
10330	9,700 c
10249	9,700 c
10331	9,400 cd
10333	9,300 de
10106	9,100 e
FA 170	7,900 f
10251	6,100 g

[a]Values followed by a common letter are not significantly different ($p < 0.05$).

Results and Discussion

Cytoplasm type. Honeydew was first observed seven days after the first inoculation. The number of infected flowers in each panicle was determined, and then analyzed by ANOVA (Table 25-1). Cytoplasm A_4 had the fewest infected flowers and cytoplasm A_1 the most. These results are consistent with those that have been previously published (*e.g.*, 2), in which cytoplasm A_4 also was the least susceptible to ergot. One of the factors related to this resistance could be the number of days to flowering. ATx398 A_4 had the fewest mean days to flowering, 97, while the more susceptible A_1 line had the most, 103.

Pollen production. The nine lines could be separated into seven groups with respect to pollen production (Table 25-2). FA163 produced the most pollen and 10251 the least. We also observed, but did not evaluate, differences in anther size.

Fungal development *in vitro*. After three days of incubation on PDA, a thick white mycelium covered the infected ovary. Normally when this fungus is cultured from field samples, the initial inoculum is from a sclerotium. Recovering the fungus from an infected ovary is not a common procedure. Our results show that the fungus can be recovered relatively easily from an infected ovary. This procedure allows us to directly culture fungal strains from infected tissue.

Conclusions. Our results confirm the hypothesis that A_4 cytoplasm is an important source of resistance to ergot. The utilization of this resistance source will depend on the interaction of these cytoplasmic genes with nuclear genes to produce superior female lines for hybrid production. For the male parent, although it is difficult to measure, pollen production should be used as a selection criterion to help decrease the risk of ergot in grain and hybrid seed production fields.

References

1. **Bandyopadhyay, R., D. E. Frederickson, N. W. McLaren, and G. N. Odvody.** 1996. Ergot, a global threat to sorghum. *Intl. Sorg. Millets Newsl.* **37:** 1-32.
2. **Bandyopadhyay, R., D. E. Frederickson, N. W. McLaren, G. N. Odvody, and M. Ryley.** 1998. Ergot, a new threat to sorghum in the Americas and Australia. *Plant Dis.* **82:** 356-367.
3. **FAO.** 1996. Agrostat-PC, 1961-1994. FAO, Rome, Italy.
4. **Thakur, R. P., V. P., Rao, and S. B. King.** 1989. Ergot susceptibility in relation to cytoplasmic male sterility in pearl millet. *Plant Dis.* **73:** 676-678.

26

Ergot and Its Impact on Hybrid Sorghum Seed Production in Venezuela

Tirso Pacheco

Sorghum ergot, or sugary disease, was first reported in Venezuela in 1996. Conidial dispersion from neighboring countries as well as the unnoticed presence of the disease in noncultivated hosts, *e.g.*, *Sorghum arundinaceum*, *Sorghum halepense*, *Sorghum verticilliflorum*, and *Panicum maximun*, are two of the means through which the fungus could have reached commercial sorghum fields (1). The disease is particularly important in hybrid seed production fields where the flowers of the male-sterile line remain receptive until fertilized.

The Impact of Ergot on the Production of Hybrid Sorghum Seed

Synchronous flowering of the two parental lines, some of which are tropical photosensitive genotypes, has always been a major goal of the hybrid sorghum seed production industry in Venezuela. With the advent of ergot disease, achieving this goal has become even more important given the competition between pollen for fertilization and conidia for infection. In addition to the production of pollen at the time of flowering, another control for ergot is to increase the amount of pollen available in the field for fertilization by reducing the usual 4:1 proportion of female:male parental lines to 3:1.

The tillering ability of male-sterile lines can be problematic, since tillers are flowering when very little pollen is left in field, and these tillers are vulnerable to fungal infection. Different plant densities were evaluated for several male-sterile lines to identify conditions under which tillering was reduced. Mena *et al.* (3) found that applications of 0.5 and 0.75 g a.i/L of paraquat effecttively controlled tillering in experimental plots. Diesel oil could be used with similar results. Neither practice has been tested in commercial fields. The manual elimination of the first plants infected in the field also has been used as a method to reduce the amount of inoculum present.

Additional control measures include moving the production of hybrid sorghum seed to areas with lower relative humidity and adjusting planting dates so that flowering occurs during the driest months of the year while simultaneously avoiding the coolest times of the year. Although these cultural practices have helped reduce ergot incidence, chemical control has been more effective. Many farmers now spray their crops three to four times at 5-7 day intervals with 0.5-1.0 L/ha of propiconazole (Tilt®, 250 g a.i./L), a triazole fungicide, beginning at the initiation of flowering (2). The additional cost of these applications has been estimated at US$ 250/ha.

References

1. **Malaguti, G., and N. Pons.** 1997. Ergot, honeydew or sugary disease of sorghum in Venezuela. *Proc. Global Conf. on Ergot of Sorghum (1-7 June, 1997, Sete Lagoas, Brazil)*: 88-96.
2. **Mena, H., F. Fuenmayor, J. Tejera, E. Georges, and R. Jimenez.** 1998. Chemical control of sorghum ergot (*Claviceps africana*) in the field. *Intl. Sorg. Millets Newsl.* **39:** 105-106.
3. **Mena, H., E. Georges, and R. Jimenez.** 1997. Tiller control in sorghum hybrid seed production as a strategy for the control of sorghum ergot (*Claviceps africana*). *Intl. Sorg. Millets Newsl.* **38:** 78.

27

The Genus *Claviceps*: Evolution at Work

Sylvie Pažoutová

Abstract

Phylogenetic trees of 19 *Claviceps* species were constructed based on the alignment of 5.8s ribosomal DNA (rDNA) and the adjacent internally transcribed spacer regions (ITS1 and ITS2). Two clades were found whose biogeography suggests that the genus originated in South America and has been influenced evolutionary by co-migration with its hosts and global climate changes. In *Claviceps purpurea*, three groups of isolates were found: G1 from fields and open meadows, G2 from shady or wet habitats, and G3 from *Spartina anglica* and *Spartina alterniflora* in salt marshes. The seemingly preferred host species are all grasses that occur in the preferred habitat. Intraspecific variation within *C. purpurea* suggests that smaller homogenous populations and chemoraces developed from a variable, basal G1 population and then specialized to fill adjoining ecological niches. A similar differentiation mechanism appears to operate in *Claviceps africana*. Here the invading populations have a clonal character, while the basal African population is more variable. Random Amplified Polymorphic DNA (RAPD) markers in African populations of *C. africana* are recombined relative to those seen in invading populations in Asia, Australia, and the Americas.

Phylogeny of *Claviceps*

At present, about 45 species of *Claviceps* have been described, and their geographic distribution is interesting and informative. For example, there is a striking difference in the number of species colonizing the chloridoid, pooid, and panicoid subfamilies of the *Poaceae*. Thirty species have been found on panicoid hosts, but only a few have been collected from sedges, rushes, and grasses of other subfamilies. Most *Claviceps* species are found in subtropical and tropical regions. Only five species are found in temperate regions, of which the most widespread is *C. purpurea* (for review, see 19). Host specificity is commonly monogeneric, or extending to perhaps a few closely related taxa, with the exceptions of *C. purpurea*, and *Claviceps phalaridis*, which both colonize all Pooidae, and *Claviceps pusilla*, which colonizes approximately 20 andropogonoid genera.

Claviceps species are characterized based on their stroma color, conidia shape and size, and dormant structure, the sclerotium. In two endemic South American species, *Claviceps flavella* and *Claviceps diadema*, simple sclerotia occur that are composed of an undifferentiated mycelial mass, which encompasses the flower parts of one, or several, florets and differentiates directly to stromata on the host (14). This type of sclerotium is considered a primitive character (11). No species with such sclerotia have been found in tropical and subtropical forests of Africa and South Asia. Langdon's (11) hypothesis, which is based on biogeography and type of sclerotia, states that *Claviceps* is of tropical origin and that species with more complex sclerotia that resist xeric conditions and temperature variation evolved as their hosts spread from a diminishing tropical forest to an initially less favorable area. Evolution of the parasitic fungus *Puccinia lagenophorae* and the clavicipitaceous *Cordyceps* and *Epichloe* species also is hypothesized to be linked with the evolution of their hosts (15, 25, 26).

I collected 16 *Claviceps* species from different grass subfamilies and studied their phylogenetic relationships by using the ITS1–5.8S–ITS2 rDNA sequences (17). The preliminary alignments of these sequences showed great intrageneric variation, especially in the ITS1 region. The *Claviceps* species in the later diverging lineages had apparent insertions in this region, approximately 60 nucleotides in length, but with differing sequences. Moreover, *Claviceps zizaniae* had a second, approximately 70 bp, insertion that appears to be a modified duplication of a portion of ITS1.

A strict consensus tree obtained by maximum parsimony analysis of the rDNA sequences had 455 steps

(Fig. 27-1). Two highly supported clades were found. The lower clade contains, in the ancestral position, *Claviceps paspali* (the only species in this clade with a short ITS1 spacer), three species adapted to cold winters (*C. zizaniae*, *Claviceps grohii*, and *C. purpurea*), and two species (*Claviceps fusiformis* and *Claviceps sulcata*) from semi-arid regions in Africa. The second clade consists of tropical species. *Claviceps citrina* and *C. phalaridis* (both with short ITS spacers) occur in the ancestral position. The remaining six species form an unresolved polytomy with a clade of five African species and the ancestral *C. pusilla*. Previously (15), *Claviceps* sp. SG and *Claviceps maximenis* (previously labeled *Claviceps* sp. PM), also with short ITS1 regions, were considered closely related to the ancestral species *C. citrina* and *C. phalaridis*. However, with the addition of new species to the data set, it became clear that the loss of the loop structure that shortens ITS1 was secondary in *Claviceps* sp. SG, *C. maximensis*, *Claviceps cynodontis*, and *Claviceps rhynchelytri*, because *C. pusilla* and other species with long ITS1 regions are ancestral to this closely related group of species.

We also made Southern hybridizations between a 1-kb portion of the *cpd1* gene (EMBL Nucleotide Sequence Database accessions AJ011963), which encodes dimethylallyltryptophan (DMAT) synthase in *C. purpurea* strain T5 (31), and DNA from strains of the other *Claviceps* species. DMAT synthase is the first enzyme specific for the alkaloid biosynthetic pathway. A sequence that hybridizes to the *C. purpurea* probe is present in all of the closely related species of the first clade (15), although the hybridization with the *C. paspali* gene was noticeably weaker. In the second clade, weak hybridization was observed only for *Claviceps gigantea*, *Claviceps africana*, and *C. pusilla*, with the remaining species showing no detectable hybridization. The size of the hybridizing bands differed by species.

Morphological features, *e.g.*, stroma color and conidial size and shape, do not correlate well with the position of the species in the phylogenetic tree. Species with yellow stromata and species with dark (purple, brown, or vinaceous) stromata were not resolved into different clades.

The rDNA tree is not consistent with the hypothesis that *Claviceps* species have directly coevolved with their grass hosts. In the first clade, species colonizing panicoid, oryzoid, and pooid grasses are interspersed. Also in this clade, species with a Paleoarctic distribution that have cold-resistant sclerotia are closely related to species from the African savanna. The similarity of rDNA sequences within this group is striking, ≥ 97% identity

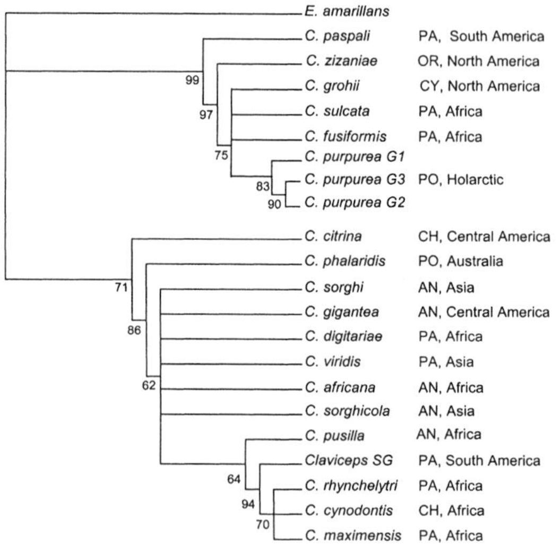

FIGURE 27-1. Phylogram of *Claviceps* species. Tree was constructed using maximum parsimony method. Numbers under branches indicate statistical support of these clades (jackknife). Clades supported by less than 50% of bootstraps are collapsed to polytomies. *Epichloe amarillans* was used as an outgroup. *Claviceps* sp. SG was found on *Setaria geniculata* in Brazil; its stroma color is unknown. Host abbreviations: PA, panicoid; AN, andropogonoid; PO, pooid; CH, chloridoid; OR, oryzoid; and CY, Cyperaceae.

between *C. purpurea* and *C. grohii*, which is consistent with relatively recent divergence.

In the second clade, species colonizing advanced andropogonoid hosts diverge later; however, their closest relatives, *Claviceps viridis* and *C. digitariae*, are specialized on panicoid grasses with many primitive characters. Also in this clade are three species that colonize the same host genus in different locations: *C. africana*, *Claviceps sorghi*, and *Claviceps sorghicola*. Parasites of the andropogonoid grasses, *e.g.*, the maize parasite *C. gigantea* from Central America and the sorghum parasites from Africa and Asia, are closely related. These relationships suggest that *Claviceps* spp. may spread with their hosts during their radiation. The *C. pusilla* clade consists of closely related species that colonize panicoid, andropogonid, and even chloridoid grasses. This clade is an example of how an invading species might have colonized other, often unrelated, grass genera in the new loca-tions, giving rise to specialized populations that later became new species.

Whyte (32) hypothesized that grasses originated in humid tropical forests with evolutionary advancement correlated with adaptation to aridity. *Claviceps* phylogeny is consistent with this hypothesis, because the primitive sclerotia, consisting of rather loosely interwoven hyphae (11, 18), would not survive in xerophytic zones. The absence of species with apparently primitive sclerotia in Africa could be due to the inability of the Gondwanan tropical forests there to adapt to the climate changes of the Tertiary (20), especially the aridization in the Oligocene and the fluctuations in the late Pliocene-Pleistocene. These fluctuations resulted in expansion/contraction cycles of savanna habitats (4), and may have selected for the differentiation of *Claviceps* species with more resistant sclerotia. In South America, similar climatic extremes did not occur in the rain forest areas (30).

The first *Claviceps* species probably occurred on the predecessors of panicoid grasses in the warm, humid climate of the South American region of the former Gondwana during the Upper Cretaceous. Panicoid grasses probably originated in that region, and South American *Claviceps* species have primitive undifferentiated sclerotia and occupy ancestral positions on the rDNA phylogeny tree. These lineages probably were preserved by the isolation of South America from the end of the Cretaceous until the end of the Tertiary (30).

The first expansion of *Claviceps* species probably gave rise to the species in the second clade, and could have occurred as the panicoid grasses spread in the early Tertiary. This *Claviceps* clade also was influenced by the spread of the andropogonoid grasses from southern Asia to Africa, southern Europe, and Central America (8). The close relationship between the Mexican maize parasite, *C. gigantea*, and parasites of andropogonoids from the Paleotropics is consistent with this hypothesis. Members of the first clade often are well adapted to semi-arid conditions, but their cold resistance is limited.

The ancestors of the second clade could have migrated from South America to North America and from there to Europe and Africa. These species developed the ability to withstand cold winters. Given the possibilities for long distance dispersal, the split between the two clades could have occurred 6-7 million years ago. The extant species of the second clade that occur in temperate climates and semi-arid Africa appear to have evolved relatively recently, to fill available ecological niches. The limited detectable hybridization observed with the DMAT synthase probe between species in different clades also is consistent with the hypothesis that these clades diverged relatively long ago.

Populations of *Claviceps purpurea*

C. purpurea has a wide host range, which includes the entire subfamily Pooideae, many Arundinoideae, and some species belonging to chloridoids and panicoids (2, 13). Its distribution is essentially Holarctic, but it has been recorded in Arctic regions (12), in the southern temperate regions, and in subtropical regions. Transfers with seed and grain make it difficult to pinpoint the center of origin for this species. The size and shape of conidia (from 5-13 μm in length, oval to cylindric), the varying coloration of stroma (pale red to purple or orange), the range in sclerotial size (5-50 mm), and the variety of peptide alkaloids produced makes *C. purpurea* one of the most variable species of the genus.

For the last century, researchers have tried to classify this variation into subtaxa, *e.g.*, varieties, special forms, or races (1), with most of the attention given to the establishment of host-specific groups. Stäger (27, 28) introduced four special forms: *secalis*, *lolii* (later joined with *secalis*), *milii* (on *Milium* and *Brachypodium* only), and *glyceriae* (suspected to be *C. wilsonii*) for *C. purpurea* sensu Tulasne. However, in an almost forgotten work (29), he reported that sclerotia found on grasses from wet habitats could float in water for hours, which improves their spreading ability. This group was defined as a new f. sp. *Phalaridis arundinaceae natans* (= floating, swimming). Unfortunately, this unconventional habitat approach was not continued, and later workers returned to subdivisions based on host specificity, even though cross inoculations between hosts and isolates belonging to different groups were successful (3). Chemoraces, based on the profile of ergot alkaloids produced, also have been described, and the qualitative composition of the alkaloid mixture produced is genetically stable (9, 10).

We recently combined all of these approaches, and characterized *C. purpurea* isolates by host plant, habitat, alkaloid composition, floating ability of sclerotia, conidial morphology, and RAPD markers (16). DNA fragments amplified by 15 RAPD primers could be used to distinguish three groups of isolates. Primer 257 (CGT GATGTCAGTGATGC) was chosen for routine typing because the resulting patterns were the most distinct (Fig. 27-2). All isolates shared only one band, the species-specific amplicon A, derived from a multicopy sequence. The 57 isolates in the G1 group all had a unique band B, although other bands in the RAPD profiles varied. The second group, G2, was formed by 41 isolates and was characterized by bands C1 and C2 with most primers. The RAPD pattern of the third group, G3, shared one band with G2 (21).

In the 5.8S rDNA of *C. purpurea* from G1 isolates Pepty 695/S and 134, and G3 isolate 511 (EMBL nos. AJ000069, AJ000070, and AJ311951, respectively), we found a conserved *Eco*RI site common in Ascomycetes. However, in the sequence of G2 isolate 434 (AJ311950), the site was missing due to a transversion that changed GAATTC to GACTTC. Restriction analysis of the 5.8 rDNA region of all isolates confirmed the presence of the *Eco*RI site in the G1 and G3 isolates and its absence in G2 isolates.

Isolates of G1 originated from fields, open meadows, and grasses along roads, *i.e.*, open sunny locations that often are dry. Under field conditions, *Alopecurus pratensis, Ammophila arenaria, Arrhenatherum elatior, Dactylis* sp., *Festuca ovina, Festuca rubra, Phleum* sp., and *Poa pratensis* were colonized by isolates of both G1 and G2. Thus, we detected no host specialization in these groups.

For the isolates of G2, there was a clear tendency towards occurrence on hosts that more commonly occurred in shady or moist habitats. *Calamagrostis, Phalaroides, Phragmites*, and *Molinia* often were the host genera. Common locations were pond and river banks, ditches, forests, and even mountain woods. G3 isolates, 10 from Great Britain and three from North America, originated from salt marshes with brackish water on *Spartina anglica* and *Spartina alterniflora*.

G1 isolates have oval to elongated conidia, usually 5-7 μm in length. G2 isolates have somewhat larger (6.5-8.5 μm in length) elongated to cylindrical conidia that overlap significantly in size with conidia of the G1 isolates. Isolates of G3 have, on average, the longest conidia

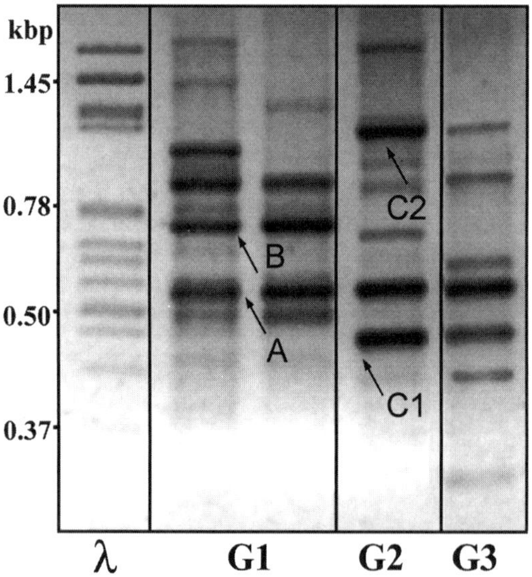

FIGURE 27-2. Typical RAPD profiles for *C. purpurea*. A, species-specific band; B, G1-specific band; and C1 and C2, bands typical of G2 isolates. Band C1 also is shared with G3 isolates.

(9.5-12 μm). Due to the overlap for G1 and G2, conidial size is of only secondary taxonomic importance.

All of the sclerotia produced by G2 and G3 isolates could float, while sclerotia produced by G1 isolates sank within the first 30 min in water.

Sclerotia produced by G1 isolates contained various peptide ergot alkaloids. However, sclerotia produced by G2 isolates contained only ergosine (from the ergotamine group), ergocristine, and small amounts of ergocryptine. Sclerotia produced by G3 isolates contained a mixture of ergocristine and ergocryptine. Therefore, G2 and G3 are distinct chemoraces, as no isolate with similar alkaloid composition was found among the G1 isolates.

The intraspecific variation in *C. purpurea* suggests that from a large, variable, basal population (G1), occasionally, smaller homogenous populations develop that perhaps should be considered ecological subspecies, or varieties, that have specialized to fill neighboring ecological niches. We identified two such ecological subspecies that both have floating sclerotia and that are found preferentially on grass species that occur in similar habitats.

Populations of *C. africana*

C. africana is endemic to Africa, but has recently invaded the Paleotropics and the Neotropics. The first invasion probably occurred in the 1980s from Africa to India, resulting in the near extinction of the local endemic species, *C. sorghi* (16). The clone continued to spread through Southeast Asia (6) and reached Australia in 1996 (23). The second invasion was more rapid, and a clone that appeared in Brazil in 1995 (22) reached Texas in two years (7). Such rapid dispersal is a preliminary condition for further species evolution and for the colonization of new hosts.

C. africana from sorghum is a quite homogenous species based on RAPD analyses (16). Most of the 100 primers tested yielded uniform patterns for isolates of different geographic origin. Our study (16) was based on ten RAPD DNA fragments amplified with one of seven primers. Invasion populations from the Americas and India-Australia were very homogeneous, almost clonal. Australian isolates were distinguished from the Indian group only by the acquisition of single band, which also was found in an isolate from Thailand. No polymorphic bands (shared by at least two isolates) differentiating isolates within these groups were found.

C. africana isolates from Africa were more variable, with the RAPD bands found in various patterns. For example, in isolates F1 and E3 from Zimbabwe, bands typical of both the American and Indo-Australian isolates were identified. Three South African isolates were identical to the American isolates, further strengthening the hypothesis that the invasive clone originated in Africa (16).

Recently, several *Claviceps* species appeared at new locations, probably due to movement with seed. The African species *C. sulcata* appeared in Brazil (5), and *C. fusiformis* was reported from Mexico (24). Such introductions may result in the extinction of endemic species that colonize the same host and distort the perceived biogeography of *Claviceps*. On the other hand, these events may promote colonization of new hosts and provide new opportunities for *Claviceps* evolution, through the geographic spread of host grasses and the subsequent colonization of neighboring alternate hosts and ecological niches.

Acknowledgements

This work was supported by the Czech Grant Agency grants 206/97/0611 and 522/99/0517, Grant Project COST 835 of Ministry of Education, and by the Sorghum and Millet Collaborative Research Support Program (INT-SORMIL) AID/DAN-1254-G-00-0021-00 from the U.S. Agency for International Development.

References

1. **Békésy, N.** 1956. Ein beitrag zur biologie des mutterkorns. *Phytopathol. Z.* **26:** 49-56.
2. **Brewer, J., and A. R. Loveless.** 1977. Ergot of *Pennisetum macrourum* in South Africa. *Kirkia* **10:** 589-600.
3. **Campbell, W. P.** 1957. Studies of ergot infection in gramineous hosts. *Can. J. Bot.* **35:** 315-320.
4. **Coughenour, M. B.** 1985. Graminoid responses to grazing by large herbivores: Adaptations, exaptations, and interacting processes. *Ann. Miss. Bot. Gard.* **72:** 852-863.
5. **Fernandes, C. L., A. T. F. Fernandes, and J. L. Bezerra.** 1995. "Mela": Uma nova doença em sementes de *Brachiaria* spp. no Brasil. *Fitopatol. Brasil.* **20:** 501-503.
6. **Frederickson, D. E., P. G. Mantle, and W. A. J. de Milliano.** 1991. *Claviceps africana* sp. nov., the distinctive ergot pathogen of sorghum in Africa. *Mycol. Res.* **95:** 1101-1107.
7. **Isakeit, T., G. N. Odvody, and R. A. Shelby.** 1998. First report of sorghum ergot in the United States. *Plant Dis.* **82:** 592.
8. **Jones, C. A.** 1991. *C4 Grasses and Cereals: Growth, Development and Stress Response*. John Wiley & Sons, New York City, New York.
9. **Kobel, H., and J. J. Sanglier.** 1978. Formation of ergotoxine alkaloids by fermentation and attempts to control their biosynthesis, p. 233-242. *In* R. Hütter, T. Leisinger, J. Nüesch, and W. Wehrli (eds.), *Antibiotics and Other Secondary Metabolites: Biosynthesis and Production*. Academic Press, New York City, New York.
10. **Kybal, J., and V. Brejcha.** 1955. Problematik der Rassen und Stämmen des Mutterkorns *Claviceps purpurea* Tul. *Pharmazie* **10:** 752-755.
11. **Langdon, R. F. N.** 1954. The origin and differentiation of *Claviceps* species. *Univ. Qld Papers (Bot.)* **3:** 61-68.
12. **Linder, D. H.** 1948. Fungi. Botany of the Canadian eastern Arctic. Part II. Thallophyta and Bryophyta. *Bull. Nat. Mus. Can.* **97:** 234-297.
13. **Loveless, A. R.** 1971. Conidial evidence for host restriction in *C. purpurea*. *Trans. Br. Mycol. Soc.* **56:** 419-434.
14. **Nikoh, N., and T. Fukatsu.** 2000. Interkingdom host jumping underground: Phylogenetic analysis of entomoparasitic fungi of the genus *Cordyceps*. *Mol. Biol. Evol.* **17:** 629-638.
15. **Pažoutová, S.** 2000. The phylogeny and evolution of the genus *Claviceps*. *Mycol. Res.* **105:** 275-283.
16. **Pažoutová, S., R. Bandyopadhyay, D. E. Frederickson, P. G. Mantle, and R. A. Frederiksen.** 2000. Relations among sorghum ergot isolates from the Americas, Africa, India, and Australia. *Plant Dis.* **84:** 437-442.
17. **Pažoutová, S., J. Olšovská, M. Linka, R. Kolínská, and M. Flieger.** 2000. *Claviceps purpurea* populations, chemoraces and habitat specialization. *Appl. Environ. Microbiol.* **66:** 5419-5425.
18. **Pažoutová, S., A. F. Raybould, A. Honzátko, and R. Kolínská.** 2002. Genotypic and phenotypic variation in *Claviceps purpurea* from *Spartina* species and the origin of the ergot infection on *S. anglica* in the British isles. *Mycol. Res.* **107:** in press.
19. **Pažoutová, S., and D. P. Parbery.** 1998. The phylogeny and taxonomy of the genus *Claviceps*, p. 57-77. *In* V. Křen. and L. Cvak (eds.), *Medicinal and Aromatic Plants — Industrial Profiles*, Vol. 6. Harwood Academic Publishers.
20. **Raven, P. H. and D. I. Axelrod.** 1974. Angiosperm biogeography and past continental movements. *Ann. Miss. Bot. Gard.* **61:** 539-673.
21. **Raybould, A. F., A. J. Gray, and R. T. Clarke.** 1998. The long-term epidemic of *Claviceps purpurea* on *Spartina anglica* in Poole Harbour: Pattern of infection, effects on seed production and the role of *Fusarium heterosporum*. *New Phytol.* **138:** 497-505.
22. **Reis, E. M., P. G. Mantle, and H. A. G. Hassan.** 1996. First report in the Americas of sorghum ergot disease, caused by a pathogen diagnosed as *Claviceps africana*. *Plant Dis.* **80:** 463.
23. **Ryley, M. J., J. L. Alcorn, J. K. Kochman, G. A. Kong, and S. M. Thompson.** 1996. Ergot on *Sorghum* spp. in Australia. *Austr. Plant Pathol.* **25:** 214.
24. **San Martín, F., P. Lavín, A. García, and G. García.** 1997. Estados anamórficos de *Claviceps africana* y *Claviceps fusiformis* (Ascomycetes, Clavicipitaceae) asociados a diferentes pastos en Tamaulipas, México. *Rev. Mex. Micol.* **13:** 52-57.
25. **Schardl, C. L., A. Leuchtmann, K. R. Chung, D. Penny, and M. R. Siegel.** 1997. Coevolution by common descent of fungal symbionts (*Epichloe* spp.) and grass hosts. *Mol. Biol. Evol.* **14:** 133-143.
26. **Scholler, M.** 1998. Obligate phytoparasitic fungi as plant taxonomists–An overview under special consideration of the rust fungus *Puccinia lagenophorae*. *Z. Pflanz. Pflanzen* **105:** 239-245.

27. **Stäger, R.** 1903. Infectionsversuche mit Gramineen-bewohnenden *Claviceps*-arten. *Bot. Ztg.* **61:** 111-118.
28. **Stäger, R.** 1905. Weitere Beiträge zur Biologie des Mutterkorns. *Zbl. Bakteriol. Parasitenk. Infektionskrank. Abt. II* **14:** 25-32.
29. **Stäger, R.** 1922. Beiträge zur Verbreitungsbiologie der *Claviceps*-Sklerotien. *Zbl. Bakteriol. Parasitenk. Infektionskrank. Abt. II* **56:** 329-339.
30. **Stebbins, G. L.** 1981. Coevolution of grasses and herbivores. *Ann. Miss. Bot. Gard.* **68:** 75-86.
31. **Tudzynski, P., K. Hölter, T. Correia, C. Arntz, N. Grammel, and U. Keller.** 1999. Evidence for an ergot alkaloid cluster in *C. purpurea*. *Mol. Gen. Genet.* **261:** 133-141.
32. **Whyte, R. O.** 1974. *Tropical Grazing Lands: Communities and Constituent Species.* W. Junk Publishing, The Hague, The Netherlands.

28

An Overview of the Biology of Sorghum Ergot

Malcolm J. Ryley, Damian J. Herde, Shamsul A. Bhuiyan, Robert G. Henzell, and David R. Jordan

Abstract

Sorghum ergot, caused by *Claviceps africana*, *Claviceps sorghi*, or *Claviceps sorghicola*, was confined to Africa and India before the mid-1990s, but is now found on all continents. *Claviceps africana* is the most widespread of these pathogens, probably due to its ability to produce airborne secondary conidia. Although more than 70 years of published literature exists on the biology of sorghum ergot, some critical processes remain poorly understood. These processes include survival mechanisms; the production, dispersal, and germination of secondary conidia of *C. africana*; and the interactions between weather parameters and pollen. A concerted, collaborative, international research effort is needed to better understand these processes.

Introduction

Sorghum ergot, a disease caused by species of *Claviceps*, is confined to the reproductive organs (gynoecia) of spikelets and is favored when pollination and fertilization are hindered by the lack of viable pollen (2, 3). Male sterile lines in seed production or breeders' nurseries are at particular risk for ergot, especially when pollination is delayed by asynchrony of flowering (2, 3). Unevenly flowering grain sorghum crops, tillers, and many forage sorghums also are at high risk (49) if ergot-conducive weather occurs at flowering (37, 58) and/or during the pre-leptotene/leptotene phase (38). Infection mimics fertilization and occurs when conidia of the causal pathogen germinate on the stigmas or, more rarely, the styles or the ovary, and hyphae invade and colonize the ovary (2, 3), which is replaced by a fungal sphacelium. A sticky liquid (honeydew) is secreted into the spikelet, and short conidiophores on the surface of the sphacelium produce macroconidia, which are released into the honeydew. Seven to eleven days after infection, the honeydew oozes from the infected spikelets, and, for *Claviceps africana* Frederickson, Mantle & de Milliano, macroconidia near the honeydew surface germinate during periods of high humidity, producing a sterigma that protrudes above the honeydew surface and upon which a single secondary conidium develops (19, 20). Macroconidia, microconidia (which also are found in honeydew), and secondary conidia can all infect gynoecia (4). Sphacelia gradually develop into sclerotia, which in *Claviceps sorghicola* Tsukiboshi, Shimanuki & Uematsu and to a lesser extent *Claviceps sorghi* Kulkarni, Seshadri & Hegde, readily germinate to produce ascostromata with immersed perithecia (2, 3, 56). Ascospores of *C. sorghi* can infect A-lines, albeit at low levels (52).

Sphacelial and sclerotial tissues are intimately associated and should be regarded as "different tissues of a single unit" (23). Sclerotia of *C. africana* produce alkaloids, predominantly dihydroergosine, which in Australia has caused agalactica in sows and cows (9) and reduced weight gain by broiler chickens (5) and beef cattle (39). Similar reports have not been made from Africa and Asia (3), which suggests that there may be differences in alkaloid production by isolates from these regions.

Four species of *Claviceps* are known to infect *Sorghum* species: *C. africana* (20), *C. sorghi* (29), *C. sorghicola* (56), and *Claviceps pusilla* Cesati (30). The first three species are pathogens of cultivated sorghum, *Sorghum bicolor* (L.) Moench, while *C. africana* also has been reported from other *Sorghum* species, *e.g.*, *Sorghum almum* Parodi and *Sorghum halepense* (L.) Pers. (49, 50). Pearl millet [*Pennisetum glaucum* (L). R. Br.] also can serve as a host for *C. africana* (18; M. J. Ryley, *unpublished*). *Claviceps africana* is the most widespread species, with *C. sorghi* and *C. sorghicola* being reported only from India and Japan, respectively (2, 56). The limited distribution of *C. africana* in Asia and Africa until the mid-1990s and its recent aggressive spread through the Americas and Australia has been well documented (2, 3). Recently (10, 45), Indian isolates assumed to be *C. sorghi* were found to be identical to *C. africana*, suggesting that the latter species has gradually replaced the former, perhaps due to *C. africana*'s ability to produce airborne secondary conidia (45). *Claviceps pusilla* has been recorded on a number of Australian native grasses, including *Sorghum leiocladum* (Hack.) Hubbard (30).

The three *Claviceps* species that infect *S. bicolor* are distinguished by differences in the morphology of the teleomorph (stipe, capitulum asci, and ascospores), sclerotia (shape), and, to a lesser extent, the anamorph (macroconidial dimensions). *C. africana* also is distinguished by the production of dihydroergosine and secondary conidia (2, 20, 56). The anamorph of both *C. africana* and *C. sorghi* is termed *Sphacelia sorghi* McRae, despite differences in conidium morphology and the production of secondary conidia by *C. africana*. The taxonomic standing of the three major sorghum ergot species as distinct species has been reinforced by recent molecular studies (28, 45, 54).

Genetic Variability

Sorghum ergot was recorded in southern Queensland, Australia in late April 1996 (48), but we suspect that it had been present in Australia for some time before then. The disease was found in northern Queensland (> 1400 km from the initial site in southern Queensland) within three weeks (3). The morphological characters of most Australian isolates are identical with those reported for *C. africana* (20), but some have features consistent with their inclusion in *C. sorghi*. The method of entry of the pathogen into Australia remains a mystery. Molecular studies (28, 45, 54, 55) have found that isolates of *C. africana* from India and Australia were very similar, supporting the hypothesis that at least some *C. africana* isolates in Australia were introduced through southeast Asia (44). In an analysis of over 50 Australian isolates, Komolong *et al.* (28) identified four genotypes of *C. africana*, including some isolates similar to those from the Americas. The available data suggest that *C. africana* was introduced into Australia several times, probably from different sources (44). Some genotypes also may have evolved locally on native sorghums (28). More research on the genetic relationships between isolates, particularly those from Asia, and on the role of sexual reproduction, is needed to understand the genetic diversity of *C. africana* and its worldwide distribution.

Survival and Inoculum Sources

In some species of *Claviceps*, *e.g.*, *Claviceps purpurea* (Fr. ex Fr.) Tul., sclerotia are important to the survival of the pathogen and for the initial infection by ascospores, which develop in the ascostromata (31). The role of sclerotia-like bodies in the life cycle of *Claviceps* species that infect sorghum is less clear. Sclerotia of *C. sorghicola* germinate readily (56) and may play an important role in the epidemiology of the disease in Japan, but few attempts to germinate the sclerotia-like bodies from either *C. africana* or *C. sorghi* have been successful (2, 3, 17). Bandyopadhyay *et al.* (2) reported that often there are locules, which contain viable macroconidia, inside the sphacelia of *C. africana*, and suggested that these sphacelia may be capable of surviving for some time. The gradual degradation of the tissues of such sphacelia could ultimately expose the macroconidia, allowing them to germinate and produce secondary conidia.

Given the lack of clear evidence for the role of sclerotia or the sexual state in the life cycle of *C. africana*, we surmise that alternative hosts, and conidia surviving on ergot-infected panicles or seed, are the modes of survival for this pathogen in Africa (2, 17, 22), Australia (49), and perhaps elsewhere. In addition, honeydew on plant parts also may play a role in survival, and ultimately infection (2). In Australia, ergot honeydew can be found on common weeds, *e.g.*, *S. halepense* and *S. almum*, and forage sorghums throughout the year in eastern Australia. This honeydew is thought to be the major source of inoculum for early infections. In addition, at least seven native Australian *Sorghum* species are susceptible to *C. africana* (Ryley and Herde, *unpublished*), and may play a role in the pathogen's survival in northern Australia.

Field research in Australia has demonstrated that macroconidia in ergot-infected panicles buried in soil survive for less than two months (8). In contrast, approximately 40% of the macroconidia in panicles 1 m above the soil could germinate after four months of winter (May-August; mean daily maximum 21.2°C, mean daily minimum 6.7°C). By mid-December (mid-summer), 7.5% of the macroconidia were still viable. During the summer months macroconidia lost their viability rapidly, with only 7.8% germinating after three months (January-early April; mean daily maximum. 29.8°C, mean daily minimum 17.8°C). Bhuiyan *et al.* (8) also found that macroconidia in honeydew on seed survived for less than two months at ambient temperatures (15-38°C) during the summer, but for 14 months on seed at 4°C.

Other research supports our hypothesis that macroconidia in honeydew can survive for long periods at low temperatures. Futrell and Webster (24) reported that macroconidia became nonviable after seven months in dried honeydew on panicles during the "dry season" (winter) in Africa. Frederickson and Odvody (22) showed that macroconidia at high temperatures (32°C) survived for only a short period, while those at 6°C still had 50% germination after 22 weeks. Jensen (27) demonstrated that macroconidia on panicles could survive temperatures as low as -26°C during the off-season (mid-November to May) in the Great Plains of the United States.

Although the evidence for alternative hosts as an important inoculum reservoir for *C. africana* is compelling, the role of macroconidia of *C. africana* in long-term survival is less clear. Despite their apparent "fragile" morphology, *i.e.*, thin-walled and hyaline, macroconidia can survive for a significant period of time in honeydew and sphacelia, particularly at low temperatures. Honeydew containing macroconidia and microconidia can be dispersed by water, and mechanically on seed, machinery, clothing, or by insects, but the relative importance of these modes of dispersal is not known (2). Frederickson and Odvody (22) showed that in Zimbabwe, a large proportion of macroconidia remain viable at the end of the dry season (winter), but that the germination of these conidia dropped rapidly as the rainy season (summer) began, probably due to high relative humidity. They hypothesized that these macroconidia might infect alternative hosts that flower in early summer. These alternative hosts then act as inoculum sources for grain sorghum, which flowers later. The contribution that macroconidia in honeydew make to intercontinental spread of the sorghum ergot pathogens, and their importance, relative to alternative hosts, in initiating infections and epidemics, however, remains to be determined.

Secondary Conidium Production, Release, and Dispersal

Environmental influences on the production, release, and dispersal of secondary conidia of *C. africana* are poorly understood. Bandyopadhyay *et al.* (2) reported that secondary conidia of "*C. sorghi*" developed from honeydew on sorghum panicles in a growth cabinet at 14-24°C and relative humidity >90%, and in a greenhouse at 23-28°C. However, the identity of the isolates used in this study is uncertain, as Pažoutova *et al.* (45) reported that these isolates appear to be *C. africana*. Data of Frederickson *et al.* (19) suggest that secondary sporulation of *C. africana* occurred on days when the temperature was 14-30°C and nocturnal relative humidity was high. Macroconidia of *C. africana* germinate at 14-35°C, with an optimum near 19°C (7, 16; Fig. 28-1).

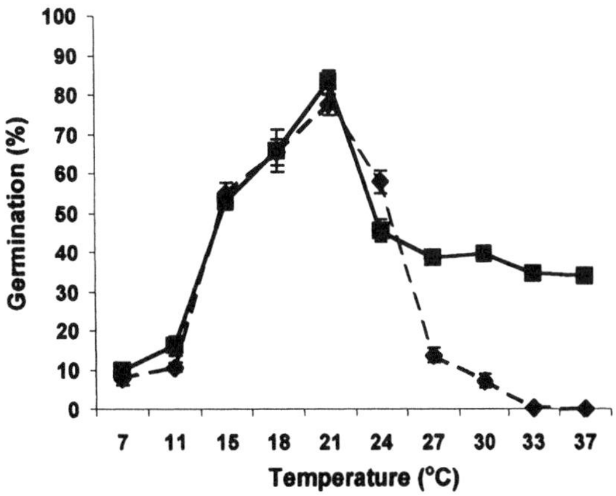

FIGURE 28-1. Effect of temperature on *in vitro* germination of secondary conidia of an Australian isolate of *Claviceps africana* on water agar (- - ◆ - -, 24 hours; –■–, 48 hours). Each data point represents the mean value of six replicates of 100 conidia each. Bars represent one standard deviation.

The concentration of sugars, particularly sucrose (20), in honeydew may influence macroconidium germination and, hence, secondary sporulation. Rainfall, which stimulates secondary sporulation (21), may have a two-fold effect–first, to increase the relative humidity around the honeydew drop, and second, to reduce the sugar concentration and, therefore, the osmotic potential, on the surface of the droplet. Frederickson (16) reported that macroconidia of *C. africana* did not germinate when the sucrose concentration of the medium was > 10% w/v. In contrast, Bhuiyan (7) found that at 20°C the highest germination of macroconidia occurred at -5 bar (= 7% sucrose), with some (10%) macroconidia germinating at -55 bar (= 44% sucrose).

The reports by Frederickson (16, 19, 21) of the spread of *C. africana* within an A-line planting are the only published studies of the factors that affect the production, release, and dispersal of secondary conidia of this pathogen. Secondary conidia of *C. africana* were present during daylight hours, but were more abundant a few hours after the relative humidity rise associated with nightfall (19). Rainfall triggered a significant increase in the number of secondary conidia in the air. Frederickson *et al.* (21) hypothesized that rainfall provided humidity, which was conducive to secondary sporulation, but that a period of "drying off," presumably at a lower humidity, was needed for high aerial concentrations of secondary conidia.

It is surprising that such critical stages in the life cycle of *C. africana* are so poorly understood. There is very good evidence that secondary conidia usually are the primary inoculum (2, 3, 21), but very little is known about the honeydew × environment × secondary sporulation interactions, the conditions that favor the release of the conidia, or the survival of secondary conidia during long-distance dispersal.

The ergot models of McLaren *et al.* (32, 34-36) and Wang *et al.* (57) are based on the influence of weather parameters during the flag leaf stage and flowering on ergot severity. Knowledge of the temporal and spatial relationships associated with sporulation and dispersal of secondary conidia is a vital missing link in these models. Greenhouse and field studies are needed to quantify these relationships.

Secondary Conidium Germination and Infection

There are few reports of the interaction between environmental parameters and the germination of secondary conidia. Most studies of *in vitro* and *in vivo* germination have been on macroconidia. In a study of Australian isolates of *C. africana* (7), macroconidia germinated on sorghum extract agar incubated at 8-31°C, with an optimum at 20°C. These results are similar to those obtained with African isolates (16). Bhuiyan (7) found that the temperature response of secondary conidia from the Australian isolates was similar to that of macroconidia after 24 hours (optimum 19.5°C), but after 48 hours, over 30% of the conidia had germinated even at 37°C (Fig. 28-1). Germination of both types of conidia began after 4 hours of incubation at 20°C, with median germination times of 5.9 hours for macroconidia and 8.8 hours for secondary conidia. In contrast, Frederickson (16) reported that *C. africana* macroconidia began germinating after 12 hours at 14°C and 35°C, and that secondary conidia began germinating after 15 hours at 24°C. Bhuiyan (7) found that macroconidia germinated to produce secondary conidia at all temperatures, while secondary conidia incubated at < 25°C germinated to produce tertiary conidia, but that these conidia produced germ tubes when germinated at temperatures > 25°C. The apparent rapid germination of *C. africana* conidia reported in the Australian study is consistent with *in vivo* studies. Only four hours of free moisture on the stigmas was required to obtain $> 50\%$ infection of panicles of a male-sterile line inoculated with a mixture of microconidia, macroconidia, and secondary conidia (6).

The mode of infection and the subsequent postinfection processes have been well documented (2, 3) and are essentially the same for *C. africana* and *C. sorghi*, although there are differences in the timing of events. Conidia (including secondary conidia) of *C. africana* require at least 15 hours for germination (2, 16), and infection hyphae require an additional 36-48 hours to reach the basal portion of the ovary. Our results suggest that conidium germination and infection processes may be much faster than previously thought.

Pollen–Conidia Interactions

The interactions between pollen and conidia determine whether the ovaries are fertilized or are invaded by hyphae from germinated conidia. The pistils usually escape or resist infection after effective fertilization of the ovary (3, 42). The timing of the arrival of pollen and conidia on stigmas are critical to this process. Bandyopadhyay *et al.* (3) reported that the highest infection levels by *C. sorghi* occurred when florets were inoculated four days before anthesis (95%). The level of infection decreased as the time between inoculation and anthesis decreased, with inoculation at anthesis resulting in 32% of the florets being infected. When the reported (2) times for pollen tube growth (2.5-12 hours) are compared with those for conidium germination and invasion of the ovary by the pathogen (up to 96 hours), it is surprising that there is such a high infection level at anthesis. This result may indicate that a significant proportion of the conidia infect the stigmas or the ovary directly, colonizing the ovary before it is fertilized by pollen.

Bandyopadhyay *et al.* (3) stated that, "In pollinated and fertilized pistils, the fungus fails to grow or grows slowly, thereby reducing the ergot severity." However, the anatomical and/or physiological bases for the postfection interactions between pollen and the germ tubes of conidia of the sorghum ergot pathogens have not been documented. Information on this aspect of the biology of sorghum ergot may be important in elucidating the resistance mechanism(s) in some sorghum genotypes.

The Role of the Environment in Disease Initiation and Development

Temperatures near 20°C during flowering appear to favor ergot development (1, 21, 37, 57), and at temperatures ≥ 28°C ergot severity is negligible (21, 37). The role of relative humidity is less clear. Sundaram (53) concluded that infection by *C. sorghi* was favored by cool, wet weather during flowering, and Futrell and Webster (24) reported that relative humidity near 100% for 24 hours was optimal for infection. Bhuiyan (6) demonstrated that only two hours at a relative humidity of 100% was necessary for infection of a male-sterile line by a mixture of macroconidia and secondary conidia of *C. africana*. Others have reported that significant infection can occur at relative humidities < 100% (1, 51). McLaren and Wehner (37) found a strong correlation between relative humidity and ergot severity–the higher the relative humidity during flowering, the greater the ergot severity.

McLaren *et al.* (35-38) have conducted the most comprehensive studies of the influence of weather on the severity of *C. africana*. In inoculated trials, they showed that the highest correlation between ergot severity and a climatic parameter was with daily maximum temperature 1-5 days after pollen shed, and that very little ergot occurred above a mean daily maximum temperature of 28°C. There also was a significant positive correlation between relative humidity and ergot severity over this time. They also identified significant correlations between maximum temperature, maximum humidity, and sunshine hours, suggesting a multivariate interaction with ergot severity. Field and greenhouse experiments have shown that cold weather (mean night temperatures < 12°C) 3-4 weeks before flowering induces male sterility (36, 38). There were strong relationships between mean minimum temperature during this period, pollen viability, and ergot disease incidence (as measured by the difference in disease incidence between male-sterile lines and male-normal genotypes). Their results were similar to those of Brooking (11), who found that low night temperatures (10°C) during the pre-leptotene/leptotene stage of microsporogenesis (18-20 days before flowering) reduced pollen viability, and that there was a linear response between pollen sterility and night temperatures. Genotypes differed in their sensitivity to low temperatures during this period (12, 38).

The parameters with the highest correlations with the mean ergot severity of 179 male-normal lines were mean minimum temperature 23-27 days before flowering and the mean values of maximum temperature and maximum relative humidity 1-5 days after flowering (36). These parameters were used to develop a model that can predict the mean ergot severity of a genetically broad-based sorghum population flowering on any day.

In Australia and Mexico, similar, but more limited, studies have been conducted (40, 41, 49, 58). In an inoculated (genotype) trial (58) there was no significant correlation between mean daily minimum temperature during the flag leaf stage (flag leaf sheath emergence to flag leaf fully expanded, corresponding to the start of the leptotene phase), pollen viability, or ergot severity. In contrast with McLaren's findings, there was a strong positive correlation between minimum and mean hourly temperatures during flowering and ergot severity (r = 0.83 and 0.85, respectively). However, this result may be misleading, because the range of temperatures encountered during this trial was small (mean hourly temperature 15.5-19.5°C). There was a strong correlation between mean relative humidity at 9:00 A.M. during flowering and ergot severity (r = 0.83).

The previously reported interactions between climatic parameters and ergot severity (35-38, 58) were determined from trials in which the entries were inoculated with suspensions of conidia at flowering. Such studies have been used to develop models for determining the susceptibility of sorghum genotypes to ergot (33, 35, 36) and to predict ergot risk periods (32, 57). However, there is a pronounced lack of published data on the interactions between environmental parameters and ergot severity under natural infection conditions.

In Australia, trials with a limited number of genotypes, including a male sterile line, were planted every 7-14 days from early September to early March for the last three years. At 50% flowering, pollen was collected from 10 panicles per five-meter row on water agar plates, and pollen viability determined by using an iodine solution (36). Pollen quantity was assessed on a qualitative 0-10 scale where 0 = no pollen and 10 = copious pollen. The portions of the panicle above and below that with bright yellow anthers were removed, and 21 days later ergot severity (mean percentage infected spikelets/panicle) was determined. Only the first planting of the male-sterile line was inoculated at flowering.

For the hybrid Buster, the highest significant correlation with ergot severity was mean hourly temperature 15-18 days before 50% flowering (r = -0.86), while for another hybrid, DK48, the corresponding period was 12-15 days before 50% flowering (r = -0.80). In similar trials in northern Mexico, Montes *et al.* (41) found that the strongest correlation between ergot incidence (number of infected panicles/plot) and maximum temperature was 13-15 days before 50% flowering for "early-maturing" hybrids and 10-12 days before 50% flowering for "medium-maturing" hybrids. They (40) also found that a minimum temperature of < 13°C 9-11 days before 50% flowering had a significant impact on the incidence of ergot. These data differ from those of McLaren *et al.* (35-38), who found that one of the critical periods for temperature effects on sorghum ergot was 23-27 days before flowering. The development of sorghum, and the time between critical stages, *e.g.*, flag leaf stage and flowering, is highly temperature dependent (58). The apparent discrepancies in these studies may be attributable to differences in the sorghum genotypes and protocols used in the trials and in the thermal units encountered during the trials.

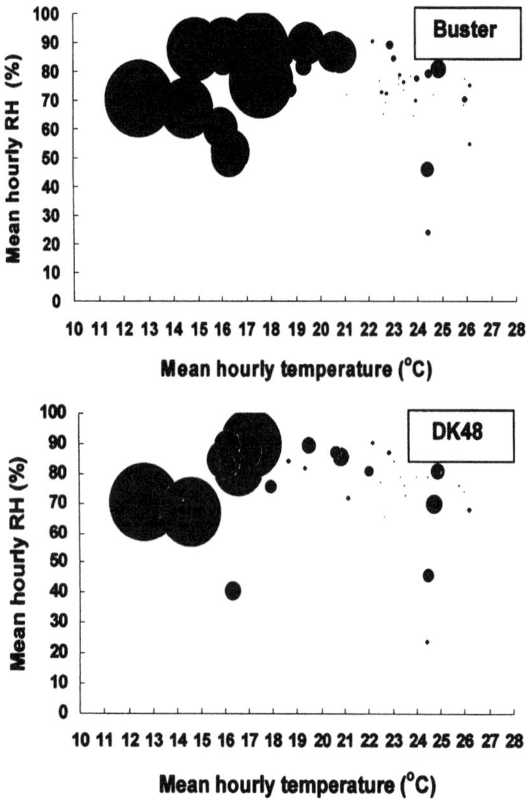

FIGURE 28-2. Relationship between mean hourly temperature and mean hourly relative humidity during flowering, and ergot severity for two commercial hybrids (Buster and DK48) in 1998/99 in Australia. Data were collected at two sites on a total of 63 flowering dates for Buster and 59 dates for DK48. The diameter of the circles represents the mean ergot severity on all panicles on each flowering date.

Environmental conditions during flowering also influence ergot severity. In Australian serial planting trials, we found highly significant correlations between the mean hourly temperatures during flowering and ergot severity ($r < -0.7$). Above an hourly mean temperature of 21 ± 1°C, a mean daily maximum of approximately 28°C, ergot severity was very low, but precise values depend upon the genotype examined (Fig. 28-2). There also was a strong correlation between the mean value of relative humidity at 9:00 A.M. during flowering and ergot severity, with severity increasing with relative humidity above 80%. The highest values of severity for both genotypes occurred when there were low temperatures and high humidity during flowering, which is experienced by grain sorghum flowering from mid-April onwards in Australia.

Montes *et al.* (41) reported that the minimum relative humidity during the period of three days before 50% flowering up to six days after 50% flowering (depending on the hybrid) was positively correlated with ergot incidence. Under Australian conditions, we found that ergot outbreaks can occur, but at low severity, during the main sorghum flowering period (December-March), when the mean hourly temperature during flowering is greater than 21°C (Fig 28-2). Outbreaks at these temperatures are strongly correlated with rainfall events (Fig. 28-3).

The availability of viable pollen is critical in disease development, and the relationship between pollen viability and ergot severity is reported to be linear (33), with severity increasing as pollen viability decreases from approximately 60% to 20%. Sorghum genotypes differ in the amount of pollen they produce (43), and this trait is influenced by environmental parameters (Henzell, *unpublished*). In our noninoculated serial planting trials, we attempted to establish a relationship between ergot severity and the amount of viable pollen produced (49). We found that ergot severity is negligible when the product of pollen viability (0-100%) and pollen quantity (0-10) (PVQ) is greater than 100, and that severity increases in a linear fashion as PVQ decreases below 100.

We also attempted to quantify the relationship between temperature and pollen production and viability. In our trials, for both Buster and DK48, the highest correlation between mean hourly temperature and pollen viability was 3-6 days before 50% flowering (r = 0.83). The highest correlation between both pollen quantity and PVQ, and mean hourly temperature, was the period from 50% flowering (the day of sampling) to 3 days later (r = 0.69 and 0.75 respectively). This latter period corresponds with flowering on the trimmed panicles. In these trials, the best correlations between ergot severity and mean hourly temperature occurred 12-18 days before 50% flowering.

Thus, although there are some data on the overall relationships between environmental parameters, pollen, and ergot severity, there is a critical lack of information on the influence of these environmental parameters on the underlying biological processes. The influences of temperature, relative humidity, and free moisture on stigmas, and on infection by conidia under controlled conditions need further study. Most infection studies use aqueous conidial suspensions sprayed onto florets, but in the case of *C. africana*, where secondary conidia are thought to be the dominant inoculum, this protocol does not simulate natural infection. A method for the "dry" inoculation of stigmas with secondary conidia needs to be developed.

In addition, the interaction between the environment and pollen is poorly understood, apart from the effects of low temperatures during the pre-leptotene/leptotene stage on pollen viability. Conditions during flowering affect the production, release, and deposition of macroconidia and secondary conidia, plant infection, and pollen production and viability. Quinby (46) reported that cloudiness and, presumably, high relative humidity during anthesis aided disease development, perhaps due to delayed anther dehiscence and pollen deposition on stigmas (2). Data from our trials suggest that low temperatures during flowering reduce pollen quantity. More research on the influences of temperature and other factors on the production and viability of pollen is needed to elucidate the complex host × environment × pathogen interactions.

Resistance (or Disease Escape)

The lack of confirmed sources of resistance to sorghum ergot is a problem that requires urgent attention (2). Genotypes that apparently are resistant at one location may be susceptible at another (2). Thus, the G × E component is of great significance in determining ergot severity. McLaren and his collaborators (33-36) have developed a system for identifying ergot resistance that uses the parameters "ergot breakdown point" (EBP) and the rate of resistance breakdown. In essence, this process determines the relationship between observed ergot incidence for a particular genotype and the ergot potential (predicted using a model based on mean minimum temperature 23-27 days before flowering, and mean maximum temperature and mean maximum relative humidity 1-5 days after flowering) for each flowering date. To ensure that a spread of ergot potentials is achieved, genotypes are sown at different planting times at multiple locations, as described in more detail elsewhere in this volume (35).

Another predictive model for ergot incidence is that of Wang *et al.* (57, 58). This model uses mean daily minimum temperature during the flag leaf stage, a temperature factor based on hourly temperatures during flowering, and the mean daily 9:00 A.M. relative humidity during flowering. The APSIM (Agricultural Production Systems SIMulator) sorghum physiology model (25) is used to predict the onset and duration of both the flag leaf stage and flowering, which vary with temperature. Neither model, however, accounts for inoculum potential (the concentration of viable conidia) on different flowering dates, a parameter that also needs to be incorporated into any successful model.

S. bicolor lines and hybrids with high levels of "resistance" to *C. africana* have been identified in recent years (14, 15, 34, 47, 49). Resistance also has been re-

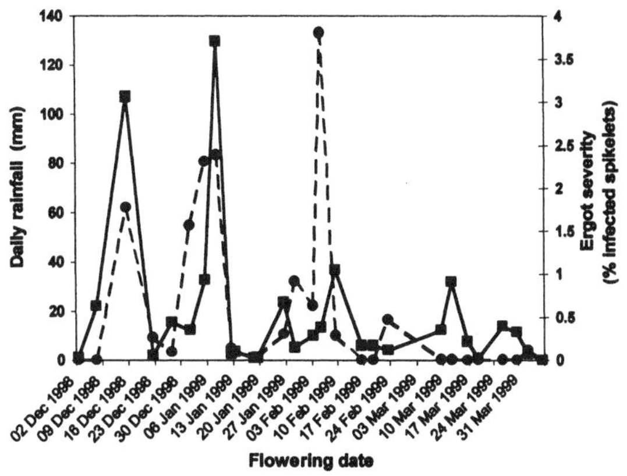

FIGURE 28-3. Relationship between daily rainfall (- - ● - -) and ergot severity (—■—) for the commercial hybrid Buster in 1998/99 at Gatton, Queensland, Australia.

ported in some accessions of *S. bicolor* spp. *Arundinaceum* and spp. *drumondii* (13). Dahlberg *et al.* (14, 15) reported that IS8525 (from Ethiopia) had the lowest level of ergot of all lines tested, and that its male sterile equivalent, A3*IS8525, was one of the most resistant A lines that they tested (15). Data from our inoculated trials over the past two seasons, and the results of others (47), support these conclusions regarding IS8525, but when IS8525 was backcrossed (one backcross) into A1 cytoplasm, an essentially male-sterile genotype that was very susceptible resulted. In Australia, we are developing methods for screening for resistance to sorghum ergot (26), and we anticipate using a system similar to that developed by McLaren *et al.* (34, 35) to evaluate ergot potentials on different flowering days.

The relationship between pollen viability and ergot severity has been established (35-38). Other characteristics, *e.g.*, a short period of stigma receptivity, rapid flower opening, and rapid fertilization, also might serve as possible mechanisms of "resistance" (2). Dahlberg *et al.* (15) noted an association between ergot resistance in IS8525 and some floral characteristics, *e.g.*, short exposure time of stigma to inoculum before pollination, rapid stigma drying before pollination, and small stigmas. Our data indicate that although IS8525 does not produce large quantities of pollen, its pollen viability remains high (> 90%), even during very cool weather. The relative importance of these various resistance mechanisms and their effects on other characters once incorporated into lines with desirable agronomic characters remains to be determined.

Conclusions

Despite over 80 years of research on sorghum ergot, there are some critical gaps in our knowledge of the biology of the causal organisms. These gaps affect our ability to manage the disease. Recent studies of genotypic diversity (*e.g.*, 28, 45, 54, and 55) are establishing the inter- and intracontinental genetic relationships of ergot isolates, and may help us to understand the apparently rapid spread of *C. africana* in the 1990s. Unfortunately, to date such studies have added little to our knowledge of the pathogenic and alkaloid production variation between isolates.

The relative importance of different potential sources of inoculum (ascospores from germinated sclerotia, secondary conidia from alternative hosts, and other sources) is still poorly understood. The host × environment × pathogen interactions involved in sorghum ergot development are complex, and there is a glaring lack of information on these interactions under natural conditions. The effects of temperature, relative humidity, and other factors on conidium production, dispersal, and germination, and on host factors, *e.g.*, pollen release, pollen germination, and stigma receptivity have not been adequately quantified. In addition, the concept of inoculum potential and its role in disease development needs to be explored, as does quantification of the amount of viable pollen needed for effective pollination and fertilization.

The appearance of *C. africana* in the Americas and Australia has, despite its negative impacts, resulted in a resurgence of research on the biology and variation of the pathogens, on fungicide efficacy, and on the search for resistance, particularly in A-lines. Improved communications and strengthened international links are needed to ensure that researchers studying sorghum ergot continue to work together to increase our basic knowledge of *C. africana*, and the other sorghum ergot pathogens, and to develop effective disease management strategies.

Acknowledgments

We thank the Australian Grains Research & Development Corporation for financial support, and Darrell Fletcher and Tony McCosker for technical assistance.

References

1. **Anahosur, K. H., and H. S. Patil.** 1982. Effect of date of sowing on the incidence of ergot of sorghum. *Ind. Phytopathol.* **35:** 507-509.
2. **Bandyopadhyay, R., D. E. Frederickson, N. W. McLaren, and G. N. Odvody.** 1996. Ergot–A global threat to sorghum. *Intl. Sorg. Millets Newsl.* **37:** 1-32.
3. **Bandyopadhyay, R., D. E. Frederickson, N. W. McLaren, G. N. Odvody, and M. J. Ryley.** 1998. Ergot: A new disease threat to sorghum in the Americas and Australia. *Plant Dis.* **82:** 356-367.
4. **Bandyopadhyay, R., L. K. Mughogho, S. K. Manohar, and M. V. Satyanarayana.** 1990. Stroma development, honeydew formation, and conidial production in *Claviceps sorghi. Phytopathology* **80:** 812-818.
5. **Bailey, C. A., J. J. Fazzino, M. S. Ziehr, M. Sattar, A. U. Haq, G. N. Odvody, and J. K. Porter.** 1999. Evaluation of sorghum ergot toxicity in broilers. *Poult. Sci.* **78:** 1391-1397.
6. **Bhuiyan, S. A.** 2002. *The Biology and Control of Ergot* (Claviceps africana) *in Sorghum.* Ph.D. thesis. Univ. of Queensland, Gatton, Queensland, Australia.
7. **Bhuiyan, S. A., V. J. Galea, M. J. Ryley, D. Tay, and A. T. Lilse.** 2002. Factors influencing the germination of macroconidia and secondary conidia of *Claviceps africana. Austral. J. Agric. Res.* **53:** in press.
8. **Bhuiyan, S. A., M. J. Ryley, V. J. Galea, D. Tay, and A. T. Lilse.** 2002. Survival of conidia of sorghum ergot (*Claviceps africana*) on panicles, seed and soil in Australia. *Australas. Plant Pathol.* **31:** in press
9. **Blaney, B. J., R. A. McKenzie, J. R. Walters, W. S. Bewg, M. J. Ryley, and R. Maryam.** 2000. Sorghum ergot (*Claviceps africana*) associated with agalactica and feed refusal in pigs and dairy cattle. *Austral. Vet. J.* **78:** 102-107.
10. **Bogo, A., and P. G. Mantle.** 1999. *Claviceps africana* in India. *Plant Dis.* **83:** 79.
11. **Brooking, I. R.** 1976. Male sterility in *Sorghum bicolor* (L.) Moench induced by low night temperature. I. Timing of the stage of sensitivity. *Austral. J. Plant Physiol.* **3:** 589-596.
12. **Brooking, I. R.** 1979. Male sterility in *Sorghum bicolor* (L.) Moench induced by low night temperature. II. Genotypic differences in sensitivity. *Austral. J. Plant Physiol.* **6:** 143-147.
13. **Claflin, L. E.** 2001. Agroecology and biotechnology of fungal pathogens of sorghum and millet, p. 11-16. INTSORMIL 2001 Annual Report. INTSORMIL, Lincoln, Nebraska.
14. **Dahlberg, J., R. Bandyopadhyay, B. Rooney, G. N. Odvody, and D. Frederickson.** 1998. Host-plant strategies within the United States. *Proc. 1998 Conf. on the Status of Sorghum Ergot in North America (24-26 June 1998, Corpus Christi, Texas).* http:/www.sorghumgrowers.com/Research/confernc/dahlberg.htm (last checked May 2002).
15. **Dahlberg, J., R. Bandyopadhyay, B. Rooney, G. Odvody, and P. Madera-Torres.** 2001. Evaluation of sorghum germplasm used in US breeding programmes for sources of sugary disease resistance. *Plant Pathol.* **50:** 681-689.
16. **Frederickson, D. E.** 1990. *Ergot Disease of Sorghum.* Ph.D. thesis, London Univ., London, U.K.
17. **Frederickson, D. E.** 1999. The life cycle, spread and survival of the sorghum ergot pathogens. *Proc. Global Conf. on Ergot of Sorghum, (1-8 June 1997, Sete Lagoas, Brazil)*: 135-140.
18. **Frederickson, D. E., and P. G. Mantle.** 1996. Pearl millet as an alternative host of the sorghum ergot pathogen, *Claviceps africana. Intl. Sorg. Millets Newsl.* **37:** 83–85.
19. **Frederickson, D. E., P. G. Mantle, and W. A. J. de Milliano.** 1989. Secondary conidiation of *Sphacelia sorghi* on

sorghum, a novel factor in the epidemiology of ergot disease. *Mycol. Res.* **93:** 497-502.

20. **Frederickson, D. E., P. G. Mantle, and W. A. J. de Milliano.** 1991. *Claviceps africana* sp. nov.; the distinctive ergot pathogen of sorghum in Africa. *Mycol. Res.* **95:** 1101-1107.
21. **Frederickson, D. E., P. G. Mantle, and W. A. J. de Milliano.** 1993. Windborne spread of ergot disease (*Claviceps africana*) in sorghum A-lines in Zimbabwe. *Plant Pathol.* **42:** 368-377.
22. **Frederickson, D. E., and G. N. Odvody.** 2002. Survival of inoculum of *Claviceps africana* in Zimbabwe: Potential sources of initial inoculum, p. 79-82. *In* J. F. Leslie (ed.), *Sorghum and Millets Diseases*. Iowa State Press, Ames, Iowa.
23. **Frederickson, D. E., G. N. Odvody, and T. Isakeit.** 1999. *Distinguishing Sphacelia and Sclerotia of* Claviceps africana *in Seed*. Texas Agricultural Extension Service Bull. L-5315, Texas A&M Univ., College Station, Texas.
24. **Futrell, M. C., and O. J. Webster.** 1966. Host range and epidemiology of the sorghum ergot organism. *Plant Dis. Rept.* **49:** 680-683.
25. **Hammer, G. L., and R. C. Munchow.** 1994. Assessing climatic risk to sorghum production in water-limited environments. I. Development and testing of a simulation model. *Field Crops Res.* **36:** 221-234.
26. **Henzell, R. G., M. Ryley, D. S. Fletcher, D. Herde, A. N. McCosker, D. Butler, and E. Wang.** 1998. Screening for resistance to sorghum ergot in Australia. In: *Proc.1998 Conf. on the Status of Sorghum Ergot in North America (24-26 June 1998, Corpus Christi, Texas).* http:/www.sorghumgrowers.com/Research/confernc/henzell.htm (last checked May 2002).
27. **Jensen, S. G.** 1998. Winter survival of *Claviceps africana* spores in the central plains of the USA. *Intl. Sorg. Millets Newsl.* **39:** 113-114.
28. **Komolong, B., S. Chakraborty, M. J. Ryley, and D. Yates.** 2002. Four genotypes of *Claviceps* spp. cause sorghum ergot in Australia, p. 87-89. *In* J. F. Leslie (ed.), *Sorghum and Millets Diseases*. Iowa State Press, Ames, Iowa.
29. **Kulkarni, B. G. P., V. S. Seshadri, and R. K. Hegde.** 1976. The perfect stage of *Sphacelia sorghi* McRae. *Mysore J. Agric. Sci.* **10:** 286-289.
30. **Langdon, R. F. N.** 1942. Ergot of native grasses in Queensland. *Proc. Roy. Soc. Qld.* **54:** 23-32.
31. **Mantle, P. G., and S. Shaw.** 1976. Role of ascospore production by *Claviceps purpurea* in aetiology of ergot disease in male sterile wheat. *Trans. Brit. Mycol. Soc.* **67:** 17-22.
32. **McLaren, N. W.** 1998. Ergot risk analysis. In: *Proc. 1998 Conf. on the Status of Sorghum Ergot in North America (June 24-26, 1998, Corpus Christi, Texas).* http://www.sorghumgrowers.com/Research/confernc/mclaren.htm (last checked May 2002).
33. **McLaren, N. W.** 1998. Breeding for ergot resistance in South Africa. In: *Proc. 1998 Conf. on the Status of Sorghum Ergot in North America (24-26 June 1998, Corpus Christi, Texas).* http://www.sorghumgrowers.com/Research/confernc/mclaren2.htm (last checked May 2002).
34. **McLaren, N. W.** 2000. Resistance of sorghum hybrids to ergot caused by *Claviceps africana*. *S. Afr. J. Plant Soil* **17:** 1-5.
35. **McLaren, N. W.** 2002. Genotype × environment effects on the response of sorghum to ergot and repercussions for disease screening, p. 95-101. *In* J. F. Leslie (ed.), *Sorghum and Millets Diseases 2000*. Iowa State Press, Ames, Iowa.
36. **McLaren, N. W., and B. C. Flett.** 1998. Use of weather variables to quantify sorghum ergot potential in South Africa. *Plant Dis.* **82:** 26-29.
37. **McLaren, N. W., and F. C. Wehner.** 1990. Relationship between climatic variables during early flowering of sorghum and the incidence of sugary disease caused by *Sphacelia sorghi*. *J. Phytopathol.* **130:** 82-88.
38. **McLaren, N. W., and F. C. Wehner.** 1992. Preflowering low temperature predisposition of sorghum to sugary disease (*Claviceps africana*). *J. Phytopath.* **135:** 328-334.
39. **McLennan, S. R., B. J. Blaney, J. F. Kidd, J. A. Connell, and V. Doogan.** 2001. Effects of sorghum ergot on the performance of cattle in feedlots. *In* A. K. Borrell and R. G. Henzell (eds.), *Proc. 4th Australian Sorghum Conf. (5-8 February 2001, The Kooralbyn Hotel Resort).* CD Rom ISBN 0724221638. Range Media Pty. Ltd., Toowoomba, Queensland, Australia.
40. **Montes, N., G. N. Odvody, and M. Marin Silva.** 2002. Effect of cold degree units on incidence of *Claviceps africana* in sorghum hybrids, p. 103-104. *In* J. F. Leslie (ed.), *Sorghum and Millets Diseases*. Iowa State Press, Ames, Iowa.
41. **Montes, N., G. N. Odvody, and H. Williams-Alanis.** 2002. Relationship between climatic variables and *Claviceps africana* incidence on sorghum hybrids in northern Mexico, p. 111-112. *In* J. F. Leslie (ed.), *Sorghum and Millets Diseases*. Iowa State Press, Ames, Iowa.
42. **Mushabyimana, T., C. Sehene, and R. Bandyopadhyay.** 1995. Ergot resistance in sorghum in relation to flowering, inoculation technique and disease development. *Plant Path.* **44:** 109-115.
43. **Ortiz-Pérez, E., F. Zavala-García, N. E. García-Trevino, and J. D. Eastin.** 1999. Pollen production in male-fertile lines of sorghum. *Intl. Sorg. Millets Newsl.* **40:** 34-35.
44. **Pažoutová, S.** 2002. The genus *Claviceps*: evolution at work, p. 135-140. *In* J. F. Leslie (ed.), *Sorghum and Millets Diseases*. Iowa State Press, Ames, Iowa.
45. **Pažoutová, S., R. Bandyopadhyay, D. E. Frederickson, P. G. Mantle, and R. A. Frederiksen.** 2000. Relations among sorghum ergot isolates from the Americas, Africa, India and Australia. *Plant Dis.* **84:** 437-442.
46. **Quinby, J. R.** 1958. *Grain Sorghum Production in Texas*. Texas Agric. Experiment Station Bull. no. 912. Texas A&M Univ., College Station, Texas.
47. **Reed, J., M. Tuinstra, and L. Claflin.** 2002. Identification of host resistance to ergot in sorghum, p. 475. *In* J. F. Leslie (ed.), *Sorghum and Millets Diseases*. Iowa State Press, Ames, Iowa.
48. **Ryley, M. J., J. L. Alcorn, J. K. Kochman, G. A. Kong, and S. M. Thompson.** 1996. Ergot on *Sorghum* spp. in Australia. *Australas. Plant Pathol.* **25:** 214.
49. **Ryley M. J., R. G. Henzell, S. Chakraborty, S. Bhuiyan, and D. Herde.** 2001. The biology and management of sorghum ergot in Australia. *In* A. K. Borrell and R. G. Hen-

zell, eds., *Proc. 4th Australian Sorghum Conf. (5-8 February 2001, The Kooralbyn Hotel Resort).* CD Rom ISBN 0724221638. Range Media Pty. Ltd., Toowoomba, Queensland, Australia.

50. **Ryley, M. J., D. Persley, D. R. Jordan, and R. G. Henzell**. 2002. Status of sorghum and pearl millet diseases in Australia, p. 441-448. *In* J. F. Leslie (ed.), *Sorghum and Millets Diseases*. Iowa State Press, Ames, Iowa.
51. **Sangitrao, C. S., and G. H. Bade.** 1979. Meteorological factors associated with honeydew development and sclerotial stage in sorghum ergot. *Sorg. Newsl.* **22:** 107-108.
52. **Sangitrao, C. S., S. Indira, and R. Bandyopadhyay.** 1999. Sorghum ergot in India. *Proc. Global Conf. on Ergot of Sorghum, (1-8 June 1997, Sete Lagoas, Brazil)*: 41-54.
53. **Sundaram, N. V.** 1971. Possible resistance to sugary disease in sorghum. *Ind. J. Gen. Pl. Breed.* **31:** 383-387 (referenced in 2).
54. **Tooley, P. W., E. D. Goley, M. M. Carras, R. D. Frederick, and E. L. Weber.** 2002. Characterization of *Claviceps* species pathogenic on sorghum by sequence analysis of the β-tubulin gene intron 3 region and EF-1α gene intron 4. *Mycologia* **93:** 541-551.
55. **Tooley, P. W., and N. R. O'Neill.** 2002. Intra-specific variation in *Claviceps africana*, p. 151-155. *In* J. F. Leslie (ed.), *Sorghum and Millets Diseases*. Iowa State Press, Ames, Iowa.
56. **Tsukiboshi, T., T. Shimanuki, and T. Uematsu.** 1999. *Claviceps sorghicola* sp. nov., a destructive ergot pathogen of sorghum in Japan. *Mycol. Res.* **103:** 1403-1408.
57. **Wang, E., H. Meinke, and M. J. Ryley.** 2000. Event frequency and severity of sorghum ergot in Australia. *Austral. J. Agric. Res.* **51:** 457-466.
58. **Wang, E., H. Meinke, M. J. Ryley, D. Herde, and R. G. Henzell.** 2000. On the relation between weather variables and sorghum ergot infection. *Austral. J. Agric. Res.* **51:** 313-324.

29

Intraspecific Variation in *Claviceps africana*

Paul W. Tooley and Nichole R. O'Neil

Abstract

We analyzed genetic variability within a global collection of the sorghum ergot pathogen *Claviceps africana* by using DNA sequence analysis, random amplified microsatellite (RAM) analysis, and amplified fragment length polymorphism (AFLP) analysis. Distinct groups of *Claviceps* species could be resolved based on the sequence of β-tubulin and elongation factor 1α (EF-1α) gene sequences, and PCR primers were designed for specific detection of *C. africana* and other species. AFLP analysis allowed assessment of intraspecific variation in *C. africana*, and high levels of genetic similarity were observed among Australian/Indian/Japanese isolates and U.S./Mexican/African isolates. Within individual populations, including the one in the United States, AFLP polymorphisms were observed, which indicates that multiple *C. africana* genotypes are present.

Introduction

Sorghum ergot, caused by *Claviceps africana*, has recently become established in the United States. The pathogen initially was restricted to Asia and Africa, but has since spread to South America, Australia, Japan, the United States, and other regions (2, 18). In 1997, ergot was found in Puerto Rico as well as the lower Rio Grande Valley of Texas (11) and spread northward as far as Nebraska and eastward to Mississippi (25). Sorghum ergot threatens sorghum production in a number of states and is of concern to the sorghum industry.

Population genetic analysis of the sorghum ergot pathogen using molecular markers could help elucidate which species may be present in the United States, enable comparisons of isolates found in the United States with those from other regions, and lead to a rapid means of pathogen detection. However, *Claviceps* is not a widely studied genus, and the number of studies of the genetics and population structure of *C. africana* is not large.

Claviceps purpurea is the most widely studied species in the genus *Claviceps* (6, 7, 12). Levels of genetic variation within *C. purpurea* (12) and *C. africana* (18) have been assessed with random amplified polymorphic DNA (RAPD). In *C. purpurea*, DNA fragments resulting from PCR amplification of genomic DNA with a single RAPD primer distinguished more than 20 *C. purpurea* isolates, with most strains from specific hosts grouping together. In 28 geographically diverse *C. africana* isolates, 65 RAPD primers amplified uniform, species-specific patterns, and seven primers could be used to distinguish four groups of isolates, but little variation was observed within groups (18). In terms of molecular taxonomy, phylogenetic analyses comparing ribosomal DNA sequences place the Clavicipitales within the unitunicate perithecial ascomycetes (8, 14, 20).

The amplified fragment length polymorphism (AFLP) technique (24) is a relatively new DNA fingerprinting technique that has become a valuable and powerful method to study population structure in fungi. The AFLP technique selectively amplifies a small portion, usually < 0.4%, of the DNA fragments resulting from a complete double restriction enzyme digest of genomic DNA. AFLPs have several advantages relative to other tools commonly used to assess genetic variation, including high reproducibility, large information content per reaction, theoretical neutrality of the markers detected, and dispersal of markers detected apparently randomly throughout the entire genome (15). AFLP analysis has become a valuable tool for identifying previously undetected levels of genetic diversity, especially among isolates within a species of fungal pathogens, and for identifying genetic markers associated with important biological, ecological, and phylogenetic characters (1, 17, 19). We analyzed strains of *C. africana* with AFLPs and found that the technique provided significant information about genetic variation within the ergot pathogen that was consistent with the results of RAPD analysis (18) in reconstructing recent patterns of pathogen migration.

Cultural Variation in *C. africana*

We have 233, mostly single conidial, isolates of *C. africana* in our collection, with 80 from the mainland United States, 33 from Puerto Rico, 51 from India, 14 from Australia, 17 from Africa, 9 from Japan, 8 from Mexico, 11 from Brazil, 5 from El Salvador, and 5 from Thailand, that we can work with in specialized containment facilities (16) not available elsewhere in the United States. Isolates are stored cryogenically at the Central Cell Repository of the National Cancer Institute at Ft. Detrick, Maryland both as dry mycelial plugs (from PDA cultures) and in 10% dimethyl-sulfoxide cryoprotectant. As outgroup species, we use *C. purpurea* (ATCC 34699, ATCC 34501), *Claviceps fusiformis* (ATCC 26019), *Claviceps sorghicola* (23), and *Claviceps paspali* from *Paspalum* grasses.

C. purpurea and *C. fusiformis* grow faster than *C. africana*, and *C. fusiformis* has "feathery" colony edges. *C. sorghicola* has a "smooth" appearance and produces a brown pigment, which is especially prominent on the underside of the colony. Cultures of *C. africana* from different geographic locations grown on potato dextrose agar at 22-24°C in darkness vary in growth rate (0.79-2.54 mm/day) and colony morphology. Cultures from Texas grow slowly, with irregular margins and lumpy colony surfaces, while those from Kansas include some with similar morphology to those from Texas as well as others with a "raised" colony morphology and/or feathery colony margins. Indian and Australian isolates generally have smooth colony surfaces and less restricted growth. Thai cultures have slow-growing, hard, dark, brown areas interspersed with smooth or lumpy, white areas of growth. We also have cultures that are intermediate between the smooth and lumpy phenoltypes and/or that appear to sector for this phenotype.

Molecular Analysis of *Claviceps africana*

We used three different molecular techniques to characterize genetic variation in *C. africana* including DNA sequence analysis, RAM analysis, and AFLP analysis. We cloned and sequenced the β-tubulin intron 3 region of 40 *Claviceps* isolates representing five species, and the EF-1α intron 4 region (21) of 11 isolates from the same five species. A very small amount of intra-specific variation was observed in the two regions, and the differences in sequence could be used to separate the five *Claviceps* species.

We also used RAM and AFLP analyses to assess intra-specific variation in *C. africana*. RAM analysis (9, 10) is a PCR-based method that uses primers containing microsatellite sequences and degenerate anchors at the 5′ ends, and has been used to evaluate variation in several fungal species (9, 10). We used the primers described by Hantula *et al.* (9) and named by the target repeat sequences: ACA, CCA, CGA, and GT. PCR was performed in a Perkin-Elmer 9600 thermal cycler with previously described reaction conditions (9). The annealing temperature used depended upon the primer used: 64°C for CCA, 61°C for CGA, 58°C for GT, and 49°C for ACA. PCR products were separated in 1.4% agarose gels in 0.5 × TBE buffer and stained with ethidium bromide. Three sets of RAM primers gave reproducible banding patterns and could be used to differentiate species. Amplified DNA fragments ranged from ~200-2000 bp in length; characteristic bands were observed for each species (22).

AFLPs were generated for *C. africana* (22) with the AFLP Analysis System II (Life Technologies Inc., Gaithersburg, MD). The AFLP reaction products were labeled with ^{32}P and separated by electrophoresis through a denaturing 6% (w/v) polyacrylamide DNA sequencing gel containing 7.5 M urea. Autoradiographs were obtained by exposing Kodak BioMax MR-2 film (Eastman Kodak Co., Rochester, NY) to the dried gel overnight at room temperature. Autoradiographs were scanned and each lane scored for the presence or absence of bands. Bands of 75-295 bp were scored by using Proscore version 2.36 (DNA Proscan Inc., Nashville, TN). Binary matrices consisting of "0"s (denoting absence of a band) and "1"s (denoting presence of the band) were analyzed to obtain simple matching coefficients among the isolates by using NTSYS-pc, version 2.0 (Exeter Biological Software, Setauket, New York). Simple matching coefficients were clustered to generate similarity trees using the SAHN clustering option and the unweighted pair-group method with arithmetic average (UPGMA) algorithm in NTSYS-pc. We analyzed 20 *C. africana* isolates plus isolates of *C. purpurea*, *C. fusiformis*, *C. sorghicola*, and *Pleospora papaveraceae*. Selective primers used were *Mse*+C and *Eco*RI+AG, and the resulting AFLP patterns were highly reproducible.

When AFLP data were analyzed using the UPGMA algorithm within NTSYS-pc (Fig. 29-1), different *Claviceps* species ranged in similarity from ≤ 60% to approximately 75%. Based on AFLP analysis, *C. sorghicola*, which causes sorghum ergot in Japan, was more closely related to *C. africana* than were either *C. purpurea* or *C. fusiformis* (22). Indian and Australian isolates also were approximately 90% similar, and shared several polymorphic bands. These findings and those from the RAM data and the RAPD data (18) all are consistent with the hypothesis that the Indian and Australian isolates are closely related. *C. africana* isolates from South Africa (Cla-56 and Cla-58) were nearly identical to those from the United States, and shared a 111 bp band not present in either the Australian (Cla-2 and Cla-3) or Indian (Cls-1 through Cls-4) isolates.

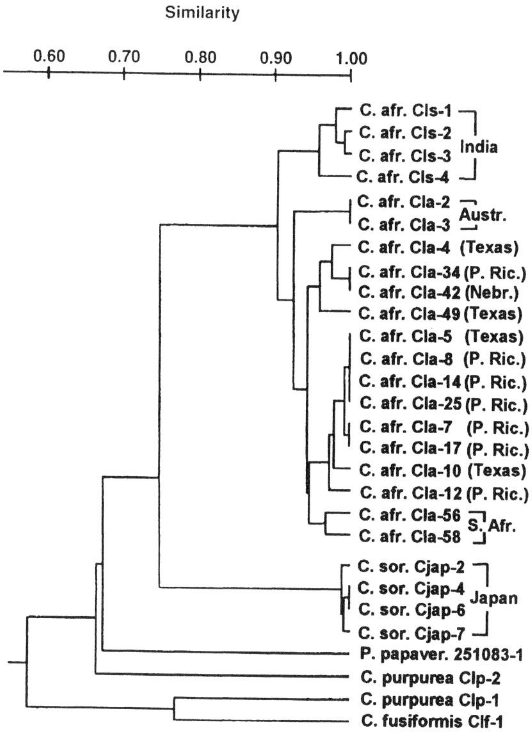

FIGURE 29-1. Dendrogram showing similarities among isolates and species of *Claviceps* based on AFLP analysis and simple matching coefficients. The tree was generated using the SAHN clustering program with the unweighted pair-group method with arithmetic average (UPGMA) algorithm in NTSYS-pc, version 2.0. PCR products between molecular sizes 75 and 295 bp were scored for the analysis. Species names are abbreviated as: C. afr., *Claviceps africana*; C. sor., *Claviceps sorghicola*; P. papaver, *Pleospora papaveracea* (5). Reprinted from (22) with the permission of the American Phytopathological Society.

The UPGMA dendrogram generated from the AFLP fingerprints has the four *C. africana* isolates from India in one cluster, which, as a group, had approximately 90% similarity to *C. africana* isolates from Australia, the United States, Puerto Rico, and South Africa (Fig. 29-1). The first isolate we obtained from infected sorghum in Nebraska in 1997 (Cla-42) was identical to Cla-34, which originated from the South Coast region (Guayanilla) of Puerto Rico in 1998. Isolates Cla-4, Cla-5, Cla-10, and Cla-49 from three different locations in Texas were 95-100% similar to isolates Cla-7, Cla-8, Cla-12, Cla-14, Cla-17, and Cla-25 from Isabela, in northwestern Puerto Rico (Fig. 29-1).

Indian *C. africana* isolates Cls-1 through Cls-4 clustered with > 95% similarity in the AFLP dendrogram, and were approximately 90% similar to other *C. africana* isolates. These Indian isolates were identical to other *C. africana* isolates based on β-tubulin and EF-1α sequences (21). Thus, we consider these isolates to be *C. africana* rather than *C. sorghi*. These Indian isolates also were identified as *C. africana* by Bogo and Mantle (3) based on morphology and the presence of the alkaloid dihydroergosine.

Some isolates had unique AFLP bands that distinguished them from other isolates from the same geographic location. Thus, a substantial amount of genetic variation exists within *C. africana*. Even amongst the isolates from the United States, there is some variation (isolates are approximately 95% similar), which indicates that several clonal genotypes of the pathogen are present.

In 1997, ergot was observed in Mexico prior to being found in the mainland United States. Puerto Rican isolates could have originated from the mainland United States following its establishment there, or from a third location, which could have been the source for the epidemics in Mexico and the mainland United States. The high (95%) similarity between isolates from South Africa and the United States implicates Africa, rather than India or Australia, as the original source of the isolates responsible for the 1997 epidemics in Puerto Rico and the mainland United States. Support for this hypothesis includes the recovery of isolates with identical RAPD patterns from both South Africa and the Americas (18). Based on RAPDs (18), isolates from the United States all represented a single clone, although the isolates tested were collected in Texas in a single year.

Conclusions and Future Research

AFLP analysis of *C. africana* isolates shows that Indian, Australian, and Japanese isolates, and U.S., Mexican, and African isolates are related to one another but fall into two groups. These results are consistent with those obtained by RAPD analysis (18) and reflect existing patterns in sorghum seed trade. Within each of these groups, AFLP variation exists, which indicates the presence of multiple genotypes in each region. Mechanisms responsible for the observed variation could include mutation, or more recently characterized modes of genetic change such as chromosome alterations, cytoplasmically transmitted genetic elements, or transposon activity (13). The parasexual cycle can occur in *Claviceps* sp. (4), but it is not known if it occurs in *C. africana*. Greater knowledge of the role of the sexual cycle, and whether it occurs in areas of recent introduction, *e.g.*, the United States, could help estimate the pathogen's ability to rearrange existing genetic variation and form new genotypes that could complicate the deployment of long-term control measures.

The AFLP method should be useful in further studies of ergot species diversity and in the monitoring of pathogen migrations. Further analysis of greater numbers of isolates from diverse regions including Africa, India, Australia, South and Central America, and the United States will allow associations of AFLP patterns with pathotype diversity, host of origin, and/or geographic origin to be made. The resulting associations may help predict future migrations and changes in the population composition of this potentially rapidly spreading pathogen. Furthermore, analysis of genotypes that are present regionally over a period of years can be used to determine whether the same genotypes cause disease every year, whether the pathogen can survive between seasons on alternative hosts, *e.g.*, johnsongrass, and whether initial inoculum is composed of immigrant genotypes from other regions. These data will provide a better overall picture of the population genetics and dynamics of *C. africana* in regions of new introduction and lead to predictions regarding the epidemiology of the disease.

Acknowledgements

We thank Richard E. Fredericksen, Debra E. Frederickson, Tom Isakeit, Jeff A. Dahlberg, Ranajit Bandyopadhyay, Gary N. Odvody, Malcolm J. Ryley, T. Tsukiboshi, Stan Jensen, and Neal McClaren for supplying infected sorghum and/or *Claviceps* cultures.

References

1. **Baayen, R. P., K. O'Donnell, P. J. M. Bonants, E. Cigelnik, L. P. N. M. Kroon, E. J. A. Roebroeck, and C. Waalwijk.** 2000. Gene geneologies and AFLP analyses in the *Fusarium oxysporum* complex identify monophyletic and nonmonophyletic formae speciales causing wilt and rot disease. *Phytopathology* **90:** 891-900.
2. **Bandyopadhyay, R., D. E. Frederickson, N. W. McLaren, G. N. Odvody, and M. J. Ryley.** 1998. Ergot: A new disease threat to sorghum in the Americas and Australia. *Plant Dis.* **82:** 356-367.
3. **Bogo, A., and P. G. Mantle.** 1999. *Claviceps africana* discovered in India. *Plant Dis.* **83:** 79.
4. **Brauer, K. L., and J. E. Robbers.** 1987. Induced parasexual processes in *Claviceps* sp. strain SD58. *Appl. Environ. Microbiol.* **53:** 70-73.
5. **Farr, D. F., N. R. O'Neill, and P. B. van Berkum.** 2000. Morphological and molecular studies on *Dendryphion penicillatum* and *Pleospora papaveracea*, pathogens of *Papaver somniferum*. *Mycologia* **92:** 145-153.
6. **Garre, V., U. Muller, and P. Tudzynski.** 1998. Cloning, characterization, and targeted disruption of *cpcat1*, coding for an *in planta* secreted catalase of *Claviceps purpurea*. *Molec. Plant-Microbe Interact.* **11:** 772-783.
7. **Giesbert, S., H. B. Lepping, K. B. Tenberge, and P. Tudzynski.** 1998. The xylanolytic system of *Claviceps purpurea*: Cytological evidence for secretion of xylanases in infected rye tissue and molecular characterization of two xylanase genes. *Phytopathology* **88:** 1020-1030.
8. **Glenn, A. E., C. W. Bacon, R. Price, and R. T. Hanlin.** 1996. Molecular phylogeny of *Acremonium* and its taxonomic implications. *Mycologia* **88:** 369-383.
9. **Hantula, J., M. Dusabenyagasani, and R. C. Hamelin.** 1996. Random amplified microsatellites (RAMs) a novel method for characterizing genetic variation in fungi. *Eur. J. For. Pathol.* **26:** 159-166.
10. **Hantula, J., E. M. Niemi, J. Kaitera, R. Jalkanen, and T. Kurkela.** 1998. Genetic variation of the resin top fungus in Finland as determined by random amplified microsatellites (RAMs). *Eur. J. For. Pathol.* **28:** 361-372.
11. **Isakeit, T., G. N. Odvody, and R. A. Shelby.** 1998. First report of sorghum ergot caused by *Claviceps africana* in the United States. *Plant Dis.* **82:** 592.
12. **Jungehulsing, U., and P. Tudzynski.** 1997. Analysis of genetic diversity in *Claviceps purpurea* by RAPD markers. *Mycol. Res.* **101:** 1-6.
13. **Kistler, H. C., and V. P. W. Miao.** 1992. New modes of genetic change in filamentous fungi. *Annu. Rev. Phytopathol.* **30:** 131-152.
14. **Kuldau, G. A., J.-S. Liu, J. F. White, Jr., M. R. Siegel and C. L. Schardl.** 1997. Molecular systematics of Clavicipitaceae supporting monophyly of genus *Epichloë* and form genus *Ephelis*. *Mycologia* **89:** 431-441.
15. **Majer, D., R. Mithen, B. G. Lewis, P. Vos, and R. P. Oliver.** 1996. The use of AFLP fingerprinting for the detec-tion of genetic variation in fungi. *Mycol. Res.* **100:** 1107-1111.
16. **Melching, J. S., K. R. Bromfield, and C. H. Kingsolver.** 1983. The plant pathogen containment facility at Frederick, Maryland. *Plant Dis.* **67:** 717-722.
17. **O'Neill, N. R., P. van Berkum, J. J. Lin, J. Kuo, G. N. Ude, W. Kenworthy, and J. A. Saunders.** 1997. Application of amplified restriction fragment length polymorphism for genetic characterization of *Colletotrichum* pathogens of alfalfa. *Phytopathology* **87:** 745-750.
18. **Pažoutová, S., R. Bandyopadhyay, D. E. Frederickson, P. G. Mantle, and R. A. Frederiksen.** 2000. Relations among sorghum ergot isolates from the Americas, Africa, India, and Australia. *Plant Dis.* **84:** 437-442.
19. **Pongam, P., T. C. Osborn, and P. H. Williams.** 1998. Genetic analysis and identification of amplified fragment length polymorphism markers linked to the *alm1* avirulence gene of *Lepotsphaeria maculans*. *Phytopathology* **88:** 1068-1072.
20. **Spatafora, J. W., and M. Blackwell.** 1993. Molecular systematics of unitunicate perithecial ascomycetes: the Clavicipitales–Hypocreales connection. *Mycologia* **85:** 912-922.
21. **Tooley, P. W., E. D. Goley, M. M. Carras, R. D. Frederick, E. L. Weber, and G. Kuldau.** 2001. Characterization of *Claviceps* species pathogenic on sorghum by sequence analysis of the β-tubulin gene intron 3 region and EF-1α gene intron 4. *Mycologia* **93:** 541-551.
22. **Tooley, P. W., N. R. O'Neill, E. D. Goley, and M. M. Carras.** 2000. Assessment of diversity in *Claviceps africana* and other *Claviceps* spp. by RAM and AFLP analyses. *Phytopathology* **90:** 1126-1130.

23. **Tsukiboshi, T., T. Shimanuki, and T. Uematsu.** 1999. *Claviceps sorghicola* sp. nov., a destructive ergot pathogen of sorghum in Japan. *Mycol. Res.* **103:** 1403-1408.
24. **Vos, P., R. Hogers, M. Bleeker, M. Reijans, T. van de Lee, M. Hornes, K. A. Frijters, J. Pot, J. Peleman, M. Kuiper, and M. Zabeau.** 1995. AFLP: a new technique for DNA fingerprinting. *Nucleic Acids Res.* **23:** 4407-4414.
25. **Zummo, N., L. M. Gourley, L. W. Trevanthan, and M. S. Gonzalez.** 1998. Occurrence of ergot (sugary disease) incited by a *Sphacelia* sp. on sorghum in Mississippi in 1997. *Plant Dis.* **82:** 590.

Part IV

Striga

30

Striga Control in Sudan: An Integrated Approach

A. G. T. Babiker

Abstract

The parasitic weed *Striga hermonthica* (Del.) Benth is a major constraint to sorghum production in the Sudan. The objective of this study was to test integrated control strategies in both high-input and low-input farming systems. The resistant sorghum cultivar SRN39 experienced less *Striga* infestation than the tolerant cultivar G/H. Intercropping with *Lablab purpureus* reduced *Striga* infestation, dry weight, and the number of capsules. Delayed planting of the intercrop reduced the efficiency of *Striga* control. Urea (95 and 190 kg/ha) and dicamba (300 g/ha) had inconsistent effects on *Striga* growing on tolerant sorghum cultivars. However, satisfactory *Striga* control was achieved with the resistant cultivar. Chlorsulfuron and its tank mix with dicamba effected excellent and persistent control of the parasite on both tolerant and resistant cultivars. Intercropping alone, chlorsulfuron, and its tank mixture with dicamba all increased sorghum yield and yield components. These treatments were more effective on the tolerant cultivar. The effectiveness of the herbicides and the intercrop on weed reduction and sorghum improvement were improved considerably by urea. At low inputs, the *Striga*-resistant cultivar (SRN39) gave the same or greater yields than a *Striga*-tolerant one (G/H); however, with high inputs, the *Striga*-tolerant cultivar outyielded the resistant one. Thus, in a low-input system where low yields are expected, *Striga*-resistant cultivars should be adopted. In high-input, high-expected-yield settings, however, the improved high-yielding *Striga*-tolerant cultivars need to be seriously considered.

Introduction

The genus *Striga* currently includes 35 species with multiple strains as well as physiological and ecological variants (4, 7). Of these species, *S. hermonthica*, a cross pollinator, is the most important in Subsaharan Africa. Two distinct strains of this species, one on millet (*Penisetum glaucum* L. R. Br.) and the other on sorghum (*Sorghum bicolor* L. *Moench*) have been identified (11).

The life cycle of *Striga* is highly coordinated with that of its host. *Striga* seeds, which are produced in large numbers, germinate only after pretreatment (conditioning) in a warm moist environment followed by subsequent reception of a chemical stimulant. Under field conditions, the stimulant is exuded by roots of hosts and some non-host plant species. The emerging *Striga* radicle forms a haustorium following contact with the host root and the reception of a second chemical signal. The haustorium, which marks the initiation of the parasitic mode, contacts and penetrates the host root and connects to the host xylem (10). Cell differentiation and division are essential for haustorium initiation and continuity of development within the host tissue (8).

Striga grows parasitically underground for 6-8 weeks prior to emergence. During this period, the parasite is the most damaging (5). Upon emergence, the parasite switches to a hemi-parasitic mode in which the flow of sap and nutrients from the host is maintained by the high rates of transpiration. The high transpiration rate in *Striga* is attributed to stomata that remain open under a variety of environmental conditions (9). When transpiretion is reduced, the *Striga* plant collapses and dies (6). *Striga* does best under conditions of low soil fertility accompanied by sporadic rains or drought conditions. Thus, the climatic and edaphic conditions in sub-Saharan Africa present an ideal environment for *Striga* spread, survival, and damage. The problem is further compounded by intensive sorghum monocropping and the replacement of the low-yielding, traditional, *Striga*-tolerant cultivars with improved higher-yielding susceptible ones.

In the Sudan, more than 10% of the five million hectares of rain-fed sorghum is heavily infested with *Striga*, and yield losses of 70-100% commonly occur. *Striga* research began in the Sudan in the early 1940s, and focused on cultural practices, including crop rotation, hand weeding, fertilizers, herbicides, herbicide/fertilizer combinations, and *Striga* bait and trap crops (1, 2, 3, 11). These practices reduce *Striga* seed reserves in the soil. The benefits are both gradual and cumulative and must be sustained for several years to have a noticeable impact.

Since the 1970s, research has focused on the development of integrated control measures that are based on *Striga*'s life cycle and the host/parasite relationship. The negative relationship between *Striga* infestation and soil fertility, seed germination, haustorium formation, host attachment, and the dependence on transpiration to obtain nutrients and possibly phytohormones from the host xylem sap all provide avenues that could lead to cost-effective control measures.

Our prime objective has been to develop integrated *Striga*-control methodologies that reduce the seed reserve in the soil, that prevent seed replenishment, and that are technically and economically feasible. These practices include *Striga*-resistant and/or *Striga*-tolerant cultivars as major components. In this report, I describe my observations of the effects of intercropping, fertilizers, and herbicides on *Striga* in irrigated sorghum.

Materials and Methods

General. Sorghum was planted in holes on ridges in artificially *Striga*-infested plots. The ridges were 80 cm apart, and the holes within a ridge were 15 cm apart. The crop was thinned to two plants/hole two weeks after plant-ing.

Intercropping. *Lablab purpureus* (L.) (cv. Brazeli) was planted in holes (three seeds/hole) 15 cm apart, each midway between sorghum planting holes. The intercrop was thinned to two seedlings/hole two weeks after planting. In the first experiment, the cultivars SRN 39, which is *Striga* resistant, and Gadam El Hamam (G/H), which is *Striga* tolerant, were used. *L. purpureus* was planted the same day as the sorghum. In the second experiment, Wad Ahmed, a high-yielding *Striga*-tolerant improved cultivar, was used. *L. purpureus* was planted on the same day as sorghum and again one or two weeks later. Urea at 95 and 190 kg/ha was placed as a side dressing, directly, beneath the sorghum plants two weeks after planting.

Herbicides and fertilizers. Urea (0 or 190 kg/ha) was applied at sowing. The herbicides chlorsulfuron (2.4 g a.i./ha) and dicamba (300 g a.i./ha) were applied separately or in a tank mix, as soil-directed sprays four weeks after planting.

Results

Effects on *Striga*. Emergence of the parasite varied with season and was more intense on G/H than on SRN 39 (Table 30-1). *L. purpureus* planted on the same day as sorghum reduced *Striga* population density by 48–93%, dry weight by 83–97%, and the number of seed capsules by 52–100% (Tables 30-2 and 30-3). *L. purpureus* planted one week after crop sowing reduced *Striga* population density, dry weight, and number of seed capsules by 46, 48, and 32%, respectively. *L. purpureus* planted two weeks after sowing had no significant effect on *Striga* population density and dry weight, but did reduce the number of seed capsules.

Urea, at 95 kg/ha, had no significant effect on *Striga* population density (Table 30-3). However, *Striga* dry weight and the number of seed capsules were reduced by 55 and 41%, respectively. Urea, at 190 kg/ha, reduced *Striga* population density,

TABLE 30-1. Effects of urea and herbicides on *Striga* plant density (plants/m^2) 60 days after sorghum sowing, and on sorghum straw and grain (tons/ha) yield.

Treatment	Sorghum Cultivar[a]	*Striga* Plant Density		Sorghum yield: Straw		Sorghum yield: Grain	
		1991	1992	1991	1992	1991	1992
Untreated	G/H	50	38	6.1	2.7	0.21	0.87
control	SRN 39	34	13	9.4	4.2	0.40	1.9
Urea	G/H	52	26	9.3	6.9	1.2	1.4
(190 kg/ha)	SRN 39	6	5	11	9.0	1.4	2.5
Chlorsulfuron	G/H	5	3	13	8.4	2.1	3.5
(2.4 g/ha)	SRN 39	0	3	9.5	7.9	1.7	1.7
Dicamba	G/H	21	3	9.0	5.1	0.92	1.5
(300 g/ha)	SRN 39	6	3	11	5.8	0.86	1.8
Urea +	G/H	2	4	17	14	4.0	4.6
Chlorsulfuron	SRN 39	0	1	12	10	2.1	2.8
Urea + Dicamba	G/H	35	3	16	7.0	2.0	3.5
	SRN 39	3	1	12	9.0	1.3	2.2
Chlorsulfuron +	G/H	2	5	13	9.8	2.0	3.1
Dicamba	SRN 39	0	0	12	7.4	1.5	3.1
Urea + Dicamba	G/H	2	3	16	13	3.9	3.4
+ Chlorsulfuron	SRN 39	0	0	13	12	1.9	3.2
S.E.				1.0	0.9	0.3	0.4

[a]G/H – Gadam El Hamam.

TABLE 30-2. Effects of *L. purpureus* intercrop on *Striga* population density (shoots/m²), dry weight (g/m²), and seed production (capsules/plant), and number of sorghum heads (× 1000/ha).

Sorghum Cultivar	Intercrop present?	*Striga*				Sorghum Heads		
		Population Density		Dry wt.[b]	Seed[b]	40 DAS	70 DAS	100 DAS
		50 DAS[a]	120 DAS					
SRN 39	No	12	9	25	7	89	102	105
SRN 39	Yes	4	1	1.8	0	90	96	102
Gadam El Hamam	No	50	46	97	15	2	13	27
Gadam El Hamam	Yes	21	3	3.6	0	43	71	71

[a]DAS – days after sowing.
[b]Determined at harvest.

dry weight, and the number of seed capsules by 57, 72, and 94%, respectively. Moreover, urea, at 190 kg/ha, increased the suppressive effects of late-planted *L. purpureus* on *Striga* population density and dry weight (Table 30-3).

Dicamba, urea at 190 kg/ha, and their combination gave satisfactory-to-excellent (62-92%) control of *Striga* on SRN 39. However, with G/H the results were inconsistent (Table 30-1). Chlorsulfuron and its tank mix with dicamba effectively controlled the parasite (77-100%) on both cultivars.

Effects on sorghum. *Flowering.* Unrestricted *Striga* parasitism delayed flowering and reduced the number of heads in G/H and Wad Ahmed, but not in SRN 39 (Tables 30-2 and 30-3). Intercropping with *L. purpureus* considerably decreased time to flowering and increased the number of heads in G/H and Wad Ahmed, but had no significant effect on SRN 39. Simultaneous planting of *L. purpureus* with sorghum was generally more effective than delayed planting (Table 30-3). Urea at 95 and 190 kg/ha applied to sorghum growing alone increased the number of heads by over 200%. The combination of urea and the intercrop did not cause a further increase in number of heads over the urea treatment alone.

Straw yield. Unrestricted *Striga* parasitism significantly reduced the straw yield of all three cultivars (Tables 30-1, 30-2, and 30-3). Untreated SRN39 outyielded G/H, albeit not significantly (Table 30-1). Intercropping alone increased straw yield by over 100%, at all planting dates (Table 30-3). Urea alone, at 95 and at 190 kg/ha, increased straw yield by over 300%. *L. purpureus* planted simultaneously with sorghum and followed by urea increased straw yield less than urea alone. Late plantings of the intercrop did not reduce yield gains due to urea.

Dicamba alone did not result in a significant increase in straw yield of either the susceptible or the resistant cultivars (Table 30-1). Dicamba plus urea outyielded the untreated control, but was not significantly different from the increase due to urea alone. Chlorsulfuron and its tank mix with dicamba consistently increased the yield of G/H significantly, but gave inconsistent results with SRN 39. In 1991, when *Striga* was less damaging and the straw yield of SRN 39 was high, only the tank mix of chlorsulfuron and dicamba, when applied following urea, significantly increased straw yield. However, in 1992 straw yield was significantly increased by all treatments containing chlorsulfuron.

Grain yield. Unrestricted *Striga* parasitism reduced the grain yield of all three cultivars (Tables 30-1 and 30-3). *L. purpureus* sown simultaneously with sorghum, or one week later, without a supplementary urea treatment, increased the grain yield of Wad Ahmed by about 9-fold (Table 30-3). Delaying planting of the intercrop to two weeks after sowing reduced the beneficial effects of the intercrop on grain yield. Urea alone at 95 kg/ha did not significantly increase yield, but urea alone at 190 kg/ha did.

With G/H, chlorsulfuron, regardless of the urea treatment, increased grain yield significantly (Table 30-1). Dicamba alone had negligible effects. All chlorsulfuron or dicamba treatments preceded by urea effected a consistent, significant yield increase (Table 30-1). SRN 39, urea, dicamba, and their combination did not significantly increase yield. Yield increases in the chlorsulfuron treatments were not consistent over time (Table 30-1).

Discussion

Unrestricted *Striga* parasitism delayed flowering and reduced straw and grain yield of all cultivars (Tables 30-1, 30-2, and 30-3). Straw and grain yields were negatively associated with *Striga* population density ($-0.89 < r < -0.56$, $p > 0.5$). At low inputs, both good control of the parasite and high yields were obtained from the *Striga*-resistant cultivar SRN 39, reducing the time needed for weeding.

In general, control of *Striga* in all cultivars by chlorsulfuron, at low rates, and by intercropping, combined with increased crop growth and yield indicate that these treatments have potential utility. Intercropping reduced *Striga* growth, its ability to produce seeds, and its ability

TABLE 30-3. Effects of *L. purpureus* intercrop and urea (kg/ha) on *Striga* population density (plants/m² 60 days after sowing), dry weight (g/m²), seed production (seed capsules/plant), sorghum (cv. Wad Ahmed) heads (× 1000/ha), straw, and grain yield (tons/ha) at harvest.

Intercrop	Urea	*Striga*			Sorghum		
		Population	Dry wt.	Seeds	Heads	Straw	Grain
None	0	35	48	18	20	2.3	0.13
	95	38	22	11	60	9.2	0.81
	190	15	13	1	66	10	1.1
At sorghum sowing	0	4	8.4	8.7	54	5.9	1.3
	95	6	7.8	2.3	61	7.5	1.4
	190	1	0.3	0	56	7.3	1.8
1 week after sorghum sowing	0	19	25	13	47	5.6	1.3
	95	20	13	3.7	67	9.1	1.3
	190	6	1	6.0	73	11	2.6
2 weeks after sorghum sowing	0	37	45	12	31	5.2	0.65
	95	30	14	8.7	73	9.4	1.7
	190	12	4.1	1.3	66	11	1.9
S.E.			4.7	1.9	6.3	0.5	0.3

to debilitate its host (Tables 30-2 and 30-3). The intercrop, perhaps by increasing the microclimate's relative humidity, reduces *Striga* transpiration and, hence, its nutrient supply. Many of the *Striga* seedlings observed shriveled and died under the intercrop. The intercrop also may interfere with pollination.

The decrease in *Striga* control and the increase in sorghum growth and yield with delayed planting of the intercrop (Table 30-3) suggest that the overall effect of the intercrop is a function of its interference with both the host and the parasite. Urea fertilization makes sorghum more competitive. Intercropping with a legume, in addition to its suppressive effects on the parasite, may increase soil fertility and thus benefit subsequent crops.

Intercropping is a valuable, cheap, and effective method for controlling the parasite in localized infestations on relatively small farms. However, its practical utility on large areas is not proven, and chemical control may be more appropriate. Use of high-yielding *Striga*-tolerant cultivars, *e.g.*, Wad Ahmed, and the herbicide chlorsulfuron, in combination with urea and/or dicamba, may be more practical. Work in my laboratory indicates that chlorsulfuron, at the rates used in the field, does not inhibit *Striga* germination. Instead, the germlings have shorter radicles, many of which fail to develop normal haustoria in response to sorghum root macerates. As chlorsulfuron also suppresses the growth of already attached subterranean *Striga* plants, hand weeding of the few emerging *Striga* plants should rapidly deplete the parasite's seed reserves.

Chlorsulfuron and related herbicides have narrow safety margins on sorghum. To attain maximum selectivity and high activity the herbicide has to be applied post-emergence to sorghum and pre-emergence to *Striga*. Post-sowing rains may preclude timely application of the herbicide and, for maximum efficiency, a tank mix of chlorsulfuron with dicamba may have to be used. Delayed rainfall after post-emergence treatments also may reduce herbicide efficacy.

Pre-emergent chlorsulfuron treatments of sorghum indicate considerable herbicide sensitivity even at low rates (Babiker, *unpublished*). However, dressing of sorghum seeds with NA (8-naphthalic anhydride), which is a general herbicide safener, reduced toxicity. A more specific safener could improve product selectivity and make pre-emergence control of *Striga* feasible. Herbicide safeners, unlike herbicide-resistant cultivars, allow more flexibility in cultivar selection, even though they increase costs.

At present, my program is focused on the development of sorghum cultivars with high tolerance to chlorsulfuron and other acetolactate synthase (ALS) inhibitors by using tissue culture techniques with emphasis on improved, high-yielding, widely adopted cultivars. We also are promoting the use of high-yielding *Striga* tolerant/resistant cultivars in combination with cultural practices that decrease *Striga* seed reserves in the soil.

In a comparative study (Babiker, *unpublished*) in 1998 and 1999 involving approximately 4,600 hectares of rain-fed sorghum revealed that the susceptible cultivars Tabat and Mr. White were associated with heavier *Striga* infestations than was the *Striga*-tolerant/resistant cultivar Wad Ahmed. Of the fields cultivated to the susceptible cultivars, 63 and 37% showed moderate and heavy infestations, respectively. Infestation of fields planted to the tolerant/resistant cultivar was low to moderate on 41 and 59% of the cultivated fields, respectively, and no heavy infestations were observed. The heavy infestations occurred on the susceptible cultivars despite fertilization with urea (95 kg/ha), suggesting that *Striga* infestation may increase significantly following repeated planting of highly susceptible cultivars without adequate *Striga* control measures.

Replacing local cultivars, with limited yield potentials, with improved high-yielding ones is urgently needed in light of the increasing population pressure. However, reactions of newly introduced cultivars to *Striga* should be determined prior to any large-scale release. Farmers and policy makers need to understand the seriousness of the problem. Extension workers and farmers should be aware of the ease of *Striga* seed dissemination and should be taught to deal quickly with new infestations.

Efforts to develop new innovative cost-effective *Striga* control measures and to improve those already available should continue. The long-term effects of control measures on *Striga* seed reserves should be included in the evaluation of treatment effects.

Acknowledgments

This work was supported in part by the U.S. Agency for International Development (USAID) through INTSORMIL, the International Sorghum and Millet Collaborative Research Support Program. I thank the conference organizers for financial support for me to attend this conference.

References

1. **Andrews, F. W.** 1945. The parasitism of *Striga hermonthica* on *Sorghum* spp. under irrigation. I. Preliminary results and the effect of heavy and light irrigation on *Striga* attack. *Ann. Appl. Biol.* **32:** 193-200.
2. **Andrews, F. W.** 1946. The parasitism of *Striga hermonthica* (Benth.) on leguminous plants. *Ann. Appl. Biol.* **34:** 267-275.
3. **Last, F. T.** 1960. Incidence of *Striga hermonthica* (Del.) Benth. on two varieties of sorghum differently manured, spaced and thinned. *Trop. Agric.* **37:** 309-319.
4. **Parker, C., and C. R. Riches.** 1993. *Parasitic Weeds of the World. Biology and Control.* CAB International, Wallingford, U.K.
5. **Piterse, A. H., and C. J. Pesch.** 1983. The witchweeds (*Striga* spp.)–A review. *Trop. Agric.* **9:** 9-37.
6. **Press, M. C., J. J. Nour, F. F. Bebawi, and G. R. Stewart.** 1989. Antitranspirant-induced heat stress in the parasitic plant *Striga hermonthica*–A novel method of control. *J. Exper. Bot.* **40:** 585-591.
7. **Rao, V. M. J., and L. J. Musselman.** 1987. Host-specificity in *Striga* and physiological strains, p. 213-250. *In* L. J. Musselman (ed.), *Parasitic Plants in Agriculture*, Vol. 1. CRC Press, Boca Raton, Florida.
8. **Riopel, J. L., and Timko, M. P.** 1995. Haustorial initiation and differentiation, p. 39-79. *In* M.C. Press and J. Graves (eds.), *Parasitic Plants*. Chapman and Hall, London, U.K.
9. **Shah, N., N. Smirnoff, and G. R. Stewart.** 1987. Photosynthesis and stomatal characteristics of *Striga hermonthica* in relation to its parasitic habit. *Physiol. Plant.* **69:** 699-703.
10. **Visser, J. H., and I. Dorr.** 1987. The haustorium, p. 91-106. *In* L. J. Musselman (ed.), *Parasitic Weeds in Agriculture*, Vol. 1. CRC Press, Boca Raton, Florida.
11. **Wilson-Jones, K.** 1952. Three experiments in witchweed control. *J. Exper. Agric.* **20:** 98-102.

31

Striga: Biological Control Strategies for a New Millennium

Dale E. Hess, J. Kroschel, Doulaye Traoré, A. E. M. Elzein, Paul S. Marley, A. A. Abbasher, and C. Diarra

Abstract

Surveys of natural enemies of *Striga* spp. in Africa and India have identified numerous insects (22 families in seven orders) and microorganisms that attack these parasitic weeds. Insects of the genera *Smicronyx* and *Eulocastra* show potential as biological control agents of the parasites. Due to their limited occurrence in Africa and India, some of these insects can be considered for release into new areas. Among fungal pathogens, *Fusarium* spp. are the most commonly associated with diseased striga plants in West Africa. Isolates of these fungi reduce the incidence and reproduction of the weed under both controlled and field conditions. There is a continued need to assess the field efficiency of the known antagonists and to develop delivery systems appropriate for subsistence farming. Bacteria have been isolated from the host rhizosphere and "striga-suppressive soils." Bacteria of the genera *Pseudomonas* and *Azospirillum* have shown bio-control potential, but available information is limited. Al-though biological control of striga appears feasible, re-search is still needed in all areas from the identification of potential biocontrol agents to the development of effi-cient formulations for application as part of an integrated pest management program.

Introduction

Striga spp. are a major biological constraint to production of cereals and grain legumes in Africa. Large portions of cereal production zones in sub-Saharan Africa are infested by these parasitic weeds, with the greatest damage occurring in the Sahelian and Savannah zones. Among the species parasitizing graminaceous crops, *Striga hermonthica* is the most important, whereas *Striga gesnerioides* is a major biotic constraint to cowpea production in the Sahel, the Sudanese Savannah, and the northern Guinea Savannah.

Striga management approaches include cultural practices, fertilizers, herbicides, and resistant crop varieties, but generally no single method provides an acceptable level of control (18, 19, 31). An integrated striga management strategy is required and would be strengthened by the use of natural enemies as biocontrol agents (2, 13).

Natural enemies of *Striga* spp. have received increased attention in recent years due to the success achieved in the biological control of nonparasitic weeds. The first surveys were initiated after the accidental introduction of *Striga asiatica* into the United States in 1956. Biological control of striga has been reviewed previously (*e.g.*, 6, 16), with more recent summaries of the potential of natural antagonists (insects and microorganisms) for parasitic weed control (2, 24, 33).

Natural Enemies of *Striga* spp.

Insects. Early reports of insects attacking striga came from India (21) and Africa (12, 43). Following the discovery of *S. asiatica* in the United States in 1956, the United States Department of Agriculture (USDA) funded an extensive survey of insects by the Commonwealth Institute of Biological Control (presently CAB International or CABI) in India (29, 32) and East Africa (17). Insect herbivores of *Striga* spp. also have been reported by Markham (26), Kroschel *et al.* (22), and Traoré *et al.* (39), among others. However many of these insects, *e.g.*, *Spodoptera* spp. and *Helicoverpa armigera*, are polyphagous species and important pests of crops (16). Oligo- and monophagous herbivores of greatest interest include *Eulocastra argentisparsa* in India, *Smicronyx* spp. in

India and Africa, *Junonia orithya* in Asia and Africa, and *Ophiomyia strigalis* in East Africa (24, 39).

Microorganisms. In the early part of this century, Wager (40) obtained *Pythium ultimum* in culture from striga seeds in South Africa. Nag Raj (29) reported the occurrence of *Alternaria* sp., *Cercospora* sp., and *Phoma* sp. on *S. asiatica* and *Striga densiflora* in India. Meister and Eplee (28) isolated five fungi in North Carolina: *Curvularia geniculata, Fusarium roseum, Fusarium solani, Rhizoctonia solani,* and *Sclerotium rolfsii.* Zummo (44) reported three diseases of *S. hermonthica* in west Africa, including a leaf spot, caused by an undescribed species of *Cercospora*; a vascular wilt caused by *Fusarium equiseti*; and a stem lesion caused by *Phoma* sp. However, none of these fungi was evaluated for its potential use as an agent to control *Striga* spp. In recent years, surveys have been conducted to identify pathogens attacking *Striga* spp. in west Africa (2, 10, 23). *Fusarium* spp. and *Curvularia* spp. were the two most common genera isolated from infected *S. hermonthica* and *S. gesnerioides.* Among the Fusaria, *F. oxysporum* was the dominant (93%) species (1).

Limited work has been done with bacteria. *Azospirillum brasilense,* isolated from a sorghum field in Mali, significantly inhibited *in vitro* germination of *S. hermonthica* seed (8). One strain also exhibited a plant growth-promoting effect, suggesting that under field conditions these bacteria might significantly reduce parasite infestation and enhance crop growth. Oswald *et al.* (30) found that inoculation of maize seedlings (Kenyan land race "Local White") with plant growth promoting rhizobacteria, significantly reduced striga germination. In contrast, this effect was not observed with the commercial maize hybrid (Pioneer 3152), suggesting that the Kenyan plant growth promoting rhizobacteria were specific for the local maize variety. A laboratory test was developed to screen *Pseudomonas syringae* pv. *glycinea* for its ability to promote suicidal germination of *Striga* spp. seed (7). Ethylene-producing bacteria also have a potential role in biological control strategies (7).

Biological Control Strategies

Natural enemies can be important factors in limiting the number and distribution of most living organisms. The basis of biological control is the calculated use of natural enemies to suppress the growth or reduce the population of a target pest species.

Three biocontrol approaches have been described for weeds (42). In the augmentative approach, the numbers of a preexisting natural enemy are increased through capture or mass rearing and release into the target zone. In classic, or inoculative, biocontrol, host-specific natural enemies are released into areas in which they did not previously occur. This approach may be appropriate if insects are used for striga biocontrol (24). The inundative, or mycoherbicidal, approach is applicable when biocontrol agents can be mass produced under controlled conditions. The goal of this strategy is to overwhelm the target weed with large numbers of a natural enemy and is well suited for phytopathogenic fungi. Under favorable climatic conditions, application of massive doses of inoculum to a weed population creates a rapid and severe epidemic. In some cases, mycoherbicides can control weeds in annual crops as well as or better than chemical herbicides. Effective mycoherbicides are highly specific, rapid, and effective. They also may control weeds in cropping situations where chemicals are insufficient.

Prospects for Biological Control of *Striga* spp.

Insects. The effects of *Ophiomyia strigalis* on the host plant are too insignificant for effective biocontrol. *Junonia orithya* occurs naturally in both India and Africa and cannot be considered for classical biocontrol. It also is unlikely that this insect can be efficiently reared for use in an inundative approach (24). CABI implemented classical biological control in Ethiopia in 1974 and 1978 with the importation (from India) and release of *Smicronyx albovariegatus* and *Eulocastra argentisparsa*. However, follow-through of the study was problematic, and to date there is no evidence of success (24).

The weevils *Smicronyx umbrinus* and *Smicronyx guineanus* and the butterfly *J. orithya* have recently been

TABLE 31-1. Effect of *Fusarium* spp. applied as foliar spray and incorporated maize meal-sand (MMS) on emergence of *Striga hermonthica* in pots.

Fungal Treatment	% Reduction in Emerged Striga/Pot[a]	
	Foliar Spray[b]	Soil Incorporation[c]
F. oxysporum (DCP)	44 (3.8)[d]	57 (4.0)
F. acuminatum	62 (4.1)	58 (4.0)
F. equiseti	58 (4.1)	59 (4.1)
F. oxysporum (PSM-197)	89 (4.5)	100 (4.6)
Fusarium sp. (PSM-297)	57 (4.0)	56 (4.0)
Control	0	0
LSD (p = 0.05)	(0.08)	(0.01)

[a]Values are the means of three replicates.
[b]Foliar spray (cfu = 1×10^6/ml) was applied to emerged striga until run-off (70 days after sowing).
[c]MMS (cfu = 1×10^6/ml) was incorporated into striga-infested soil before sowing sorghum.
[d]Values in parentheses are natural logarithmic transformed data.

evaluated for potential as biocontrol agents in Burkina Faso and Ghana (20, 22, 39). Good synchrony exists between the active stages of the weevils and the period of striga growth in Burkina Faso (38) and Ghana (20). Galls were found in every field visited, with a mean percentage galling of 11 to 52%. Larvae were located in the upper stratum of the striga plant, where they destroyed large numbers of seed (37). Since *S. guineanus* spends almost half (47%) of its time on the striga inflorescence (bud, corolla and calyx) and ≥ 80% of its time in the upper portion of the plant, these larvae are prone to feed, copulate, and lay eggs on striga reproductive organs (39).

However, the populations of these insect herbivores in their natural habitats are very low and do not significantly reduce the overall striga population (20). Smith *et al.* (34) evaluated the biocontrol potential of *S. umbrinus* by utilizing a simulation model for *S. hermonthica* population dynamics. They concluded that when *S. umbrinus* is used as the sole control agent, it would have to destroy approximately 95% of the *S. hermonthica* seeds produced each year to reduce the density of emerging plants by 50%. In a more recent publication, Smith and Webb (35) corrected some of these calculations and reduced their estimate of the number of seeds that need to be destroyed from 95% to 70-80%. In northern Ghana, Jost *et al.* (20) estimated an average reduction in seed production of only 17% occurred, a value unlikely to significantly reduce established striga infestations.

Microorganisms. Fungi have received more attention than other microorganisms isolated from infected striga plants because they can be aggressive, host-specific parasites (9, 36). In addition, it is usually simple to grow them under *in vitro* conditions, and large amounts of inoculum can be prepared relatively quickly and easily.

During the last decade, surveys have been conducted in Burkina Faso, Ghana, Mali, Niger, Nigeria, and the Sudan (1, 3-5, 10), resulting in the isolation and identification of numerous fungi and several bacteria. Numerous soilborne *Fusarium* spp. are potential biocontrol agents: *F. nygamai* (4), *F. semitectum* var. *majus* (3), *F. oxysporum* (1, 10, 14, 23, 27), *F. equiseti* (3, 27), *F. acuminatum* (27), and *F. verticillioides* (1).

TABLE 31-2. Effect of temperature on percentage viability of *Fusarium oxysporum* propagules in the "Pesta" granular formation after 12 months of storage.

Propagule	Temperature	
	4°C	23°C
Microconidia	34[a] ± 2.7 b	4.7 ± 0.47 c
Microconidia and mycelia	66 ± 8.8 b	8.2 ± 0.74 c
Fresh chlamydospores	95 ± 1.9 a	58 ± 6.1 b
Dry chlamydospores	100 ± 0 a	67 ± 11 b

[a]Mean ± 1 S.E. Values followed by the same letter are not significantly different at $p = 0.05$.

Host specificity is of utmost importance when considering a plant pathogen for biological control. Irrespective of potential benefits, the safety of non-target native and cultivated plants must be ensured prior to use. In host range tests, indigenous *F. oxysporum* from Burkina Faso, Ghana, Mali, and Niger infected only *Striga* spp. and no crops or vegetables (1, 10; Elzein, *unpublished*). An important constraint to the use of *Fusarium* spp. as biological control agents is their ability to produce mycotoxins that may lead to cancer or toxicoses in humans and domestic animals (25). *F. oxysporum* does not produce any of the major *Fusarium* mycotoxins, *e.g.*, fumonisins, deoxynivalenol, nivalenol, T-2 toxin, and zearalenone, making it a potential candidate for the biological control of *Striga* spp. (10, 23).

Similar application techniques apply to both mycoherbicides and chemical herbicides. Delivery systems must be easy, effective, and compatible with available

TABLE 31-3. Reduction of *Striga hermonthica* and increase in sorghum dry weight as influenced by different *Fusarium* spp. applied presowing using different inoculum substrates.

Fusarium isolate[a]	Striga shoots/pot		Infestation reduction (%)	Sorghum shoot dry weight (g)		Biomass increase (%)
	Control	Treated		Control	Treated	
F. nygamai[b]	17	1.2	93	1.0	18	1700
F. semitectum var. majus[b]	21	1.2	94	1.8	45	2400
F. oxysporum M12-4A[c]	2.3	0	100	0.3	45	400
F. oxysporum isolate 1[d]	23	2.5	88	15	50	240
F. oxysporum isolate 2[d]	23	0.2	98	15	57	290
F. solani[d]	23	5.3	76	15	48	230

[a]*Fusarium nygamai*, *F. semitectum*, and *F. oxysporum* (1 and 2) were cultured on autoclaved sorghum grain and applied presowing at a rate of 10 g per kg soil; *F. oxysporum* (M12-4A) was cultured on autoclaved sorghum straw and applied presowing at a rate of 3 g per kg of soil; all *Fusarium* isolates significantly reduced striga emergence and increased sorghum shoot dry weight at $p \leq 0.05$.

[b]Isolated from diseased *S. hermonthica* from the Sudan (3).

[c]Isolated from diseased *S. hermonthica* from Mali (10).

[d]Isolated from diseased *S. hermonthica* in Ghana (23).

agricultural equipment. The simplest mycoherbicide delivery system is an aqueous solution of fungal spores, usually applied at a rate of 900-1000 L/ha with 10^5-10^7 spores/ml. With few exceptions, liquid formulations of bioherbicides are best suited for use as post-emergence sprays and are used primarily to incite leaf and stem diseases (2).

Pathogens that infect at or below the soil line are probably best delivered in a granular formulation applied to the soil. A simple approach is in the form of infested cereal grains (Table 31-1), but formulated granules have the advantage of a longer shelf-life and a reduction in the quantity of inoculum applied to the field. Wheat flour-kaolin granules, "Pesta," prepared as described by Connick *et al.* (11), and alginate granules, prepared by a method modified from that of Walker and Connick (41), were tested with a chlamydospore-rich biomass of *F. oxysporum* as inoculum. The application of 2 g/pot (4 kg of soil) of both fresh and dried biomass in "Pesta" significantly reduced the number of emerged *S. hermonthica* by 78% and 57%, respectively. As a result, the growth of the host sorghum was significantly improved. Chlamydospore-rich biomass also is the preferred inoculum for long-term viable and storable "Pesta" granules, especially when stored at low temperatures. Up to 100% of the initial colony-forming units were recovered after 12 months of storage at 4°C (15; Table 31-2). The shelf-life of Pesta granules containing chlamydospores should enhance the acceptance and commercial potential of the product.

Research with *Fusarium* spp. has demonstrated that, applied pre-sowing, they can reduce *Striga* spp. incidence under controlled or field conditions (1, 4, 5, 10, 14, 23, 27). These fungi attacked all *Striga* growth stages from ungerminated seeds to flowering stalks. Thus, *Fusarium* spp. can be very effective in reducing the seed bank of the parasite by destruction of the seeds and prevention of reproduction. Since significant damage to host plants occurs while the parasite is still underground, use of *Fusarium* spp. in biocontrol can contribute to improving crop yield (Table 31-3).

Pathogenicity tests have shown that one *Fusarium* species can control more than one *Striga* species (2). *Fusarium nygamai*, *F. oxysporum*, and *F. solani* isolated from infected *S. hermonthica* plants from Sudan and Ghana all reduced emergence of *S. asiatica* and *S. gesnerioides* and improved growth of the host crops (maize and cowpea). A frequent limitation of bioherbicides, the inability to control more than one weed species, is not a constraint for *Fusarium* spp.

Recommendations for Future Work

- Systematic field surveys to identify microorganisms, including bacteria, associated with *Striga* spp. that might be suitable as biological control agents.
- Continued development of effective, economical, environmentally friendly, and technically feasible bioherbicide formulations.
- Continued development/improvement of delivery systems, *e.g.*, the "Pesta" granular mycoherbicide (*F. oxysporum*), through amendments to increase spore germination and improve pathogen stability and virulence.
- Field trials to evaluate indigenous *Fusarium* isolates in collaboration with national research programs.
- Implementation of classical biological control of striga in Africa through field releases of oligo- and monophagous herbivores collected from India.

Although biocontrol approaches have the potential to significantly contribute to the control of parasitic weeds, successful striga management will result from the development of integrated control systems that utilize multiple control technologies and that are carefully applied by farmers in targeted agroecological zones.

References

1. **Abbasher, A. A., D. E. Hess, and J. Sauerborn.** 1998. Fungal pathogens for biological control of *Striga hermonthica* on sorghum and pearl millet in west Africa. *Afr. Crop Sci. J.* **6:** 179-188.
2. **Abbasher, A. A., D. E. Hess, J. Sauerborn, and J. Kroschel.** 1999. Biological control of *Striga*, p. 47-61. *In* D. E. Hess and J. M. Lenné (eds.), *Report on the ICRISAT Sector Review for* Striga *Control in Sorghum and Millet.* ICRISAT, Bamako, Mali.
3. **Abbasher, A. A., J. Kroschel, and J. Sauerborn.** 1995. Microorganisms of *Striga hermonthica* in northern Ghana with potential as biocontrol agents. *Biocontrol Sci. Technol.* **5:** 157-161.
4. **Abbasher, A. A., and J. Sauerborn.** 1992. *Fusarium nygamai*, a potential bioherbicide for *Striga hermonthica* control in sorghum. *Biocontrol* **2:** 291-296.
5. **Adeoti, A. A.** 1993. Fungi associated with *Striga hermonthica*. *Striga Newsl.* **4:** 9.
6. **Bashir, M. O.** 1987. The potential for biocontrol of witchweeds, p. 183-206. *In* L. J. Musselman (ed.), *Parasitic Weeds in Agriculture.* Vol. 1 *Striga*. CRC Press, Boca Raton, Florida.
7. **Berner, D. K., N. W. Schaad, and B. Völksch.** 1998. Ethylene-producing bacteria for *Striga* spp. control. *Phytopathology* **88:** S7.
8. **Bouillant, M. L., M. Kabir, G. Alexandre, C. Jacoud, L. Miche, R. Bally, G. Sallé, and O. Ouédraogo.** 1996.

Inhibition of *Striga* seed germination by bacteria isolated from soil under sorghum field, p. 900-904. *In* M. T. Moreno J. I. Cubero, D. Berner, D. Joel, L. J. Musselman, and C. Parker (eds.), *Advances in Parasitic Plants Research.* Dirección General de Investigación Agraria, Sevilla, Spain.

9. **Charudattan, R.** 1985. The use of natural and genetically altered strains of pathogens for weed control, p. 347-372. *In* M. A. Hoy and D. C. Herzog (eds.), *Biological Control in Agriculture IPM Systems.* Academic Press, Orlando, Florida.
10. **Ciotola, M., A. K. Watson, and S. G. Hallett.** 1995. Discovery of an isolate of *Fusarium oxysporum* with potential to control *Striga hermonthica* in Africa. *Weed Res.* **35:** 303-309.
11. **Connick, W. J., Jr., C. D. Boyette, and J. R. McAlpine.** 1991. Formulation of mycoherbicides using a pasta-like process. *Biol. Control* **1:** 281-287.
12. **Davidson, A.** 1963. Insects attacking *Striga* in Kenya. *Nature* **197:** 923.
13. **Diarra, C.** 1999. Biological control of *Striga hermonthica* in the Sahel, p. 103-115. *In* J. Kroschel, H. Mercer-Quarshie, and J. Sauerborn (eds.), *Advances in Parasitic Weed Control at On-Farm Level. Vol. 1. Joint Action to Control* Striga *in Africa.* Margraf Verlag, Weikersheim, Germany.
14. **Diarra, C., M. Ciotola, S. G. Hallett, D. E. Hess, and A. K. Watson.** 1996. Field efficacy of *Fusarium oxysporum* for the control of *Striga hermonthica. Nuisibles-Pests-Pragas* **4:** 257-263.
15. **Elzein, A., J. Sauerborn, D. Müller-Stöver, and J. Kroschel.** 2000. Effect of storage temperature, granule size, and inoculum type on the viability of *Fusarium oxysporum*, a pathogen of *Striga hermonthica*, encapsulated in wheat-kaolin ("Pesta") granules, p. 179-180. In *Book of Abstracts, IIIrd Internatl. Weed Science Cong. (6-11 June 2000, Foz do Iguassu, Brazil).* Intl. Weed Science Society, Corvallis, Oregon.
16. **Greathead, D. J.** 1984. The natural enemies of *Striga* spp. and their utilization as biological control agents, p. 133-166. *In* E. S. Ayensu, H. Doggett, R.D. Keynes, J. Marton-Lefevre, L. J. Musselman, C. Parker, and A. Pickering, (eds.), Striga *Biology and Control.* IRL Press Ltd., Oxford, UK.
17. **Greathead, D. J., and J. E. D. Milner.** 1971. A survey of *Striga* spp. and their natural enemies in east Africa with a discussion on the possibilities of biological control. *Trop. Agric.* **48:** 111-124.
18. **Haussmann, B. I. G., D. E. Hess, H. G. Welz, and H. H. Geiger.** 2000. Improved methodologies for breeding striga-resistant sorghums. *Field Crops Res.* **66:** 195-211.
19. **Hess, D. E., and B. I. G. Haussmann.** 1999. Status quo of *Striga* control: Prevention, mechanical and biological control methods & host plant resistance, p. 75-87. *In* J. Kroschel, H. Mercer-Quarshie, and J. Sauerborn, (eds.), *Advances in Parasitic Weed Control at On-Farm Level. Vol. 1. Joint Action to Control* Striga *in Africa.* Margraf Verlag, Weikersheim, Germany.
20. **Jost, A., J. Kroschel, and J. Sauerborn.** 1996. Studies on *Smicronyx* spp. and *Junonia orithya* and their potential for biocontrol of *Striga hermonthica* in northern Ghana, p. 888-897. *In* M. T. Moreno J. I. Cubero, D. Berner, D. Joel, L. J. Musselman, and C. Parker (eds.), *Advances in Parasitic Plants Research.* Dirección General de Investigación Agraria, Sevilla, Spain.
21. **Khan, M. Q., and D. V. Murthy.** 1955. *Smicronyx albovariegatus* Faust (Curculioidae) Coleoptera) on *Striga* spp. *Indian J. Entomol.* **17:** 362.
22. **Kroschel, J., A. A. Abbasher, and J. Sauerborn.** 1995. Herbivores of *Striga hermonthica* in northern Ghana and approaches of their use as biological control agents. *Biocontrol Sci. Technol.* **5:** 163-164.
23. **Kroschel, J., A. Hundt, A. A. Abbasher, and J. Sauerborn.** 1996. Pathogenicity of fungi collected in northern Ghana to *Striga hermonthica. Weed Res.* **36:** 515-520.
24. **Kroschel, J., A. Jost, and J. Sauerborn.** 1999. Insects for *Striga* control–Possibilities and constraints, p. 117-132. *In* J. Kroschel, H. Mercer-Quarshie, and J. Sauerborn, (eds.), *Advances in Parasitic Weed Control at On-Farm Level. Vol. 1. Joint Action to Control* Striga *in Africa.* Margraf Verlag, Weikersheim, Germany.
25. **Marasas, W. F. O., P. E. Nelson, and T. A. Toussoun.** 1984. *Toxigenic* Fusarium *Species: Identity and Mycotoxicology.* Pennsylvania State Univ. Press, University Park, Pennsylvania.
26. **Markham, R. H.** 1985. *Possibilities for the Biological Control of* Striga *Species in the Sahel.* Commonwealth Institute of Biological Control, Nairobi, Kenya, 16 pp.
27. **Marley, P. S., S. M. Ahmed, J. A. Y. Shebayan, and S. T. O. Lagoke.** 1999. Isolation of *Fusarium oxysporum* with potential for biocontrol of the witchweed in the Nigerian savanna. *Biocontrol Sci. Technol.* **9:** 159-163.
28. **Meister, C. W., and R. E. Eplee.** 1971. Five new fungal pathogens of witchweed (*Striga lutea*). *Plant Dis. Rept.* **55:** 861-863.
29. **Nag Raj, T. R.** 1966. Fungi occurring on witchweed in India. *Commonwealth Inst. Biol. Control Tech. Comm.* **7:** 75-80.
30. **Oswald, M., J. Sauerborn, A. Oswald, and P. Mbogo Okoth.** 2000. Interaction of plant growth promoting rhizobacteria (PGPR) with maize and *Striga hermonthica* (Del.) Benth. seeds, p. 262. In *Book of Abstracts, IIIrd Intl. Weed Science Cong. (6-11 June 2000, Foz do Iguassu, Brazil).* Intl. Weed Science Society, Corvallis, Oregon.
31. **Ransom, J.** 1999. The status quo of *Striga* control: Cultural, chemical and integrated aspects, p. 133-143. *In* J. Kroschel, H. Mercer-Quarshie, and J. Sauerborn, (eds.), *Advances in Parasitic Weed Control at On-Farm Level. Vol. 1. Joint Action to Control* Striga *in Africa.* Margraf Verlag, Weikersheim, Germany.
32. **Sankaran, T., and V. P. Rao.** 1966. Insects attacking witchweed (*Striga*) in India. *Commonwealth Inst. Biol. Control Tech. Bull.* **7:** 63-73.
33. **Sauerborn, J., A. A. Abbasher, and J. Kroschel.** 1994. Biological control of parasitic weeds by phytopathogenic fungi, p. 545-549. *In* A. H. Pieterse, J. A. C. Verkleij, and S. J. ter Borg (eds.), *Proc. 3rd Intl. Workshop on* Orobanche *and Related* Striga *Research.* The Royal Tropical Institute, Amsterdam, The Netherlands.
34. **Smith, M. C., J. Holt, and M. Webb.** 1993. Population model of the parasitic weed *Striga hermonthica* (Scrophulariaceae) to investigate the potential of *Smicronyx umbri-*

nus (Coleoptera: Curculionidae) for biological control in Mali. *Crop Protect.* **12:** 470- 476.

35. **Smith, M. C., and M. Webb.** 1996. Estimation of the seed bank of *Striga* spp. (Scrophulariaceae) in Malian fields and the implications for a model of biocontrol of *S. hermonthica*. *Weed Res.* **36:** 85-92.
36. **Templeton, G. E.** 1982. Status of weed control with plant pathogens, p. 29-34. *In* R. Charudattan and H. L. Walker (eds.), *Biological Control of Weeds with Plant Pathogens*. John Wiley and Sons, New York City, New York.
37. **Traoré, D., C. Vincent, and R. K. Stewart.** 1994. Dispersion pattern of adults and vertical distribution of larvae of *Smicronyx guineanus* Voss and *Sm. umbrinus* Hustache (Coleoptera: Curculionidae) on *Striga hermonthica* (Del.) Benth (Scrophulariaceae). *Insect Sci.* **15:** 479-485.
38. **Traoré, D., C. Vincent, and R. K. Stewart.** 1996. Association and synchrony of *Smicronyx guineanus* Voss, *Sm. umbrinus* Hustache (Coleoptera: Curculionidae) and *Striga hermonthica* (Del.) Benth. (Scrophulariaceae). *Biol. Control* **7:** 307-315.
39. **Traoré, D., C. Vincent, and R. K. Stewart.** 1999. *Smicronyx guineanus* Voss and *S. umbrinus* Hustache (Coleoptera: Curculionidae): Potential biocontrol agents of *Striga hermonthica* (Del.) Benth. (Scrophulariaceae), p. 105-115. *In* D. E. Hess and J. M. Lenné (eds.), *Report on the ICRISAT Sector Review for* Striga *Control in Sorghum and Millet.* ICRISAT, Bamako, Mali.
40. **Wager, V. A.** 1931. Diseases of plants in South Africa due to members of Pythiaceae. *U. So. Afr. Dept. Agric. Bull.* **105:** 1-43.
41. **Walker, H. L., and W. J. Connick, Jr.** 1983. Sodium alginate for production and formulation of mycoherbicides. *Weed Sci.* **33:** 333-338.
42. **Wapshere, A. J.** 1982. Biological control of weeds, p. 47-56. *In* W. Holzner, and N. Numata (eds.), *Biology and Ecology of Weeds*. W. Junk, The Hague, The Netherlands.
43. **Williams, C. N., and G. H. Caswell.** 1959. An insect attacking *Striga*. *Nature* **184:** 1668.
44. **Zummo, N.** 1977. Diseases of giant witchweed, *Striga hermonthica* in West Africa. *Plant Dis. Rept.* **61:** 428-430.

Part V

Pathogen Variability

32

Sorghum Grain Mold: Through the 1990s into the New Millennium

Ranajit Bandyopadhyay, Christopher R. Little, Ralph D. Waniska, and David R. Butler

Abstract

We review progress in research on sorghum grain molds during the past ten years. The complexity of the disease is accentuated by variability within the major mold-causing organisms and the predisposition of grain attacked by head bugs to infection by mold fungi. Mold-damaged grains, some containing mycotoxins, pose health hazards to both humans and animals. Recent studies on epidemiology have emphasized weather-mold relationships. Major research efforts to manage grain molds still rely on host plant resistance. Significant progress has been made in understanding the mechanisms of resistance, and new studies are underway to further enhance these understandings at the molecular level. Other approaches for mold management, *e.g.*, timely harvest and grain processing, also are possible. We think that grain mold management programs should consider the socioeconomic, marketing, organizational, and institutional framework within which the farmers operate. The limitations imposed by grain molds must be overcome if increased production and utilization of sorghum is to occur in warm, humid environments.

Introduction

Grain mold, caused by a complex of fungi (46), continues to be one of the most intractable biotic constraints to sorghum improvement, production, and utilization worldwide. Globally, grain mold causes an estimated loss of $130 million annually (19). Considerable knowledge has been generated since the disease gained importance with the widespread use of high-yielding, short-season cultivars. Research during the 1970s and 1980s focused on etiology, biology of the pathogens, infection pathways of major causal fungi, host plant resistance, and other control methods such as chemical control (13, 72). However, currently recommended control methods, including host plant resistance, have had limited success in significantly reducing the disease in farmers' fields. Research in the 1990s emphasized etiology, epidemiology, and mechanisms of resistance, which were areas for future research that were identified in the last global review (13). We focus on new knowledge generated, its potential application(s), and research needs in these three areas. We also suggest other technical and institutional options for grain mold management.

Etiology

Pathogen variability. *Gibberella fujikuroi (Fusarium* Section *Liseola)* is an important grain mold fungus infecting sorghum grain, and considerable genetic diversity is known in this fungus (31, 32). Seven of the eight known mating populations, or biological species, of the fungus have been isolated from sorghum from different parts of the world. Among these, mating population F (also termed *Fusarium thapsinum* or *Gibberella thapsina*), is the most common on sorghum, and mating population A (also termed *Fusarium moniliforme, Fusarium verticillioides*, or *Gibberella moniliformis*) is the most common on maize. However, these two mating populations do not differ in their ability to cause stalk rot on either maize or sorghum (25, 26). Resistance appears to be general across the species and not strain specific (37). Differences in pathogenicity have been reported in *Curvularia lunata* isolated from different geographical areas in India (64). There is a need to determine the frequency of occurrence of the different *G. fujikuroi* mating populations in India, primarily with respect to mycotoxin and virulence.

Head bug × grain mold interaction. Head bugs feed *(Calocoris angustatus* in India) or oviposite *(Eurystylus oldi* in west Africa) on developing grains. Head-bug injury provides a site through which mold fungi can enter and colonize the grain. Sap exuded from head-bug injury sites provides a substrate for fungal growth on the grain surface. Head-bug injured grains are more prone to damage by mold fungi. Even slight head-bug injury renders most mold-resistant genotypes susceptible by breaching their resistance. Insect damage significantly increases the number of fungal colonies associated with grain and alters their relative frequency. Following insect damage, *G. fujikuroi*, *Phoma sorghina*, and *Cochliobolus lunatus* were the most common fungi in decreasing order of frequency in comparison to the relative frequency on undamaged grain in which the positions of *P. sorghina* and *G. fujikuroi* were reversed (39). Some factors, *e.g.*, grain hardness, that are associated with grain mold resistance also confer resistance to head bugs. Thus, it should be possible to develop combined resistance to both grain mold and head bugs (56).

Sorghum-Mold Interactions at the Molecular Level

Mechanisms of resistance and defense. The first line of defense to fungal conidia is the morphological or physical characteristics of the seed, glume, and/or panicle. These preformed attributes primarily contribute to grain weathering resistance. If infection occurs during earlier stages of seed development, *i.e.*, prior to black layer formation, then the disease is more properly defined as grain mold (13).

When a conidium makes contact with the lodicule, lemmal tissue, and/or ovary base of the sorghum spikelet, the presence of the fungus may be detected by signaling mechanisms within the affected and surrounding cells even before penetration has occurred (8, 16). When the cell wall and plasmalemma are breached by facultatively necrotrophic pathogens, *e.g.*, *F. thapsinum* or *C. lunata*, the resulting stress may be recognized as attempted infection and appropriate physiological modifications may result, *i.e.*, a level of resistance, or not, *i.e.*, a level of susceptibility. If the plant activates responses such as H_2O_2 generation, OH^- radical formation (changes redox status and cross-linking in the cell wall), influx of various anions, efflux of Ca^{++}, lipid peroxidation, or enhanced phospholipase activity, then cell death and/or the accumulation of salicylic acid, ethylene, or jasmonate may result. These processes are generally associated with dying cells, whether undergoing a hypersensitive response or being killed by a necrotrophic pathogen (16, 30).

Signal transduction mechanisms, which elicit expression of defense response genes, activate defense mRNAs. The defense mRNAs are translated into protein products that may function in metabolic pathways yielding fungistatic or fungitoxic products, or that directly affect the integrity and viability of the invading pathogen. These gene products may have observable *in vitro* effects against fungi. However, it is difficult to demonstrate a significant correlation between the induction of any single defense gene and fungistasis, as resistance commonly requires many different defense elements working in concert (27, 30). This concept of multiple synergistic interactions should guide our perception of resistance to grain mold.

Genes involved in phenylpropanoid and flavonoid metabolism. Genes associated with the phenylpropanoid and flavonoid pathways that are coordinately regulated are phenylalanine ammonia-lyase (PAL), chalcone synthase (CHS), chalcone isomerase, and dihydroflavonol reductase. Expression of these genes is normally regulated at the transcriptional level via ultraviolet and visible light. In many plant-fungal interactions, inoculation of "resistant" and "susceptible" plants with appropriate fungal pathogens, and nonspecific and/or specific elicitors changes the expression of defense-related genes (9, 27, 47).

Secondary metabolites, whether preformed or induced, may be most important during early infection (50). Examples of preformed secondary metabolites include flavan-4-ols and pro-anthocyanins (tannins) (22). Flavonoid compounds induced in sorghum include the 3-deoxyanthocyanins. These compounds, *e.g.*, the sorghum phytoalexin apigeninidin, often are red pigments (62, 63). Snyder *et al.* (62) showed that red-pigmented compounds accumulated below infection structures of *Colletotrichum graminicola.* The concentration of luteolinidin in these inclusions was about 150 μM; much more than 9 μM required for *in vitro* fungitoxic effects. Thus, sorghum plants produce secondary metabolites that can directly inhibit or retard further infection, growth, and ramification of fungal hyphae.

Schutt and Netzley (54) tested apigeninidin, bound and unbound apiforol (a flavan-4-ol), and naringenin chalcone against several saprophytic and facultatively pathogenic fungi. They found that apiforol, apiforol-7-rhamnoglucoside, and naringenin had no significant activity against any of the fungi tested. Apigeninidin (synthesized *in vitro)* had no observable effects on *Sclerotium rolfsii, Phytophthora* spp., *Rhizoctonia solani*, or *Rhizo-*

pus stolonifer, but, at 3.6 μM, inhibited the growth of *Alternaria solani*, *Fusarium oxysporum*, *Gibberella zeae*, and *Gliocladium roseum*. Unfortunately their experiments did not include any of the major sorghum grain-mold fungi.

PAL is a multisubunit enzyme that catalyzes the first step in phenylpropanoid biosynthesis, the conversion of L-phenylalanine to *t*-cinnamic acid via a deamination (10). PAL is highly inducible at the transcriptional level, and induction of PAL genes indicates the potential for increased biosynthetic flux through the pathway (17). In the soybean/*Phytophthora megasperma* f. sp. *glycinea* system, PAL activity increased more than four-fold after bean cell cultures were treated with a β-glucan elicitor. CHS increased more than five-fold, but activity levels of other enzymes in the pathway, *e.g.*, *p*-coumaryl:CoA ligase were not significantly altered (11). Ultraviolet or visible light, wounding, nutrient and salt stress, nitrogen starvation, fungal elicitors, and fungal infection all can induce PAL activity relatively quickly (10, 17, 47). Sorghum seedlings of lines resistant to downy mildew inoculated with *Peronosclerospora sorghi* synthesized higher levels of PAL mRNA than did seedlings of susceptible lines. *Sporisorium reilianum* did not induce the accumulation of PAL mRNA in either meristematic or nonmeristematic resistant lines (9). PAL mRNA accumulated in sorghum within a few hours after inoculation with *Bipolaris maydis*, a nonpathogen of sorghum (9). A partial clone of the sorghum PAL gene (*PAL1-1*) has 70% homology to 40 gene/cDNA sequences from 24 plant species, and was most similar (91% similarity in nucleotide sequence) to the GP28 gene for PAL from *Oryza sativa* (9).

PAL1-1 is being used to analyze the early sorghum-grain mold interaction. Initial results suggest that PAL activity is higher in spikelets of the sorghum genotype Tx430 inoculated at anthesis with *F. thapsinum* than in those inoculated just with water. In reverse transcriptase polymerase chain reactions (RT-PCR), several bands, one of which hybridizes to *PAL1-1*, appear consistently in inoculated and control treatments (Little, *unpublished*). Other PAL genes amplified from sorghum cDNA are being cloned.

Response to infection appears to be quite rapid at the level of phenolic and flavonoid compound accumulation. In response to infection and colonization by *C. lunata*, Tx2911, a red-seeded genotype, produces large amounts of red pigments best visualized five days after inoculation. The phenol layer contained what is most likely 3-deoxyanthocyanins, *e.g.*, apigeninidin and luteolinindin, whereas similar extracts from water-treated control plants typically are a light green color (Little, *unpublished*). The downstream products of the flavonoid pathway include the flavones, flavanols, flavanones, anthocyanins, and deoxyanthocyanidins.

The phenylpropanoid pathway is "switched" to flavonoid biosynthesis via the condensation of three molecules of malonyl CoA with one molecule of *p*-coumaryl CoA to yield chalcone. Cui *et al.* (9) found that CHS accumulated in BTx635 and RTx7078 sorghum seedlings within six hours of inoculation with *B. maydis*. *S. reilianum* did not induce the accumulation of CHS in either meristematic or nonmeristematic resistant lines. *P. sorghi* induced higher levels of CHS in downy mildew resistant lines than in susceptible ones. Cui *et al.* (9) obtained a 620-bp clone, *CHS2G*, that had 70% similarity to CHS genes from 27 plant species. This clone's highest similarity was with *C2* from *Zea mays*. In both Sureno (moderately resistant to grain mold) and Tx430 (susceptible) at 3-120 hours postinoculation, CHS transcription appears relatively constant, with RT-PCR amplifying equally a single DNA fragment, approximately 600 bp in length, from control plants or from those challenged with either *F. thapsinum* or *C. lunata* (Little, *unpublished*).

Antifungal Protein Genes. The combined effects of hydrolytic enzymes with antifungal proteins and seed-storage proteins also may play an important role at early stages of the host/pathogen interaction, see Waniska *et al.* (71) in this volume for more details.

Mycotoxins

Several fungi in the grain mold complex produce mycotoxins and other secondary metabolites. In particular, fungi belonging to *Fusarium* spp. in the *Liseola* section produce chemically and biosynthetically diverse mycotoxins such as the fumonisins, moniliformin, fusaproliferin, fusaric acid, fusarins, beauvericin, and gibberellic acids (32), with additional toxins still remaining to be identified (34). Fumonisins B_1, B_2, and B_3 are part of a family of mycotoxins produced by the molds *Fusarium verticillioides* and *Fusarium proliferatum* (15, 68). Zearalenone is primarily produced by the mold *Fusarium graminearum* (= *Gibberella zeae*), a fungus that also may produce nivalenol and deoxynivalenol (sometimes called vomitoxin) (38). Visible grain mold may indicate contamination by these mycotoxins; however, absence of visible mold growth does not necessarily ensure grain that is mycotoxin free (6). Ochratoxin A, most commonly produced by *Aspergillus ochraceus* and *Penicillium viridicatum*, can be recovered from maize, barley, and sorghum. In a recent preliminary study at ICRISAT, ochratoxin A levels exceeding 300 μg/kg were detected in

6/14 sorghum grain samples intended for incorporation into commercial chicken feed (Reddy, personal communication). In general, sorghum is not considered as good a substrate for the production of aflatoxin as is maize, peanuts, or other oil-rich seeds. In Australia, extremely moldy sorghum containing *Alternaria* and *Fusarium* mycotoxins decreased feed conversion in pigs and chickens by up to 14% (5). Normally, sorghums that mature during the rainy season are more prone to mycotoxin contamination than are those that mature after the end of the rainy season (53).

An outbreak of poisoning, characterized by abdominal pain and diarrhea, attributed to the ingestion of fumonisin-contaminated maize and sorghum, has been reported from several villages in India (4). Poisoning was observed only in adults who consumed moldy grain. The affected people were primarily marginal farmers and landless agricultural laborers who did not have access to other foods. *F. thapsinum*, which is isolated frequently from sorghum, produces much less fumonisin B_1 than does *F. verticillioides*, which is more common on maize (33), and on which fumonisin contamination is more severe. There is a need to determine the relative frequency of *Fusarium* spp. on sorghum, primarily from a mycotoxicological point of view. The best protection against mycotoxins is to monitor for their presence in feeds and foods by testing materials from initial harvest to the finished product. Rapid and inexpensive diagnostic assays for mycotoxins and institutional response mechanisms are needed to ensure sorghum quality and to expand their utilization in the commercial food and feed sectors.

Relationships Between Grain Mold and Weather

Although grain mold is associated with sorghum that matures under wet conditions, published reports on quantitative relationships between weather and grain mold are scarce. This lack of data may be attributed, at least in part, to the complex nature of grain mold, which may be caused by various fungal species that infect the grain at different growth stages.

Recent studies at ICRISAT used three approaches to investigate the effect of weather on the grain mold complex. One was to identify empirical relationships by comparing naturally occurring grain mold at various locations and in different growing seasons, and then analyzing the variation in grain mold severity in relation to the observed weather. A second approach was to conduct experiments in controlled environments to determine response curves for processes such as infection and sporulation with a range of values for individual weather variables. The third approach was to manipulate the physical environment in field experiments through the use of shelters and controlled wetting of the crop during critical periods in the growing season.

Empirical relationships. Empirical relationships were established from an international grain mold trial in west and central Africa, which involved 15 sites in ten countries (49). The trials were undertaken in two years, providing 26 site/season combinations, with daily weather data for the complete growing season. Grain mold was assessed visually at the time of harvest in plots with and without the control of head bugs. Correlations between disease and weather variables were obtained with a "window" computer program (7) that calculates mean or totals of each variable for different periods, or windows, of time. Windows that began on each day of the growing season and that varied in length (2, 5, 10, 15, 25, 30, 35, and 40 days) were chosen, so that weather data from the entire growing season were examined. The weather variables included precipitation frequency, total precipitation, the number of consecutive days with precipitation, the number of consecutive days without precipitation, maximum temperature, minimum temperature, temperature range, and minimum and maximum relative humidity.

Very few highly significant correlations ($p < 0.001$) were obtained between rainfall-derived variables and grain mold scores. However, a large number of highly significant correlations were found with the mean maximum relative humidity, especially in the case of susceptible sorghum genotypes. No significant correlations were found with grain-mold-resistant genotypes, *e.g.*, ICSV 1079. When head bugs were controlled, there was a dramatic decrease in the number of highly significant correlations.

Scatter diagrams of grain mold score versus maximum relative humidity revealed nonlinear relationships. When the maximum relative humidity (RH_{max}) was less than 90%, the mean grain mold score for 21 genotypes was consistently low, but a marked increase in grain mold was found at around 95% RH_{max} and above.

Controlled environment experiments. Experiments were conducted to determine the effects of wetness duration on infection by various grain mold fungi, *C. lunata, Drechslera australiensis, P. sorghina, F. moniliforme sensu lato*, and *Cladosporium oxysporum*. Sorghum panicles were inoculated at flowering, milk, soft dough, hard dough, and physiological maturity, and, immediately after inoculation, were placed in a dew chamber for various time periods (0, 16, 24, 40, 48, and 72 hours) to keep them wet. Subsequently, the plants were kept in a greenhouse for a week and tested for infection by plating surface-sterilized grains on culture medium.

Results with all fungal species were generally similar. For *C. lunata* and *P. sorghina*, infection was much less at flowering, than at any other growth stage. At the milk stage, a longer wetness period (72 hours) was needed for maximum infection than at soft dough, hard dough, or maturity. At hard dough and maturity, some infection occurred with *F. moniliforme sensu lato* and *P. sorghina* even with no wetness treatment.

Mist and shelter experiments. Field experiments have been carried out at ICRISAT in recent years in which small areas of a sorghum crop have been covered from milk stage until harvest. Under some of the shelters, the crop was kept wet each night by applying mist from 18:00 to 06:00 hours. In different treatments, mist was applied at different growth stages.

Maximum grain mold was observed in the natural control (no shelter or mist) in 1998 (a wet year), but not in 1999 (a drier year). In the control treatment of 1998 season, the dominant grain mold pathogen was *C. graminicola*. Very little grain mold occurred in the dry treatment (with shelter and no mist). Under the shelter, mold severity was more when misting was provided from physiological maturity to harvest two weeks later, compared to misting from milk stage to physiological maturity. Therefore, predetermined wetness regimes under rain-free conditions during the rainy season can induce different levels of grain mold.

Future prospects. The results of the work outlined here indicate that there is good potential for the development of risk assessment models for sorghum grain mold. These models could be used to identify suitable sorghum varieties for planting in areas with different levels of mold risk. Information on the relationship(s) between weather and grain mold could be used to improve grain mold resistance screening techniques. The identification of partial resistance (tolerance) may require moderate disease pressure, which could be achieved by manipulating the environment. More information is needed on sorghum grain mold epidemiology. Effects of temperature on grain mold have not been investigated, and controlled environment studies of sporulation, which determines the grain mold symptoms, are urgently needed.

Improved knowledge of disease epidemiology, coupled with information on tolerant varieties, will be a valuable tool for developing management options for sorghum production and marketing. A grain-mold risk model based on weather could forecast the occurrence and severity of mold in different sorghum-growing areas. Such a model could indicate the availability of sorghum of different quality standards in specific areas, reducing the cost of a search for sorghum grain of a predetermined standard.

Management of Grain Mold

Research on grain mold usually has focused on only a single method of control, *e.g.*, chemical control, avoidance, resistance, or timely harvest. In most cases, avoidance or chemical control in farmers' fields is impractical. Almost all research on control of grain mold has been conducted under on-station conditions, away from the socioeconomic, marketing, organizational, and institutional framework under which the farmers operate. Therefore, most methods for control of grain mold on farms are far from satisfactory.

Resistance. Host plant resistance is the preferred method of control and is the backbone of integrated disease management strategies. For this reason, major research efforts have focused on development of resistant cultivars. Efforts to produce sorghum genotypes with tolerance to grain mold by conventional breeding methods have been partially successful, but to date, high levels of resistance have not been incorporated into white-grain genotypes with high yield potential.

Screening techniques. Screening at ICRISAT is currently done without artificial inoculation (2) since natural inocula are present during the rainy season (3). High moisture levels are assured by sprinkling on rain-free days from flowering to harvest. This methodology results in very high disease pressure, obliterates minor differences in resistance, and is unsuitable for the identification of cultivars with moderate levels of grain-mold resistance. Ideally, the screening technique should result in variable levels of mold pressure on a set of standard sorghum lines against which the test lines are evaluated. The degree of resistance of a genotype depends on the grain mold pressure it can withstand. The breakdown point beyond which grain mold resistance of a genotype is overcome is a good indicator of the level of the disease pressure that the line can withstand. Research at ICRISAT is currently aimed at inducing predetermined mold pressures by manipulating wetness duration in the field under rainout shelters normally used for drought research. Wet conditions are provided by a wetness sensor-linked misting system.

A laboratory-based screening technique has been developed to evaluate photoperiod-sensitive sorghum genotypes that are difficult to test under field conditions. The method involves dip-inoculating mature seeds with either individual or a mixture of mold fungi and incubating the grain for five days in a moist chamber. The advantages of this method are that grain can be evaluated against individual fungi and that the evaluation is not dependent on the season (59).

Sources of resistance. Searches for sources of grain-mold resistance have been the topic of several research papers. The screening methodology used varies widely and has included: (i) evaluation under natural conditions (45) or artificial wetness (48), (ii) multilocational tests (20), and (iii) *in vitro* techniques (60). Resistance has been found in different grain types; however, the level of resistance is invariably highest in colored-grain genotypes with tannins, followed by colored-grain genotypes without tannin, and finally white and hard-grain sorghums. Four white, grain-mold-resistant lines were identified among 66 guinea-based sorghum accessions (61), suggesting that it is possible to obtain resistance in white-grain lines. However, the factors associated with mold resistance in guinea sorghums are difficult to incorporate into white-grain lines without adversely affecting yield (66). White-grain, agronomically elite, mold-resistant lines in non-guinea background have been described (1, 42, 61). There is a need to assemble all the reported sources of resistance and to evaluate them together under common conditions to confirm the reported resistance.

Plant traits associated with mold resistance. There are conflicting reports on the role of panicle morphology and flower structure in grain mold resistance. Some authors (36, 70) concluded that the extent of glume coverage, open panicle shape, and tall plant height could reduce grain mold severity. A compact panicle provides a better habitat for head bugs and increases head-bug damage (55), and, therefore, could increase grain mold. However, other research that includes a relatively large number of test genotypes has found no significant relationship between either the extent of glume cover on the grain or panicle compactness and grain-mold resistance (1, 44). The latter conclusion also is reinforced in earlier reviews of grain mold (72). Glume length was less related to grain mold resistance than darker glume color (1). However, lighter-colored glumes and plant colors are preferred in food-type sorghums. The challenge is to enhance mold resistance in white-grain sorghum by incorporating the resistance associated with the colored-glume character without permitting the pigments from the glumes to leach and stain the grains.

Grain biochemistry and structure associated with mold resistance. Caryopsis traits, *e.g.*, a pigmented testa layer, red pericarp color, and hard endosperm, are associated with grain-mold resistance (70). Sorghums with a thin pericarp appear less molded; however, when genotypes varying in pericarp (mesocarp) thickness and grain-mold resistance were evaluated, pericarp thickness was not related to grain mold ratings (12).

Several studies concluded that tannins or 3-deoxy-anthocyanidins in the pigmented testa layer of sorghum were the primary mechanism of resistance to grain mold (1, 12, 43). Improved grain-mold resistance occurs in sorghums with a red pericarp and a pigmented testa; however, some sorghums with pigmented testa and tannins are susceptible to grain mold. Therefore, the characterization of the specific bioactive compounds, enzymes, and mechanisms in the resistant sorghums needs to go beyond the results of previous studies of tannin fractionation and bioactivity.

Flavan-4-ols are similar in structure to the tannin monomer. Leaves from grain-mold-resistant sorghum genotypes contain 3-fold more flavan-4-ols than do susceptible genotypes (21). Red-pericarp caryopses contain similar levels of flavan-4-ols at the milk stage of development, but susceptible red-pericarp sorghums have much lower levels at maturity than did resistant red-pericarp sorghums (24). Significant, but weak, correlations were observed between grain mold resistance and flavan-4-ol contents in the caryopsis and in the glumes (1). The caryopses of grain mold-resistant sorghum with red pericarp contain flavan-4-ols, but sorghum caryopses with white pericarp do not (43). The absence of flavan-4-ols is probably one reason why genotypes with white pericarp have less resistance to grain mold than do grain-mold-resistant sorghums with red pericarps. With the exception of Schutt and Netzley (54), flavan-4-ols have not been isolated, nor have their bioactivities against grain-mold-inducing fungi been determined.

Improved sorghums with a red pericarp exhibit more grain-mold tolerance or resistance than those with a white pericarp (1, 12, 43). Several of these improved red sorghums appear much darker due to the dominant intensifier gene, *I*. The phenolic composition, their bioactivities, the enzymes activated by the intensifier gene, and the mechanism of resistance of red pericarp sorghums all need further investigation. The differences in the quality and quantity of the phenolic and other bioactive compounds in red- and white-pericarp sorghums also needs to be determined.

Sorghums varying in pericarp color and grain mold resistance can be correctly classified by identifying the free and bound phenolic acids extracted from the caryopsis (52). The change in phenolic acids in response to inoculation with *F. thapsinum* also can be used to classify sorghums with respect to their resistance to grain molding (52). More extensive studies of phenolic acids and other phenolic compounds in caryopses and glumes are needed.

Endosperm hardness is related to grain-mold resistance (1, 12, 23). Sorghums vary in the proportion of outer, vitreous, translucent area and inner, soft, opaque, floury area. Grain-mold-resistant sorghums had significantly harder endosperms than did susceptible sorghums when grown in environments that were conducive to grain molds (1, 23, 51). However, in dry environments that are not conducive to grain mold, the susceptible sorghums had much harder endosperms that were not statistically different from those of the resistant sorghums (1, 52). This result suggests that the ability to retain grain hardness under grain-mold-inducing conditions is crucial for the expression of gain-mold resistance.

In sorghum grain, vitreous areas are enriched in kafirins and, in particular, in γ-kafirins (41, 67). γ-kafirins are rich in cysteine and form extensive intrachain disulfide bonds that may contribute to vitreous texture, and increase resistance to proteolysis and to fungal colonization (40). The cell wall composition of the outer and inner endosperm also differs in the hard and soft grains (28, 29).

None of the traits mentioned in this section is sufficient by itself to explain the variation in grain-mold resistance. Pyramiding these defense mechanisms with one another and other attributes that contribute to grain-mold resistance is a good strategy to increase grain-mold resistance.

Significance of the pericarp in grain-mold resistance. Mold growth is less easily seen in grain with red pericarp than it is in grain with white pericarp. Grain-mold-resistant cultivars with red pericarp contain phenolics that help reduce fungal growth on grain surfaces. White grain with corneous, hard endosperm resists fungal colonization internally, but cannot suppress late infection and sporulation by fungi such as species of *Cladosporium*, *Alternaria*, and *Curvularia* on the pericarp. Since it is difficult to inhibit late fungal infection and sporulation of the pericarp, physical methods, *e.g.*, polishing to remove mold growth from the grain surface, may be of value. Red-grained cultivars are not consumed as food in India. Removal of the red pericarp from testa-less cultivars by dehulling might allow red sorghum to be used as food since the dehulled sorghum can be milled into white flour. However, the greatest potential use for mold-resistant red-grain sorghums is as animal food.

Preharvest sprouting and mold resistance. Internal colonization of grain often is accompanied by sprouting of the grain before harvest. Preharvest sprouting can occur as early as 15 days after pollination (65), suggesting that significant internal grain colonization, predisposed by preharvest sprouting, can begin before the hard dough stage. The extent of internal colonization of grain can be minimized in genotypes that have seed dormancy which extends for a month after maturity. Grain dormancy in sorghum could be a valuable trait if there is a strong relationship between resistance to preharvest sprouting and resistance to grain mold. Molecular approaches can aid in the incorporation of resistance to preharvest sprouting into elite lines since quantitative trait loci (QTLs) that determine resistance to preharvest sprouting have been mapped (35).

Time of harvest. Incidence and severity of grain mold increases after physiological maturity of the grain, particularly when harvesting is delayed and/or wet conditions persist. Therefore, reduction in mold damage is possible if the crop is harvested at physiological maturity when moisture levels range from 15-18%. However, grain should be dried to 10-12% moisture to avoid molding during storage and further processing

Processing of sorghum and management of grain mold. Moldy grain, usually being cheaper is often consumed by the poor in spite of the inferior quality of such grain for food preparation and the increased health hazards due to mycotoxin contamination. Moldy grain also may be used as a raw material in some industrial processes. Industrial utilization of moldy grain is important not only because it generates cash income for farmers, but also because such utilization reduces the amount of moldy sorghum in the diets of the poor. Research to increase industrial use of moldy grain and to encourage the adoption of other postharvest methods to reduce grain mold should consider the socioeconomic, operational, and institutional framework of the target group in addition to the technical aspects.

Drying. The problem of grain mold is most acute when the mature crop is exposed to high humidity and rain. Harvesting at physiological maturity is the best method for reducing grain mold severity and is feasible if farmers have access to grain-drying technology. Grain harvested at physiological maturity or under wet conditions should be dried, which commonly is done by solarization. Artificial grain drying is possible but may not be economical for the farmer unless drying is performed in community-owned dryers or via simple indigenous methods. It also may be possible to use mold inhibitors, *e.g.*, propionic acids, during the drying process to reduce the risk of mold development (58). Although considerable research has been done on sorghum grain drying, no information is available on the technical and economic feasibility of drying sorghum grain in order to reduce grain mold. Grain drying as a mold management tool must be economical to be practical. Commercial drying of sorghum will become important only if the demand for this service increases and a premium is paid for clean grain.

Polishing. The result of early fungal infection usually is considerable colonization of internal grain tissues resulting in soft grains that often are unsuitable for food preparation. Fungal growth resulting from late infection under moderate disease pressure often is restricted to the pericarp with little internal colonization of the endosperm. These grains superficially appear moldy and sell for a low price in the market. If the pericarp is removed by mechanical dehulling, however, then such grains appear almost as clean as nonmoldy grains (18), and may contain fewer mycotoxins, as has been seen with rice (69). Village-level dehulling has been suggested as a means to increase the quality and safety of molded grain (66). The commercial value of dehullers and dehulled products has been demonstrated with clean sorghum grain (14). Further research is required in terms of value-added sorghum products, and linkages with sorghum processors and consumers are necessary. Future research areas include: (i) the identification of suitable cultivars with adequate grain mold resistance and grain quality characters, (ii) optimization of the dehulling procedure and design of equipment for use with grain with different levels of mold contamination and grain structure, (iii) identification of the limits of grain-mold contamination for which the technology is economical, and (iv) the effects of the technology on consumer acceptance, nutrient composition, keeping quality, and safety of the dehulled grain. Before extensive research begins, consumer acceptability and marketability of, and market opportunities for the dehulled products need to be assessed.

Industrial use. The primary use of sorghum is as animal feed in most of the world other than India and Africa. Demand for poultry feed continues to grow in India due to high annual growth rates in layer hens (10%) and broilers (15%). Currently, 30-35% of this diet is maize. The demand for finished poultry feed is nearly 12 million tons/year, which suggests that maize demand will rise from current levels of 3.5 million tons/year to 31 million tons/year by 2020. The feed industry anticipates that sorghum will be needed to replace much of the maize currently being used. Moldy sorghum may be detrimental to the growth of chickens. The reduction or removal of contaminating mycotoxins may need to be addressed prior to the use of moldy sorghum in poultry feed (69), as these birds are quite sensitive to moniliformin, which is produced at relatively high levels by many strains of *F. thapsinum* (34). Grain-mold-resistant, red grain sorghum is a safe component in poultry feed. Significant utilization of moldy sorghum also occurs in the grain alcohol sector (57). The brewing industry also is interested in using sorghum as an adjunct in the brewing process. Appropriate institutional arrangements are necessary to catalyze the use of sorghum in various industrial sectors.

Recommendations for Future Research

- Determine the relative distribution of fungi belonging to *Fusarium* Section *Liseola* on sorghum and assess the risk posed for mycotoxin production.
- Determine the extent of mycotoxin contamination of sorghum grain and identify factors that are associated with mycotoxin contamination.
- Increase baseline knowledge of grain-mold epidemiology to enable forecasts of disease occurrence and severity.
- Develop a screening system to evaluate sorghum genotypes under variable grain-mold disease pressure.
- Determine the utility of limited postharvest grain dormancy as a means to reduce postmaturity grain mold in the field and in storage.
- Pyramid defense mechanisms with other attributes that contribute to grain-mold resistance.
- Increase long-term grain-mold-management research on host-plant resistance, with emphasis on the use of molecular biological tools to understand and integrate mechanisms of resistance in grain-mold-resistant cultivars.
- Increase medium- and short-term grain mold management research to determine the technical feasibility, consumer acceptability, and economic viability of alternate mold management methodologies, and to encourage the use of sorghum grain for animal feed and other industrial processes that are more tolerant of grain mold.

References

1. **Audilakshmi, S., J. W. Stenhouse, T. P. Reddy, and M. V. R. Prasad.** 1999. Grain mould resistance and associated characters of sorghum genotypes. *Euphytica* **107:** 91-103.
2. **Bandyopadhyay, R., and L. K. Mughogho.** 1988. Evaluation of field screening techniques for resistance of sorghum to grain molds. *Plant Dis.* **72:** 500-503.
3. **Bandyopadhyay, R., L. K. Mughogho, M. V. Satyanarayana, and M. E. Kalisz.** 1991. Occurrence of airborne spores of fungi causing grain mold over a sorghum crop. *Mycol. Res.* **95:** 1315-1320.
4. **Bhat, R. V., P. H. Shetty, R. P. Amrith, and R. V. Sudershan.** 1997. A foodborne disease outbreak due to the consumption of moldy sorghum and maize containing

fumonisin mycotoxins. *J. Toxicol. Clin. Toxicol.* **35:** 249-255.

5. **Blaney, B. J., and K. C. Williams.** 1991. Effective use in livestock feeds of moldy and weather-damaged grain containing mycotoxins: Case histories and economic assessments pertaining to pig and poultry industries of Queensland. *Austr. J. Agric. Res.* **42:** 993-1012.
6. **Bowman, D. T., and W. M. Hagler, Jr.** 1991. Potential use of visual mold ratings to predict mycotoxin contamination of grain sorghum. *J. Prod. Agric.* **4:** 132-134.
7. **Coakley, S. M., L. R. McDaniel, and R. F. Line.** 1988. Quantifying how climatic factors affect variation in plant disease severity: A general method using a new way to analyze meteorological data. *Climate Change* **12:** 157-175.
8. **Collings, D. B., and A. J. Slusarenko.** 1987. Plant gene expression in response to pathogens. *Plant Mol. Biol.* **9:** 389-410.
9. **Cui, Y., J. Magill, R. A. Frederiksen, and C. Magill.** 1996. Chalcone synthase and phenylalanine ammonia lyase mRNA levels following exposure of sorghum seedlings to three fungal pathogens. *Physiol. Mol. Plant Pathol.* **49:** 187-199.
10. **Dixon, R. A., C. A. Maxwell, W. Ni, A. Oommen, and N. L. Paiva.** 1994. Genetic manipulation of lignin and phenylpropanoid compounds involved in interactions with microorganisms, p. 153-179. *In* B. E. Ellis, G. W. Kuroki, and H. A. Stafford (eds.), *Genetic Engineering of Plant Secondary Metabolism.* Plenum Press, New York City, New York.
11. **Ebel, J., M. R. Stab, and W. E. Schmidt.** 1985. Induction of enzymes of phytoalexin synthesis in soybean cells by fungal elicitor, p. 247-254. *In* K. H. Neumann, W. Barz, and E. Reinhard (eds.), *Primary and Secondary Metabolism in Plant Cell Cultures.* Springer-Verlag, London, U.K.
12. **Esele, J. P., R. A. Frederiksen, and F. R. Miller.** 1993. The association of genes controlling caryopsis traits with grain mold resistance in sorghum. *Phytopathology* **83:** 490-495.
13. **Forbes, G. A., R. Bandyopadhyay, and G. Garcia.** 1992. A review of sorghum grain mold, p. 253-264. *In* W. A. J. de Milliano, R. A. Frederiksen, and G. D. Bengston (eds.), *Sorghum and Millets Diseases: A Second World Review.* ICRISAT, Patancheru (A.P.), India.
14. **Geervani, P., and V. Vimala.** 1993. *Rural Food Enterprises, Operational Research on Development of Sorghum Food Enterprises for Alternate Uses and Supplementary Feeding.* Acharya N. G. Ranga Agric. Univ., Rajendranagar, Hyderabad, India.
15. **Gelderblom, W. C. A., K. Jaskiewicz, W. F. O. Marasas, P. G. Thiel, R. M. Horak, R. Vleggaar, and N. P. J. Kriek.** 1988. Fumonisins. Novel mycotoxins with cancer-promoting activity produced by *Fusarium moniliforme. Appl. Environ. Microbiol.* **54:** 1801-1811.
16. **Graham, T. L., and M. Y. Graham.** 1991. Cellular coordination of molecular response in plant defense. *Molec. Plant-Microbe Interact.* **4:** 415-427.
17. **Hahlbrock, K., and D. Scheel.** 1989. Physiology and molecular biology of phenylpropanoid metabolism. *Annu. Rev. Plant Physiol. Plant Molec. Biol.* **40:** 347-369.
18. **ICRISAT (International Crops Research Institute for the Semi-Arid Tropics).** 1986. *Annual Report 1985.* ICRISAT, Patancheru (A.P.), India.
19. **ICRISAT.** 1992. *Medium Term Plan 1994-1998. Research Theme Datasets, Vol. 3.* ICRISAT, Patancheru (A.P.), India.
20. **Indira, S., J. Mohan, and B. S. Rana.** 1991. Genotype × environmental interaction for grain mold resistance and seed weight in sorghum. *Ind. Phytopathol.* **44:** 523-525.
21. **Jambunathan, R., and M. S. Kherdekar.** 1991. Flavan-4-ol concentration in leaf tissues of grain mold susceptible and resistant sorghum plants at different stages of leaf development. *J. Agric. Food Chem.* **39:** 1163-1165.
22. **Jambunathan, R., M. S. Kherdekar, and R. Bandyopadhyay.** 1990. Flavan-4-ols concentration in mold-susceptible and mold-resistant sorghum at different stages of grain development. *J. Agric. Food Chem.* **38:** 545-548.
23. **Jambunathan, R., M. S. Kherdekar, and J. W. Stenhouse.** 1992. Sorghum grain hardness and its relationship to mold susceptibility and mold resistance. *J. Agric. Food Chem.* **40:** 1403-1408.
24. **Jambunathan, R., M. S. Kherdekar, and P. Vaidya.** 1991. Ergosterol concentration in mold-susceptible and mold-resistant sorghum at different stages of grain development and its relationship with flavan-4-ols. *J. Agric. Food Chem.* **39:** 1866-1870.
25. **Jardine, D. J., and J. F. Leslie.** 1992. Aggressiveness of isolates of *Gibberella fujikuroi* (*Fusarium moniliforme*) isolates to grain sorghum under greenhouse conditions. *Plant Dis.* **76:** 897-900.
26. **Jardine, D. J., and J. F. Leslie.** 1999. Aggressiveness to mature maize plants of *Fusarium* strains differing in the ability to produce fumonisin. *Plant Dis.* **83**: 690-693.
27. **Joost, O.** 1993. *Early Genetic Events in the Interaction of* Verticillium dahliae *and* Gossypium *Species.* Ph.D. dissertation. Texas A&M Univ., College Station, Texas.
28. **Kavitha, R., and A. Chandrashekar.** 1992. Content and composition of nonstarchy polysaccharides in endosperm of sorghum varying in hardness. *Cereal Chem.* **69:** 440-443.
29. **Kavitha, R., and A. Chandrashekar**. 1993. Isolation and characterization of cell-wall components from the endosperm of sorghum varieties varying in hardness. *Carbo. Polymers* **22:** 107-115.
30. **Knogge, W.** 1996. Fungal infection of plants. *Plant Cell* **8:** 1711-1722.
31. **Leslie, J. F.** 1995. *Gibberella fujikuroi*: Available populations and variable traits. *Can. J. Bot.* **73**: S282-S291.
32. **Leslie, J. F.** 1999. Genetic status of the *Gibberella fujikuroi* species complex. *Plant Pathol. J.* **15:** 259-269.
33. **Leslie, J. F., and A. S. B. Mansuetus.** 1995. Biological species and vegetative compatibility group as population descriptors in *Fusarium*, p. 277-288. *In* J. F. Leslie and R. A. Frederiksen (eds.), *Disease Analysis Through Biotechnology: Interdisciplinary Bridges to Improved Sorghum and Millet Crops.* Iowa State University Press, Ames, Iowa.
34. **Leslie, J. F., W. F. O. Marasas, G. S. Shephard, E. W. Sydenham, S. Stockenström, and P. G. Thiel.** 1996. Duckling toxicity and the production of fumonisin and

moniliformin by isolates in the A and F mating populations of *Gibberella fujikuroi* (*Fusarium moniliforme*). *Appl. Environ. Microbiol.* **62:** 1182-1187.

35. **Lijavetzky, D., M. C. Martínez, F. Carrari, and H. E. Hopp.** 2000. QTL analysis and mapping of pre-harvest sprouting resistance in sorghum. *Euphytica* **112:** 125-135.
36. **Mansuetus, A. S. B., R. A. Frederiksen, R. D. Waniska, G. N. Odvody, J. Craig, and D. T. Rosenow.** 1988. The effects of glume and caryopses characteristics of sorghum on infection by *Fusarium moniliforme* Sheldon. *Sorghum Newsl.* **31:** 100.
37. **Mansuetus, A. S. B., G. N. Odvody, R. A. Frederiksen, and J. F. Leslie.** 1997. Biological species of *Gibberella fujikuroi* (*Fusarium* section *Liseola*) recovered from sorghum in Tanzania. *Mycol. Res.* **101:** 815-820.
38. **Marasas, W. F. O., P. E. Nelson, and T. A. Toussoun.** 1984. *Toxigenic* Fusarium *Species: Identity and Mycotoxicology*. Pennsylvania State University Press, University Park, Pennsylvania.
39. **Marley, P. S., and A. M. Malgwi.** 1999. Influence of headbugs (*Eurystylus* sp.) on sorghum grain mold in the Nigerian savanna. *J. Agric. Sci.* **132:** 71-75.
40. **Mazhar, H., and A. Chandrashekar.** 1993. Differences in kafirins composition during endosperm development and germination in sorghum cultivars varying in hardness. *Cereal Chem.* **70:** 667-671.
41. **Mazhar, H., and A. Chandrashekar.** 1995. Quantification and distribution of kafirins in the kernels of sorghum cultivars varying in endosperm hardness. *J. Cereal Sci.* **21:** 155-162.
42. **Meckenstock, D. H., F. Gomez, D. T. Rosenow, and V. Guiragossian.** 1993. Registration of Sureno sorghum. *Crop Sci.* **33:** 213.
43. **Melake-Berhan, A., L. G. Butler, G. Ejeta, and A. Menkir.** 1996. Grain mold resistance and polyphenol accumulation in sorghum. *J. Agric. Food Chem.* **44:** 2428-2434.
44. **Menkir, A., G. Ejeta, L. Butler, and A. Melakeberhan.** 1996. Physical and chemical kernel properties associated with resistance to grain mold in sorghum. *Cereal Chem.* **73:** 613-617.
45. **Menkir, A., G. Ejeta, L. G. Butler, A. Melakeberhan, and H. L. Warren.** 1996. Fungal invasion of kernel and grain mold damage assessment in diverse sorghum germ plasm. *Plant Dis.* **80:** 1399-1402.
46. **Navi, S. S., R. Bandyopadhyay, A. Hall, and P. J. Bramel-Cox.** 1999. *A Pictorial Guide for the Identification of Mold Fungi on Sorghum Grain.* Information Bulletin No. 59. ICRISAT, Patancheru (A.P.), India.
47. **Orczyk, W., J. Hipskind, E. de Neergaard, P. Goldsbrough, and R. L. Nicholson.** 1996. Stimulation of phenylalanine ammonia-lyase in sorghum in response to inoculation with *Bipolaris maydis*. *Physiol. Mol. Plant Pathol.* **48:** 55-64.
48. **Rao, K. E. P., S. D. Singh, and J. W. Stenhouse.** 1995. Grain mold resistance in guinea sorghum germplasm. *Intl. Sorg. Millets Newsl.* **36:** 94-95.
49. **Ratnadass, A., D. R. Butler, P. S. Marley, O. Ajayi, R. Bandyopadhyay, M. A. Hamada, D. E. Hess, F. Assamoi, I. D. K. Atokple, J. Beyo, O. Cisse, D. Dakouo, M. Diakite, S., Dossou-Yovo, B. Le Diambo, I. Sissoko, M. B. Vopeyande, and I. Akintayo.** 1999. Final report of the West and Central Africa Sorghum Research Network (WCASRN) 1996 and 1997 regional sorghum head bug and GM trial. Paper presented at the West African Sorghum Research Network meeting, 8-12 March 1999, Lome, Togo.
50. **Rhodes, M. J. C.** 1994. Physiological roles for secondary metabolites in plants: Some progress, many outstanding problems. *Plant Mol. Biol.* **24:** 1-20.
51. **Rodriguez, R., W. L. Rooney, R. A. Frederiksen, D. T. Rosenow, and R. D. Waniska.** 1997. Stability analysis of grain mold resistance in sorghum using the AMMI model. *Intl. Sorg. Millets Newsl.* **38:** 73-76.
52. **Rodriguez-Ballesteros, O. R., A. S. B. Mansuetus, R. D. Waniska, R. A. Frederiksen, G. N. Odvody, and D. T. Rosenow.** 2002. Free and bound phenolic acids in mature sorghum caryopsis as affected by inoculation with *Fusarium thapsinum*, p. 269-272. *In* J. F. Leslie (ed.), *Sorghum and Millets Diseases*. Iowa State Press, Ames, Iowa.
53. **Salifu, A.** 1981. Mycotoxins in short season sorghums in northern Nigeria. *Samaru J. Agric. Res.* **1:** 83-88.
54. **Schutt, C., and D. Netzly.** 1991. Effect of apiforol and apigeninidin on growth of selected fungi. *J. Chem. Ecol.* **17:** 2261-2266.
55. **Sharma, H. C., Y. O. Doumbia, J. F. Scheuring, K. V. Ramaiah, N. F. Beninati, and M. Haidra.** 1994. Genotypic resistance and mechanisms of resistance to the sorghum head bug, *Eurystylus immaculatus* Odh. in west Africa. *Insect Sci. Appl.* **15:** 39-48.
56. **Sharma, H. C., M. V. Satyanarayana, S. D. Singh, and J. W. Stenhouse.** 2000. Inheritance of resistance to head bugs and its interactions with grain molds in *Sorghum bicolor*. *Euphytica* **112:** 167-173.
57. **Sheorain, V., R. Banka, and M. Chavan.** 2000. Ethanol production from sorghum, p. 228-239. *In* A. Chandrashekar, R. Bandyopadhyay, and A. J. Hall (eds.), *Technical and Institutional Options for Sorghum Grain Mold Management: Proc. Intl. Consultation (18-19 May 2000, Patancheru, India)*. ICRISAT, Patancheru (A.P.), India.
58. **Shetty, H. S., K. L. Patkar, C. M. Usha, N. Paster, R. Kennedy, and J. Lacey.** 1995. Effect of propionic acid on the incidence of storage fungi in stored rice, sorghum, and groundnut under tropical conditions. *Trop. Sci.* **35:** 40-48.
59. **Singh, S. D., and S. S. Navi.** 2000. An *in vitro* screening technique for the identification of grain mold resistance in sorghum. *Ind. Phytopathol.* **53:** 249-253.
60. **Singh, S. D., S. S. Navi, K. E. P. Rao, and M. V. Satyanarayana.** 1993. Identification of resistance to grain mold in white grain sorghum. *Intl. Sorg. Millets Newsl.* **34:** 24.
61. **Singh, S. D., S. S. Navi, J. W. Stenhouse, and K. E. P. Rao.** 1995. Grain mold resistance in white grain sorghum. *Intl. Sorg. Millets Newsl.* **36:** 95-96.
62. **Snyder, B. A., B. Leite, J. Hipskind, L. G. Butler, and R. L. Nicholson.** 1991. Accumulation of sorghum phytoalexins induced by *Colletotrichum graminicola* at the infection site. *Physiol. Molec. Plant. Pathol.* **39:** 463-470.

63. **Snyder, B. A., and R. L. Nicholson.** 1990. Synthesis of phytoalexin in sorghum as a site specific response to fungal ingress. *Science* **248:** 1637-1639.
64. **Somani, R. B., R. W. Ingle, and S. S. Wanjari.** 1994. Variability in *Curvularia lunata* (Wakker) Boedijn causing grain mold in sorghum. *Intl. Sorg. Millets Newsl.* **35:** 106-107.
65. **Steinbach, H. S., R. L. Benech-Arnold, G. Kristof, R. A. Sanchez, and S. Marcucci-Poltri.** 1995. Physiologic basis of pre-harvest sprouting in *Sorghum bicolor* (L.) Moench. ABA levels and sensitivity in developing embryos of sprouting-resistant and -susceptible varieties. *J. Exper. Bot.* **46:** 701-709.
66. **Stenhouse, J. W., R. Bandyopadhyay, S. D. Singh, and V. Subramanian.** 1998. Breeding for grain mold resistance in sorghum. *Proc. Intl. Conf. on Genetic Improvement of Sorghum and Pearl Millet (1-8 June 1997, Lubbock, Texas)*: 326-336.
67. **Sunitha, R. K., and A. Chandrashekar.** 1994. Relationships between grain hardness and the content of prolamins and three antifungal proteins in sorghum. *J. Cereal Sci.* **20:** 93-99.
68. **Thiel, P. G., W. F. O. Marasas, E. W. Sydenham, G. S. Shephard, W. C. A. Gelderblom, and J. J. Nieuwenhuis.** 1991. Survey of fumonisin production by *Fusarium* species. *Appl. Environ. Microbiol.* **57:** 1089-1093.
69. **Vasanthi, S., and R. V. Bhat.** 1998. Mycotoxins in food: Occurrence, health and economic significance and food control measures. *Ind. J. Med. Res.* **108:** 212-222.
70. **Waniska, R. D., G. A. Forbes, R. Bandyopadhyay, R. A. Frederiksen, and L. W. Rooney.** 1992. Cereal chemistry and grain mold resistance, p. 265-272. *In* W. A. J. de Milliano, R. A. Frederiksen, and G. D. Bengston (eds.), *Sorghum and Millets Disease: A Second World Review.* ICRISAT, Patancheru (A.P.), India.
71. **Waniska, R. D., R. T. Venkatesha, A. Chandrashekar, S. Krishnaveni, F. P. Bejosano, J. Jeoung, J. Jayaraj, S. Muthukrishnan, and G. H. Liang.** 2002. Antifungal proteins and other mechanisms in the control of sorghum stalk rot and grain mold, p. 287-297. *In* J. F. Leslie (ed.), *Sorghum and Millets Diseases.* Iowa State Press, Ames, Iowa.
72. **Williams, R. J., and K. N. Rao, K. N.** 1981. A review of sorghum grain molds. *Trop. Pest Manag.* **27:** 200-211.

33

Stalk Rots of Sorghum

Larry E. Claflin and Laura M. Giorda

Abstract

Stalk rots remain a constraint to yields in many sorghum-growing areas. Significant progress has been made in revising the taxonomy of *Fusarium* spp. Biotechnology has advanced knowledge of both the pathogens and the host plant. *Macrophomina phaseolina* and *Fusarium* spp., the causal agents of charcoal rot and Fusarium stalk rot, respectively, have been examined with molecular techniques, *e.g.*, randomly amplified polymorphic DNA (RAPD) and amplified fragment length polymorphisms (AFLPs). Similar techniques and the identification of quantitative trait loci (QTL) have been used to evaluate host plants for resistance to these stalk rot pathogens. We expect more progress in these areas to be made in the future and to reduce losses attributed to *Fusarium* spp. and *M. phaseolina.*

Introduction

Stalk rots of sorghum often are a constraint to yields in the tropics, subtropics, and temperate regions under stressful environmental conditions (32, 42). Lodging may result from anatomical weakness, climatic stresses, and/or a complex of plant pathogens. Yield or fodder losses are attributable to (i) loss of quality or quantity of grain, (ii) inability to harvest grain with mechanical harvesters, (iii) other pests that attack grain in lodged plants, and (iv) diminished quality or quantity of fodder.

Major stalk rot diseases of sorghum and their causal agents include: Charcoal rot (*Macrophomina phaseolina*), Fusarium stalk rot (*Fusarium proliferatum*, *Fusarium thapsinum*, and *Fusarium verticilliodes*), and Anthracnose stalk rot (*Colletotrichum graminicola*; syn. *Colletotrichum sublineolum*). Stalk rot diseases of lesser importance or with a more limited distribution include Pokkah boeng (*F. proliferatum*, *Fusarium subglutinans*, and *F. thapsinum*), bacterial stalk rot (*Erwinia* spp.), and Acremonium wilt (*Acremonium strictum*).

Stalk rots have been reported in sorghum for over 100 years from every geographical area in which the crop is grown. The problem persists even though it has been greatly diminished in the last 30 years. Acceptance of hybrid types containing three to four dwarfing genes has been very important in reducing losses attributed to lodging.

Milo disease (*Periconia circinata)* nearly devastated the sorghum industry in the 1930s in the central Great Plains of the United States. A single resistance gene was incorporated into breeding materials and has provided immunity to milo disease for over 60 years (39, 42). In recent years, milo disease has been reported only from the Republic of South Africa (4). At present, germplasm is screened for resistance to milo disease only at Chillicothe, Texas.

Other noteworthy achievements in the last 30 years include (i) identification and incorporation of the "stay green" trait that imparts drought tolerance to sorghum and provides resistance to various stalk rots, most notably to charcoal rot; and (ii) reduction of Fusarium stalk rot incidence. Revision of the taxonomy of relevant *Fusarium* spp. will enhance resistance-screening methodology through clearer identification of the causal agents of the disease.

Fusarium Stalk Rot

Fusarium stalk rot may be caused by several sibling species of the genus *Fusarium*, which are sometimes termed mating populations. Tremendous strides have been made in the last ten years in identifying the genetic relationships that exist in this complex genus. For nearly 75 years, the causal agent was listed as *Fusarium moniliforme*, a species name that has been retired for conformity with the International Code for Botanical Nomenclature, and that has been subdivided into a number of species (19, 21, 24). The teleomorph, *Gibberella fujikuroi*, contains at least eight different mating populations, or biological species, commonly designated with the letters A thru H. For example, *G. fujikuroi* mating population F also is known as *F. thapsinum* or *Gibberella thapsina* (Table 33-1). The common maize fungus known as *G. fujikuroi* mating population A is now called either *F. verticilliodes* or *Gibberella moniliformis*. The changes resulted from new information on molecular relatedness and sexual cross-fertility, and the need to make the nomenclature consistent with the rules in the International Code of Botanical nomenclature. DNA sequence-based tests are commonly used to make these distinctions, but isozyme analysis, sensitivity to benomyl and hygromycin, host range, and cross-fertility with a known standard strain also can be used (18, 21-23). Previously, cultures were identified to species based solely on morphological characteristics, *e.g.*, the production of microconidia in chains or in false heads (29, 40).

It is imperative in conducting research with Fusarium stalk rot that known pathogenic isolates within the section *Liseola* be utilized. Twelve *Fusarium* spp. are reportedly recovered from sorghum, although pathogenicity has not been conclusively proven for all of these species (21). No significant differences were noted in aggressiveness between *F. verticillioides* and *F. thapsinum* on either maize (16) or sorghum (15) under greenhouse conditions. There also is considerable difference in the frequency of recovery of these species from field-grown maize and sorghum, which suggests that significant differences may occur under field conditions that are not easily duplicated in the greenhouse.

Fusarium spp. occur in both temperate and tropical areas. *Fusarium* spp. incite seed rots, seedling blights, head blights, and root and stalk rots of numerous crops such as maize, millet, sudangrass, and sorghum (32, 42). Some strains also produce toxic metabolites including moniliformin, fusaric acid, fusarin C, and fumonisins. Yield reductions depend on the growth stage of the host, environmental conditions, and variables such as fertility, soil drainage, cultural practices, insects, nematodes, and other diseases. Losses usually result from poor grain fill, lodged peduncles, or stalk breakage that hinders harvesting.

Symptoms of Fusarium stalk rot may range from inconspicuous lesions, *e.g.*, small circular spots on leaves in the plant whorl, to lodged plants. Lesion color may be associated with plant type–purple, tan, or red. Early season symptoms may not be apparent except on the roots of seedlings, usually as a red discoloration of the cortex. As the plant matures, internodes at the base of the plant may be infected as indicated by near total abnormal coloration, and the internal pith may be cream or pink to red-

TABLE 33-1. Mating populations and species in the *Gibberella fujikuroi* species complex (after 37).

Mating Population	*Fusarium* anamorph[a]	*Gibberella* teleomorph	Common Plant Host
A	*F. verticillioides* (*F. moniliforme*)	*G. moniliformis*	Maize
B	*F. sacchari*	*G. sacchara*	Sugarcane
C	*F. fujikuroi*	*G. fujikuroi*	Rice
D	*F. proliferatum*	*G. intermedia*	Maize, Sorghum, Prairie grasses, Asparagus
E	*F. subglutinans* (*F. sacchari* var. *subglutinans*)	*G. subglutinans*	Maize
F	*F. thapsinum*	*G. thapsina*	Sorghum
G	*F. nygamai*	*G. nygamai*	Millets, Sorghum
H	*F. circinatum* (*F. subglutinans* f. sp. *pini*)	*G. circinata*	Pine

[a]Preferred names given first; alternate names in parentheses.

dish. The uppermost internodes and the peduncle may contain discolored (red to brown) vascular bundles caused by *Fusarium* spp. Symptoms often are more pronounced if mechanical injury or insect damage has occurred. Lodging of plants normally occurs after anthesis and can occur at locations along the stalk from soil level to the peduncle. Secondary invaders are commonly associated with stalk rot infections. Leaves on plants severely infected with stalk rot may suddenly acquire a bluish-gray color that resembles frost damage. Heads on affected plants are noticeably smaller, with numerous sterile florets, and the kernels that do develop usually are smaller than healthy ones.

Primary inoculum consists of conidia and mycelia that overwinter in crop debris. The fungus also is seedborne. Fungal propagules could not survive for more than three months in the winter in the absence of plant debris in Kansas (25), but can survive the milder winters in Argentina (Giorda, *unpublished*). Chlamydospores were not detected in grain sorghum debris. No loss in conidial viability occurred when *F. moniliforme sensu lato* was stored in host tissue at -16°C for six months (25), and spore suspensions stored in 15% glycerol at -70°C survive for at least ten years (J. F. Leslie, personal communication). The fungus enters roots and stalks through root wounds resulting from mechanical, insect, or other types of injury.

Stalk rot is reduced considerably when sorghum is grown under ecofallow conditions rather than conventional tillage (8). High nitrogen coupled with low potassium increased stalk rot severity. A ratio of 1:1 N:K is optimal for reducing stalk rot. Manipulation of irrigation practices to avoid water stress, particularly at anthesis, can reduce lodging (7, 10).

Protease inhibitors and hydrolytic enzymes, *e.g.*, chitinases and β-1,3-glucanases, are synthesized in response to insect damage, *Fusarium* infection, and mechanical injury (20). These pathogenesis-related proteins can hydrolyze the cell walls of many fungi. These enzymes are induced in both resistant and susceptible inbred lines in response to *Fusarium* infection, so the role of pathogenesis-related proteins in the disease interaction is not clear (20).

Characteristics associated with Fusarium stalk rot resistance include lodging resistance, nonsenescence (stay green), and greenbug resistance (13). Nonsenescence and lodging resistance are considered phenotypic expressions of postflowering drought tolerance. Nonsenescence was identified as the most important plant character related to stalk rot, lodging, and postfloral drought resistance (14). The efficacy of research and breeding efforts would be increased if QTLs associated with Fusarium stalk rot under optimum environmental conditions were identified. If traits such as postflowering drought tolerance and Fusarium root and stalk rot resistance were genetically marked, then molecular markers could be used to more rapidly combine desirable resistance genes with other agronomically important traits.

Charcoal Rot

Macrophomina phaseolina (Tassi) Goid. [syn. *Tiarosporella phaseolina* (Tassi) vam der Aa] is a soilborne fungus that causes charcoal rot and ashy stem blight of over 500 plant species in nearly 100 families (26). Other synonyms include *M. phaseoli, Macrophoma corchori*, *Macrophoma cajani*, *Sclerotium bataticola*, and *Rhizoctonia bataticola* (26). This fungus produces two anamorphic structures–pycnidiospores and microsclerotia. A teleomorphic stage has not been reported. *M. phaseolina*, at present, is the only species in this diverse genus and is likely to be divided into more species as DNA sequencing and genetic fingerprinting techniques, *e.g.*, amplified fragment length polymorphisms (AFLPs), are brought to bear (44). Isolates of *M. phaseolina* may vary in host preference (6). A general trend for isolates of *M. phaseolina* to be more aggressive towards the host species from which they were recovered than towards other hosts has been reported (5). Isolates from a single sorghum host varied in pathogenicity when inoculated on other cultivars (6). Strains may differ in growth on minimal media containing potassium chlorate (26, 33, 34), and considerable phenotypic variation exists for pycnidia production, pathogenicity, and colony characteristics on media. Growth of *M. phaseolina* was similar on media adjusted for osmotic water potential by KCl, NaCl, or sucrose, but growth was best on media with sucrose without additional salts (26). *M. phaseolina* produced a large number of sclerotia when cultured on media with low water potential (6, 26).

M. phaseolina occurs in all sorghum-growing areas and is most severe in environments with persistent, hot, dry climates. The fungus may not be the sole incitant of charcoal rot as other plant parasitic fungi usually are found in association with *M. phaseolina* in plants with this disease. Nematodes also may intensify the disease. Temperature also plays an important role, as charcoal rot is more common when soil temperatures are high (> 32°C), whereas Fusarium stalk rot is more common at lower soil temperatures. Colonization of root and stalk tissue usually occurs when the plants are drought stressed during grain development (9, 31). Lodging of plants at maturity is a useful diagnostic characteristic, but other symptoms include decayed roots, plant death, and abnormal panicles with poor-quality grain. Diseases caused by *M. phaseolina* often are misdiagnosed as drought damage because the symptoms are similar. The small, black microsclerotia in roots and stems also are diagnostic for charcoal rot. *M. phaseolina* can result in yield losses of up to 64% (27).

M. phaseolina, in the absence of host plants, overseasons as sclerotia in debris or in soil. The sclerotia apparently germinate when they contact root exudates, and pathogenesis begins when the host plant is infected. Lesion nematodes, particularly *Pratylenchus hexincisus*, can be important in providing a portal of entry (30). Stunt (*Tylenchlorhynchus annulatus*) and ring (*Criconemella xenoplax*) nematodes commonly are involved in a disease complex with *M. phaseolina* (45). Root and panicle dry weights were lower, and reproduction of both nematode species was suppressed by *M. phaseolina*. These changes can be explained by a reduction in the number of feeding sites, as both nematodes are ectoparasites.

Various protocols have been tried as controls for charcoal rot. Biocontrol microorganisms, *e.g.*, *Trichoderma harzianum* and *Pseudomonas fluorescens*, reduced germination of sclerotia of *M. phaseolina* by 60-63% (41), which is probably insufficient for effective charcoal rot control. A natural oil product containing actinidine extracted from *Nepeta clarkei* is highly effective against *M. phaseolina* (38). Isolates containing double-stranded RNA are less virulent and grew more slowly than dsRNA-free isolates, which suggests an association between the presence of dsRNA and hypovirulence (35). Use of hypovirulent isolates may provide an alternative approach for control of *M. phaseolina*.

At present, crop management and germplasm tolerant to *M. phaseolina* offer the only practical means of reducing losses (2, 43). Crop management systems aim to reduce fungal inoculum, primarily sclerotia, through crop rotation and moisture conservation practices (27, 42). In sorghum fields grown under conventional tillage (moldboard plow), stalk rot incidence was reduced from 39% with conventional tillage to 23% and 11% with minimum tillage and no tillage, respectively (8). Sorghum grown in a mixed cropping system had less stalk rot than when sorghum was grown as a monoculture (8).

Breeding for nonsenescence (stay green) is a promising strategy for the simultaneous selection of resistance to *M. phaseolina* (43). This methodology is based on the nonsenescence of plants subjected to drought stress during grain development and is correlated with reduced lodging. Charcoal rot resistance accrued from both additive and nonadditive genes, and heterosis was not important, as the parents often were more resistant than their hybrids (36). General combining ability effects were more important than specific combining ability, which suggests that nonadditive effects are likely the result of dominant and recessive epistatic interactions between resistant loci (36). General combining ability effects also were more important than specific combining ability in dryland environments (2). Resistance to *M. phaseolina* appears to be multigenic, with expression heavily influenced by the environment, and neither consistently dominant nor recessive across all of the host resistance loci (2, 3).

Bacterial Stalk Rot

Bacteria are opportunistic microorganisms that under optimum moisture and temperature conditions may cause stalk rots. The principal incitant in bacterial stalk and top rot is *Erwinia chrysanthemi* Burkholder, McFadden and Dimock (17). Other potential synonyms and causal agents include *Erwinia carotovora* ssp. *carotovora*, *Erwinia chrysanthemi* pv. *zeae*, *Erwinia chrysanthemi* pv. *chrysanthemi*, and *Erwinia dissolvens*. This organism is a gram-negative rod, and a facultative anaerobe with a wide host range. Bacteria probably are disseminated by splashing rain or in reclaimed water from wastewater holding ponds used for irrigation.

Symptoms of bacterial stalk rot vary from plant death to affecting only a few leaves in a whorl. Typically, symptoms are observed at or near boot stage when the top four to five leaves appear to be drought stressed and then desicate within the next week. The leaves and stalk tissue are easily removed from the whorl; the tissue usually is soft rotted and often has a putrid odor. Infection is greatest when air temperatures are > 30°C. Bacterial soft rot is more common in plants in low-lying areas of the field. Resistance mechanisms have not been identified.

Reports of stalk rots of sorghum attributed to plant pathogenic bacteria are not common. Only a limited number of sorghum pathologists are trained as bacteriologists, and laboratory conditions and equipment often are insufficient for an accurate diagnosis. *Acidovorax avenae* (=*Pseudomonas avenae*), causal agent of bacterial leaf stripe, resulted in stalk rots and death to the growing point after plants were artificially inoculated in a screening program (Claflin, *unpublished*). *Burkholderia andropogonis* (*Pseudomonas andropogonis*) caused stalk rots and the premature death of plants of very susceptible genotypes. Other *Erwinia* spp. besides *E. chrysanthemi* also cause similar top and stalk rot symptoms of maize and sorghum (Claflin, *unpublished*). Bacterial stalk rot has been reported in Mauritania, Puerto Rico, the Philippines, India, and Venezuela, and probably occurs wherever sorghum is grown under environmental conditions that favor disease development.

Acremonium Stalk Rot

The causal agent of Acremonium stalk rot, *Acremonium strictum* W. Gams, is a cosmopolitan fungus that is pathogenic to numerous and diverse monocotyledonous and dicotyledonous crops including *Acacia*, *Alnus*, *Ficus*, *Glycine*, *Gossypium*, *Triticum*, and *Zea*. Considerable nomenclatural confusion exists, as *Cephalosporium acremonium* Corda was previously identified as the causal agent. A different fungal species has been described in Egypt as the causal agent of a disease there that is similar to Acremonium wilt (11).

Acremonium wilt symptoms are initially observed when leaves of plants approaching maturity desicate. The dessication is coupled with leaf veins and the midrib assuming a distinct coloration that is commonly the plant type color, *i.e.*, red, tan, or purple. These symptoms generally occur only on one side of the midrib. As Acremonium wilt progresses, the entire leaf becomes desicated and the apex dies, although lower leaves usually remain alive. If Acremonium wilt is present, then a transverse cut of the stalk near the soil line usually reveals discolored vascular bundles of orange, red, and brown (28).

Acremonium wilt has been reported from Argentina, Honduras, India, and the United States (1, 28). The disease probably occurs wherever sorghum is produced and has been observed in numerous African and Central American countries, and in Mexico (1; Claflin, *unpublished*). The fungus is systemic and may be isolated from all tissues of infected plants including seeds. *A. strictum* is probably disseminated via infected seeds (1).

Conclusions

Research involving root and stalk rots of any crop is difficult and complex. Numerous interactions probably occur, and biotic and abiotic factors, *e.g.*, other fungi, bacteria, insects, cultural practices, and, in general, any soil inhabitant could be involved. It is impossible to identify all of the potential interactions, but observations from a single parameter may be misleading and incorrect. For stalk rot of sorghum we should remember that, "Investigation with one microorganism kept pure and free from contamination with any other has been the classical procedure ever since Koch and others perfected the pure-culture methods that facilitate so greatly the separation of microorganisms. Students in our laboratories have been thoroughly inbred with the idea that cultures must be pure for a single organism. This necessary insistence on pure cultures of single organisms has perhaps led unconsciously to a feeling that to allow the use of a mixture in plant pathological work is extremely unscientific, as if it is actually a deadly plant pathological sin. Nature does not work with pure cultures alone but most frequently with associations" (12).

Acknowledgments

Contribution No. 01-197-B from the Kansas Agricultural Experiment Station, Manhattan.

References

1. **Bandyopadhyay, R., L. K. Mughogho, and M. V. Satyanarayana.** 1987. Systemic infection of sorghum by *Acremonium strictum* and its transmission through seed. *Plant Dis.* **71:** 647-650.
2. **Bramel-Cox, P. J., and L. E. Claflin.** 1989. Selection for resistance to *Macrophomina phaseolina* and *Fusarium moniliforme* in sorghum. *Crop Sci.* **29:** 1468-1472.
3. **Bramel-Cox, P. J., I. S. Stein, D. M. Rodgers, and L. E. Claflin.** 1988. Inheritance of resistance to *Macrophomina phaseolina* (Tassi) Goid. and *Fusarium moniliforme* Sheldon in sorghum. *Crop Sci.* **28:** 37-40.
4. **de Milliano, W. A. J.** 1992. Sorghum diseases in southern Africa, p. 9-19. *In* W. A. J. de Milliano, R. A. Frederiksen, and G. D. Bengston (eds.), *Sorghum and Millets Diseases: A Second World Review.* ICRISAT, Patancheru (A.P.), India.
5. **Diourte, M.** 1987. *Pathogenic Variation and Morphological Studies of* Macrophomina phaseolina *(Tassi) Goid.* M.S. Thesis, Texas A & M Univ., College Station, Texas.
6. **Diourte, M., J. L. Starr, M. J. Jeger, J. P. Stack, and D. T. Rosenow.** 1995. Charcoal rot (*Macrophomina phaseolina*) resistance and the effects of water stress on disease development in sorghum. *Plant Pathol.* **44:** 196-202.
7. **Doupnik, B., Jr.** 1984. Cultural and biological control of root and stalk rot disease of sorghum, p. 201-208. *In* L. K. Mughogho and G. Rosenberg (eds.), *Sorghum Root and Stalk Rots: A Critical Review.* ICRISAT, Patancheru (A.P.), India.
8. **Doupnik, B., M. Boosalis, G. Wicks, and D. Smika.** 1975. Ecofallow reduced stalk rot in grain sorghum. *Phytopathology* **65:** 1021-1022.
9. **Edmunds, L. K.** 1964. Combined relation of plant maturity, temperature, and soil moisture to charcoal stalk rot development in grain sorghum. *Phytopathology* **54:** 514-517.
10. **El-Meleigi, M. A., L. E. Claflin, and R. J. Raney.** 1983. Effect of seedborne *Fusarium moniliforme* and irrigation scheduling on colonization of root and stalk tissue, stalk rot incidence and grain yields. *Crop Sci.* **23:** 1025-1028.
11. **El-Shafey, H. A., M. F. Abdel-Rahim, and M. M. Refaat.** 1979. A new Cephalosporium wilt of grain sorghum in Egypt. *Proc. 3rd Egypt. Phytopathol. Congr.*: 514-532.
12. **Fawcett, H. S.** 1931. The importance of investigations on the effects of known mixtures of microorganisms. *Phytopathology* **21:** 545-550.
13. **Giorda, L. M., and M. J. Martinez.** 1995. Opportunities for biotechnology research on sorghum stalk rot and virus diseases. *Afr. Crop Sci. J.* **3:** 191-201.
14. **Henzell, R. G., R. L. Dodman, A. A. Done, K. L. Brengman, and P. E. Mayers.** 1984. Lodging, stalk rot, and root rot in sorghum in Australia, p. 225-236. *In* L. K. Mughogho and G. Rosenberg (eds.), *Sorghum Root and Stalk Rots: A Critical Review.* ICRISAT, Patancheru (A.P.), India.
15. **Jardine, D. J., and J. F. Leslie.** 1992. Aggressiveness of *Gibberella fujikuroi* (*Fusarium moniliforme*) isolates to grain sorghum under greenhouse conditions. *Plant Dis.* **76:** 897-900.
16. **Jardine, D. J., and J. F. Leslie.** 1999. Aggressiveness to mature maize plants of *Fusarium* strains differing in the ability to produce fumonisin. *Plant Dis.* **83:** 690-693.
17. **Jensen, S. G., W. R. Mayberry, and J. A. Obrigawitch.** 1986. Identification of *Erwinia chrysanthemi* as a soft-rot-inducing pathogen of grain sorghum. *Plant Dis.* **70:** 593-596.
18. **Kerènyi, Z., K. A. Zeller, L. Hornok, and J. F. Leslie.** 1999. Molecular standardization of mating type terminology in the *Gibberella fujikuroi* species complex. *Appl. Environ. Microbiol.* **65:** 4071-4076.
19. **Klittich, C. J. R., J. F. Leslie, P. E. Nelson, and W. F. O. Marasas.** 1997. *Fusarium thapsinum* (*Gibberella thapsina*): a new species in section *Liseola* from sorghum. *Mycologia* **89:** 643-652.
20. **Krishnaveni, S., S. Muthukrishnan, G. H. Liang, G. Wilde, and A. Manickam.** 1999. Induction of chitinases and β-1,3-glucanases in resistant and susceptible cultivars of sorghum in response to insect attack, fungal infection and wounding. *Plant Sci.* **144:** 9-16.

21. **Leslie, J. F.** 1995. *Gibberella fujikuroi*: available populations and variable traits. *Can. J. Bot.* **73(Suppl. 1):** S282-S291.
22. **Leslie, J. F.** 1999. Genetic status of the *Gibberella fujikuroi* species complex. *Plant Pathol. J.* **15:** 259-269.
23. **Leslie, J. F.** 2000. *Fusarium* species associated with sorghum, p. 30-31. *In* R. F. Frederiksen and G. N. Odvody (eds.), *Compendium of Sorghum Diseases*, 2nd ed. APS Press, St. Paul, Minnesota.
24. **Leslie, J. F., and W. F. O. Marasas.** 2002. Will the real *Fusarium moniliforme* please stand up!, p. 201-209. *In* J. F. Leslie (ed.), *Sorghum and Millets Diseases*. Iowa State Press, Ames, Iowa
25. **Manzo, S. K., and L. E. Claflin.** 1984. Survival of *Fusarium moniliforme* hyphae and conidia in grain sorghum stalks. *Plant Dis.* **68:** 866-867.
26. **Mihail, J. D., and S. J. Taylor.** 1995. Interpreting variability among isolates of *Macrophomina phaseolina* in pathogenicity, pycnidium production, and chlorate utilization. *Can. J. Bot.* **73:** 1596-1603.
27. **Mughogho, L. K., and S. Pande.** 1983. Charcoal rot of sorghum, p. 11-24. *In* L. K. Mughogho and G. Rosenberg (eds.), *Sorghum Root and Stalk Rots: A Critical Review.* ICRISAT, Patancheru (A.P.), India.
28. **Natural, M. P., R. A. Frederiksen, and D. T. Rosenow.** 1982. Acremonium wilt of sorghum. *Plant Dis.* **66:** 863-865.
29. **Nelson, P. E., T. A. Toussoun, and W. F. O. Marasas.** 1983. Fusarium *Species: An Illustrated Guide for Identification.* Pennsylvania State University Press, University Park, Pennsylvania.
30. **Norton, D. C.** 1958. The association of *Pratylenchus hexincisus* with charcoal rot of sorghum. *Phytopathology* **48:** 355-358.
31. **Odvody, G. N., and L. D. Dunkle**. 1979. Charcoal stalk rot of sorghum: Effect of environment in host-parasite relations. *Phytopathology* **69:** 250-254.
32. **Partridge, J. E., J. E. Reed, S. G. Jensen, and G. S. Sidhu.** 1984. Spatial and temporal succession of fungal species in sorghum stalks as affected by environment, p. 59-78. *In* L. K. Mughogho and G. Rosenberg (eds.), *Sorghum Root and Stalk Rots: A Critical Review.* ICRISAT, Patancheru (A.P.), India.
33. **Pearson, C. A. S., F. W. Schwenk, and J. F. Leslie.** 1986. Variable chlorate resistance in *Macrophomina phaseolina* from corn, soybean and soil. *Phytopathology* **76:** 646-649.
34. **Pearson, C. A. S., F. W. Schwenk, and J. F. Leslie.** 1987. Host preference correlated with chlorate resistance in *Macrophomina phaseolina. Plant Dis.* **71:** 828-831.
35. **Pecina-Quintero, V., M. J. Alvarado, H. Williams-Alanis, R. T. Almaraz, and G. J. Vandemark.** 2000. Detection of double-stranded RNA in *Macrophomina phaseolina. Mycologia* **92:** 900-907.
36. **Pecina-Quintero, V., H. Williams-Alanis, and G. J. Vandemark.** 1999. Diallel analysis of resistance to *Macrophomina phaseolina* in sorghum. *Cereal Res. Commun.* **27:** 99-106.
37. **Samuels, G. J., H. I. Nirenberg, and K. A. Seifert.** 2001. Perithecial species of *Fusarium*, p. 1-14. *In* B. A. Summerell, J. F. Leslie, D. Backhouse, W. L. Bryden, and L. W. Burgess (eds.), Fusarium*: Paul E. Nelson Memorial Symposium*. APS Press, St. Paul, Minnesota.
38. **Saxena, J., and C. S. Mathela.** 1996. Antifungal activity of new compounds from *Nepeta leucophylla* and *Nepeta clarkei. Appl. Environ. Microbiol.* **62:** 702-704.
39. **Schertz, K. F., and Y. P. Tai.** 1969. Inheritance of reaction of *Sorghum bicolor* (L.) Moench to toxin produced by *Periconia circinata* (L. Mangin) Sacc. *Crop Sci.* **9:** 621-624.
40. **Snyder, W. C., and H. N. Hansen.** 1945. The species concept in *Fusarium* with reference to discolor and other sections. *Amer. J. Bot.* **32:** 657-666.
41. **Srivastava, A. K., D. K. Arora, S. Gupta, R. R. Pandey, and M. W. Lee.** 1996. Diversity of potential microbial parasites colonizing sclerotia of *Macrophomina phaseolina* in soil. *Biol. Fertil. Soils* **22:** 136-140.
42. **Tarr, S. A. J.** 1962. *Diseases of Sorghum, Sudangrass and Broom Corn.* Commonwealth Mycological Institute, Kew, Surrey, UK.
43. **Tuinstra, M. R., T. Teferra, L. E. Claflin, R. G. Henzell, A. Borrell, N. Seetharama, G. Ejeta, and D. T. Rosenow.** 2002. Breeding for resistance to root and stalk rots in sorghum, p. 281-286. *In* J. F. Leslie (ed.), *Sorghum and Millets Diseases 2000.* Iowa State Press, Ames, Iowa.
44. **Vandemark, G., O. Martinez, V. Pecina, and M. J. Alvarado.** 2000. Assessment of genetic relationships among isolates of *Macrophomina phaseolina* using a simplified AFLP technique and two different methods of analysis. *Mycologia* **92:** 656-664.
45. **Wenefrida, I., E. C. McGawley, and J. S. Russin.** 1997. Interrelationships among *M. phaseolina*, *Criconemella xenoplax* and *Tylenchorhynchus annulatus* on grain sorghum. *J. Nematol.* **29:** 199-208.

34

A Population Genetic Approach to Variation in *Colletotrichum graminicola*, the Causal Agent of Sorghum Anthracnose

Liane Rosewich Gale

Abstract

Colletotrichum graminicola, the causal agent of sorghum anthracnose, is known as a variable pathogen. This variability has been assessed by using virulence and, more recently, random amplified polymorphic DNA (RAPDs) as phenotypic and molecular markers. I review the relevant literature on these approaches and describe results in which nuclear restriction fragment length polymorphisms (RFLPs) are used in a population genetic context. I analyzed 1278 isolates of *Colletotrichum graminicola* from diverse ecological settings, including commercial grain sorghum fields, maicillos (a land race sorghum from Central America), and johnsongrass (*Sorghum halepense*) to generate a broader picture of the genetic variation in *C. graminicola*.

Current Understanding of Genetic Variation in *Colletotrichum graminicola*

Colletotrichum graminicola (Ces.) Wils. is a plant pathogenic fungus that causes anthracnose of a wide variety of cereals and grasses. In grain sorghum [*Sorghum bicolor* (L.) Moench], *C. graminicola* is one of the more important pathogens.

***Colletotrichum graminicola* or *Colletotrichum sublineolum*?** Members of the genus *Colletotrichum* Cda. belong to the subdivision Deuteromycotina, the class Coelomycetes and the order Melanconiales (conidia produced in acervuli) (19). Perfect stages associated with the Coelomycetes belong to the subdivision Ascomycotina, with *Glomerella* considered the teleomorph of most or all *Colletotrichum* species (1, 34). Nevertheless, the perfect stage for *Colletotrichum* spp. is not commonly found in nature and has not been described for some species. For example, the sexual stage of sorghum isolates of *C. graminicola*, *Glomerella graminicola* Politis, has been produced in the laboratory (38) but has never been found in the field. Thus, most researchers assume that this fungus reproduces asexually under field conditions.

Most taxonomic approaches to establish relationships within and between *Colletotrichum* spp. have relied on the morphology of the commonly observed asexual stage. Unfortunately, the anamorph has relatively few morphological characters that are suitable for taxonomic purposes, and the expression of these characters is dependent on culture conditions and media. The problem is aggravated further since a standard set of culture conditions for identification purposes has not been established for these fungi. Consequently, the species concept in *Colletotrichum* is broad and often unreliable and is based on a combination of criteria such as conidial shape and size, presence and morphology of setae, presence of sclerotia and appressoria, and symptom expression on hosts (35).

Wilson (39) and von Arx (4) united most of the falcate-spored *Colletotrichum* known from the Gramineae into the group species *C. graminicola*, which previously had been separated into 36 taxa. In 1968, Sutton (33) described primarily qualitative differences between the appressoria of isolates from sorghum, maize (*Zea mays* L.) and sugarcane (*Saccharum officinarum* L.), recognized *Colletotrichum falcatum* Went (from sugarcane) as a separate species, and suggested that *C. graminicola* from maize and sorghum should be recognized as different taxa. In 1980, Sutton (34) stated that the species *C. graminicola* should include only isolates from *Zea*, as an isolate from sorghum from Togo could be distinguished from maize isolates by conidial shape and appressorial size and morphology, in addition to other cultural characters, *e.g.*, colony color and the color of conidial masses. He suggested the name *Colletotrichum sublineolum* P. Henn apud Kabát & Bub. for isolates from *Sorghum*, and also indicated that it may be necessary to further subdivide the species if isolates from other grass hosts differ from those that have previously been examined.

Molecular techniques have been applied only recently to species of *Colletotrichum* (*e.g.*, 31). There are relatively few published studies on the genetic relationships within the genus and within the group species *C. graminicola.* An early study by Huguenin *et al.* (21) found distinct isozyme patterns in isolates of *C. graminicola* from maize and sorghum. Ali *et al.* (3) described the use of filter aminopeptidase, gel electrophoretic patterns of soluble proteins, and a laser-enhanced aminopeptidase technique to characterize isolates of *C. graminicola* from sorghum, maize, and barnyard grass [*Echinochloa cruzgalli* (L.) P. Beauv.]. All three methods separated isolates based on their host origin. The laser-aminopeptidase profiles further separated the sorghum isolates into pathotypes.

Vaillancourt and Hanau (38) examined mitochondrial (mt-)DNA and RAPD patterns of six maize and seven sorghum isolates, collected from different geographic regions, and compared molecular data with morphological characters. They also included an isolate of *C. falcatum* from sugarcane. RFLP analysis of DNA digested with *Hin*dIII yielded mtDNA patterns that correlated with the host from which the isolates were collected, even though some polymorphisms were present within each group. Analysis of RAPD banding patterns revealed that maize and sorghum isolates were about 45% similar but that, within a cluster, the level of similarity was at least 85%. The results are consistent with the hypothesis that isolates of *Colletotrichum* from maize and sorghum were as different from one another as either group was from the isolate of *C. falcatum* (38). The degree of difference suggests that isolates from maize and sorghum have been genetically isolated for an extended period of time.

Comparisons between the maize and sorghum isolates and the isolate from sugarcane also led to the identification of morphological differences in colony growth rate, mean size of falcate conidia, and shape and size of mycelial appressoria. Although the authors believed these differences to be consistent, the range of measurements within each group was broad and sometimes overlapped with that of the other groups. Thus, these morphological criteria were considered neither sufficient nor appropriate for distinguishing *Colletotrichum* isolates that originated from maize and sorghum.

In 1993, Guthrie (15) described taxonomic relationships between *Colletotrichum* isolates from maize, sorghum, and johnsongrass. With respect to cultural characters, primarily nonsignificant differences were noted in colony diameters and sporulation characters. Significant differences between all three groups were found only in area and maximum diameter of conidial appressoria. The groups also could be differentiated by using combined banding patterns generated by three RAPD primers, when similarities between groups were estimated from the number of RAPD bands shared. Again, the maize isolates constituted a genetically distinct group. The johnsongrass isolates might represent a second distinct group, based on appressorial dimensions, but the johnsongrass isolates were relatively few and from a limited geographical area. Some cultural characters, *e.g.*, colony color and the color of conidial masses, of the sorghum isolates were not consistent with Sutton's (34) description of *C. sublineolum*, the proposed species name for *Colletotrichum* isolates from *Sorghum* spp. Guthrie (15) therefore stated that he thought "... that the suggested specific names were based on a faulty premise. Since there is a real need for clarification of the taxonomy of this pathogen, the adoption of a new nomenclature system based on more extensive research may be the most satisfactory solution." As the species name *C. sublineolum* has not been widely accepted for *Colletotrichum* isolates from sorghum, and since there may be problems with the present description, I recommend that the designation *C. graminicola* be retained for isolates from sorghum, until these isolates and those from a range of grass hosts and agricultural crops are systematically reexamined with both molecular and morphological markers.

Pathogen variability on *Sorghum bicolor*. In sorghum pathology, *C. graminicola* is regarded a "variable pathogen." As early as 1967, Harris and Johnson (18) suggested the existence of races in *C. graminicola.* Differential cultivar reactions that depend on location have been observed for many years (*e.g.*, 13, 22, 28). A change in pathogen virulence over time also has been reported for some locations (*e.g.*, 10, 17).

The presumed variability in virulence characteristics found within the species has been confirmed and somewhat quantified in controlled experiments in the greenhouse and in field studies by using isolates collected either from a single location or from a few widely separated locations. In these studies, virulence was used to assess differential sorghum cultivars for their reactions to selected fungal strains (2, 6-8, 11, 15, 27). In all of these studies, isolates recovered following the subculture of a single spore could be grouped into different virulence phenotypes (Table 34-1).

TABLE 34-1. Pathotypes identified in isolates of *Colletotrichum graminicola* from *Sorghum bicolor.*

Isolates	Sorghum Differentials	Pathotypes	Geographic Origin	Ref.
7	12	7	Brazil	11
9	6	3	Florida, Georgia, Puerto Rico, Indiana	2
9	30	9	India	27
7	8	6	Puerto Rico, Texas, Georgia	6
78	8	21	Brazil, Texas, Georgia	7
72	6	12	Global	15

Though these studies have increased our knowledge of pathogen variability, there has not been a standardization of sorghum differentials, inoculation procedures, or evaluation criteria. Therefore, no overall picture of virulence variability in *C. graminicola* can be assembled at present, although some general conclusions can be drawn. The same virulence phenotype may be assigned to isolates derived from different areas, countries, or even continents. For example, in Brazil many pathotypes are widespread, a result interpreted to mean that there are no major geographic restrictions on pathotype distribution within the country (9). Yet it also is possible for a single microgeographic location to harbor several virulence phenotypes (*e.g.*, 7, 15).

The related question of variation in aggressiveness for various isolates of *C. graminicola* has been addressed in only one study (27). Aggressiveness was defined as the severity of lesions on six susceptible sorghum genotypes, and nine sorghum isolates from India were placed in three aggressiveness categories. An observed differential relationship between virulence and aggressiveness was ascribed to specific isolate × host genotype interactions (27).

In summary, it seems that virulence in *C. graminicola* from sorghum is not yet well defined. The differentials have been historically derived, based primarily on data from the International Sorghum Anthracnose Virulence Nursery (ISAVN). A criterion for the selection of a differential is based on a clear reaction type and the assumption that these cultivars have a vertical resistance phenotype (12). Neither the genetic basis of resistance towards *C. graminicola* nor the number of resistance genes that distinguish any of the differentials has been critically examined, so if host differentials have more than one resistance gene, maximum pathogen diversity is not detected (14). Additionally, a formal analysis of genes for resistance and avirulence has not been conducted, and gene-for-gene interactions, while assumed, have not been proven. The fact that only six to eight genetically undefined differentials are commonly used suggests that a relatively large number of virulence genotypes might contribute to an individual virulence phenotype, rendering virulence phenotype a rather imprecise marker for the study of genetic variation in *C. graminicola*.

RAPDs. Recently, the data from virulence testing has been supplemented with studies that used RAPDs as molecular markers (7, 15, 16). These studies demonstrated that variation within *C. graminicola* could be detected at the molecular level. As with virulence studies, the same or very similar RAPD patterns were observed in isolates originating from different locations.

In one of these studies (7, but also see 9), 78 isolates from various sorghum growing areas in Brazil and the United States were analyzed with eight arbitrary primers. From the 78 isolates, 38 RAPD haplotypes were identified. The average similarity among haplotypes was 93% and the genetically most divergent isolate was 89% similar to the other isolates. This observation led Casela (7) to the conclusion that "populations of this pathogen are not geographically defined," implying that gene flow occurs. This suggested gene flow may keep populations of this pathogen from diverging or, if gene flow occurred only in the past, it may have prevented a substantial substructuring within the species as a whole. The hypothesis of no substantive population substructure has been extended (9) to suggest that *C. graminicola* from a widespread geographic area is all part of a single phylogenetic lineage.

A Population Genetics Approach Using Nuclear RFLPs

The virulence phenotype and RAPD studies demonstrated the need for a study of a relatively large number of fungal isolates, from diverse hosts and locations, with codominant markers, *e.g.*, RFLPs (29, 30). The seven objectives of this study were to:

- Develop a robust and efficient system to study genetic diversity in *C. graminicola*.
- Examine and increase knowledge of the genetic structure of and genetic diversity within populations of this fungal species. Assuming solely asexual reproduction, do specific populations consist of one dominant clone or several clonal lineages?
- Determine the degree of genetic differentiation between populations. Is there evidence for gene flow between populations and, if so, at what spatial scale? Are clones widely dispersed geographically (direct

movement of clones), or is each geographic area characterized by unique clonal lineages?

- Compare the amount and distribution of genetic diversity of fungal populations from different hosts and ecological settings. Is genetic variation higher in more complex environments, *e.g.*, disease nurseries, than in monoculture commercial fields? How does this genetic diversity compare with that seen in fungal populations in fields cultivated with sorghum land races or in samples from noncultivated hosts?
- Determine if populations of *C. graminicola* from a non-cultivated host (*Sorghum halepense*) and a cultivated one (*S. bicolor*) are genetically isolated. Can isolates from johnsongrass provide the initial inoculum for a cultivated sorghum crop?
- Assess the genetic stability of *C. graminicola* populations from *S. bicolor*. Do detectable changes occur in genetic structure or genetic diversity with time?
- Determine if there is evidence for sexual reproducetion by this presumed asexual fungal species.

Isolates examined. To address these questions, *C. graminicola* isolates were collected from:

Experimental sorghum nursery in Griffin, Georgia (30). Fungal populations from this disease nursery, which is part of the Georgia Experiment Station, University of Georgia, were sampled for three consecutive years: 1991 (105 isolates), 1992 (98 isolates), and 1993 (208 isolates).

Honduras. Anthracnose samples were collected from two agricultural settings, commercial fields, and subsistence farmers' fields. In December 1993, I collected 110 diseased leaves from each of three commercial sorghum hybrid fields in two sorghum-growing areas, following a hierarchical sampling strategy. RFLP data was gathered from 303 single-spored isolates. Some samples of *C. graminicola* also were obtained from fields of subsistence farmers in the department of Choluteca, which is near the Pacific Ocean and borders El Salvador and Nicaragua. Anthracnose on maicillos, the commonly grown sorghum land race cultivars in the area, is very rare, with symptoms similar to those on johnsongrass; *i.e.*, small (1-2 mm), eye-spot type lesions with a defined margin. Sporulation occurs in the center of the lesion and is limited to one or a few acervuli. A total of 73 leaf pieces with anthracnose symptoms taken from five fields in Choluteca yielded 64 single-spored isolates. The bulk of the samples were from three fields, with only one leaf sample each from the other two locations.

Zambia. The disease nursery at Mansa is generally regarded as a "hot spot" for *C. graminicola*. Though the sorghum-growing countries in Southern Africa contain many ecological zones, anthracnose is not reported to be a major disease problem in this region. The station at Mansa is the Southern African Development Cooperation's (SADC) anthracnose testing site for the Introductory Disease Resistance Screening Nursery. In March 1994, Dr. Gary N. Odvody collected 110 leaf samples from a part of this nursery.

Texas. Anthracnose on sorghum is sporadically observed in the upper coastal plains of Texas. In August 1993, Delroy Collins collected 100 infected sorghum stalks from a sorghum field near La Ward. In that same year, 45 leaf samples were obtained from a disease nursery in Beeville and 25 infected stalks from a second disease nursery in Corpus Christi. The Beeville nursery was the only Texas collection site that had sorghum leaf infection in 1993. Twenty-nine additional stalk samples from other locations in Texas also were obtained in 1994.

Johnsongrass. *C. graminicola* on johnsongrass is difficult to detect, as the disease is expressed as small lesions (1-2 mm), usually characterized by an eye-spot with a defined margin. Sporulation occurs in the center of necrotic lesions, is limited to one or a few acervuli, and is similar to the reaction found in the maicillos. Samples were taken from johnsongrass in two locations near La Ward in August 1993. One location was adjacent to the grain sorghum field from which the stalk samples were taken. The other location was about 3 km north of the first site and bordered a second sorghum field. Samples were taken along a 25-m line parallel to but approximately 5 m from the edge of the neighboring sorghum field. This sampling strategy yielded 33 samples from the first site and 62 samples from the second. Another johnsongrass-infested area, approximately 10 km south of La Grange, also was sampled in 1993. There was no commercial sorghum field within at least 2 km of this site. Disease incidence at this site was higher, and 160 leaf samples were collected.

Diversity within/between populations. Seven single- or low-copy RFLP probes were selected from a genomic DNA library of *C. graminicola* (30) and used to construct multilocus haplotypes for the 1278 isolates analyzed.

Gene and genotypic diversity were determined for all populations (30). Variation in allele frequencies at RFLP loci, *i.e.*, the number and frequency of alleles at a single locus, between populations, even from the same ecological background, limited general conclusions. Collections from sorghum disease nurseries usually had more variation at RFLP loci than did collections from commercial sorghum production fields. Collections from johnsongrass resembled those from the nurseries in terms of the variation observed at RFLP loci, even though the incidence of *C. graminicola* on this host, and therefore the total population size, was generally low. In contrast, isolates collected from maicillo land race cultivars had low levels of variability at the RFLP loci.

Despite obvious differences in variation at RFLP loci between populations, the genotypic diversity measure of Stoddart and Taylor (32) was low in all populations, due to the dominance of one or a few haplotypes at each location. In many instances, the same haplotype was found at more than one sampling location. Finding the same haplotypes at different sites is consistent with hypotheses that include gene flow or long distance dispersal. This conclusion is strengthened as the dominant haplotypes were relatively widespread and recovered from all hosts. In grain sorghum, haplotype "A" (alleles 1111111) dominated in the United States (Georgia), Puerto Rico, Zambia, and Honduras, while haplotype "TB" (alleles 2621322) dominated all of the populations collected from the Texas Gulf coast. Also, haplotypes were shared by isolates collected from several maicillo fields in Honduras. Long-distance dispersal of usually splash-dispersed plant pathogens has traditionally been correlated with agricultural practices. In the present study, haplotype "JA" (alleles 4946934) was identified in collections from johnsongrass separated by more than 100 km and requires an alternative explanation for its dispersal.

Regional dispersal is common. Samples from a specific region, *e.g.*, grain sorghum isolates from Texas and Honduras, maicillo isolates from Honduras, and johnsongrass isolates from populations near La Ward, Texas, shared common frequent haplotypes that resulted in similar population structure and low genetic distance values for collections within each of these three regions. Therefore, genetic cohesion of *C. graminicola* populations can occur across extended geographic areas.

Analysis of the data with maximum parsimony techniques did not group haplotypes within the grain sorghum collection. Most of the alleles at all seven RFLP loci were found in most of the populations, with only a few private alleles detected. Therefore, populations of *C. graminicola* from grain sorghum may not be extensively subdivided. This conclusion is similar to those of previous studies in which macrogeographic sampling strategies were employed. Each regional population may have its own dynamics and be evolutionarily independent, and these regional differences can be detected in terms of haplotype composition, but the analysis of the degree of differentiation between populations from different regions often failed to mathematically track these differences. Within the collection of isolates from grain sorghum, only the isolates from Texas, as a group, were sufficiently well-differentiated to be considered a specific subpopulation.

The "A" haplotype was found at intermediate to high frequencies in all populations from modern sorghum cultivars, except for the Texas collections, and may have been introduced into many sorghum-growing areas. Distinguishing between the introduction and local origin of other *C. graminicola* haplotypes is more difficult to assess. With the exception of these relatively few immigration events, the populations apparently evolve in relative isolation even though significant differentiation between populations has not yet occurred, as indicated by the low frequency of private alleles.

In the grain sorghum collection from Honduras, for example, haplotype "A" coexists with many derivative types. These derivatives were not commonly observed in the collections of *C. graminicola* from Griffin, Georgia and Mansa, Zambia. This difference could result from a larger initial introduction of haplotype "A" into Honduras, or if haplotype "A" has been present in Honduras for a longer time than at the other locations.

Testing these conclusions will require samples from more populations of *C. graminicola.* The distribution of haplotype "A" in other sorghum-growing regions and countries will be of particular interest, as will the detection of additional private alleles and/or haplotypes that might help to pinpoint the center(s) of origin or diversity for *C. graminicola.*

Diversity over time. The three-year study of the population of *C. graminicola* from the experimental sorghum disease nursery at Griffin, Georgia identified little variation in allele and haplotype frequencies over years, although some rare haplotypes were not recovered every year (30). Thus, genetic drift does not appear to have a significant impact on population structure within a three-year time frame. As the population of *C. graminicola* in Griffin is provided with optimum conditions for overseasoning, these observations probably can not be extrapolated to populations at other locations.

Selection pressure in this nursery may be intense since haplotype "A" dominates and is presumed to have a selective advantage relative to the other haplotypes. As no apparent changes in haplotype frequency were detected for any of the identified haplotypes, this population currently may be at equilibrium.

Primary inoculum. In 1993 and 1994, sorghum anthracnose was observed in the upper coastal plains of Texas after many years of a virtual absence of the disease. The environmental conditions may have been more conducive for disease development in those years, but the source of the primary inoculum is unknown. Microsclerotia, seed transmission, and alternative hosts, *e.g.*, johnsongrass, all have been implicated as sources for the initial inoculum. Seed transmission and johnsongrass are not sources for the primary inoculum for *C. graminicola* populations from Texas grain sorghum. Seed transmission as a regular event does not occur, due to the geographic area-specific occurrence of haplotype "TB" in Texas. Sorghum seed production is highly concentrated, and, if regular seed transmission were occurring, then the same or similar haplotypes should occur in all areas sown with commercial sorghum seed. Johnsongrass isolates are genetically distinct from grain sorghum isolates and do not serve as the inoculum for anthracnose infection in sorghum. Thus, microsclerotia are left as both a means for survival across seasons and as a source for the primary inoculum. It is unlikely that every Texas location carries numerous microsclerotia across many years, and I think some regional dispersal occurs within seasons that are conducive for the disease. This hypothesis also would explain the commonly observed regional similarities in population structure, but requires a mechanism for relatively rapid and widespread dispersal of what is generally regarded as a splash-dispersed plant pathogen. Airborne movement of conidia, as described by Nicholson and Moraes (26), could provide such a dispersal mechanism.

Clonality/gametic disequilibrium. The influence of reproductive systems on population structure can be studied in various ways (37). In diploid organisms, such studies usually determine whether recombination and segregation of alleles occurs. Data consistent with asexual reproduction include fixed heterozygosity, absence of segregating genotypes, and deviation from Hardy-Weinberg expectations (37). Hardy-Weinberg proportions cannot be determined for haploid organisms. In these organisms, including *C. graminicola*, tests for sexual reproduction are based on the null-hypothesis of random mating. If sexual reproduction and random mating occur, then alleles at independent loci are randomly associated.

Four characteristics of haploid organisms can be exploited in determining whether sexuality impacts population structure (37). These characters include: (i) overrepresented, ubiquitous multilocus genotypes, (ii) the absence of recombinant genotypes, (iii) strong gametic disequilibrium, and (iv) correlations between two independent sets of genetic markers. All these characteristics are consequences of a lack of independent assortment in the genomes of asexually reproducing organisms. Of these indicators, significant gametic disequilibrium is commonly viewed as statistically the most powerful method to ascertain asexuality. Both physical linkage and clonal reproduction can cause these deviations, which can be measured in several ways (20). Though generally regarded as statistically powerful, these methods also have some significant drawbacks, of which the most important is a need for relatively common alleles. This requirement is especially pronounced when sample sizes are small (5, 25).

Among several available statistics for measuring gametic disequilibria, D_{ij}, the gametic disequilibrium parameter (25) is commonly used. The rationale for selecting this test method was: (i) it results in a χ^2 test for association between two loci so that statistical significance is easily ascertained, and (ii) a normalized measure, $D'_{ij} = D_{ij}/D_{max}$, can be computed and used to determine the extent of gametic disequilibrium. The range of the normalized measure D'_{ij} is independent of allele frequencies and varies between -1 and +1.

Two of the four characteristics of asexual reproducetion identified by Tibayrenc *et al.* (37), *i.e.*, overrepresented, ubiquitous multilocus genotypes and the absence of recombinant genotypes, were typical for all populations of *C. graminicola*. Over-represented, ubiquitous genotypes were haplotypes "A," "TB," and "JA." Within all populations examined in this study, multilocus haplotypes could be identified as members of clonal lineages, and the null hypothesis that sexual reproduction is common was not consistent with the data.

Low genotypic diversity is characteristic of asexual reproduction (populations consist of clones). Even with sample sizes as large as those in this study, the number of haplotypes was limited. Clone "censored" values, *i.e.*, each clone is used only once, must be used to measure gametic disequilibrium and to prevent skewing. The power of the statistical tests for D_{ij} depends on sample size (5), which makes it difficult to reject the null hypothesis that $D_{ij} = 0$, *i.e.*, that the population is at gametic equilibrium for the two loci examined, when the sample size is small. Unfortunately, small sample sizes, when clone-corrected values are used, are typical for asexually reproducing organisms. This problem leads to the apparent paradox that even when asexuality is obvious, it may be difficult to detect significant gametic disequilibrium.

To circumvent this problem, the sample size was artificially increased by using allelic information from all of the haplotypes of *C. graminicola* from grain sorghum identified in this study (51 total). This mixing of genetic information from haplotypes originating from different populations is not normally done; however, I thought this analysis was justified because most alleles were shared by the populations sampled.

The D'_{ij} values for pairs of loci in *C. graminicola* from grain sorghum (Table 34-2) average 0.4 for the 21 pairwise comparisons among the seven RFLP loci. Of 345 comparisons with individual allele pairs, only 42 were significant at ($p < 0.05$). These data probably should

TABLE 34-2. Measures of gametic disequilibrium among pairs of RFLP loci in *Colletotrichum graminicola* isolated from grain sorghum.

	pSCg 3	pSCg10	pSCg 17	pSCg 23	pSCg28	pSCg31
pSC g1	2/24* 0.454	5/9*** 0.596	2/18 0.462	1/15 0.325	2/6 0.42	0/6 0.031
pSC g3		3/24 0.600	5/48 0.485	1/40 0.410	3/16 0.511	2/16* 0.449
pSC g10			0/18 0.338	2/15 0.441	4/6** 0.643	0/6 0.138
pSC g17				3/30 0.138	2/12 0.251	4/12 0.405
pSC g23					0/10 0.531	1/10 0.480
pSC g28						0/4 0.309

Data are measures of disequilibrium based on a single representative of each clone identified from grain sorghum (51 haplotypes). The top row is the number of significant ($p < 0.05$) χ^2 tests between individual alleles at different RFLP loci over the total number of tests made, with the level of significance (*, $p < 0.05$; **, $p < 0.01$; and ***, $p < 0.001$) noted for χ^2 tests for association between all alleles at the two loci. Numbers in the second row are values for D'_{ij}.

not be interpreted as evidence for sexual reproduction in *C. graminicola* but, rather, that asexuality has not been conclusively demonstrated.

Thus, sexual reproduction in *C. graminicola* populations cannot be excluded, but if it does occur, then it is probably too infrequent to substantially contribute to the microevolutionary dynamics of individual populations. As noted by Kohn (23), "... clonality, year-to-year carry-over of clones, spatial mixing via immigration, and possibly selection pressure on clone frequencies may all be associated with monoculture, agricultural practices and the movement of agricultural material, such as seed and machinery. Invasion of agricultural hosts by pathogen genotypes from wild plant populations may be expected to follow epidemic, clonal patterns as single adaptive genotypes arise from within the wild populations."

Virulence/selection. Although no virulence studies were conducted with the 1200+ isolates, and the range or variability of virulence characteristics within the haplotypes is unknown, the genetic results obtained can be used to direct research related to virulence and selection. Priority should be placed on determining the range of vir-ulence of clones that are widespread and common. Thus, future studies should focus first on haplotypes "A" (ubiquitous), "B" and "C" (both from Griffin, Georgia), "TB" (Texas), and "ZB", "ZC" and "ZD" (Mansa, Zambia). To make these studies of virulence even more robust, efficient, and easier to interpret, a set of sorghum isolines, or at least differentials based on single resistance genes, should be developed. Another major drawback of current virulence testing methodology is its labor and time intensity, mainly due to juvenile resistance in sorghum. This procedural limitation severely reduces the number of isolates that can be tested at any one time. Thus, alternative methods of virulence testing need to be evaluated and developed. Tests could be conducted, for example, to determine the correlation between the reaction of detached leaves and intact plants to infection with *C. graminicola*, as a detached leaf test would reduce virulence testing time enormously.

At this time it cannot be concluded that genetically less variable populations respond less to selection pressures, as the variation for virulence or aggressiveness within haplotype groups is not known. Research with the rice blast fungus, *Magnaporthe grisea*, indicates that a single clonal lineage may comprise many diverse pathotypes and that the mutation/selection rate for specific virulence genes is greater than that for RFLP markers (24, 40). An interesting question in this context is, "Why are populations from johnsongrass and sorghum genetically differentiated, even though both greenhouse and field studies have shown that isolates from johnsongrass are pathogenic on cultivated sorghum and may even show differential virulence?" The present results indicate that it can be dangerous to extrapolate data from greenhouse or even field experiments to describe population parameters.

Host specificity/co-evolution/taxonomy. A maximum parsimony analysis was performed with PAUP version 3.1 (36), with all 79 haplotypes detected in this study. Bootstrap analysis resulted in a high level of support (96%) for separation of the grain sorghum haplotypes from those from johnsongrass and maicillos. These observations suggest that *C. graminicola* displays host specificity, which probably results in substantial genetic substructuring within the species. Genetic relationships between these groups may not be easily predicted. For example, isolates from maicillos and grain sorghum, were expected to cluster and to be distinct from the johnsongrass isolates. The phylogenetic analysis indicates that the situation is not this simple and that more (or different) data are needed to enable reliable predictions.

Collections of *C. graminicola* from land race sorghums, wild grass species, and domesticated or wild sorghum hybrids, especially from Africa, need to be examined to determine a more appropriate basis for taxonomic differentiation between entities within the *C. graminicola* species complex. Further studies of land race cultivars and noncultivated hosts also may provide more insight into co-evolutionary processes within the group. Additional markers, *e.g.*, rDNA regions (31) and mtDNA polymorphisms, should be included as part of any taxonomic reevaluation. Co-evolutionary processes play an important role in the evolutionary dynamics of *C. graminicola* and need to be included in the discussion of future species concepts. The substantial genetic substructuring and sophistication observed in this fungal species is somewhat contradictory to the common belief that pathogens with a wide host range are much less refined and specialized in their interactions with their hosts than are biotrophic plant pathogens.

References

1. **Agrios, G. N.** 1988. *Plant Pathology*, 3rd ed. Academic Press, New York City, New York.
2. **Ali, M. E. K., and H. L. Warren.** 1987. Physiological races of *Colletotrichum graminicola. Plant Dis.* **71:** 402-404.
3. **Ali, M. E. K., H. L. Warren, D. M. Huber, F. E. Lytle, and K. Hughes.** 1989. Characterization of *Colletotrichum graminicola* isolates with aminopeptidase profiles and polyacrylamide gel electrophoresis of soluble proteins. *Phytopathology* **79:** 1148.
4. **von Arx, J. A.** 1957. Die arten der gattung *Colletotrichum* Cda. *Phytopath. Z.* **29:** 413-468.
5. **Brown, A. H. D.** 1975. Sample sizes required to detect linkage disequilibrium between two or three loci. *Theor. Pop. Biol.* **8:** 184-201.
6. **Cardwell, K. F., P. R. Hepperly, and R. A. Frederiksen.** 1989. Pathotypes of *Colletotrichum graminicola* and seed transmission of sorghum anthracnose. *Plant Dis.* **73:** 255-257.
7. **Casela, C. R.** 1992. *Investigations on the variability of the sorghum anthracnose fungus* Colletotrichum graminicola. Ph.D. Dissertation, Texas A&M Univ., College Station, Texas.
8. **Casela, C. R., and A. S. Ferreira.** 1988. Proposta para um sistema de classificação de raças de *Colletotrichum graminicola* em sorgo (*Sorghum bicolor*). *Fitopatol. Brasil.* **12:** 337-344.
9. **Casela, C. R., A. S. Ferreira, K. A. Zeller, and M. Levy.** 1995. Pathotype variation in the sorghum anthracnose fungus: a phylogenetic perspective for resistance breeding, p. 257-276. *In* J. F. Leslie and R. A. Frederiksen (eds), *Disease Analysis Through Genetics and Molecular Biology: Interdisciplinary Bridges to Improved Sorghum and Millet Crops.* Iowa State University Press, Ames, Iowa.
10. **Duncan, R. R.** 1985. Reaction of selected sorghum genotypes to four pathotypes of *Colletotrichum graminicola. Sorghum Newsl.* **28:** 97-99.
11. **Ferreira, A. S. and C. R. Casela.** 1986. Raças patogenicas de *Colletotrichum graminicola*, agente causal da antracnose em sorgo (*Sorghum bicolor*). *Fitopatol. Brasil.* **11:** 83-87.
12. **Frederiksen, R. A., D. Collins, and K. F. Cardwell.** 1987. Changes in the International Sorghum Anthracnose Virulence Nursery. *Sorghum Newsl.* **30:** 89.
13. **Frederiksen, R. A., and D. Rosenow.** 1971. Disease resistance in sorghum. *Proc. Annu. Corn Sorghum Res. Conf. Amer. Seed Trade Assoc.* **26:** 71-82.
14. **Groth, J. V., and A. P. Roelfs.** 1987. Analysis of virulence diversity in populations of plant pathogens, p. 63-74. *In* M. S. Wolfe and C. E. Caten (eds.), *Populations of Plant Pathogens: Their Dynamics and Genetics.* Blackwell Scientific Publications, Oxford, U.K.
15. **Guthrie, P. A. I.** 1993. *RAPD Markers as a System for Differentiating Among Isolates of the Anthracnose Fungus* Colletotrichum graminicola. Ph.D. Dissertation, Texas A & M Univ., College Station, Texas.
16. **Guthrie, P. A. I., C. W. Magill, R. A. Frederiksen, and G. N. Odvody.** 1992. RAPD markers: a system for identifying, and differentiating among, isolates of *Colletotrichum graminicola. Phytopathology* **82:** 832-835.
17. **Harris, H. B., and B. M. Cunfer.** 1976. Observations on sorghum anthracnose in Georgia. *Sorghum Newsl.* **19:** 100-101.
18. **Harris, H. B., and B. J. Johnson.** 1967. Sorghum anthracnose–Symptoms, importance, resistance. *Proc. Bienn. Grain Sorghum Res. Utiliz. Conf.* **5:** 49-52.
19. **Hawskworth, D. L., B. C. Sutton, and G. C. Ainsworth.** 1983. *Ainsworth & Bisby's Dictionary of the Fungi*, 7th ed. Commonwealth Mycological Institute, Kew, U.K.
20. **Hedrick, P. W.** 1987. Gametic disequilibrium measures: proceed with caution. *Genetics* **117:** 331-341.
21. **Huguenin, B., M. Lourd, and J. P. Geiger.** 1982. Comparaison entre isolats de *Colletotrichum falcatum* et *Colletotrichum graminicola* sur la base de leur characteristique morphologiques, physiologiques et pathologiques. *Phytopathol. Z.* **105:** 293-304.
22. **King, S. B., and R. A. Frederiksen.** 1976. Report on the International Sorghum Anthracnose Virulence Nursery. *Sorghum Newsl.* **19:** 105-106.
23. **Kohn, L. M.** 1995. The clonal dynamic in wild and agricultural plant pathogen populations. *Can. J. Bot.* **73 (Suppl. 1):** S1231-S1240.
24. **Levy, M., F. J. Correa-Victoria, R. S. Zeigler, S. Xu, and J. E. Hamer.** 1993. Genetic diversity of the rice blast fungus in a disease nursery in Colombia. *Phytopathology* **83:** 1427-1433.
25. **Lewontin, R. C.** 1988. On measures of gametic disequilibrium. *Genetics* **120:** 849-852.
26. **Nicholson, R. L., and W. B. C. Moraes.** 1980. Survival of *Colletotrichum graminicola*: importance of the spore matrix. *Phytopathology* **70:** 255-261.
27. **Pande, S., L. K. Mughogho, R. Bandhyopadhyay, and R. I. Karunakar.** 1991. Variation in pathogenicity and cultural characteristics of sorghum isolates of *Colletotrichum graminicola* in India. *Plant Dis.* **75:** 778-783.
28. **Pastor-Corrales, M. A., and R. A. Frederiksen.** 1979. Sorghum reaction to anthracnose in the United States, Guatemala, and Brazil. *Sorghum Newsl.* **22:** 127-128.
29. **Rosewich, U. L.** 1996. *The Population Genetics of* Colletotrichum graminicola *from Different Ecosystems of Sorghum.* Ph.D. Dissertation, Texas A & M Univ., College Station, Texas.
30. **Rosewich, U. L., R. E. Pettway, B. A. McDonald, R. R. Duncan, and R. A. Frederiksen.** 1998. Genetic structure and temporal dynamics of a *Colletotrichum graminicola* population in a sorghum disease nursery. *Phytopathology* **88:** 1087-1093.
31. **Sherriff, C., M. J. Whelan, G. M. Arnold, J.-F. Lafay, Y. Brygoo, and J. A. Bailey.** 1994. Ribosomal DNA se-

quence analysis reveals new species groupings in *Colletotrichum. Exp. Mycol.* **18:** 11-138.
32. **Stoddart, J. A., and J. F. Taylor.** 1988. Genotypic diversity: estimation and prediction in samples. *Genetics* **118:** 705-711.
33. **Sutton, B. C.** 1968. The appressoria of *Colletotrichum graminicola* and *Colletotrichum falcatum. Can. J. Bot.* **46:** 873-876.
34. **Sutton, B. C.** 1980. *The Coelomycetes*. Commonwealth Mycological Institute, Kew, U.K.
35. **Sutton, B. C.** 1992. The genus *Glomerella* and its anamorph *Colletotrichum*, p. 1-26. *In* J. A. Bailey and M. J. Jeger, (eds.), Colletotrichum*: Biology, Pathology and Control*. CAB International, Wallingford, U.K.
36. **Swofford, D. L.** 1993. PAUP: Phylogenetic Analysis Using Parsimony, Version 3.1. Illinois Natural History Survey, Champaign, Illinois.
37. **Tibayrenc, M., F. Kjellberg, and F. J. Ayala.** 1990. A clonal theory of parasitic protozoa: the population structures of *Entamoeba*, *Giardia*, *Leishmania*, *Naegleria*, *Plasmodium*, *Trichomonas*, and *Trypanosoma* and their medical and taxonomical consequences. *Proc. Natl. Acad. Sci. USA* **87:** 2414-2418.
38. **Vaillancourt, L. J., and R. M. Hanau.** 1992. Genetic and morphological comparisons of *Glomerella* (*Colletotrichum*) isolates from maize and from sorghum. *Exp. Mycol.* **16:** 219-229.
39. **Wilson, G. W.** 1914. The identity of the anthracnose of grasses. *Phytopathology* **4:** 106-112.
40. **Zeigler, R. S., L. X. Cuoc, R. P. Scott, M. A. Bernardo, D. H. Chen, B. Valent, and R. J. Nelson.** 1995. The relationship between lineage and virulence in *Pyricularia grisea* in the Philippines. *Phytopathology* **85:** 443-451.

35

Will the Real "*Fusarium moniliforme*" Please Stand Up!

John F. Leslie and Walter F. O. Marasas

Fusarium moniliforme Sheldon is the name that has traditionally been used for various isolates of *Fusarium* from the *Liseola* section of the genus recovered from sorghum stalks and grain and associated with stalk rot and grain mold and was established by Wollenweber and Reinking in 1935 (63). Snyder and Hansen (58) proposed that the entire *Liseola* section be reduced to *F. moniliforme* Sheldon emend Snyder & Hansen. The variety *Fusarium moniliforme* var. *subglutinans* Wollenweber & Reinking was separated from *F. moniliforme* in the Section *Liseola* by Booth (2) based on the absence of microconidial chains and the presence of polyphialides. A list of the currently recognized species that might have at one time been termed *F. moniliforme* and their authorities is given in Table 35-1. Booth (2) also stated that *F. moniliforme* occurs on rice and sugarcane and is a "major parasite" of several Gramineae, including maize and sorghum, and that it occurs on at least 31 other families of plants. The purported host range of *F. moniliforme* was extended even further by Bacon *et al.* (1) who calculated that more than 11,000 species of plants may serve as hosts for this fungus.

Nelson *et al.* (45) expanded the number of *Fusarium* species to six, adding another chain-forming species, *Fusarium proliferatum*, separated because of the presence of polyphialides, and elevating *F. moniliforme* var. *subglutinans* to species rank as *Fusarium subglutinans*. Gerlach and Nirenberg (13) used much the same material as did Nelson *et al.*, but identified ten taxa instead of six. They also noted that, based on priority, the correct name for *F. moniliforme* is *Fusarium verticillioides* as descrybed by Nirenberg (46). These differences did not signifycantly impact the sorghum research community, as both Gerlach and Nirenberg (13) and Nelson *et al.* (45) agreed that a single species was responsible for the sorghum disease problems, even though they differed in the name

TABLE 35-1. *Fusarium* species that form chains and might be identified as *Fusarium moniliforme*; preferred name given first, followed by synonyms (indented).

Anamorph (Asexual) *Fusarium*	Teleomorph (Sexual) *Gibberella*
F. andiyazi Marasas *et al.* (39)	None
F. annulatum Bugnicourt (4)	None
F. brevicatenulatum Nirenberg *et al.* (48)	None
F. fujikuroi	*G. fujikuroi* *G. fujikuroi* mating population C (28)
F. globosum Rheeder *et al.* (55)	None
F. lactis (Pirotta & Riboni) Nirenberg & O'Donnell (47)	None
F. napiforme Marasas *et al.* (38)	None
F. nygamai Burgess & Trimboli (5)	*G. nygamai* Klaasen & Nelson (22) *G. fujikuroi* mating population G (29)
F. phyllophilum Nirenberg & O'Donnell (47)	None
F. proliferatum (Matsushima) Nirenberg (46)	*G. intermedia* (Kuhlman) Samuels *et al.* (57) *G. fujikuroi* mating population D (28)
F. pseudocircinatum O'Donnell & Nirenberg (47)	None
F. pseudonygamai O'Donnell & Nirenberg (47)	None
F. thapsinum Klittich *et al.* (25)	*G. thapsina* Klittich *et al.* (25) *G. fujikuroi* mating population F (24)
F. verticillioides (Sacc.) Nirenberg (46)	*G. moniliformis* Wineland (62) *G. fujikuroi* mating population A (27)

proposed for it–*F. verticillioides* or *F. moniliforme*. For sorghum pathologists, this kept the identification task relatively simple, as most of the *Fusarium* isolates could be quickly assigned to this species. This simplicity, however, had its price, for the same fungus was thought to infect both maize and sorghum, and researchers who were working with different strains of "*F. moniliforme*" could obtain results that were nearly polar opposites.

Beginning in 1977, Hsieh *et al.* (15) began to work with the sexual stage of the fungi in this section. They differentiated three different heterothallic mating groups. Within a group, strains were generally cross-fertile, but between groups, no successful crosses could occur. These mating groups were a *de facto* application of the biological species concept to these fungi. Kathariou (19) would later extend this work to a fourth mating population, and Kuhlman (27) described each of these four populations as a variety of *Gibberella fujikuroi*, even though some had been assigned species names by Wineland (62) and Edwards (10). To further complicate matters, Leslie (28) and Klittich and Leslie (24) described two additional mating populations (bringing the total to six, A-F). The E and the F mating populations were the first to seriously threaten the relatively small number of species in this section, as they resulted in both the *F. moniliforme* and *F. subglutinans* anamorphs each containing two biological species, A and F, and B and E, respectively. Further studies with isozymes (16) and karyotypes (65) confirmed the differences that were detected through crossing behavior leading first to the separation of the A and F mating populations as separate species (25) and later to the separation of B and E (29, 57). The concept of sexual cross-fertility to define species has led to the description of two additional mating populations, G (22) and H (3, 47), and the identification of a possible third (8).

Molecular technology also has been applied to study these problems, and two general approaches have been taken. Sequencing of relatively conserved genes, *e.g.* β-tubulin, nuclear 28S rDNA, mitochondrial small subunit rDNA (50), histone (59), and *MAT-2* (60), has been used to construct phylogenies and as the primary basis for the delineation of groups of strains that were then described as new species (47). As an alternative, Marasas *et al.* (39) used Amplified Fragment Length Polymorphisms (AFLPs) to group similar cultures. They found that strains within the same species usually shared 65% or more of the bands produced, and that those in different species usually shared 40% or less. The phytopathological importance of these new species has not been firmly established, and too few strains have been identified to determine their range. It would be surprising, however, if none of these species were important plant pathogens. As many of the cultures are originating from collections from sorghum and millet in Africa, the terminology used to describe *Fusarium* at the fourth global conference a decade from now will undoubtedly be quite different from what was used at this meeting.

A central question is whether these species differences matter in any practical sense. Why not just find a name of some sort that can be applied to all of them as a group and continue with diagnostic processes as in the past? We would encourage such a practice if we thought these differences unimportant. Unfortunately, these differences may be critical. For example, Leslie and Plattner (32) found that the different mating populations differed in the frequency with which they colonized maize and sorghum (Table 35-2). They also differ in the mycotoxins that they produce (Table 35-3) and in their relative aggressiveness to sorghum (Table 35-4). To continue to lump these quite different fungi under the same name does not do the sorghum research community a useful service.

Host preference (Table 35-2) can be an important tool in identifying these fungi, as there are some important differences. Strains from maize, for example, usually are *F. verticillioides*, while those from sorghum are more commonly *Fusarium thapsinum*. The relatively small number of strains clearly identified to other species prevents clear-cut conclusions, although *Fusarium andiyazi* is almost certainly important for sorghum (33, 39) and *Fusarium pseudonygamai* is almost certainly important on millet (33). There is some evidence that environment also plays an important role, since, for example, *Gibberella zeae* is only found in climates that are relatively cool and wet for sorghum (42). The general importance of geographic location as an indicator of species

TABLE 35-2. Relative frequency of recovery of strains belonging to different *Fusarium* species from the *Liseola* section from maize and sorghum (after 32 and 36).

Sample	*F. verticillioides* (%)	*F. proliferatum* (%)	*F. thapsinum* (%)	Other *G. fujikuroi* MPs (%)	Sterile (%)	Total no. of isolates
Maize–USA (32)	40	25	5	9	21	311
Sorghum–USA (32)	6	10	55	4	25	303
Sorghum–Tanzania (36)	21	7	49	8	15	354

present has not been adequately evaluated due to the general lack of population-sized samples.

Mansuetus *et al.* (36) found in a study of *Fusarium* spp. from sorghum grown in Tanzania that the distribution of strains in different mating populations was similar to that seen in the United States by Leslie and Plattner (32). Of most importance is that in both studies (Table 35-2), a significant number of sterile strains were identified. These sterile strains are likely to belong to the species that are just now being differentiated and formally described. Some of the reports of strains belonging to the "B" and "E" mating populations also may be erroneous, as the tester strains for these species are known to occasionally produce perithecia homothallically, and could lead to reactions that are scored as false positives (3). There also are reports of strains that can not be identified to any known species that require further attention, *e.g.*, Klaasen and Nelson (23), Leslie and Plattner (32), Marasas *et al.* (41), Onyike *et al.* (53), and Onyike and Nelson (51, 52). Re-evaluation of the identity of strains previously classified as "*Fusarium moniliforme*" is an essential effort that will be needed to understand, at least retrospectively, the results of much previous research.

Secondary Metabolites

The most important toxin produced by these fungi is fumonisin (Table 35-3). Fumonisin B_1 is known to cause leukoencephalomalacia in horses (20, 37), pulmonary edema in swine (14), and liver cancer in rats (12). It is associated with esophageal cancer in humans (56). Of the species associated with sorghum, *F. verticillioides* and *F. proliferatum* are the most prolific fumonisin producers. Although common on maize, these species are much less common on sorghum. While fumonisins undoubtedly are produced on sorghum, the contamination level is likely to be much lower than would a sample of maize that was naturally infested at a similar level.

Of the *Fusarium* species commonly found on sorghum, *F. thapsinum* produces much more moniliformin than fumonisin (33, 40). Moniliformin is toxic to poultry (26), but has not been studied as extensively as have the fumonisins. The relative risk to humans and domesticated animals who eat material contaminated with moniliformin is not well understood. Numerous additional seconddary metabolites are known to be produced by some of the species in this section, including beauvericin (11, 34, 43), fusaproliferin (11, 35, 43), fusarins (61), fusaric acid (1), and gibberellic acid (7, 54). Sorghum grain has not been routinely subjected to analysis for these metabolites, and baseline data on these toxins in sorghum grain is generally missing. In addition, there is reason to believe

TABLE 35-3. Relative fumonisin and moniliformin levels produced by strains of 12 *Fusarium* spp. that produce microconidia in chains.

Species	Fumonisin	Moniliformin
F. andiyazi	-	-
F. brevicatenulatum	±	-
F. globosum	+	-
F. lactis	+	-
F. napiforme	+	+
F. nygamai	+	+
F. phyllophilum	+	+
F. pseudocircinatum	+	+
F. proliferatum	++	+
F. pseudonygamai	-	+++
F. thapsinum	+	++
F. verticillioides	+++	-

that additional toxins remain to be identified, based on studies of duckling toxicity (31). The relative levels of toxin present could be extremely important for subsistence farmers who rely primarily on sorghum in their diet. The potential differences in the amount of fumonisins present could have significant human health implications if these farmers switch from a traditional African diet, composed primarily of sorghum, to a diet that is now composed primarily of maize.

Differences in relative aggressiveness of the different species towards sorghum also need to be thoroughly investigated. Leslie *et al.* (33) (Table 35-4) found that there were significant differences between the five species they tested, with the two most commonly recovered from sorghum, *F. thapsinum* and *F. andiyazi*, being the most aggressive in a seedling assay. In a test using toothpick inoculations of mature maize or sorghum plants, Jardine and Leslie (17, 18) found that both *F. verticillioides* and *F. thapsinum* produced lesions of similar length. Once the dominant species have been identified from sorghum, it will be necessary to determine the degree of resistance present in breeding materials and the inheritance patterns of the resistance that is present. More exhaustive tests of aggressiveness to both seedlings and to mature plants are necessary to determine the degree of correlation of these phenotypes and whether their genetic bases are similar.

Intraspecific Variation

No analysis of differences between species is complete if it does not include an analysis of variation within the species. Without such an analysis it is not possible to conclude if the differences observed are limited to those of a single strain, or whether they are the property of the species being examined. When conducted properly, these studies require a substantial number of strains, as the goal is to be ≥ 95% certain that all alleles found in at least 5% (or 1%) of the individuals in a population have been detected. For detecting all alleles present at a frequency ≥ 5%, the number of strains to be examined is 59 (95% confidence), 90 (99% confidence), or 135 (99.9% confidence). If all alleles present at a frequency ≥ 1% are to be detected, then the number of strains to be examined is 299 (95% confidence), 459 (99% confidence), or 688 (99.9% confidence). In practice, the 95% confidence level for alleles present at a level of at least 5% is usually targeted. For population collections, usually approximately 100 strains must be recovered to ensure that at least 59 of the most common species have been recovered. If samples of a less-frequent species are desired, then the total number of samples recovered must be increased sufficiently for there to be ≥ 59 isolates of that species in the sample. Most studies of genetic variation have relied on tests of fertility or vegetative compatibility (VCGs), or on molecular markers.

TABLE 35-4. Relative aggressiveness of five different species of *Fusarium* towards sorghum (after 33).

Species	Disease[a]	Shoot		Root	
		Mass[b] (mg)	Length[b] (mm)	Mass[b] (mg)	Length[b] (mm)
F. andiyazi	370 d	47 d	250 c	17 a	110 b
F. nygamai	170 b	55 bc	280 b	18 a	110 b
F. pseudonygamai	160 b	52 c	280 b	17 a	74 c
F. thapsinum	440 e	40 e	210 d	13 b	67 d
F. verticillioides	210 c	56 b	290 a	18 a	110 b
Control	43 a	61 a	290 a	17 a	120 a

[a]Rank mean.
[b]Values within a column followed by the same letter are not significantly different (p = 0.05).

Tests of female fertility can be used to estimate the relative levels of sexual and asexual reproduction occurring within a population. For *F. verticillioides*, *F. proliferatum*, and *F. thapsinum*, Leslie and Klein (30) found that 49, 30, and 10% of the strains from a global collection were self-sterile hermaphrodites (could serve as either the male or female parent in a cross), respectively, and that the remainder of the strains could function only as the male parent. Mansuetus *et al.* (36) found comparable results in strains isolated from sorghum in Tanzania (*F. verticillioides*, 51%; and *F. thapsinum*, 19%), as did Leslie and Salleh (*unpublished*) from maize in Malaysia (*F. verticillioides*, 79%). These values can be used to estimate N_e, the effective population number, which is a standard population genetics statistic first derived by Sewall Wright (6, 63), that permits the comparison of a population with a non-random mating structure to one in which random mating occurs.

A side benefit from these calculations for heterothallic ascomycete fungi, such as these *Fusarium* spp., is the ability to estimate the relative number of sexual and asexual generations, see Leslie and Klein (30) for details of these calculations. In the *F. verticillioides* populations, the global population had N_e of 89% of the count, the Tanzanian population 69% of the count, and the Malaysian population 97% of the count. The number of sexual generations is less than the number of asexual generations, and was 1:35-1:70 (sexual–asexual) for the global population, similar for the Tanzanian population, and less for the Malaysian population. In the *F. thapsinum* populations, the global population had N_e of 32% of the count and the Tanzanian population 53% of the count, with corresponding frequencies of sexual reproduction of 1:110–1:230 (global) and 1:35–1:70 (Tanzania). The frequency of sexual reproduction in the *F. thapsinum* populations is low enough that many strains will appear to be clones, and that a relatively few multi-locus genotypes may dominate in the population. Under these conditions, effective resistance breeding probably will require the identification of the local dominant strain type(s), and the use of representative strains to test the material in the program. The frequency of sexual reproduction in *F. verticillioides* (1:35–1:70 asexual generations), while not high, is high enough so that most of the strains isolated are genotypically distinct, and that techniques such as VCG analysis usually gives little information other than every strain is different from either most or all of the rest. In this case, breeding for resistance to a particular strain type is likely to be unsuccessful, and either strains carrying different pathogenicity genes will need to be identified, or resistance to the species as a whole will need to be selected. At this time, little is known of the mechanisms governing resistance to these fungi, and successful breeding is more likely to involve an empirical approach than one guided by a knowledge of the underlying mechanisms.

Data on molecular variation and N_e are not available for most of the newly described species; indeed, in some cases there are no more than two strains of these species available. More detailed study of these other fungi with respect to sorghum and millet is needed to determine which are phytopathological threats and which are mycological curiosities.

The "Real" *Fusarium moniliforme*?

The fungus to which the name *F. moniliforme* is applied certainly depends upon the context in which the fungus was collected and the host and or toxins it is thought to produce. From sorghum, we think that most "*F. moniliforme*" isolates would belong to one of five species: *F. thapsinum*, *F. andiyazi*, *F. nygamai*, *F. proliferatum*, and perhaps *F. verticillioides*. The two most frequently recovered of these fungi produce little or no fumonisin but can synthesize high levels of moniliformin, which may be a particular problem for poultry. This situation is quite different from that in maize, where *F. verticillioides* is by far and away the dominant species, followed by *F. proliferatum* and *F. subglutinans*. The more common of these species both can produce high levels of fumonisins, and the secondary metabolite profiles of maize and sorghum grain should reflect these differences. In terms of other hosts, "*F. moniliforme*" is not adequately defined, generally due to the limited number of samples. "*F. moniliforme*" from millet is probably *F. pseudonygamai*, from rice probably *F. fujikuroi* or *F. proliferatum*, and from mango *F. proliferatum*, *F. pseudocircinatum* (Leslie, *unpublished*), or one of at least two other undescribed taxa.

Distinguishing the different species using solely morphological characters is not easy, and even experienced mycologists can have difficulty in locating and identifying the critical distinguishing features. Culture conditions can be extremely important, *e.g.*, morphologically based diagnoses of cultures grown on PDA (potato-dextrose agar) plates are likely to be wrong, and the critical characters extremely subtle. A dichotomous key to the most prominent species can be found in Table 35-5, but this key is useful only if the cultures have been grown on carnation leaf agar, or another natural substrate, with adequate light and temperature (see 13 and 45 for details).

Sexual cross-fertility remains the "gold" standard for identification within this group. Fertile tester strains are available from the Fungal Genetics Stock Center, Department of Microbiology, University of Kansas Medical Center, Kansas City, Kansas. Strains that are cross-fertile with these testers at a high level can be considered to be conclusively identified. Crosses in which only a few perithecia are formed must be evaluated with greater care, as some of the tester strains occasionally produce perithecia homothallically (3). The ascospores in these perithecia are all identical to the parent, and if questions about fertility arise, then the identity can be quickly checked with a VCG analysis of the parent and a few of the progeny.

There are several problems related to crossing, however. First, the process is lengthy. Even though the effort required to make the crosses is not particularly high, the fertilized plates usually incubate for 2-4 weeks before a definitive diagnosis can be made. Second, interpretation of no-cross results is not definitive. At most, one can say that a strain does not belong to a particular species. A positive cross with one of the testers is required if the species diagnosis is to be definitive. Finally, fertile tester strains for most of the newly described species have not been developed. Given the relatively low number of female-fertile strains found in the species with identified teleomorphs and the effort required to construct reliable tester strains, usually a few generations of backcrossing or sib mating, the necessary tester strains are unlikely to become available for many of the more recently described species in the near future. The recent development of PCR tests (21, 60) that can be used to identify the mating type alleles (*MAT-1* and *MAT-2*) in these fungi also should help

TABLE 35-5. Dichotomous key for morphological identification of eight *Fusarium* species forming microconidia in chains that may be associated with sorghum or millet.

Chlamydospores or Pseudochlamydospores absent	
Polyphialides absent	
Yellow pigment absent	*F. verticillioides*[a]
Yellow pigment present	*F. thapsinum*
Polyphialides present	
Globose microconidia absent	*F. proliferatum*
Globose microconidia present	*F. globosum*
Chlamydospores or Pseudochlamydospores present	
Chlamydospores present in aerial mycelium	
Polyphialides absent, napiform microconidia present	*F. napiforme*
Polyphialides present, napiform microconidia absent	*F. nygamai*
Chlamydospores absent in aerial mycelium	
Pseudochlamydospres present in carnation leaves on CLA	
Polyphialides present	*F. pseudonygamai*
Polyphialides absent	*F. andiyazi*

[a]Some *F. thapsinum* strains also fail to produce yellow pigment.

to simplify the tester development process, by enabling *MAT* allele identification in the absence of a positive cross.

Isozymes can be used to differentiate some of the groups, but little information is available on the alleles present in many of the new species. Population level studies employing this technique also are limited. The single most useful isozyme appears to be triose phosphate isomerase, for which *F. verticillioides* appears to carry a unique allele. Identification of any of the other species examined requires a combination of two, or more, isozymes to be conclusive. If isozymes are not a technology already in use in a laboratory, then we recommend that one of the DNA-based technologies be adopted instead.

DNA-based technologies are of three basic types, none of which can be addressed in great detail here. At this time, these technologies are the only means by which all of the species in Table 35-1 can be repeatedly and reliably distinguished on every occasion. The simplest relies on amplification via PCR of a fragment specific for a given *Fusarium* species such as those developed by Möller *et al.* (44) for *Fusarium* from maize. Such assays are undoubtedly the future of *Fusarium* diagnostics, but these assays are not yet generally available for *Fusarium* species common on sorghum. Hopefully, diagnostic primers combined with a simplified DNA extraction (9) will be routine by the time of the Fourth Global Conference on sorghum and millet pathology, and the focus will be on the biology of these strains and their interactions with the host rather than simply their identification.

AFLPs provide a good method for grouping strains that belong in the same species (39). Generally, strains that share ≈ 65% or more of the AFLP bands are in the same species, while strains in different species rarely share more than 40% of the bands. AFLP patterns also can be used to determine genetic relationships between strains in the same species and are, therefore, useful for studies of migration and hybridization. The utility of AFLP banding patterns for phylogenetic studies is less clear. Once strains are grouped, then the strains can be examined for morphological similarities that might not have been immediately obvious. Alternatively, the existing *MAT* primers (21, 60) can be used to determine mating type with the strains of opposite mating type then checked for cross-fertility.

DNA sequencing, commonly of the β-tubulin, mitochondrial small subunit RNA (49), the 5 portion of the nuclear 28S rDNA (50), histones (59), and *MAT-2* (60) also can be used. Note that the ITS2 region of the nuclearly encoded ribosomal repeat should generally *not* be used for this purpose in these fungi (49). DNA sequences can be used to construct phylogenetic trees and estimate distances between species, which is more difficult to do with AFLP data. DNA sequences, on the other hand, sample a much more limited portion of the genome than do AFLPs, and sequences with an appropriate level of variation for determining species relationships often are too conserved to be of use in determining relationships between strains in the same species. As with AFLPs, once similar strains have been grouped, then the search for morphological characters and strains that form fertile perithecia can begin.

Recommendations

Given the known genetic variability within these fungi and the fact that the host plant may be colonized by several species or genetically distinct isolates of the same species, single-spore subcultures of any strains to be examined should be essential. Mixed cultures will give confusing results regardless of the technique used to analyze them. Similarly, the practice of putting a strain through a plant to "restore its pathogenicity" should not be done unless a multigenic marker is assayed in both the introduced and the recovered strain to ensure that the strains really are the same. Developers of species-specific diagnostic markers should be particularly sensitive to the need to account for all of the significant variation that can occur within a species, and they also should analyze a sample that is both large enough and diverse enough to include this variation.

We recommend that if the only criteria available are morphological characters, then the strain should be assigned to one of the species listed in Table 35-5. If such an assignment is not possible, then we recommend that the strain be termed *Fusarium moniliforme sensu lato*. This diagnosis will be as accurate as many presently being done, but this diagnosis is more intellectually honest as it indicates that an accurate diagnosis to species level was not made and alerts others that the strains described may belong to one of several different species. If access to PCR-based technology is available, then the diagnostic assays should be adopted as soon as they become practical. Until that time, AFLPs and DNA sequencing probably will remain the most reliable tools for the identification and differentiation of these *Fusarium* spp.

Conclusions

There are numerous candidates for the "real" *Fusarium moniliforme*. These candidate fungi differ significantly in their host preference, phytopathological capability, toxin production, and, probably, geographic and climatic distribution. Those species common on sorghum are relatively new, in terms of the length of time since they were described, and their relative importance, in terms of plant pathogenicity and toxin production, remains to be determined.

Acknowledgments

This research was supported in part by the Kansas Agricultural Experiment Station and by the Sorghum and Millet Collaborative Research Support Program (INTSORMIL) AID/DAN-1254-G-00-0021-00 from the U. S. Agency for International Development. Contribution no. 02-195-B from the Kansas Agricultural Experiment Station, Manhattan.

References

1. **Bacon, C. W., J. K. Porter, W. P. Norred, and J. F. Leslie.** 1996. Production of fusaric acid by *Fusarium*. *Appl. Environ. Microbiol.* **62:** 4039-4043.
2. **Booth, C.** 1971. *The Genus* Fusarium. Commonwealth Agricultural Bureau, Bucks, UK.
3. **Britz, H., T. A. Coutinho, M. J. Wingfield, W. F. O. Marasas, T. R. Gordon, and J. F. Leslie**. 1999. *Fusarium subglutinans* f. sp. *pini* represents a distinct mating population in the *Gibberella fujikuroi* species complex. *Appl. Environ. Microbiol.* **65**: 1198-1201.
4. **Bugnicourt, F.** 1952. Une espèce fusaruenne nouvelle, parasite du riz. *Rev. Gén. Bot.* **59:** 13-18.
5. **Burgess, L. W., and D. Trimboli.** 1986. Characterization and distribution of *Fusarium nygamai*, sp. nov. *Mycologia* **78:** 223-229.
6. **Caballero, A.** 1994. Developments in the prediction of effective population size. *Heredity* **73:** 657-679.
7. **Cerdá-Olmedo, E., M. R. Fernández, and J. Avalos.** 1994. Genetics and gibberellin production in *Gibberella fujikuroi*. *Antonie van Leeuwenhoek* **65:** 217-225.
8. **Desjardins, A. E., R. D. Plattner, and T. R. Gordon.** 2000. *Gibberella fujikuroi* mating population A and *Fusarium subglutinans* from teosinte species and maize from Mexico and Central America. *Mycol. Res.* **104:** 865-872.
9. **DuTeau, N. M., and J. F. Leslie.** 1991. A simple, rapid procedure for the isolation of DNA for PCR from *Gibberella fujikuroi* (*Fusarium* section *Liseola*). *Fung. Genet. Newsl.* **38**: 72.
10. **Edwards, E. T.** 1941. Internal grain infection in maize due to *Gibberella fujikuroi* and *Gibberella fujikuroi* var. *subglutinans*. *J. Austr. Inst. Agric. Sci.* **7:** 74-82.
11. **Fotso, J., J. F. Leslie, and J. S. Smith.** 2002. Production of beauvericin, moniliformin, fusaproliferin, and fumonisins B_1, B_2 and B_3 by fifteen ex-type strains of *Fusarium* species. *Appl. Environ. Microbiol.* **68:** 5195-5197.
12. **Gelderblom, W. C. A., N. P. J. Kriek, W. F. O. Marasas, and P. G. Thiel.** 1991. Toxicity and carcinogenicity of the *Fusarium moniliforme* metabolite fumonisin B_1 in rats. *Carcinogenesis* **12:** 1247-1251.
13. **Gerlach, W., and H. Nirenberg.** 1982. The genus *Fusarium*–A pictorial atlas. *Mitt. Biol. Bundesanst. Land-Forstwirtsch, Berlin-Dahlem* **209:** 1-406.
14. **Harrison, L. R., B. M. Colvin, J. T. Green, L. E. Newman, and J. R. Cole.** 1990. Pulmonary edema and hydrothorax in swine produced by fumonisin B_1, a toxic metabolite of *Fusarium moniliforme*. *J. Vet. Diagn. Investig.* **2:** 217-221.
15. **Hsieh, W. H., S. N. Smith and W. C. Snyder.** 1977. Mating groups in *Fusarium moniliforme*. *Phytopathology* **67:** 1041-1043.
16. **Huss, M. J., C. L. Campbell, D. B. Jennings, and J. F. Leslie.** 1996. Isozyme variation among biological species in the *Gibberella fujikuroi* species complex (*Fusarium* section *Liseola*). *Appl. Environ. Microbiol.* **62:** 3750-3756.
17. **Jardine, D. J., and J. F. Leslie.** 1992. Aggressiveness of *Gibberella fujikuroi* (*Fusarium moniliforme*) isolates to grain sorghum under greenhouse conditions. *Plant Dis.* **76:** 897-900.
18. **Jardine, D. J., and J. F. Leslie.** 1999. Aggressiveness to mature maize plants of *Fusarium* strains differing in ability to produce fumonisin. *Plant Dis.* **83:** 690-693.
19. **Kathariou, S.** 1981. *Gene pool organization in* Fusarium moniliforme. Ph.D. dissertation. Univ. of California, Berkeley, California.
20. **Kellerman, T. S., W. F. O. Marasas, P. G. Thiel, W. C. A. Gelderblom, M. Cawood, and J. A. W. Coetzer.** 1990. Leukoencephalomalacia in two horses induced by oral dosing of fumonisin B_1. *Onderstepoort J. Vet. Res.* **57:** 269-275.
21. **Kerényi, Z., K. Zeller, L. Hornok, and J. F. Leslie.** 1999. Standardization of mating-type terminology in the *Gibberella fujikuroi* species complex. *Appl. Environ. Microbiol.* **65**: 4071-4076.
22. **Klaasen, J. A., and P. E. Nelson.** 1996. Identification of a mating population, *Gibberella nygamai* sp. nov., within the *Fusarium nygamai* anamorph. *Mycologia* **88:** 965-969.
23. **Klaasen, J. A., and P. E. Nelson.** 1998. Identity of *Fusarium nygamai* isolates with long and short microconidial chains from millet, sorghum and soil in Africa. *Mycopathologia* **140:** 171-176.
24. **Klittich, C. J. R., and J. F. Leslie.** 1992. Identification of a second mating population within the *Fusarium moniliforme* anamorph of *Gibberella fujikuroi*. *Mycologia* **84:** 541-547.
25. **Klittich, C. J. R., J. F. Leslie, P. E. Nelson, and W. F. O. Marasas.** 1997. *Fusarium thapsinum* (*Gibberella thapsina*): A new species in *Liseola* from sorghum. *Mycologia* **89:** 643-652.
26. **Kriek, N. P. J., W. F. O. Marasas, P. S. Steyn, S. J. van Rensburg, and M. Steyn.** 1977. Toxicity of a moniliformin-producing strain of *Fusarium moniliforme* var. *subglutinans* isolated from maize. *Food and Cosm. Toxicol.* **15:** 579-587.
27. **Kuhlman, E. G.** 1982. Varieties of *Gibberella fujikuroi* with anamorphs in *Fusarium* section *Liseola*. *Mycologia* **74:** 759-768.
28. **Leslie, J. F.** 1991. Mating populations in *Gibberella fujikuroi* (*Fusarium* section *Liseola*). *Phytopathology* **81:** 1058-1060.
29. **Leslie, J. F.** 1995. *Gibberella fujikuroi*: Available populations and variable traits. *Can. J. Bot.* **73(Suppl. 1):** S282-S291.

30. **Leslie, J. F., and K. K. Klein.** 1996. Female fertility and mating type effects on effective population size and evolution in filamentous fungi. *Genetics* **144:** 557-567.
31. **Leslie, J. F., W. F. O. Marasas, G. S. Shephard, E. W. Sydenham, S. Stockenström, and P. G. Thiel.** 1996. Toxicological differences between isolates in the A and F mating populations of *Gibberella fujikuroi* (*Fusarium moniliforme*). *Appl. Environ. Microbiol.* **62:** 1182-1187.
32. **Leslie, J. F., and R. D. Plattner.** 1991. Fertility and fumonisin B_1 production by strains of *Fusarium moniliforme* (*Gibberella fujikuroi*). *Proc. 17th Sorg. Improve. Conf. No. Amer. (Lubbock, Texas)*: 80-84.
33. **Leslie, J. F., K. A. Zeller, W. F. O. Marasas, S. C. Lamprecht, and J. P. Rheeder.** 2003. Toxicity, pathogenicity and genetic differentiation of five species of *Fusarium* from sorghum and millet. (Submitted.)
34. **Logrieco, A., F. Fornelli, G. Castella, M. Kostecki, P. Golinski, A. Ritieni, and J. Chelkowski.** 1998. Beauvericin production by *Fusarium* species. *Appl. Environ. Microbiol.* **64:** 3084-3088.
35. **Logrieco, A., A. Moretti, F. Fornelli, V. Fogliano, A. Ritieni, M. F. Caiaffa, G. Randazzo, A. Bottalico, and L. Macchia.** 1996. Fusaproliferin production by *Fusarium subglutinans* and its toxicity to *Artemia salina*, SF-9 insect cells, and IARC/LCL 171 human B lymphocytes. *Appl. Environ. Microbiol.* **62:** 3378-3384.
36. **Mansuetus, A. S. B., G. N. Odvody, R. A. Frederiksen, and J. F. Leslie.** 1997. Biological species in the *Gibberella fujikuroi* species complex (*Fusarium* section *Liseola*) recovered from sorghum in Tanzania. *Mycol. Res.* **101:** 815-820.
37. **Marasas, W. F. O., T. S. Kellerman, W. C. A. Gelderblom, J. A. W. Coetzer, P. G. Thiel, and J. J. van der Lugt.** 1988. Leukoencephalomalacia in a horse induced by fumonisin B_1 isolated from *Fusarium moniliforme*. *Onderstepoort J. Vet. Res.* **55:** 197-203.
38. **Marasas, W. F. O., C. J. Rabie, A. Lübben, P. E. Nelson, T. A. Toussoun, and P. S. van Wyk.** 1987. *Fusarium napiforme*, a new species from millet and sorghum in southern Africa. *Mycologia* **79:** 910-914.
39. **Marasas, W. F. O., J. P. Rheeder, S. C. Lamprecht, K. A. Zeller, and J. F. Leslie.** 2001. *Fusarium andiyazi* sp. nov., a new species from sorghum. *Mycologia* **93:** 1203-1210.
40. **Marasas, W. F. O., P. G. Thiel, C. J. Rabie, P. E. Nelson, and T. A. Toussoun.** 1986. Moniliformin production in *Fusarium* section *Liseola*. *Mycologia* **78:** 242-247.
41. **Marasas, W. F. O., P. G. Thiel, E. W. Sydenham, C. J. Rabie, A. Lübben, and P. E. Nelson.** 1991. Toxicity and moniliformin production by four recently described species of *Fusarium* and two uncertain taxa. *Mycopathologia* **113:** 191-197.
42. **Menkir, A., G. Ejeta, L. G. Butler, A. Melakeberhan, and H. L. Warren.** 1996. Fungal invasion of kernels and grain mold damage assessment in diverse sorghum germplasm. *Plant Dis.* **80:** 1399-1402.
43. **Moretti, A., A. Logrieco, A. Bottalico, A. Ritieni, V. Fogliano, and G. Randazzo.** 1996. Diversity in beauvericin and fusaproliferin production by different populations of *Gibberella fujikuroi* (*Fusarium* Section *Liseola*). *Sydowia* **48:** 44-56.
44. **Möller, E. M., J. Chełkowski, and H. H. Geiger.** 1999. Species-specific PCR assays for the fungal pathogens *Fusarium moniliforme* and *Fusarium subglutinans* and their application to diagnose maize ear rot disease. *J. Phytopathol.* **147:** 497-508.
45. **Nelson, P. E., T. A. Toussoun, and W. F. O. Marasas.** 1983. Fusarium *Species: An Illustrated Manual for Identification*. Pennsylvania State University Press, University Park, Pennsylvania.
46. **Nirenberg, H. I.** 1976. Untersuchungen über die morphologische und biologische Differenzierung in der *Fusarium*-Sektion *Liseola*. *Mitt. Biol. Bundesanst. Land-u. Forstwirtsch. Berlin-Dahlem* **169:** 1-117.
47. **Nirenberg, H. I., and K. O'Donnell.** 1998. New *Fusarium* species and combinations within the *Gibberella fujikuroi* species complex. *Mycologia* **90:** 434-458.
48. **Nirenberg, H. I., K. O'Donnell, J. Kroschel, A. P. Andrianaivo, J. M. Frank, and W. Mubatanhema.** 1998. Two new species of *Fusarium*: *Fusarium brevicatenulatum* from the noxious weed *Striga asiatica* in Madagascar and *Fusarium pseudoanthophilum* from *Zea mays* in Zimbabwe. *Mycologia* **90:** 459-464.
49. **O'Donnell, K., and E. Cigelnik.** 1997. Two divergent intragenomic rDNA ITS2 types within a monophyletic lineage of the fungus *Fusarium* are nonorthologous. *Molec. Phylogen. Evol.* **7:** 103-116.
50. **O'Donnell, K., E. Cigelnik, and H. I. Nirenberg.** 1998. Molecular systematics and phylogeography of the *Gibberella fujikuroi* species complex. *Mycologia* **90:** 465-493.
51. **Onyike, N. B. N., and P. E. Nelson.** 1992. *Fusarium* species associated with sorghum grain from Nigeria, Lesotho and Zimbabwe. *Mycologia* **84:** 452-458.
52. **Oniyke, N. B. N., and P. E. Nelson.** 1993. The distribution of *Fusarium* species in soils planted to millet and sorghum in Lesotho, Nigeria and Zimbabwe. *Mycopathologia* **121:** 105-114.
53. **Onyike, N. B. N., P. E. Nelson and W. F. O. Marasas.** 1991. *Fusarium* species associated with millet grain from Nigeria, Lesotho and Zimbabwe. **83:** 708-712.
54. **Phinney, B. O., and C. A. West.** 1960. Gibberellins as native plant growth regulators. *Annu. Rev. Plant Physiol.* **11:** 411-436.
55. **Rheeder, J. P., W. F. O. Marasas, and P. E. Nelson.** 1996. *Fusarium globosum*, a new species from corn in southern Africa. *Mycologia* **88:** 509-513.
56. **Rheeder, J. P., W. F. O. Marasas, P. G. Thiel, E. W. Sydenham, G. S. Shephard, D. J. van Schalkwyk.** 1992. *Fusarium moniliforme* and fumonisins in corn in relation to human esophageal cancer in Transkei. *Phytopathology* **82:** 353-357.
57. **Samuels, G. J., H. I. Nirenberg, and K. A. Seifert.** 2001. Perithecial species of *Fusarium*, p. 1-14. *In* B. A. Summerell J. F. Leslie, D. Backhouse, W. L. Bryden, and L. W. Burgess (eds.), Fusarium: *Paul E. Nelson Memorial Symposium*. APS Press: St. Paul, Minnesota.
58. **Snyder, W. C., and H. M. Hansen.** 1945. The species concept in *Fusarium* with reference to *Discolor* and other sections. *Amer. J. Bot.* **32:** 657-666.
59. **Steenkamp, E. T., B. D. Wingfield, T. A. Coutinho, M. J. Wingfield, and W. F. O. Marasas.** 1999. Differentia-

tion of *Fusarium subglutinans* f. sp. *pini* by histone gene sequence data. *Appl. Environ. Microbiol.* **65:** 3401-3406.

60. **Steenkamp, E. T., B. D. Wingfield, T. A. Coutinho, K. A. Zeller, M. J. Wingfield, W. F. O. Marasas, and J. F. Leslie**. 2000. PCR-based identification of *MAT-1* and *MAT-2* in the *Gibberella fujikuroi* species complex. *Appl. Environ. Microbiol.* **66**: 4378-4382.
61. **Wiebe, L. A., and L. F. Bjeldanes.** 1981. Fusarin C, a mutagen from *Fusarium moniliforme*. *J. Food Sci.* **46:** 1424-1426.
62. **Wineland, G. O.** 1924. An ascigerous stage and synonomy for *Fusarium moniliforme*. *J. Agric. Res.* **28:** 909-922.
63. **Wollenweber, H. W., and O. A. Reinking.** 1935. *Die Fusarien, ihre Beschreibung, Schadwirkung und Bekämpfung*. Paul Parey, Berlin, Germany.
64. **Wright, S.** 1931. Evolution of Mendelian populations. *Genetics* **16:** 97-159.
65. **Xu, J.-R., K. Yan, M. B. Dickman, and J. F. Leslie.** 1995. Electrophoretic karyotypes distinguish the biological species of *Gibberella fujikuroi* (*Fusarium* section *Liseola*). *Molec. Plant-Microbe Interact.* **8:** 74-84.

36

Sorghum Anthracnose–Problem and Management Strategies

Kusum Mathur, Ram P. Thakur, Adama Neya, Paul S. Marley, and Carlos R. Casela

Abstract

We review the economic importance and geographic distribution of sorghum anthracnose (caused by *Colletotrichum graminicola*) and advances made in sorghum anthracnose research, including pathogen population boilogy, pathogenic and genotypic diversity, epidemiology, the association of anthracnose with other diseases, host and organ specificity, and disease management strategies. Future research thrusts include documentation and description of the geographical and ecological distribution of species and subspecies of the pathogen, development of a comprehensive system, including DNA-marker technology, for characterizing races, virulence pattern, organ specificity and symptom types, identification of R genes, and strategic utilization and deployment of R genes in combination with other management options for economic and efficient management of this disease.

Introduction

The recommendations committee for sorghum anthracnose [*Colletotrichum graminicola* (Ces.) Wils.] during the Second World Review in 1988 (18) suggested research studies of host-parasite-environment interaction, pathogen variability, host range, the relationship of grain and foliar anthracnose to stalk rot, taxonomy of the pathogen *C. graminicola* from different graminaceous hosts, epidemiology, significance of seedborne inoculum in disease perennation, development of effective and reliable screening methods for identifying resistant material, mechanism of resistance, identification and utilization of resistance by classical and molecular (DNA-marker technology) approaches, genetics and inheritance of resistance, and disease management strategy involving chemicals, cultural practices, and biological methods. The committee also recommended the establishment of a working group on "Anthracnose on Gramineae" to promote cooperation among interested researchers worldwide. In this chapter, we evaluate progress made in these areas in the last decade and identify problems that remain to be addressed.

Geographic Distribution

Sorghum anthracnose is prevalent whenever sorghum is grown in a warm, humid environment. The disease occurs widely and is considered of primary importance in Asia–China and India; Central America and the Caribbean basin–Argentina, Brazil, El Salvador, Guatemala, Honduras, Hawaii, Mexico, Panama, Puerto Rico, and Venezuela; and Africa–Botswana, Ethiopia, Kenya, Mali, Niger, Mozambique, Somalia, South Africa, Sudan, Tanzania, Uganda, Zambia and Zimbabwe. The disease is of minor importance in Japan, the Philippines, and Thailand (25, 50, 69). Anthracnose stalk rot in the form of irregularly mottled or marbled red patterns on leaf, mid-vein, panicle and rachis branches, and stalks is prevalent in the United States (25, 26). Anthracnose caused by *Colletotrichum gloeosporioides* also has been reported on grain from and foliage of local land race sorghums in Thailand and India (5, 40).

Economic Importance

The disease causes both direct and indirect yield losses. The extent of direct losses varies with region, cultivar, and prevalent climatic conditions. The reduction in 1000-seed mass, 1000-seed density, and early abortion of seeds are the most important factors in yield reduction. The premature drying of leaves and defoliation due to foliar anthracnose can reduce the yield of sorghum grain and fodder by 30-50% in susceptible cultivars during severe epidemics (2). Stalk infection (stalk rot or red rot) also results in lodging and considerable reduction in grain and fodder yield (23, 46). Secondary infection of the stalk by red rot may cause yield losses proportional to the severity of anthracnose and stalk rot (49). In a study conducted in Mali (69, 73), yield loss, as calculated from fungicide-sprayed and unsprayed plots, of the susceptible inbred line was up to 69%. A moderately resistant inbred line, with almost the same area under the disease progress curve (AUDPC), had no apparent loss in grain yield due to anthracnose for two years and a reduction of ≤ 4% in the third year. Kernel weight losses were 16-36% in the susceptible line and 15% in the moderately resistant line (69, 73).

Indirect losses also result from grain anthracnose, due to reduced seed germination, and the transmission of the disease to new geographic regions (3). For example, five Brazilian seed samples contained 18-23% seed infected by *C. graminicola* (76). In a second example, a seed sample of sorghum cultivar QL3 from Puerto Rico had visible acervuli, and 23% of the seedlings resulting from this seed lot developed anthracnose within 1-2 weeks of germination with lesions developing on mesocotyledonary sheaths, primary leaf lamina, and primary roots (8). In a third example, 5/37 seed samples tested in India had anthracnose infection. Severely infected seeds did not germinate, while seeds with lesser infections had lower test weight, germinated poorly, and had a low vigor index (58).

Anthracnose does not usually occur in isolation in sorghum. In India, for example, several foliar fungal diseases are common (59), *e.g.*, zonate leaf spot (caused by *Gloeocercospora sorghi*), gray leaf spot (caused by *Cercospora sorghi*), rust (caused by *Puccinia purpurea*), leaf blight (caused by *Exserohilum turcicum*), and rough leaf spot (caused by *Ascochyta sorghi*). Environmental conditions that favor infection by one of these fungi also may favor infection by other fungi, and, consequently, multiple infections may be more common than generally perceived (53). There have been no systematic studies to assess the extent of damage due to mixed infection by two or more diseases or to determine whether the interactions between different diseases are additive or synergistic in terms of the amount of debilitation to the plant. Anthracnose stalk and Fusarium stalk rots are associated in the United States, where these pathogens interact to cause stalk rot, and are more prevalent when short periods of rain and drought occur (19).

The Pathogen

The taxonomic status of the sorghum anthracnose pathogen *Colletotrichum graminicola* (Ces.) Wils (= *C. sublineolum* P. Henn., Kabát & Bub.) has been controversial, probably due to the wide host range and the morphological and pathogenic variability associated with this fungus (53). Isolates recovered from one host will not necessarily infect other hosts. Strains infecting maize were designated *C. graminicola* and those infecting sorghum *C. sublineolum*, based on their appressorial morphology (64), rDNA sequence (60), and host specificity (32, 36). The perfect state, *Glomerella graminicola* Politis, is known for both maize and sorghum isolates of *C. graminicola* (76), but the conditions for development of the perfect state by sorghum isolates are not well defined (71). Sorghum and johnsongrass isolates from the upper coastal plains of Texas were genetically isolated from each other, while the sorghum isolates from the land race (maicillos) sorghums in Honduras were closely related to those from the Texas johnsongrass (57).

C. graminicola can infect all above-ground plant parts, including stem, leaves, peduncle, inflorescence, and grain. The most common form of this disease is leaf or foliar anthracnose. The first signs of foliar infection are circular to elliptical red spots 0.5-5.0 mm in diameter. These lesions may coalesce to occupy most or all of an infected leaf and may even kill the infected leaf. Acervuli of *C. graminicola* develop as brown-grayish to black specks usually arranged in concentric rings.

Differences in foliage and grain symptoms may be due to host reaction, physiological status of the host, the environment (22, 53), and pathogenic variability (23, 52). The large range in symptoms, cultural characters and isolate morphology led to the description of two *forma speciales* based on morphological characters and pathogenicity: *C. graminicola* var. *zonatum* from India, which forms diffuse spots (up to 50 mm in diameter) with numerous acervuli in concentric zones (55), and *C. graminicola* var. *isolatum* from Nigeria (1).

It is not clear whether fungal isolates from foliage, grain, and stalks are different, or whether variation in symptoms on different plant parts is due to the age of the plant, organ specificity of the particular isolate, environmental conditions, or some combination of these factors. Sorghum lines resistant to foliar anthracnose can develop heavy grain infection, and there is a weak, but significant, negative correlation between foliar and grain anthracnose ratings of hybrids, suggesting that resistance to foliar and grain anthracnose are independent characters (39) and that some isolates may be specific for foliage or grain (74). However, leaf and stalk isolates from sorghum in Texas were indistinguishable when analyzed with molecular markers, suggesting that there is no genetic basis for organ specificity (57). Clearly, much remains to be done to determine if the disease symptoms observed in the host have a genetic basis in the pathogen.

Cultural and morphological variability. Isolates of *C. graminicola* from different locations, lesions, or even the single-conidial derivatives from single-lesion cultures can vary considerably in morphology (15, 31, 41, 52, 56, 71). Isolates from an infected plant may be mixed in terms of genotypes and pathotypes present (39, 42). These mixtures may persist through mass mycelial subculturing, but also can be resolved by subculturing single spores. Some isolates exhibit conidial dimorphism and produce oval conidia when cultured in liquid media (70). Acervuli may range from not well differentiated with occasional setae (indistinct) to well developed (distinct) with 2 to > 20 setae per acervulus. Other variable characters include size of acervuli and production of appressoria and sclerotia (31, 56, 74). Significant intrapopulation variation in colony growth, pigmentation, and conidial size can occur within isolates subcultured from a colony initiated from a single conidium. Some cultures from a single conidium form distinct mycelial sectors that differ genetically and pathogenically from the original isolate (38). Morphological variation, however, is independent of pathogenic or genetic variation (39, 56).

Population biology. In the absence of sexual reproduction by *C. graminicola* on sorghum under field or laboratory conditions, it is not known how genetic information is exchanged within or between populations of *C. graminicola*. In putatively asexually reproducing fungi, genetic exchange via heterokaryosis and the parasexual cycle can be studied by using auxotrophic mutants (35, 54). Under field conditions, strains in the same vegetative compatibility group (VCG) can form heterokaryons and exchange genetic information in the absence of sexual reproduction. VCGs can be identified by using nitrate nonutilizing (*nit*) mutants to force heterokaryons under laboratory conditions, and have been used to characterize genetic relatedness of isolates from several *Colletotrichum* spp. (4, 17). In preliminary studies with 11 isolates of *C. graminicola* and two of *C. gloeosporioides* from anthracnose-infected leaves, *nit1* (nitrate reductase holoenzyme) mutants were readily generated, but a NitM (molybdenum cofactor for nitrate reductase complex) was recovered from only one of the strains. Thus, the pairings possible were limited, and additional *nit* mutants and strains need to be examined for the utility of the VCG protocol in evaluating genetic diversity in *C. graminicola* to be adequately assessed (41).

Epidemiology. There are conflicting reports about host specificity and host range of *C. graminicola* isolates from different hosts. *Sorghum halepense*, *Echinocloa colonum*, *Digitaria sanguinalis*, and *Dactyloctenium aegyptum* are all putative collateral hosts of *C. graminicola* infecting sorghum (33). However, Rosewich (57) found that *C. graminicola* isolates from *S. halepense* in Texas were genetically distinct from those of sorghum, and she concluded that *S. halepense* does not provide inoculum for the sorghum crop. *C. graminicola* can survive in infected leaf debris buried in the soil for up to 10 months (45). The survival period is influenced by temperature, and soil temperature ≥ 45°C greatly decreased infectivity of the lesions (44). Sporogenic sclerotia of *C. graminicola* can survive in sorghum stalk debris on the soil surface for up to 18 months. Sclerotial germination decreases faster if the sorghum stalk residue is buried 10-20 cm deep in soil than if the residue is on the soil surface (14). Thus, sclerotia in debris on the soil surface may act as a primary source of inoculum for anthracnose epidemics (14). However, not all fungal isolates produce sclerotia (31, 56). Further research is needed to determine whether sclerotium-forming isolates have a relative survival advantage. In Nigerian sandy loam soil, infective conidia could be recovered from soil nine months after harvest (75). As conidia are thin walled and prone to desiccation and fungistasis in the soil, the origin of these infective conidia is thought to be sporogenic sclerotia in the soil. Use of specific techniques to identify sclerotia in soil may provide more information on the survival of *C. graminicola*. Infected seeds also are potential sources of infection in fields first planted to sorghum and may give rise to infected seedlings (3, 8, 58, 77). The pathogen can survive in infected seeds stored at room temperature (20-35°C) for up to 30 months (45).

Pathogenic variability. The use of virulence to assess genetic variability provides direct information on the effect of host selection. Within *C. graminicola*, variation is known for virulence (disease reaction) and for aggressiveness (disease severity). This variation usually is detected on the basis of differential foliage reaction amongst a set of sorghum cultivars (12, 20, 21, 50, 53). In sorghum anthracnose, approximately 40 races/pathotypes have been identified amongst geographically diverse strains that have been tested under both greenhouse and field conditions. In many cases, host lines, pathogen strains, and/or disease scoring systems are unique to the research group. For example, eight pathotypes were identified amongst 12 isolates collected in Griffin, Georgia; Isabella, Puerto Rico; and central and southern Texas (Catspring, Telferner, Port Lavaca, College Station, and Orange Grove) based on reactions to eight differential grain sorghum lines (9).

Physiological races were differentiated by the reaction of sorghum cultivars tested in the United States, Guatemala, and Brazil (48, 53), and a virulence shift in the Brazilian populations occurred in response to host-guided selection. Anthracnose-resistant sweet sorghum varieties–Brandes, Theis, Rio, and Wray–were resistant to anthracnose when introduced into Brazil in 1987 but were susceptible to new fungal races that developed after 1987 (79). Subsequently, several physiological races of *C. graminicola*, differentiated by reactions of twelve sorghum cultivars, were identified in Brazil. The sorghum cultivars Redland (Tx378), SC 326-6, and SC 283 can be used to distinguish eight important groups of races, designated A to H. Tx623, Brandes, SC 112-14, Tx398, Tx2536, and Theis can be used to distinguish 32 races in each group. Based on this system, 16 races were identified in Brazil in 1984-1985, including three that had been detected previously (10, 12, 20, 21).

In India, researchers have identified various numbers of pathotypes. Nine isolates collected from nine Indian states were designated as nine races on the basis of their reactions towards 32 sorghum lines (52). In a later study, however, 33 isolates collected from eight states of India (including those in first study) were grouped into six distinct phenotypic clusters (56). Isolates from the same state could belong to separate clusters. Forty-six foliage and grain isolates collected from seven districts of Andhra Pradesh during 1992-1997 varied in colony growth and were classified into seven clusters based on a virulence index, with variation within each cluster (39). Twenty single-spore purified isolates of *C. graminicola* from Kovilpatti, India, varied in pathogenicity and genotype, suggesting that some isolates are highly heterogeneous. Virulence towards improved popular hybrids was detected in populations of *C. graminicola* from Andhra Pradesh and Maharashtra. Populations from a single hybrid, CSH 5, collected from different locations in Maharashtra varied considerably in pathogenicity and could be differentiated into to five pathotypes (42). In another study, 12 isolates from four states in India were placed in three pathogenic and four genotypic groups (32).

Molecular markers have been useful in identifying genetic variation among isolates of *C. graminicola* infecting maize and sorghum (60) and also in distinguishing sorghum isolates from different countries (29). Genetic and pathogenic variability has been reported among isolates from the Sudan and North Guinea zones of western Africa (72), from Texas and Griffin, Georgia (57), and from Brazil and the United States (13). In all of these studies, which compared variation generated by using random amplified polymorphic DNA (RAPD) markers and virulence phenotypes, the RAPD markers were more variable than the virulence phenotypes. The cluster composition for RAPDs and virulence in *C. graminicola* varied in all of these studies, indicating that these traits are independent of one another.

An International Sorghum Anthracnose Virulence Nursery (ISAVN) was established by the Texas Agricultural Experiment Station in 1975 (37), which was later coordinated by ICRISAT, to help standardize pathotypes of *C. graminicola* (34). The nursery consists of a local resistant and a local susceptible check and 15-18 sorghum lines, with diverse geographic origins and differential reactions to known pathogen populations identified in earlier nurseries. Until 1998, the ISAVN was evaluated at 9-19 locations in 12 countries in Asia, Latin America, and southeastern, western and central Africa. Anthracnose severity was scored at the soft dough stage, on a standard disease rating scale of 1-9, based on the percentage of leaf area covered with necrotic lesions (65, 67). The basic assumption underlying the nursery is that each environment/location represents a unique pathogen population that is being evaluated on a set of differential sorghum lines. If a pathogen caused a susceptible reaction (necrotic lesions with acervuli), it was considered 'virulent', otherwise it was considered 'avirulent'. Virulence was determined from the mean severity score on a host differential line over several years. Isolates (or locations) with a mean severity score ≤ 5 were considered avirulent, and those > 5 were virulent. Correspondingly, host differential lines with a mean score ≤ 5 were considered resistant, and those > 5, susceptible.

Disease severity scores of sorghum lines varied by location and by year (Table 36-1). Based on the mean severity score across the entries, the population at Griffin, Georgia was the most aggressive with a mean severity of 7.1, followed by those at Pantnagar (India), Mansa (Zambia), Baguada (Nigeria), Surat (India), Bako (Ethiopia), Samaru (Nigeria), Kovilpatti (India), Maidiguri (Nigeria), Alupe (Kenya), and Udaipur (India). The least aggressive populations were in Bangkok (Thailand) and Indore (India). The Griffin and Pantnagar populations caused an S reaction on 11/14 test lines, while those from Bangkok and Dharwad (India) did not cause S reactions on any of the 14 test lines. Populations differed in their reaction types on the host lines, and the 19 populations could be divided into 16 putative pathotypes. Each population appeared unique, except Bangkok/Dharwad (Pathotype I) and Indore/Kovilpatti/Udaipur (Pathotype II), which had similar reactions on all 14 sorghum lines (Table 36-1). Sorghum line IS 18758 was resistant at all 19 locations. Several other lines (IS 12467, IS 18760, IS 6928, IS 2508, IS 3758, and IS 6958) had R reactions at all locations and are potential resistance sources for resistance breeding programs (67).

TABLE 36-1. Anthracnose reactions (R and S) based on severity scores on 15 sorghum lines for 2-7 years, 1992-1998.

Location	Sorghum Line														No. "S" Lines	Patho-type	Mean Virulence
	IS 8354	IS 3758	IS 3552	IS 2508	IRAT 204	IS 6928	IS 854	IS 1006	IS 6958	IS 12467	IS 17141	IS 18760	ICSV 247	IS 18442			
Bangkok/ Dharwad	R	R	R	R	R	R	R	R	R	R	R	R	R	R	0	I	2.0
Indore/ Kovilpatti/ Udaipur	R	R	R	R	R	R	R	R	R	R	R	R	R	S	1	II	2.9
Serere	R	R	R	R	R	R	R	R	S	R	R	R	S	R	2	III	2.8
Samanko	R	R	R	R	S	R	R	R	R	R	R	R	R	S	2	IV	3.9
College Station	R	R	R	R	R	R	R	S	R	R	R	R	R	S	2	V	3.0
Patancheru	S	R	R	R	R	R	S	R	R	R	R	R	R	S	3	VI	3.1
Bako	R	S	S	R	R	R	R	S	R	R	R	R	R	S	4	VII	4.1
Longorola	R	R	R	R	S	R	R	S	R	R	R	R	S	S	4	VIII	3.1
Maiduguri	R	R	S	R	R	R	R	S	R	R	S	R	R	S	4	IX	3.4
Alupe	R	S	R	R	S	R	R	R	S	R	R	R	S	R	4	X	3.4
Surat	R	S	R	S	R	R	R	S	R	R	S	R	R	S	5	XI	4.1
Bagauda	S	R	R	R	R	R	S	S	R	R	S	R	R	S	6	XII	4.4
Samaru	R	R	S	R	S	S	R	S	S	R	S	R	S	R	7	XIII	3.9
Mansa	S	R	R	R	S	S	S	S	S	R	S	R	S	R	8	XIV	5.2
Pantnagar	S	S	S	S	S	R	S	S	R	S	S	S	R	S	11	XV	5.3
Griffin	S	R	S	S	S	S	S	S	R	S	R	S	S	S	11	XVI	7.1
No. "R" Locations	14	15	14	16	12	16	14	9	15	17	13	17	13	6			
Mean Resistance	3.9	3.7	3.5	3.2	4.0	3.3	3.4	5.1	3.3	3.0	3.7	3.2	3.6	6.0			

R = resistant (≤ 5 severity score), S = susceptible (> 5 severity score); IS 8758 was resistant at all locations.

Disease Management

Host plant resistance. Host plant resistance has been the most reliable and economical method of disease management for diseases such as anthracnose. Although breeding for anthracnose resistance appears simple, the numerous physiological races of the pathogen often result in breakdown of resistance. Identification of durable resistance requires screening sorghum genotypes against multiple races of the pathogen at multiple locations.

Screening techniques for anthracnose resistance have been standardized for both field and greenhouse conditions. In the field, a "disease" garden technique that involves all of the important/prevalent races has been successful. Normally, infector/indicator/spreader rows are used, such that tested genotypes are exposed to the maximum epidemic that a potential pathogen can produce under prevailing climatic conditions during the crop season. Four inoculation methods–mixing infected debris in soil, spreader rows, conidial spray, and whorl inoculation with infected leaf bits–were compared on two sorghum cultivars under two irrigation systems–furrow irrigation and sprinklers. Whorl inoculation was the most effective for sorghum anthracnose (62). In the greenhouse, plant age is critical for anthracnose symptom development, with 60-day-old plants developing symptoms more readily than 30- and 45-day-old ones (47). Spraying 21- to 30-day-old plants with a conidial suspension (1×10^5 spores/ml), followed by high humidity ($\geq$ 95% RH) for 36 hours has been suggested as an effective method of inoculation under greenhouse conditions (9, 66). For grain anthracnose, inoculation at the green grain stage, just after grain setting, was effective (41, 74).

Anthracnose severity is best scored at the soft dough stage for field evaluations. In greenhouse tests, it is best scored 14 days after foliar inoculations or 30 days after grain inoculations. Various disease-scoring scales for foliar anthracnose severity, usually based on the percentage of the leaf area covered with necrotic lesions, have been used. A 1-9 scale is now used with the ISAVN, where 1 represents no lesions; 2, 1-5% lesions; 3, 6-10% lesions; 4, 11-20% lesions; 5, 21-30% lesions; 6, 31-40% lesions; 7, 41-50% lesions; 8, 51-75% lesions, and 9 = > 75% of the leaf area covered with lesions (65). In greenhouse inoculations it is possible to record latent period (time in days to development of first necrotic/chlorotic lesion), disease severity on the 1-9 scale, and disease reaction: R, resistant (no symptoms or chlorotic flecking); MR, moderately resistant (hypersensitive lesions, red spots or necrotic spots without acervuli); and S, susceptible (lesions with acervuli). Disease severity of grain anthracnose can be recorded 30 days after inoculation by using a similar 1-9 scale and based on the percentage of the head area infected (24). Disease ratings can be converted to AUDPC, or to a virulence index: 1 + (virulence × aggressiveness) ÷ latent period (38, 56).

Mechanism of resistance. Young sorghum leaves accumulate a complex of phenols in response to invasion by *C. graminicola*. Four of these compounds, 3-deoxyanthocyanidin and the phytoalexins apigeninidin, luteolinidin, and the caffeic acid ester of arabinosyl-5-O-apigeninidin, accumulated in resistant lines and their F_1 progeny to significantly higher levels than in susceptible lines in response to attempted infection by *C. graminicola* (51). Phytoalexin concentrations were considerably lower in plants that had grown beyond the juvenile stage (68). Compatible interactions were characterized by an early biotrophic phase, lasting approximately 24 hours, during which time intracellular infection vesicles were formed and primary hyphae colonized epidermal cells. In incompatible interactions, infection vesicles form inside the epidermal cells within 42 hours of inoculation, but there was no evidence of biotrophic interactions, as penetrated cells appeared dead and fungal development was restricted to single epidermal cells (78).

Genetics of resistance. Different genes probably control resistance to leaf and stalk anthracnose (30). Resistance to grain and foliar anthracnose also is probably governed by different genes, as several lines that are highly resistant to foliar anthracnose develop high levels of grain anthracnose in both field and greenhouse screens (32, 41, 74). Resistance to foliar anthracnose is inherited as a single gene recessive trait in a cross between resistant SC 326-6 and susceptible BTx623. Two RAPD bands co-segregated with the resistant allele, while a third was linked to the susceptible allele (6, 7). Combinations of five male-sterile (A) lines and five restorer (R) lines were evaluated against three Brazilian populations of *C. graminicola*. Fifteen host combinations were identified as stable sources of resistance (11). Dilatory resistance in sorghum against *C. graminicola* was evaluated with nine sorghum genotypes and six Brazilian races of the pathogen (16). A differential interaction between pathogen race and sorghum genotype was observed both in the field and in the greenhouse experiments, suggesting the presence of incomplete vertical resistance.

Sources of resistance. Stable sources of resistance (IS 18758, IS 18760, IS 2508, IS 6928, and IS 6958) to diverse races were identified through the ISAVN (65). Some lines, *e.g.*, ICSV 247, IS 8354, and IS 2508, showed slow disease development in the field, indicating dilatory resistance (67). Dilatory resistance also was observed in sorghum hybrids inoculated with six isolates of *C. graminicola* in the greenhouse, with the hybrids having a significantly longer latent period than the susceptible check (41). Phenotypic stability concepts were used to interpret and express dilatory resistance. Genotypes CMS × S221B, CMS × S207R, CMS × 173R, and CMS × S101B had the highest level of dilatory resistance, moderate phenotypic stability, and predictability (22, 26, 28). Varieties S263, SPV669, SPV707, and hybrid CSH5 from the All India Coordinated Sorghum Improvement Program were highly resistant to anthracnose at Pantnagar (61). Cultivars SC36, Tx2737, SC224, SC418-14, and 79HW207 were highly resistant to four Brazilian races in greenhouse tests (9). None of the sorghum cultivars tested in Burkina Faso showed complete resistance to anthracnose, but Gnonfing, Sariaro 10, Siripe 1, and ICSV 1002 were moderately resistant (50).

Sources of resistance to foliar anthracnose are known, but resistance sources for grain and stalk anthracnose have not yet been identified. A positive correlation between grain and foliar anthracnose has been reported (30), while other studies report that lines resistant to foliar anthracnose are susceptible to grain anthracnose (31, 41). Different pathotypes of *C. graminicola* also cause diverse symptoms, *e.g.*, red spots, raised scattered black dots, and black streaks on the grain. The ability of isolates to infect grain, as well as the type of symptoms produced on the grain, both provide useful characters for differentiating pathotypes of *C. graminicola* (74).

Other management methods. Seed treatment with vitavax (carboxin) followed by a spray with zineb was the most effective anthracnose control (63). However, this control method is not economical for farmers. *In vitro* antagonism by some bacterial isolates to *C. graminicola* has been reported (43), but to date there are no studies directed towards developing an effective biocontrol of sorghum anthracnose.

Areas for Future Research

- Identification and description of the *Colletotrichum* species and subspecies associated with sorghum anthracnose in various regions with molecular tools.
- Evaluation of known sources of foliar anthracnose resistance against grain and stalk anthracnose isolates to identify stable sources of resistance for anthracnose disease management.
- Identification and characterization of genes for resistance to foliar, grain, and stalk anthracnose for use in breeding for anthracnose resistance.
- Develop near-isogenic lines with different R genes/alleles for characterization of pathotypes/races.
- Determine genetic relatedness of fungal isolates, and mechanism(s) by which pathogenic variation arises.
- Develop a comprehensive system for race characterization that includes virulence, organ specificity, symptom type, molecular genetic characters, and vegetative compatibility phenotype.
- Evaluation of population dynamics of the pathogen (type and density of propagules over time), especially through comparisons of populations surviving in crop debris with those causing foliar, grain, and stalk diseases of sorghum.
- Determination of roles and relative significance of sclerotia, chlamydospores, infected seeds, and collateral hosts in disease perpetuation and perennation.
- Elucidation of the epidemiology of the disease.
- Monitor populations for virulence shifts through differential host lines and with molecular techniques.
- Evaluate and develop resistance gene management strategies for the development of durable resistance.
- Develop an integrated management strategy that combines host plant resistance with cultural control, to reduce the number of surviving propagules, fungicides (seed treatment), and/or systemic acquired resistance.

References

1. **Alawode, D. A., S. K. Manzo, and N. V. Sundaram.** 1983. Anthracnose of sorghum in northern Nigeria caused by *Colletotrichum graminicola*. Ahmadu Bello University Sorghum Pathology Report. 18 pp.
2. **Ali, M. E. K., and H. L. Warren.** 1992. Anthracnose of sorghum, p. 203-208. *In* W. A. J. de Milliano, R. A. Frederiksen, and G. D. Bengston (eds.), *Sorghum and Millet Diseases: A Second World Review*. ICRISAT, Patancheru (A.P.), India.
3. **Basu Choudhary, K. C., and S. B. Mathur.** 1979. Infection of sorghum seeds by *Colletotrichum graminicola*. 1. Survey, location in seed, and transmission of the pathogen. *Seed Sci. Technol.* **7:** 87-92.
4. **Brooker, N. L., J. F. Leslie, and M. B. Dickman.** 1991. Nitrate non-utilizing mutants of *Colletotrichum* and their use in studies of vegetative compatibility and genetic relatedness. *Phytopathology* **81:** 672-677.
5. **Boon-Long, T.** 1992. Sorghum diseases in Thailand, p. 41-43. *In* W. A. J. de Milliano, R. A. Frederiksen, and G. D. Bengston (eds.), *Sorghum and Millet Diseases: A Second World Review*. ICRISAT, Patancheru (A.P.), India.
6. **Boora, K. S., R. A. Frederiksen, and C. W. Magill.** 1997. Identification of RAPD markers tightly linked to gene for resistance to anthracnose in sorghum. *Proc. Internatl. Conf. Genet. Improve. Sorghum & Pearl Millet (22-27 September 1996, Lubbock, TX)*: 643-644.
7. **Boora, K. S., R. A. Frederiksen, and C. W. Magill.** 1998. DNA-based markers for a recessive gene conferring anthracnose resistance in sorghum. *Crop Sci.* **38:** 1708-1709.
8. **Cardwell, K., P. R. Fhepperly, and R. A. Frederiksen.** 1989. Pathotypes of *Colletotrichum graminicola* and seed

transmission of sorghum anthracnose. *Plant Dis.* **73:** 255-257.

9. **Casela, C. R., and A. S. Ferreira.** 1988. Identification of sources of resistance to *Colletotrichum graminicola*, causal agent of sorghum anthracnose. *Fitopatol. Brasil.* **13:** 351-353.
10. **Casela, C. R., A. S. Ferreira, and N. Brancao.** 1996. Virulence structure and variability in sorghum anthracnose fungus *Colletotrichum graminicola. Fitopatol. Brasil.* **21:** 357-361.
11. **Casela, C. R., A. S. Ferreira, and F. G. Santos.** 1998. Virulence association with genetic resistance in the sorghum anthracnose fungus *Colletotrichum graminicola. Fitopatol. Brasil.* **23:** 143-146.
12. **Casela, C. R., A. S. Ferreira, and R. E. Schaffert.** 1992. Physiological races of *Colletotrichum graminicola* in Brazil, p. 209-212. *In* W. A. J. de Milliano, R. A. Frederiksen, and G. D. Bengston (eds.), *Sorghum and Millet Diseases: A Second World Review.* ICRISAT, Patancheru (A.P.), India.
13. **Casela, C. R., and R. A. Frederiksen.** 1993. Variability in the sorghum anthracnose fungus *Colletotrichum graminicola* in Brazil and USA, p. 310. *In* T. Jacobs and J. E. Parlevleit (eds.), *Durability of Disease Resistance.* Kluwer Academic Publishers, Dordrecht, The Netherlands.
14. **Casela, C. R., and R. A. Frederiksen.** 1993. Survival of *Colletotrichum graminicola* sclerotia in sorghum stalk residue. *Plant Dis.* **77:** 825-827.
15. **Casela, C. R., and R. A. Frederiksen.** 1994. Pathogenic variability in mono-conidial isolates of the sorghum anthracnose fungus *Colletotrichum graminicola* from single lesions and from mono-conidial cultures. *Fitopatol. Brasil.* **19:** 149-153
16. **Casela, C. R., and R. A. Frederiksen.** 1995. Evidence of dilatory resistance to anthracnose in sorghum. *Plant Dis.* **79:** 908-911.
17. **Correll, J. C., T. E. Morelock, and J. C. Guerber.** 1993. Vegetative compatibility and virulence of the spinach anthracnose pathogen, *Colletotrichum dematium. Plant Dis.* **77:** 688-691.
18. **de Milliano, W. A. J., R. A. Frederiksen, and G. D. Bengston, eds.** 1992. *Sorghum and Millet Diseases: A Second World Review.* ICRISAT, Patancheru (A.P.), India.
19. **Duncan, R. R.** 1983. The anthracnose-*Fusarium* disease complex on sorghum in the Southeastern USA. *Sorg. Newsl.* **26:** 121-122.
20. **Ferreira, A. S., and C. R. Casela.** 1986. Races pathogenices *de Colletotrichum graminicola*, agent causal de anthracnose em sorgho [*Sorghum bicolor* (L.) Moench]. *Phytopatol. Brasil.* **11:** 833-837.
21. **Ferreira, A. S., R. A. Frederiksen, H. L. Warren, and K. C. de Castillo.** 1985. Identification of races of *Colletotrichum graminicola* in Brazil. *Sorg. Newsl.* **28:** 80-83.
22. **Ferreira, A. S., and H. L. Warren.** 1982. Resistance of sorghum to *Colletotrichum graminicola. Plant Dis.* **66:** 773-775.
23. **Frederiksen, R. A.** 1984. Anthracnose stalk rot, p. 37-42. *In* L. K. Mughogho (ed.), *Sorghum Root and Stalk Rots: A Critical Review.* ICRISAT, Patancheru (A.P.), India.
24. **Frederiksen, R. A., L. L. Castor, and D. T. Rosenow.** 1982. Grain mold, small seeds and head blight: The *Fusarium* connection. *Proc. Annu. Corn Sorg. Res. Conf.* **37:** 26-36.
25. **Frederiksen, R. A., and R. R. Duncan.** 1992. Sorghum diseases in North America, p. 85-88. *In* W. A. J. de Milliano, R. A. Frederiksen, and G. D. Bengston (eds.), *Sorghum and Millet Diseases: A Second World Review.* ICRISAT, Patancheru (A.P.), India.
26. **Frederiksen, R. A., and D. T. Rosenow.** 1971. Disease resistance in sorghum. *Proc. Annu. Corn Sorg. Res. Conf.* **26:** 71-82.
27. **Gumaraes, F. B., C. R. Casela, F. G. Santos, and A. S. Ferreira.** 1998. Control of sorghum anthracnose by the use of cultivar mixture. *Summa Phytopathol.* **24:** 131-135.
28. **Gumaraes, F. B., C. R. Casela, F. X. R. Vale, L. Zombolin, and F. G. Santos.** 1998. Dilatory resistance of sorghum genotypes to different races of *Colletotrichum graminicola. Summa Phytopathol.* **24:** 136-140.
29. **Guthrie, P. A., C. W. Magill, R. A. Frederiksen, and G. N. Odvody.** 1992. Random Amplified Polymorphic DNA markers: A system for identifying and differentiating isolates of *Colletotrichum graminicola. Phytopathology* **82:** 832-835.
30. **Harris, H. B., and B. J. Johnson.** 1967. Sorghum anthracnose, symptoms, importance and resistance. *Proc. Bien. Sorg. Improve. Conf. No. Amer.* **5:** 48-52.
31. **Hazra, S.** 1998. *Pathogenic Variability in Sorghum Anthracnose Incited by* Colletotrichum graminicola *(Ces.) Wils.* M. Sc. Thesis, Acharya N. G. Ranga Agric. Univ., Rajendranagar, Hyderabad (A.P.), India.
32. **Hazra, S., R. P. Thakur, G. Umadevi, and K. Mathur.** 1999. Pathogenic and molecular variability among twelve isolates of *Colletotrichum graminicola* from sorghum. *J. Mycol. Plant Pathol.* **29:** 176-183.
33. **Holliday, P.** 1980. *Fungal Diseases of Tropical Crops.* Cambridge University Press, Cambridge, UK.
34. **ICRISAT.** 1983. Annual Report 1982. ICRISAT, Patancheru (A.P.), India.
35. **Jacobson, D. J., and T. R. Gordon.** 1988. Vegetative compatibility and self-incompatibility within *Fusarium oxysporum* f. sp. *melonis. Phytopathology* **78:** 668-672.
36. **Jamil, F. F., and R. L. Nicholson.** 1989. Susceptibility of corn to isolates of *Colletotrichum graminicola* pathogenic to other grasses. *Plant Dis.* **71:** 809-810.
37. **King, S. B., and R. A. Frederiksen.** 1976. Report on the International Sorghum Anthracnose Virulence Nursery. *Sorghum Newsl.* **19:** 105-106
38. **Mathur, K., V. P. Rao, and R. P. Thakur.** 1997. Intra-population variability in *Colletotrichum sublineolum* infecting sorghum. *Indian J. Mycol. Plant Pathol.* **27:** 302-310.
39. **Mathur, K., and R. P. Thakur.** 1998. *A Research Report On Pathogenic Variability in Sorghum Anthracnose.* ICRISAT, Patancheru (A.P.), India. 172 pp.
40. **Mathur, K., R. P. Thakur, and V. P. Rao.** 1998. A strain of *Colletotrichum gloeosporioides* causing anthracnose of yellow sorghum. *Intl. Sorg. Millet Newsl.* **37:** 69-70.
41. **Mathur, K., R. P. Thakur, and V. P. Rao.** 2000. Pathogenic variability and vegetative compatibility among isolates of *Colletotrichum graminicola* and *C. gloeosporioides* causing foliar and grain anthracnose of sorghum. *Ind. Phytopathol.* **53**: 407-414.

42. **Mathur, K., R. P. Thakur, and V. P. Rao.** 2001. Characterization of *Colletotrichum graminicola* populations from a sorghum hybrid CSH 9 for morphological and pathogenic variability. *Ind. Phytopathol.* **54**: 299-303
43. **Michereff, S. J., N. S. S. Silveira, and R. L. R. Mariano.** 1994. Antagonism of bacteria to *Colletotrichum graminicola* and potential for biocontrol of sorghum anthracnose. *Fitopatol. Brasil.* **19:** 541-545.
44. **Mishra, A. N., and A. P. Sinha.** 1996. Studies on the survival of sorghum anthracnose (*Colletotrichum graminicola*) pathogen on crop debris. *Ind. Phytopathol.* **49:** 381-383.
45. **Mishra, A., and B. S. Siradhana.** 1979. Studies on the survival of sorghum anthracnose (*Colletotrichum graminicola*) pathogen. *Phil. Agric.* **62**: 141-152
46. **Mishra, A., and B. S. Siradhana.** 1979. Estimation of losses due to anthracnose of sorghum. *J. Mycol. Plant Pathol.* **9:** 257.
47. **Mishra, A., and B. S. Siradhana.** 1992. Effect of host age on sorghum anthracnose development. *Gujarat Agric. Univ. Res. J.* **18:** 135-136.
48. **Nakamura, K.** 1982. *Especializacao Fisiologic em* Colletotrichum graminicola *(Ces.) Wils. (Sensu Arx., 1957) Agente Causal da Antracnose em Sorgho*. Ph.D. Thesis, Univ. Estadual Paulista, Jaboticabal, Brazil.
49. **Neya, A., and K. B. Kabore.** 1987. Evaluating the incidence of anthracnose and red stalk rot caused by *Colletotrichum graminicola* on sorghum. *Phytoprotection* **68:** 121-126.
50. **Neya, A., and M. Normad.** 1998. Response of sorghum genotypes to leaf anthracnose (*Colletotrichum graminicola*) under field conditions in Burkina Faso. *Crop Protect.* **17:** 47-53.
51. **Nicholson, R. L., F. Jamil, B. A. Snyder, W. I. Lue, and J. Hipskind.** 1988. Phytoalexin synthesis in the juvenile sorghum leaf. *Physiol. Mol. Plant Path.* **33:** 271-278.
52. **Pande, S., L. K. Mughogho, R. Bandyopadhyay, and R. I. Karunakar.** 1991. Variation in pathogenicity and cultural characteristics of sorghum isolates of *Colletotrichum graminicola* in India. *Plant Dis.* **75:** 778-783.
53. **Pastor-Corrales, M. A., and R. A. Frederiksen.** 1980. Sorghum anthracnose, p. 289-294. *In* R. J. Williams, R. A. Frederiksen, L. K. Mughogho, and G. D. Bengston (eds.), *Sorghum Diseases: A World Review*. ICRISAT, Patancheru (A.P.), India.
54. **Puhalla, J. E.** 1985. Classification of strains of *Fusarium oxysporum* on the basis of vegetative compatibility. *Can. J. Bot.* **63:** 179-183.
55. **Rajasab, A. H., and A. Ramalimgam.** 1981. Zonate anthracnose, a new disease of sorghum caused by *Colletotrichum graminicola* var. *zonatum* var. nov. *Curr. Sci.* **50:** 34-36.
56. **Rao, V. P., R. P. Thakur, and K. Mathur.** 1998. Morphological and pathogenic diversity among grain sorghum isolates of *Colletotrichum graminicola* in India. *Ind. Phytopathol.* **51:** 64-174.
57. **Rosewich, U. L.** 1996. *The Population Genetics of* Colletotrichum graminicola *from Different Ecosystems of Sorghum.* Ph.D. Thesis, Texas A&M Univ., College Station, Texas.
58. **Saifula, M., and K. G. Rangnathiah.** 1989. Seed health testing of sorghum with special reference to *Colletotrichum graminicola. Ind. Phytopathol.* **42:** 73-78.
59. **Sharma, H. C.** 1980. Screening of sorghum for leaf disease resistance in India, p. 249-264. *In* R. J. Williams, R. A. Frederiksen, L. K. Mughogho, and G. D. Bengston (eds.), *Sorghum Diseases: A World Review.* ICRISAT, Patancheru (A.P.), India.
60. **Sherriff, C., M. J. Whelan, G. M. Arnold, and J. Bailey.** 1995. rDNA sequence analysis confirms the distinction between *Colletotrichum graminicola* and *C. sublineolum. Mycol. Res.* **99:** 475-478.
61. **Sinha, A. P., R. Singh, and P. K. Shiotria.** 1986. Screening of breeder's material in advance varietal trials for resistance to sorghum anthracnose, zonate leaf spot and grain mold. *Sorg. Newsl.* **29:** 12-13.
62. **Singh, S. D.** 1992. Screening techniques and series of resistance to important leaf diseases of sorghum. *Proc. XXII*nd *Annual Sorghum Workshop (2-4 April 1992, Surat, Gujarat, India)*: 27-28.
63. **Singh, M., P. Shukla, and S. K. Sharma.** 1989. Chemical control of anthracnose disease of jowar (*Sorghum vulgare* Pers.). *Farm Sci. J.* **4:** 121-123.
64. **Sutton, B. C.** 1968. The appressoria of *Colletotrichum graminicola* and *C. falcatum. Can. J. Bot.* **46:** 873-876.
65. **Thakur, R. P.** 1995. Status of International Sorghum Anthracnose and Pearl Millet Downy Mildew Virulence Nurseries, p. 75-92. *In* J. F. Leslie and R. A. Frederiksen (eds.), *Disease Analysis through Genetics and Biotechnology: Interdisciplinary Bridges to Improved Sorghum and Millet Crops.* Iowa State University Press, Ames, Iowa.
66. **Thakur, R. P., K. Mathur, K., V. P. Rao, V. R. Hiremath, R. S. Kushwaha, R. R. Dwivedi, H. C. Tailor, and S. Indira.** 1999. Pathogenic and genetic variability among Indian isolates of *Colletotrichum graminicola. Ind. Phytopathol.* **51:** 220-228.
67. **Thakur, R. P., and V. P. Rao.** 1999. *International Sorghum Anthracnose Virulence Nursery (ISAVN): A Summary Report 1992-1998.* ICRISAT, Patancheru (A.P.), India, 30 pp.
68. **Tenkouano, A., F. Miller, G. E. Hart, R. A. Frederiksen, and R. L. Nicholson.** 1993. Phytoalexin assay in juvenile sorghum: an aid to breeding for anthracnose resistance. *Crop Sci.* **33:** 243-248.
69. **Thomas, M. D.** 1995. Sorghum anthracnose research in West Africa: A look at the present and the future, p. 127-136. *In* J. F. Leslie and R. A. Frederiksen (eds.), *Disease Analysis through Genetics and Biotechnology: Interdisciplinary Bridges to Improved Sorghum and Millet Crops.* Iowa State University Press, Ames, Iowa.
70. **Thomas, M. D., and R. A. Frederiksen.** 1995. Dynamics of oval and falcate conidium production of *Colletotrichum graminicola* from sorghum. *Mycologia* **87:** 87-89.
71. **Thomas, M. D., and R. A. Frederiksen.** 1995. Fertility tests on sorghum isolates of *Colletotrichum graminicola* from West Africa. *Intl. Sorg. Millet Newsl.* **36:** 79.
72. **Thomas, M. D, U. L. Rosewich, and R. A. Frederiksen.** 1995. Preliminary studies on total DNA restriction fragment length polymorphism (RFLP) of West African isolates of *Colletotrichum graminicola* from sorghum. *Intl. Sorg. Millet Newsl.* **36:** 90-93.

73. **Thomas, M. D., I. Sissoko, and M. Sacko.** 1996. Development of leaf anthracnose and its effect on yield and grain weight of sorghum in West Africa. *Plant Dis.* **80:** 151-153.
74. **Totla, K. G.** 1990. *Studies on Foliar, Grain and Stalk Anthracnose of Sorghum [*Sorghum bicolor *(L.) Moench]*. M. Sc. Thesis, Rajasthan Agric. Univ., RCA, Udaipur, India.
75. **Traore, B., and K. B. Kabore.** 1992. Survival of *Colletotrichum graminicola* in the soil in Burkina Faso, p. 293. *In* W. A. J. de Milliano, R. A. Frederiksen, and G. D. Bengston (eds.), *Sorghum and Millet Diseases: A Second World Review.* ICRISAT, Patancheru (A.P.), India.
76. **Vaillancourt, L. S., and R. M. Hanau.** 1992. Genetic and morphological comparisons of *Glomerella* (*Colletotrichum*) isolates from maize and sorghum. *Exp. Mycol.* **16:** 219-229.
77. **Vechiato, M. H., P. J. Valarini, and C. C. Lasca.** 1986. Evaluation of methods for the detection of *Colletotrichum graminicola* (Ces.) Wils. on sorghum (*Sorghum* spp.) seeds. *Summa Phytopathol.* **12:** 195-201.
78. **Wharton, P. S., and A. M. Julian.** 1996. A cytological study of compatible and incompatible interactions between *Sorghum bicolor* and *Colletotrichum sublineolum. New Phytol.* **134:** 25-34.
79. **Zummo, N.** 1992. Sweet sorghum varieties resistant to anthracnose in USA are highly susceptible to the disease in Brazil, p. 289. *In* W. A. J. de Milliano, R. A. Frederiksen, and G. D. Bengston (eds.), *Sorghum and Millet Diseases: A Second World Review.* ICRISAT, Patancheru (A.P.), India.

37

Genetic Variability Among and Within Host-Specialized Isolates of *Sporisorium reilianum*

G. Naidoo and J. Heriberto Torres-Montalvo

Abstract

We used molecular markers to examine genetic variation and diversity among populations of *Sporisorium reilianum*. Karyotypes were stable and correlated with host-specificity of the isolate. A large number of differences were detected between sorghum and maize isolates with 473 anonymous random probes in a RFLP screen, but only 8 and 3% of these probes were polymorphic among the sorghum and maize isolates, respectively. Based on RFLP analysis of 459 sorghum isolates from Mexico, Niger, and the United States, the level of gene diversity in each of these populations was low.

Introduction

Sporisorium reilianum (Kuhn) Langdon and Fullerton, the causal agent of sorghum head smut, is dimorphic and can grow saprophytically, but requires sexual reproduction and passage through a compatible host to complete its life cycle. Pathogenicity in populations of *S. reilianum* varies. The fungus attacks both sorghum and maize worldwide, and two distinct populations can be distinguished based on host specificity (1, 2, 10), but not on the basis of morphology or isozymes (4). While there are no recognized pathotypes in maize head smut populations, the sorghum pathotype is more variable. Four different races or pathotypes of sorghum head smut are recognized in the United States (2, 10) and Mexico (8). Each population of sorghum head smut is a mixture of pathotypes that is differentiated by the smut reaction on a set of four differential varieties (10). The two pathotypes observed most frequently are races 1,3 and 1,3,4, with races 1,2,3,4 and 1,4 being recovered occasionally. In general, very little race 2 exists in Texas (2).

Characterization of sorghum head smut populations requires the use of uniform disease nurseries grown in areas known to have high levels of smut infestation (2). However, field evaluation techniques can be unreliable because of escapes (2). Under field conditions, maize head smut attacks only maize, and sorghum head smut attacks only sorghum (1, 2, 10). Yet, concerns remain that the same pathogen may attack both hosts or that genetic exchange may occur between maize head smut and sorghum head smut in areas where maize and sorghum are grown side by side. These concerns are of particular import for the breeding programs, since populations of sorghum head smut have overcome host plant resistance in the past, and similar events could occur in maize head smut populations.

Over the last few decades, sources of resistance to sorghum head smut have been overcome, and the disease has again become economically important on sorghum. A population differing in pathogenicity from others was observed in central Texas in 1993 (11) for the first time since the 1970s. The amount of variation within a pathogen population and the ability of subpopulations to exchange genetic information can be used to make projecttions about pathogen population responses to new sources of resistance. These projections can be used to evaluate potential breeding lines, to evaluate hybrids, and in the rational design of plant breeding and other disease control strategies.

Population structure discerned from selectable markers, *e.g.*, virulence or fungicide resistance, probably reflects the selective pressures operating within a population. The use of selectively neutral genetic markers is a more precise method to study changes in the genetic structure of field populations and may provide information on other processes, *e.g.*, gene flow, drift, and the frequency of sexual reproduction. Our objectives in this study were: (i) to correlate molecular genetic markers with host-specific pathotypes of *S. reilianum*, and (ii) to evaluate the genetic structure of nine geographically diverse field populations of sorghum isolates of *S. reilianum*.

Distinguishing Sorghum and Maize Pathotypes of *S. reilianum*

Electrophoretic karyotypes. We used contour-clamped homogenous field (CHEF) electrophoresis, to identify chromosome length polymorphisms between the sorghum and maize pathotypes of *S. reilianum*. We used 21 sorghum head smut isolates collected over a four-year period from across Texas and seven isolates of maize head smut from Halfway, Texas. Sporidia were derived from single teliospores, converted to protoplasts as described by Kinscherf and Leong (3), and suspended in low melting point (LMP) agarose. Agarose plugs were treated with protease (Type XIV) and stored in 0.5 M EDTA at 5-8°C. Alternatively, concentrated, washed sporidia were imbedded directly in the agarose, treated with pr-tease in 0.5 M EDTA (5), and used for electrophoresis with similar results. Chromosomal preparations of *Saccharomyces cerevisiae*, *Schizosaccharomyces pombe*, and *Ustilago maydis*, the causal agent of common corn smut, were used as size standards.

Following CHEF gel electrophoresis, 13-15 bands could be distinguished for the *S. reilianum* isolates, and 14 bands each for *S. cerevisiae* (strain YNN295) and *U. maydis* (isolates from College Station, TX). Electrophoretic conditions that permitted this separation were a switching interval of 60 sec for 15 hr and 90 sec for 8 hr with 200 v across the electrodes. The band sizes in the *S. reilianum* samples ranged from approximately 270-2,300 kb. By increasing the switching interval to 90 sec for 16 hr, and then to 240 sec for 32.8 hr while decreasing the voltage from 200 v to 150 v, an additional band of about 1,900 kb could be seen in *S. reilianum*, and a 2,100 kb band could be seen in *U. maydis*, but the resolution of some of the other bands was then lost. All sorghum head smut isolates had the same pattern with bands estimated at 2,300, 1,360, 940, 900, 750, 660, 630, 560, 490, 470, 430, 380, and 270 kb. Based on relative staining intensity, the 560-, 630-, and 750-kb bands are at least doublets. The maize head smut isolates all had an identical electrophor-etic karyotype with 15 bands. The estimated band sizes were the same as those for the sorghum head smut iso-lates, except that instead of a single intense band near 750 kb there were three bands between 700 and 800 kb. Also, a band near 490 kb from the maize head smut iso-lates was slightly larger than its counterpart from the sor-ghum head smut isolates. The chromosome sizes and numbers we observed for *S. reilianum* are similar to those seen in other smut fungi, *e.g.*, 13 in *U. hordei* and 16-21 in *U. tritici*. The total genome size for *S. reilianum* is 12-15 Mb, which is similar to that reported for the other smut fungi and the yeasts (5).

Random Amplified Polymorphic DNA (RAPD) fingerprints. Decameric nucleotide primers of defined sequence and greater than 50% GC were used to prime PCR amplifications of DNA from four maize and five sorghum isolates of *S. reilianum*. Amplification reactions, as described by Williams *et al.* (13), each contained 25 ng of template DNA. After initially denaturing DNA for 3 min at 94°C, there were 45 cycles of 1 min at 94°C, 1 min at 36°C for primer annealing, and 1 min at 72°C for primer extension, which was extended to 7 min following the last cycle. PCR products were resolved by electrophoresis through a 1% agarose gel.

For most primers, maize head smut and sorghum head smut isolates yielded multiband patterns following agarose gel electrophoresis. While some bands in these patterns were shared by both the maize and the sorghum head smut isolates, other bands were pathotype specific. One maize head smut isolate, however, often had a

TABLE 37-1. Putative multilocus genotypes of *Sporisorium reilianum* isolates from sorghum[a].

Isolate	RFLP Probe-Enzyme Combination						
	126 *Pst*I	130 *Hind*III	143B *Pst*I	210 *Hind*III	239 *Hind*III	250A[b] *Pst*I	250B[b] *Pst*I
China 101	2 2	2 2	1 1	2 2	1 1	2 2	2 2
CoCr 3-13 & SfMx 46	2 2	1 1	2 2	2 2	1 1	1 1	1 1
Dane 3-3	1 1	1 1	1 2	2 2	1 1	1 1	1 1
Mali 3-91	1 1	1 1	1 1	1 1	2 2	1 1	2 2
MtMx 50	1 1	1 1	2 2	2 2	1 1	1 1	1 1
RBMx 3	1 1	1 1	1 1	2 2	1 1	1 1	1 1
Tayl 4-18	1 1	1 1	1 1	1 1	1 1	1 1	1 1
Ugan 2-93	2 2	1 1	1 1	1 1	2 2	2 2	2 2
Vict 101-4	1 1	1 1	1 1	1 1	1 1	1 1	1 2

[a]Each digit represents a putative allele at a putative nuclear RFLP locus defined by a particular probe-enzyme combination (two putative alleles per locus). A set of different polymorphic bands that hybridized to the same probe was named A, B, C, etc. The most frequent band or banding pattern in the sample was termed "allele" 1.

[b]Common probes followed by different letters represent different bands generated by the same probe. These probes may, but need not, represent distinct loci.

RAPD profile that was more typical of that seen in sorghum head smut isolates than that seen in the isolates from maize. Thus, while RAPDs are quick and relatively simple, their reliability in distinguishing host-specialized isolates of *S. reilianum* is unproven, as is their utility in population analyses.

Restriction fragment length polymorphisms (RFLPs). RFLP analysis (6) was conducted with anonymous probes obtained from two sorghum head smut genomic libraries. Single-locus or low-copy probes were used in combination with restriction enzyme digests of DNA from *S. reilianum* isolates obtained from maize and sorghum. Probe-enzyme combinations that identified RFLPs between the maize and the sorghum head smut isolates and within each host type were used to construct multilocus genotypes. Two hundred and sixty-five (55%) of the probe-enzyme combinations resulted in polymorphisms between the maize and the sorghum head smut isolates. For these probe-enzyme combinations, we focused on DNA fragments that were shared by all members of the same pathotype, *i.e.*, all of the maize isolates were the same as one another as were all of the sorghum isolates, but these DNA fragments were not shared by strains of different pathotype.

RFLP Variation Within Sorghum and Maize Pathotypes of *S. reilanum*

Variation within isolates obtained from sorghum. A subset of ten sorghum head smut isolates from China, Mali, Uganda, and the United States was examined with 505 probe-enzyme combinations. Only 41 (8%) of these probe-enzyme combinations detected RFLPs. For 19 of the 505 probe-enzyme combinations (4%), the probe hybridized to ≥ 10 fragments for every enzyme digest. Of these 19 probe-enzyme combinations, five detected polymorphisms among the 10 sorghum head smut isolates. Fifty-nine percent (297 probe-enzyme combinations) were single copy, and 37% were low copy (< 10 fragments) copy. Two, and only two, electrophoretic variants were found among the 10 isolates for each polymorphic probe-enzyme combination. Six single- or low-copy probe-enzyme combinations, which identified 11 putative polymorphic loci, were used to construct multilocus genotypes (Table 37-1). A repetitive probe-enzyme combina-tion, pSR117-*Pst*I, identified nine different banding pat-terns in 10 international isolates.

Variation within maize head smut. Most of the clones that hybridized to the sorghum head smut isolates also hybridized to a set of maize head smut isolates, but only 14 of 473 (3%) probe-enzyme combinations identified polymorphisms. Four single- or low-copy probe-enzyme combinations, which identified six bands, could be used to distinguish eight putative multilocus genotypes from nine isolates (Table 37-2). Only two repetitive probe-enzyme combinations detected polymorphisms among the maize head smut isolates. The repetitive probe-enzyme combination, pSR 117-*Pst*I, which identified multiple banding patterns amongst the sorghum head smut isolates, was monomorphic when hybridized to DNA from any of the maize head smut isolates digested with any of the five restriction enzymes tested.

Genetic Structure of *S. reilianum* Populations from Mexico, Niger, and the United States

We examined 459 isolates of sorghum head smut from field populations in Mexico, Niger, and the United States for polymorphisms at nine bands, based on nine probe-enzyme combinations that were polymorphic in our preliminary survey. Low levels of genetic diversity were detected in all nine populations. Only two variants, treated as putative alleles in the genetic analysis, were present for each polymorphic band. The variant for two bands that was the most common for the seven populations from the Western Hemisphere was the least frequent in the two African populations. Genotypic diversity was moderately high across the nine populations. On average, 12 putative multilocus genotypes were found in a sample of 51 isolates from each population, with 54 putative multilocus genotypes identified across all nine populations. The populations from Africa and the Americas do not share any putative multilocus genotypes. These results are consistent with the hypothesis that the sampled populations are not from the centers of diversity for sorghum head smut. For example, the migrations from the center of origin, perhaps eastern Africa, that established the American and west African populations could be separate events.

There was a significant excess of homozygotes in all nine populations. This pattern could result from high levels of mating between sister sporidia that germinate and fuse to form dikaryons beneath the soil surface. Thus, there are relatively few opportunities for fusions between sporidia that originate from different teliospores. Thus, even though the sexual cycle occurs frequently, the genetic structure of *S. reilianum* populations is not consistent with a great deal of outbreeding.

The genetic differentiation among the nine populations was large (F_{ST} = 0.48). Within the Americas, genetic differentiation was moderate (F_{ST} = 0.14). Thus, most of the genetic differences observed were due to differences between the African and American popula-

TABLE 37-2. Putative multilocus genotypes of *Sporisorium reilianum* isolates from maize[a].

Isolate	RFLP Probe-Enzyme Combination				
	143C *Hind*III	233 *Pst*I	250F[b] *Hind*III	250H[b] *Xho*I	251 *Xho*I
CoCr 88-1 & Lubk 6-5	2 2	1 2	1 2	1 2	1 1
Lubk 3-5 & TxMx 9-9	2 2	1 2	1 1	1 1	1 1
Lubk 89-1	2 2	2 2	1 1	1 1	1 1
Lubk 89-2	2 2	2 2	1 2	1 1	1 1
QrMx 4-15	1 1	1 1	1 2	1 1	2 2
TxMx 8-14	1 1	1 1	1 2	1 2	2 2
ZpMx 2-7	2 2	1 2	1 1	1 2	2 2

[a]Each digit represents a putative allele at a putative nuclear RFLP locus defined by a particular probe-enzyme combination (two putative alleles per locus). The most frequent band or banding pattern in the sample was termed "allele" 1.
[b]Common probes followed by different letters represent different bands generated by the same probe. These probes may, but need not, represent distinct loci.

tions (F_{ST} between 0.49 and 0.72) and not to differences that occurred within a continent. The F_{ST} between the Koni and Tillabery populations from Niger was 0.13. Pairwise comparisons between populations showed that the average number of migrants per generation (N_m) was ≤ 0.26 between Konni or Tillabery and any of the populations from the Americas. In contrast, there was evidence for significant gene flow among populations from Texas and Mexico ($N_m = 3.1$). We hypothesize that gene flow is occurring on a regional scale through grain movement and or wind-blown spore dispersal within the region.

Conclusions

Karyotype stability. Electrophoretic karyotyping provides a phenotype that correlates with host specificity. The electrophoretic karyotypes of 17 *S. reilianum* isolates from sorghum collected from Texas, China, and Mali over six growing seasons were uniform, which suggests that this character is not changing rapidly. Similarly, 16 maize head smut isolates from Texas and Mexico collected over eight years also had a common karyotype that was different from that of the sorghum isolates. These results are different from those for other plant pathogens, *e.g., U. maydis* (3), and may reflect the relative importance of the sexual cycle in *S. reilianum*.

The lack of karyotype polymorphism in populations of *S. reilianum* suggests that if crosses occur between the maize and the sorghum head smut isolates, then progeny are either rare with or at a selective disadvantage. Note, however, that many of the chromosomes in the karyotypes are similar in size and that detecting a mixed karyotype is not easy.

Population structure. The low level of genetic diversity within the sorghum head smut and the maize head smut strain sets could have numerous causes, including relatively recent population establishment, non-random mating and selection (7, 9, 12). The level of divergence between sorghum head smut and maize head smut isolates, however, was high with 55% of the probe-enzyme combinations examined yielding fixed polymorphisms between the maize and the sorghum isolates. The high genetic diversity between the sorghum and the maize head smut isolates, and the limited genetic diversity within isolates recovered from each host, suggests that these smuts have evolved independently and that little genetic exchange has occurred between them recently.

Acknowledgments

This work was supported in part by The Nicholson Foundation, Mexican Council of Science and Technology (CONACyT), and Pioneer Hi-Bred International. We thank R. A. Frederiksen, C. W. Magill, and B. A. MacDonald for guidance in the design, execution, and interpretation of these experiments.

References

1. **Al-Sohaily, I. A., C. I. Mankin, and G. Semeniuk.** 1963. Physiological specialization of *Sphacelotheca reiliana* to sorghum and corn. *Phytopathology* **53:** 723-726.
2. **Frederiksen, R. A., and L. Reyes.** 1980. The head smut program at Texas A&M, p. 367-372. *In* R. J. Williams, R. A. Frederiksen, and L. K. Mughogho (eds.), *Proc. Intl. Workshop on Sorghum Diseases (11-15 December 1978, Hyderabad, India).* ICRISAT, Patancheru (A.P.), India.
3. **Kinscherf, T. C., and S. A. Leong.** 1988. Molecular analysis of the karyotype of *Ustilago maydis. Chromosoma* **96:** 427-433.
4. **Kubicek, Q. B.** 1986. *Isozyme Analysis of* Sporisorium reilianum *Populations from Sorghum Fields in Texas.* Ph.D. dissertation, Texas A & M Univ., College Station, Texas.
5. **McCluskey, K., and D. Mills.** 1990. Identification and characterization of chromosome length polymorphisms among strains representing fourteen races of *Ustilago hordei. Molec. Plant-Microbe Interact.* **3:** 366-373
6. **McDonald, B. A., and J. P. Martinez.** 1990. DNA restriction fragment length polymorphisms among *Mycosphaerella graminicola* (anamorph *Septoria tritici*) isolates collected from a single wheat field. *Phytopathology* **80:** 1368-1373.
7. **Naidoo, G.** 1992. *Identification and Differentiation of Host-Specialized Isolates of* Sporisorium reilianum *Using Molecular Markers.* Ph.D. dissertation. Texas A & M Univ., College Station, Texas.
8. **Narro, J., V. A. Betancourt, and J. I. Aguirre.** 1992. Sorghum diseases in Mexico, p. 75-84. *In* W. A. J. de Milliano, R. A. Frederiksen, and G. D. Bengston (eds.),

Sorghum and Millets Diseases: A Second World Review. ICRISAT, Patancheru (A.P.), India.

9. **Nei, M.** 1973. Analysis of gene diversity in subdivided populations. *Proc. Natl. Acad. Sci. USA* **70:** 3321-3323.
10. **Omer, E.-H., and R. A. Frederiksen.** 1992. Sorghum smuts, p. 245-252. *In* W. A. J. de Milliano, R. A. Frederiksen, and G. D. Bengston (eds.), *Sorghum and Millets Diseases: A Second World Review.* ICRISAT, Patancheru (A.P.), India.
11. **Palma, A.** 1993. *Evaluation of Horizontal Resistance to Head Smut* Sporisorium reilianum *(Kuhn) Langdon and Fullerton in* Sorghum bicolor *(L.) Moench.* M.S. thesis. Texas A & M Univ., College Station, Texas.
12. **St. Louis, V. L., and J. C. Barlow.** 1988. Genetic differentiation among ancestral and introduced populations of the Eurasian tree sparrow (*Passer montanus*). *Evolution* **42:** 266-276.
13. **Williams, J. G. K., A. R. Kubelik, K. J. Livak, J. A. Rafalski, and S. V. Tingey.** 1990. DNA polymorphisms amplified by arbitrary primers are useful as genetic markers. *Nucleic Acids Res.* **18:** 6531-6535

38

Grain Mold Fungi from Sorghum in Ghana

Stephen K. Nutsugah, Vibeke Leth, Ibrahim D. K. Atokple, and Francis K. Tsigbey

A sorghum disease survey was conducted in Ghana during the 1997 cropping season. Grey leaf spot (caused by *Cercospora sorghi*), zonate leaf spot (caused by *Gloeocercospora sorghi*), leaf blight (caused by *Exserohilum turcicum*), and grain mold were common in all of the sorghum-growing regions of the country (1, 6). Grain molds and weathering effects are major problems with introduced genotypes of sorghum maturing during relatively high rainfall in the Southern and Northern Guinea savannah zones (5). Under these conditions, grain yields and quality often are reduced. Grain mold is particularly serious on improved varieties of sorghum in the savannah zones of Ghana (4). The fungi causing grain mold have been neither isolated nor identified, and their respective roles in the disease are not known. The objective of this study was to isolate and identify the pathogens of grain mold of sorghum in Ghana.

Materials and Methods

We conducted a field experiment during the rainy seasons (June-October) of 1996 and 1997 at the Savannah Agricultural Research Institute (SARI), Nyankpala, Ghana. We used 21 sorghum genotypes to study the interactions between sorghum head bug population dynamics and the grain mold disease complex under a minimal insecticide application management strategy (4). We used natural infestation of head bugs and infection by grain molds, *i.e.*, no field inoculations, in a split plot design with three replications.

Twenty-one molded and healthy seed samples from the 1996 and 1997 cropping seasons were saved in plastic bags and stored at 4°C for study in 1998. The samples were taken to the Danish Institute of Seed Pathology, Copenhagen, Denmark, where they were examined for the presence of fungi by the blotter method following the recommendations of the International Seed Testing Association (3). Two hundred seeds of each sample were placed on three layers of moistened blotters in sterilized Petri dishes at the rate of 25 seeds per dish. The Petri dishes were incubated at 21°C under alternating cycles of 12 hours near ultraviolet light and 12 hours darkness for 7 days. Seeds were examined under a stereo microscope for the presence of seed-borne fungi.

Results and Discussion

Some of the sorghum genotypes produced normal heads,

TABLE 38-1. Percentage infection of fungi recorded in 21 samples of sorghum from Ghana tested by the blotter method.

Fungus	Number of Infected Samples	Incidence (%)	
		Mean	Range
Acremonium strictum	4	0.7	0.2-1.7
Bipolaris bicolor	4	0.4	0.3-0.5
Bipolaris maydis	13	0.5	0.2-1.3
Cercospora sorghi	1	0.5	0.5
Colletotrichum gloeosporioides	8	0.8	0.3-2.3
Colletotrichum sublineolum	19	11	0.3-45
Curvularia lunata	1	1.9	1.9
Exserohilum monoceras	1	0.3	0.3
Exserohilum rostratum	7	0.3	0.2-0.5
Fusarium equiseti	15	0.9	0.2-2.8
Fusarium moniliforme sensu lato	21	30	15-56
Fusarium pallido-roseum	18	2.0	0.5-3.0
Gloeocercospora sorghi	8	0.5	0.3-1.3
Nigrospora sp.	1	0.3	0.3
Phoma sorghina	21	69	31-93
Phomopsis sp.	6	0.4	0.2-0.8
Verticillium sp.	4	0.6	0.2-1.5

while others were completely spoiled as the grain became covered with molds as the season progressed. The germination of moldy grains was reduced to 0-12%, and they were of no commercial value and could not be used for seed (4). The most frequently isolated seed-borne fungi (Table 38-1) were *Phoma sorghina* (range of infection 31-93%, mean 69%, causal agent of *Phyllostica* seed disease and stalk rot) and *Fusarium moniliforme sensu lato* (range of infection 15-56%, mean 30%, reported to cause seedling blight, root and stalk rot, Pokkah boeng, grain mold, and head blight) (2, 7-9). Other fungi recorded were *Acremonium strictum* (mean infection 0.7%), *Bipolaris bicolor* (mean infection 0.4%), *Bipolaris maydis* (mean infection 0.5%), *Cercospora sorghi* (mean infection 0.5%), *Colletotrichum gloeosporioides* (mean infection 0.8%), *Colletotrichum sublineolum* (mean infection 11%), *Curvularia lunata* (mean infection 1.9%), *Exser-ohilum monoceras* (mean infection 0.3%), *Exserohilum rostratum* (mean infection 0.3%), *Fusarium equiseti* (mean infection 0.9%; *Fusarium pallido-roseum* (mean infection 2.0%), *Gloeocercospora sorghi* (mean infection 0.5%), *Nigrospora* sp. (mean infection 0.3%), *Phomopsis* sp. (mean infection 0.4%), and *Verticillium* sp. (mean in-fection 0.6%).

Seedborne diseases in Ghana do not yet pose a major constraint to agricultural development, but some of the fungi reported in this study can cause seed rot and other sorghum diseases. Sorghum yields in Ghana are currently very low, which is at least partially due to the use of poor quality seed. There is a lack of awareness among resource-poor farmers of seed health and its rela-tionship to crop yields. Attention should be paid to avoid the build-up of inoculum of seed-borne pathogens.

Two fungi, *Phoma sorghina* and *Fusarium moniliforme sensu lato*, constitute more than 30% of the fungi isolated. Seed testing of sorghum for these pathogens, if introduced into the seed quality control system of production, could significantly improve plant and seed health. Low infection of *B. maydis* (0.2-1.3%) in thirteen samples indicates the importance of inspecting seed lots of sorghum for this pathogen to prevent spread to maize. The build-up of seedborne inoculum of pathogens such as *Colletotrichum*, *Curvularia*, and *Gloeocercospora* in certified sorghum seed also should be avoided.

Acknowledgements

We thank Anna-Lise Hynkemejer for technical assistance and the Danish Government Institute of Seed Pathology, Copenhagen, Denmark, for providing facilities for the mycoflora studies.

References

1. **Bandyopadhyay, R., D. E. Hess, I. Sissoko, S. K. Nutsugah, and I. D. K. Atokple**. 1997. Sorghum disease survey in Ghana. Summary report, October 20-24, 1997. ICRISAT, Patancheru (A.P.), India.
2. **Frederiksen, R. A., and G. N. Odvody, eds**. 1999. *Compendium of Sorghum Diseases*, 2nd ed. APS Press, St. Paul, Minnesota.
3. **International Seed Testing Association.** 1976. International rules for seed testing. Annexes 1976. *Seed Sci. Technol.* **4**:51-177.
4. **Nutsugah, S. K.** 1996. *Annual Report: Plant Pathology*. Savannah Agric. Res. Inst., Nyankpala, Tamale, Ghana.
5. **Nutsugah, S. K.** 1997. *Annual Report: Plant Pathology*. Savannah Agric. Res. Inst., Nyankpala, Tamale, Ghana.
6. **Nutsugah, S. K., F. K. Ayensu, and F. K. Tsigbey.** 1999. Survey of seed-borne diseases in Ghana: Initiation of cooperative research programme for seed-borne disease management. *Proc. 1999 World Seed Conf. (Cambridge, UK)*: 50.
7. **Tarr, S. A. J.** 1962. *Diseases of Sorghum, Sudangrass, and Broom Corn*. Commonwealth Mycological Inst., Kew, Surrey, UK.
8. **Tullis, E. C.** 1951. *Fusarium moniliforme*, the cause of stalk rot of sorghum in Texas. *Phytopathology* **41**:529-535.
9. **Zummo, N.** 1980. Fusarium disease complex of sorghum in West Africa, p. 297-299. *In* R. J. Williams, R. A. Frederiksen, and L. K. Mughogho (eds.), *Sorghum Diseases, A World Review*. ICRISAT, Patancheru (A.P.), India.

39

Mycotoxins from Fungal-Infected Sorghum: *Claviceps*, *Fusarium*, and the *Striga* Connection

James K. Porter, Charles W. Bacon, William P. Norred, Emma M. Wray, Gretchen A. Kuldau, Anthony E. Glenn, and John F. Leslie

Sorghum (*Sorghum bicolor*) is a globally important cereal and forage crop, whose production is reduced by both pathogenic and parasitic fungi and by parasitic plants. *Claviceps africana* is the latest global disease to challenge the sorghum industry (7, 22, 33, 34). *Claviceps* and *Fusarium* species, however, not only threaten sorghum yield but also can produce mycotoxins in the grain that pose health concerns for domesticated animals and humans (7, 11-13, 16, 21, 25, 29, 31, 36-38, 45-47, 61). Sorghum production also is reduced by a parasitic angiosperm known as *Striga*, or witchweed (43, 58-60). Defending against these problems, which may also affect other agricultural commodities such as corn, wheat, rye, millet, and rice, requires an understanding of the ecology of host-pathogen interactions, of the conditions required for mycotoxin production, and of the effects of the mycotoxins on livestock, poultry, and humans. In this chapter we focus first on the primary mycotoxins, *i.e.*, the ergot alkaloids, produced by *C. africana* and results from some of the toxicology investigations in poultry and swine associated with *C. africana*-infected sorghum. We then shift our focus to the analysis of *Fusarium* toxins associated with *C. africana*-infected sorghum, and the utilization of these toxins, and the strains that produce them, as natural biocontrol agents for *Striga*.

Claviceps

The three primary species of *Claviceps* that infect sorghum are: *Claviceps africana*, found in Africa, Southeast Asia, Australia, and South and North America; *C. sorghi*, found in India and Southeast Asia; and *C. sorghicola*, found in Japan (7, 25, 36). *Claviceps* causes an ovarian tissue replacement that impacts production by reducing grain yields and hybrid seed production. This fungus also limits sorghum yield by interfering with mechanical harvest due to the honeydew exudate produced during the early stages of fungal infection that leads to clogged harvesting equipment and difficulties in separating the grains (7, 25, 36). *Claviceps africana* further reduces sorghum production by predisposing the plant to infection by other pests and pathogens. *C. africana* also reduces the quality of any resulting sorghum grain by producing mycotoxins that belong to the ergot alkaloid group. Historically, ergot alkaloids have posed major health concerns for livestock, poultry, and humans consuming contaminated products made from the infected grain (11, 12, 31, 37, 45-47).

The major ergot alkaloids identified from *C. africana* infected sorghum are dihydroergosine, dihydroelymoclavine, festuclavine, and pyroclavine (Fig. 39-1; 36, 37, 47). These alkaloids are structurally related, but represent two distinct classes of ergot alkaloids: the ergopeptine alkaloids (dihydroergosine) and the clavine alkaloids (dihydroelymoclavine, festuclavine, and pyroclavine). These different alkaloid classes have different toxic effects on livestock, poultry, and humans (11, 12, 47). The ergopeptines are associated with the vasoconstrictive properties of the ergot alkaloids; *i.e.*, these compounds lead to dry gangrene of the peripheral extremities of animals and humans that consume contaminated feed and foods (11, 12). This group of alkaloids is produced primarily by *Claviceps purpurea*, the "ergot of wheat and rye" that swept through Europe during the Middle Ages and is responsible for "gangrenous ergotism." They also are produced by *Neotyphodium coenophialum*, the endophyte of toxic tall fescue that causes fescue foot in cattle (3, 45-47, 50).

The only structural differences in the ergot alkaloids produced by *C. purpurea*, *N. coenophialum*, and *C. africana* is that *C. africana* produces the dihydro-ergot alkaloids; *i.e.*, the double bond is reduced at the 9,10-position of the ergoline nucleus (11). Fortuitously, the ergot alkaloids produced by *C. africana* do not appear to be as toxic as those produced by *C. purpurea* or *N. coenophialum*, nor do they have the vasoconstrictive properties. Nevertheless, current toxicological data in poultry and swine associated with consumption of the ergot alkaloids produced by *C. africana* on infected sorghum have not resulted in consensus conclusions.

For example, in poultry broilers (hatch to 3 weeks of age), Bailey *et al.* (8) reported that sorghum ergot (*C. africana*) mixed in a basal diet (with total alkaloid concentration, 11 ppm; dihydroergosine, 2.1 ppm; and dihydroelymoclavine, 9.1 ppm) did not significantly affect growth or feed efficiency, but increased liver weight 12% in these birds. In broilers raised to 6 weeks in a three phase feeding program using formulated starter, finisher, and withdrawal diets, to approximate broiler production regimens, sorghum ergot increased the feed/gain ratio by 12% at 3 weeks, decreased body weight gain by 4.1% at 4 weeks, and increased the feed/gain ratio by 14% and 12% at 5 and 6 weeks, respectively. The alkaloid concentrations in the feed were the same as for the previous 3-week study. In a second set of experiments with day-old broilers fed for 3 weeks with mature sorghum-ergot sclerotia at 2.5%, 5.0%, and 10% of the formulated diets and a more concentrated and different ratio of alkaloids, *i.e.*, 235 ppm total ergot alkaloids–dihydroergosine 181 ppm, dihydroelymoclavine 39 ppm, festuclavine 13 ppm, and pyroclavine 1.9 ppm. Bailey *et al.* (8) reported increased feed/gain ratios of 5.9% and 8.7% at 2 and 3 weeks, respectively. Thus, in the first set of experiments, they reported reduced feed efficiency in poultry that consumed more dihydroelymoclavine than dihydroergosine, while in the succeeding experiments with more mature sclerotia, dihydroergosine was at a higher level than dihydroelymoclavine and the other minor alkaloids.

In a 14-day growth study of day-old broilers fed uncontaminated sorghum and ergot-contaminated sorghum with dihydroergosine concentrations as low as 1.8 ppm of the diet, Dean *et al.* (20) reported that average daily gain and feed intake were reduced 5.9% and 4.5%, respectively. In a 28-day growth assay, with nursery pigs, Dean *et al.* (19) reported that average daily gains and average feed intake were reduced 5.5% and 5.8%, respectively, with dihydroergosine at 1.7 ppm of the diet, and reduced 17% and 18%, respectively, at 3.4-ppm. In these studies only trace amounts of dihydroelymoclavine, festuclavine, pyroclavine, and other unidentified ergot alkaloids were present in the diets. Thus, these investigators (19, 20) observed more dramatic effects in both poultry and swine with much lower concentrations of dihydroergosine and only trace quantities of the minor alkaloids than did Bailey *et al.* (8).

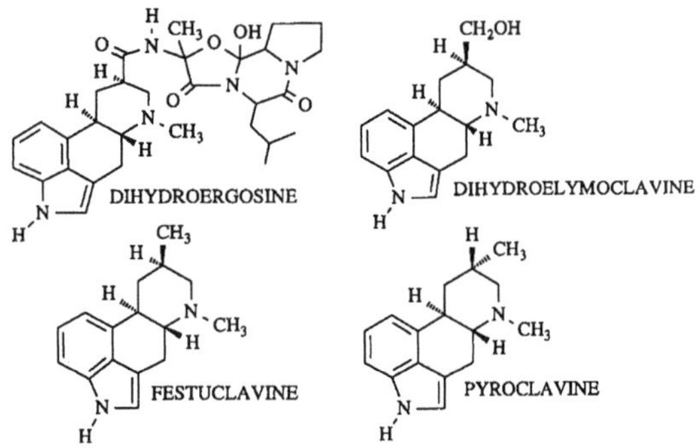

FIGURE 39-1. Chemical structures of four ergot alkaloids.

Although dihydroergosine is the primary alkaloid found in mature sclerotia of *C. africana* and belongs to that class of ergot alkaloids usually associated with toxicity problems in livestock and humans, *i.e.*, the ergopeptine alkaloids (11, 12, 31, 45-47, 50), the data to date suggest that a combination of the ergot alkaloids produced by *C. africana* may be more toxicologically significant than any single component. Even though mortality was not significant in any of the above experiments, the significant reduction in feed conversion, feed intake, and reduced body weight gains indicate that *C. africana* infected sorghum containing the ergot alkaloids can have an economically significant effect on production in poultry and swine. Moreover, diets containing as little as 1-20 ppm *C. africana* alkaloids should be viewed with caution when fed to either poultry or swine and, especially, to nursing pigs. These levels reduced feed efficiency by 5-12% and weight gains by > 5%, and would have a significant negative impact on production. These data are somewhat consistent with other investigations in which much higher concentrations of alkaloids in the diets of livestock produced more dramatic effects (7). Nonetheless, due to the diversity in alkaloid concentrations produced at different stages of fungal growth and the less-than-uniform results outlined above, further studies are needed to assess the long-term effects of these compounds on the livestock that consume them, and to ensure that all of the mycotoxigenic compounds have been identified.

Fusarium

In addition to *Claviceps*, *Fusarium* species also represent a major threat to the sorghum industry and to animal and

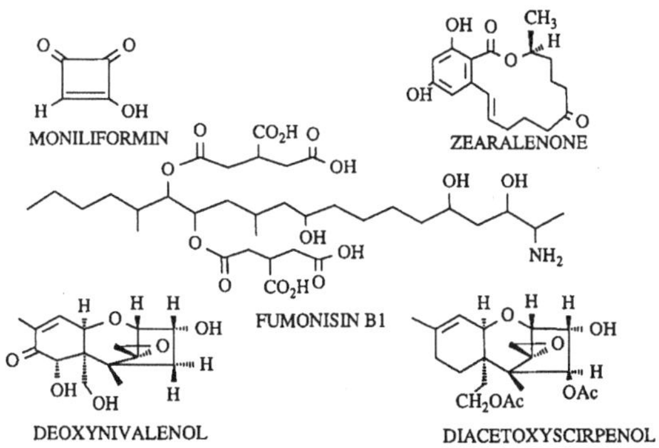

FIGURE 39-2. Chemical structures of five commonly occurring *Fusarium* mycotoxins.

human health. Unlike *Claviceps*, *Fusarium* species produce a diverse group of structurally unrelated mycotoxins with equally diverse biological activities (13, 18, 21, 39); most notably, the fumonisins, zearalenone, moniliformin, fusaric acid, and trichothecenes, *e.g.*, deoxynivalenol and diacetoxyscirpenol (Figs. 39-2 and 39-3; for other examples of *Fusarium* toxins see 13, 18, 21, 39). Fusaric acid (5-butyl-2-pyridinecarboxylic acid; Fig. 39-3) is a common metabolite produced by most *Fusarium* species. We have found that the presence of this mycotoxin is a good indicator of *Fusarium* infection (4-6, 32, 47, 48, 51, 52).

Fusaric acid and other alkyl-pyridinecarboxylic acids (APAs, Fig. 39-3) have been reported either singly or in combination from *Fusarium verticillioides* (previously known as *Fusarium moniliforme*), *Fusarium proliferatum*, *Fusarium oxysporum*, *Fusarium subglutinans*, *Fusarium solani*, *Fusarium heterosporum*, *Fusarium sambucinum*, *Fusarium nygamai*, *Fusarium fujikuroi*, *Fusarium crookwellense*, *Fusarium napiforme*, *Fusarium lateritium*, *Fusarium semitectum*, and *Fusarium thapsinum* (6, 9, 14, 15, 23, 35, 41, 44, 45, 48). Thus, fusaric acid is widely distributed in nature since these taxa probably occur on several thousand plant species, many of which are agronomically important. *F. thapsinum* (teleomorph, *Gibberella thapsina*) is a relatively newly described species with a global distribution (29, 33) that is commonly isolated from sorghum stalks, surfaced-sterilized sorghum seeds, sorghum plants with stalk rot, and asymptomatic sorghum. Porter *et al.* (48, 49) reported that this species produces numerous structural variations of the fusaric acid, with 10-OHFA as the major compound (Fig. 39-3). The collective production of such diverse fusaric acid analogues may be unique to this *Fusarium* species.

Fusaric acid also has been isolated from *Claviceps*-infected sorghum, which indicates not only the co-occurrence of mycotoxins, but also a double infection by a *Fusarium* species and *C. africana*. Subsequent studies of *C. africana*-infected sorghum confirmed a *Fusarium* species (Kuldau, *unpublished*). Thus, the co-occurrence of both *Claviceps* and *Fusarium* mycotoxins should be considered in chemical and toxicological evaluations for either synergistic and/or antagonistic effects (5, 22, 48, 51, 61). In addition to interactions with plant hosts, some *Fusarium* species have been reported as mycoparasites on *Claviceps* (40), which suggests yet another level of interaction. The co-occurrence of mycotoxins in a single foodstuff is relatively common (13, 21, 22, 45, 47, 53), which may indicate that the producing organisms are competing with each other for a niche or that they colonized the host at different times or locations.

The *Striga* Connection

Originally, fusaric acid was identified as a wilt toxin (phytotoxin) and thought to play a role in the pathogenesis of *Fusarium* toward certain plants (26). Fusaric acid, some of the fusaric acid derivatives, and some isolates of *F. oxysporum* and *F. nygamai* have natural herbicidal activity against the parasitic weed, *Striga hermonthica* (1, 2, 64, 68). In addition, *Fusarium brevicatenulatum*, a relatively newly identified species of *Fusarium* isolated from *Striga asiatica* (42), may have use as a *Striga* biocontrol on sorghum. The hydroxylated fusaric acid derivatives produced by *F. thapsinum* may be translocated throughout the plant, thereby providing natural systemic protection against *Striga*. However, many questions remain, and the efficacy of these fusaric acid derivatives as bioherbicides is unknown. Similarly, modifications that increase or decrease the action spectrum, that alter the solubility, or that enhance the chemical stability of the molecule remain to be defined. The discovery that *F. thapsinum* produces these unusual derivatives of fusaric acid should provide an opportunity to explore some of these questions in more detail. Some of

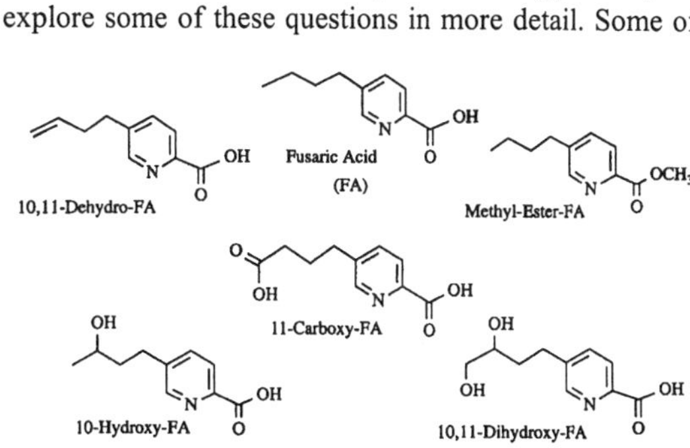

FIGURE 39-3. Chemical structure of fusaric acid and five derivatives synthesized by *Fusarium* spp.

our current research is directed at screening various combinations of the APAs produced by *F. thapsinum* for their activity against *Striga* (45).

Isolates of *F. thapsinum* from sorghum and *F. verticillioides* from maize have been described as symptomless, endophytes of their plant host (4, 29). In this regard, these two species have rather unique associations with plants. Similar associations, *i.e.*, endophytic infections, are commonly mutualistic, as the endophytic fungus produces metabolites that are toxic to mammals, insects, and/or nonhost plants (3, 45, 46). These associations and activities affect species survival and impart drought tolerance, which gives infected plants an ecological advantage (3, 45). Indeed, these associations sometimes are referred to as defensive mutualisms (17). Consequently, the production of fusaric acid and its derivatives by *Fusarium* spp. and the phytotoxicity of these compounds to other plant species suggests that the interaction between *F. thapsinum* and sorghum may be both unique and complex. Initial studies have demonstrated positive interacttions in *F. verticillioides*-infected maize plants in terms of photosynthesis (57) and growth (67). Further studies are needed with both *F. verticillioides*-infected maize and *F. thapsinum*-infected sorghum plants to better define these interactive-associations.

Species survival often depends on the development of intricate biochemical relationships with and defense mechanisms towards their environment (3, 45). Within these ecological constraints, co-evolutionary processes can occur in which the biochemical properties of one species are either exploited by or dependent upon the other. Usually both the fungus and the host have co-evolved in these associations. The host has developed resistance to the fungus's pathogenicity mechanism, and the fungus can elude or overcome its host plant's defense mechanisms. In some endophyte-host associations, the defense mechanism of one partner also becomes a defense mechanism for the other (3, 17, 45). Whether there is a parallel mutualism that has co-evolved between some *Fusarium* species and sorghum and their defense mechanisms against *Striga*, *e.g.*, fusaric acid and derivatives, is currently an open question.

Applications

Exploiting the host plant's natural defense mechanisms to control the devastating effects of *Claviceps*, *Fusarium*, and *Striga* is intriguing, but not without controversy. The use of fungi that produce mycotoxins or the use of mycotoxins individually or collectively on food crops as natural biocontrol agents against other fungi, weeds or insects raises both environmental and major health concerns for domesticated animals and humans. These concerns are amplified because *Fusarium* toxins are associated with some of the most pernicious syndromes affecting domesticated animals and humans (13, 16, 21, 53), not to mention the possibility of synergistic interactions with other toxins (5, 22, 51, 52, 61, 66). Fusaric acid has some of the most diverse activities of any of the *Fusarium* toxins (5, 6, 24, 48, 51, 52, 55, 56, 61-63), and although it alone is not acutely toxic to neonate, juvenile, or adult rats, it may enhance the effects of other toxins produced by both *Fusarium* and *Claviceps* (5, 13, 21, 22, 45, 48, 50, 55, 56). Furthermore, the lactational transfer of fusaric acid from the feed of nursing dams to suckling neonates along with the neurochemical effects of this compound (48, 51, 52, 55, 56) suggests that its use on food crops should be viewed with caution. At the same time, fusaric acid and the methyl-ester of fusaric acid dissipate in the soil within a few days of their application (65), and these compounds have few, if any, phytotoxic effects on sorghum or corn (Kittle, Bacon, Porter and Kuldau, *unpublished*). Direct correlations have been reported between resistance to fusaric acid and that to certain *Fusarium* species in gladioli and potatoes (54, 64). Thus, the judicious application of a select group of fusaric acid derivatives may be a useful adjunct to biocontrol practices against *Striga* and/or other weedy parasites on food and feed crops and ornamental plants. Similarly, genetic manipulation of the plant and its endophyte could maximize their joint natural defenses against other parasites–animal, plant, or microbial.

Summary and Conclusions

Sorghum is the fifth most important world cereal crop, second only to maize in portions of South America and first in some portions of Africa. Historically, *Claviceps* and *Fusarium* are two of the most important groups of fungi associated with economic losses to sorghum production, while the mycotoxins these fungi produce pose significant health concerns for livestock, poultry, and humans. These two groups of fungi affect not only sorghum, but also maize, wheat, rice, millet, barley, rye, and pasture grasses consumed by grazing ungulates.

The major ergot alkaloids produced by *C. africana* on infected sorghum–dihydroergosine, dihydroelymoclavine, festuclavine, and pyroclavine–at a level of 1-20 ppm in the diet reduce feed efficiency in poultry and average daily gains and feed intake in nursery pigs, with nursery pigs apparently more susceptible than poultry. Furthermore, *Fusarium* species and mycotoxins that they produce, *e.g.*, fusaric acid, are known to co-occur on *C. africana* infected sorghum. Therefore, the combined effects of these co-occurring mycotoxins need to be considered in both chemical and toxicological studies, especially since their disparate biological activities could result in synergistic toxicities. *F. thapsinum*, the endophyte of sorghum, produces several fusaric acid derivatives and also differs from its sibling species *F. verticillioides*, an endophyte of maize, in the ability to produce moniliformin and fumonisins. Yet both species can infect either sorghum or maize (27-30, 33, 34).

Originally, mycotoxins were defined as fungal metabolites toxic to mammals and birds and were thought to be unimportant secondary metabolites or dead-end biosynthetic pathways associated with primitive evolutionary processes. However, bioassays now define mycotoxins as fungal metabolites that affect bacteria (antibiotics), other fungi (natural antifungal agents), insects (natural insecticides), plants (phytotoxins or natural herbicides) (45, 46), and animals. Some of these activities are simultaneously invasive and defensive mechanisms, which provides an ecological rationale for mycotoxin production (3, 17).

Claviceps species produce one of the most biologically active and structurally unique classes of compounds known to the medical sciences: the ergot alkaloids (11, 12). Historically, these compounds have been a major scourge to both livestock and humans consuming contaminated cereals grains, yet they also have been a boon to the medical sciences (11, 12, 45, 46). *Fusarium* species not only produce the structurally most diverse secondary metabolites isolated from any group of fungi, but also produce compounds with the most divergent biological activities isolated from one genus (13, 21, 45). Understanding the biological and toxicological mechanisms of the mycotoxins produced by *Fusarium* has generated new awareness, concepts, and practices in handling *Fusarium*-associated toxicities in domesticated animals and humans (4, 6, 10, 16, 39, 53). Further, some *Fusarium* species not only have been described as mycoparasites on *Claviceps*, but also have been suggested as a possible biological control mechanism for these fungi (40). *Fusarium* species, fusaric acid produced by *Fusarium* spp., and derivatives of fusaric acid also have been suggested as a biocontrol agent for *Striga*, a parasitic weed. Even though *Claviceps* and *Fusarium* are two of the most important fungal genera associated with negative influences on agricultural production and human and animal health, major benefits have been derived from the "mycotoxins" produced by these two groups of fungi. Only future investigations will provide us with the knowledge we need to capitalize on their beneficial aspects and to control and manage the devastating effects they can wreak throughout the food production chain.

Acknowledgments

This research was supported in part by the Kansas Agricultural Experiment Station, and by the Sorghum and Millet Collaborative Research Support Program (INTSORMIL) AID/DAN-1254-G-00-0021-00 from the U.S. Agency for International Development. Contribution no. 02-240-B from the Kansas Agricultural Experiment Station, Manhattan.

References

1. **Abbasher, A. A., J. Kroschel, and J. Sauerborn.** 1995. Microorganisms of *Striga hermonthica* in Northern Ghana with the potential as biocontrol agents. *Biocontrol Sci. Technol.* **5:** 157-161.
2. **Abbasher, A. A., and J. Sauerborn.** 1992. *Fusarium nygamai*, a potential bioherbicide for *Striga hermonthica* control of sorghum. *Biol. Control* **2:** 291-296.
3. **Bacon, C. W., and N. S. Hill, eds.** 1997. Neotyphodium-*Grass Interactions–Proc. 3rd Intl. Symp. on* Acremonium/ *Grass Interactions (28-31 May 1997, Athens, Georgia).* Plenum Press, New York City, New York.
4. **Bacon, C. W., and D. M. Hinton.** 1996. Symptomless endophytic colonization of maize by *Fusarium moniliforme. Can. J. Bot.* **74:** 1195-1202.
5. **Bacon, C. W., J. K. Porter, and W. P. Norred.** 1995. Toxic interaction of fumonisin B_1 and fusaric acid measured by injection into fertile chicken eggs. *Mycopathologia* **129:** 29-35.
6. **Bacon, C. W., J. K. Porter, W. P. Norred, and J. F. Leslie.** 1996. Production of fusaric acid by *Fusarium* species. *Appl. Environ. Microbiol.* **62:** 4039-4043.
7. **Bandyopadhyay, R., D. E. Frederickson, N. W. McLaren, and M. J. Ryley.** 1998. Ergot: A new disease threat to sorghum in the Americas and Australia. *Plant Dis.* **82:** 356-367.
8. **Bailey, C. A., J. J. Fazzino, Jr., M. S. Ziehr, M. Sattar, A. U. Haq, G. N. Odvody, and J. K. Porter.** 1999. Effects of sorghum ergot on broiler performance. *Poult. Sci.* **78:** 1391-1397.
9. **Ballio, A.** 1981. Structure-activity relationships, p. 395-441. *In* R. D. Durbin (ed.), *Toxins in Plant Diseases*. Academic Press, New York City, New York.
10. **Beardall, J. M., and J. D. Miller.** 1994. Diseases in humans with mycotoxins as possible causes, p. 487-539. *In* J. D. Miller and H. L. Trenholm (eds.), *Mycotoxins in Grain: Compounds Other than Aflatoxin*. Egan Press, St. Paul, Minnesota.
11. **Berde B., and H. O. Schild.** 1978. *Ergot Alkaloids and Related Compounds. Handbook of Experimental Pharmacology*, Vol. 49. Springer-Verlag, New York City, New York.
12. **Bove, F. J.** 1970. *The Story of Ergot.* S. Karger, New York City, New York.
13. **Bryden, W. L., A. Logrieco, H. K. Abbas, J. K. Porter, R. F. Vesonder, J. L. Richard, and R. J. Cole.** 2001. Other significant *Fusarium* toxins, p. 360-392. *In* B. A. Summerell, J. F. Leslie, D. Backhouse, W. L. Bryden, and L. W. Burgess (eds.), Fusarium*: Paul E. Nelson Memorial Symposium*. APS Press, St. Paul, Minnesota.

14. **Burmeister, H. R., M. D. Grove, R. E. Peterson, D. Weisleder, and R. D. Plattner.** 1985. Isolation and characterization of two new fusaric acid analogs from *Fusarium moniliforme* NRRL 13,163. *Appl. Environ. Microbiol.* **50:** 311-314.
15. **Capasso, R., A. Evidente, A. Cutignano, M. Vurro, M. C. Zonno, and A. Bottalico.** 1996. Fusaric acid and 9,10-dehydrofusaric acid and their methyl esters from *Fusarium nygamai*. *Phytochemistry* **41:** 1035-1039.
16. **Charmley, L. L., A. Rosenberg, and H. L. Trenholm.** 1994. Factors responsible for economic losses due to *Fusarium* mycotoxin contamination of grains, foods, and feedstuffs, p. 471-486. *In* J. D. Miller and H. L. Trenholm (eds.), *Mycotoxins in Grain: Compounds Other than Aflatoxin*. Egan Press, St. Paul, Minnesota.
17. **Clay, K., and J. Holah**. 1999. Fungal endophyte symbiosis and plant diversity in successional fields. *Science* **285:** 1742-1744.
18. **Cole, R. J., and R. H. Cox.** 1981. *Handbook of Toxic Fungal Metabolites*. Academic Press, New York City, New York.
19. **Dean, D. W., R. H. Hines, J. D. Hancock, G. A. Kennedy, J. K. Porter, C. A. Maloney, J. M. DeRouchey, D. J. Lee, J. S. Park, and H. Cao.** 1999. Effects of sorghum ergot in diets for nursery pigs. *J. Anim. Sci.* **77(Suppl. 1):** 337.
20. **Dean, D. W., R. H. Hines, J. D. Hancock, J. K. Porter, and C. L. Jones.** 2000. Effects of sorghum ergot in diets for broiler chicks. *J. Poult. Sci.* **70(Suppl. 1):** 302.
21. **D'Mello, J. P. F., J. K. Porter, A. M. C. MacDonald, and C. M. Placinta.** 1997. *Fusarium* toxins, p. 287-301. *In* J. P. F. D'Mello (ed.), *Handbook of Plant and Fungal Toxicants*. CRC Press, Boca Raton, Florida.
22. **Dowd, P. F.** 1988. Toxicology and biochemical interactions of fungal metabolites fusaric acid and kojic acid with xenobiotics in *Heliothis zea* (F.) and *Spodoptera frugiperda* (J. E. Smith). *Pestic. Biochem. Physiol.* **32:** 123-134.
23. **Drysdale, R. B.** 1984. The production and significance in phytopathology of toxins produced by species of *Fusarium*, p. 95-105. *In* M. O. Moss and J. E. Smith (eds.), *The Applied Mycology of* Fusarium. Cambridge University Press, New York City, New York.
24. **Fernandez-Pol, J.A., D. J. Klos, and P. D. Hamilton.** 1993. Cytotoxic activity of fusaric acid on human adenocarcinoma cells in tissue cultures. *Anticancer Res.* **13:** 57-64.
25. **Frederickson, D. E., P. G. Mantle, and W. A. J. de Milliano.** 1991. *Claviceps africana* sp. nov.: the distinctive ergot pathogen of sorghum in Africa. *Mycol. Res.* **95:** 1101-1107.
26. **Gaumann, F.** 1957. Fusaric acid as a wilt toxin. *Phytopathology* **47:** 342-357.
27. **Jardine, D. J., and J. F. Leslie.** 1992. Aggressiveness of *Gibberella fujikuroi* (*Fusarium moniliforme*) isolates to grain sorghum under greenhouse conditions. *Plant Dis.* **76**: 897-900.
28. **Jardine, D. J., and J. F. Leslie.** 1999. Aggressiveness to mature maize plants for *Fusarium* strains differing in ability to produce fumonisin. *Plant Dis.* **83:** 690-693.
29. **Klittich, C. J. R., J. F. Leslie, P. E. Nelson, and W. F. O. Marasas.** 1997. *Fusarium thapsinum* (*Gibberella thapsina*): A new species in section *Liseola* from sorghum. *Mycologia* **89:** 643-652.
30. **Klittich, C. J. R., and J. F. Leslie.** 1992. Identification of a second mating population within the *Fusarium moniliforme* anamorph of *Gibberella fujikuroi*. *Mycologia* **84:** 541-547.
31. **Kuldau, G. A., and C. W. Bacon.** 2000. Biology and toxicity of *Claviceps* and related fungi, p. 503-534. *In* Y. H. Hui, R. A. Smith, and D. G. Spoerke, Jr. (eds.), *Food-borne Disease Handbook: Vol. 3. Plant Toxicants*. Marcel Dekker, New York City, New York.
32. **Kuldau, G. A., J. K. Porter, E. M. Wray, R. Bennett, A. E. Glenn, and C. W. Bacon.** 1999. Isolation of *Fusarium moniliforme* mutants with reduced fusaric acid production. *Fung. Genet. Newsl.* **46(Suppl):** 108.
33. **Leslie, J. F., W. F. O. Marasas, G. S. Shephard, E. W. Sydenham, S. Stockenstrom, and P. G. Thiel.** 1996. Duckling toxicity and the production of fumonisin and moniliformin by isolates of the A and F mating populations of *Gibberella fujikuroi* (*Fusarium moniliforme*). *Appl. Environ. Microbiol.* **62:** 1182-1187.
34. **Leslie, J. F., R. D. Plattner, A. E. Desjardins, and C. J. R. Klittich.** 1992. Fumonisin B_1 productions by strains from different mating populations of *Gibberella fujikuroi* (*Fusarium* selection *Liseola*). *Phytopathology* **82:** 341-345.
35. **Luz, J. M., R. R. M. Paterson, and D. Brayford.** 1990. Fusaric acid and other metabolite production in *Fusarium oxysporum* f. sp. *vasinfectum*. *Lett. Appl. Microbiol.* **11:** 141-144.
36. **Mantle, P. G.** 1999. Sorghum ergot: an overview. *Proc. Global Conf. on Ergot of Sorghum (1-8 June 1997, Sete Lagoas, Brazil)*: 3-6.
37. **Mantle, P. G., and E. S. Waight.** 1968. Dihydroergosine: A new naturally occurring alkaloid from *Spacelia sorghi* (McRae). *Nature* **218:** 581-582.
38. **Mansuetus, A. S. B., G. N. Odvody, R. A. Frederiksen, and J. F. Leslie**. 1997. Biological species in the *Gibberella fujikuroi* species complex (*Fusarium* section *Liseola*) recovered from sorghum in Tanzania. *Mycol. Res.* **101:** 815-820.
39. **Miller, J. D., and H. L. Trenholm, eds.** 1994. *Mycotoxins in Grain: Compounds Other than Aflatoxin*. Egan Press, St. Paul, Minn.
40. **Mower, R. L., W. C. Snyder, and J. G. Hancock.** 1975. Biological control of ergot by *Fusarium*. *Phytopathology* **65:** 5-10.
41. **Mutert, W. U., H. Lutfring, and W. Barz.** 1990. Formation of fusaric acid by fungi of the genus *Fusarium*. *Z. Naturforsch. (B)* **11:** 141-144.
42. **Nirenberg, H. I., K. O'Donnell, J. Kroschel, J. M. Frank, and W. Mubatanhema.** 1998. Two new species of *Fusarium*: *Fusarium brevicatenulatum* from the noxious weed *Striga asiatica* in Madagascar and *Fusarium pseudoanthophilum* from *Zea mays* in Zimbabwe. *Mycologia* **90:** 459-464.
43. **Pare, J., B. Dembele, O. Ouedraogo, A. Raynal-Roquws, C. Tuquet, and G. Salle.** 1997. Dynamics of flowering and embryology of *Striga hermonthica* (Del.) Benth. (Scrophulariaceae): implications for chemical control. *Internat. J. Pest Manag.* **43:** 285-290.

44. **Pitel, D. W., and L. C. Vining**. 1970. Accumulation of dehydrofusaric acid and its conversion to fusaric acid and 10-hydroxyfusaric acid in cultures of *Gibberella fujikuroi*. *Can. J. Biochem.* **48:** 623-630.
45. **Porter, J. K.** 1994. Chemical constituents of grass endophytes, p. 103-123. *In* C. W. Bacon and J. F. White, Jr. (eds.), *Biotechnology of Endophytic Fungi of Grasses*. CRC Press, Boca Raton, Florida.
46. **Porter, J. K.** 1997. Endophyte alkaloids, p. 51-62. *In* J. P. F. D'Mello (ed.), *Handbook of Plant and Fungal Toxicants*. CRC Press, Boca Raton, Florida.
47. **Porter, J. K.** 2000. Analytical methodology for mycotoxins, p. 653-682. *In* Y. H. Hui, R. A. Smith, D. G. Spoerke, Jr. (eds.), *Foodborne Disease Handbook: Vol. 3. Plant Toxicants*. Marcel Dekker, New York City, New York.
48. **Porter, J. K., C. W. Bacon, E. M. Wray, and W. M. Hagler, Jr.** 1995. Fusaric acid in *Fusarium moniliforme* cultures, corn, and feeds toxic to livestock and the neurochemical effects in the brain and pineal gland of rats. *Nat. Tox.* **3:** 91-100.
49. **Porter, J. K., C. W. Bacon, E. M. Wray, G. A. Kuldau, and J. F. Leslie.** 2000. Fusaric acid (FA), other alkylpyridinecarboxylic acids (APAs), and fumonisin (FB_1) by *Fusarium thapsinum* and *F. moniliforme*: GC/MS analysis, p. 41-47. *In* W. J. de Koe, R. A. Samson, H. P. van Egmond, J. Gilbert, and M. Sabino, (eds.), *Proc. of the Xth IUPAC Symp. on Mycotoxins and Phycotoxins (21-25 May 2000, Sao Paulo, Guaruja, Brazil). Mycotoxins and Phycotoxins in Perspective at the Turn of the Millennium.* W. J. de Koe, Wageningen, The Netherlands.
50. **Porter, J. K., and F. N. Thompson, Jr.** 1992. Effects of fescue toxicosis on reproduction in livestock. *J. Anim. Sci.* **70:** 1594-1603.
51. **Porter, J. K., E. M. Wray, R. M. Eppley, and W. M. Hagler, Jr.** 1998. Zearalenone and fusaric acid in the diet of nursing dams enhances both mycotoxins' lactational transfer to the suckling neonate rat. *Abstr. 112th Annu. Meet. Assoc. Anal. Chem. Internatl. (12-17 Sept. 1998, Montreal, Canada)*, abstract no. A-107.
52. **Porter, J. K., E. M. Wray, A. M., Rimando, P. C. Stancel, C. W. Bacon, and K. A. Voss.** 1996. Lactational passage of fusaric acid from the feed of nursing dams to the neonate rat and effects on pineal neurochemistry in the F_1 and F_2 generations at weaning. *J. Toxicol. Environ. Health* **49:** 161-175.
53. **Prelusky, D. B., B. A. Rotter, and R, G. Rotter.** 1994. Toxicology of mycotoxins, p. 359-403. *In* J. D. Miller and H. L. Trenholm (eds.), *Mycotoxins in Grain: Compounds Other than Aflatoxin*. Egan Press, St. Paul, Minnesota.
54. **Remottii, P. C.** 1996. The role of fusaric acid in the *Fusarium-Gladiolus* interaction and its application in *in vitro* selection for resistance and breeding. Ph.D. thesis, Katholieke Universiteit, Nijmegen, The Netherlands.
55. **Rimando, A. M., and J. K. Porter**. 1997. Fusaric acid increases melatonin levels in the weanling rat and in pineal cell cultures. *J. Toxicol. Environ. Health* **50:** 275-284.
56. **Rimando, A. M., and J. K. Porter.** 1999. Effects of *Fusarium* mycotoxins on levels of serotonin, melatonin and 5-hydroxytryptophan in pineal cell cultures. *Adv. Exp. Med. Biol.* **467:** 425-432.
57. **Rodrigues, L. S., J. L. Azevedo, J. O. Pereira, M. L. C. Vieira, and C. A. Labate.** 2000. Symptomless infection of banana and maize by endophytic fungi impairs photosynthetic efficiency. *New Phytol.* **147:** 609-616.
58. **Salle, G., A. Raynal-Roquws, and C. Tuquet.** 1995. Un fleau en Afrique, les Striga. *C. R. Acad. Sci. (Paris), La Vie Sci.* **12:** 27-46.
59. **Salle, G., C. Tuquet, and A. Raynal-Roquws.** 1998. Biologie of *Phanerogames* parasites. *C. R. Soc. Biol.* **192:** 9-36.
60. **Sauerborn, J.** 1991. The economic importance of the phytoparasites *Orobanche* and *Striga*, p. 137-143. *In* J. K. Ransom, L. J. Musselman, A. D. Worsham, and C. Parker (eds.), *Proc. 5th Internatl. Symp. on Parasitic Weeds*. Survey, Nairobi, Kenya.
61. **Smith, T. K.** 1992. Recent advances in the understanding of *Fusarium* trichothecene mycotoxicoses. *J. Anim. Sci.* **70:** 3989-3993.
62. **Smith, T. K., E. G. McMillian, and J. B. Castillo.** 1997. Effects of feeding blends of *Fusarium*-contaminated grains containing deoxynivalenol and fusaric acid on growth and feed consumption of immature swine. *J. Anim. Sci.* **69:** 2184-2191.
63. **Smith T. K., and M. G. Sousadias.** 1993. Fusaric acid content in swine feedstuffs. *J. Agric. Food Chem.* **41:** 2296-2298.
64. **Venter, S. L., and P. J. Steyn**. 1998. Correlation between fusaric acid production and virulence of isolates of *Fusarium oxysporum* that causes potato dry rot in South Africa. *Potato Res. Wageningen: Eur. Assn. Potato Res.* **41:** 289-294.
65. **Vischetti, C., and A. Esposito.** 1999. Degradation and transformation of a potential natural herbicide in three soils. *J. Agric. Food Chem.* **47:** 3901-3904.
66. **Voss, K. A., J. K. Porter, C. W. Bacon, F. I. Meredith, and W. P. Norred.** 1999. Fusaric acid and modification of subchronic toxicity to rats of fumonisins in *F. moniliforme* culture material. *Food Chem. Toxicol.* **37:** 853-861.
67. **Yates, I. E., C. W. Bacon, and D. M. Hinton.** 1997. Effects of endophytic infection by *Fusarium moniliforme* on corn growth and cellular morphology. *Plant Dis.* **81:** 723-728.
68. **Zonno, M. C., M. Vurro, R. Capasso, A. Evidente, A. Cutignano, J. Sauerborn, and H. Thomas.** 1996. Phytotoxic metabolites produced by *Fusarium nygamai* from *Striga hermonthica*. *Proc. IXth Intl. Symp. Biol. Control of Weeds (21-26 January 1996, Cape Town, South Africa)*: 223-226.

Part VI

Molecular Biology, Genome Mapping, and Host Plant Resistance

40

Identification of Molecular Markers for an Oval Leaf Spot Resistance Gene in Sorghum

Khazan S. Boora, A. Sindhu, P. Boora, and Clint W. Magill

Abstract

Resistance to oval leaf spot in sorghum [*Sorghum bicolor* (L.) Moench] accession IS 4725 segregates as a simple dominant trait controlled by a single gene in a cross to the susceptible cultivar HC136. Bulk segregant analysis coupled with random amplified polymorphic DNA (RAPD) analysis was used to identify DNA polymorphisms linked to a resistance gene. F_3 progeny tests following self-pollination of 105 F_2 individuals identified homozygous resistant and susceptible F_2 plants for use in bulk segregant analysis. DNA from the parental lines and the homozygous resistant and susceptible bulks was analyzed by RAPD using 100 arbitrary primers. Primer OPC 04 amplified a RAPD fragment only in the susceptible parent, HC 136, and the susceptible bulk, and primer OPC 16 amplified a RAPD fragment only in the resistant parent, IS 4725, and resistant homozygous bulk. The marker OPH 7, which is linked to an oval leaf spot resistance gene in sorghum accession G48, was not linked to the gene in IS 4725. Cosegregation of OPC 16 was verified in an F_2 population segregating for oval leaf spot resistance. Use of this RAPD marker could facilitate breeding for oval leaf spot resistance with marker-assisted selection.

Introduction

Oval leaf spot of sorghum, *Sorghum bicolor* (L.) Moench, caused by *Ramulispora sorghicola* Harris, is a foliar disease characterized by initial small, water-soaked spots that develop into small, roughly circular (2-4 mm in diameter), lesions with dark red to brown margins and lighter centers in which small black sclerotia may be produced (5). Grain losses caused by this disease are usually low, and the disease is usually of greatest concern on forage sorghums.

The search for closely linked markers has two primary goals. The first is to identify gene tags suitable for marker-assisted selection; in this case, the ability to identify the presence of a resistance allele in the progeny of selected crosses even in the absence of the disease. The second goal is map-based cloning, *i.e.*, to clone the gene(s) responsible for a discernible phenotype(s) based on the gene's map position. Map-based gene cloning generally requires physical mapping, chromosome walking, and gene identification, in addition to the identification of very closely linked markers (11). This strategy has been used successfully to isolate a number of important genes, including some genes for disease resistance (4, 6, 7).

In this report, our objective was to tag an oval-leaf-spot resistance gene using Random Amplified Polymorphic DNA (RAPD) markers. These markers can then be used to facilitate early selection of resistant lines in a sorghum breeding program.

Materials and Methods

Fungal cultures. Cultures of *Ramulispora sorghicola* collected from diseased plants from the field were grown on oatmeal agar (HiMedia Laboratories Pvt. Ltd., Mumbai, India) supplemented with ampicillin and streptomycin (25 ng/ml of each final concentration) at 22°C under constant fluorescent light, which limits mycelial growth and induces sporulation. Two-week-old agar-plate cultures were flooded with sterile distilled water supplemented with Tween 20® (1 µl/ml) and conidia were scraped from the plates. Spore density was adjusted to 10^6/ml based on hemacytometer counts.

Plant material, plant inoculations, and scoring for disease. Sorghum line IS4725 is resistant to oval leaf spot caused by *R. sorghicola*, and variety HC136 is agronomically superior but susceptible to this pathogen. The F_1 from HC136 × IS4725 was grown at ICRISAT, Patancheru, Hyderabad, India, in the rabi season. One hundred and five F_2 progeny were grown in a standard organic soil mix [one part of farmyard manure mixed with two parts of sandy soil supplemented with 10 g each of urea, muriate of potash, and superphosphate per kg of soil mix (1)] in a net house in April 1998, and resistance to artificial inoculation was determined. Plants were inoculated by injecting 1 ml of the conidial suspension into the whorl of 6-week-old plants; all of the F_2 plants and 15 plants of each parental genotype were inoculated. Resistance was scored 10, 20, and 30 days after inoculation. Twenty plants of each of the F_2-derived F_3 plants were scored for disease resistance to identify F_2 accessions homozygous for the target locus.

DNA extraction. Leaves were collected from 3-4-week-old seedlings of each of the parental accessions, and F_2 individuals were used for DNA extraction by the method of Murray and Thompson (8) as modified by Saghai-Maroof *et al.* (10). DNA concentration was measured with a DNA fluorometer (Dyna Quant 200, Hoefer Pharmacia Biotech, Piscataway, NJ).

RAPD Analysis. We screened 100 arbitrary primers (10 bp) from Operon Technologies (Alameda, CA) for their ability to amplify polymorphic DNA fragments following PCR. The RAPD marker, OPH07, linked to an oval leaf spot resistance gene from another resistance source, G48, also was screened (2). Amplification reactions were performed in volumes of 20 μl containing 2 μl of 10× buffer, 100 μM of each dNTP, 2 μM $MgCl_2$, 0.5 units of *Taq* DNA polymerase, 0.2 μM primer, and 50 ng genomic DNA. Amplifications were performed in a MJ Research (Watertown, MA) thermal cycler PTC100 programmed for 45 cycles of 1 min. (5 min. for first cycle) at 94°C, 1 min. at 40°C, and 2 min. at 72°C followed by a final extension of 15 min. at 72°C. Amplified products were resolved in 1.6% agarose gels and visualized by staining with ethidium bromide for viewing on a UV transilluminator. Fragments amplified only in the resistant or susceptible parent and the respective bulks were considered linked to the locus.

Results and Discussion

Genetic analysis of resistance. The F_1 of HC136 × IS4725 was resistant to a local biotype of oval leaf spot, indicating that disease resistance is controlled by a dominant gene(s). Of the 105 F_2 plants screened, 78 were resistant and 27 were susceptible to oval leaf spot, a ratio consistent with the Mendelian ratio of 3:1 for segregation of a single dominant gene. All 27 of the susceptible F_2s were homozygous for susceptibility, and 28/78 resistant F_2s were homozygous for resistance to the disease, confirming that a single gene controlled the disease resistance phenotype. These results are consistent with previous reports on foliar diseases of sorghum (3), sorghum leaf blight (2), and rice blast (4).

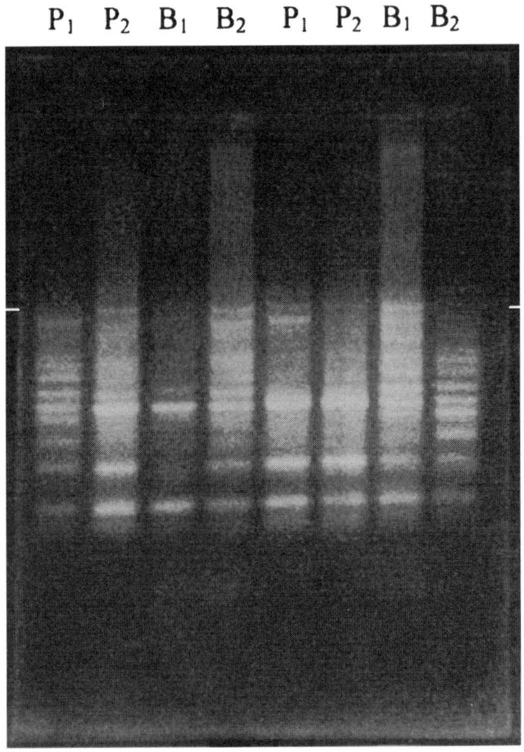

FIGURE 40-1. Susceptible parent, HC136 (P_1); resistant parent, IS4725 (P_2); resistant bulk (B_1), and susceptible bulk (B_2) amplified with arbitrary primers, OPC04 (lanes 1-4) and OPC16 (lanes 5-8). The uppermost band in lanes 1 and 4, and next to the top band in lanes 5 and 7 are unique polymorphic bands from OPC04 and OPC16, respectively.

RAPD Analysis. Of the 100 primers screened, 88 amplified one or more DNA fragments (890 fragments total). Sixteen primers did not amplify any DNA fragments, while 12 primers could be used to amplify fragments from one of the parents but not both. Sixty primers amplified a total of 144 polymorphic DNA fragments, but 58/60 of these primers produced indistinguishable patterns in the homozygous resistant and homozygous susceptible bulks. Primers OPC 04 and OPC 16 each amplified a fragment that could differentiate the two parents. OPC 04 amplified a DNA fragment specific for HC136 and the susceptible bulk of homozygous F_2 lines. OPC 16 amplified a fragment specific for IS4725 and the resistant bulk (Fig. 40-1). OPH7, a marker linked to oval leaf spot resistance gene in resistant line G48, was not linked to oval leaf blight resistance in HC136 × IS4725. This result shows that the resistance gene for oval leaf spot from strain IS4725 may be different from the gene from strain G48 (2). Since RAPD markers are dominant, these polymorphisms must be converted to co-dominant sequenced characterized amplified regions marker (9) before they can be used in a marker-assisted selection breeding program.

Acknowledgements

KSB thanks the Rockefeller Foundation for providing funds to conduct this research. We thank Dr. S. D. Singh for helping us grow F_1 plants at ICRISAT, Patancheru, India.

References

1. **Anonymous.** 1987. *Handbook of Agriculture.* Indian Council of Agricultural Research, New Delhi, India.
2. **Boora, K. S., A. Sindhu, and P. Boora.** 2000. Identification of a molecular marker linked to oval leaf spot disease resistance gene in sorghum. *Forage Res.* **26:** 1-5.
3. **Boora, K. S., R. A. Frederiksen, and C. W. Magill.** 1999. A molecular marker that segregates with sorghum leaf blight resistance in one cross is maternally inherited in another. *Mol. Gen Genet.* **261:** 317-322.
4. **Chao, C. T., K. A. K. Moldenhauer, and A. H. Ellingboe.** 1999. Genetic analysis of resistance/susceptibility in individual F_3 families of rice against strains of *Magnaporthe grisea* containing different genes for avirulence. *Euphytica* **109:** 183-190.
5. **Girard, G. C.** 1980. A review of sooty stripe and rough, zonate and oval leaf spots, p. 229-239. *In* R. J. Williams, R. A. Frederiksen and L. K. Mughogho (eds.), *Sorghum Diseases, a World Review.* ICRISAT, Patancheru (A.P.), India.
6. **Giraudat, J., B. M. Hauge, C. Valon, J. Smalle, F. Parcy, and H. Goodman.** 1992. Isolation of the *Arabidopsis ABI3* gene by positional cloning. *Plant Cell* **4:** 1251-1261.
7. **Grant, M. R., L. Godiard, E. Straube, T. Ashfield, J. Lewald, A. Sattler, R. W. Innes, and J. L. Dangl.** 1995. Structure of the *Arabidopsis RPM1* gene enabling dual specificity disease resistance. *Science* **269:** 843-846.
8. **Murray, M. G., and W. F. Thompson.** 1980. Rapid isolation of high molecular weight plant DNA. *Nucleic Acids Res.* **8:** 4321-4325.
9. **Paran, I., and R. W. Michelmore.** 1993. Development of reliable PCR-based markers linked to downy mildew resistance genes in lettuce. *Theor. Appl. Genet.* **85:** 985-993.
10. **Saghai-Maroof, M. A., K. M. S., R. A. Jorgensen, and R. W. Allard.** 1984. Ribosomal DNA spacer-length polymorphisms in barley: Mendelian inheritance, chromosomal location, and population dynamics. *Proc. Natl. Acad. Sci. USA* **81:** 8014-8018.
11. **Woo, S. S., J. Jiang, B. S. Gill, A. H. Paterson, and R. A. Wing.** 1994. Construction and characterization of a bacterial artificial chromosome library for *Sorghum bicolor. Theor. Appl. Genet.* **89:** 139-145.

41

Pearl Millet Genomics and Breeding for Resistance to Downy Mildew

Wendy A. Breese, C. Tom Hash, Katrien M. Devos, and Catherine J. Howarth

Abstract

Our objective is to summarize work in pearl millet genomics as it relates to the mapping of disease resistance. Since 1990 there have been three major achievements: (i) construction of a genetic map for pearl millet, (ii) the mapping of downy mildew resistance genes, and (iii) the development of marker-assisted selection for breeding for resistance to *Sclerospora graminicola*, the causal agent of pearl millet downy mildew.

Introduction

In 1990 the Plant Sciences Research Program of the UK Overseas Development Agency (ODA), now known as the Department for International Development (DFID), began funding a collaborative research project on pearl millet. The collaboration was to use molecular plant breeding methods to address long-standing problems in pearl millet production (7, 22). Specifically, the project targeted downy mildew epidemics [caused by *Sclerospora graminicola* (Sacc.) J. Schröt.] and yield losses due to drought.

There have been three major milestones in this work: (i) Creation of the molecular-marker-based genetic linkage map of pearl millet (14) with updates (5, 6). (ii) Successful mapping of downy mildew resistance genes (11). (iii) Incorporation of marker-assisted selection (MAS) into the ongoing resistance breeding program (8, 10).

Pearl Millet Genetic Map

The first step was to construct a genetic map of pearl millet. We used polymorphic DNA markers that are assumed to be selectively neutral and that do not affect the phenotype. The types of changes detected include single nucleotide differences, small insertions to or deletions from the DNA sequence, or changes in the number of copies of subunits of a repetitive DNA element.

DNA markers often are referred to by their acronyms: RFLPs (restriction fragment length polymorphisms), AFLPs (amplified fragment length polymerphisms), and SSRs (simple sequence repeats or microsatellites). Since 1990, more than 600 pearl millet markers have been identified and mapped (Fig. 41-1). RFLPs were used initially, but these and AFLPs are now being superceded by SSRs. RFLPs are based on hybridization methodology, and although they are simple and reproducible, they require relatively large quantities of DNA, are time consuming to conduct, and are expensive in terms of materials and reagents. By comparison, AFLPs and SSRs are both PCR-based, so that only small initial amounts of DNA are needed. They also are much quicker, cheaper, and more easily transferred to developing country laboratories. So far, approximately 100 pearl millet SSRs have been developed (1, 17), of which about 50 have been mapped.

The original RFLP map of pearl millet was produced

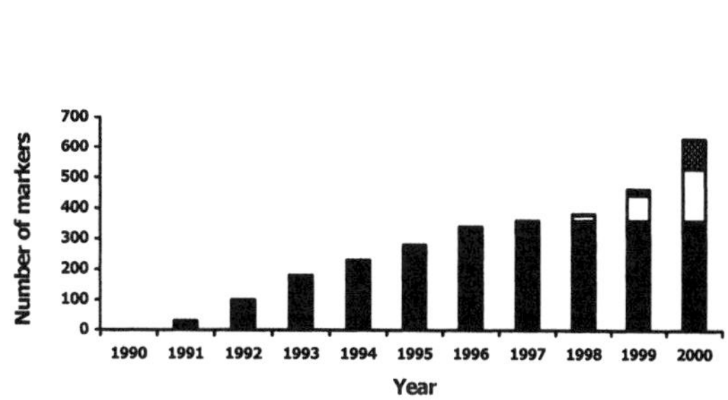

FIGURE 41-1. Development of pearl millet molecular markers from 1990 to 2000. RFLP, ■; AFLP, □; and SSR, ░.

by analyzing F_2 progeny from two single crosses, involving four different parents (Fig. 41-2). The F_2 progeny segregate for the markers from each parent. By comparing the RFLP banding pattern profiles of the F_2 individuals with those of the parents, it is possible to determine the parental source of the alleles present in each of the progeny. By inference, the parental origin of various portions of the individual's chromosome complement also can be determined, as can the heterozygous/homozygous status of any particular region.

Genetic mapping is based on the statistical likelihood that markers will be inherited together, and this probability is used to order the markers. The closer the markers are on the chromosome, the less likely are crossovers, and the more likely the linked markers will be inherited together. Computer programs, *e.g.*, Map Maker (12, 13), Map Manager QT (15), and JoinMap (18, 19), are available to perform these analyses. Mapping also indicates relative distances between markers [in centiMorgans (cM)] and places markers on linkage groups. Pearl millet has seven linkage groups and seven chromosomes ($2n = 14$). The Millet Genes database contains information on probes, RFLP polymorphisms, gel images, genetic maps, and Quantitative Trait Loci (QTL) data for pearl millet, and is publicly available (http://jic-bioinfo.bbsrc.ac.uk/cereals/millet.html; last checked January, 2002).

Map Utilization Strategy

Grasses have a great deal of genomic colinearity (3, 4), *i.e.*, gene content and order are highly conserved in rice, foxtail millet, sorghum, pearl millet, and other cereals (6). Knowledge of the relationship between the pearl millet genome and that of other grass species often enables the utilization of information gleaned from one species in understanding another. For example, a gene found on rice chromosome 1 is likely to be found on pearl millet chromosome 6, and loci associated with the rice gene can be used in pearl millet mapping populations to find the counterpart.

Once the genetic map is established, it can be searched for traits of interest. Downy mildew resistance is a quantitative trait, with the phenotype a composite that depends on the segregation of several to many naturally occurring polymorphic genes. The effects of individual allelic differences on the phenotype are generally small compared with environmental effects, and it is not always possible to deduce the genotype from the phenotype. An area or region on the genome that affects a target trait is known as a quantitative trait locus (QTL). In a cross between resistant and susceptible parents, in which resis-

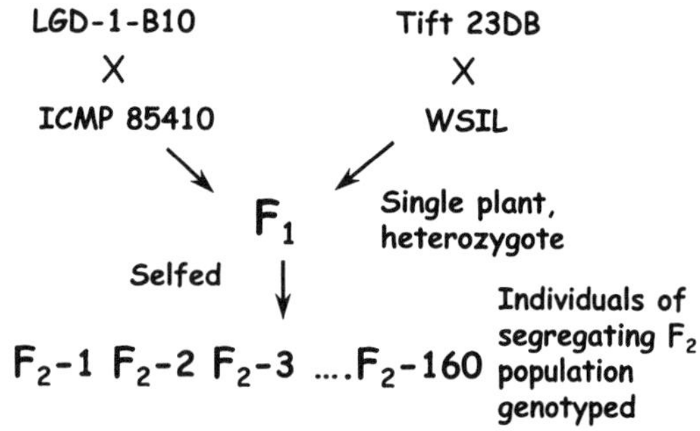

FIGURE 41-2. Crosses used to develop the original RFLP-based map of pearl millet. F_2 progeny of two single crosses involving four parents (LGD-1-B10 × ICMP 85410 and Tift 23DB × WSIL).

tance is entirely quantitative, few or none of the progeny are as resistant as the resistant parent, and some may be more susceptible than the susceptible parent.

To map QTLs in any particular cross, a skeleton map is developed with anchor loci. These loci should be spaced approximately every 10 cM and must be polymerphic between the parents of the cross. Sets of pearl millet RFLP probes and SSR primers for such studies are available from the John Innes Centre. We used this approach to develop a skeleton map for the cross of ICMP 451, which is downy mildew resistant, with H 77/833-2, which is susceptible to downy mildew but is the elite male parent of grain hybrids in India with characteristics of seedling thermotolerance, high tillering capacity, and earliness (Fig. 41-3).

We used a standard strategy to identify resistance QTLs (9). Highly inbred, homozygous parents were crossed to produce heterozygous F_1 seed. An F_1 plant is grown from a single seed, selfed, and the resultant F_2 seed is sown to produce a segregating F_2 population. A tissue sample is taken from each F_2 plant for marker genotyping, and the individual segregating plants are selfed to produce $F_{2:3}$ progenies. These plants also are selfed to provide $F_{2:4}$ bulks for disease screening. From the F_2 genotyping each progeny can be scored as homozygous for the resistant parent allele, homozygous for the susceptible parent allele, or heterozygous. The $F_{2:4}$ bulk progenies are screened for downy mildew resistance either in a greenhouse or in field sick plots. In the greenhouse, we screen pots of $F_{2:4}$ seed corresponding to each marker-genotyped F_2 plant, their parents, and the controls. Five-day-old seedlings are spray-inoculated with downy mildew sporangia and scored for disease incidence 14 days later. At Bangor, we screened ten mapping populations against downy mildew isolates from India and five African countries, for a total of 38 mapping-population ×

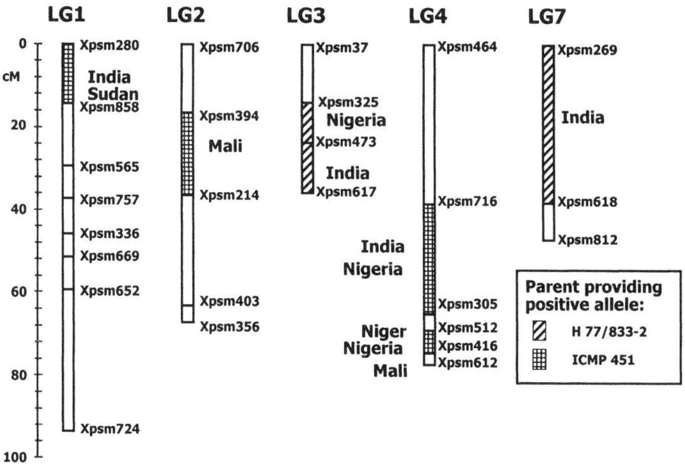

FIGURE 41-3. Skeleton genetic linkage map of the pearl millet cross ICMP 451 (resistant) with H 77/833-2 (susceptible) indicating positions of QTLs associated with resistance to downy mildew and countries in which the resistance is effective.

pathogen-isolate combinations. An additional ten mapping populations have been developed at ICRISAT. These populations consist of skeleton-mapped F_2 plants, with sets of $F_{2:4}$ self-bulk progeny and $F_{2:3}$ testcrosses. Parents of these mapping populations include genetically diverse inbred lines of Asian, African, and American origin.

To go from disease scores to QTL mapping, the genotype and disease screening data are combined and analyzed with a computer program, *e.g.*, MAPMAKER/QTL (16), PLABQTL (20), MapQTL (21), or QTL Cartographer (2). These programs use different algorithms to determine the likelihood that the trait is (i) influenced by the genomic region associated with each marker (single-marker regression), (ii) between an adjacent pair of markers disregarding the possible effects of other genomic regions (simple interval mapping), or (iii) between an adjacent pair of markers while considering the effects of other genomic regions (composite interval mapping). The likelihood or logarithm of odds (LOD) of a QTL being within a marker interval is calculated and then plotted. The peaks in the LOD profile are the most likely positions of putative QTLs. Different pearl millet cultivars contain different downy mildew resistance QTLs effective against various isolates of the pathogen.

Marker Assisted Selection in Pearl Millet

For the cross of ICMP 451 × H 77/833-2, resistance QTLs from the resistant parent map to linkage groups 1, 2, and 4 (Fig 41-3). These QTLs are effective against pathogen isolates from India, Sudan, Mali, Nigeria, and Niger. The susceptible parent also carries resistance QTLs on linkage groups 3 and 7 that are effective against isolates from Nigeria and India. A similar approach has been used to identify QTLs for the maintenance of grain yield during drought stress (23).

Once the QTLs are mapped, MAS can be used to breed for resistance (Fig. 41-4). Molecular markers are used as tags to identify pieces of the donor parent linkage group with the putative resistance QTL. The F_1 is backcrossed to the agriculturally important variety several times. At each stage, the molecular markers are used to identify individual plants with as much of the recurrent parent genome as possible, but that are heterozygous for the region carrying the resistance QTL. Selected plants are then selfed to obtain homozygous progeny.

MAS has important advantages in breeding for resistance, as compared to conventional breeding, including (i) multiple resistance genes can be introduced, also termed gene pyramiding; (ii) genes for resistance to local strains of the pathogen can be more rapidly introgressed into existing, popular, well-adapted hybrids that lack these resistance genes; (iii) resistance to African isolates of the disease can be introduced into hybrids bred in India; and (iv) breeding progress is more rapid, and expensive, time-consuming and often unreliable multilocation trials often are not needed. Products of breeding supple-

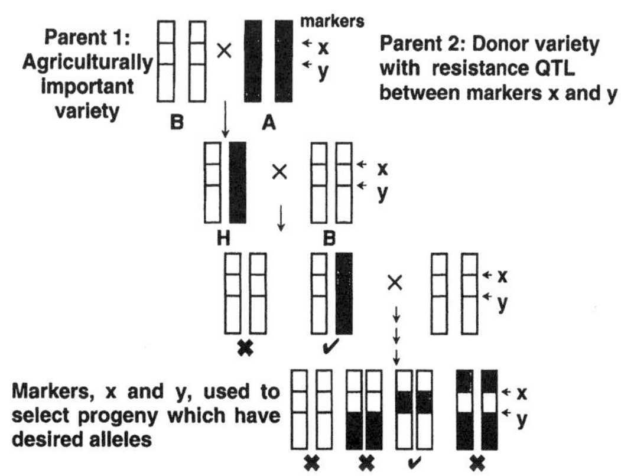

FIGURE 41-4. Schematic diagram for the marker-assisted selection of desirable traits, such as resistance to pearl millet downy mildew.

mented with MAS are beginning to become available in India.

New technologies, *e.g.*, molecular markers, are expensive to develop, and returns from the initial research accrue slowly. However, once the knowledge reaches a critical level, the rate of return accelerates rapidly. The information acquired since 1990 will be useable by breeders for decades of plant breeding in the incorporation and pyramiding of resistance genes into pearl millet. Products of breeding supplemented with MAS are just now beginning to become available in India, and work is continuing to transfer downy mildew QTLs into at least seven economically important hybrid parental lines at ICRISAT. The next decade will hopefully see this technology benefiting pearl millet farmers world-wide.

Acknowledgements

This work was supported in part by the United Kingdom Department for International Development (DFID) Plant Sciences Research Program and managed by the Centre for Arid Zone Studies, University of Wales, Bangor for the benefit of developing countries. The views expressed are not necessarily those of DFID.

References

1. **Allouis, S., X. Qi, S. Lindup, M. D. Gale, and K. M. Devos.** 2001. Construction of a BAC library of pearl millet, *Pennisetum glaucum*. *Theor. Appl. Genet.* **102:** 1200-1205.
2. **Basten, C. J., Z. B. Zeng, and B. S. Weir**. 1997. *QTL Cartographer: A Reference Manual and Tutorial for QTL Mapping.* Department of Statistics, North Carolina State University: Raleigh, North Carolina.
3. **Devos, K. M., and M. D. Gale.** 1997. Comparative genetics in grasses. *Plant Mol. Biol.* **35:** 3-15.
4. **Devos, K. M., and M. D. Gale.** 2000. Genomic relationships: the grass model in current research. *Plant Cell* **12:** 637-646.
5. **Devos, K. M., T. S. Pittaway, C. S. Busso, M. D. Gale, J. R. Witcombe, and C. T. Hash.** 1995. Molecular tools for the pearl millet nuclear genome. *Intl. Sorg. Mill. Newsl.* **36:** 64-66.
6. **Devos, K. M., T. S. Pittaway, A. Reynolds, and M. D. Gale.** 2000. Comparative mapping reveals a complex relationship between the pearl millet genome and those of foxtail millet and rice. *Theor. Appl. Genet.* **100:** 190-198.
7. **Gale, M. D., and J. R. Witcombe.** 1992. DNA markers and marker-mediated applications in plant breeding, with particular reference to pearl millet breeding, p. 323-332. *In* J. P. Moss (ed.), *Biotechnology and Crop Improvement in Asia.* ICRISAT, Patancheru (A.P.), India.
8. **Hash, C. T., and P. J. Bramel-Cox.** 2000. Marker applications in pearl millet, p. 112-127. *In* B. I. G. Haussmann, H. H. Geiger, D. E. Hess, C. T. Hash, and P. J. Bramel-Cox (eds.), *Application of Molecular Markers in Plant Breeding. Training Manual for a Seminar Held at IITA, Kano, Nigeria, (16-17 August 1999).* ICRISAT, Patancheru (A.P.), India. http://grep.icrisat.cgiar.org/mol/S2_5 Hash.pdf (last checked January 2002).
9. **Hash, C. T., and J. R. Witcombe.** 1994. Pearl millet mapping populations at ICRISAT, p. 69-75. *In* J. R. Witcombe, and R. R. Duncan (eds.), *Use of Molecular Markers in Sorghum and Pearl Millet Breeding for Developing Countries.* Overseas Development Agency (ODA), London, UK.
10. **Hash, C. T., and J. R. Witcombe.** 2002. Gene management and breeding for downy mildew resistance, p. 27-36. *In* J. F. Leslie (ed.), *Sorghum and Millets Pathology 2000.* Iowa State Press, Ames, Iowa.
11. **Jones, E. S., C. J. Liu, M. D. Gale, C. T. Hash, and J. R. Witcombe.** 1995. Mapping quantitative trait loci for downy mildew resistance in pearl millet. *Theor. Appl. Genet.* **91:** 448-456.
12. **Lander, E., and D. Botstein.** 1989. Mapping Mendelian factors underlying quantitative traits using RFLP linkage maps. *Genetics* **121:** 185-199.
13. **Lander, E., P. Green, J. Abrahamson, A. Barlow, M. Daley, S. Lincoln, and L. Newburg.** 1987. MAPMAKER: An interactive computer package for constructing primary genetic linkage maps of experimental and natural populations. *Genomics* **1:** 174-181.
14. **Liu, C. J., J. R. Witcombe, T. S. Pittaway, M. Nash, C. T. Hash, C. S. Busso, and M. D. Gale.** 1994. An RFLP-based genetic map of pearl millet (*Pennisetum glaucum*). *Theor. Appl. Genet.* **89:** 481-487.
15. **Manly, K. F. and J. M. Olson.** 1999. Overview of QTL mapping software and introduction to Map Manager QT. *Mamm. Genome* **10:** 327-334.
16. **Paterson, A., E. Lander, S. Lincoln, J. Hewitt, S. Peterson, and S. Tanksley.** 1988. Resolution of quantitative traits into Mendelian factors using a complete RFLP linkage map. *Nature* **335:** 721-726.
17. **Qi, X., T. S. Pittaway, H. Liu, and K. M. Devos.** 2000. Towards developing and mapping microsatellite markers in pearl millet. *Intl. Sorg. Mill. Newsl.* **41:** 73-74.
18. **Stam, P.** 1993. Construction of integrated genetic linkage maps by means of a new computer package: JoinMap. *Plant J.* **3:** 739-744.
19. **Stam, P., and J. W. van Ooijen.** 1995. *JoinMap™ Version 2.0: Software for the Calculation of Genetic Linkage Maps.* CPRO-DLO, Wageningen, The Netherlands.
20. **Utz, H. F., and A. E. Melchinger.** 1995. *PLABQTL. A Computer Program to Map QTL, Version 1.0.* Institute of Plant Breeding, Seed Science, and Population Genetics, Univ. of Hohenheim, Stuttgart, Germany.
21. **Van Ooijen, J. W., and C. Maliepaard.** 1996. *MapQTL™ Version 3.0: Software for the Calculation of QTL Positions on Genetic Maps.* CPRO-DLO, Wageningen, The Netherlands.
22. **Witcombe, J. R., and R. R. Duncan, eds.** 1994. *Use of Molecular Markers in Sorghum and Pearl Millet Breeding for Developing Countries.* Overseas Development Agency (ODA), London, UK.
23. **Yadav R.S., C. T. Hash, F. R. Bidinger, G. P. Cavan, and C. J. Howarth.** 2002. Quantitative trait loci associated with traits determining grain and stover yield in pearl millet under terminal drought stress conditions. *Theor. Appl. Genet.* **104:** 67-83

42

Molecular Tags for Disease Resistance Genes in Sorghum: Improved Prospects for Mapping

Clint W. Magill, Richard A. Frederiksen, Khazan Boora, Ramasamy Perumal, and S. Sivaramakrishnan

Abstract

Additional molecular markers linked to genes for resistance to leaf blight, anthracnose, oval leaf spot, and head smut have been identified. As with most previously described tags, these markers did not segregate in a recombinant inbred mapping population. The ability to identify and utilize single-nucleotide polymorphisms should alleviate this problem. Even if no differences are found within the Random Amplified Polymorphic DNA (RAPD) or Amplified Fragment Length Polymorphism (AFLP) sequence-tagged-sites, polymorphisms usually can be identified by expanding the sequences to include flanking regions. Conserved sequences in cloned resistance genes from other crop species, including soybeans, tomato, tobacco, and rice, are being used to amplify analogous sequences in sorghum, resulting in a quicker and more efficient search for resistance gene tags.

Introduction

In the 1996 conference in Lubbock, we reported the tagging of eight genes for resistance to five sorghum diseases (19). All of the tags were linked to single genes segregating in the F_2 progeny of specific crosses between resistant and susceptible parents. Many of the genes were tagged by amplification with single 10-base primers (RAPDs) by using the bulk-segregant analysis method, and in most cases we identified tags that were no more than seven map units from the resistance locus. However, we also reported a problem that greatly limited the prospects for using these markers for breeding purposes, *i.e.*, that these markers seldom segregated in other crosses. Especially disappointing was our consistent failure to place the markers on a restriction fragment length polymorphism (RFLP)-based map, which would help to identify other potentially useful markers. Likewise, our inability to identify tags for specific bacterial artificial chromosome (BAC) clones prevented the use of the BACs as a tool for genome-walking to initiate cloning of resistance genes (R-genes). Even sequencing the amplified fragment to create longer primers, referred to as sequence characterized amplified region (SCAR) primers, generally did not help. In many cases, use of the longer primers led to loss of the original polymorphism. Many of these problems remain unresolved. The objective of this chapter is to update progress in our program and to review the prospects for application of more sophisticated approaches to gene tagging in sorghum and pearl millet in light of the many problems that remain unresolved.

One Sequence–Two Map Locations?

An unexpected problem was identified by Boora *et al.* (2), who identified a 10-base primer that yielded a difference in the bands amplified by the polymerase chain reaction (PCR) in a cross segregating for leaf blight resistance. PCR amplification of DNA from SC326-6, the resistant parent, resulted in a 323-bp band that was missing in the susceptible parent, BTx623. BTx623 also is the male parent of a mapping cross. Both the RAPD and SCAR primers amplified a product from the maternal mapping parent, IS3620C, that appeared to be identical to the band amplified from SC326-6. However, when the recombinant inbred mapping population was tested, the band was present in all 137 progeny. Once cloned and sequenced, the bands from SC326-6 and IS32620C were found to differ slightly in length (323 vs. 344 bp). Both have the correct primer sites at each end and differ only slightly in sequence outside the small insertion/deletion, but these sequences are inherited in very different patterns. The 323-bp sequence in SC326-6 is inherited as a nuclear gene linked to leaf blight resistance, while that in IS3620C has the maternal inheritance pattern expected for mitochondrial or chloroplast DNA. These results failed to identify the map location of the tagged R-gene, and underscore the danger of assuming that PCR products of similar size are homologous and map to the same site.

More RAPD-Tagged Genes

We also tagged a dominant gene for resistance to oval leaf spot (3) and a recessive gene for resistance to anthracnose in SC326-6 (1). Tags for recessive genes are extremely useful as breeding tools since potential heterozygotes can be identified in each generation. Primer OPHC-16 amplifies a band that segregates with a single gene for resistance in IS 4725, as detected in the F_2 progeny of a cross to the susceptible cultivar, HC 136. Markers for the equivalent allele in the susceptible parent also were identified. The resistance in IS-4725 appears to be a different gene than that in G48, which was linked to a polymorphism amplified with primer OPH-7 (Boora, *unpublished*).

Head Smut Resistance-Tagging with Single Nucleotide Polymorphisms (SNPs)

Tags for a single-gene, race-specific form of resistance to head smut, caused by *Sporisorium reilianum*, which appears to function in the meristem, since resistant plants do not become infected despite injection of fungal spores, were reported previously (19), but none were located on the sorghum map. Markers for head smut resistance genes would be very useful, because the disease is not scored until heading, and because there is typically a very high rate of escapes. For example, less than 16% of the plants of cultivar B3, which is completely susceptible to needle inoculation, showed any symptoms in a head smut disease nursery. These and other observations have led to a hypothesis that a nonmeristematic, nonrace-specific form of resistance also exists.

We searched for genes that contribute to nonmeristematic resistance in an F_2 population that also was segregating for a meristematic-resistance gene. All the F_3 plants from 110 individual F_2 heads were scored as resistant in the field, so 10-12 additional plants were grown for greenhouse inoculation. Some F_3 field-resistant families gave evidence for segregation in these tests, while others (only eight) were scored as having only nonmeristematic resistance. Any F_3 family scored as possibly susceptible in field tests (often < 10% of the plants/row) contained 100% susceptible plants in needle inoculations in the greenhouse (17 families). Bulks of resistant and susceptible genotypes were made by mixing DNA extracted from leaves of scored plants. Standard AFLP procedures (32) were used to amplify fragments from both parents and bulks. Genomic DNA was digested with restriction endonucleases *Eco*RI and *Mse*I, leaving AATT and TA tails, respectively. After ligation of unique adapters to each of the tails, the fragments were amplified via PCR using primers complementary to both adapters. Subsets of the total fragment set were then amplified by adding one or more bases to each primer. (All enzymes, adapters and buffers are available from Invitrogen™, San Diego, CA.) Amplified products were detected by silver staining following polyacrylamide gel electrophoresis.

Several AFLP fragments were identified that differed in the resistant and susceptible bulks. For these combinations, the AFLP process was repeated for the individuals that constituted each of the bulks. None of the markers were especially close, *i.e.*, closer than four map units, to the target gene(s). To simplify the search for closer markers and the identification of flanking markers, these AFLP fragments have now been converted into single nucleotide polymorphism (SNP) markers.

Four AFLP products that appeared to cosegregate with the R or S pools were identified. Each AFLP band was derived from a different combination of primers, all of which were made using three-base extensions. The bands, ranging from 675 to 950 bp, were eluted, amplified first by asymmetric PCR with a single primer (the *Eco*RI-based primer) for 95 cycles, and then reamplified with the original primer pair for verification that the product was the expected size. Next, each AFLP product was sequenced using ABI Prism® Big-Dye™ terminators (Applied Biosystems, Foster City, CA). The sequences were examined using Oligo®, a primer analysis program for desktop computers (Molecular Biology Insights, Cascade, CO), and primer pairs identified for specific amplification of nearly full-length copies of the genomic portion of each AFLP band, converting them to sequence-tagged sites (STS). Two of the primer pairs amplified more than one band, several of which also were cloned and sequenced. No primer pair resulted in a length polymorphism, and, when DNA from the recombinant inbred mapping parents was amplified and sequenced, no differences were detected.

We next performed a modified inverse-PCR protocol (24) with *Eco*RI-digested genomic DNA. This modification involves the addition of *Eco*RI adapters so that a combination of the internal primers and a standard primer could be used to extend each AFLP sequence in both directions. To simplify sequencing, the inverse-PCR products from four sets of STS primers were cloned into the *Eco*RI site of the TOPO cloning vector (Invitrogen).

We sequenced an extended AFLP fragment generated by the primer pair E/ACA-M/CTC that contained a useful SNP. This difference results in an *Ava*I site in only one of the mapping parents. New primers that flank the site were developed, and digestion of amplified DNA was confirmed to occur for only one parent. Even if the CAPS (cleaved amplified polymorphic sequence) site had not been detected, the dCAPS (derived CAPS) program we used to screen for useful SNPs often will identify one or two base changes that can be built into a primer to create a new restriction site in the product of one of the alleles (23). Given the number and variety of restriction endonuclease target sites that can be cleaved by commercially available enzymes, most SNPs can be detected by this relatively simple assay. More sophisticated procedures also are being developed for detecting SNPs in humans. For example, it is possible to create one-base extensions of a primer in the presence of variably labeled dideoxy-A, C, G, and T nucleotides for automated fluorometric detection. In general, these protocols are still too expensive to consider for screening plant populations.

Shortcuts to Tagging R Genes?

The very best molecular tag for a resistance gene is a tag based on the DNA sequence from within the resistance gene itself. Since the R and S alleles must differ, once R genes are cloned, it is theoretically possible to design PCR-based tests to determine which allele(s) is(are) present. A number of resistance genes have been cloned, and some have been confirmed to confer resistance by transformation into a previously susceptible cultivar or by segregation into resistant progeny. Information on the sequence of these genes will simplify the search for equivalent genes in sorghum and pearl millet.

Sequences of cloned R genes provide only hints of potential functional mechanisms. To date, most of the cloned R genes encode proteins that contain C-terminal Leucine Rich Repeats (LRRs; 35). Many also have an N-terminal conserved sequence that can bind DNA, *i.e.*, a Nucleotide Binding Sequence (NBS). At least five "classes" of R genes (Table 42-1) have been defined based on the inclusion of these and other conserved sequences (20). The DNA sequence of *Arabidopsis* contains approximately 200 NBS-LRR genes or pseudogenes, which represent about 1% of the genome. Studies from organisms as diverse as fruit flies and pigs (12) have demonstrated that the LRR elements of such proteins can function as a recognition site for binding to other proteins or ligands. Thus, it is hypothesized that LRR proteins serve as specific receptors for elicitors derived from expression of *avr* genes in a potential pathogen. This hypothesis has been proven for interactions between an *Arabidopsis* resistance gene (*RPS2*) with a *Pseudomonas* avirulence factor (*AvrRps2*) (15). It also is hypothesized that ligand-bound and free forms of the R-gene protein differentially affect transcription of other defense-response genes by altering binding of the NBS to promoter sequences.

TABLE 42-1. Resistance gene classes based on conserved motifs.

Class	Gene	Host	Pathogen	NBS[a]	LRR[b]	TME[c]	PK[d]	TIL[e]	Comment	Ref.
I	*RPP* (3)	*Arabidopsis thaliana*	Downy Mildew (*Phytophthora parasitica*)	+	+			+	Three genes in a four-gene complex account for resistance to three of four races	4
II	*Cf-2, 4, 5 & 9*	*Lycopersicon esculentum* (Tomato)	Leaf Mold (*Cladosporium fulvum*)		+	+			Tandem repeats, > 90% amino acid identity	30
III	*Pib*	*Oryza sativa* (Rice)	Blast (*Magnaporthe grisea*)	+	+		+		Expression affected by light and temperature	33
IV	*RPS2* *RPM1*	*A. thaliana*	*Pseudomonas syringae*	+	+				Conflicting reports of protein expression in cells or in membrane	5, 14
	Rp1-D	*Zea mays* (Maize)	*Puccinia sorghi*						No evidence induced by rust	6

[a]NBS, nucleotide binding site.
[b]LRR, leucine rich repeat.
[c]TME, transmembrane expression.
[d]PK, protein kinase.
[e]TIL, Toll-interleukin receptor.

TABLE 42-2. Resistance gene analogs and their putative function.

Host	Number	Type[a]	Potential Function	Clusters	Ref.
Oryza sativa (Rice)	25	NBS-LRR	Linked to resistance to three rice blast races and three bacterial blight races	Four mixed, one with five RGAs in 200 kb	16
Hordeum vulgare (Barley)	25	NBS-LRR	Powdery mildew and rust		16
H. vulgare	?	NBS-LRR	Cereal cyst nematode and corn leaf aphid	Two on Chromosome 2H, two scattered	27
Capsicum annuum (Peppers)		NBS-LRR	Homology to virus resistance	Seven in one region	25
C. annuum	3	PK	Homology to *Cf, Pto, Fen*	None	25
C. annuum	1	NBS	QTL for CMV	–	25
Phaseolus vulgaris (Bean)	3	NBS-LRR	May be *C0-2* (anthracnose resistance)	Yes	7
Latuca sativa (Lettuce)	4	NBS	Two near *RPM-1* and *RPM-2*, respectively	Four families	28
Brassica oleracea	36/103		Mapped	Colinear with *Arabidopsis*	29
Tagetes erecta (Marigold)	1	TIL	*Striga* resistance	Small family	10

[a]Abbreviations: NBS, nucleotide binding site; LRR, leucine rich repeat; PK, protein kinase; and TIL, Toll-interleukin receptor.

Other R genes are known to include a kinase domain, which indicates a possible function in a signal transduction pathway. Signaling pathways also are likely to be involved in the induction of active host defense mechanisms, *e.g.*, phytoalexin production and hypersensitive responses. The R genes that fall into these groups contain highly conserved DNA sequence motifs. For example, known LRR-containing genes have from 14 to > 40 repeats of the xxLxLxx motif. Differences in the exposed amino acids (x) presumably provide unique binding surfaces (22). Thus, race specificity is assumed to derive from changes in the amino acids on the surface of the protein. These sites vary, both in length (number of repeats) and in amino acid sequence (22). Single amino acid substitution (missense) mutations in one LRR repeat of the *Arabidopsis RPS5* gene lowers resistance to several pathogens, suggesting that this element may interact with a shared signal transduction component rather than a single specific elicitor (34).

With respect to gene tagging, the most significant aspect of the cloning of R genes is that the nucleotide sequences of several key domains are relatively well conserved across species. This sequence conservation means that degenerate PCR primers can be designed and used to amplify resistance gene analogs (RGAs, sometimes also called RGCs for resistance gene candidates), even from species where no R genes have been cloned. Such studies are already in progress, including studies that utilize sorghum (21). Searches for RGAs have detected clusters of genes involved in gene-for-gene type interactions and QTLs that contribute to disease resistance (Table 42-2).

We attempted to map five sorghum RGAs (21) onto the latest version of the sorghum RI map (13). As expected for clones made by amplifying conserved sequences, multiple bands were detected when the clones were labeled and used to probe blots of restriction-digested DNA from the RI parents, BTx623 and IS3620C. We sequenced each of the clones and used the sequence to develop specific PCR primers. In this case, one of the primer pairs resulted in a distinct polymorphism, allowing us to genotype all 137 RI progeny. The bands present in the parents were not allelic, *i.e.*, the bands originated from different loci and were each segregating with a null allele. Null alleles are less informative than allele pairs with codominant banding patterns, since heterozygotes cannot be distinguished from homozygotes that carry the allele for the amplified band. We mapped both loci assuming that there would be very few heterozygotes in the RI population. The best-defined band mapped to a cluster of RFLP and SSR markers on sorghum Linkage Group B. The bands detected in the RI parents are being sequenced based on the hypothesis that the CAPS or d-CAPS procedures can then be used to map other RGAs.

Defense response genes that act downstream of the early recognition function proposed for R genes also may serve as tags for resistance. For example, in wheat, a cluster of genes that includes catalase-, chitinase-, and thaumatin-coding sequences on chromosome 7B is associated with leaf rust resistance in adult plants. Other defense response genes in the same study (8) were mapped to locations previously identified as major QTLs for tan spot, leaf rust, Karnal bunt, and stem rust resistance.

Since the speed and level of activation of these genes can be critical in determining resistance or susceptibility, a common hypothesis is that differences in promoter affinity for the NBS region of R genes or for subsequent transcription factors also affect disease response.

Several defense response genes in sorghum have been cloned and mapped (Kumari and Magill, *unpublished*). Since very few disease resistance genes and only a few R-gene tags have been mapped in sorghum, it is not surprising that no significant associations have yet been detected.

The use of RGAs to define resistance loci in marker-assisted selection has both positive and negative aspects. There usually is sufficient diversity in the less highly conserved segments of related R genes to permit each cloned RGA to be followed individually through segregation by the use of PCR amplification with specific primers or by DNA hybridization (31). Given the conservation of gene order across cereal species (9), it is possible that map locations of R genes determined in other grasses could be used to locate similar genes in sorghum and pearl millet. Rice could be an especially good source of such information since much of its genome has been sequenced (26). Unfortunately, comparisons of the map locations of PCR-generated RGAs in rice, barley, and millet provided little evidence for syntenic locations or colinearity with flanking markers, even though many of the RGAs mapped to known resistance gene locations (17). However, there are still relatively few RGAs that have been mapped, and several of the cloned RGAs hybridize to multiple fragments, some of which are not yet mapped. Given the relative frequency of NBS-LRR sequences in plant genomes, it is not surprising that RGAs from different locations sometimes were tagged. Prospects for cross-species colinearity of R gene locations seem more promising in a recent comparison of *Solanaceous* plants, *i.e.*, tomato, potato, and pepper (11, 18). In this case, the RGA probes were generated by hybridization to previously cloned R genes rather than by PCR amplification with a degenerate primer. It was rare that genes with specificity for a particular pathogen had conserved linkage relationships, but RGAs often were found in the same relative map locations. In a sense these results are very encouraging, since they suggest that LRR regions at different loci evolved independently to provide protection against specific pathogens.

Resistance genes and RGAs commonly are found in clusters. The clusters typically include more than one type of R gene, and also may include genes for resistance to different pathogens as well as to different races of a pathogen with which the plant has gene-for-gene specificity (22). Clustering can cause major problems in the use of marker-assisted selection to pyramid R genes. If parents that carry different R genes located in the same cluster are crossed, the likelihood of finding recombineants that carry both R alleles is diminished by their close proximity, even though the chance for gene conversion and unequal crossing over that can lead to changed specificities is simultaneously increased. If a doubly resistant line is identified, however, this resistance combination will be more stable than one in which the resistance loci are not linked, although if the desired alleles map to different clusters it will be easier to synthesize the multiply resistant lines.

The potential for using closely linked molecular tags for marker assisted selection to combine desirable genes has long been recognized. However, direct application to practical plant breeding projects has been limited by problems in developing markers that can be used with more than one set of parental lines. The methods described here for identifying single nucleotide polymorphisms and for taking advantage of conserved sequences within resistance genes to develop accurate and easily scored PCR-based markers should greatly enhance prospects for practical applications in sorghum in the future.

References

1. **Boora, K. S., R. A. Frederiksen, and C. W. Magill.** 1998. DNA-based markers for a recessive gene conferring anthracnose resistance in sorghum. *Crop Sci.* **38:** 1708-1709.
2. **Boora, K., R. A. Frederiksen, and C. W. Magill.** 1999. A molecular marker that segregates with sorghum leaf blight resistance in one cross is maternally inherited in another. *Mol. Gen. Genet.* **261:** 317-322.
3. **Boora, K. S., A. Sindhu, P. Boora, and C. W. Magill.** 2002. Identification of molecular markers for oval leaf spot resistance in sorghum, p. 239-241. *In* J. F. Leslie, (ed.), *Sorghum and Millets Pathology 2000.* Iowa State Press, Ames, IA.
4. **Botella, M. A., J. E. Parker, L. N. Frost, P. D. Bittner-Eddy, J. L. Beynon, M. J. Daniels, E. B. Holub, and J. D. Jones.** 1988. Three genes of the Arabidopsis *RPP1* complex resistance locus recognize distinct *Peronospora parasitica* avirulence determinants. *Plant Cell* **10:** 1847-1860.
5. **Boyes, D. C., J. Nam, and J. L. Dangl.** 1988. The *Arabidopsis thaliana RPM1* disease resistance gene product is a peripheral plasma membrane protein that is degraded coincident with the hypersensitive response. *Proc. Natl. Acad. Sci. USA* **95:** 15849-15854.
6. **Collins, N., J. Drake, M. Ayliffe, Q. Sun, J. Ellis, S. Hulbert, and T. Pryor.** 1999. Molecular characterization of the maize *Rp1*-D rust resistance haplotype and its mutants. *Plant Cell* **11:** 1365-1376.
7. **Creusot, F., C. Macadre, E. Ferrier Cana, C. Riou, V. Geffroy, M. Sevignac, M. Dron, and T. Langin.** 1999. Cloning and molecular characterization of three members of the NBS-LRR subfamily located in the vicinity of the

Co-2 locus for anthracnose resistance in *Phaseolus vulgaris*. *Genome Res.* **42:** 254-264.

8. **Faris, J. D., W. L. Li, D. J. Liu, P. D. Chen, and B. S. Gill.** 1999. Candidate gene analysis of quantitative disease resistance in wheat. *Theor. Appl. Genet.* **98:** 219-215.
9. **Gale, M. D., and K. M. Devos.** 1998. Comparative genetics in the grasses. *Proc. Natl. Acad. Sci. USA* **95:** 1971-1974.
10. **Gowda, B. S., J. L. Riopel, and M. P. Timko.** 1999. *NRSA-1*: a resistance gene homolog expressed in roots of non-host plants following parasitism by *Striga asiatica* (witchweed). *Plant J.* **20:** 217-230.
11. **Grube, R. C., E. R. Radwanski, and M. Jahn.** 2000. Comparative genetics of disease resistance within the *Solanaceae*. *Genetics* **155:** 873-887.
12. **Kobe, B., and J. Deisenhofer.** 1994. The leucine-rich repeat: A versatile binding motif. *Trends Biochem. Sci.* **19:** 415-421.
13. **Kong, L., J. Dong, and G. E. Hart.** 2000. Characteristics, linkage-map positions, and allelic differentiation of *Sorghum bicolor* (L.) Moench DNA simple-sequence repeats (SSRs). *Theor. Appl. Genet.* **101:** 438-448.
14. **Leister, R. T., F. M. Ausubel, and F. Katagiri.** 1996. Molecular recognition of pathogen attack occurs inside of plant cells in plant disease resistance specified by the Arabidopsis genes *RPS2* and *RPM1*. *Proc. Natl. Acad. Sci. USA* **93:** 15497-15502.
15. **Leister, R. T., and F. Katagiri.** 2000. A resistance gene product of the nucleotide binding site-leucine rich repeats class can form a complex with bacterial avirulence proteins *in vivo*. *Plant J.* **22:** 345-354.
16. **Leister, D., J. Kurth, D. A. Laurie, M. Yano, T. Sasaki, A. Graner, and P. Schulze-Lefert.** 1999. RFLP and physical mapping of resistance gene homologues in rice (*O. sativa*) and barley (*H. vulgare*). *Theor. Appl. Genet.* **98:** 509-520.
17. **Leister, D., J. Kurth, D. A. Laurie, M. Yano, S. Takugi, K. Devos, A. Graner, and P. Schulze-Lefert.** 1998. Rapid reorganization of resistance gene homologues in cereal genomes. *Proc. Natl. Acad. Sci. USA* **95:** 370-375.
18. **Lenné, J., and R. Ortiz.** 2002. Agrobiodiversity in pest management, p. 309-319. *In* J. F. Leslie, (ed.), *Sorghum and Millets Diseases*. Iowa State Press, Ames, Iowa.
19. **Magill, C. W., K. Boora, S. Kumari, J. Osario, B. J. Oh, B. Gowda, Y. Cui, and R. A. Frederiksen.** 1996. Tagging sorghum genes for disease resistance: Expectations and reality. *Proc. Intl. Conf. on Genetic Improvement of Sorghum and Pearl Millet (22-27 Sept. 1996, Lubbock, Texas)*: 316-325.
20. **Martin, G. B.** 1999. Functional analysis of plant disease resistance genes and their downstream effectors. *Curr. Opin. Plant Biol.* **2:** 273-299.
21. **Meyers, B. C., A. W. Dickerman, R. W. Michelmore, S. Sivaramakrishnan, B. W. Sobral, and N. D. Young.** 1999. Plant disease resistance genes encode members of an ancient and diverse protein family within the nucleotide-binding superfamily. *Plant J.* **20:** 317-332.
22. **Michelmore, R. W., and B. C. Meyers.** 1998. Clusters of resistance genes in plants evolve by divergent selection and a birth-and-death process. *Genome Res.* **8:** 1113-1130.
23. **Neff, M. M., J. D. Neff, J. Chory, and A. E. Pepper.** 1998. dCAPS, a simple technique for the genetic analysis of single nucleotide polymorphisms: Experimental applications in *Arabidopsis thaliana* genetics. *Plant J.* **14:** 387-392.
24. **Offringa, R., and F. van der Lee.** 1995. Isolation and characterization of plant genomic DNA sequences via (Inverse) PCR amplification, p. 181-195. *In* H. Jones (ed.), *Methods in Molecular Biology: Plant Gene Transfer and Expression Protocols*. Humana Press, Totowa, New Jersey.
25. **Pflierger, S., V. Lefebvre, C. Caranta, A. Blattes, B. Goffinet, and A. Palloix.** 1999. Disease resistance gene analogs as candidates for QTLs involved in pepper-pathogen interactions. *Genome Res.* **42:** 1100-1110.
26. **Sasaki, T., and B. Burr.** 2000. International rice genome sequencing project: the effort to completely sequence the rice genome. *Curr. Opin. Plant Biol.* **3:** 138-141.
27. **Seah, S., K. Sivasithamparam, A. Karakousis, and E. S. Lagudah.** 1997. Cloning and characterization of a family of disease resistance gene analogs from wheat and barley. *Theor. Appl. Genet.* **97:** 937-945.
28. **Shen, K. A., B. C. Meyers, N. Islam-Faridi, D. M. Stelly, and R. W. Michelmore.** 1998. Resistance gene candidates identified using PCR with degenerate oligonucleotide primers map to resistance gene clusters in lettuce. *Molec. Plant–Microbe Interact.* **11:** 815-823.
29. **Sillito, D., I. Parkin, R. Mayerhofer, D. Lydiate, and A. G. Good.** 2000. *Arabidopsis thaliana*: a source of candidate disease-resistance genes for *Brassica napus*. *Genome* **43:** 452-460.
30. **Thomas, C. M., M. S. Dixon, M. Parniske, C. Golstein, and J. Jones.** 1998. Genetic and molecular analysis of tomato *Cf* genes for resistance to *Cladosporium fulvum*. *Trans. Roy. Soc. Lond. Ser. B, Biol. Sci.* **353:** 1413-1424.
31. **Vicente, J., and G. King.** 2001. Characterization of disease resistance gene-like sequences in *Brassica olaracea* L. *Theor. Appl. Genet.* **102:** 555-563.
32. **Vos, P., R. Hogers, M. Bleeker, M. Reijans, T. van de Lee, M. Hornes, A. Frijters, J. Pot, J. Peleman, M. Kuiper, and M. Zabeau.** 1995. AFLP: a new technique for DNA fingerprinting. *Nucleic Acids Res.* **23:** 4407-4414.
33. **Wang, Z. X., M. Yano, U. Yamanouchi, M. Iwamoto, L. Monna, H. Hayasaka, Y. Katayos, and T. Sasak.** 1999. The *Pib* gene for rice blast resistance belongs to the nucleotide binding and leucine-rich repeat class of plant disease resistance genes. *Plant J.* **19:** 55-64.
34. **Warren, R. F., A. Henk, P. Mowery, E. Holub, and R. W. Innes.** 1998. A mutation within the leucine-rich repeat domain of the *Arabidopsis* gene *RPS5* partially suppresses multiple bacterial and downy mildew resistance genes. *Plant Cell* **19:** 1439-1452.
35. **Young, N.** 2000. The genetic architecture of resistance. *Curr. Opin. Plant Biol.* **3:** 285-290.

Part VII

Sorghum Breeding and Disease Physiology

43

Developmental Stages of *Sorghum* Caryopses, With Emphasis on the Aleurone Transfer Cell and Placental Sac

Kirsten Engell, Lise Bolt Jørgensen, Jette Dahl Møller, Eigil de Neergaard, and Elisabeth Wester

Abstract

We examined the development of the caryopses in *Sorghum bicolor* (L.) Moench, variety SC 972 from 1 to 44 days after pollination. Aleurone transfer cells develop in the outer layer of the endosperm adjacent to the integumentary gap. A few days after pollination, small invaginations occur on the cell walls, and, concomitantly, the placental sac develops by degeneration of nucellar cells. The wall invaginations become more distinct until they fill most of the cell lumen in the mature embryo.

Introduction

The development of sorghum embryos from zygote to maturity, 1-44 days after pollination (DAP), has been divided into proembryonal stages and five embryo stages, the fifth being the mature embryo, which occurs approximately 23 DAP. In this chapter, we describe the development of the aleurone transfer cells, which differentiate into a distinct region of the outer layer of the endosperm.

Plasmodesmata between the different generations, the sporophyte (the nucellus) and the gametophyte, have not been identified in the ovule (1), so symplast transport is unlikely. Within the gametophyte (the embryo sac), the embryo and endosperm develop, so that the endosperm is adjacent to the nucellus. In *Sorghum*, a placental sac is formed (13, 10, 9) between the nucellus and the outer surface of the endosperm. Within this apoplastic space, assimilates accumulate prior to being absorbed by the endosperm. In some genera of Poaceae, *e.g.*, *Zea* (6, 8) and *Sorghum* (13), early investigations identified cells with wall invaginations in a morphologically distinct region of the aleurone layer adjacent to the placental sac. Later, these cells were described as aleurone transfer cells in several genera, including *Setaria* (12), *Echinochloa* (15), *Paspalum* (5), *Pennisetum* (4), *Pappophorum* (11), *Sorghum* (9, 10), and *Zea* (2).

Materials and Methods

We used *Sorghum bicolor* (L.) Moench, variety SC 972 for our studies. We used previously described methods for transmission electron microscopy (TEM; 7), and glycolmethacrylate embedding for light microscopy. Cell walls were stained with periodic acid schiff (PAS).

Results and Discussion

Sorghum endosperm development depends on the nuclear endosperm type and begins before the first zygotic division with rapid mitoses of the triploid endosperm nucleus resulting in a single layer of nuclei located at the periphery of the embryo sac surrounding a large central vacuole. Anticlinal divisions follow, and after elongation, these cells are divided by periclinal cell walls. The outermost cell layer forms the initials of the aleurone layer, and the remaining cells form the starchy endosperm. The aleurone cells adjacent to the placental sac develop into transfer cells while the embryo is at a proembryonal stage. The outer periclinal and the anticlinal walls increase in thickness and, subsequently, cell wall invaginations appear both from the outer periclinal and the anticlinal walls. These wall invaginations eventually become very prominent and occupy most of the outer part of the cell lumen (Figs. 43-1 and 43-2) when the embryo is mature, at which time the cytoplasm is degenerating (Fig. 43-2).

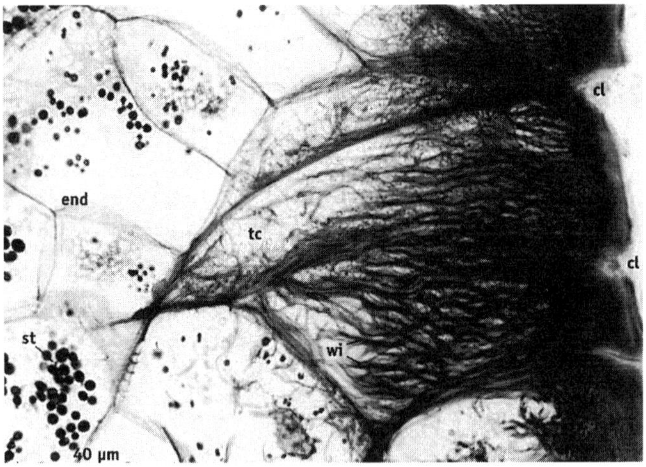

FIGURE 43-1. Light microscopy of aleurone transfer cells at the mature embryo stage stained with periodic acid schiff. A late developmental stage of the aleurone transfer cells. The cell wall invaginations are prominent and extend deeply into the cell lumen. The endosperm cells contain starch grains. Note the cleavage in the anticlinal wall of the transfer cells. (cl, cleft in anticlinal wall; end, endosperm; pl, placental sac; st, starch grain; tc, aleurone transfer cell; and wi, wall invagination).

Adjacent to the aleurone transfer cells, the apoplastic cavity, the placental sac, appears to result from the degeneration of the innermost cells of the nucellus in the integumentary gap region. The placental sac is formed early in embryo formation (at the symmetrically club-shaped stage) and extends along the aleurone transfer cells. The outermost part of the anticlinal cell wall between adjacent aleurone transfer cells is cleaved, thereby increasing the cell wall surface exposed to the placental sac (Fig. 43-1) and, consequently, the accessibility of nutrients to the endosperm is enhanced. In this study, we observed a placental sac at about 8 DAP, although a cavity was present at 5 DAP (10).

The placental sac contains nutrients and connects, through the integumentary gap, the aleurone transfer cells and the vascular cells (9). Thus, the placental sac provides an intermediate apoplastic sink for assimilate accumulation and a source for nutrients absorbed by the endosperm transfer cells. These results are similar to those seen in *Zea* (3) and *Triticum* (14), where anatomical considerations indicate that solutes enter the apoplast at the integumentary gap and are transferred into the symplast by aleurone transfer cell absorption.

To understand transport processes, it is important to understand the structure of the transport cells. In this presentation we described the development of the transfer cells and the placental sac as they correlate with embryo developmental changes. At the late pro-embryo stage the transfer cells form small invaginations. These invaginations are well-established when the placental sac is present at the early embryo stages. The cell wall invaginations fill most of the transfer cell lumen at the mature embryo stage as the cytoplasm degenerates.

References

1. **Carr, D. J.** 1976. Plasmodesmata in growth and development, p. 243-289. *In* B. E. S. Gunning, and A. W. Robards (eds.), *Intercellular Communication in Plants: Studies on Plasmodesmata*. Springer, Berlin, Germany.

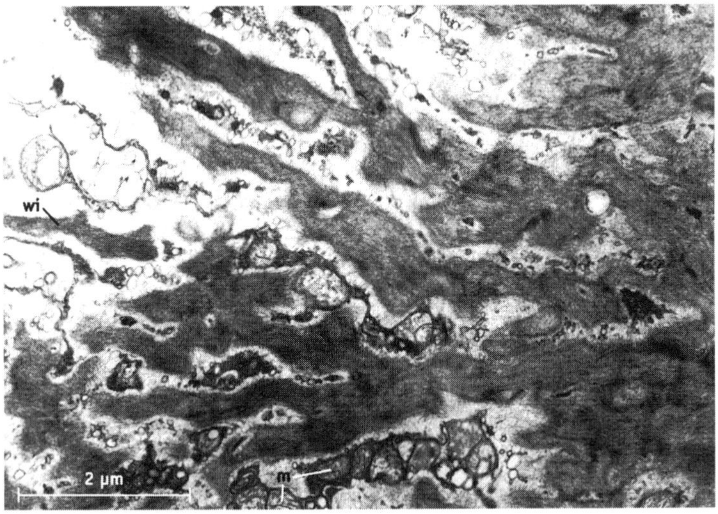

Figure 43-2. Transmission electron micrograph of aleurone transfer cells at the mature embryo stage. Ultrastructure of an aleurone transfer cell at a late developmental stage showing prominent cell wall invaginations. The cytoplasm is degenerating, but the mitochondria are still distinguishable. m, mitochondrion; wi, wall invagination.

2. **Davis, R. W., J. D. Smith, and B. G. Cobb.** 1990. A light and electron microscope investigation of the transfer cell region of maize caryopses. *Can. J. Bot.* **68:** 471-479.
3. **Felker F. C., and J. C. Shannon.** 1980. Movement of ^{14}C-labeled assimilates into kernels of *Zea mays* L. III. An anatomical examination and microautoradiographic study of assimilate transfer. *Plant Physiol.* **65:** 864-870.
4. **Fussell, L. K., and D. M. Dwarte.** 1980. Structural changes of the grain associated with black region formation in *Pennisetum americanum*. *J. Exper. Bot.* **31:** 645-654.
5. **Izaguirre de Artucio, P., and G. Ziliani.** 1975. La region placentaria en la ontogenia del caryopse de *Paspalum dilatatum* (Gramineae). *Rev. Biol. Uruguay* **3:** 161-167.
6. **Johann, H.** 1935. Histology of the caryopsis of yellow dent corn, with reference to resistance and susceptibility to kernel rots. *J. Agric. Res.* **51:** 855-883.
7. **Jørgensen, L. B.** 1995. Stomatal myrosin cells in Caricaceae: Taxonomic implications for a glucosinolate-containing family. *Nord. J. Bot.* **15:** 523-540.
8. **Kiesselbach, T. A., and E. R. Walker.** 1952. Structure of certain specialized tissues in the kernel of corn. *Amer. J. Bot.* **39:** 561-569.
9. **Maness, N. O., and G. G. McBee.** 1986. Role of placental sac in endosperm carbohydrate import in *Sorghum* caryopses. *Crop Sci.* **26:** 1201-1207.
10. **Maness, N. O., G. G. McBee, and F. R. Miller.** 1982. Placental sac influence on grain sorghum caryopsis development. Anatomical view. *Abstracts Amer. Soc. Agron. Annu. Meet.*: 104.
11. **Rost, T. L., P. Izaguirre de Artucio, and E. B. Risley.** 1984. Transfer cells in the placental pad and caryopsis coat of *Pappophorum subbulbosum* Arech. (Poaceae). *Amer. J. Bot.* **71:** 948-957.
12. **Rost, T. L., and N. R. Lersten.** 1970. Transfer aleurone cells in *Setaria lutescens* (Gramineae). *Protoplasma* **71:** 403-408.
13. **Sanders, H. E.** 1955. Developmental morphology of the kernel in grain sorghum. *Cereal Chem.* **32:** 12-25.
14. **Wang, H. L., J. W. Patrick, C. E. Offler, and V.-D. Wang.** 1995. The cellular pathway of photosynthate transfer in the developing wheat grain. III. A structural analysis and physiological studies of the pathway from the endosperm cavity to the starchy endosperm. *Plant, Cell Environ.* **18:** 389-407.
15. **Zee, S.-Y., and T. P. O'Brien.** 1971. Aleurone transfer cells and other structural features of the spikelets of millet. *Austr. J. Biol. Sci.* **24:** 391-395.

44

Plant-Parasitic Nematodes of Sorghum and Pearl Millet: Emphasis On Africa

Issoufou Abdourhamane Kollo

Abstract

Sorghum and pearl millet are host to a large number of parasitic nematodes. Species of *Tylenchorhynchus*, *Trichodorus*, *Longidorus*, and *Xiphinema* all are highly pathogenic to sorghum. Pearl millet also is susceptible to most of these nematodes. Yield loss data are lacking for the majority of these nematodes, especially with respect to pearl millet. More research also is needed on the interacttions between nematodes and other soilborne pathogens and on methods for culturing nematodes. Breeding for resistance to nematodes has become essential.

Introduction

Sorghum (*Sorghum bicolor*) and pearl millet (*Pennisetum glaucum*) are important food crops in semi-arid areas of the world. Sorghum is adapted to a wide variety of environments. It is grown in areas located at the margins of the Sahara desert, in temperate countries, and in high-rainfall areas of the world (38). Sorghum ranks fifth among the world's cereal food crops, followed by pearl millet. Pearl millet is extremely tolerant of drought, low soil fertility, and heat stress. It can grow and yield under conditions in which no other cereal can (38). In developed countries, both pearl millet and sorghum are grown almost exclusively for animal feed. In developing countries, particularly in India and Africa, these crops constitute the staple food for millions of subsistence farmers (24). Like all other food crops, pearl millet and sorghum are attacked by various pathogens. In contrast to diseases caused by fungi, bacteria, and viruses, plant parasitic nematodes affecting sorghum have received little attention. Nematode damage often mimics drought, nutrient deficiencies, or insect damage and is easily overlooked or misidentified. This problem occurs more readily in Africa, where pearl millet and sorghum usually are grown on poor soils prone to drought. This review deals with plant parasitic nematodes affecting sorghum, with an emphasis on Africa.

Plant parasitic nematodes are classified based on their feeding behaviors into two main groups: the ectoparasites and the endoparasites. Ectoparasites feed on the root surface but do not enter the host tissue. In contrast, endoparasites enter and feed in the host tissue. Some endoparasites migrate within the host tissue; others attach themselves to the host tissue and never migrate. The latter are called sedentary endoparasites, and the former migratory endoparasites.

A nematode species is considered a parasite of a plant species if it can grow and reproduce on that plant, and such plants are considered hosts of the nematode. Conversely, a plant species is a nonhost for a nematode if the nematode cannot reproduce on that plant. A nematode is pathogenic to a plant species if it adversely affects the growth or yield of that plant, whether or not it is a parasite of that plant.

Ectoparasites

Ectoparasitic nematodes have been reported both as parasites and pathogens of sorghum and millet. Johnson and Burton (25, 26) compared the growth of millet and sorghum-sudangrass in nematicide-treated and -untreated plots infested with several species of nematodes. Based on yield and growth data and nematode population counts, *Belonolaimus longicaudatus* was considered the species primarily responsible for yield reduction. This nematode is so destructive to maize that one individual per 100 cm^3 of soil is the damage threshold (9).

Among the Belonolaimidae, *Tylenchorhynchus* spp. are the most prevalent on sorghum and millet. At least 10 species of *Tylenchorhynchus* have been reported as path-

TABLE 44-1. Effect of nematodes on the growth of sorghum and millet in West Africa as relative reduction of root and shoot fresh weight (% of control) [data from Beaujard (4) and Beaujard *et al.* (5)].

Nematode Species	Inoculum[a]	Pearl Millet		Sorghum	
		Root	Shoot	Root	Shoot
Ectoparasites					
Aorelamus macbeti	600	60	11	-[b]	-
Aphasmatylenchus straturatus	2000	-	-	-	-
A. variabilis	2000	0	0	0	0
Cepahlenchus lobus	1700	21	0	6.6	0
Criconemella curvata	150	0	0	0	0
C. sphaerocephala	100	0	0	0	0
Helicotylenchus dihystera	1000	35	15	-	-
Hoplolaimus pararobustus	1000	29	0	0	0
Paralongidorus bullatus	500	27	3.1	59	31
P. duncani	200	36	1.8	37	7.6
Paratrichodorus minor	1000	0	0	52	24
P. nanus	240	7.1	7.3	59	37
Paratylenchus pernoxius	1900	15	4.8	19	27
Scutellonema cavenessi	5000	0	0	0	0
S. clathrycaudatum	600	0	0	0	0
Trichotylenchus falciformis	700	25	4.5	49	14
T. palustris	500	39	26	37	11
Triversus annulatus	1100	0	0	53	0
Tylenchorhynchus annulatus	1000	16	4.9	31	12
T. germanii	500	30	7.8	48	23
T. gladiolatus	700	-	-	46	10
T. mashoodi	2000	17	3.6	39	26
T. phaseoli	900	29	0	36	12
T. sulcatus	1100	26	7.1	53	19
T. ventralis	400	5.0	4.5	33	1.7
T. vulgaris	1000	15	26	36	7.2
Xiphenima elongatum	200	5.1	14	8.1	9.4
X. savanicola	150	27	25	62	22
Migratory endoparasites					
Pratylenchus brachyrus	700	10	0	18	0
P. loosi	500	37	22	0	0
P. sefaensis	-	-	-	15	0
P. zeae	300	0	0	0	0
Sedentary endoparasites					
Senegaloma sorghi	1000	39	23	29	14

[a]Inoculum–number of individuals/tube (4.5 cm diameter × 17.5 cm high).
[b]No data.

ogenic on sorghum. Smolik (44) reported that in fertilized and unfertilized plots *Tylenchorhynchus nudus* reduced sorghum growth by 10% and 56% respectively. He also reported that in the greenhouse, *T. nudus,* at 400 individuals/100 cm^3 of soil, significantly reduced sorghum growth. Claflin *et al.* (10) reported a significant sorghum yield increase after nematicides were applied to a field infested with *Tylenchorhynchus martini*. Hafez and Claflin (22) reported that in a field test, *T. martini* at 800-2000 individuals/100 cm^3 soil, significantly reduced sorghum growth. Curezam-Teran and Trevatan (14, 15) reported that *Quinisulcius acutus* and *Pratylenchus zeae* were the two most common nematodes on sorghum in Mississippi. *Q. acutus* alone reduced sorghum growth, but *P. zeae* and *Q. acutus* act synergistically to reduce sorghum root and top growth.

Perhaps, the most extensive host status and pathogenicity tests concerning nematodes on pearl millet and sorghum were conducted by the French team working for the Office of Scientific Research for Overseas Territories (ORSTOM), based in Dakar, Senegal. For years, this team has been measuring the response of several crops to the application of fumigant and non-fumigant nematicides. Although dramatic yield responses could sometimes be correlated with nematode population levels, it was difficult to identify the particular nematodes responsible for yield reduction. One reason for this difficulty was that the soils were infested with more than one nematode species. In addition, little was known about the pathogenicity of most of the nematodes to the crops grown in West Africa. More than 30 nematodes species were tested against pearl millet, and 29 against sorghum, by Beaujard (4) and Beaujard *et al.* (5) (Tables 44-1 and 44-2). Five of these nematodes were endoparasites, and the rest were ectoparsites. Sorghum and millet were compared with groundnut and cowpea as hosts for these nematodes. All Belonolaimidae, including eight *Tylenchorhynchus* species (*T. annulatus*, *T. germanii*, *T. gladiolus*, *T. mash-*

oodi, *T. phaseoli*, *T. sulcatus*, *T. ventralis*, and *T. vulgaris*), two *Trichotylenchus* species (*T. falciformis* and *T. palustris*) and *Triversus annulatus*, are highly pathogenic to sorghum, causing more than 30% reduction in sorghum root fresh weight (Table 44-1). Based on the severity of root damage, pearl millet appeared less susceptible than sorghum to these nematode species.

In both pearl millet and sorghum, top growth was less affected than root growth. In addition, all of the Belonolaimidae species reproduced on sorghum and pearl millet. Sorghum was the better host for the Belonolaimidae, with the exception of *T. vulgaris* (Table 44-2).

In contrast, none of the seven ectoparistic Hoplolaimidae (*Hoplolaimus pararobustus, Helicotylenchus dihystera, Aorolaimus macbethi*, *Aphasmatylenchus straturatus, A. variabilis*, *Scutelonema cavenessi*, and *S. clathricaudatum*) were pathogenic on sorghum (Table 44-1), although pearl millet was susceptible to *H. pararobustus*, *H. dihystera*, and *A. macbethi.* At 1000 individuals/tube, *H. pararobustus* and *H. dihystera* reduced millet fresh root weight by 29% and 35% respectively. *A. macbethi*, at 600 individuals/tube, reduced millet root growth by more than 60% and the top weight by 11%. *A. straturatus* did not reproduce on sorghum or pearl millet. Sorghum was the best host for *A. variabilis*, while pearl millet was the best host for *S. cavenessi, H. pararobustus, H. dihystera*, and *A. macbethi* (Table 44-2).

Two criconematidae, *Criconemella curvata* and *C. sphaerocephala*, also were tested for pathogenicity (4, 5). Sorghum is a good host for *C. curvata* and *C. sphaerocephala*; however, pearl millet is a host only for *C. curvata* (Table 44-2). Neither *C. curvata* nor *C. sphaerocephala* was pathogenic to sorghum or pearl millet (Table 44-1). These results are not consistent with those of Kinlock (28), who reported that *C. sphaerocephala* is pathogenic to sorghum. In a field test, Kinlock (28), found a significant negative correlation between initial populations of *C. sphaerocephala* and sorghum grain yield and height. The difference in the results may be attributable to differences in the sorghum cultivars used as the host. *Criconemoides ornatus* is another ring nematode reported to be a parasite of millet and of a sorghum-sudangrass hybrid (25, 26), but this nematode has not been reported from West Africa. More recently Wenefri-

TABLE 44-2. Effect of host plant on the multiplication rate of nematodes in West Africa [data from Beaujard *et al.* (5)].

Nematode Species	Host			
	Peanut	Pearl Millet	Sorghum	Cowpea
Ectoparasites				
Aorelamus macbeti	-[a]	++[b]	+[c]	+
Aphasmatylenchus straturatus	+	-	-	++
A. variabilis	+	+	++	+
Cepahlenchus lobus	+	++	-	-
Criconemella curvata	-	+	++	++
C. sphaerocephala	-	-	+	++
Helicotylenchus dihystera	-	++	+	+
Hoplolaimus pararobustus	-	++	+	+
Paralongidorus bullatus	++	+	+	+
P. duncani	-	++	+	-
Paratrichodorus minor	-	-	++	+
P. nanus	-	+	++	+
Paratylenchus pernoxius	-	+	++	+
Scutellonema cavenessi	+	++	+	+
S. clathrycaudatum	+	+	+	++
Trichodorus eburneus	[d]ND	ND	++	ND
Trichotylenchus falciformis	-	+	++	-
T. palustris	-	+	++	+
Tylenchorhynchus annulatus	+	++	++	+
T. germanii	+	+	++	+
T. gladiolatus	+	+	++	+
T. indicus	-	+	++	+
T. mashoodi	+	+	++	+
T. phaseoli	+	+	++	+
T. sulcatus	+	+	++	+
T. ventralis	-	+	++	-
T. vulgaris	-	+	+	-
Xiphenima elongatum	-	+	++	+
X. savanicola	+	+	++	+
Migratory endoparasites				
Pratylenchus brachyrus	+	+	+	++
P. loosi	-	+	+	++
P. sefaensis	-	+	+	++
P. zeae	-	+	++	-
Sedentary endoparasites				
Senegaloma sorghi	-	+	++	-

[a]Nematode multiplication rate < 1.
[b]Maximum nematode multiplication rate.
[c]Nematode multiplication rate > 1.
[d]Nematode multiplication rate not determined.

da *et al.* (49) reported that when *C. xenoplax* and *Tylenchorhynchus annulatus* were both present, they reduced sorghum root growth and head weight. The interaction between these two nematodes and *Macrophomina phaseolina*, the causal agent of charcoal rot, was antagonistic. Since the nematodes were not tested alone, it is impossible to attribute the reduction of sorghum yield and root growth to a single nematode species.

Cephalenchulus lobus and *Paratylenchus pernoxius* are pathogenic on pearl millet and sorghum (4, 5). In west Africa, these nematodes have been found only in Senegal. Nothing is known about their importance in the field.

The second most important nematode group on sorghum in west Africa belongs to the order Dorylamida (Table 44-1). Sorghum, but not pearl millet is highly susceptible to *Paratrichodurus minor*, *Paratrichodurus nanus*, and *Paratrichodurus rhodesiensis*. These Trichodorids caused more than 50% reduction in sorghum root fresh weight, and reduced top growth by more than 20%. Sorghum also is susceptible to *Paralongidorus bullatus*, *Paralongidorus duncani*, *Xiphinema savanicola*, and, to a lesser extent, *Xiphinema elongatum*. Similarly, these nematodes also are pathogenic to pearl millet. *Xiphinema parasetaria* was not tested against pearl millet and sorghum, and there are no data on the status of this nematode as a pathogen of these crops. Sorghum was the better host for the three *Xiphinema* species, *P. duncani*, *P. minor*, *P. nanus*, *P. rhodesiensis*, and *Trichodorus eburneus*. The importance of the parasitic Dorylamids probably has been underestimated in Africa because these nematodes are seldom sampled (4, 5). During the dry season, these nematodes migrate deep in the soil, usually below 50 cm, to avoid desiccation. These levels are beneath the normal sampling depth.

Pathogenic Dorylamids also are important on sorghum in other parts of the world. *P. minor* is common on sorghum in South Africa (16) and Florida (33). Smolik (44) reported that *Trichodorus allius*, at 125 individuals/100 cm^3, reduced sorghum growth. Moreover, the effect of the *T. allius* was temperature dependent. Sorghum growth reduction occurred at 15-25°C, but not at 10°C or 30-35°C. In greenhouse tests, *Longidorus africanus* at a low level, 50 individuals/100 cm^3 soil, reduced sorghum growth (30).

Sedentary Endoparasites

Three species of *Meloidogyne* have been reported as parasites of sorghum–*M. acronea*, *Meloidogyne naasi*, and *M. incognita*. *Meloidogyne acronea* has been reported on both sorghum (11) and pearl millet (12), but there is no evidence to indicate that this nematode is pathogenic to either sorghum or pearl millet. Orr (40) reported that sorghum was seriously damaged by *M. incognita* when planted after cotton in a rotation sequence in Texas. Plants in infested areas were chlorotic and stunted, their flowering was delayed, and their yield reduced by one third, with galls on sorghum roots that appeared as discrete knobs or swelling (41). Birchfield (7) tested nine sorghum cultivars against *M. incognita* in the greenhouse and found that all of the tested cultivars were susceptible. However, the race of *M. incognita* used in these experiments was not specified. Thomas and Murray (46) reported that in a microplot experiment, *M. incognita* race 3 reduced grain sorghum yield by 45% and early plant growth by one third. On the other hand, Fortnum and Currin (19) tested 10 sorghum cultivars against *M. incognita* race 3, *M. arenaria* races 1 and 2, and *M. javanica*. They found that sorghum was a poor host for *Meloidogyne* spp., and they also suggested that the reaction of sorghum to *M. incognita* might depend upon the nematode isolate, the host cultivar(s), or the geographic location. Babatola and Idowu (3) found a highly significant correlation between inoculum levels of *M. incognita* and grain yield of three sorghum cultivars in a greenhouse test in Nigeria. Again, the race of *M. incognita* was not specified. They also reported the reaction of six sorghum cultivars to *M. incognita*. While the local cultivar showed some resistance, all others were fully susceptible.

M. naasi has been reported on sorghum only in Kansas, where five races of this nematode were identified (35). Only race 5 could infect and reproduce on sorghum. In the field, symptoms of *M. naasi* consist of irregularly distributed chlorotic and stunted plants (2). In a greenhouse test, treatment of infested soil with 1,3-dichloropropene or 1,2-dichlorpropane improved plant growth (2). Evidence of pathogenicity of *M. naasi* was generated by Ediz and Dickerson (17), who found that in a greenhouse test *M. naasi* reduced sorghum root and top growth. *M. naasi* and *M. incognita* are the only root-knot nematodes that have been conclusively shown to be pathogenic to sorghum. Their importance on pearl millet is unknown.

Another sedentary nematode, *Senegaloma sorghi*, was identified in Senegal (21) and is pathogenic to sorghum in greenhouse tests but has not been implicated in yield losses in the field (4, 5). *Heterodora gambiensis* was reported on sorghum from Gambia (34) and also is of unknown importance.

Heterodora sorghi has been identified from sorghum in India (29), where there is a significant negative linear relationship between the inoculum levels of *H. sorghi* and several sorghum growth parameters (fresh root and shoot weight, dry weight of shoot and top), but not root length (36).

Migratory Endoparasites

All migratory endoparasites reported on sorghum belong to the genus *Pratylenchus*. Norton (39) reported that *P. hexincius*, at 65 individuals/100 cm^3 of soil, reduced sorghum growth by 20% and that the reduction increased if the plants were drought stressed. He also reported an additive interaction between *P. hexincius* and *Macrophomina phaseolina*, and that there was no synergistic disease complex. However, lodging occurred only when both organisms were present.

Endo (18) studied the host range of *P. zeae* and *Pratylenchus brachyrus* and found that sorghum is a good host for *P. zeae*, but not for *P. brachyrus*. Johnson and Burton (25, 26) reported a significant yield increase from pearl millet and sorghum-sudangrass hybrids after nematicides were applied to a field plot infested with several species of nematodes. Based on nematode population dynamics, *P. zeae* and *P. brachyrus* both were implicated in the yield reduction. Curezma-Teran and Trevathan (14) reported that *P. zeae* was common on sorghum in Mississippi. In addition, *P. zeae* alone at 500 individuals/pot or when combined with *Q. acutus* significantly reduced sorghum growth. In Puerto Rico, the death of 2-3-week-old sorghum seedlings was attributed to *P. zeae* (1). Bee-Rodriguez and Ayala (6) reported that in the greenhouse *P. zeae* at 1500 individuals/20-cm-pot reduced sorghum top and root growth. Moreover, *P. zeae* interacts with *M. phaseolina*, *Curvularia* spp., *Fusarium moniliforme sensu lato*, and *Rhizoctonia solani* to reduce root and top growth of sorghum.

The host status and susceptibility of sorghum to four *Pratylenchus* species was studied by Motalaote *et al.* (37), but they found that only *P. zeae* was pathogenic to sorghum. In greenhouse tests, *P. zeae*, at 600 individuals/pot, reduced plant height, shoot weight, and root weight. In microplots, however, *P. zeae*, even at high population levels (8000 individuals/microplot), did not reduce sorghum growth. *P. brachyrus* had no effect on sorghum even at levels as high as 8000 individuals/pot, and probably is not pathogenic to sorghum. Since the reproductive index of *P. crenatus* on sorghum was less than one, the authors concluded that sorghum is a nonhost for *P. crenatus*. Motalaote *et al.* (37) also reported that sorghum was a poor host for *P. coffea* and *P. brachyrus*.

There are conflicting reports on pathogenicity of *Pratylenchus* species to sorghum. Sharma and de Souza Meideros (43) reported that of 16 sweet sorghum cultivars tested against *P. brachyrus*, only one was moderately resistant, four were tolerant, and the others were susceptible. McDonald and van den Berg (32) reported that *P. zeae* at 9000 individuals/pot had no effect on sorghum growth, even when the plants were drought stressed. Beaujard (4) and Beaujard *et al.* (5) reported that in west Africa sorghum and millet are susceptible to *P. brachyrus* and *P. loosi*, but not to *P. zeae* (Table 44-1). Of all the *Pratylenchus* species described on sorghum, Starr (45) considered only three (*P. zeae*, *P. thornei*, and *P. hexincius*) to have been conclusively shown to be pathogenic on sorghum. Luc *et al.* (31) stated that only two *Pratylenchus* species (*P. zeae* and *P. hexincius*) are important on sorghum. Much of this controversy probably can be attributed to differences in cultivars, inoculum levels, and environment.

Pratylenchus species have been implicated as components of several disease complexes with wilt-causing fungi (20). In Niger, *P. zeae* was associated with *Acremonium strictum* (29; Fig. 44-1) the causal agent of Acremonium wilt. In greenhouse tests, *Pratylenchus zeae* interacts synergistically with *A. strictum* to cause severe

TABLE 44-3. Interaction between *Acremonium strictum* and *Pratylenchus zeae* on two sorghum genotypes: Mota, which is resistant to *A. strictum*, and NAD-1, which is a susceptible hybrid (ATx623 × MR732).[a]

Treatment	Wilt Severity[b]		Root Necrosis[c]		Top Weight[d] (g)	
	Mota	NAD-1	Mota	NAD-1	Mota	NAD-1
A. strictum	1.6 b	3.4 b	1.6 c	3.8 b	60 ab	44 b
P. zeae	1.2 b	1.1 c	2.4 b	3.1 b	54 b	51 b
A. strictum + *P. zeae*	3.7 a	4.6 a	5.1 a	5.5 a	32 c	25 c
Control A[e]	1.0 b	1.1 c	1.3 c	2.1 bc	62 a	64 a
Control B[f]	4.2 a	5.0 a	5.4 a	6.0 a	25 c	5.1 d

[a]Soil was taken from a farmer's field with a known history of Acremonium wilt and infested with *P. zeae*. Part of the soil was steam pasteurized, then infested with *A. strictum* and *P. zeae*, either alone or in combination. The experiment was conducted in a screen house in Niger (after 29). Within the same cultivar, numbers followed by the same letters are not significantly different.

[b]Wilt severity was visually estimated on 1 to 5 scale. 1 = no symptoms, 2 = limited vascular discoloration, 3 = vascular discoloration and large portions of the leaves necrotic, 4 = general leaf and leaf sheath wilting, 5 = whole plant wilting and dying. CV = 25%.

[c]Root necrosis was visually estimated on a scale of 1 to 6. 1 = no necrosis, 2 = 1-10% root necrosis, 3 = 11-25% necrosis, 4 = 26-50% necrosis, 5 = 51-75% necrosis, and 6 = > 75% necrosis. CV = 27%.

[d]Top weight is the dry weight of the top growth of sorghum plant in grams. CV = 25%

[e]Noninfested pasteurized soil

[f]Nonpasteurized and noninfested soil from the farmer's field.

wilt and root necrosis and to reduce top growth (Table 44-3). On the resistant cultivar(s), disease develops only when *P. zeae* and *A. strictum* are combined. On the susceptible cultivar, *P. zeae* increases wilt severity (Table 44-3). *P. zeae* may not be the only nematode that interacts with *A. strictum*. In a case of severe Acremonium wilt in Niger in 2000, only *Helicoylenchus* spp. were found.

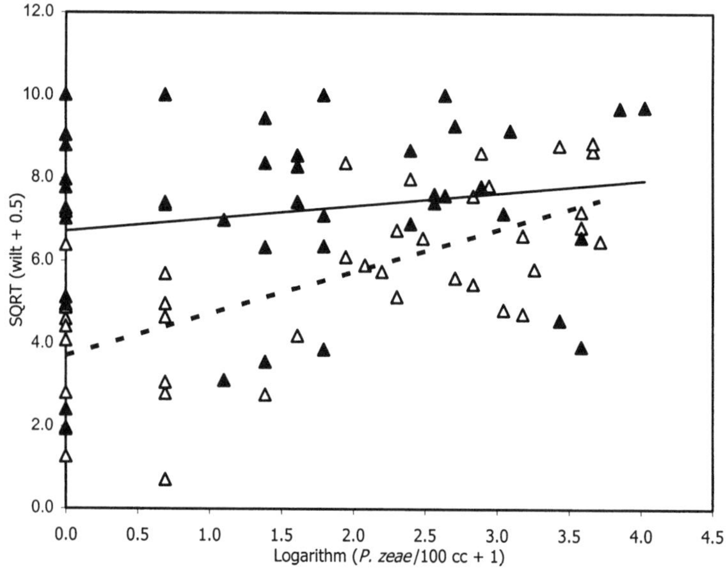

FIGURE 44-1. Relationship between the natural logarithm of the number of *Pratylenchus zeae* at planting time and the incidence of Acremonium wilt in sorghum cultivars in a farmer's field around Konni (Niger) in 1997. Incidence of wilt was transformed to $\sqrt{(x + 0.5)}$. ▲, —: Mota (resistant); $y = 3.72 + 1.01\ x$; $r^2 = 0.44$. △, - - -: NAD-1 (susceptible); $y = 6.72 + 0.31\ x$; $r^2 = 0.03$.

Other Nematodes

Rotylenchus devonensis (16), *Rotylenchulus borealis* (4, 48), and *Rotylenchulus parvus* (4) have been reported as parasites of either sorghum or pearl millet, but there are no published data supporting the pathogenicity of these nematodes to either host.

Management of Plant-Parasitic Nematodes

The control of plant parasitic nematodes on sorghum and pearl millet is a difficult task. Pearl millet and sorghum are both low-value crops, so the use of nematicides may not be cost effective (23, 50). Nematicides also are highly toxic, and the release of large quantities of toxic chemicals may pose serious environmental and human health hazards. If nematicides are applied as a seed dressing, as we have done with Apron Plus® (42), then the environmental threat is lessened and nematicide use may be cost effective.

The use of host resistance is very cost effective when it exists. Resistance against specialized sedentary endoparasites exists in many crop species (13, 50). Resistance probably exists in pearl millet and sorghum since sorghum cultivars respond differently to *M. incognita* (3) and to some ectoparasites (9). However, resistance against one species of nematodes may not be effective against another, which makes breeding for nematode resistance difficult, as most soils are infested by several nematode species.

Nematode damage is usually more severe in poor soils than in fertile ones. Therefore, improving soil fertility may reduce nematode damage to crops (8, 47). This solution may be cost effective where fertilizer is not expensive, but in many African countries the use of mineral fertilizers on pearl millet and sorghum is not cost effective, even if no pests or pathogens are present. Increasing soil organic matter by applying large quantities of farmyard manure or by leaving crop residue in the field also can be beneficial. In addition to increasing fertility, such practices can increase the soil microbial population and the water-holding capacity of the soil. Competitive or predatory soil microorganisms also may help control nematodes.

Many fungi, bacteria, and protozoa feed on nematodes (27), but the use of biological control agents to manage plant-parasitic nematodes has not yet been very successful under field conditions (27, 50). Perhaps the biggest problem is how to modify the soil environment so that the biological control agent, once introduced, can persist for an extended period of time.

Crop rotation has been widely used to manage plant-parasitic nematodes (50). This practice is feasible if the nematode has a narrow host range and if the alternative crop has some value. In situations where the soil is infested with several species of parasitic nematodes, as is the case in many parts of west Africa, rotation may not be an effective means to reduce the nematode population.

Research Needs

Nematodes on sorghum and millets have not been intensively or extensively studied relative to most other pathogens. Some of the most important research needs include:

- Determine threshold levels at which control methods should be deployed.
- Obtain reliable field data on losses attributable to nematodes.
- Develop better descriptions and characterizations of nematodes with fungal and bacterial pathogens.
- Determine the role of nematodes in seedling diseases and poor stand establishment.
- Determine how nematodes interact with one another.
- Develop *in vitro* culturing methods for important nematode species.

Acknowledgements

The research work on which this manuscript is based was supported by the Sorghum and Millet Collaborative Research Support Program (INTSORMIL) AID/DAN-1254-G-00-0021-00 from the U.S. Agency for International Development. I thank Dr. Richard Frederiksen for his assistance in developing a sorghum pathology program for Niger, and my colleagues at INRAN, especially Drs. Haougui Adamou and Issoufou Kapran, for their collaborative efforts in these studies.

References

1. **Ayala, A., and D. Bee-Rodriguez.** 1978. Control of phytoparasitic nematodes attacking sorghum in Puerto Rico. *J. Agric. Univ. PR* **62:** 119-132.
2. **Ayatan, S., and O. J. Dickerson.** 1969. *Meloidogyne naasi* on sorghum in Kansas. *Plant Dis. Rept.* **53:** 737.
3. **Babatola, J. O., and K. A. Idowu.** 1990. Responses of some sorghum cultivars to the root-knot nematode *Meloidogyne incognita* in Nigeria. *Nematol. Mediter.* **18:** 65-66.
4. **Beaujard, P.** 1994. Nématicides, nématodes phytoparasitaires et rendements des cultures pluviales dans la zone sahélienne de l' Afrique de l'ouest. *Afro-Asian J. Nematol.* **4:** 129-146.
5. **Beaujard, P., E. Bour, and B. Martiny.** 1995. Incidence des nématodes phytoparasitaires sur la culture du sorgho dans la zone sahélienne du Sénégal, Afrique de l'Ouest. *Afro-Asian J. Nematol.* **5:** 1-10.
6. **Bee-Rodriguez, D., and A. Ayala.** 1977. Interaction of *Pratylenchus zeae* with four soil fungi on sorghum. *J. Agric. Univ. PR* **61:** 501-506.
7. **Birchfield, W.** 1983. Wheat and grain sorghum varietal reaction to *Meloidogyne incognit*a and *Rotylenchulus reniformis*. *Plant Dis.* **67:** 41-42
8. **Bridge, J.** 1987. Control strategies in subsistence agriculture, p. 389-418. *In* R. H. Brown and B. R. Kerry (eds.), *Principles and Practice of Nematode Control*. Academic Press, Orlando, Florida.
9. **Claflin, L. E.** 1984. Plant-parasitic nematodes affecting sorghum, p. 53-58. *In* L. K. Mughogho and G. Rosenberg (eds.), *Sorghum Root and Stalk Rots: A Critical Review*. ICRISAT, Patancheru (A.P.), India.
10. **Claflin, L. E., S. L. Hafez, and T. C. Todd.** 1983. Effects of insecticides-nematicides on stunt nematode populations and grain sorghum yields. *Abstr. 4th Intl. Cong. of Plant Pathol. (17-24 August 1983, Melbourne, Australia)*: 145.
11. **Coetzee, V.** 1956. *Meloidogyne acronea*, a new species of root-knot nematode. *Nature* **177:** 899-900.
12. **Coetzee, V., and H. J. Botha.** 1965. A redescription of *Hyposerine acronea* (Coetzee, 1956) Sledge and Golden 1964. (Nematoda: Heteroderidae), with a note on its biology and host specifity. *Nematologica* **11:** 480-485.
13. **Cook, R., and K. Evans.** 1987. Resistance and tolerance, p. 179-221. *In* R. H. Brown and B. R. Kerry (eds.), *Principles and Practice of Nematode Control*. Academic Press, Orlando, Florida.
14. **Cuarezma-Teran, J. A., and L. E. Trevathan.** 1984. Nematodes associated with sorghum in Mississippi. *Plant. Dis.* **68:** 1083-1085.
15. **Cuarezma-Teran, J. A., and L. E. Trevathan.** 1985. Effects of *Pratylenchus zeae* and *Quinissulcius acutus* alone and in combination on sorghum. *J. Nematol.* **17:** 169-174.
16. **de Waele, D., and M. Jordan.** 1988. Plant-parasitic nematodes on field crops in South Africa. 2. Sorghum. *Rev. Nématol.* **11:** 203-212.
17. **Ediz, S. A., and O. J. Dickerson.** 1976. Life cycle, pathogenicity, histopathology, and host range of race 5 of the barley root-knot nematode. *J. Nematol.* **8:** 228-232.
18. **Endo, B. Y.** 1959. Responses of root-lesion nematodes, *Pratylenchus*, *Brachyurus* and *P. zeae* to various plants and soil types. *Phytopathology* **49:** 417-421.
19. **Fortnum, B. A, and R. E. Currin III.** 1988. Host suitability of grain sorghum cultivars to *Meloidogyne* spp. *Ann. Appl. Nematol.* **2:** 61-64.
20. **Francl, L. J., and T. A. Wheeler.** 1993. Interactions of plant-parasitic nematodes with wilt-inducing fungi, p. 79-103. *In* W. Khan (ed.), *Nematode Interactions*. Chapman and Hall, London, UK.
21. **Germani, G., M. Luc, and J. G. Baldwin.** 1984. A new Rotylenchulinae: *Senegaloma sorghi* n. gen., n. sp. (Nematoda: Tylenchina). *Rev. Nématol.* **7:** 49-56.
22. **Hafez, S. L., and L. E. Claflin.** 1982. Control of plant parasitic nematodes on grain sorghum and yield response. *Fung. Nemat. Tests* **37:** 1198.

23. **Hague, N. G. M., and S. R. Gowen.** 1987. Chemical control of nematodes, p. 131-173. *In* R. H. Brown, and B. R. Kerry (ed.), *Principles and Practice of Nematode Control.* Academic Press, Orlando, Florida.
24. **House, L. R.** 1995. Sorghum: one of the world's greatest cereals. *Afr. Crop Sci. J.* **3:** 135-142.
25. **Johnson, A. W., and G. W. Burton.** 1973. Comparison of millet and sorghum-sudangrass hybrids grown in untreated soil and soil treated with two nematicides. *J. Nematol.* **1:** 54-59.
26. **Johnson, A. W., and G. W. Burton.** 1977. Influence of nematicides on nematodes and yield of sorghum-sudangrass hybrids and millets. *Plant. Dis. Rept.* **61:** 1013-1017.
27. **Kerry, B. R.** 1987. Biological control, p. 233-257. *In* R. H. Brown, and B. R. Kerry (ed.), *Principles and Practice of Nematode Control.* Academic Press, Orlando, Florida.
28. **Kinlock, R. A.** 1987. The population dynamics of *Criconemella sphaerocephala* and the nematode's relationship to sorghum yield. *J. Nematol.* **19:** 536.
29. **Kollo, A. I.** 2000. *Acremonium Wilt Complex of Sorghum in Niger, Etiology and Control.* Ph.D. Dissertation. Texas A&M Univ., College Station, Texas.
30. **Lamberti, F.** 1969. Pathogenicity of *Longidorus africanus* on selected field crops. *Plant Dis. Rept.* **53:** 421-424.
31. **Luc, M., R. A. Sikora, and J. Bridge.** 1990. *Plant Parasitic Nematodes in Subtropical Agriculture.* CAB International, Oxford, UK.
32. **McDonald, A. H., and E. H. van den Berg.** 1993. Effect of watering regimen on injury to corn and grain sorghum by *Pratylenchus* species. *J. Nematol.* **25:** 654-658.
33. **McSorly, R., and R. Gallaher.** 1992. Comparison of nematode population densities on six summer crops at seven sites in North Florida. *J. Nematol.* **24:** 699-706.
34. **Merny, G., and C. Netscher.** 1976. *Heterodera gambiensis* n. sp. (Nematoda: Tylenchida) parasite du mil et du sorgho en Gambie. *Cahiers Office de la Recher. Scientif. et Tech. d'Outre-Mer Série Biol.* **11:** 209-218.
35. **Michell, R. E., R. B. Malek, D. P. Taylor, and D. I. Edwards.** 1973. Races of the barley root-knot nematode, *Meloidogyne naasi.* I. Characterization by host preference. *J. Nematol.* **54:** 41-44.
36. **Mittal, A., and S. C. Dhawan.** 1991. Effect of varying densities of sorghum cyst nematode, *Heterodera sorghi*, on plant growth, and nematode multiplication on sorghum. *Ind. Phytopathol.* **44:** 429-432.
37. **Motalaote, B., J. L. Starr, R. A. Frederiksen, and F. R. Miller.** 1987. Host status and susceptibility of sorghum to *Pratylenchus* species. *Rev. Nématol.* **10:** 81-86.
38. **National Research Council.** 1996. *The Lost Crops of Africa. Volume 1: Grains.* National Academy Press, Washington, D.C.
39. **Norton, D. C.** 1958. The association of *Pratylenchus hexincius* with charcoal rot. *Phytopathology* **48:** 355-358.
40. **Orr, C. C.** 1967. Observations on cotton root-knot nematode in grain sorghum in West Texas. *Plant Dis. Rept.* **51:** 29.
41. **Orr, C. C., and E. D. Morey.** 1978. Anatomical response of grain sorghum roots to *Meloidogyne incognita acrita. J. Nematol.* **10:** 48-52.
42. **Scheuring, J. F., S. O. Katilé, and I. A. Kollo.** 2002. Boosting pearl millet yields with Apron Plus® and Apron Star® Seed treatments, p. 47-49. *In* J. F. Leslie (ed.), *Sorghum and Millets Diseases.* Iowa State Press, Ames, Iowa.
43. **Sharma, R. D., and A. C. de Souza Medeiros.** 1982. Reaçoes de Alguins de genotipos de sorgho sacarino. Aos nematoides, *Meloidogyne javanica* e *Pratylenchus brachyrus. Pesq. Agropec. Bras. (Brasilia)* **17:** 697-701.
44. **Smolik, J. D.** 1977. Effects of *Trichodorus allius* and *Tylenchorynchus nudus* on growth of sorghum. *Plant Dis. Rept.* **61:** 855-858.
45. **Starr, J. L.** 1992. Nematode pathogens of sorghum, p. 179-185. *In* W. A. J. de Milliano, R. A. Frederiksen, and G. D. Bengston (eds.), *Sorghum and Millets Diseases: A Second World Review.* ICRISAT, Patancheru (A.P.), India.
46. **Thomas, S. H., and L. Murray.** 1987. Yield reductions in grain sorghum associated with injury by *Meloidogyne incognita* race 3. *J. Nematol.* **19:** 559.
47. **Thomason, I. J., and E. P. Caswell.** 1987. Principles of nematode control, p. 87-124. *In* R. H. Brown, and B. R. Kerry (ed.), *Principles and Practice of Nematode Control.* Academic Press, Orlando, Florida.
48. **Volvas, N., and I. N. Inserra.** 1982. Biological relationships of *Rotylenchus borealis* on several plant cultivars. *J. Nematol.* **14:** 373-377.
49. **Wenefrida, I., E. C. McGawley, and J. S. Russin.** 1997. Interrelationships among *Macrophomina phaseolina, Criconemella xenoplax*, and *Tylenchorhynchus annulatus* on grain sorghum. *J. Nematol.* **29:** 199-208.
50. **Whitehead, A. G.** 1998. *Plant Nematode Control.* CAB International, Wallingford, Oxon, U.K.

45

Evaluation of Reduced Sorghum Seed Germination

Neal W. McLaren, Jeanetta Saayman, J. Benade, and Marlene van der Walt

Abstract

We quantified the effects of fungal colonization of sorghum seed embryos in terms of germination and abnormal seedling growth. The fungi most commonly recovered from infected embryos were *Alternaria* spp., *Curvularia lunata*, and *Phoma sorghina*. Germination was reduced and abnormal seedling growth increased regardless of the species that colonized the embryos. In pathogenicity studies, germination was reduced only under high humidity and, primarily, when heads were inoculated at the soft dough stage.

Introduction

Early infection of sorghum [*Sorghum bicolor* (L.) Moench)] grain by fungal pathogens generally results in reduced grain fill (4) and premature black layer formation (1) with concomitant losses in seed mass (2, 9) and grain density (1). Colonization of the embryo may result in abnormal or reduced germination (1, 6). *Alternaria* spp., *Fusarium moniliforme sensu lato, Curvularia lunata, Fusarium semitectum*, and *Phoma sorghina* are the most important early infection pathogens (3, 8).

McLaren (7) could attribute neither reduced germination nor seedling disease severity to visible fungal colonization on seed surfaces. The objective of this study was to determine the relationship between embryo colonization by seedborne fungi and seed quality.

Materials and Methods

Thirteen sorghum seed samples of the hybrid NK283 from commercial seed production fields were visually assessed for grain mold and visible seed cracking by using a stereo microscope. Seed mass (1000 kernels), germination, and seedling abnormalities were determined according to ISTA methodologies (5). Fifty embryos per sample were excised, surface sterilized, and plated onto half-strength malt extract agar (25 g malt extract, 10 g agar, and 1 L H_2O). After four days, fungal colonies were counted and purified for identification.

Pathogenicity of selected isolates towards NK 283 was evaluated in the greenhouse. Seeds were planted in 5-L plastic pots. Pots were overseeded and thinned after emergence to two plants per pot. Pots were maintained under a 26/22°C day/night regime with 14 hr daylight. At the commencement of flowering, heads of ten plants each were sprayed with an aqueous spore suspension of *Alternaria* sp., *C. lunata, Fusarium thapsinum*, and *P. sorghina* from 10-day-old cultures. Five heads were covered with a transparent plastic bag to ensure high humidity for five days. The procedure was repeated on a second group of sorghum heads at the soft dough stage of grain development. At maturity (approximately 12% moisture), grain was harvested and assessed for germination, seedling abnormalities, and embryo infection.

Results and Discussion

Seed germination in NK 283 collected from commercial seed production fields ranged from 21-81%, with high percentages of either dormant or dead seed. Abnormal seedlings ranged from 8.8-28% of those seed that germinated. The most commonly recovered fungi from infected grain were *Alternaria* spp., *P. sorghina*, and *C. lunata*, which constituted 65% of the isolates. The high incidence of *Alternaria* spp. is similar to the results of Menkir *et al.* (8). Other isolates included unidentified bacteria (3.2%), *Drechslera* spp. (7.0%), *F. thapsinum* (6.8%), and other *Fusarium* spp. (5.9%), with *Chaetomium* spp., *Cladosporium* spp., *Penicillium* spp., and unidentified fungi making up the remainder.

Germination and 1000 kernel mass were reduced and seed dormancy increased in colonized ovaries, irrespective of the pathogen involved (Table 45-1). The relationship between abnormal seedling growth and total isolation frequency was not significant; a result that might be attributable to pathogen specificity or to the site of infection. We also observed no statistically significant relationship between seed cracking or visible colonization of grain and either seed germination or seedling vigor. The lack of a significant relationship between these variables suggests that visual assessment of seed is not always an appropriate criterion for assessing seed quality (7).

TABLE 45-1. Correlations between fungal-colonized embryos and sorghum seed quality.

	Germination				Normal Seedling Growth	1000 Kernel Mass
	Normal	Dormant	Abnormal	Dead		
Colonized Embryos	-0.85***	0.71**	-0.01	0.89***	-0.42	-0.67**
Grain Discoloration	-0.39	0.34	0.46	0.26	0.08	-0.07

** $p < 0.01$; *** $p < 0.001$.

In the greenhouse, with the exception of the control treatments, bagging the heads after artificial inoculation to increase humidity decreased germination by up to 60% and increased the number of abnormal seedlings observed. Germination of seed from unbagged heads was not significantly reduced, indicating that high humidity is essential for infection. Reductions in germination were more severe when the heads were inoculated at soft dough stage. Under bagged heads at soft dough stage, *Alternaria* sp. was the most pathogenic of the tested isolates, reducing germination by 88%, followed by *P. sorghina* (61%), *F. thapsinum* (55%), and *C. lunata* (54.6 %). Menkir *et al.* (8) reported that despite *Alternaria* spp. being the most commonly isolated fungus from moldy grain in Indiana, it was not significantly associated with grain discoloration. They did not, however determine the effect of fungi on germination. No reports attributing reduced germination of sorghum to *Alternaria* spp. could be found, and the above result, therefore, warrants further investigation.

We conclude that fungal colonization of embryos is an important cause of reduced seed germination in sorghum seed lots during some seasons in South Africa.

References

1. **Castor, L. L.** 1981. *Grain Mold Histopathology, Damage Assessment, and Resistance Screening Within* Sorghum bicolor *(L.) Moench Lines.* Ph.D. Thesis, Texas A&M Univ., College Station, Texas.
2. **Castor, L. L., and R. A. Frederiksen.** 1981. *Fusarium* head blight occurrence and effects on sorghum yield and grain characteristics in Texas. *Plant Dis.* **64:** 1017-1019.
3. **Forbes, G. A., R. Bandyopadhyay, and G. Garcia.** 1992. A review of sorghum grain mold, p. 253-264. *In* W. A. J. de Milliano, R. A. Frederiksen, and G. D. Bengston (eds.), *Sorghum and Millets Diseases: A Second World Review.* ICRISAT, Patancheru (A.P.), India.
4. **Frederiksen, R. A., L. L. Castor, and D. T. Rosenow.** 1982. Grain mold, small seed and head blight: the *Fusarium* connection in sorghum. *Proc. Annu. Corn Sorg. Indus. Res. Conf.* **37:** 26-36.
5. **International Seed Testing Association (ISTA).** 1999. International rules for seed testing. *Seed Sci. Technol.* **27:** 155-199.
6. **Maiti, R. K., P. S. Raju, and F. R. Bidinger.** 1985. Studies on germinability and some aspects of pre-harvest physiology of sorghum grain. *Seed Sci. Technol.* **13:** 27-35.
7. **McLaren, N. W.** 1987. *Studies on Pre- and Post-emergence Damping-off and Seedling Blight of Sorghum.* M.Sc. Thesis, Univ. of Natal, Pietermaritzburg, South Africa.
8. **Menkir, A., G. Ejeta, L. G. Butler, A. Melakeberhan, and H. L. Warren.** 1996. Fungal invasion of kernels and grain mold damage assessment in diverse sorghum germplasm. *Plant Dis.* **80:** 1399-1402.
9. **Singh, A. R., and V. G. Makne.** 1985. Correlation studies on seed viability and seedling vigor in relation to seed size in sorghum (*Sorghum bicolor*). *Seed Sci. Technol.* **13:** 139-142.

46

Free and Bound Phenolic Acids in Mature Sorghum Caryopses as Affected by Inoculation with *Fusarium thapsinum*

Oscar R. Rodriguez-Ballesteros, Anaclet S. B. Mansuetus, Richard A. Frederiksen, Gary N. Odvody, Ralph D. Waniska, and Darrell T. Rosenow

Sorghum [*Sorghum bicolor* (L.) Moench] is grown worldwide for food, fodder, and fuel. The worldwide market demands acceptable grain color, with milling ability and absence of microbial deterioration. Sorghum grain mold, caused by *Fusarium thapsinum*, is a major disease of sorghum that reduces grain and food quality and presents a significant challenge to sorghum improvement programs (12). Various control measures for grain mold have been devised, but the use of resistant cultivars has so far provided the most effective control (13). Germplasm sources with resistance to grain mold have been identified, as have several factors associated with resistance (1-4, 8, 11, 12).

Sorghum cultivars vary in the amount of free and bound phenolic acids that they contain. The level of these compounds may be related to resistance to grain mold (6), and cultivars that are highly resistant to grain weathering have high levels of phenolic acids (5). Resistant sorghum varieties contain higher levels of and more diverse types of phenolic acids than do susceptible varieties, but other factors, *e.g.*, the response of glume tissue to attacks by phytopathogenic fungi, also contribute to grain mold resistance (8).

Many attempts have been made to elucidate the role that phenolic acids and other phenolic compounds play in resistance to grain weathering. The results have, in some cases, been contradictory (6, 11). In this study, we measured levels of free and bound phenolic acids and other phenolic compounds in inoculated and noninoculated caryopses and found that phenolic acids were involved in the mechanism of resistance to grain weathering. We also developed prediction and classification equations that can be used as part of a grain mold resistance breeding program.

TABLE 46-1. Sorghum plant characteristics, and race classification of the 13 cultivars used in this study.

Cultivar	Mold Rating[a]	Resistance Level[b]	Pericarp Color[c]	Plant Color[c]	Glume Color[c]	Race
SC279-14E	1.8	R	R	R	R	Guinea
SC630-11E	2.5	I	R	R	R	Kafir
84BH5629,31	2.5	I	R	T	Dark T	Caudatum-Kafir
BTx626	2.5	I	R	R	Dark P	Kafir
6BH6077	2.3	I	R	P	Dark P	Caudatum-Kafir
80B2892	2.0	R	R	P	B	Kafir
IS9530	3.3	I	R	P	R-P	Caudatum
Sureno	2.0	R	W	T	Dark T	Caudatum
84BH5734	1.8	R	W	T	T	Caudatum
CS3541	2.3	I	W	T	T	Caudatum
Malisor 7	3.0	S	W	T	Dark T	Caudatum
SC170-6-17	4.0	S	W	R	R	Caudatum
Tx2536	4.5	S	W	P	R-P	Caudatum

[a]Mold rating system after Rodriguez-Herrera *et al.* (9). Data from Texas Agricultural Experiment Station tests.
[b]R, resistant; I, intermediate; and S, susceptible.
[c]P, purple; R, red; T, tan; B, black; and W, white.

TABLE 46-2. Phenolic acids selected by the stepwise discriminant analysis, and listed in the order of importance as they enter the classification model.[a]

Bound Control	Bound Inoculated	BI-BC Composite	Free Control	Free Inoculated	FI-FC Composite
Fer	Syr	Pro	PC-glumes	PC-glumes	PC-glumes
Syr	Gal	Gen	Gal	Gal	Pro
POH	POH	POH	Pro	Pro	Gen
Caf	Caf	Caf	Gen	Gen	POH
Cin	Gen	Syr	Van	Van	Van
Gal	Pro	Fer	Caf	Caf	Cou
Pro	Cin	Cin	Fer	Syr	
Cou			Sal	Cou	
Van				Fer	
Gen				Sal	
				Cin	

[a]Only phenolic acids that make a statistically significant contribution to each model (Wilk's λ, $p > 0.05$) are listed.
Abbreviations: BC-BI, difference between bound control and bound inoculated phenolic acids; Caf, caffeic acid; Cin, cinnamic acid; Cou, *p*-coumaric acid; FC-FI, difference between free control and free inoculated phenolic acids; Fer, ferulic acid; Gal, gallic acid; Gen, gentisic acid; PC-glumes, phenolic compounds from the glumes; POH – parahydroxybenzoic; Pro, protocatechuic acid; Sal, salicylic acid; Syr, syringic acid; and Van, vanillic acid.

Materials and Methods

We used 13 sorghum cultivars (Table 46-1) and one isolate of *F. thapsinum*. Plants were inoculated at anthesis with a spore suspension (5 ml/per head of a 1×10^6 spores/ml suspension); control plants were inoculated similarly with sterile distilled water.

Panicles of control and inoculated plants were harvested, threshed, cleaned, and lyophilized prior to chemical analysis. A composite sample from the inoculated and noninoculated plants of each cultivar (26 total samples) was made and four subsamples were made from each sample. Cultivars were planted in a randomized complete block design, replicated four times, identifying subsamples as blocks, free and bound phenolic acids as types, and inoculated and noninoculated plants as treatments.

Free and bound phenolic acids were assayed with a high performance liquid chromatography (HPLC) method (6, 7). Retention times and peaks for 11 phenolic acid standards–caffeic, cinnamic, ferulic, gallic, gentisic, *p*-coumaric, *p*-OH-benzoic, protocatechuic, salicylic, syringic, and vanillic–were used to quantitate phenolic acids (μg/g). Phenolic compounds in glumes and grain were measured by using an automated Folin-Ciocalteu method with gallic acid as a standard (7).

We conducted a stepwise discriminant analysis (10) of the 11 phenolic acids and analyzed each type × treatment combination independently (Table 46-2). Discriminant analysis is used to predict group membership from a set of variables. The purpose of this type of analysis is to identify discriminant functions that provide the maximum ratio of difference between a pair of group multivariate means to the multivariate variance within the groups. These predictors enable the user to test the validity of the groups, based on actual data, relative to created test groups. Only free phenolic acids were analyzed from glumes and caryopses. Grain mold ratings (Table 46-1) were used as the grouping variable in the analysis. Canonical discriminant functions, standardized by within variances, were estimated for each type × treatment combination.

Results

The phenolic acids were not all equally important in the discriminant analysis (Table 46-2), with caffeic, cinnamic, ferulic, gentisic, *p*-OH-benzoic, protocatechuic, and syringic acids being the most important bound acids, and gentisic, *p*-coumaric, *p*-OH-benzoic, protocatechuic, and vanillic acids, and the phenolic compounds from the glumes the most important free phenolic acids, as related to grain mold resistance. We compared the *a priori* classification (intermediate, resistant, and susceptible) of each cultivar with the *a posteriori* grouping based on the analysis of the phenolic acids (Table 46-3), and we found the best correlations occurred between the free control and free inoculated plants. The phenolic acids included in each type × treatment combination (Table 46-2) represent a discriminant function (available from authors) that depicts the order and contribution of each phenolic acid to the overall model. Amongst the free phenolic acids, the discriminant analysis accounts for 93% of the dispersion observed, while the bound phenolic acids account for 55% of the observed dispersion.

Discussion

We analyzed individual phenolic acids by subtracting the values of the control treatments from those for the inoculated plants and identifying the acids that increased (ferulic, salicylic, caffeic, syringic, and *p*-coumaric) or decreased (gallic) in response to inoculation. These results are consistent with previous reports (12), which indicated that caffeic, ferulic, and *p*-coumaric acids were involved in grain mold resistance. Levels of some of the phenolic acids changed due to the inoculation process but did not make a significant contribution to any of the analyzed models. We used discriminant analysis to distinguish between control and inoculated treatments and to identify the phenolic acids that are present constitutively and those whose production is induced by inoculation (BI-BC and FI-FC models). Caffeic, cinnamic, ferulic, gentisic, *p*-OH-benzoic, protocatechuic, and syringic acids are the most important bound phenolic acids, and gentisic, *p*-OH-benzoic, *p*-coumaric, protocatechuic and vanillic acids, and the phenolic compounds in the glumes are the most important free phenolic acids. Although the relative importance of the phenolic acids varies in the different models, in both the BI-BC and FI-FC models the protocatechuic, gentisic, and *p*-OH-benzoic acids made the greatest contributions to their respective models.

TABLE 46-3. Percentage of correct observations based on a Jackknife classification matrix for each treatment × type combination of a stepwise discriminant analysis.

Grain Mold Resistance	Bound Control	Bound Inoculated	BI-BC Composite	Free Control	Free Inoculated	FI-FC Composite
Resistant	91	86	83	100	100	82
Intermediate	98	94	50	100	92	86
Susceptible	100	90	75	100	100	80
Total	94	90	71	100	97	83

Abbreviations: BC-BI, composite sample of bound phenolics control and bound phenolics inoculated samples; and FC-FI, composite sample of free acid control and free acid inoculated samples.

The results from a Jackknife classification matrix (Table 46-3) are particularly important, since the best fit was obtained with the free control treatment, in which all of the cultivars were classified correctly. This result indicates that the *a priori* classification based on grain mold ratings, was similar to the variation present in the phenolic acids as weighted by the discriminant function (Table 46-2). Amongst the free phenolic acids, the discriminant analysis accounts for 93% of the dispersion observed, while the bound phenolic acids account for 55% of the observed dispersion. Thus, only free phenolic acids, which are easier to quantitate than are the bound phenolic acids, in noninoculated plants need to be analyzed as part of a grain mold resistance breeding program.

Conclusions

Caffeic, cinnamic, ferulic, gentisic, *p*-OH-benzoic, protocatechuic, and syringic acids are the most important bound phenolic acids, and gentisic, *p*-OH-benzoic, *p*-coumaric, protocatechuic, and vanillic acids, and the phenolic compounds in the glumes are the most important free phenolic acids in determining resistance to grain mold. The results from this study are consistent with previous reports that that caffeic, ferulic, and *p*-coumaric acids are important in grain mold resistance. The use of discriminant analysis on the residual models (BI-BC and FI-FC) identified additional phenolic acids, *e.g.*, protocatechuic, gentisic, and *p*-OH-benzoic, which are elicited by the presence of the pathogen and make an important contribution to grain mold resistance. The variation explained by the analysis of free phenolic acids indicates that analysis of these phenolic acids could be used as part of a grain mold resistance breeding program.

References

1. **Bandyopadhay, R., and L. K. Mughogho.** 1988. Evaluation of field screening techniques for resistance to sorghum grain mold. *Plant Dis.* **72:** 500-503.
2. **Castor, L. L.** 1981. *Grain Mold Histopathology, Damage Assessment and Resistance Screening with* Sorghum bicolor *(L.) Moench Lines.* Ph.D. Dissertation, Texas A&M Univ., College Station, Texas.
3. **Doherty, C. A., R. D. Waniska, L. W. Rooney, and J. H. Poe.** 1987. Free phenolic compounds and tannins in sorghum caryopsis and glumes during development. *Cer. Chem.* **64:** 42-46.
4. **Forbes, G. A.** 1986. *Characterization of Grain Mold Resistance in Sorghum [*Sorghum bicolor *(L.) Moench].* Ph.D. Dissertation, Texas A&M Univ., College Station, Texas.
5. **Hann, D. H.** 1981. *Qualitative and Quantitative Analysis of Low Molecular Weight Phenols by High Pressure Liquid Chromatography.* M.S. Thesis, Texas A&M Univ., College Station, Texas.
6. **Hann, D. H., L. W. Rooney, and J. M. Fubion.** 1983. Sorghum phenolic acids, their HPLC separation and their relationship to fungal resistance. *Cer. Chem.* **60:** 255-259.
7. **Kaluza, W. Z., R. M. McGrath, R. C. Roberts, and H. H. Schroeder.** 1980. Separation of phenolics of *Sorghum bicolor* (L.) Moench grain. *J. Agric. Food Chem.* **25:** 1191-1196.
8. **Mansuetus, A. B. S.** 1990. *The Effect of Glume and Caryopsis Characteristics of Sorghum [*Sorghum bicolor *(L.) Moench] on Infection by* Fusarium moniliforme *Sheldon.* M.S. Thesis, Texas A&M Univ., College Station, Texas.
9. **Rodriguez-Herrera, R., W. L. Rooney, D. T. Rosenow, and R. A. Frederiksen.** 2000. Inheritance of grain mold resistance in grain sorghum without a pigmented testa. *Crop Sci.* **40:** 1573-1578.
10. **SAS Institute, Inc.** 1988. *SAS/STAT Users Guide: Statistics, 1988 Edition, Release 6.03.* SAS Institute, Cary, North Carolina.

11. **Waniska, R. D., G. A. Forbes, R. Bandyopadhay, R. A. Frederiksen, and L. W. Rooney**. 1992. Cereal chemistry and grain mold resistance, p. 265-272. *In* W. A. J. de Milliano, R. A. Frederiksen, and G. D. Bengston (eds.), *Sorghum and Millets Diseases: A Second World Review.* ICRISAT, Patancheru (A.P.), India.
12. **Waniska, R. D., J. H. Poe, and R. Bandyopadhay.** 1989. Effects of growth conditions on grain molding and phenols in sorghum caryopsis. *J. Cer. Sci.* **10:** 217-225
13. **Williams, F. J., and K. N. Rao.** 1981. A review of sorghum grain molds. *Trop. Pest Manag.* **27:** 200-211

47

Breeding Sorghum for Resistance to Anthracnose, Grain Mold, Downy Mildew, and Head Smuts

William L. Rooney, S. Delroy Collins, R. R. Klein, P. J. Mehta, Richard A. Frederiksen, and R. Rodriguez-Herrera

Sorghum [*Sorghum bicolor* (L.) Moench] is a staple food and feed grain in the semiarid tropics that was cultivated on 45 million hectares in 1999 (15). Sorghum is a host plant for an array of plant pathogens including viruses, bacteria, and fungi. Because the environmental conditions in sorghum production areas vary widely, the diseases of economic importance also vary. Some of the most damaging diseases of sorghum are anthracnose, grain mold, root and stalk rots, ergot, head smut, and downy mildew. For most of these diseases, breeding for host-plant resistance remains the most cost-effective method of disease control. Sorghum breeders and pathologists continue to screen sorghum germplasm to identify new sources of genetic resistance, and breeders utilize these materials to produce improved varieties and hybrids and to reduce losses due to disease susceptibility. In the last ten years, molecular analysis of the sorghum genome has led to the mapping of many different resistance genes. Breeders, pathologists and molecular geneticists are now utilizing these techniques to produce the next generation of disease-resistant hybrids and cultivars.

In terms of acreage, sorghum is the fifth most important cereal crop of the world (after wheat, maize, rice, and barley). Approximately 45 million hectares of sorghum are harvested annually (15). Because of its superior drought and heat tolerance, sorghum production occurs primarily in the semiarid tropics. Sorghum production often is constrained by plant diseases such as anthracnose, grain mold, downy mildew, head smut, and maize dwarf mosaic virus. Breeding for host-plant resistance is the single most important and feasible mechanism for defending yield potential against these diseases. To effectively control a disease with host plant resistance, four requirements must be met (41, 47): (i) The disease must cause sufficient economic loss to warrant a breeding program for resistance; (ii) there must be a mechanism to screen and identify sources of resistance to the pathogen; (iii) the resistance must be heritable and transferable to other

TABLE 47-1. IS designations, country of origin and working group classification for converted sorghum accessions reported to have resistance to anthracnose (23, 29, 40, 41, 47, and 50).

Number[a]	IS Designation	Country of Origin	Working Group Classification
SC84	12593C	Uganda	Durra-Nigricans
SC120	2816C	Zimbabwe	Zerazera
SC137	12628C	Ethiopia	Durra-Dochna
SC155	12646C	Ethiopia	Subglabrascens-Durra-Dochna
SC166	12657C	Ethiopia	Durra-Bicolor
SC275	7361	Nigeria	Conspicuum
SC325	2462	USA	Nigricans
SC326	3758C	Ethiopia	Nigricans
SC393	7362	Nigeria	Caudatum-Kaura
SC647	2395C	South Africa	Caffrorum-Birdproof
SC689	2729C	Uganda	Nigricans-Feterita
SC701	3462C	Sudan	Caudatum
SC748	3552C	Sudan	Caudatum-Guineense
SC991	12219C	Uganda	Bicolor
SC1033	11894	Ethiopia	Subglabrescens-Durra-Dochna
SC1160	12018	Ethiopia	Durra

[a]SC number is the sorghum conversion number assigned to the line in the Texas Agricultural Experiment Station-USDA sorghum conversion program.

breeding sources; and (iv) the resistance must be sufficiently stable across environments and over time. Whenever these criteria are met, host plant resistance usually is the best method available to control a disease. There are numerous diseases that cause economic damage in sorghum production systems (47). Our objective in this chapter is to review the current status of efforts to improve the resistance of sorghum to anthracnose, grain mold, downy mildew, head smut, and some important foliar diseases.

Anthracnose

Anthracnose, caused by the fungus *Colletotrichum graminicola* (Ces.) Wils., is one of the most destructive diseases of sorghum in warm, humid regions. The pathogen can infect above-ground portions of the plant including stalk, foliage, panicle, and grain, and individual genotypes often differ in their reaction to foliar, stalk, or panicle infection (24). Yield losses may be as high as 50% in susceptible cultivars (22) and are greatest during extended periods of cloudy, warm, humid, and wet weather, especially when such weather occurs in the early grain-filling phase (48).

Sources of Resistance. Sorghum researchers from Asia, Africa, and the Americas have identified sources of resistance to many important sorghum diseases. Accessions from the ICRISAT world collection, the USDA collection in Griffin, Georgia, and the USDA/TAES sorghum conversion program have been scored for resistance to numerous sorghum diseases (Table 47-1).

Sorghum breeders have incorporated anthracnose resistance from some of the identified resistance sources into their breeding programs, but many other sources of resistance have not been utilized. Given the highly variable nature of the anthracnose pathogen, sorghum breeders need to have access to as many different sources of resistance as possible. These different resistance sources can be used to pyramid resistance genes to make durable varieties or to respond to changes in the composition of the fungal population (8). Wiltse (51) intercrossed 13 different sources of anthracnose resistance to establish genetic relationships between the germplasms. Segregation in the progeny of resistant × resistant crosses indicates that different genes control anthracnose resistance in the parents. At least five different sources of resistance were present among 13 sorghum conversion lines with anthracnose resistance. More recently, Mehta *et al.* (29) evaluated progeny from additional resistant × resistant crosses and expanded the number of unique sources of anthracnose resistance to at least six.

Resistance Stability. Sorghum cultivars can react differently to anthracnose in Texas, Georgia, Brazil, and Africa (42; Table 47-2). The differential nature of the reaction usually depends upon the races of *C. graminicola* that are present, and whose variation is well-documented (1, 6, 23). In addition to geographic variation, Rosewich *et al.* (43) found that asexual reproduction at a location may result in a predominance of one race that is only occasionally influenced by genetic drift and gene flow. Regardless of the mechanism generating the variation within *C. graminicola*, breeders must screen potentially resistant germplasm in as many environments as possible to insure that the resistance incorporated into their hybrids and cultivars is stable across environment and pathogen race.

Heritability of Resistance. The heritability of anthracnose resistance in sorghum has been studied for over 50 years. In most cases, researchers identify resistance genes that protect against a single isolate of the pathogen and have described several different simply inherited forms of genetic resistance to anthracnose (41, 47; Table 47-3). In several cases, resistance is the conferred by a recessive allele. Mehta *et al.* (29) identified three dominant and three recessive sources of resistance. Regardless of whether resistance is dominant or recessive, the simple inheritance of these resistance genes suggests that breeding to improve anthracnose resistance will succeed, although the stability and durability of the resulting resistance will be subject to changes in the dominant pathogen race(s).

Breeding Approaches. Parental lines and improved sorghum germplasm with enhanced anthracnose resistance have been released (13, 32). Some of these lines are used directly in commercial hybrid or cultivar production, while others are used as breeding sources, but most were developed by using traditional breeding methodology. To facilitate selection for anthracnose resistance, Tenkuano *et al.* (46) proposed selection at the seedling stage based on phytoalexin content. More recently, markers linked to anthracnose resistance have been reported (4, 29), and additional studies with other resistance sources are in progress. The molecular markers should make pyramiding of genes for anthracnose resistance feasible, and these genes, at least theoretically, should be more stable across environments.

Most public and private sorghum breeding programs have anthracnose resistance as an important target and evaluate new materials in multiple environments that have consistent anthracnose infections. In some cases, trials are inoculated to increase the amount of disease pressure. Marker-assisted selection will enhance current breeding efforts by enabling the introgression of resistance from exotic germplasm and permitting the pyramiding of multiple resistance genes in the same line or germplasm.

TABLE 47-2. Converted sorghum accessions with differential reactions to anthracnose populations in different environments.

Number[a]	IS Designation	Arkansas	Georgia	Texas	Brazil	India
SC2	12552C	R[b]	S[b]	R	R	S
SC60	2569C	R	R	R	R	S
SC103	2403C	R	S	R	S	S
SC110	12610C	S	S	S	S	S
SC112	12612C	S	S	R	S	S
SC126	12617C	R	R	R	S	S
SC155	12646C	S	R	R	R	S
SC322	1309C	S	R	R	R	S
SC326	3758	R	R	R	R	R
SC574	8337C	S	R	R	S	S
SC647	2395C	S	R	R	R	R
SC655	2862C	S	S	R	R	R
SC810	957C	S	S	R	S	R

[a]SC number is the sorghum conversion number assigned to the line in the Texas Agricultural Experiment Station-USDA sorghum conversion program.
[b]R, resistant; S, susceptible as described by Metha *et al.* (29) and Wiltse (51).

Grain Mold

Grain mold of sorghum is a major problem when grain development coincides with warm, wet weather. Grain mold describes the diseased appearance of sorghum grain resulting from infection by one or more parasitic fungi (50). Grain mold is most commonly caused by *Fusarium moniliforme sensu lato* (27) and *Curvularia lunata* (Walker) Boedjin (9), although many other species also cause grain mold. An array of screening and selection methodologies have been developed for grain mold, and a significant amount of information has been published on the identification of resistance, the mechanism of resistance, the inheritance of resistance, and the potential for future improvements.

Sources and Mechanisms of Resistance. Once screening methodologies and appropriate screening environments were developed, numerous sources of grain mold resistance in sorghum were identified and described (2, 9, 16, 19). Numerous traits have been associated with increased grain mold resistance, but none of them confer complete resistance to the disease. Initially, resistant sorghums usually had a testa layer and were high in tannins; however, additional and more thorough screening revealed that many guinea-type sorghums and zerazera sorghums had higher levels of grain mold resistance without a testa (21). Many of these original sources of resistance have been used to develop new germplasm with even better grain mold resistance and parental lines that are used to produce hybrids with a higher level of grain mold resistance. Corneous endosperm types are more resistant to grain mold than floury endosperm types (17), and grain hardness also is associated with grain mold resistance (25). Other structural-based kernel traits that have been reported to improve grain mold resistance are a thin mesocarp, the presence of surface wax, red pericarp color (due to the presence of flavon-4-ols), and a loose panicle structure (14, 17, 30). With all of the reported associations, however, there are numerous exceptions to any general rule.

Attempts to identify the biochemical basis for grain mold resistance have been similarly difficult. Most cultivars with a testa resist weathering and produce flavan-4-ols and high levels of other phenolic compounds that may enhance mold resistance (49). Esele *et al.* (14) found that red pericarp genotypes were more resistant to weathering than genetically similar white pericarp genotypes. Genotypes with a red pericarp typically contain higher levels of flavan-4-ols, and this compound may increase grain mold resistance (49). Seetharaman *et al.* (44) identified four proteins from sorghum endosperm with antifungal activity, but there was no association between antifungal protein content and grain mold incidence in the lines that they evaluated. Rodriguez-Herrera *et al.* (38) found a positive correlation between antifungal protein content and grain mold resistance. These results imply that antifungal proteins *per se* do not confer grain mold resistance, but that these proteins do have an important role in grain mold resistance.

Inheritance of Resistance. Many of the traits described in the previous section are qualitatively inherited (40), but their contribution to total grain mold resistance usually accounts for only a small portion of the total variation for the trait. Thus, grain mold resistance commonly is considered a quantitative trait, and general and specific combining ability components of variation for grain mold resistance are highly significant (12). Murty and House (33) used generation means analysis to evaluate resistance to sorghum grain mold. They found that the F_1 hybrid was more resistant to *C. lunata* and *F. moniliforme sensu lato* than the midparent, suggesting that grain mold resistance has dominance effects.

More recently, Rodriguez-Herrera *et al.* (39) conducted a generation means analysis of a Sureno × Tx430

TABLE 47-3. Inheritance of disease resistance in sorghum.

Disease	Inheritance Pattern	Reference
Anthracnose	Dominant and recessive qualitative loci	4, 29, 46
Downy Mildew	Mostly qualitative with dominant and partially dominant loci	11, 18, 34, 37, 45
Grain Mold	Mostly quantitative with additive, dominant, and some epistatic effects	12, 33, 39
Head Smut	Quantitative with dominant and additive effects	7
	Qualitative with recessive resistance	28
Leaf Blight	Qualitative - dominant	5, 28
MDMV	Dominant and recessive loci	42
Rust	Qualitative - dominant	31
Zonate Leaf Spot	Quantitative with additive and dominant effects	20

population for grain mold resistance in eight Texas environments. They detected additive effects in all eight environments and dominance effects in seven of the eight. Epistatic effects were detected in only two of eight environments, but a combined analysis indicated that higher order interactions were important when evaluated across environments. Rodriguez-Herrera *et al.* (39) concluded that selection in specific environments is useful for enhancing resistance to mold in those environments, but that this process may not provide grain mold resistance across a wide range of environments.

Further improvements in grain mold resistance are needed to maintain sorghum yield and grain quality. Several unlinked quantitative trait loci (QTLs) that affect grain quality and grain mold incidence in west Africa (35) and Texas (26) have been identified. In one study (26), each QTL individually accounted for 10-24% of the phenotypic variation, and, collectively, these QTLs account for approximately 45% of the phenotypic variation for grain mold. These QTLs are not consistently expressed in every environment, however, and genotype × environment interaction is critical in grain mold resistance. None of the five grain mold QTLs were detected in all five environments. These results are consistent with prior observations that breeding for stable resistance to grain mold is difficult, and that grain mold resistance mechanisms differ by environment and that these resistance mechanisms are controlled by different genetic loci (39).

Breeding Efforts. Although the mechanism(s) underlying grain mold resistance require more research, sorghum breeding programs have made substantial improvements in grain mold resistance by using the currently available information. The traditional strategy of hybridizing two parents with high levels of grain mold resistance (due to different resistance mechanisms) to produce derived lines and hybrids with higher levels of grain mold resistance has been successful. The use of overhead sprinkler systems to insure high levels of disease pressure for screening also has made scoring grain mold resistance more reliable. Continued use of these technologies should further increase grain mold resistance, but the improvements are likely to be small in effect.

Combining molecular genetic technology with sorghum breeding probably will yield the most substantial improvements in grain mold resistance. For example, the first association of grain mold resistance with antifungal proteins in sorghum was in a grain mold QTL mapping population (38). While not yet technically feasible, genetic transformation of sorghum may allow sorghum researchers to increase grain mold resistance by incorporating appropriate genes from other plant species.

Downy Mildew

Downy mildew, caused by the fungus *Peronosclerospora sorghii*, is a disease that occurs throughout the world. This fungus can infect sorghum systemically and/or through local lesions. Outbreaks of sorghum downy mildew are sporadic and depend upon inoculum load and environmental conditions. Downy mildew outbreaks have been significantly reduced in Mexico and the United States by the use of metalaxyl seed treatments (10).

Resistance and Its Inheritance. Resistance to downy mildew has been identified from several sources (10, 45; Table 47-3). These sources are race specific, and changes in race often result in lines losing their resistance (10). In the United States, the development of the metalaxyl seed treatment fungicides has minimized the incidence of sorghum downy mildew and probably reduced the overall effort in breeding for downy mildew resistance. Nevertheless, sorghum breeding programs have continued to breed for resistance to downy mildew through the use of downy mildew nurseries in which susceptible hybrids or cultivars are grown to maintain high inoculum levels.

The inheritance of downy mildew resistance in sorghum is dependent on the environment and the source of resistance used. In most cases, the inheritance of downy mildew resistance has been oligogenic (Table 47-3), but the number of genes involved varies from 1-6 (3, 36, 45). In some cases (*e.g.*, 18, 34, 52), molecular markers have been linked to downy mildew resistance, but none of these markers has yet been placed on a map of the sorghum genome.

Head Smut

Of the four smuts that infect sorghum, head smut (caused by *Sporisorium relianum*) is the most economically damaging. Several good sources of resistance to head smut have been identified, characterized, and utilized in breeding programs (Table 47-3 and 47-4). Inheritance of resistance to head smut is variable, depending upon the source of resistance and the environment in which the resistance is evaluated (7, 28, 40). Efforts to map the genetic factors conditioning resistance to head smut are in progress, and molecular markers linked to head smut resistance should be available in the near future.

Most sorghum breeding programs to reduce head smut still utilize field evaluation and select against progeny that show any level of infection. Screening for head smut is done in fields with high inoculum levels, and rows of highly susceptible checks are included in a systematic pattern as controls and to ensure suitable levels of inoculum for the following year's evaluation.

Foliar Diseases

Numerous diseases affect the leaves of sorghum, but the specific foliar disease of importance depends on location. The inheritance of resistance to some foliar diseases is known. The inheritance pattern for resistance often depends on the source of resistance and the environment in which the trait is evaluated. Breeders have utilized the available sources of resistance in their breeding programs as needed.

Conclusions

Disease resistance breeding has been very effective in protecting yield and quality in sorghum for many years, and breeding for disease resistance in sorghum remains an important objective in almost every sorghum breeding program. There are numerous sorghum diseases not yet controlled by host plant resistance, and the ability to effectively develop resistant germplasm is contingent on the availability and heritability of resistance and the availability of resources and time to move the resistance into elite germplasm. In the past ten years, there has been a significant reduction in the number of personnel and the amount of resources devoted to sorghum breeding and pathology. These reductions have reduced the capacity of sorghum improvement programs to respond to the needs of the crop. Further reductions will reduce the number of effective programs even further and diminish the ability of those that do remain to effectively address the disease resistance needs of the crop.

TABLE 47-4. Converted sorghum accessions designated with a "✓" possess resistance to grain mold, downy mildew, head smut, and/or anthracnose of sorghum based on evaluations conducted in the Texas Agricultural Experiment Station sorghum improvement program disease screening nurseries from 1995-1999.[a]

Number[b]	Grain Mold[c]	Downy Mildew[d]	Head Smut[e]	Anthracnose[f]
SC103-11E	✓		✓	
SC120, SC137, SC155, SC166, SC647, SC689			✓	✓
SC277, SC414, SC423, SC425, SC575		✓	✓	
SC325		✓	✓	✓
SC630-11E, SC650-11E, SC719-11E	✓			
SC748-5	✓			✓

[a]Reactions for grain mold, downy mildew, and head smut are based on natural infection, while anthracnose reactions are based on inoculation with *Colletotrichum gramincola* isolate TX430-BB-5.

[b]SC number is the sorghum conversion number assigned to the line in the Texas Agricultural Experiment Station-USDA sorghum conversion program.

[c]Sorghums with grain mold resistance had grain mold ratings (39) that averaged less than 2.0 [on a scale of 1 (resistant) to 5 (susceptible)] over multiple environments.

[d]Sorghums had less than 1% systemic infection and no local lesion infection caused by pathotypes 1 and 3 of *Peronosclerospora sorghi* over several environments.

[e]Sorghums had less than 5% mean systemic head smut infection caused by *Sporisorium reilianum* over several environments.

[f]Sorghums with anthracnose disease rating of less than 1.2 are included in this listing. The rating scale (29, 51) is 1 (no disease) to 5 (plant death).

References

1. **Ali, M. E. K., and H. L. Warren.** 1987. Physiological races of *Colletotrichum graminicola* on sorghum. *Plant Dis.* **71:** 402-404.
2. **Bandyopadhyay, R., and L. K. Mughogho.** 1988. Evaluation of field screening techniques for resistance to sorghum grain molds. *Plant Dis.* **72:** 500-503.
3. **Bhat, M. G., B. T. S. Gowda, K. H. Anahosur, and J. V. Goud.** 1982. Inheritance of plant pigmentation and downy mildew resistance in sorghum. *SABRAO J.* **14:** 53-59.
4. **Boora, K. S., R. A. Frederiksen, and C. W. Magill.** 1998. DNA-based markers for a recessive gene conferring anthracnose resistance in sorghum. *Crop Sci.* **38:** 1708-1709.
5. **Boora, K. S., R. A. Frederiksen, and C. W. Magill.** 1999. A molecular marker that segregates with sorghum leaf blight resistance in one cross is maternally inherited in another. *Mol. Gen. Genet.* **261:** 317-322.

6. **Cardwell, K. F., P. R. Hepperly, and R. A. Frederiksen.** 1989. Pathotypes of *Colletotrichum graminicola* and seed transmission of sorghum anthracnose. *Plant Dis.* **73:** 255-257.
7. **Cao, R. H., X. L. Wang, J. H. Ren, and C. H. Nan.** 1988. The resistance of sorghum to head smut and its inheritance, p. 121-124. *In* S. Suzuki (ed.), *Crop Genetic Resources of East Asia.* International Board for Plant Genetic Resources, Rome, Italy.
8. **Casela C. R., A. S. Ferreira, and R. E. Schaffert.** 1992. Physiological races of *Colletotrichum graminicola* in Brazil, p. 209-212. *In* W. A. J. de Milliano, R. A. Frederiksen, and G. D. Bengston (eds.), *Sorghum and Millets Diseases: A Second World Review.* ICRISAT, Patancheru (A.P.), India.
9. **Castor, L. L., and R. A. Frederiksen.** 1980. *Fusarium* and *Curvularia* grain molds in Texas, p. 93-102. *In* R. J. Williams, R. A. Frederiksen, L. K. Mughogho, and G. D. Bengston (eds.), *Sorghum Diseases, A World Review.* ICRISAT, Patancheru (A.P.), India.
10. **Craig, J., and G. N. Odvody.** 1992. Current status of sorghum downy mildew control, p. 213-217. *In* W. A. J. de Milliano, R. A. Frederiksen, and G. D. Bengston (eds.), *Sorghum and Millets Diseases: A Second World Review.* ICRISAT, Patancheru (A.P.), India.
11. **Craig, J., and K. F. Schertz.** 1985. Inheritance of resistance in sorghum to three pathotypes of *Peronosclerospora sorghi. Phytopathology* **75:** 1077-1078.
12. **Dabholkar, A. R., and S. S. Baghel.** 1980. Inheritance resistance to grain mould of sorghum. *Ind. J. Genet. Plant Breed.* **40:** 472-476.
13. **Duncan, R. R., F. R. Miller, D. T. Rosenow, A. Sotomayor-Rios, and S. Torres-Cardona.** 1990. Registration of GTPP7R(H)C5 white seeded, tan plant, and foliar disease resistant sorghum germplasm population. *Crop Sci.* **30:** 239-240.
14. **Esele, J. P., R. A. Frederiksen, and F. R. Miller.** 1993. The association of genes controlling caryopsis traits with grain mold resistance in sorghum. *Phytopathology* **83:** 490-495.
15. **Food and Agricultural Organization of the United Nations.** 1999. FAOSTAT agriculture data. FAO, Rome, Italy. http://apps.fao.org
16. **Glueck, J. A., and L. W. Rooney.** 1976. Physical and chemical characterization of sorghum lines with resistance to grain deterioration. *Cer. Foods World* **21:** 436-437.
17. **Glueck, J. A., and L. W. Rooney.** 1980. Chemistry and structure of grain in relation to mold resistance, p. 119-140. *In* R. J. Williams, R. A. Frederiksen, L. K. Mughogho, and G. D. Bengston (eds.), *Sorghum Diseases, A World Review.* ICRISAT, Patancheru (A.P.), India.
18. **Gowda, P. S., G. W. Xu, R. A. Frederiksen, and C. W. Magill.** 1995. DNA markers for downy mildew resistance genes in sorghum. *Genome* **38:** 823-826.
19. **Gray, E., G. D. Lacefield, and G. D. Lowe.** 1971. Head mold on grain sorghum. *Plant Dis. Rept.* **55:** 337-339.
20. **Grewal, R. P. S.** 1988. Genetic basis of resistance to zonate leaf spot disease in forage sorghum. *Theor. Appl. Genet.* **76:** 550-554.
21. **Hahn, D. H., and L. W. Rooney.** 1986. Effect of genotype on tannins and phenols of sorghum. *Cer. Chem.* **63:** 4-8.
22. **Harris, H. B., B. J. Johnson, J. W. Dobson, and E. S. Luttrell.** 1964. Evaluation of anthracnose on grain sorghum. *Crop Sci.* **4:** 460-462.
23. **Hazra, S., R. P. Thakur, and R. P. Devi.** 1999. Pathogenic and molecular variability among twelve isolates of *Colletrotrichum graminicola* from sorghum. *J. Mycol. Plant Pathol.* **29:** 176-183.
24. **Hess, D. E., R. Bandyopadhyay, and I. Sissoko.** 2001. Reactions of sorghum genotypes to leaf, panicle and grain anthracnose (*Colletotrichum graminicola*) under field conditions in Mali, p. 163-177. *In* I. Akintayo and J. Sedgo (eds.), *Towards Sustainable Sorghum Production, Utilization, and Commercialization in West and Central Africa.* ICRISAT, Patancheru (A.P.), India.
25. **Jambunathan, R., M. S. Kherdekar, and J. W. Stenhouse.** 1992. Sorghum grain hardness and its relationship to mold susceptibility and mold resistance. *J. Agric. Food Chem.* **40:** 1403-1408.
26. **Klein, R. R., R. Rodriguez-Herrera, J. A. Scheulter, P. E. Klein, Z. H. Yu, and W. L. Rooney.** 2000. Identification of genomic regions that affect grain mold incidence and other traits of agronomic importance in sorghum. *Theor. Appl. Genet.* **102:** 307-319.
27. **Leslie, J. F., and W. F. O. Marasas.** 2002. Will the real *Fusarium moniliforme* please stand up!, p. 201-209. *In* J. F. Leslie (ed.), *Sorghum and Millets Diseases.* Iowa State Press, Ames, Iowa.
28. **Magill, C. W., K. S. Boora, R. S. Kumari, J. Osorio, B. J. Oh, B. S. Gowda, Y. Cui, and R. A. Frederiksen.** 1997. Tagging sorghum genes for disease resistance: expectations and reality. *Proc. Intl. Conf. on Genetic Improvement of Sorghum and Pearl Millet (22-27 September 1996, Lubbock, Texas)*: 316-325.
29. **Mehta, P. J., S. D. Collins, W. L. Rooney, R. A. Frederiksen, and R. R. Klein.** 2000. Identification of different sources of genetic resistance to anthracnose in sorghum. *Intl. Sorg. Millets Newsl.* **41:** 51-54.
30. **Menkir, A., G. Ejeta, L. Butler, and A. Melakeberhan.** 1996. Physical and chemical kernel properties associated with resistance to grain mold in sorghum. *Cer. Chem.* **73:** 613-617.
31. **Miller, F. R., and H. J. Cruzado.** 1969. Allelic interactions at the *Pu* locus in *Sorghum bicolor* (L.) Moench. *Crop Sci.* **9:** 333-339.
32. **Miller, F. R., T. F. Dusek, K. L. Prihoda, and L. W. Rooney.** 1992. Registration of RTx436 sorghum parental line. *Crop Sci.* **32:** 1518.
33. **Murty, D. S., and L. R. House.** 1984. Components of generation means for resistance to grain mold-causing fungi *Curvularia* and *Fusarium* in sorghum. *Cer. Res. Comm.* **12:** 237-244.
34. **Oh, B. J., R. A. Frederiksen, and C. W. Magill.** 1996. Identification of RFLP markers linked to a gene for downy mildew resistance in sorghum. *Can. J. Bot.* **74:** 315-317.
35. **Rami, J. F., P. Dufour, G. Trouche, G. Fliedel, C. Mestres, F. Davrieux, P. Blanchard, P. Hamon.** 1998. Quantitative trait loci for grain quality, productivity, morphological, and agronomical traits in sorghum (*Sorghum bicolor* L. Moench). *Theor. Appl. Genet.* **97:** 605-616.
36. **Rana, B. S., K. H. Anahosur, M. J. V. Rao, V. J. M. Rao, R. Parameshwarappa, and N. G. P. Rao.** 1982.

Inheritance of field resistance to sorghum downy mildew. *Ind. J. Genet. Plant Breed.* **42:** 70-74.

37. **Reddy, B. S., S. Reddy, L. K. Mughogho, Y. D. Narayana, K. D. Nicodemus, and J. W. Stenhouse.** 1992. Inheritance patterns of downy mildew resistance in advanced generations of sorghum. *Ann. Appl. Biol.* **121:** 249-255.
38. **Rodriguez-Herrera, R., R. D. Waniska, W. L. Rooney.** 1999. Antifungal proteins and grain mold resistance in sorghum with a non-pigmented testa. *J. Agric. Food Chem.* **47:** 4802-4806.
39. **Rodriguez-Herrera, R., W. L. Rooney, D. T. Rosenow, and R. A. Frederiksen.** 2000. Inheritance of grain mold resistance in grain sorghum without a pigmented testa. *Crop Sci.* **40:** 1573-1578.
40. **Rooney, W. L.** 2000. Sorghum genetics and cytogenetics, p. 261-307. *In* C. W. Smith and R. A. Frederiksen (eds.), *Sorghum: Evolution, History, Production and Technology.* John Wiley, New York City, New York.
41. **Rosenow, D. T.** 1992. Using germplasm from the world collection in breeding for disease resistance, p. 319-324. *In* W. A. J. de Milliano, R. A. Frederiksen, and G. D. Bengston (eds.), *Sorghum and Millets Diseases: A Second World Review.* ICRISAT, Patancheru (A.P.), India.
42. **Rosenow, D. T., and R. A. Frederiksen.** 1982. Breeding for disease resistance in sorghum. p. 447-455. *In* L. R. House, L. K. Mughogho, and J. M. Peacock (eds.), *Sorghum in the Eighties*, Vol. 1. ICRISAT, Patancheru (A.P.), India.
43. **Rosewich, U. L., R. E. Pettway, B. A. McDonald, R. R. Duncan, and R. A. Frederiksen.** 1998. Genetic structure and temporal dynamics of a *Colletotrichum graminicola* population in a sorghum disease nursery. *Phytopathology* **88:** 1087-1093.
44. **Seetharaman, K. E., R. D. Waniska, and L. W. Rooney.** 1996. Physiological changes in sorghum antifungal proteins. *J. Agric. Food Chem.* **44:** 2435-2441.
45. **Sifuentes, J., and R. A. Frederiksen.** 1988. Inheritance of resistance to pathotypes 1, 2, and 3 of *Peronosclerospora sorghi* in sorghum. *Plant Dis.* **72:** 332-333.
46. **Tenkuano, A., F. R. Miller, G. E. Hart, R. A. Frederiksen, and R. L. Nicholson.** 1993. Phytoalexin assay in juvenile sorghum: An aid to breeding for anthracnose resistance. *Crop Sci.* **33:** 243-248.
47. **Thakur, R. P., R. A. Frederiksen, D. S. Murty, B. V. S. Reddy, R. Bandyopadhyay, L. M. Giorda, G. N. Odvody, and L. E. Claflin.** 1997. Breeding for disease resistance in sorghum. *Proc. Intl. Conf. on Genetic Improvement of Sorghum and Pearl Millet (22-27 September 1996, Lubbock, Texas)*: 303-315.
48. **Thakur, R. P., and K. Mathur.** 2000. Anthracnose, p. 40. *In* R. A. Frederiksen and G. N. Odvody (eds.), *Compendium of Sorghum Diseases*, 2nd ed. APS Press, St. Paul, Minnesota.
49. **Waniska, R. D., G. A. Forbes, R. Bandyopadhyay, R. A. Frederiksen, and L. W. Rooney.** 1992. Cereal chem.-istry and grain mold resistance, p. 265-272. *In* W. A. J. de Milliano, R. A. Frederiksen, and G. D. Bengston (eds.), *Sorghum and Millets Diseases: A Second World Review.* ICRISAT, Patancheru (A.P.), India.
50. **Williams, R. J., and K. N. Rao.** 1981. A review of sorghum grain molds. *Trop. Pest Manag.* **27:** 200-211.
51. **Wiltse, C. C.** 1998. *A Survey of Anthracnose Resistant Sorghum Germplasm Lines to Identify Additional Resistance Genes.* M.S. Thesis. Texas A & M Univ., College Station, Texas.
52. **Xu, G. W., C. W. Magill, K. F. Schertz, and G. E. Hart.** 1994. A RFLP linkage map of *Sorghum bicolor* (L.) Moench. *Theor. Appl. Genet.* **89:** 139-145.

48

Breeding for Resistance to Root and Stalk Rots in Sorghum

Mitchell R. Tuinstra, Tesfaye T. Teferra, Larry E. Claflin, Robert G. Henzell, Andrew Borrell, N. Seetharama, Gebeisa Ejeta, and Darrell T. Rosenow

Abstract

Stalk rots are one of the most prevalent diseases of sorghum. Charcoal rot is the most aggressive and widespread stalk rot disease in tropical and more arid environments, but Fusarium stalk rot has greater impact in temperate environments, including the central Great Plains of the United States. Stalk rot resistance is controlled by multiple genes. Most breeding efforts focus on integrating resistance genes into elite parent lines through classical breeding and selection. Genes for charcoal rot resistance in sorghum are being mapped to enable marker-assisted selection and to enhance breeding efficiency.

Introduction

Root and stalk rots are one of the most prevalent diseases of sorghum (31), and breeding for resistance to these diseases remains a research priority in most sorghum improvement programs. Stalk rots directly impact yield potential and feed quality by reducing seed size and the quality of grain and stover. Indirectly, stalk rots impact yield through increased incidence of stalk and root lodging. Significant yield losses attributed to stalk rots have been reported in most sorghum-producing regions.

Root and stalk rots are caused by a complex of diverse fungal pathogens. Infection is enhanced under conditions of environmental stress (5, 11, 12, 20, 43, 50). Pathogen genera associated with these diseases in sorghum include *Macrophomina*, *Fusarium*, *Acremonium*, *Alternaria*, *Colletotrichum*, *Nigrospora*, *Trichoderma*, and *Epicocum* (31). Stalk rots in sorghum usually are caused by *Macrophomina phaseolina*, the causal agent of charcoal rot, and *Fusarium moniliforme sensu lato*, the pathogen responsible for Fusarium stalk rot (12, 26, 43, 50). Charcoal rot is the most aggressive and widespread stalk rot disease of sorghum in tropical and more arid environments (16, 27). Fusarium stalk rot generally has greater impact in more temperate environments, including the central Great Plains of the United States (19).

Fusarium and charcoal rots are generally characterized by stress-induced senescence of stalk pith cells followed by the degradation of pith tissue at or near the base of the stalk (31). Consequently, stalk rot infections often lead to poor grain development, premature plant death, and stalk lodging. Researchers have identified a number of complex factors including cultural practice, environmental stress, and anatomical weakness of the stalk that are associated with increased stalk rot disease incidence (11, 12, 20, 43, 50). The environment in which the host plant is grown also can predispose the plant to stalk rot. Factors that influence the expression of stalk rot include high plant populations that increase competition for scarce resources (7), planting date (1), water deficit (37), high temperature (11), and an imbalance of mineral nutrients (7). Not all of these factors, however, are conducive for all types of stalk rot.

The use of cultivars that are resistant to both biotic and abiotic stresses is generally considered the most effective and environmentally sound approach for addressing stalk rot disease problems. The development of resistant cultivars involves germplasm characterization to identify sources of resistance, population development to derive segregating progenies, and the identification of superior genotypes to bring together the required traits into a single plant. The steps in this process are more difficult when the phenotype observed is dependent upon environmental conditions. Charcoal rot incidence, for example, is much higher when the plants are exposed to prolonged drought stress and high temperature during the grain filling period (11). The conditions that favor Fusarium stalk rot are not as well understood as the moisture stress relationship with charcoal rot; however, the disease is generally most severe in later growth stages when cool, wet conditions follow hot, dry weather (41, 50). These results suggest that stalk rots should be less severe on postflowering drought-tolerant genotypes.

Historically, most research effort has focused on addressing problems with charcoal rot because it is generally more aggressive than Fusarium stalk rot and impacts a larger geographic area. However, Fusarium stalk rot is a major problem in more temperate and highly productive environments. Our objective in this review is to provide a systematic overview of current research efforts in sorghum breeding directed towards resistance to charcoal rot and Fusarium stalk rot.

Charcoal Rot

Charcoal rot is the most aggressive stalk rot disease of sorghum (16, 27). The disease is caused by the fungus *Macrophomina phaseolina* (Tassi) Goid. *M. phaseolina* infects sorghum through roots and continues to the base of the plant, resulting in disintegration of stalk pith cells, premature leaf and plant senescence, and crop lodging. Charcoal rot usually appears late in the season when grain filling coincides with drought stress and higher soil temperatures. Edmunds (11) reported total death of sorghum plants five days after inoculation at soil temperature of 35°C and < 25% available soil moisture. Seetharama *et al.* (37) also demonstrated the impact of water deficit on charcoal rot incidence by using a line-source sprinkler irrigation technique that resulted in increased charcoal rot incidence as the distance to the sprinkler line, and the water deficit, increased. Other factors, *e.g.*, nitrogen fertilization, plant population, and planting date, also can influence charcoal rot incidence (29).

Symptoms. Charcoal rot infections are enhanced by rapid onset of drought and by high-temperature stress. Although charcoal rot infections lead to soft stalks, premature senescence, and reduced grain size, the most dramatic and easily recognized symptom of infection is an increase in stalk lodging. Inspection of lodged plants generally reveals vascular shredding and masses of small, black fungal structures, termed sclerotia (14). The presence of these sclerotia discolor the pith of the stalk, giving it a charcoal black or dark gray color.

Economic Importance and Control. Charcoal rot is the most widely distributed stalk rot pathogen of sorghum. Charcoal rot epidemics are common in west Africa, extending from Senegal to Chad (16), and in east Africa, including Sudan, Uganda, Kenya, Somalia, and the semi-arid Rift Valley areas of Ethiopia (17). It also is a major disease of sorghum in India, particularly during the post-rainy season (37), and Australia (40). In south and central Texas, stress-induced plant death and stalk lodging are almost always associated with charcoal rot; however, this disease is less common in sorghum-producing areas further north (19, 31).

Most recommendations for controlling charcoal rot in sorghum integrate cultivar selection for stalk rot resistance with crop management strategies that focus on reducing the risk of drought stress. In some cases, stalk rots are less of a problem in conservation tillage and no-till fields where soil moisture is conserved (13), or if lower plant populations are cultivated (7). Variation in soil fertility also is associated with stalk rot incidence (8).

Breeding for Host Plant Resistance to Charcoal Rot Infection. Effective selection for host plant resistance requires quantitative measurement and scoring of the disease and/or its affects. Unfortunately, scoring charcoal rot resistance often requires an environmental stress at an appropriate growth stage for adequate disease expression. Despite this problem, progress has been made in identifying genotypes that are resistant to charcoal rot (16, 31, 33).

Inoculation of sorghum plants with *M. phaseolina*-infested toothpicks is widely used to induce infection artificially and to permit comparisons and to identify resistant genotypes (11). Although inoculation does not exactly mimic natural infection, it can be useful for induction of consistent infection in different environments, a prerequisite for an effective selection program (5).

Several different scoring techniques are available to quantify stalk infection following artificial inoculation or natural infection. Direct measurement of the disease lesion, the number of nodes crossed by the pathogen, and the percentage of internode tissue infected all have been used to quantify disease severity (5). Frederickson *et al.* (15) classified disease reaction on a 1-5 scale depending on the number of nodes and internodes crossed. Patil *et al.* (30) proposed a similar system based on the percenttage of infected plants. They identified highly resistant (when the plant is completely free from the disease), resistant (< 10% infection), moderately resistant (10-30% infection), susceptible (30-50% infection), and highly susceptible (> 50% infection) varieties. Bramel-Cox and Claflin (4) rated stalk disintegration on a 1-6 scale that integrated the number of internodes crossed and the degree of discoloration.

Genetic variation in stalk rot resistance in sorghum is well-documented (5, 18, 30, 32, 34). Most breeding for charcoal rot resistance has focused on selection for the staygreen trait under terminal moisture stress conditions. This strategy is based on the association between staygreen and charcoal rot resistance (18, 32). Staygreen results in reduced senescence during the grain filling period and leads to healthier plants even under late-season drought stress. Several studies have shown that the staygreen trait is significantly and positively correlated with reduced lodging and increased charcoal rot resistance (10, 18, 32, 33).

Rosenow (32) demonstrated that indirect selection for charcoal rot resistance by using the staygreen trait was effective in applied plant breeding efforts. The genotypes were scored visually scored on a 1-5 scale based on premature leaf and stem death, where 1 = completely green and 5 = dead, a method that is both reliable and quick (47, 48). Rosenow *et al.* (35) recently compiled a list of exotic sorghums resistant to postflowering drought stress under field conditions (Table 48-1). Most of these sorghums are staygreen types that were identified in observations of converted sorghum lines and photo-period-insensitive accessions from the Sudan Collection.

There is a direct relationship between senescing tissue and susceptibility to stalk rots (21) and physiological studies (2, 37), which suggest that the association between staygreen and decreased charcoal rot and lodging is causal. A reduction in available carbohydrate may lead to tissue senescence and provide an opportunity for root and stalk rot infection (7). The rate of leaf sensecence and the amount of stem reserves mobilized are correlated (3). Staygreen lines maintained a greater proportion of recently produced assimilates in their stems at maturity (38). Collectively, these reports suggest that staygreen hybrids are less reliant on nonstructural stem carbohydrate during the grain-filling period, which increases stem strength and reduces lodging.

Inheritance of Charcoal Rot Resistance. Charcoal rot resistance is controlled by multiple genes with both recessive and additive modes of action, and the general combining ability (GCA) effects for *M. phaseolina* resistance in some crosses is significant (5). The staygreen and charcoal rot resistance derived from SC35 is controlled by dominant and recessive epistatic interactions with no maternal effects (28, 46). In B35, staygreen and charcoal rot resistance were controlled by one major dominant gene, a recessive gene, and a third locus with modifying effects (42). The analysis of staygreen and charcoal rot resistance in SC599 indicated a similar pattern of inheritance for charcoal rot resistance, but the expression of staygreen was recessive in this genetic background (42). Interestingly, charcoal rot resistance and staygreen did not cosegregate in populations derived from SC599, suggesting that the charcoal rot resistance mechanism in this accession is not staygreen (42). A broad evaluation of staygreen parent lines indicated that numerous inheritance patterns, from completely dominant to completely recessive, were operating for this trait (36).

Sorghum improvement programs at Texas A&M University, Purdue University and the Queensland Department of Primary Industries (QDPI) have focused on selection for staygreen derived from B35 (34). This line originated from the first backcross generation of Martin with IS 12555, a durra from Ethiopia. Quantitative trait locus (QTL) mapping is now in progress to identify genes associated with staygreen and charcoal rot resis-

TABLE 48-1. Drought-tolerant sorghum accessions with resistance to postflowering drought stress under field conditions in Lubbock, Texas.

Accession	IS Designation	Geographic Origin	Race
SC22	IS 12542C	Ethiopia	Durra
SC23	IS 12543C	Ethiopia	Durra
SC33	IS 12553C	Ethiopia	Durra
SC35	IS 12555C	Ethiopia	Durra
SC38	IS 12558C	Ethiopia	Durra
SC56	IS12568C	Sudan	Caudatum
SC134	IS 12625C	Ethiopia	Durra-Bicolor
SC146	IS 12637C	Ethiopia	Caudatum
SC237	IS 3071C	Sudan	Caudatum
SC265	IS 6705C	Burkina Faso	Guinea
SC279	IS 7419C	Nigeria	Guinea
SC284	IS 6729C	Burkina Faso	Guinea
SC452	IS 5145C	India	Dochna
SC564	IS 3909C	Uganda	Kafir-Caudatum
SC574	IS 8337C	Pakistan	Caudatum
SC599	IS 17459C	Sudan	Caudatum
SC689	IS 2729C	South Africa	Kafir-Caudatum
SC693	IS 7132C	Uganda	Caudatum
SC694	IS 7193C	Nigeria	Caudatum
SC701	IS 3462C	Sudan	Caudatum
SC1014	IS 11443C	Ethiopia	Durra-Bicolor
SC1017	IS 11549C	Ethiopia	Durra-Bicolor
SC1031	IS 11885C	Ethiopia	Durra-Bicolor
SC1039	IS 12171C	Ethiopia	Durra-Bicolor
SC1154	IS 11814C	Ethiopia	Durra
SC1158	IS 11930C	Sudan	Durra
SU439	IS 8688	Sudan	Kafir-Caudatum
SU629	IS 9782	Sudan	Caudatum
SU1672	IS 22364	Sudan	Caudatum
CSM-63		Mali	Guinea
P898012		Unknown	Caudatum
SRN39		Unknown	Caudatum

tance and to enhance these breeding efforts (6, 39, 40, 45, 49). In each study B35, or one of its derivatives, was crossed to a charcoal-rot susceptible parent to develop a population of recombinant inbred lines. These populations were evaluated for differences in staygreen under postflowering drought conditions at multiple locations for several years. Molecular genetic markers scored included restriction fragment length polymorphisms (RFLP), random amplified polymorphic DNAs (RAPDs), and simple sequence repeats (SSRs). In each population, two to four major QTLs were identified that explained up to 50% of the variation in staygreen across locations. At least three of these loci, *stg1, stg2, stg3*, appear to be expressed in different mapping populations (39). Genes for charcoal rot resistance *per se* were not mapped in these studies.

Researchers at Purdue, QDPI, Commonwealth Scientific and Industrial Research Organization (CSIRO), and ICRISAT are currently mapping genes for staygreen and charcoal rot resistance under postflowering drought conditions. Preliminary analyses indicate that a subset of the major QTL markers for staygreen identified in the United States and Mexico map to the same region as do those identified in India.

Fusarium Stalk Rot Resistance in Sorghum

Fusarium (*Fusarium* sp.) stalk rot is another important disease of grain sorghum. *F. moniliforme sensu lato* is generally perceived to be less aggressive than *M. phaseolina*, and less research has been conducted to address this problem. *Fusarium thapsinum* is the most common fungus associated with sorghum stalk rot in more temperate environments including much of the central Great Plains of the United States (5, 22-26). Although *Fusarium* often is regarded as a weak pathogen or a secondary invader, *Fusarium* infections become more aggressive under the appropriate environmental conditions.

Symptoms. *Fusarium* persists in the soil on crop debris and weedy hosts. Fusarium stalk rot, unlike charcoal rot, is most damaging when cool, wet conditions follow hot, dry weather (44, 50). *Fusarium* infection usually is accompanied by extensive root damage, which may migrate to the above-ground stalk (7). The cortical tissues of the roots are attacked first, followed by the root vascular tissues. Although Fusarium stalk rot is relatively less aggressive than charcoal rot, it can spread quickly and result in extensive lodging and yield loss. Symptoms of Fusarium stalk rot include red or purple disease lesions that become visible in the pith of the stalk following infection (14). Premature plant death and stalk lodging are the most dramatic symptoms of infection by Fusarium stalk rot. Analysis of genetically diverse Fusarium isolates with respect to virulence under greenhouse conditions did not identify any significant differences between the fungal isolates tested (19).

Economic Importance and Control. In the United States, Fusarium stalk rot is particularly important on the high plains from Texas to Nebraska. It is reported that *F. moniliforme sensu lato* is the primary pathogen associated with sorghum stalk rot in Kansas, one of the main sorghum-producing areas in the United States. Although the average estimate of grain yield reduction caused by this pathogen is only about 4%, disease incidence at some locations can be very high, with yield losses of more than 90%. Reports from Nebraska (31) and the southeastern United States (9) indicate that *F. moniliforme sensu lato*, which includes *Fusarium verticillioides*, *F. thapsinum*, *Fusarium proliferatum*, *Fusarium andiyazi*, and *Fusarium pseudonygamai* among others, is the major causal organism of sorghum stalk rot in those areas as well.

Most recommendations for controlling Fusarium stalk rot integrate variety selection for stalk rot resistance with crop management strategies that focus on reducing the risk of crop water deficits. As with other stalk rot pathogens, the incidence of Fusarium stalk rot increases in response to environmental stresses, *e.g.*, fertilizer imbalance, high plant populations, and root damage.

Breeding for Host Plant Resistance to Fusarium Stalk Rot. The development of sorghum hybrids that are resistant to Fusarium stalk rot has been difficult because the disease is caused by a complex of pathogens and disease incidence is confounded by environmental stress. Thus, year and location interaction effects often complicate germplasm characterization studies that evaluate differences in Fusarium stalk rot resistance (5).

Most sources of germplasm for stalk rot resistance were identified based on resistance to *M. phaseolina*. Although some reports suggest that genotypes that are resistant to charcoal rot also are resistant to Fusarium stalk rot (5), anecdotal observations indicate that some charcoal rot resistant lines and their hybrids are susceptible to Fusarium stalk rot. Thus, it is important to identify sources of germplasm with superior resistance to Fusarium stalk rot.

Some progress has been made in identifying genotypes resistant to Fusarium stalk rot (5, 31, 41). Most of these studies used a toothpick inoculation technique similar to those used to study charcoal rot resistance; however, liquid inoculation procedures also have been developed at Kansas State University to identify and compare sources of genetic resistance.

Inheritance of Fusarium Stalk Rot Resistance. Several germplasm characterization studies evaluated genetic variation for Fusarium stalk rot resistance and assessed the feasibility of obtaining resistant varieties or hybrids through breeding (5, 30). In most studies, resistance is controlled by dominant and additive genes and there is a significant general combining ability (GCA) effect (5). Thus, Fusarium stalk rot resistance is attributed to many loci. The best presently available source of Fusarium stalk rot resistance is SC599 (5). Resistance in test cross hybrids developed from this line was excellent, and both general and specific combining ability were important. Teferra *et al.* (41) confirmed Fusarium stalk rot resistance in SC599 and in its hybrid derivatives in multilocation experiments conducted in Kansas. SC599 was the only dominant source of Fusarium stalk rot resistance identified in these experiments.

Bramel-Cox and Claflin (4) found a positive correlation between resistance to charcoal rot and Fusarium stalk rot in half-sib progeny derived from KP8B, suggesting some overlap in the resistance mechanism for these pathogens. In contrast, combining ability studies of the best sources of charcoal rot resistance, including SC 33 and SC 35, for sensitivity to Fusarium infection indicated that these two sources of charcoal rot resistance were highly susceptible to *Fusarium* cultures (41). The correlation reported by Bramel-Cox and Claflin (4) may be explained by the fact that SC 599 and its derivatives constituted a significant proportion of the germplasm base of the sorghum population used in their experiments. This source of resistance appears to confer resistance to both charcoal rot and Fusarium stalk rot.

Conclusions

Considerable effort has focused on breeding for root and stalk rot resistance in sorghum. Sources of genetic resistance have been identified for the two most important stalk and root rot pathogens, *M. phaseolina* and *F. thapsinum*, and these resistance genes have been deployed in some commercial sorghum varieties and hybrids. The complexity of these resistance traits, however, has slowed crop improvement efforts. Ongoing research efforts have focused on mapping and identifying resistance genes to improve the efficiency of crop improvement through marker-assisted selection. If these genes are identified and cloned, future manipulation of these resistance traits in sorghum and other crops could occur through deployment as transgenes.

References

1. **Anahosur, K. H., and S. H. Patil.** 1981. Date of planting and incidence of charcoal rot of sorghum. *Sorg. Newsl.* **24:** 118-119.
2. **Borrell, A. K., G. L. Hammer, and R. G. Henzell.** 2000. Does maintaining green leaf area in sorghum improve yield under drought? I. Leaf growth and senescence. *Crop Sci.* **40:** 1026-1037.
3. **Borrell, A. K., G. L. Hammer, and R. G. Henzell.** 2000. Does maintaining green leaf area in sorghum improve yield under drought? II. Dry matter production and yield. *Crop Sci.* **40:** 1037-1048.
4. **Bramel-Cox, P. J., and L. E. Claflin.** 1989. Selection for resistance to *Macrophomina phaseolina* and *Fusarium moniliforme* in sorghum. *Crop Sci.* **29:** 1468-1472.
5. **Bramel-Cox, P. J., J. S. Stein, D. M. Rodgers, and L. E. Claflin.** 1988. Inheritance of resistance to *Macrophomina phaseolina* (Tassi) Goid and *Fusarium moniliforme* Sheldon in sorghum. *Crop Sci.* **28:** 37-40.
6. **Crasta, O. R., W. W. Xu, D. T. Rosenow, J. Mullet, and H. T. Nguyen.** 1999. Mapping of post-flowering drought resistance traits in grain sorghum: Association between QTLs influencing premature senescence and maturity. *Mol. Gen. Genet.* **262:** 579-588.
7. **Dodd, J. L.** 1980. The photosynthetic stress-translocation balance concept of sorghum stalk rots, p. 300-305. *In* R. J. Williams, R. A. Frederiksen, L. K. Mughogho, and G. D. Bengston (eds.), *Sorghum Diseases: A World Review.* ICRISAT, Patancheru (A.P.), India.
8. **Doupnik, B., M. G. Boosalis, G. Wicks, and D. Smika.** 1975. Ecofallow reduces stalk rot in grain sorghum. *Phytopathology* **65:** 1021-1022.
9. **Duncan, R. R.** 1983. Anthracnose-Fusarium disease complex on sorghum in southeastern USA. *Sorg. Newsl.* **26:** 121-122.
10. **Duncan R. R., A. J. Bockholt, and F. R. Miller.** 1981. Descriptive comparison of senescent and non-senescent sorghum genotypes. *Agron. J.* **73:** 849-853.
11. **Edmunds, L. K.** 1964. Combined relation of plant maturity, temperature and soil moisture to charcoal stalk rot development in grain sorghum. *Phytopathology* **54:** 514-517.
12. **Esechie, H. E., J. W. Maranville, and W. M. Ross.** 1977. Relationship of stalk morphology and chemical composition to lodging resistance in sorghum. *Crop Sci.* **17:** 609-612.
13. **Flett, B. C.** 1996. Crop rotation and tillage effects on yield and the incidence of root and stalk rot in sorghum (*Sorghum bicolor*). *S. Afr. J. Plant and Soil* **13:** 136-138.
14. **Frederiksen, R. A., and G. N. Odvody, eds.** 2000. *Compendium of Sorghum Diseases*, 2nd ed. APS Press, St. Paul, Minnesota.
15. **Frederiksen, R. A., D. T. Rosenow, G. L. Teetes, and J. W. Johnson.** 1976. Disease and insect ratings. *Sorg. Newsl.* **19:** 118-123.
16. **Frowd, J. A.** 1980. Sorghum stalk rots in West Africa, p. 322-324. *In* R. J. Williams, R. A. Frederiksen, L. K. Mughogho, and G. D. Bengston (eds.), *Sorghum Diseases: A World Review.* ICRISAT, Patancheru (A.P.), India.
17. **Gebrekidan, B., and Y. Kebede.** 1979. Highlights of sorghum improvement in Ethiopia. *Sorg. Newsl.* **22:** 2-5.
18. **Henzell, R. G., R. L. Dodman, A. A. Done, R. L. Brengman, and P. E. Mayers.** 1984. Lodging, stalk rot, and root rot in sorghum in Australia, p. 225-235. *In* L. K. Mughogho (ed.), *Sorghum Root and Stalk Diseases, a Critical Review.* ICRISAT, Patancheru (A.P.), India.

19. **Jardine, D. J., and J. F. Leslie.** 1992. Aggressiveness of *Gibberella fujikuroi* (*Fusarium moniliforme*) isolates to grain sorghum under greenhouse conditions. *Plant Dis.* **76:** 897-900.
20. **Jordan, W. R., R. B. Clark, and N. Seetharama.** 1984. The role of edaphic factors in disease development, p. 209-217. *In* L. K. Mughogho and G. Rosenberg (eds.), *Sorghum Root and Stalk Rots: A Critical Review.* ICRISAT, Patancheru (A.P.), India.
21. **Katsanos, R. A., and A. J. Pappelis.** 1965. Seasonal trends in density and cell death in sorghum stalk tissue. *Phytopathology* **55:** 97-99.
22. **Klittich, C. J. R., J. F. Leslie, P. E. Nelson, and W. F. O. Marasas.** 1997. *Fusarium thapsinum* (*Gibberella thapsina*): A new species in section *Liseola* from sorghum. *Mycologia* **89:** 643-652.
23. **Leslie, J. F.** 1995. *Gibberella fujikuroi*: Available populations and variable traits. *Can. J. Bot.* **73**: S282-S291.
24. **Leslie, J. F., and W. F. O. Marasas.** 2002. Will the real *Fusarium moniliforme* please stand up!, p. 201-209. *In* J. F. Leslie (ed.), *Sorghum and Millets Diseases.* Iowa State Press, Ames, Iowa.
25. **Leslie, J. F., C. A. S. Pearson, P. E. Nelson, and T. A. Toussoun.** 1990. *Fusarium* species from corn, sorghum, and soybean fields in the central and eastern United States. *Phytopathology* **80:** 343-350.
26. **Marasas, W. F. O., J. P. Rheeder, S. C. Lamprecht, K. A. Zeller, and J. F. Leslie.** 2001. *Fusarium andiyazi* sp. nov., a new species from sorghum. *Mycologia* **93:** 1203-1210.
27. **Odvody, G. N., and L. D. Dunkle.** 1979. Charcoal stalk rot of sorghum: Effect of environment on host parasite relations. *Phytopathology* **69:** 250-254.
28. **Oosterom, E. J., R. Jaychandran, and F. R. Bidinger.** 1996. Diallel analysis of stay green trait and its components in sorghum. *Crop Sci.* **36:** 549-555.
29. **Pande, S., L. K. Mughogho, N. Seetharama, and R. I. Karunakar.** 1989. Effects of nitrogen, plant density, moisture stress and artificial inoculation with *Macrophomina phaseolina* on charcoal rot incidence in grain sorghum. *Phytopathology* **126:** 343-352.
30. **Patil, R. C., N. B. Deshmane, A. P. Chaven, and A. R. Bangar.** 1980. Testing of sorghum varieties for reaction to charcoal rot. *Sorg. Newsl.* **23:** 125.
31. **Reed, J. E., J. E. Partridge, and P. T. Nordquist.** 1983. Fungal colonization of stalks and roots of grain sorghum during the growing season. *Plant Dis.* **64:** 417-420.
32. **Rosenow, D. T.** 1977. Breeding for lodging resistance in sorghum. *Proc. Annu. Corn Sorg. Res. Conf. Am. Seed Trade Assoc.* **32:** 171-185.
33. **Rosenow, D. T.** 1984. Breeding for resistance to root and stalk rots in Texas, p. 209-217. *In* L. K. Mughogho and G. Rosenberg (eds.), *Sorghum Root and Stalk Rots: A Critical Review.* ICRISAT, Patancheru (A.P.), India.
34. **Rosenow, D. T., J. E. Quisenberry, C. W. Wendt, and L. E. Clark.** 1983. Drought tolerant sorghum and cotton germplasm. *Agric. Water Manag.* **7:** 207-222.
35. **Rosenow, D. T., C. A. Woodfin, L. E. Clark, and J. W. Sij.** 1999. Drought resistance in exotic sorghums. *1999 Agron. Abstr.*: 166.
36. **Rosenow, D. T., C. A. Woodfin, G. Gebeyehu, M. S. Beder, A. N. McCosker, and H. T. Nguyen.** 1994. Evaluation of parental sorghum lines for staygreen in F_1 hybrids. *1994 Agron. Abstr.*: 130.
37. **Seetharama, N., F. R. Bidinger, K. N. Rao, K. S. Gill, and M. Mulgund.** 1987. Effect of pattern and severity of moisture deficit stress on stalk rot incidence in sorghum. I. Use of line source irrigation technique, and the effect of time of inoculation. *Field Crops Res.* **15:** 289-308.
38. **Seetharama, N., R. C. Sachan, A. K. Huda, K. S. Gill, K. N. Rao, and F. R. Bidinger.** 1991. Effect of pattern and severity of moisture deficit stress on stalk rot incidence in sorghum. II. Effect of source/sink relationships. *Field Crops Res.* **26:** 355-374.
39. **Subudhi, P. K., D. T. Rosenow, and H. T. Nguyen.** 2000. Quantitative trait loci for the stay green trait in sorghum (*Sorghum bicolor* L. Moench): Consistency across genetic backgrounds and environments. *Theor. Appl. Genet.* **101:** 733-741.
40. **Tao, Y. Z., R. G. Henzell, D. R. Jordan, and C. L. McIntyre.** 2000. Identification of genomic regions associated with stay green in sorghum by testing RILs in multiple environments. *Theor. Appl. Genet.* **100:** 1225-1232.
41. **Teferra, T. T., M. R. Tuinstra, N. W. Ochanda, and L. E. Claflin**. 2001. Analysis of genetic diversity for Fusarium stalk rot resistance in sorghum. *2001 Agron. Abstr.* (CD ROM computer file).
42. **Tenkuano, A., F. R. Miller, R. A. Frederiksen, and D. T. Rosenow.** 1993. Genetics of nonsenescence and charcoal rot resistance in sorghum. *Theor. Appl. Genet.* **85:** 644-648.
43. **Thompson, D. L.** 1963. Stalk strength of corn as measured by crushing strength and rind thickness. *Crop Sci.* **3:** 323-329
44. **Trimboli, D. S., and L. W. Burgess.** 1983. Reproduction of *Fusarium moniliforme* basal stalk rot and root rot of grain sorghum in the greenhouse. *Plant Dis.* **67:** 891-894.
45. **Tuinstra, M. R., E. M. Grote, P. B. Goldsbrough, and G. Ejeta.** 1997. Genetic analysis of post-flowering drought tolerance and components of grain development in sorghum. *Mol. Breeding* **3:** 439-448.
46. **Walulu, R. S., D. T. Rosenow, D. B. Wester, and H. T. Nguyen.** 1994. Inheritance of stay green trait in sorghum. *Crop Sci.* **34:** 970-972.
47. **Wanous, M. K., F. R. Miller, and D. T. Rosenow.** 1991. Evaluation of visual rating scales for green leaf retention in sorghum. *Crop Sci.* **31:** 1691-1694.
48. **Xu, W., D. T. Rosenow, and H. T. Nguyen.** 2000. Stay-green trait in sorghum: relationship between visual ration and leaf chlorophyll concentration. *Plant Breeding* **119:** 365-367.
49. **Xu, W, P. K. Subudihi, O. R. Crasta, D. T. Rosenow, J. E. Mullet, and H. T. Nguyen.** 2000. Molecular mapping of QTLs conferring stay-green in grain sorghum (*Sorghum bicolor* L. Moench). *Genome* **43:** 461-469.
50. **Zummo, N.** 1980. Fusarium disease complex of sorghum in West Africa, p. 297-299. *In* R. J. Williams, R. A. Frederiksen, L. K. Mughogho, and G. D. Bengston (eds.), *Sorghum Diseases: A World Review.* ICRISAT, Patancheru (A.P.), India.

49

Antifungal Proteins and Other Mechanisms in the Control of Sorghum Stalk Rot and Grain Mold

Ralph D. Waniska, R. T. Venkatesha, A. Chandrashekar, S. Krishnaveni, F. P. Bejosano, Jeoung-Mee Jeoung, Jayaraj Jayaraman, Subbaratnam Muthukrishnan, and George H. Liang

Abstract

We review progress in research on antifungal proteins and other mechanisms that provide the biochemical basis for host plant resistance to stalk rot and grain molds. Stalk rot, caused by *Fusarium* spp., leads to substantial yield loss due to poor grain filling and/or lodging. Transgenic sorghum plants that express high levels of chitinase have less stalk rot when exposed to conidia of *Fusarium thapsinum*. Grain mold of sorghum is associated with warm humid environments and results from colonization of the caryopsis by several fungi, *e.g.*, *F. thapsinum*, *Curvularia lunata*, and *Alternaria alternata*. Biochemical mechanisms that may contribute to grain mold resistance include tannins, phenolic compounds, red pericarp, hard endosperm, and antifungal proteins. The effects attributed to these characters appear to be additive, and pyramiding of genes is a feasible approach to limit grain deterioration.

Fungal Root and Stalk Rot of Sorghum

Fungi cause many severe diseases of sorghum [*Sorghum bicolor* (L.) Moench], including root and stalk rots that result from infection by *Fusarium moniliforme sensu lato*, *F. thapsinum*, or *Colletotrichum graminicola*; seedling diseases induced by *Pythium* sp.; foliar diseases, *e.g.* sooty stripe, incited by *Ramulispora sorghi*; and head smut caused by *Sporisorium reilianum*. In Kansas, *Fusarium* stalk rot is considered to be the most devastating disease, and losses attributed to this disease exceeded $80 million in 1996 (D. J. Jardine, personal communication). Yield losses due to stalk rot result from lower kernel weight, weakened peduncles, plant lodging, and stalk breakage. *Fusarium* spp. are spread by wind, rain, machinery, and insects. The fungus may penetrate the plant directly or enter via natural openings, *e.g.*, stomata. The fungus survives as spores or resting structures in soil or plant debris, or as hyphae within living plants.

Pathogenesis-related proteins. Pathogenesis-related (PR) proteins are inducible proteins that restrict the spread of the pathogen when an incompatible interaction occurs and that form the basis of systemic acquired resistance. Most research with PR proteins has been carried out with leaves and stalks in which protein synthesis is induced. Many PR proteins have *in vitro* antifungal activity and specific subcellular targets in the pathogen. For example, chitinases and β-1,3-glucanases attack fun-gal cell walls, and thaumatin-like proteins affect the permeability of fungal membranes (66). Genes or cDNAs for these proteins have been isolated from the cereals rice, barley, and wheat (34, 36, 53). Constitutive overexpression of these genes in transgenic rice and wheat plants (66) results in improved resistance to fungal pathogens (15, 54).

A gene coding for a rice chitinase was incorporated into an elite sorghum inbred line, Tx430, by using a biolistic transformation protocol, and six primary transgenic plants were shown to contain the transgene via Southern blot analysis (105). The chitinase transgene was inherited in a Mendelian manner by the progeny of these primary transgenic plants and expressed in the first, second, and third generation progeny of the primary transgenic plants

(T_1, T_2, and T_3 generations). One transgenic line that did not segregate for the transgene also was identified. Plants from the T_2 and T_3 generations were tested for resistance to infection by inoculating the stalks of mature plants with 1.0 ml of a conidial suspension of *F. thapsinum* (10^6 conidia/ml). Stalk rot symptoms were significantly reduced in transgenic sorghum plants that expressed relatively high levels of rice chitinase (0.05-0.1% of total protein), while plants with little or no expression of the rice chitinase suffered normal disease progression (47). Thus, constitutive overexpression of rice chitinase in sorghum appears to protect against this stalk rot pathogen.

Fungal Caryopsis Diseases of Sorghum

Grain mold of sorghum results from colonization by fungi, *e.g.*, *F. thapsinum*, *C. lunata*, and *A. alternata*, of developing grain, and is associated with warm, humid environments during caryopsis development. The major colonizing fungi also produce mycotoxins, which are harmful to humans and livestock (1, 9, 26). Fungal growth initially causes surface discoloration and then breaks down the components of the grain. These changes decrease milling and processing yields and the quality of sorghum for feed or food. Breeding efforts to obtain host plant resistance have been only partially successful (88), while control of grain mold with fungicides and crop management strategies is either too expensive or too complex for routine practical applications.

Plant traits, *e.g.*, panicle shape, plant height, and glume structure, are associated with grain mold resistance (14, 73), as are caryopsis traits, *e.g.*, endosperm hardness, a pigmented testa layer, and red pericarp color (5, 24, 29, 38). Although many sorghums with tannins and higher levels of phenol-based pigments are resistant to molding, these compounds cause dark colors, astringency, and/or decreased nutritional value in foods or feeds (22, 32). Accordingly, sorghums for food use are either grown in environments that are not conducive to grain mold deterioration, with a correspondingly lower yield, or cultivars resistant to grain mold are selected which yield less desirable food. Hence, grain mold resistance is necessary for high-yielding food-type sorghums.

Sorghum grain has several defenses that reduce infestation by fungi including: caryopsis structure, phenolic compounds, and antifungal proteins.

Fungal invasion. Forbes *et al.* (26) reviewed colonization events in mold-resistant and mold-susceptible cultivars. During the initial period of flower invasion, *C. lunata* can infect the apical part of the ovary wall from the colonized lemma, palaea, lodicules, filaments, pollen grain, and the decaying style (26). Hyphae penetrate the pericarp and ramify through the cross and tube cells within 5-10 days, but the pericarp generally is not colonized. The placental sac provides a niche for fungal growth that may carry over to the endosperm and the embryo as well. *F. thapsinum* appears later and grows beneath the pericarp and then attacks the floury endosperm. The significance of these differences in infection patterns is currently being studied.

Mycotoxins. Mycotoxins are less a problem in sorghum than in maize. Lower levels of fumonisin occur in sorghum (0.7-36 ppm) than in maize (5-4000 ppm) (67, 99). Normal sorghum used to prepare Indian foods contains very little fumonisin (96). Leslie *et al.* (52) reported that *F. thapsinum*, which dominates on sorghum, produces little or no fumonisin, while *Fusarium verticillioides*, which dominates on maize, produces much more. Klittich *et al.* (46) observed that *F. thapsinum* does not produce fumonisin.

Aflatoxins are not found at significant levels in sorghum brought directly in from the field (G. N. Odvody, personal communication). High levels of mycotoxins, including aflatoxins, are associated with high-moisture storage of sorghum and other cereals. Maize may have dangerous levels of aflatoxins even though sorghum cultivated in similar drought-prone environments is aflatoxin free. Some sorghums, as well as wheat and maize, may contain another type of toxin, deoxynivalenol, which is an emetic and toxic to rats (72, 77, 97). *Alternaria* spp. also produce toxins, which can be confused with metabolites from other fungi (83). Ergosterol levels, corresponding to nonspecific mold colonization of sorghum, have been used to monitor grain molding, but they have not been linked to animal health problems (82).

Sorghum ergot, *Claviceps africana*, has recently infested sorghum in the Western Hemisphere (37). The sclerotia produced by this fungus on sorghum contain alkaloids, such as dihydroergosine (10). A diet containing 5% sclerotia (70 mg alkaloids/kg diet) decreased young Duroc pig growth by 30% via decreased feed intake and feed conversion (10). A nutrient balanced diet containing 235 mg alkaloids/kg of the diet caused no or inconsistent affects on weight gain, feed efficiency, and liver weight in poultry, and did not affect mortality or produce obvious symptoms or ergot toxicity (6).

Grain Structure and Development in Sorghum

The structure of the mature grain of sorghum, like all cereals, comprises the triploid storage endosperm and diploid zygotic embryo surrounded by the pericarp, a maternal tissue, and testa. The testa in some sorghum species is pigmented and contains condensed tannins. The thickness of the testa varies from 8-40 μm (23), with the thickest area being below the style and the thinnest on the side of the kernel. Some sorghums containing tannins have a high degree of resistance to molds (64), but most sorghums are virtually tannin free (71). The pigmented testa appears as a dark layer between the light endosperm and the pericarp when the caryopsis is scraped to remove the pericarp.

The endosperm, the bulk of the caryopsis, is inside the pericarp and testa (62, 86). The cells in the first layer, the rectangular aleurone cells, have thick walls and contain oil and protein bodies. The peripheral area is inside the aleurone cells and comprises several layers of dense cells that are rich in protein but contain only small starch granules. Inside this area is the corneous (horny) area, which contains angular starch granules embedded in a continuous protein matrix. In the center is a white, opaque, floury area in which protein is packed loosely with air spaces and starch granules are more spherical in shape (84).

McDonough and Rooney (62) reported that the pericarp is completely formed 6-9 days after anthesis and that thereafter, it begins to compress. Cells in the testa layer were apparent 3-6 days after anthesis, but the aleurone layer took longer, 6-12 days after anthesis, to develop. Ovary walls contain simple and compound starch granules at anthesis. Simple starch granules and protein bodies begin to develop in endosperm cells by 3-6 days after anthesis. Protein bodies at this stage are cov-ered by thin filamentous webbing, which later develops into a distinctive protein matrix. Shull *et al.* (86) noted that grains that are more resistant to fungi and those that produce hard grains deposit protein much earlier than do grains that are destined to be floury.

Grain texture and resistance to grain molds. The relative proportions of the outer translucent layer (also called glassy, horny, vitreous, or corneous) and the inner opaque, white area (also called soft or floury) vary among cultivars (44). Grain hardness is related to the relative amounts of corneous and floury endosperm. Endosperm hardness attenuates the growth of grain mold (5, 24, 29, 38, 92). Molds more quickly degrade the endosperm structure of susceptible cultivars than they do that of resistant cultivars, even though both lines may have similar endosperm hardness in a mold-free environment (76).

Studies of mutant lines of maize provide clues to the resistance mechanism. The *opaque-2* mutation increases lysine but has a soft texture, reduced levels of α-zeins and ribosome inactivating protein (RIP) (8), and is very susceptible to infection by molds (85). A DNA-binding

TABLE 49-1. Categorization of cultivars, plant maturity, caryopsis properties, and amounts of sormatin and chitinase in sorghums that vary in resistance to grain mold in 1997.

Cultivar	AFP[a]	GMR[b]	Mature[c]	Pericarp	Testa Pigment	Spreader[d]	Endosperm[e]	Sormatin[f]	Chitinase[g]	Molding[h]
SC719-11E	Low	High	Early	Red	Yes	Yes	Medium	13	100	1.4
Malisor 84-7	High	Mod.	Medium	White	No	No	High	110	400	1.2
R9025	High	High	Medium	Red	No	No	Low	83	170	1.2
Sureno	High	Mod.	Medium	White	No	No	High	50	390	1.2
Hegari* Dobbs	High	Mod.	Medium	White	Yes	No	Medium	220	560	1.8
IS2319	Low	Low	Late	White	Yes	No	Low	83	460	5.0
E35-1	Low	Low	Late	White	No	No	Medium	17	270	4.8
RTX2536	Low	Low	Medium	White	No	No	Low	32	250	3.8
RTX430	High	Low	Medium	White	No	No	Low	150	530	3.0
BTX638	High	Low	Medium	Red	No	No	Medium	270	450	1.6

[a]Antifungal protein content and grain mold resistance (after 81).
[b]Grain mold resistance (after 81).
[c]Early maturity, < 60 days after planting; medium maturity, 61-80 days after planting; late maturity, > 81 days after planting.
[d]When dominant spreader gene (*S*) is present, tannins accumulate in pericarp and testa.
[e]Hard, < 30 % abraded during decortication; medium, 30-50% abraded; soft, > 51% abraded.
[f]μg/caryopsis. Coefficient of variation varies from 11-23%. Values are averages of duplicate analysis of replicated treatments.
[g]μg/caryopsis. Coefficient of variation varies from 11-23%. Values are averages of duplicate analysis of replicated treatments.
[h]Mold rating (1 = no mold to 5 = > 50% of surface molded; 26) at 50 days after anthesis at College Station, Texas.

protein encoded at *opaque-2* (56, 78, 79) regulates the levels of both α-zein and RIP. The homologous RIP from barley has antifungal properties (50), while Dowd *et al.* (21) suggest that the maize RIP may be insecticidal. Decreased RIP levels may increase susceptibility to molds. The hard-textured types of *opaque-2* maize contain the same lower levels of α-zein, as do the unmodified *opaque-2* lines while their γ-zein levels are increased (27, 100). The corneous endosperm of sorghum and normal maize contain similar levels of γ-zein (70).

The corneous endosperm of sorghum is enriched in kafirins, especially γ-kafirins (49, 60, 90). The γ-kafirins contain more cysteine and form extensive intrachain disulfide bonds, which may contribute both to hard texture and to resistance to fungal infection (59, 61). The prolamins in the protein bodies are remarkably resilient to proteolysis by a fungal protease until their disulfide bonds are reduced (Mazhar and Chandrashekar, *unpublished*).

Cell wall composition also differs between the corneous and floury endosperm and between the hard and the soft grains (42, 43). Since endosperm structure is modified during fungal colonization of the caryopsis, lines with altered cell wall structure and composition might have increased resistance to grain molds.

A combination of factors probably is responsible for grain mold resistance (12). Sorghums with red pericarp and/or tannins (Table 49-1) that lack a hard endosperm may exhibit resistance, but white pericarp sorghums require a hard endosperm for observable resistance. Otherwise, sorghums with low and medium endosperm hardness incur the most deterioration. Harder grain also has lower mold ratings when plants are stressed by sprinkling or by inoculation with fungal pathogens (12).

Tannins and phenolic compounds. Most sorghums do not contain tannins, but all sorghums contain phenols and most contain flavonoids. Sorghums are classified as type I, no tannins; type II, tannins in pigmented testa; or type III, tannins in pigmented testa and pericarp. Types II and III sorghums have dominant *B1* and *B2* genes and a pigmented testa (seed coat). Type III sorghums have a dominant spreader (*S*) gene and higher levels of tannins than do type II sorghums (84).

Phenolic compounds are divided into three major categories: phenolic acids, flavonoids, and tannins (17, 32). Phenolic acids are derivatives of benzoic or cinnamic acids. Flavonoids consist of two units: a C6-C3 fragment from cinnamic acid and a C6 fragment from malonyl-CoA. The major groups of flavonoids in sorghum are the flavans, flavan-3-en-3-ols with a double bond between C3 and C4, and the C3-hydroxylated anthocyanidins. Tannins are polymers of 5-7 flavan-3-ol units (catechin) linked through acid labile carbon-carbon bonds (13, 32, 63). Tannin sorghums contain proanthocyanidins as part of their phenolic compounds but do not contain tannic acid or hydrolyzable tannins. Phenolic acids and compounds increase during caryopsis development with a maximum at physiological maturity and a decrease afterwards (102).

Assabgui *et al.* (4) reported a correlation ($r = 0.68$, $p = 0.002$, $n = 57$) between the content of ferulic acid in maize kernels with resistance to *F. graminearum*, while Waniska *et al.* (102) found higher levels of *p*-coumaric acid in some white pericarp, nontannin sorghums that were susceptible to molding. The ability to convert *p*-coumaric acid to ferulic acid may be reduced or absent in susceptible cultivars. Free or bound phenolic acids in control and inoculated caryopses could be used to classify sorghums into resistant, intermediate, and susceptible groups (75).

Sorghum cultivars resistant to fungal attack contain both a greater variety and larger amounts of free phenolic acids, especially tannin sorghums (31, 102). Red pericarp sorghums are more resistant to mold than are white pericarp sorghums and contain more flavon-4-ols (7, 24, 26, 38, 39, 64). A cause-and-effect relationship has not been established for flavan-4-ols.

No phytoalexin has been detected in the caryopsis of sorghum. Lo *et al.* (55) isolated cDNA clones from a sorghum mesocotyl library after infection with *Colletotrichum sublineolum*. The level of chalcone synthase, an enzyme that is involved in the synthesis of the non-caryopsis phytoalexins, was greater when mesocotyls were treated with *Cochliobolus heterostophus*, which does not infect sorghum, than with *C. sublineolum*, which does (55).

Antifungal Proteins in Sorghum Caryopsis

Several proteins are constitutively expressed in developing cereal seeds and have antifungal properties either *in vivo* or *in vitro*. Sorghum grain has not been studied in as much detail for antifungal proteins as have more widely cultivated cereals (19, 35, 36, 58, 74, 98, 103). However, homologs of many of these proteins, particularly those present in maize, probably also are present in sorghum. Several types of bioactive and antifungal proteins have been identified and characterized in sorghum, maize, and millet (Table 49-2).

Sunitha and Chandrashekhar (89, 90) identified three proteins of 18, 26 and 30 kDa that affected hyphal growth of *F. moniliforme sensu lato*. The 18-kDa protein resulted in sloughing of cell wall polysaccharides, while the other two proteins resulted in leakage of cytoplasmic

TABLE 49-2. Collated information on antifungal proteins of sorghum, maize, and millets.

Species	Protein Size/ Isoform	Fungal Species Inhibited	Effective/ Inhibitory dose	Reference
Chitinase				
Sorghum	24, 28, and 33 kDa	*Trichoderma viridae*	1 μg/ml	48
		Fusarium moniliforme sensu lato	5 μg/ml	
		Rhizoctonia solani	1 μg/ml	
Maize	pI 5.9/4.7	ND[a]	ND	19, 35, 103
Pearl Millet[b]	24 kDa	*Trichoderma reesi*	250-800 μg/ml	40
	pI 9.8	*F. moniliforme sensu lato*		
Job's Tears[c]	26.4 kDa	ND	ND	2
	pI 8.5-9.0			
Ribosome Inactivating Protein				
Maize	30 kDa	*In vitro* protein synthesis	0.7 μg/ml	8, 33
Sorghum	32-34 kDa	*In vitro* protein synthesis	ND	33
Thaumatin-Like Protein				
Maize	22 kDa	*T. reesi, R. solani*	0.1 μg/ml	35, 74, 98
	pI 9.1		10 μg/ml	58
Sorghum	26 and 30 kDa	*F. moniliforme sensu lato*	6.8 μg/ml	89
		Fusarium oxysporum		
		Curvularia lunata		
General Proteins				
Sorghum	12, 16, and 20 kDa	*Aspergillus flavus*	15 μg/ml	28
	18 kDa	*F. moniliforme sensu lato*	15 μg/ml	89, 90

[a]Not determined.
[b]The pearl millet protein is a cysteine proteinase inhibitor.
[c]The Job's tears protein is an α-amylase/chitinase double inhibitor protein (2).

contents. The authors concluded that the 18-kDa protein could be an enzyme acting on cell walls, while the 26 and 30-kDa components could be related to permatins. Antifungal protein levels were highest in hard endosperm, grain mold resistant sorghums (49, 90).

Vigers *et al.* (98) and Seetharaman *et al.* (81) identified positively charged, water-soluble proteins belonging to the permeatin (which they called sormatin), chitinase, β-glucanase, and RIP classes. The recoverable levels of several antifungal proteins increased until physiological maturity, then decreased as the grain dried (81, 91).

The average level of antifungal proteins in physiologically mature sorghum was 7-9 μg/caryopsis (80, 89), with 6.8-15 μg/dose required to inhibit hyphal growth by *Fusarium* sp. (89). An antifungal protein fraction mixture (3.6 μg/dose) containing sormatin, chitinase, β-glucanase, and RIP inhibited spore germination of *F. thapsinum*, *C. lunata* and *A. flavus* (80). Hyphal rupture at the growing tips was observed for *F. thapsinum* at 7.0 μg/dose and at 7-36 μg/dose for *C. lunata*, but was not observed for *A. flavus*. The amount of antifungal proteins in physiologically mature sorghum caryopses could inhibit germination and growth of *F. thapsinum* and perhaps *C. lunata*, depending upon how the proteins are concentrated in the pericarp during imbibition (80).

Ribosome inactivating proteins. Walsh *et al.* (101) showed that the maize RIP was synthesized as a proprotein with a region excised during germination. Hey *et al.* (33) demonstrated that the connecting hinge between two domains of the maize RIP was removed during activation and was involved in the inactivation of ribosomal RNA. The location of RIP in maize is in the aleurone and the scuetellum (30). Sorghum RIP cross reacts with the maize RIP antibody, and both RIPs have similar molecular weights (95). The proprotein was deposited in the sorghum endosperm during the later part of caryopsis development, and the proprotein was split as it was in maize during extraction. Two cross-reacting RIP bands of about 30 kDa varied in intensity, depending upon the age of the caryopsis and the sorghum line (95). Seetharaman *et al.* (81) observed that the RIP was more easily extracted from caryopses at 15 days after anthesis.

Thionins. A new family of low molecular weight proteins from sorghum caryopses was identified that inhibits insect α-amylase (11). These proteins, subsequently named γ-thionins and defensins, each contain 47 amino acid residues and have four disulfide bonds (68).

Chitinase and β-glucanase. Several chitinases (three between 21 and 24 kDa and one of 28 kDa) and one β-glucanase (30 kDa) occur in sorghum (20). Three chitinases were identified from sorghum caryopses (24, 28, and 33 kDa) that inhibited the growth of *Trichoderma viride* and *F. thapsinum* (47). The maximum accumulation of chitinase occurred 30 days after anthesis (81; Venkatesha *et al., unpublished*). The level of the 28-kDa chitinase in the caryopsis decreases 24 hours after imbibition, during germination, but these proteins were detectable for at least another 10 days. Several additional chitinase isoforms appeared during germination (81; Venkatesha *et al., unpublished*). Putative chitinase clones have been derived from genomic DNA and from a cDNA library of developing caryopses.

Variability and relative importance of antifungal proteins. A significant inverse correlation between sormatin in caryopses at 30 days after anthesis and sorghum grain mold rating at harvest time was first reported by Seetharaman *et al.* (81) based on observations of 32 sorghum cultivars. Extractable antifungal proteins increased during caryopsis development and decreased prior to grain drying (81, 91).

Sorghum antifungal proteins leached from immature caryopses but were retained in the pericarp of mature caryopses during water imbibition (81). If antifungal proteins are mobile and "bound" to the pericarp during imbibition, then all of the antifungal proteins could be concentrated in less than 10% of the caryopsis. Such localization would increase antifungal protein concentrations about 10-fold, thereby increasing their antifungal potential. Moreover, increased quantities of these proteins were detected after treatment of flour with protease (89).

Variations in antifungal protein levels in the endosperm also were observed, with the largest amount of antifungal proteins localized to the corneous endosperm of resistant sorghums, which contain more protein bodies (90, 92, 93). Sorghums with harder endosperms contained fewer nonstarch polysaccharides, and these polysaccharides contained relatively more hexose and relatively fewer pentose residues (42, 43). Protein bodies and antifungal proteins appear to be concentrated near cell walls in the corneous endosperm.

The distribution of antifungal proteins within the caryopsis also varies by line (Waniska *et al., unpublished*). Sormatin was most common in the endosperm of Dorado (moderately resistant, hard endosperm sorghum) and in the pericarp of Tx2536 (susceptible, intermediate endosperm sorghum). For both sorghum accessions, chitinase was most common in the endosperm, and β-glucanase was common in the pericarp. Soaking seeds in water for 24 hours increased extractability of both chitinase and glucanase from the endosperm, germ, and pericarp of physiologically mature caryopses.

Induction of antifungal proteins. Sormatin-like proteins are induced after fungal infection (12, 91). Periodic sprinkling of panicles in the field, *e.g.*, from rain or artificial irrigation, similar to *in vitro* imbibition, decreased either the level or the extractability of antifungal proteins in susceptible sorghums at physiological maturity (76). Puncturing the germ induced chitinase synthesis in mature Dorado caryopses, but not in Tx2536 caryopses (Waniska *et al., unpublished*). Puncturing and soaking induced sormatin mobility and accumulation in caryopses of both Dorado and Tx2536.

Rodriguez-Herrera *et al.* (76) compared the levels of four antifungal proteins in eight mold-resistant and eight mold-susceptible lines derived from a susceptible × resistant cross, grown in eight environments over three years. Infection by grain mold fungi led to the induction and/or retention of more antifungal proteins in the resistant lines (Table 49-3), suggesting that the co-expression of multiple antifungal proteins is needed for resistance in lines with a nonpigmented testa. β-1,3-glucanase levels in the resistant lines increased by an average of 110%, while chitinase, sormatin, and RIP concentrations in resistant lines were 120-460% higher than in the grain mold susceptible lines.

Genetic markers associated with grain mold. In other cereals, biochemical or molecular markers that can be used to follow resistance in breeding programs have been identified. For example, Mingeot and Jacquemin (65) found polymorphism for a thaumatin-like protein, and Farris *et al.* (25) found that the oxalate oxidase and peroxidase genes were both good markers for tan spot

TABLE 49-3. Antifungal protein levels[a] (μg/g) in mature caryopses of eight grain-mold resistant (GMR) and eight grain-mold susceptible (GMS) lines grown in seven environments in 1995-1997. Resistant lines have more antifungal proteins than susceptible lines, except for glucanase.

Environment	Sormatin		Chitinase		Glucanase		RIP[b]	
	GMR	GMS	GMR	GMS	GMR	GMS	GMR	GMS
Molding								
95B - dry	13	1.3	230	59	200	21	91	40
95C - wet	20	1.9	230	90	170	90	230	37
96C - dry	23	12	360	190	110	52	200	43
96C - wet	49	2.1	110	7	50	0	150	12
97B - dry	6.4	2.0	42	14	6	0	200	16
97C - dry	32	20	220	170	260	230	260	27
97C - wet	11	2.0	240	110	54	0	230	66
Non-Molding								
96H - dry	30	15	190	130	250	320	250	180

[a]Coefficient of variation ranges from 9-20%. Values are averages of duplicate analysis of replicated treatments.
[b]Ribosome inhibiting protein.

resistance, that phenylalanine ammonium lyase gene and a thaumatin gene were linked to the resistance genes Lr27 and Lr31, and that disease resistance genes were clustered on chromosome 7BL in wheat. Mutations in resistance genes also may be responsible for differences between susceptible and resistant varieties. In sorghum, Klein *et al.* (45) identified five QTLs associated with grain mold, and work is in progress to map the genes for antifungal protein content in sorghum and to determine their location relative to mapped QTLs for grain mold resistance (45).

Resistance genes. Resistance (R) genes present in the host plant may be activated following infection by pathogenic fungi and lead to the production of phytoalexins and antifungal proteins. The relationship(s) between R genes and pathogens may be extremely specific, resulting in responses only to specific species or races of pathogens. Synthetic changes in the endoplasmic reticulum, prolamins, antifungal proteins, and the cell wall all may be coordinated by R-genes.

In recent years, R genes have been cloned from a number of plant species. R proteins share common features, the most prominent of which is the presence of leucine-rich repeats, which may be combined with protein kinase domains. Moreover, resistance genes often are found in clusters (3, 41), which has facilitated the isolation of more R genes.

Some R genes have been isolated and partially sequenced from sorghum and completely sequenced from other species, *e.g.*, lettuce (85), maize (18), potatoes (51), rice (57), soybeans (41, 104), and tomato (69). Conserved R gene sequences can be used to design PCR primers or oligonucleotides that can, in turn, be used to isolate part or all of these presumptive disease resistance genes from sorghum (16). Once the genes, or fragments of them, are available, their relationship to grain mold resistance can be determined by measuring their expression in developing seed and their linkage with known resistance loci and QTLs.

Studies of other host-pathogen systems have shown that overexpression of R genes may result in broad-spectrum resistance with high levels of antifungal proteins and other defense-related compounds (94). Transgenic plants have been made with antifungal proteins and with the kinases and leucine-rich repeat proteins that are involved in signal reception and transduction. Transgenic plants that carry specific antifungal proteins often are more resistant to particular fungi, while a more broad-spectrum resistance is expected when kinases are used. Overexpression of such genes in a developing panicle of sorghum when it is susceptible to infection by grain mold fungi could lead to increased resistance.

Future Prospects

At present, current sources of resistance provide only partial protection against stalk rot and grain mold fungi, the most important of which are caused by *Fusarium*, *Alternaria*, and *Curvularia* spp. Grain mold resistance can be based on a combination of characters (pigmented testa, phenolic compounds, red pericarp, endosperm texture, cell wall, and antifungal proteins) that act additively or synergistically. The expression of some of these characters may be under the control of the R genes. Some of the resistance mechanisms may reduce the uses for the grain produced, and there may be an economically important trade-off between disease resistance and quality.

Recommendations

- **Pigmented testa and red pericarp.** Since both red pericarp and tannins increase grain mold resistance, the utilization of red pericarp sorghums (grown on a tan plant) and tannin sorghums are reasonable short-term solutions to decrease grain molding. The chemical compounds in these sorghums, *e.g.*, non-pigmented phenols and tannins, responsible for the antifungal phenotype remain to be identified and their mode of action determined.
- **Grain hardness.** This trait corresponds to high levels of γ-kafirin. The biochemical basis for this effect is not understood, and the existence of modifier genes, *e.g.*, as in QPM, remains to be confirmed. The effect on grain mold resistance of more copies of γ-kafirin-encoding genes needs to be determined. In addition, hardness is currently associated with high levels of antifungal proteins, and this linkage needs to be understood and then, if possible, broken.
- **Antifungal proteins.** A link between antifungal proteins and resistance is known for both stalk rot and grain molding, although only partial resistance results. In the short term, selecting for high levels of characterized antifungal proteins, *e.g.*, sormatin, RIP, β-1,3-glucanase, and endochitinase, may help in-crease resistance levels. More information also is needed on endogenous sorghum antifungal proteins to identify new antifungal proteins, to determine the activity levels and synergistic interactions of known antifungal proteins, to determine the location and amounts of antifungal proteins in developing and mature tissues, to determine how synthesis is initiated, and to relate these characters to disease resistance.
- **Marker-assisted selection.** The development of molecular markers for the sorghum genome is essen-

tial for constructing genetic maps that will enable the selection of resistant plants and for the combination of the resistance phenotype with those for yield and end-use quality. Molecular markers also will enable the separation of the various aspects of resistance, *e.g.*, those due to hardness and those due to antifungal proteins. A more directed approach also could be adopted by using markers based on known antifungal proteins, *e.g.*, thaumatin, and putative R genes (based on other species) to identify putative resistant lines.

❖ **Transformation.** Transformation is an essential prerequisite for long-term improvement of sorghum, and efforts in this area need continued support. Transformants that express sormatin, β-1,3-glucanase, and endochitinase in the endosperm under the control of a strong endosperm-specific promoter are needed to enhance molding resistance. Other transformants that control the level, tissue specificity, and duration of antifungal protein expression in the florets, ovary walls, and glumes, should further enhance the resistance of sorghum to grain mold.

Acknowledgements

We thank the Texas and Kansas Agricultural Experiment Stations, private industry, INTSORMIL, and CFTRI for their long-term support, and the Director of CFTRI for permission to make this presentation.

References

1. **Anasari, A. A., and A. K. Shrivastava.** 1990. Natural occurrence of *Alternaria* mycotoxins in sorghum and ragi from North Bihar, India. *Food Add. Contam.* **7:** 815-820.
2. **Ary, M. B., M. Richardson, and P. R. Shewry.** 1989. Purification and characterization of an insect α-amylase inhibitor/endochitinase from seeds of Job's tears (*Coix lachryma-jobi*). *Bioch. Bioph. Acta* **999:** 260-266.
3. **Ashfield, T., J. R. Danzer, D. Held, D. Clayton, P. Keim, M. A. S. Maroof-Saghai, D. M. Webb, and R. W. Innes.** 1998. *RPg1* a soybean gene effective against races of bacterial blight maps to a cluster of previously identified disease resistance genes. *Theor. Appl. Gen.* **96:** 1013-1021.
4. **Assabgui, R. A., L. M. Reid, R. I. Hamilton, and T. Arnason.** 1993. Correlation of kernel (E)-ferulic acid content of maize with resistance to *F. graminearum*. *Phytopathology* **83:** 949-953.
5. **Audilakshmi, S., J. W. Stenhouse, T. P. Reddy, and M. V. R. Prasad.** 1999. Grain mould resistance and associated characters of sorghum genotypes. *Euphytica* **107:** 91-103.
6. **Bailey, C. A., J. J. Fazzino, M. S. Ziehr, M. Sattar, A. U. Haq, G. Odvody, and J. K. Porter.** 1999. Evaluation of sorghum ergot toxicity in broilers. *Poult. Sci.* **78:** 1391-1397.
7. **Bandyopadyay, R., L. K. D. Mughogho, and K. E. P. Rao.** 1988. Sources of resistance to sorghum grain molds. *Plant Dis.* **72:** 504-508.
8. **Bass, H. W., C. Webster, G. R. O. Brian, J. K. M. Roberts, and R. S. Boston.** 1992. A maize ribosome-inactivating protein is controlled by the transcriptional activator opaque-2. *Plant Cell* **4:** 225-234.
9. **Bhat, R. V., P. H. Shetty, R. P. Amruth, and R. V. Sudershan.** 1997. A foodborne disease outbreak due to consumption of moldy sorghum and maize containing fumonisin mycotoxin. *J. Toxicol. Clin. Toxicol.* **35:** 249-255.
10. **Blaney, B. J., J. S. Kopinski, M. H. Magee, R. A. McKenzie, G. W. Blight, R. Maryam, and J. A. Downing.** 2000. Blood prolactin depression in growing pigs fed sorghum ergot (*Claviceps africana*). *Austr. J. Agric. Res.* **51:** 785-791.
11. **Bloch, C., and M. Richardson.** 1991. A new family of small (5 kDa) protein inhibitors of insect α-amylase from seeds of sorghum (*Sorghum bicolor* L. Moench) has sequence homologies with wheat γ-purothionins. *FEBS Lett.* **379:** 101-104.
12. **Bueso, F. J., R. D. Waniska, W. L. Rooney, and F. P. Bejosano.** 2000. Activity of antifungal proteins against mold in sorghum caryopsis in the field. *J. Agric. Food Chem.* **48:** 810-816.
13. **Butler, L. G.** 1990. The nature and amelioration of the anti-nutritional effects of tannins in sorghum grain, p. 191-205. *In* E. T. Mertz, G. Ejeta, L. W. Rooney, R. Schaffert, and J. Yohe (eds.), *Sorghum Nutritional Quality*. INTSORMIL, Lincoln, Nebraska.
14. **Castor, L. L., and R. A. Frederiksen.** 1980. *Fusarium* and *Curvularia* grain mold in Texas, p. 93-102. *In* R. J. Williams, R. A. Frederiksen, L. K. Mughogho, and G. D. Bengston (eds.), *Sorghum Diseases: A World Review*. ICRISAT, Patancheru (A.P.), India.
15. **Chen, W. P., P. D. Chen, D. J. Liu, R. J. Kynast, B. Friebe, R. Velazhahan, S. Muthukrishnan, and B. S. Gill.** 1999. Development of wheat scab symptoms is delayed in transgenic wheat plants that constitutively express a rice thaumatin-like protein gene. *Theor. Appl. Genet.* **99:** 755-760.
16. **Chen, X. M., R. F. Line, and H. Leung.** 1998. Genome scanning for resistance-gene analogs in rice, barley and wheat by high-resolution electrophoresis. *Theor. Appl. Genet.* **97:** 345-355.
17. **Chung, K. T., T. Y. Wong, C. I. Wei, Y. W. Huang, and Y. Lin.** 1998. Tannins and human health: A review. *Crit. Rev. Food Sci. Nutr.* **38:** 421-464.
18. **Collins, N. C., C. A. Webb, A. Seah, J. G. Ellis, S. H. Hulbert, and A. Pryor.** 1998. The isolation and mapping of disease resistance gene analogs in maize. *Molec. Plant-Microbe Interact.* **11:** 968-978.
19. **Cordero, M. J., D. Raventos, and B. San Segundo.** 1994. Differential expression and induction of chitinases and 1-3 glucanases in response to fungal infection during germination of maize seeds. *Molec. Plant-Microbe Interact.* **7:** 23-30.
20. **Darnetty, J. F. Leslie, S. Muthukrishanan, M. Swegele, A. J. Vigers, and C. P. Selitrenikoff.** 1993. Variability in antifungal proteins in the grains of maize, sorghum and wheat. *Physiol. Plant.* **88:** 339-349.

21. **Dowd, P. F., A. D. Mehta, and R. S. Boston.** 1998. Relative toxicity of the maize endosperm ribosome-inactivating protein to insects. *J. Agric. Food Chem.* **46:** 3775-3779.
22. **Earp, C. F., C. A. Doherty, and L. W. Rooney.** 1983. Fluorescence microscopy of the pericarp, aleurone layer and endosperm cell-walls of 3 sorghum cultivars. *Cer. Chem.* **60:** 408-410.
23. **Earp, C. F., and L. W. Rooney.** 1982. Scanning electron microscopy of the pericarp and testa of several sorghum varieties. *Food Microstruc.* **1:** 125-134.
24. **Esele, J. P., R. A. Frederiksen, and F. R. Miller.** 1993. The association of genes controlling caryopsis traits with grain mold resistance in sorghum. *Phytopathology* **83:** 490-494.
25. **Farris, J. D., W. L. Li, D. J. Liu, P. D. Chen, and B. S. Gill.** 1999. Candidate gene analysis of quantitative disease resistance in wheat. *Theor. Appl. Genet.* **98:** 219-225.
26. **Forbes, G. A., R. Bandyopadhyay, and G. Garcia.** 1992. A review of sorghum grain mold, p. 253-264. *In* W. A. J. de Milliano, R. A. Frederiksen, and G. D. Bengston (eds.), *Sorghum and Millets Diseases: A Second World Review.* ICRISAT, Patancheru (A.P.), India.
27. **Geetha, K. B., C. R. Lending, M. A. Lopes, J. C. Wallace, and B. A. Larkins.** 1991. Opaque-2 modifiers increase γ-zein synthesis and alter its spatial distribution in maize endosperm. *Plant Cell* **3:** 1207-1219.
28. **Ghosh, M. M., and K. Ulaganathan.** 1996. Mature seeds of sorghum contain proteins toxic to aflatoxin producing *Aspergillus flavus*. *J. Stored Prod. Res.* **32:** 339-343.
29. **Glueck, J. A., and L. W. Rooney.** 1980. Chemistry and structure of grain in relation to mold resistance, p. 119-140. R. J. Williams, R. A. Frederiksen, L. K. Mughogho, and G. D. Bengston (eds.), *Sorghum Diseases: A World Review.* ICRISAT, Patancheru (A.P.), India.
30. **Guo, B. Z., T. E. Cleveland, R. L. Brown, N. W. Widstorm, R. E. Lynch, and J. S. Russin.** 1999. Distribution of antifungal proteins in maize kernel tissues using immunochemistry. *J. Food Prot.* **62:** 295-299.
31. **Hahn, D. H., J. M. Faubion, and L. W. Rooney.** 1983. Sorghum phenolic acids, their performance liquid chromatography separation and their relation to fungal resistance. *Cer. Chem.* **60:** 255-259.
32. **Hahn, D. H., L. W. Rooney, and C. F. Earp.** 1984. Tannins and phenols of sorghum. *Cer. Foods World* **29:** 776-779.
33. **Hey, T. D., M. Hartley, and T. A. Walsh.** 1995. Maize ribosome inactivating protein (b-32): Homologs in related species, effects on maize ribosomes and modulation of activity by pro-peptide deletions. *Plant Physiol.* **107:** 1323-1332.
34. **Huang, J.-K., L. Wen, M. Swegle, H.-C. Tran, T. H. Thin, H. M. Naylor, S. Muthukrishnan, and G. R. Reeck.** 1991. Nucleotide sequence of a rice genomic clone that encodes a class I endochitinase. *Plant Molec. Biol.* **16:** 479-480.
35. **Huynh, Q. K., J. R. Borgmeyer, and J. E. Zobel.** 1992. Isolation and characterization of a 22 kDa protein with antifungal properties from maize seeds. *Biochem. Biophys. Res. Comm.* **182:** 1-5.
36. **Ignatius, M. J. S., J.-K. Huang, R. K. Chopra, and S. Muthukrishnan.** 1994. Isolation and characterization of a barley chitinase genomic clone: Expression in powdery-mildew infected barley. *J. Plant Biochem. Biotech.* **3:** 91-95.
37. **Isakeit, T., G. N. Odovody, and R. A. Shelby.** 1998. First report of sorghum ergot caused by *Claviceps africana* in the United States. *Plant Dis.* **82:** 592.
38. **Jambunathan, R., M. S. Kherdekar, and J. W. Stenhouse.** 1992. Sorghum grain hardness and its relationship to mold susceptibility and mold resistance. *J. Agric. Food Chem.* **40:** 1403-1408.
39. **Jambunathan, R., M. S. Kherdekar, and P. Vardya.** 1991. Ergosterol concentration in mold-susceptible and mold-resistant sorghum at different stages of grain development and its relationship to flavan-4-ols. *J. Agric. Food Chem.* **39:** 1866-1870.
40. **Joshi, B. N., M. N. Sainani, K. B. Bastawade, V. S. Gupta, and P. K. Ranjekar.** 1998. Cysteine protease inhibitor from pearl millet: a new class of antifungal protein. *Bioch. Biophys. Res. Comm.* **246:** 382-387.
41. **Kanazin, V., L. F. Marek, and R. C. Shoemaker.** 1996. Resistance gene analogs are conserved and clustered in soybean. *Proc. Nat. Acad. Sci. USA* **93:** 11746-11750.
42. **Kavitha, R, and A. Chandrashekar.** 1993. Isolation and characterization of cell-wall components from the endosperm of sorghum varieties varying in hardness. *Carb. Polymers* **22:** 107-115.
43. **Kavitha, R., and A. Chandrashekar.** 1997. Content and composition of nonstarchy polysaccharides in endosperm of sorghum varying in hardness during four developmental stages. *Cer. Chem.* **74:** 22-24.
44. **Kirleis, A. W., K. D. Crosby, and T. L. A. Housley.** 1984. A method for quantitatively measuring vitreous endosperm in sectioned sorghum grain. *Cer. Chem.* **61:** 556-558.
45. **Klein, R. R., R. Rodriguez-Herrera, J. A. Schlueter, P. E. Klein, Z. H. Yu, and W. L. Rooney.** 2000. Identification of genomic regions that affect grain mold incidence and other traits of agronomic importance in sorghum. *Theor. Appl. Genet.* **102:** 307-319.
46. **Klittich, C. J. R., J. F. Leslie, P. E. Nelson, and W. F. O. Marasas.** 1997. *Fusarium thapsinum* (*Gibberella thapsina*): A new species in section *Liseola* from sorghum. *Mycologia* **89:** 643-652.
47. **Krishnaveni, S., J.-M. Jeoung, S. Muthukrishnan, and G. H. Liang.** 2001. Transgenic sorghum plants constitutively expressing a rice chitinase gene show improved resistance to stalk rot. *J. Genet. Breed.* **55:** 151-158
48. **Krishnaveni, S., G. H. Liang, S. Muthukrishanan, and A. Manickam.** 1999. Purification and partial characterization of chitinases from sorghum seeds. *Plant Sci.* **144:** 1-7.
49. **Kumari, S. R., and A. Chandrashekar.** 1994. Relationships between grain hardness and the content of prolamins and three antifungal proteins in sorghum. *J. Cer. Sci.* **20:** 93-99.
50. **Leah, R., H. Tommerup, I. B. Svendsen, and J. Mundy.** 1991. Biochemical and molecular characterization of three barely seed proteins with antifungal properties. *J. Biol. Chem.* **226:** 1564-1573.
51. **Leister, D., A. Ballvora, F. Salamini, and C. Gebhardt.** 1996. A PCR-based approach for isolating pathogen resistance genes from potato with potential for wide applications in plants. *Nature Gen.* **14:** 421-429.

52. **Leslie, J. F., R. D. Plattner, D. H. Shackelford, and A. E. Desjardins.** 1992. Fumonisin B_1 production by strains from different mating populations of *Gibberella fujikuroi* (*Fusarium* section *Liseola*). *Phytopathology* **82:** 341-345.
53. **Li, W. L., J. D. Farris, S. Muthukrishnan, D. J. Liu, P. D. Chen, and B. S. Gill.** 2000. Isolation and characterization of novel cDNA clones of acidic chitinases and β-1,3-glucanases from wheat spikes infected by *Fusarium graminearum*. *Theor. Appl. Genet.* **102:** 353-362.
54. **Lin, W., C. S. Anuratha, K. Datta, I. Potrykus, S. Muthukrishnan, and S. K. Datta.** 1995. Genetic engineering of rice for resistance to sheath blight. *Bio/Technology* **13:** 686-691.
55. **Lo, S.-C., J. D. Hipskind, and R. L. Nicholson.** 1999. cDNA cloning of a sorghum pathogenesis-related protein (PR-10) and differential expression of defense related genes following inoculation with *Cochilobolus heterostrophus* or *Colletotrichum sublineolum*. *Molec. Plant-Microbe Interact.* **12:** 479-489.
56. **Lohmer, S., M. Maddaloni, M. Motto, N. DiFonzo, H. Hartings, F. Salamini, and R. D. Thompson.** 1991. The maize regulatory locus *opaque-2* encodes a DNA-binding protein which activates the transcription of the *b-32* gene. *EMBO J.* **10:** 617-624.
57. **Mago, R., S. Nair, and M. Mohan.** 1999. Resistance gene analogues from rice: cloning, seqeuncing and mapping. *Theor. Appl. Genet.* **99:** 50-57.
58. **Malehorn, D. E., J. R. Borgemeyer, C. E. Smith, and D. M. Shah.** 1994. Characterization of an antifungal zeamatin-like protein (Zip) gene from *Zea mays*. *Plant Physiol.* **106:** 1471-1481.
59. **Mazhar, H., and A. Chandrashekar.** 1993. Differences in kafirins composition during endosperm development and germination in sorghum cultivars varying in hardness. *Cer. Chem.* **70:** 667-671.
60. **Mazhar, H., and A. Chandrashekar.** 1995. Quantification and distribution of kafirins in the kernels of sorghum cultivars varying in endosperm hardness. *J. Cer. Sci.* **21:** 155-162.
61. **Mazhar, H., A. Chandrashekar, and H. S. Shetty.** 1993. Isolation and immunochemical characterization of the alcohol extractable proteins (kafirins) of *Sorghum bicolor* (L.) Moench. *J. Cereal Sci.* **17:** 83-93.
62. **McDonough, C. M., and L. W. Rooney.** 1990. Developmental study of six varieties of sorghum. *Cer. Foods World* **35:** 836.
63. **Mehansho, H., L. G. Butler, and D. M. Carlson.** 1987. Dietary tannins and salivary proline-rich proteins: Interactions, induction and defense mechanisms. *Annu. Rev. Nutr.* **7:** 423-440.
64. **Menkir, A., G. Ejeta G., L. G. Butler, and A. Melake-berhan.** 1996. Physical and chemical kernel properties associated with resistance to grain mold in sorghum. *Cer. Chem.* **73:** 613-617.
65. **Mingeot, D., and J. M. Jacquemin.** 1997. A wheat cDNA coding for a thaumatin-like protein reveals a high level of RFLP in wheat. *Theor. Appl. Genet.* **95:** 822-827.
66. **Muthukrishnan, S., G. H. Liang, H. N. Trick, and B. S. Gill.** 2001. Pathogenesis-related proteins and their genes in cereals. *Plant Cell, Tiss. Org. Cult.* **64:** 93-114.
67. **Nelson, P. E., R. D. Plattner, D. D. Shackelford, and A. E. Desjardins.** 1991. Production of fumonisin by *Fusarium moniliforme* strains from various substrates and geographic areas. *Appl. Environ. Microbiol.* **57:** 2410-2412.
68. **Nitti, G., S. Orru, C. Block, L. Morhy, G. Marina, and P. Pucci.** 1995. Amino acid and disulfide-bridge pattern of the γ-thionins from *Sorghum bicolor*. *Eur. J. Biochem.* **228:** 250-256.
69. **Ohmori, T., M. Murata, and F. Motoyoshi.** 1998. Characterization of disease resistance gene-like sequence in near-isogenic lines of tomato. *Theor. Appl. Genet.* **96:** 331-338.
70. **Pratt, R. C., J. W. Pauli, K. Miller, T. Nelsen, and J. A. Bietz.** 1995. Association of zein classes with maize kernel hardness. *Cer. Chem.* **72:** 162-167.
71. **Radhakrishan, M. R., and M. R. Sivaprasad.** 1989. Tannin content of sorghum varieties and their role in iron availability. *J. Agric. Chem.* **28:** 55-57.
72. **Ramakrishna, Y., R. V. Bhat and V. Ravindranath.** 1989. Production of deoxynivalenol by *Fusarium* isolates from samples of wheat associated with a human mycotoxicosis outbreak and from sorghum cultivars. *Appl. Environ. Microbiol.* **55:** 2619-2620.
73. **Rao, C. H., and B. S. Rana.** 1989. The genetics of characters related to grain deterioration in sorghum. *J. Maharastra Agric. Univ.* **4:** 356-357.
74. **Roberts, W. K. and C. P. Selitrennikoff.** 1986. Zeamatin, an antifungal protein from maize with membrane permeablizing activity. *J. Gen. Microbiol.* **136:** 1771-1778.
75. **Rodriguez-Ballesteros, O. R., A. S. B. Mansuetus, R. A. Frederiksen, G. N. Odvody, R. D. Waniska, and D. T. Rosenow.** 2002. Free and bound phenolic acids in mature sorghum caryopsis as affected by inoculation with *Fusarium thapsinum*, p. 269-272. *In* J. F. Leslie (ed.), *Sorghum and Millets Diseases*. Iowa State Press, Ames, Iowa.
76. **Rodriguez-Herrera, R., R. D. Waniska, and W. L. Rooney.** 1999. Antifungal proteins and grain mold resistance in sorghum with non-pigmented testa. *J. Agric. Food Chem.* **47:** 4802-4806.
77. **Rukimini, C., and R. V. Bhat.** 1978. Occurrence of F2 toxin in *Fusarium* infected sorghum from India. *J. Agric. Food Chem.* **26:** 647-649.
78. **Schmidt, R. J., R. A. Burr, and B. A. Larkins.** 1990. Maize regulatory locus *opaque2* encodes a protein with a "leucine-zipper" motif that binds a zein DNA. *Proc. Natl. Acad. Sci. USA* **87:** 46-50.
79. **Schmidt, R. J., M. Ketudat, M. J. Aukerman, and G. Hoschek.** 1992. Opaque2 is a transcriptional activator that recognizes a specific target site in the 22 kDa zein genes. *Plant Cell* **4:** 689-700.
80. **Seetharaman, K.** 1997. *In vitro* activity of sorghum seed antifungal proteins against grain mold pathogens. *J. Agric. Food Chem.* **45:** 3666-3671.
81. **Seetharaman, K., R. D. Waniska, and L. W. Rooney.** 1996. Physiological changes in sorghum antifungal proteins. *J. Agric. Food Chem.* **44:** 2435-2441.
82. **Seitz, L. M., D. B. Sauer, R. Burroughs, H. E. Mohr, and J. D. Hubbard.** 1979. Ergosterol as a measure of fungal growth. *Phytopathology* **69:** 1202-1203.
83. **Seitz, L. M., D. B. Sauer, H. E. Mohr, R. Burroughs, and J. V. Paukstelis.** 1975. Metabolites of *Alternaria* in grain sorghum: Compounds which could be mistaken for zearalenone and aflatoxin. *J. Agric. Food Chem.* **23:** 1-4.

84. **Serna-Saldivar, S., and L. W. Rooney.** 1995. Structure and chemistry of sorghum and millets, p. 69-124. *In* D. A. V. Dendy (ed.), *Sorghum and Millets: Chemistry and Technology*. American Association of Cereal Chemists, Inc., St. Paul, Minnesota.
85. **Shen, K. A., B. C. Meyers, M. N. Islam-Fraidi, D. B. Chin, D. M. Stelly, and R. W. Michelmore.** 1998. Resistance gene candidates identified by PCR with degenerate oligonucleotide primers map to clusters of resistance genes in lettuce. *Molec. Plant-Microbe Interact.* **11:** 815-823.
86. **Shull, J. M., A. Chandrashekar, A. W. Kirleis, and G. Ejeta.** 1990. Development of sorghum (*Sorghum bicolor* (L.) *Moench*) endosperm in varieties of varying hardness. *Food Struct.* **9:** 253-267.
87. **Singh, J., and V. L. Asnani.** 1975. Present status and future prospects of breeding better protein quality in maize through *opaque-2*, p. 86-101. *In* E. T. Mertz (ed.), *High Quality Protein Maize*. Dowden, Hutchinson and Ross, Stroudsburg, Pennsylvania.
88. **Stenhouse, J. W., R. Bandyopadhyay, S. D. Singh, and V. S. Subramanian.** 1998. Breeding for grain mold resistance in sorghum. *Proc. Internatl. Conf. on Genetic Improvement of Sorghum and Pearl Millet (23-27 September 1996, Lubbock, Texas)*: 326-336.
89. **Sunitha, R. K., and A. Chandrashekar.** 1994. Isolation and purification of antifungal proteins from sorghum endosperm. *J. Sci. Food Agric.* **64:** 357-364.
90. **Sunitha, R. K., and A. Chandrashekar.** 1994. Relationships between grain hardness and the content of prolamins and three antifungal proteins in sorghum. *J. Cer. Sci.* **20:** 93-99.
91. **Sunitha, R. K., A. Chandrashekar, and R. A. Frederiksen.** 1996. Levels of three antifungal proteins during development, germination and in response to fungal infection in grain *Sorghum bicolor* (L) Moench. *Afr. J. Crop Sci.* **4:** 79-88.
92. **Sunitha, R. K., A. Chandrashekar, and H. S. Shetty.** 1992. Proteins in developing sorghum endosperm of sorghum that may be involved in resistance to grain molds. *J. Sci. Food Agric.* **60:** 275-282.
93. **Sunitha, R. K., A. Chandrashekar, and H. S. Shetty.** 1994. Antifungal proteins from sorghum endosperm and their effect on fungal mycelia. *J. Sci. Food Agric.* **64:** 121-127.
94. **Tang, X., M. Xie, Y. J. Kim, J. Zhou, D. F. Klessig, and G. B. Martin.** 1999. Overexpression of *Pto* activates defense responses and confers broad resistance. *Plant Cell* **11:** 15-30.
95. **Uma, M. V.** 1999. *Molecular Cloning of Ribosome Inactivating Proteins from Sorghum.* Ph.D. Thesis, Univ. of Mysore, Mysore, India.
96. **Vasanthi, S, and R. V. Bhat.** 1998. Mycotoxins in food: Occurrence, health and economic significance and food control measures. *Ind. J. Med. Res.* **108:** 212-222.
97. **Vesonder, R. F., A. Ciegeler, and A. H. Jensen.** 1973. Isolation of the emetic principle from *Fusarium* infected corn. *Appl. Environ. Microbiol.* **26:** 1008-1010.
98. **Vigers, A. J., W. K. Roberts, and C. P. Selitrenikoff.** 1991. A new family of plant antifungal proteins. *Molec. Plant-Microbe Interact.* **4:** 315-323.
99. **Visconti, A., and M. B. Doko.** 1994. Survey of fumonisin production by *Fusarium* isolated from cereals in Europe. *J. Amer. Oil Anal. Chem. Intl.* **77:** 546-550.
100. **Wallace, J. C., M. A. Lopes, E. Paiva, E. and B. A. Larkins.** 1990. New methods for extraction and quantitation of zeins reveal a high content of γ-zein in modified *opaque-2* maize. *Plant Physiol.* **92:** 191-196.
101. **Walsh, T. A., A. E. Morgan, and T. D. Hey.** 1991. Characterization and molecular cloning of a proenzyme form of a ribosome-inactivating protein from maize. *J. Biol. Chem.* **266:** 23422-23426.
102. **Waniska, R. D., J. H. Poe, and R. Bandyopadhyay.** 1989. Effects of weathering on phenolic acids in sorghum caryopses. *J. Cereal Sci.* **10:** 217-225.
103. **Wu, S., A. L. Kriz, and J. M. Widholm.** 1994. Molecular analysis of two cDNA clones encoding acidic class 1 chitinase in maize. *Plant Physiol.* **105:** 1097-1105.
104. **Yu, Y. G., G. R. Bus, and M. A. Saghai-Maroof.** 1996. Isolation of a superfamily of candidate disease resistance genes from soybean based on a conserved nucleotide-binding site. *Proc. Natl. Acad. Sci. USA* **93:** 11751-11756.
105. **Zhu, H., S. Krishnaveni, G. H. Liang, and S. Muthukrishnan.** 1998. Biolistic transformation of sorghum using a rice chitinase gene. *J. Genet. Breed.* **52:** 243-252.

Part VIII

Technology and Approaches to Disease Management

50

Role of Sorghum in Overseasoning of *Gibberella zeae*

Lester W. Burgess, Brett A. Summerell, Greg Giblett, David Backhouse, Megan L. Blake, Jillian Smith-White, and Melanie Colville

Gibberella zeae (Schw.) Petch (anamorph *Fusarium graminearum* Schwabe) caused a localized epidemic of head blight in durum and bread wheats in the Liverpool Plains, Bellata, and Biniguy areas of the northern grain belt of New South Wales (Australia) in October 1999. Although *G. zeae* is usually the dominant pathogen associated with the disease, we also isolated the crown rot of wheat pathogen *Fusarium pseudograminearum* Aoki & O'Donnell and *F. culmorum* (W.G. Smith) Sacc. from blighted heads. The 1999 epidemic caused grain quantity and quality, especially in durum wheat, to decrease. It was the first significant outbreak of head blight caused by *G. zeae* in the main wheat belt of Australia, although the fungus is commonly associated with maize in the higher-rainfall coastal and adjacent tablelands of eastern Australia (2, 4, 5). The high levels of inoculum of *G. zeae* in 1999 were, presumably, a consequence of the recent significant increase in the acreage of durum wheat, which is very susceptible to this fungus; no-tillage cropping practices, which involve retention of crop residues; and double cropping. Head blight was particularly severe on farms where durum wheat was grown in rotations that included maize.

Sorghum also is grown in these rotations, and its potential as a carryover host of *G. zeae* was questioned following the discovery of fertile perithecia of *G. zeae* on sorghum stubble on two farms. However, there are few reports of *G. zeae* associated with sorghum stalks. Trimboli and Burgess (9) occasionally isolated *G. zeae* during systematic studies of the fungi associated with sorghum stalk rot in the northern grain belt of New South Wales, Australia. Blaney and Dodman (1) evaluated mycotoxins produced by *G. zeae* and included a strain isolated from sorghum. Reed *et al.* (8) isolated *G. zeae* from sorghum stalks from Nebraska although it was not common. Ivaschenko and Nazarovskaya (6) reported that *G. zeae* was present on sorghum residues from the Krasnodar region of the Northern Caucasus, RSFSR. Menkir *et al.* (7) also found significant levels of *G. zeae* in sorghum in northern Indiana, which they attributed to the relatively cool climate from which most of their samples originated.

Our objective in this study was (i) to determine the frequency of *G. zeae* on old sorghum stalks, and (ii) to clarify the role such crop residues might play in the overseasoning of this pathogen.

Materials and Methods

Samples of 10-12-month-old sorghum stalks were collected in January 2000 from 21 fields in the northern grain belt of New South Wales, including fields on farms where head blight was severe in October 1999. Thirty stalks from each sample were selected arbitrarily in the laboratory, washed gently to remove encrusted soil, and damp-dried. A segment of stalk tissue from the first internode above the nodal roots of each stalk was removed after the stalk surface was swabbed with 75% ethanol, and then plated on a selective medium, peptone-PCNB agar (3). *Fusarium* colonies were subcultured to carnation-leaf-piece agar, CLA (3), for identification. The stalks were examined for the presence of fertile perithecia of *G. zeae* following incubation at 100% RH and 20°C under continuous lights for 7 days. Representative perithecia were removed from stalks and examined microscopically to confirm their identity. The identity and viability of the perithecia was further confirmed by subculturing single ascospores on CLA.

Results

Of 11 samples from areas with high incidence of wheat head blight, perithecia were found in seven samples, and

G. zeae was isolated from 1-15 of the 30 stalks per sample. For the 10 samples from areas with low incidence of wheat head blight, perithecia were observed on stalks of two samples, but *G. zeae* was not recovered. We did isolate *G. zeae* from a stalk from each of two samples, but no perithecia were observed on these stalks. This result was not surprising, as tissue was taken from a single node before the stalks were examined for the presence of perithecia. Ten of the samples had neither perithecia nor recoverable *G. zeae*. Six samples were from areas with low incidence of wheat head blight, and four were from areas with high incidence of wheat head blight.

We frequently isolated of members of the *Gibberella fujikuroi* complex and detected perithecia of at least one member of this complex in samples from the Spring Ridge area of the Liverpool Plains. These fungi may have reduced the recovery of *G. zeae* from some samples.

Discussion

The coincidence of favorable weather conditions, both pre- and post-anthesis, and high levels of inoculum contributed to the development of the 1999 epidemic of head blight. *G. zeae* was generally found on sorghum stalks only on farms where head blight was severe in 1999 and where there were high levels of inoculum of *G. zeae* on durum residues. *G. zeae* generally was not recovered from sorghum stalks from farms where little or no wheat head blight was reported, even though most of these farms have a long-term history of sorghum production.

Double-cropping also has resulted in a significant increase in residue levels, and it is common to find residues of two or three crop species present in no-tilled fields. When *G. zeae* colonizes these residues, then a dramatic increase in inoculum levels can result. Sorghum had been grown in rotation partly as a break-crop for diseases of maize and wheat.

On one farm where head blight was extremely severe, perithecia were found on sorghum stalks at levels ranging from 50-100% of the stalks in the field. Maize was grown regularly in rotation with wheat and sorghum on this farm, and no-tillage practices had been followed for many years. In some fields on this farm, we identified residues of three crop species, maize, sorghum, and durum wheat, all of which carried viable perithecia of *G. zeae*. The simultaneous presence of these three types of residues presumably results from double cropping and the slow rates of decomposition resulting from extended dry periods.

The finding that *G. zeae* can overseason on sorghum residue has implications for the management of *G. zeae*. Some farmers responded to the wheat head blight epidemic by burning wheat residue and incorporating the remainder into the soil, while other farmers only incorporated the residues. No-tillage cultivation is critically important, however, to minimize soil erosion caused by severe summer storms. Consequently, it will be necessary to alter the crop rotation sequence to regularly include a crop on which *G. zeae* cannot overseason, *e.g.*, dry land cotton. The value of chopping crop residues to encourage decomposition as a process for reducing inoculum levels also is being evaluated.

G. zeae was rarely associated with sorghum stalks from areas where wheat head blight was not common, even though sorghum had been grown in regular rotation for up to 30 years on some of these farms. This finding may be a consequence of the lower rainfall in these areas compared to, for example, the Liverpool Plains. It also is possible that colonization of sorghum stalks by *G. zeae* is dependent on exposure to a high level of inoculum, such as occurs on durum wheat residues. Further research is needed to clarify these issues and to determine whether *G. zeae* colonizes sorghum stalks as an endophyte, a pathogen, a saprophyte, or some combination of the three. The environmental and agronomic factors favoring colonization also need to be clarified. This information is important, as sorghum has been used as a break crop in rotations in this region for disease control. It also will be necessary to clarify the role, if any, of potential alternate crops in the overseasoning of *G. zeae* before these crops are incorporated into crop rotations for no-tillage systems in this region where *G. zeae* is now established.

Acknowledgements

This work was supported in part by the Grains Research and Development Corporation, Australia.

References

1. **Blaney, B. J., and R. L. Dodman.** 1988. Production of the mycotoxins zearalenone, 4-deoxynivalenol and nivalenol by isolates of *Fusarium graminearum* Groups 1 and 2 from cereals in Queensland. *Austr. J. Agric. Res.* **39:** 21-29.
2. **Burgess, L. W., R. L. Dodman, W. Pont, and P. Mayers.** 1981. *Fusarium* diseases of wheat, maize and grain sorghum in eastern Australia, p. 64-76. *In* P. E. Nelson, T. A. Toussoun, and R.J. Cook (eds.), Fusarium*: Diseases, Biology and Taxonomy*. Pennsylvania State University Press, University Park, Pennsylvania.
3. **Burgess, L. W., B. A. Summerell, S. Bullock, K. Gott, and D. Backhouse.** 1994. *Laboratory Manual for* Fusarium *Research*, 3rd ed. The University of Sydney, Sydney, Australia.
4. **Francis, R. G., and L. W. Burgess.** 1975. Surveys of Fusaria and other fungi associated with stalk rot of maize in eastern Australia. *Austr. J. Agric. Res.* **26:** 801-807.

5. **Francis, R. G., and L. W. Burgess.** 1977. Characteristics of two populations of *Fusarium roseum* 'Graminearum' in eastern Australia. *Trans. Brit. Mycol. Soc.* **68:** 421-427.
6. **Ivashchenko, V. G., and L. A. Nazarouskaya.** 1990. Characteristics of the ascomycetous stage of the causal agent of ear fusariosis in various cultivars from Krasmodarskii Krai. *Doklady-Vsesoyuznoi-Ordena-Lenina-i-Ordena-Trudovogo-Krasnogo-Znameni-Akademii-Sel'skokhozyaistvennykh-Nauk-Imeni-V.I.-Lenina* **12:** 11-14.
7. **Menkir, A., G. Ejeta, L. G. Butler, A. Melakeberhan, and H. L. Warren.** 1996. Fungal invasion of kernel and grain mold damage assessment in diverse sorghum germplasm. *Plant Dis.* **80:** 1399-1402.
8. **Reed, J. E., J. E. Partridge, and P. T. Nordquist.** 1983. Fungal colonization of stalks and roots of grain sorghum during the growing season. *Plant Dis.* **67:** 417-420.
9. **Trimboli, D. S., and L. W. Burgess.** 1985. Fungi associated with basal stalk rot and root rot of dryland grain sorghum in New South Wales. *Plant Protec. Quart.* **1:** 3-9.

51

Rapid Information Dissemination on the World Wide Web

Jeffrey A. Dahlberg

In 1957, the USSR launched the first artificial earth satellite, called Sputnik. The United States responded by creating the Advanced Research Projects Agency (ARPA), within the Department of Defense, whose primary responsibility was to advance science and technology in the field of military applications. In 1962, Paul Baran of the RAND Corporation was asked by the U.S. Air Force to research how it could maintain command and control over its missiles and bombers after a nuclear strike. His work, along with that of several others who were working on similar issues at the Massachusetts Institute of Technology and the National Physical Laboratory (NPL), led to what was later to be called "packet switching." In short, scientists could send "packets" of data, with origin and destination labels attached to them, from one computer to another via various routes. This development was a major step towards what we now know as computer networking. In 1965, Thomas Merrill and Lawrence Roberts connected the TX-2 computer in Massachusetts to the Q-32 computer in California with a low-speed dial-up telephone to create the first computer network.

ARPA's name was changed to ARPANET in 1968, and by 1969 four universities–University of California (Los Angeles), SRI at Stanford University, University of California (Santa Barbara), and University of Utah–were linked together to create a physical network. In 1972, the first e-mail program was developed by Ray Tomlinson of Bolt Beranek and Newman, Inc. (BBN), and in 1974 the term "Internet" was used in a paper by Cerf and Kahn (3). These early years saw research on new data transfer protocols and the development and merger of several groups working to optimize the fledgling network. In 1992, the Corporation for Research and Education Networking (CREN) released the World Wide Web [for a much more detailed history of these early years, see Kristula (4) or Leiner *et al.* (5)].

In the beginning of what is now known as the Internet, usage was primarily restricted to research scientists who used it to communicate with each other. The system was rather slow and cumbersome, and easy access to e-mail did not arrive until the early 1990s. The 1990s saw an explosion in the growth and use of the Internet as a form of communication. Services such as CompuServe and America Online began to offer limited uses of the Internet, and people began to explore the Internet through use of e-mail and visits to sites created by various service providers. With the advent of HTML language and various programs that allowed for the creation of Web pages, the Internet saw an explosion of use and utility.

Ergot in the Americas

Sorghum ergot caused by *Claviceps africana* Frederickson, Mantle & de Milliano was first reported on sorghum in Brazil in 1995 (6). The pathogen spread quickly throughout South America, and then moved northwards to Central America, the Caribbean, and the continental United States (1).

In 1995, a group of scientists began communicating, via e-mail, about the spread of ergot in the Americas. We used this medium to communicate about the spread of the disease, and a serious discussion began about the need for a Global Conference to address this problem. In 1996, I established the Sorghum Research Homepage, located on the US Department of Agriculture's Agricultural Research Service's Web site at Mayaguez, Puerto Rico; this material was moved from the ARS site in 1999 to http://www.sorghumgrowers.com/research/sorghum.html. This site was one of the first to deal with sorghum research issues. At the time, searches with the Excite (www.excite.com) and Yahoo (www.yahoo.com) search engines identified only seven sites that contained the word sorghum. The Sorghum Research Homepage was used initially to outline the research activities of the US Sorghum Curator, but information concerning ergot also was posted. It was on this site that news concerning ergot's movement throughout the United States was first reported.

Based on e-mail exchanges, The Global Conference on Ergot of Sorghum was organized. The meeting was held 1-8 June 1997 in Sete Lagoas, Brazil. One of the recommendations from the conference (2) was that the World Wide Web site for the dissemination of information on sorghum ergot be maintained and expanded. In 1998, the *Proceedings of the 1998 Conference on the Status of Sorghum Ergot in North America* held 24-26 June 1998 in Corpus Christi, Texas were published on this site as PDF files and PowerPoint presentations. Since 1999, the Sorghum Research Home page site has been maintained by the National Grain Sorghum Producers at www.sorghumgrowers.com/Research/sorghum.html.

Pros and Cons of Using the Internet

While the Internet has been an excellent tool for the dissemination of information, there are both advantages and disadvantages to using the World Wide Web. Some of the major disadvantages include:

- Rapid dissemination of information.
- User base is relatively narrow.
- Accuracy of information presented is not necessarily known.
- Too much available information may make finding relevant information difficult.
- Cost in time and money to maintain and update a site.

Dissemination of information has never been easier than with the World Wide Web. Traditionally, scientific and nonscientific information was distributed through various channels, *e.g.*, peer-reviewed journals, newspapers, radio, television, and/or popular magazines. However, in most cases such dissemination took from days to months to accomplish. Peer-reviewed journals, for example, may have processing times of a year or more from the receipt of a manuscript to its publication. The use of the Web shortens the time required for the dissemination of information and can lead to rapid high-volume circulation of misleading or conflicting reports worldwide.

Those who use the Internet often like to think that everyone is using it, but the use of the Internet is still limited in agriculture. There are several reasons for this lack of use; however, the primary reason is a lack of reliable connections in rural areas of both developing and developed countries. Less than 4% of the members of the National Grain Sorghum Producers routinely use the World Wide Web for gathering information. This limited usage makes it imperative that traditional forms of information dissemination are not discarded in favor of postings on the World Wide Web.

One of the most difficult things to prevent on the Internet is the spread of false rumors or facts. There is no organization that screens sites for accuracy and content, which makes it extremely easy to send information thought to be true and accurate all over the world. The World Wide Web is based on a trust system that most people honor, but the system has been severely tested in recent years. It is, therefore, critical that if information is posted on the World Wide Web, then it should be as accurate as possible.

Too much information is often thought to be a good thing. From seven sites identified by the Excite search engine in 1996 as containing "sorghum," the number of sites has increased to 6250 in 2000, and if the Google search engine (www.google.com) is used, then over 107,000 sites are identified. A similar search with the term "pearl millet" leads to 550 sites through Excite and 7600 sites through Google. Even narrowing the search to include the words "sorghum" and "diseases" still leads to 10,200 sites with Google, while "pearl millet diseases" identifies 1600 sites. It is virtually impossible to manually search all these sites, and time quickly becomes a critical constraint in identifying relevant available information.

A major disadvantage to the World Wide Web that is often overlooked is cost. To develop and maintain a Web site requires a diligent person with considerable computer skills. Maintenance of a Web site, in particular, often is overlooked or neglected, but is required if a Web page is to remain fully functional and accurate.

The World Wide Web also has some major advantages for information dissemination, including:

- Rapid dissemination of information.
- Virtually unlimited available information.
- Access to information from remote parts of the world.
- Possible increased and efficient collaboration and cooperation.

While rapid dissemination of information has its drawbacks, the advantages usually outweigh the disadvantages. Used correctly, the World Wide Web is an important tool for sharing information and "spreading the word" in a rapid and efficient manner. For example, consider the rapid response to the outbreak of ergot in the Americas and how the scientific community responded to inquiries and requests in a time and fashion not even conceived of several years ago. Through rapid communication, the spread of the disease was documented as it moved through the Americas, and scientists with knowledge of this disease from other continents were brought into the discussion. In years past, this interaction and information exchange would have taken much more time and effort to establish.

Though too much information makes for time-consuming searches, the amount of information available is now measured on a scale without precedence in human history. Research, data, news, and other such items are currently available for almost any topic. It is amazing to think that we have gone from seven hits to over 120,000 hits with the word "sorghum" in less than four years. On the Web, one now can find information on such diverse topics as making molasses from sorghum, food uses of pearl millet, growing the crops or ameliorating their diseases, and production figures for past and current crops in both the United States and elsewhere.

Another advantage of the World Wide Web is the ability for those in traditionally knowledge-starved areas of the world to access the latest data and research. The National Grain Sorghum Producers has received inquiries not only from Africa, but also from the former Soviet Union and Asia on topics ranging from plant disease to how to grow sorghum. The major drawback has been the limited availability of technologies to access the World Wide Web, although access should increase as digital technologies expand into more remote areas of the world. Thus, communication via the Internet enables collaboration and cooperation between people and places that was logistically very difficult in the past. Again, the collaboration between scientists that was established when ergot was reported in Brazil serves as a prime example. Scientists from around the world worked together closely via the Internet to evaluate this situation and to exchange information and research on this extremely important disease. The use of the Internet allowed for much more rapid development of collaboration and cooperation than was possible before.

Conclusions

The World Wide Web has developed from the idea of "packet switching" started by a group of scientists to one of the fastest growing forms of communication ever seen. The World Wide Web is a powerful "tool" from which information can be sent rapidly to every corner of the globe. As more people obtain routine access to the Internet, it will become an even more important communication tool. Until digital and other technologies penetrate rural and developing areas more effectively, however, traditional forms of communication, such as research pamphlets, pocket handbooks, and other "slower" printed forms of communication, will continue to play a vital role in both agricultural research and education.

References

1. **Bandyopadhyay, R., D. E. Fredrickson, N. W. McLaren, G. N. Odvody, and M. J. Ryley.** 1998. Ergot: a new disease threat to sorghum in the Americas and Australia. *Plant Dis.* **82:** 356-367.
2. **Casela, C. R., and J. Dahlberg.** 1999. Recommendations report. *Proc. Global Conf. on Ergot of Sorghum (1-8 June 1997, Sete Lagoas, Brazil)*: 191-194.
3. **Cerf, V. G., and R. E. Kahn.** 1974. A Protocol for Packet Network Interconnection, IEEE Transactions of Communications.
4. **Kristula, D.** 1997. The history of the internet. Available at http://www.davesite.com/webstation/net-history.shtml (posted March 1997; last verified October 2001).
5. **Leiner, B. M., V. G. Cerf, D. D. Clark, R. E. Kahn, L. Kleinrock, D. C. Lynch, J. Postel, L. G. Roberts, and S. Wolf.** 2000. A brief history of the internet. Available at http://www.isoc.org/internet-history/brief.html (last verified October 2001).
6. **Reis, E. M., P. G. Mantle, and H. A.-G. Hassan.** 1996. First report in the Americas of sorghum ergot disease, caused by a pathogen diagnosed as *Claviceps africana*. *Plant Dis.* **80:** 463.

52

Agrobiodiversity in Pest Management

Jillian M. Lenné and Rodomiro Ortiz

Agrobiodiversity is essential for survival, for adaptation to specific environments, and to counter unknown future challenges as they arise. Crop biodiversity represents diversity at species, individual, and gene levels. It consists of land races, modern cultivars, breeders' populations, and the individuals and genes that they contain. This diversity may be in a farmers' field, on a research station, or in gene banks. For crops, such as sorghum and millets, the biodiversity and conservation of "intraspecific" diversity–the within species diversity or genetic diversity–is the most important.

Biodiversity for disease resistance in crops is intrinsically linked with the variability of their pests and pathogens (10, 38). Pathogens and pests have played, and continue to play, crucial roles in determining biodiversity for disease and pest resistance in economic plants through processes such as host-pest/pathogen co-evolution, insect seed predation, or weed-crop introgression, among others. It is somewhat incongruous that the continued generation of biodiversity for disease resistance for future use in crop improvement depends on active, dynamic populations of pests and pathogens that reduce food crop production in the process (48). It is not surprising that in agroecosystems, pest and pathogen diversity is seen as harmful–to be managed, and even eliminated–rather than as the means to generate useful resistances for future use.

Management strategies deployed by farmers may have profound effects on agrobiodiversity (including pests, pathogens, weeds, parasites, or predators) in agro-ecosystems. Our objective in this chapter is to examine biodiversity for pest management in the broadest sense possible; to include diversity within and between crops in single fields, between fields, and within the landscape or agroecosystem, but with emphasis on examples related to pathogens. We also consider pathogen management in crop monocultures, the source of most of our plant food, but often perceived to be unstable, unsustainable, ecologically dysfunctional, and highly vulnerable to pests. We will try to answer the question, "How much biodiversity is needed for successful pest management?" Examples will be taken from research-for-development with sorghum and millets in tropical regions.

Nature and Origin of Biodiversity for Disease Resistance

Origin of biodiversity for disease resistance. There is a continuum linking wild species with present-day cultivars through natural selection over millions of years, domestication, human and natural selection of domesticates and often wide dispersal over 10,000 years, and about 100 years of modern plant selection and breeding (18). During this process there has been continuous reduction in biodiversity, including that for disease resistance. This ongoing process is driven by the utilitarian need to produce food to feed more people (17, 27). Millions of years of natural selection of wild relatives of crops has occurred in response to both abiotic and biotic influences. The widespread occurrence of plant defense compounds and batteries of resistance genes in wild relatives of crops are evidence of powerful natural selective forces (24).

Discovery and incorporation of new genes for resistance from wild species provides a means for sustaining crop improvement. Although durability of resistance cannot be predicted, genetic diversity through preemptive breeding may provide buffering against crop losses arising due to changes in the pathogen. In this way, new breeding stocks can be generated to combat more virulent pathotypes when they become important. Germplasm enhancement with wild species has seldom resulted in direct cultivar release, but many parents with "wild" genes have become available. The genetic

TABLE 52-1. Wild species of sorghum and millet held in the ICRISAT gene bank (after 35).

Sorghum Section or Millet Genus	Number of Species	Number of Accessions
Sorghum – *Chaetosorghum*	1	3
Sorghum – *Heterosorghum*	1	12
Sorghum – *Parasorghum*	5	36
Sorghum – *Sorghastrum*	1	8
Sorghum – *Sorghum*	7	362
Sorghum – *Stiposorghum*	10	106
Sorghum – Unknown	-	17
Millet – *Cenchrus*	1	11
Millet – *Pennisetum*	24	739

resources of sorghum and pearl millet held in trust at the gene bank of the International Crops Research Institute for the Semi-Arid Tropics (ICRISAT) include 544 and 750 accessions of wild species and weedy relatives of sorghum and pearl millet (37, 41; Table 52-1), respectively. Screening of this germplasm has identified good sources of resistance to important pests and diseases (Table 52-2).

Domestication is the foundation stone of crop biodiversity (18). Early farmers selected from a limited range of plant families (51), primarily from the grass and legume families. In many cases, the new crop was quickly isolated genetically from its wild relatives, *e.g.*, wheat and many pulses. In some crops, *e.g.*, sorghum and pearl millet, sporadic recombination with wild relatives in the primary gene pool continued with ongoing diversification (23, 30).

Did domestication reduce diversity? Yes, profoundly so. But much of the diversity that was lost was not needed by the "crop version" of the species (52). There was a loss of killer toxins, head shattering, hairiness, dormancy, and perennial habit, among others. The fact that we often return to the wild species germplasm for disease resistance genes (1, 28) is an inevitable consequence of domestication and ongoing selection processes.

10,000 years of natural selection, farmer selection, and dispersal. Domestication was followed by 10,000 years of natural selection through exposure to diverse climates, pests and pathogens, and competition with weeds (18). Domesticated agriculture encouraged the spread of crops far beyond the range of their wild ancestors. Crop biodiversity also was strongly influenced by human selection for specific plant traits, *e.g.*, seed size, color, flavor, and growth habit, as well as dietary and market needs. This combination of natural and human selection accounts for the remarkable variability found in land races and their extraordinary ranges of adaptation (52).

Land races became the foundation stock for modern cultivars. The interaction between the crop and its pathogens has no doubt resulted in evolutionary changes in both the crop and the resident pathogen population. We still know very little about diversity for disease resistance within land races still grown in traditional systems or in gene banks. Most of the attention has focused on host-specific biotrophic fungi in co-evolved host-pathogen systems in centers of origin, as they are easier to study. The greatest diversity for disease resistance to coevolved pathogens is expected at such centers (22). Secondary centers of diversity, especially if the pathogen moved with the crop or if the crop encountered different populations of the same pathogen in a new location, also may offer useful diversity for disease resistance.

Ecosystem management by humans also has played a major role in the evolution of pathogens and disease resistances in crops. Farmers can substantially slow the evolution for disease resistance in crops through management practices. Direct disease control practices through roguing, seed selection, application of natural and chemical pesticides, and indirect control through planting density, rotation, weeding, intercropping, and green manures are among the most important (42).

Farmer selection for one character may result in selection against disease resistance (51). Farmers' inability to recognize a valuable character, *e.g.*, race nonspecific, incomplete resistance to a pathogen or race-specific resistances present at low levels in a crop population, could result in loss of such characters from the crop. Also, in many crop populations, *e.g.*, land races or mixed crop species populations, the selection pressure may be too low to cause rapid changes, and the evolution of resistance may be very slow (32).

Intercontinental crop introduction is a relatively recent process for many crops and was especially intense in the post-Columbian period. Major staple food crops such

TABLE 52-2. Wild species accessions in the ICRISAT gene bank with disease or pest resistance, and quality traits (after 35).

Crop	Pest or Disease	Screened	Selected
Sorghum	Grain mold	308	29
	Downy mildew	565	94
	Midge	20	15
	Shoot Fly	268	13[a]
	Stem Borer	50	15[a]
Pearl Millet	Downy mildew	534	220[b]

[a]*Parasorghum* are immune.
[b]All *Pennisetum pedicellatum* and 66/68 *Pennisetum polystachyon* accessions are immune.

as maize, cassava, groundnut, bean and potato, moved from the Americas to Africa and Asia; rice moved from Asia to the Americas and Africa; and sorghum and millets moved from Africa to Asia and the Americas. Continued selection of crops by new farmers further diversified land races in secondary centers of diversity for useful characters, including disease resistance.

In summary, the development of agriculture appears to have been associated with successive waves of crop introduction and dispersal that include the creation of new diversity and the loss of old diversity, which is no longer needed by farmers because new and more productive agrobiodiversity evolves or has been selected for. The development and continued deployment of this more productive crop diversity is driven by the basic need to produce more food.

Up to 100 years of modern plant breeding. The development of modern plant breeding in the late nineteenth century initiated a process of plant introduction and evaluation, and the assemblage of collections of crop diversity for future use as never seen before. Improved high-yielding varieties were productive and profitable, and farmers increasingly abandoned land races to increase food production. Maintaining diversity was not a high priority for these farmers (44).

Nature of Diversity for Disease Resistance

The diversity and durability of the resistance in a crop for any specific pathogen is determined by many complex factors. Plant pathogens are extremely heterogeneous, possessing diverse ecological requirements and modes of parasitism, dispersal, reproduction and survival (11). Much of the diversity for disease resistance present in crop plants today represents the ghosts of diseases past (24).

It is difficult to summarize the nature of diversity for disease resistance in crops. Race-specific and non-specific resistances fluctuate in response to the prevalence and racial composition of the pathogen. Race-specific resistances are the easiest to characterize and manipulate. Thus, more is known about the incidence and distribution of the genes governing this type of resistance than for non-specific resistance (*e.g.*, 16, 47).

Use of Diversity for Disease Resistance in Sorghum and Millets

Sources of disease resistance. Disease resistance is of direct interest to breeders for developing disease resistant varieties and to farmers who want to grow productive, resistant varieties. Resistance sources include breeders' lines, land races, wild relatives, distantly related plants, and other sources accessed through biotechnological means, *e.g.*, Ortiz (35). Resistances from wild species have made major contributions in combating the ravages of pests and diseases (1, 28), and add to the diversity in improved crops. The genetic diversity of land races is the most immediately useful and economically valuable part of global biodiversity and is of paramount importance for future world crop production (1, 51). A major impediment to the utilization and conservation of crop biodiversity is the lack of both *in situ* and *ex situ* evaluation of disease resistance.

Core collections are evolving tools for quantifying genetic variation, and are the entry point for "mining" biodiversity in gene banks. A large collection, such as the sorghum land race collection held at ICRISAT, represents a challenge for the maintenance of both the accessions and the collection information. The core collection concept is used to define a limited set of accessions derived from an existing germplasm collection. Entries in this core collection are chosen to represent the genetic spectrum of the entire collection (ecologically or genetically) and to maximize genetic diversity. The core collection serves as an entry point to the entire collection and improves the accessibility of the conserved germplasm by plant breeders and other researchers.

Different sampling strategies, either random or non-random, were proposed to obtain subsets of reduced size (core collection) of the sorghum accessions at ICRISAT (19, 20). Three subsets were established following (i) a random logarithmic sampling strategy within a stratified collection (L), (ii) a principal component score strategy based upon morpho-agronomic diversity (PCS), and (iii) a sample based upon empirical taxonomic knowledge of sorghum (T). Morpho-agronomic characterization and passport information from these three samples were assessed to determine their impact on phenotypic diversity (19). For overall diversity, the three subsets were equivalent, as seen from two-dimensional representations of the morpho-agronomic diversity and the Shannon-Weaver diversity indices (19). When comparisons for morpho-agronomic and passport data were considered, the PCS subset looked similar to the entire land race collection. The L subset showed differences from the main collection for characters associated with the photoperiod reaction that was considered in the stratification of the collection. The T subset was the most distinct from the entire land race collection because it overrepresents the landraces selected by farmers for specific uses and covered the widest range of geographical adaptation and morpho-agronomic characteristics.

TABLE 52-3. Partial resistance screening against fungi in the sor-ghum entire collection and core subsets held at ICRISAT.

Disease		Entire	Core Subsets		
Name	Severity	Collection	PCS[a]	SRLS[b]	TS[c]
Grain mold[d]	None	32	-	1	1
(*N* = 8,034)	≤ 10%	271	-	1	4
	11-25%	493	3	7	4
	26-50%	1,246	8	12	11
	> 50%	5,998	16	31	55
Anthracnose[e]	≤ 1%	77	1	-	-
(*N* = 6,953)	2-5%	1,594	5	6	14
	6-20%	475	-	-	8
	21-40%	1,336	3	11	10
	> 40%	3,471	14	24	48
Leaf blight[f]	≤ 1%	111	-	-	2
(*N* = 8,583)	2-5%	1,394	2	4	17
	6-20%	3,267	14	17	48
	21-40%	1,621	8	13	25
	>40%	2,190	3	12	20
Rust[g]	≤ 1%	64	-	1	1
(*N* = 7,264)	2-5%	646	2	1	6
	6-20%	1,627	6	8	18
	21-40%	1,091	3	3	17
	> 40%	3,836	21	8	38
Downy	0-10%	1,550	11	8	14
mildew[h]	11-20%	2,225	11	13	26
(*N* = 13,372)	21-30%	1,899	12	8	18
	31-40%	1,472	5	9	12
	41-50%	1,125	7	9	14
	51-60%	905	4	4	9
	61-70%	884	3	9	13
	71-80%	875	4	4	15
	81-90%	1,005	3	8	23
	91-100%	1,432	3	5	14

[a]PCS, Principal Component Scoring strategy.
[b]SRLS, Stratified Random Logarithmic Sampling.
[c]TS, Taxonomy Selection.
[d]Homogeneity of entire collection and core subsets, $\chi^2 = 20$, $p = 0.066$; Entire vs. PCS, $\chi^2 = 6.41$, $p = 0.17$; Entire vs. SLRS, $\chi^2 = 11.3$, $p = 0.024$; Entire vs. TS, $\chi^2 = 2.63$, $p = 0.62$.
[e]Homogeneity of entire collection and core subsets, $\chi^2 = 18$, $p = 0.13$.
[f]Homogeneity of entire collection and core subsets, $\chi^2 = 15$, $p = 0.23$.
[g]Homogeneity of entire collection and core subsets, $\chi^2 = 12$, $p = 0.28$.
[h]Results based on field screening. Homogeneity of entire collection and core subsets, $\chi^2 = 32$, $p = 0.22$.

The genetic diversity in each of the sampling procedures (PCS, L, T) was quantified by using the polymorphisms at 15 microsatellite loci (20). The average allelic richness for each subset was equivalent (16.1, 16.3, and 15.4 alleles per locus for the subsets PCS, L, and T, respectively). The average genetic diversity also was comparable for the three subsets (0.81, 0.77, and 0.80, respectively). Allelic frequency distribution for each subset was compared with a χ^2 test, but few significant differences were observed. A high percentage of the rare alleles (71-76% of 206 total rare alleles) were maintained in all three subsets (20). Thus, the global molecular diversity retained in each subset was not affected by a sampling procedure based upon phenotypic characters.

Partial assessment of host response to five sorghum diseases may provide a means to quantify the importance of agrobiodiversity in disease management (Table 52-3). Frequency distributions were the same between the entire collection and core subsets for all diseases, except between the entire collection and the L core subset for grain mold. This difference was not surprising because the sampling strategy for this core subset and the material included in the screening for this disease did not match. The L core subset has the widest range of adaptation to photoperiod, whereas only photoperiod insensitive germplasm was screened for grain mold. The lack of accessions with the highest resistance class for some diseases in the core subsets are the result of sampling statistics, but the χ^2 tests for homogeneity clearly confirmed that the entire collection and core subset included the same range and distribution of variation, with the exception of grain mold in the L core subset. New accessions with high resistance to specific diseases are likely to be identified by completing the screening of the core subsets. This rational, targeted approach also may be cost-effective and more precise than long-term screening of the entire collection.

Many of the accessions with specific and high disease resistance that were not included in one or more of the core subsets did not originate from a country with a large number of accessions exhibiting high resistance (Table 52-4). Since accessions included in the core subsets should be unique, these results suggest that areas with large numbers of accessions showing high disease resistance should not necessarily be regarded as pockets of variation for these characteristics because they may be genetically redundant. Pyramiding genes for disease resistance, after confirming that unique and distant accessions included in the core subset(s) have distinct genes for disease resistance, could be a profitable means to enhance existing lines and germplasm. DNA markers may facilitate this approach for enhancing host plant resistance.

As with morphological and molecular data, the entire collection and the core subcollections do not differ statistically in genetic diversity, and the diversity observed was just above intermediate (0.541-0.599) (Table 52-5). The genetic diversity within a collection depends on the character, being largest for host response to downy mildew and smallest for grain mold, as the sampling for screening was biased.

This analysis shows that large sample sizes are not always associated with the capture of useful variation for disease resistance (*i.e.*, entire collection vs. core subcollections), neither when the sampling was defined by breeding objectives, *e.g.*, the L subset; a mirror of the entire collection, *e.g.*, the PCS subset; or by maximizing farmers' land races, *e.g.*, the T subset. In short, agrobiodiversity *per se* does not seem to be insurance for disease control. Useful genetic variation may be acquired from only a few germplasm sources.

With the development of molecular techniques, disease resistance may be sought from unrelated organisms including distantly related plants and microorganisms. Further refinements in genome analysis and manipulation, *e.g.*, marker technology to enhance efficient access to disease resistance sources (50) and synteny studies (33), will foster access to genes from more distantly related plants. Understanding the biochemical bases of infection processes will open up more opportunities, *e.g.*, antifungal genes for sorghum grain mold. Have we reached the era of "designer" resistance genes?

TABLE 52-4. Disease resistance sources in the entire collection and the core subsets.

Disease	Core Subset	Core Location(s)	Entire Collection - Locations
Grain mold	SRLS[a] TS[b]	United States United States	Argentina, Botswana, China, India, Japan, Kenya, Malawi, South Africa, Sudan, former USSR, United States
Anthracnose	PCS[c]	Cameroon	Cameroon, Central African Republic, Chad, Ghana, India, Japan, Kenya, Mexico, Niger, Nigeria, Senegal, South Africa, Sudan, Tanzania, Uganda, United States, Zimbabwe
Leaf blight	TS	India, Mali	Australia, Burkina Faso, Cameroon, Central African Republic, Chad, Coté d'Ivoire, Egypt, Ethiopia, India, Mali, Niger, Nigeria, Philippines, South Africa, Sudan, Taiwan, Tanzania, Thailand, Uganda, United States
Rust	SRLS TS	Uganda Sri Lanka	Australia, Cameroon, Chad, Ethiopia, India, Kenya, Nigeria, Senegal, Sri Lanka, Sudan, Swaziland, Tanzania, Uganda, United States, Zimbabwe
Downy mildew	PCS SRLS TS	Cameroon, Chad, Guatemala, Japan, South Africa Cameroon, Chad, Ethiopia, India, Nigeria, South Africa Cameroon, China, Ethiopia, India, Malawi, Nigeria, South Africa, Zambia	Angola, Argentina, Australia, Bangladesh, Botswana, Burkina Faso, Cameroon, Central African Republic, Chad, China, Congo, Coté d'Ivoire, Dominican Republic, El Salvador, Ethiopia, Ghana, Guatemala, Hungary, India, Israel, Jamaica, Japan, Kenya, Lebanon, Lesotho, Malawi, Mali, Mexico, Nepal, Nicaragua, Niger, Nigeria, Philippines, Portugal, Senegal, South Africa, Sudan, Swaziland, Tanzania, Thailand, Turkey, Uganda, former USSR, United States, Zambia, Zimbabwe

[a]SRLS, Stratified Random Logarithmic Sampling.
[b]TS, Taxonomy Selection.
[c]PCS, Principal Component Scoring strategy.

Is there a finite pool of diversity for disease resistance in crops? If the potential pool of resistance sources is limited, *sensu* Frankel *et al.* (18), the rate at which resistances are overcome through pathogen evolution should be of great concern for the future management of pathogen diversity and crop productivity. Are we squandering resistance genes in agriculture in a manner similar to that for antibiotics in medicine? If there is a finite pool of resistance genes in crops, we think that the limits of pool size are still largely unknown in most crops. We think it is unlikely that the diversity of resistance genes in wild relatives for coevolved diseases will ever be exhausted, especially since additional sources of resistance are increasingly being identified in distantly related plants and microorganisms.

Composite crosses have been important tools for population geneticists to examine the evolution of adaptation in cultivated plants or the genetics of host-pathogen interactions. Based on his investigation on the barley-*Rhynchosporium secalis* pathosystem, Allard (2) suggested that genetically variable evolving populations are important sources in breeding for disease resistance and other characteristics that affect adaptation to specific environments. For example, Ibrahim and Barret (25) reported that inbred lines derived from successsive generations of a composite cross population of barley showed significant resistance to powdery mildew. Hence, composite crosses are considered, by these authors, as dynamic reservoirs of genetic diversity that may keep pace with the evolution of pathogen populations.

TABLE 52-5. Shannon-Weaver index for five sorghum diseases in three core subsets and the entire collection of sorghum held at ICRISAT.

Disease	Entire Collection	Core Subset			Mean ± S.E.
		PCS[a]	SRLS[b]	TS[c]	
Grain mold	0.397	0.464	0.382	0.354	0.399 ± 0.023
Anthracnose	0.450	0.412	0.478	0.536	0.469 ± 0.026
Leaf blight	0.494	0.559	0.592	0.600	0.562 ± 0.024
Rust	0.428	0.566	0.550	0.527	0.518 ± 0.031
Downy mildew	0.936	0.967	0.980	0.977	0.965 ± 0.10
Entire collection	0.541 ± 0.100	0.594 ± 0.098	0.597 ± 0.102	0.599 ± 0.103	

[a]PCS, Principal Component Scoring strategy.
[b]SRLS, Stratified Random Logarithmic Sampling.
[c]TS, Taxonomy Selection.

Use of diversity for disease resistance in improvement of sorghum and millets. There are many questions in this area that need answers, including:

- Does more diversity for disease resistance mean greater crop stability?
- Does less diversity for disease resistance mean greater vulnerability to diseases?
- Does crop uniformity mean genetic identity or phenotypic identity?

Answers to questions such as these should provide a better understanding and definition of "durable resistance" and the best ways in which it can be obtained and maintained.

In crops, covert genetic diversity often is masked by morphological uniformity, and phenotypic diversity may mask genetic uniformity for reaction to major diseases. A mythology has arisen that is based on false interpretation of the value of morphological diversity (51). We must be careful how we assess crop uniformity. Replacement of land races by modern varieties, for example, may mean less varietal diversity but more genetic diversity (37, 41).

The introduction and widespread adoption of modern, high-yielding, less heterogeneous, varieties of major staple food crops has been strongly criticized as a major cause of crop vulnerability, especially to diseases and pests, and genetic erosion (3, 35). This process was driven by the need to dramatically increase food production. The alternative was starvation for millions of people. Single-gene resistances are the easiest to manipulate in breeding programs and remain the principle defenses against many pathogens that affect our most important food crops. Complex, polygenic resistances have generally been difficult to understand, identify, quantify, sample, screen for, and utilize in breeding programs (18). However, the development of marker-assisted selection is enabling more efficient and effective use of such traits (7, 50).

It is not homogeneity across a crop's genome that is the problem but, rather, homozygosity for one, or a few, genes that condition susceptibility to a particular pathogen (9). Thus, a crop variety may be very diverse genetically but uniformly susceptible–or the reverse, uniform genetically but resistant to a given pathogen. The problem then becomes determining when crop homogeneity is at risk in relation to the development of an epidemic (9).

Both vertical and horizontal resistance systems have provided examples of long-term stability and rapid breakdown. Single-gene resistances in uniform cultivars constitute the most vulnerable system, but there are examples of long-term stability, *e.g.*, a gene for resistance to Bean Common Mosaic Virus in beans (4). Other resistance traits based on single genes have proved transient, *e.g.*, single genes for resistance to rice blast, wheat rust, barley powdery mildew, and sorghum downy mildew. Nonspecific, polygenic resistance may be less vulnerable, but there also are examples of its rapid and gradual decline of efficacy (31).

Breeding for single-gene resistance to variable pathogens may be misguided and can seriously affect crop diversity. Unfortunately, the popular concept of modern plant breeding is that only uniformity is desired. Thirty years after the 1970-1971 Southern Corn Leaf Blight Epidemic, the stigma remains. Much has been learnt since then about the vulnerability of single-gene resistances for variable pathogens, durability of resistance, and the need for stability of resistance whether it is based on single or multiple genes. Indeed, modern plant breeding probably has not greatly reduced diversity for disease resistance in major crops, *e.g.*, Witcombe (49).

Genetic erosion in perspective. The continuing development of agriculture has been associated with genetic erosion, and successive waves of new crops and new varieties, crop introduction and dispersal, and replacement by newer and better varieties. However, the recognition that erosion is a potential constraint to future world food production is relatively recent. Much of the concern dates from the Southern Corn Leaf Blight epidemic of 1970-1971 (45). This single event impressed upon the world the dangers of genetic uniformity and was the catalyst for the global interest in the collection and conservation of plant genetic resources. However, almost 30 years after the genetic erosion alarm was raised, little is known about the actual rate of genetic erosion of crops and their wild progenitors (13), and the extent of genetic erosion has not been measured quantitatively (8). Continuous efforts are required to maintain and broaden the genetic base of crops for disease resistance, as this remains the most practical and environmentally sound means for controlling major crop diseases (1, 34).

The Crop in the Landscape

The field in the landscape. Agricultural management of a field is superimposed on the complexities of the natural biota in the landscape. The distribution of cultivated land in a matrix of natural and seminatural vegetation gives abundant opportunities for the movement of pests and pathogens into crops. In contrast to the crop field, where some control is possible, native habitats may be beyond management. As an extreme example, consider the annual movement of virulent rust strains from wild relatives of wheat in the Himalayan foothills to cultivated wheat in India and Pakistan with resultant epidemics. Wild relatives of both sorghum and millets in sub-Saharan Africa often are affected by diseases spreading from wild relatives in and around crop fields.

Coevolution between the crop and its pests and pathogens has created useful sources of resistance to past and current variants. Recent studies by Burdon and Thrall (12), however, suggest that interactions between hosts and pathogens in native plant communities may be erratic and unpredictable and that the time scale for successful coevolution may be very long. This conclusion challenges the justification for maintaining vegetational diversity to foster useful coevolution between hosts and pathogens in the short term, and argues for *in situ* conservation of the wild relatives of crops, and the crops themselves, in agro-ecosystems.

The largest-scale impact of agricultural management on biological diversity is intercontinental crop and domestic animal introduction, which has major implications for the sustainability of crop production and for effective pest control. If crops can escape their co-evolved pests, they can flourish, but there is always a chance of new pests moving onto the crop, or of old pests catching up by escaping quarantine. For example, sorghum ergot recently has spread rapidly throughout the Americas and Australia causing serious losses to susceptible hybrids (6).

Weeds are a major problem with intercontinental transfers as well, since they also can escape pests and become a major problem in fields and pastures. Examples include *Sorghum halepense* from the Old World, which is a serious weed in North America. We still do not know enough to predict the likely impacts of intercontinental movement of crops and their associates.

Sorghum is a good example of extensive and continuous gene flow between the crop and ancestor, both belonging to *Sorghum bicolor* (15). Although the process could have added useful genes to the crop, it also created aggressive global weeds such as johnsongrass (21).

The complexity of the interrelationships between pests and crops in agro-ecosystems has generated contrasting views on pest management. Janzen (26) suggested that the agricultural potential of many parts of the tropics, especially the seasonally dry tropics, might be improved by destroying surrounding native vegetation. This process would destroy tremendous numbers of pests and pathogens that are supported by this perennial vegetation. In contrast, Altieri (5) stressed the importance of utilizing the full potential of pest/pathogen controls in the native vegetation by designing diverse, pest-stable, self-sustained, environmentally sound agro-ecosystems through the long-term maintenance of vegetational and landscape diversity. However, the understanding of the impacts of natural enemies on pests in such agro-ecosystems is at best incomplete.

The Crop

In addition to vegetational diversity in the landscape, there are many opportunities for partitioning diversity in the crop itself that affects pests and their management.

Spatial diversity. Crop diversity in space may be achieved both within the same field and between fields. Within-field diversity can include: (i) monocultures (single species) of varieties with complex genetic makeup, multilines, or varietal mixtures; (ii) intercrops–structured associations of two crops, *e.g.*, a cereal and a pulse; and (iii) mixed crops or polycultures, where two (or more) crops are mixed in an unstructured association.

Varietal mixtures. Varietal mixtures of many crops have been used by farmers for millennia and are still used in many developing countries. Studies of the contribution of varietal mixtures to disease management found that improved stability and decreased disease severity were common features of mixtures relative to their components in monoculture, but in most cases the yield advantage was small, *e.g.* 4-9% (40). In a study of the effect of hybrid mixtures in controlling sorghum leaf blight, Sifuentes-Barrera and Frederiksen (39) found that the addition of 1% resistant plants reduced disease development by 0.4% and increased the yield of susceptible hybrids by 0.55%.

Most studies on management of diseases and pests in varietal mixtures have targeted specialized cereal systems in temperate, developed regions and not subsistence agriculture in tropical regions. We expect the same principles to apply; however, the gains in production from spatial diversity within fields may be countered by the extended and overlapping seasons. Also, the close proximity of neighboring fields could exacerbate disease and pest problems. Clearly this is an area where further research is needed for sorghum and millet in tropical areas.

Intercrops. Most studies of pathogens and insect pests in intercrops have focused on cereal-legume associations. The most commonly reported effect of intercropping is reduced incidence and severity of diseases and reduced populations of insect pests, *e.g.*, the reduced disease incidence in beans intercropped with maize (3). In the majority of studies, however, the effect of reduced biotic stress on crop yield has not been measured. Wall *et al.* (46) found that sorghum-maize intercropping in Honduras reduced oval leaf spot, caused by *Ramulispora sorghicola*, disease severity compared to sole sorghum cultivation but sorghum-bean intercropping did not have any effect on sorghum disease severities.

Multiple cropping. Multiple cropping is largely practiced in tropical subsistence systems. The socioeconomic reasons for farming mixed cropping systems are better understood and far more compelling than the biological reasons (14). Although the choice of crops grown in mixed systems is usually planned, the spatial and temporal associations may be opportunistic or driven by factors unrelated to sound pest management principles. Few studies have been done to understand the multiplicity of interactions occurring between crops, pests, and beneficial organisms in such systems. This is another area in which additional research is needed (29, 36).

Until our understanding of the combined impact of the negative and positive effects of diversity within crops is greatly improved, there will always be risks of crop loss associated with maintaining weeds in crops. It is somewhat paradoxical that spatial diversity in crops is most common in agro-ecosystems where the least research has been done to understand the functional role of spatial diversity in pest management (29).

Deployment of diversity between fields may be planned by farmers and communities, but in areas where climatic and edaphic factors as well as economic considerations strongly influence a farmer's choices, the same crop, and even the same variety, may be grown widely in spite of potentially increased risks from pests. Economic influences often are the major drivers for practicing between-field diversity. If pest management strategies based on vegetational diversity are sufficiently robust within fields, then there should be less need for further strategies between fields.

Temporal diversity. Crop diversity in time is dictated primarily by climatic factors such as rainfall and temperature, and through cultural practices such as rotation (including fallows), sowing time, use of crops and varieties of different maturity, and planned deployment of specific crops and varieties over time. Temperature plays a critical role in determining temporal diversity in crops and agro-ecosystems. During the cold and/or dry season, pathogens overseason, insects enter diapause, and weed seed remains dormant in the soil. Depending on the severity of the conditions, inoculum levels usually are reduced to the benefit of the next season's crop. Year-round warmth and moisture in the tropics allows pests to multiply continuously either on the crop or on alternative hosts making pest management a never ending task (29).

Farmer practices such as fallowing, crop rotation, alteration of sowing date, and the use of cultivars with different maturities all rely on seasonality for effectively managing pests above and below ground by manipulating the diversity of components of the agro-ecosystem (42). Fallowing reduces the viability of the soil seed bank and, when associated with burning before planting, destroys insect pests, pathogens, and weed seed. Reducing the length of the fallow period in traditional systems results in significant yield reductions in rice due to increased weed pressure and declining soil fertility.

Monocultures. Monocultures, or single-species stands cultivated as uniform varieties, multiline or composite varieties, land races, and varietal mixtures, are the most common type of cropping system and produce most of our staple plant food, including sorghum and millet. In spite of becoming progressively more productive and resistant to pests through ongoing advances in agricultural science, species monocultures often are perceived to be unstable, unsustainable, ecologically dysfunctional, and highly vulnerable to pests.

From *ca.* 1950-1970, plant breeding was dominated by the search for single-gene resistances because it is easier to breed for single-gene resistance than for multiple-gene resistance. Although some single-gene resistances have proved stable and durable, there have been problems with the use of single-gene resistance for managing variable pathogens, *e.g.*, sorghum anthracnose. During the 1980s and 1990s, however, plant breeders contributed significantly to the durability of resistance to major pests by breeding crops with multiple resistances to variable pests and by introducing resistance from a wider range of sources.

Whatever the effects of within- and between-field diversity on pests in agro-ecosystems, the stability of monoculture irrigated rice–practiced for several thousand years–is the best example of the durability of monoculture agriculture. This system also produces more food than any other cropping system. For the foreseeable future, monocultures, especially for staple cereals, will continue to feed most of the world's present six billion inhabitants (17). We think that research resources would be better spent on understanding how to make monocultures more productive and resilient, rather than on how to make them more visually diverse with largely unproven benefits (29).

In conclusion, agrobiodiversity within the agricultural landscape must be planned–whether within the crop itself, within the crop in the same field, within the same field, between fields, or within the landscape–if it is to make a useful contribution to pest management (29). This contribution can be increased by combining it with cultural controls, biocontrols, and the judicious use of pesticides. A myopic focus on diversity would be a poor management strategy, as agro-ecosystem processes are influenced by more than diversity (43). How much agrobiodiversity is needed for productive agriculture? Each agricultural situation must be assessed individually for the crops, pests and associated biodiversity present. Additional diversity beyond that needed for a productive agro-ecosystem may contribute positively or negatively to pest management, but until it is tested, the outcome usually will be unpredictable.

References

1. **Akem, C., S. Ceccarelli, W. Erskine, and J. Lenné.** 2000. Using genetic diversity for disease resistance in agricultural production. *Outlook Agric.* **29:** 25-30.
2. **Allard, R. W.** 1990. The genetics of host-pathogen coevolution: Implications for genetic resource conservation. *J. Hered.* **81:** 1-6.
3. **Allen, D. J., R. Buruchara, and J. B. Smithson.** 1997. Diseases of common beans, p. 179-265. *In* D. J. Allen and J. M. Lenné (eds.), *The Pathology of Food and Pasture Legumes*. CABI Publishing, Wallingford, UK.
4. **Allen, D. J., J. M. Lenné, and J. M. Waller.** 1999. Pathogen biodiversity: Its nature, characterization and consequences, p. 123-153. *In* D. Wood, and J. M. Lenné (eds.), *Agrobiodiversity: Characterization, Utilization and Management.* CABI, Wallingford, UK.
5. **Altieri, M. A.** 1994. *Biodiversity and Pest Management in Agroecosystems.* Food Products Press, New York City, New York.
6. **Bandyopadhyay, R., and R. A. Frederiksen.** 1999. Contemporary global movement of emerging plant diseases. *Ann. New York Acad. Sci.* **894:** 28-36.
7. **Breese, W. A., C. T. Hash, K. M. Devos, and C. J. Howarth.** 2002. Pearl millet genomics and breeding for resistance to downy mildew, p. 243-246. *In* J. F. Leslie (ed.), *Sorghum and Millets Pathology 2000.* Iowa State Press, Ames, Iowa.
8. **Brush, S. B.** 1995. *In situ* conservation of land races in centers of crop diversity. *Crop Sci.* **35:** 346-54.
9. **Buddenhagen, I. W.** 1977. Resistance and vulnerability of tropical crops in relation to their evolution and breeding. *Ann. New York Acad. Sci.* **287:** 309-326
10. **Burdon, J. J.** 1987. *Diseases and Plant Population Biology.* Cambridge University Press, Cambridge, UK.
11. **Burdon, J. J.** 1992. The growth and regulation of pathogenic fungal populations, p. 173-181. *In* G. C. Carroll and D. T. Wicklow (eds.), *The Fungal Community: Its Organization and Role in the Ecosystem.* Marcel Dekker, New York City, New York.
12. **Burdon, J. J., and P. H. Thrall.** 1999. Spatial and temporal patterns in coevolving plant and pathogen associations. *Amer. Nat.* **153(Suppl.):** S15-S33.
13. **Ceccarelli, S., J. Valkoun, W. Erskine, S. Weigand, R. Miller, and J. A. G. van Leur.** 1992. Plant genetic resources and plant improvement as tools to develop sustainable agriculture. *Exper. Agric.* **28:** 89-98.
14. **Clawson, D. L.** 1985. Harvest security and intraspecific diversity in traditional tropical agriculture. *Econ. Bot.* **39:** 56-67.
15. **Cox, T. S., and D. Wood.** 1999. The nature and role of crop biodiversity, p. 35-58. *In* D. Wood and J. M. Lenné (eds.), *Agrobiodiversity: Characterization, Utilization and Management.* CABI, Wallingford, UK.

16. **de Milliano, W. A. J., R. A. Frederiksen, and G. D. Bengston, eds.** 1992. *Sorghum and Millets Diseases: A Second World Review.* ICRISAT, Patancheru (A.P.), India.
17. **Evans, L. T.** 1998. *Feeding the Ten Billion: Plants and Population Growth.* Cambridge University Press, Cambridge, UK.
18. **Frankel, O. H., A. H. D. Brown, and J. J. Burdon.** 1995. *The Conservation of Plant Biodiversity.* Cambridge University Press, Cambridge, UK.
19. **Grenier, C., P. J. Bramel-Cox, M. Noirot, K. E. P. Rao, and P. Hamon.** 2000. Assessment of genetic diversity in three subsets constituted from the ICRISAT sorghum collection using random vs. non-random sampling procedures. A. Using morpho-agronomical and passport data. *Theor. Appl. Genet.* **101:** 190-196.
20. **Grenier, C., M. Deu, S. Kresovich, P. J. Bramel-Cox, and P. Hamon.** 2000. Assessment of genetic diversity in three subsets constituted from the ICRISAT sorghum collection using random vs. non-random sampling procedures. B. Using molecular markers. *Theor. Appl. Genet.* **101:** 197-202.
21. **Harlan, J. R.** 1965. The possible role of weed races in the evolution of cultivated plants. *Euphytica* **14:** 173-6.
22. **Harlan, J. R.** 1976. Diseases as a factor in plant evolution. *Annu. Rev. Phytopathol.* **13:** 31-51.
23. **Harlan, J. R., and J. M. J. de Wet.** 1971. Towards a rational classification of cultivated plants. *Taxon* **20:** 509-17.
24. **Harper, J. L.** 1990. Pests, pathogens and plant communities: An introduction, p. 2-14. *In* J. J. Burdon and S. R. Leather (eds.), *Pests, Pathogens and Plant Communities.* Blackwell Scientific Publications, Oxford, UK.
25. **Ibrahim, K. M., and J. A. Barrett.** 1991. Evolution of mildew resistance in a hybrid bulk population of barley. *Heredity* **67:** 247-256.
26. **Janzen, D.** 1973. Tropical agroecosystems. *Science* **182:** 1212-1219.
27. **Lenné, J. M.** 1998. The biodiversity and conservation of crops for disease resistance. *7th Intl. Cong. Plant Pathol. (Edinburgh, Scotland).* Abstract 4.1.1S.
28. **Lenné, J. M., and D. Wood.** 1991. Plant diseases and the use of wild germplasm. *Annu. Rev. Phytopathol.* **29:** 35-63.
29. **Lenné, J. M., and D. Wood.** 1999. Vegetational diversity in agroecosystems: a mixed blessing for successful pest management? *BCPC Symp. Proc.* **73:** 75-98.
30. **Marchais, L.** 1994. Wild pearl millet population (*Pennisetum glaucum*, Poaceae) integrity in agricultural Sahelian areas. An example from Kieta (Niger). *Plant System. Evol.* **189:** 233-245.
31. **Marshall, D. R.** 1977. The advantages and hazards of genetic homogeneity. *Ann. New York Acad. Sci.* **287:** 1-20.
32. **Maxted, N., B. V. Ford-Lloyd, and J. G. Hawkes, eds.** 1997. *Plant Genetic Conservation: The* in situ *Approach.* Chapman and Hall, London, UK.
33. **Moore, G., K. M. Devos, Z. Wang, and M. D. Gale.** 1995. Grasses, line up and form a circle. *Curr. Biol.* **5:** 737-739.
34. **Ortiz, R.** 1999. Genetic enhancement and base broadening efforts, p. 191-203. *In* T. Gass, L. Frese, F. Begemann, and E. Lipman (eds.), *Conservation and Sustainable Utilization of Plant Genetic Resources for Food and Agriculture–Implementation of the Global Plan of Action in Europe.* International Plant Genetic Resources Institute (IPGRI), Rome, Italy.
35. **Ortiz, R.** 2002. Germplasm enhancement to sustain genetic gains in crop improvement, p. 275-290. *In* J. M. M. Engels, V. R. Rao, A. H. D. Brown, and M. Jackson (eds.), *Managing Plant Genetic Diversity.* IPGRI, Rome, Italy and CABI, Wallingford, UK.
36. **Polaszek, A., C. Riches, and J. M. Lenné.** 1999. The effects of pest management strategies on biodiversity in agroecosystems, p. 273-303. *In* D. Wood and J. M. Lenné (eds.), *Agrobiodiversity: Characterization, Utilization and Management.* CABI, Wallingford, UK.
37. **Rai, K. N., S. A. Rao, and K. N. Reddy.** 1997. Pearl millet, p. 243-258. *In* D. Fuccillo, P. L. Sears, and P. Stapleton (eds.), *Biodiversity in Trust: Conservation and Use of Plant Genetic Resources in CGIAR Centers.* Cambridge University Press, Cambridge, UK.
38. **Ridley, H. N.** 1930. *The Dispersal of Plants Throughout the World.* L. Reeve, Ashford, UK.
39. **Sifuentes-Barrera, J. A., and R. A. Fredericksen.** 1994. Evaluation of sorghum hybrid mixtures for controlling sorghum leaf blight. *Plant Dis.* **78:** 499-503.
40. **Smithson, J. B., and J. M. Lenné.** 1996. Varietal mixtures: A viable strategy for sustainable productivity in subsistence agriculture. *Ann. Appl. Biol.* **128:** 127-158.
41. **Stenhouse, J. W., K. E. P. Rao, V. G. Reddy, and S. A. Rao.** 1997. Sorghum, p. 292-308. *In* D. Fuccillo, P. L. Sears, and P. Stapleton (eds.), *Biodiversity in Trust: Conservation and Use of Plant Genetic Resources in CGIAR Centers.* Cambridge University Press, Cambridge, UK.
42. **Thurston, H. D.** 1992. *Sustainable Practices for Plant Disease Management in Traditional Systems.* Westview Press, Boulder, Colorado.
43. **Tilman, D.** 1999. The ecological consequences of changes in biodiversity: A search for general principles. *Ecology* **80:** 1455-1474.
44. **Tripp, R.** 1996. Biodiversity and modern crop varieties: sharpening the debate. *Agric. and Human Values* **13:** 48-63.
45. **Ullstrup, A. J.** 1972. The impacts of the southern corn leaf blight epidemic of 1970-1971. *Annu. Rev. Phytopathol.* **10:** 37-50.
46. **Wall, G. C., R. A. Frederiksen, J. Craig, and M. J. Jege.** 1992. Epidemiology of sorghum diseases in Central America: a case study, p. 303-317. *In* W. A. J. de Milliano, R. A. Frederiksen, and G. D. Bengston (eds.), *Sorghum and Millets Diseases: A Second World Review.* ICRISAT, Patancheru (A.P.), India.
47. **Williams, R. J., and R. A. Frederiksen, eds.** 1980. *Sorghum Diseases: A World Review.* ICRISAT, Patancheru (A.P.), India.
48. **Wills, C.** 1996. Safety in diversity. *New Scient.* **2002:** 38-42.

49. **Witcombe, J. R.** 1999. Does plant breeding lead to a loss of genetic diversity?, p. 245-273. *In* D. Wood and J. M. Lenné (eds.), *Agrobiodiversity: Characterization, Utilization and Management*. CABI, Wallingford, UK.

50. **Witcombe, J. R., and C. T. Hash.** 2000. Resistance gene deployment strategies in cereal hybrids using marker-assisted selection: Gene pyramiding, three-way hybrids, and synthetic parent populations. *Euphytica* **112:** 175-186.

51. **Wood, D., and J. M. Lenné.** 1997. The conservation of agrobiodiversity on-farm: Questioning the emerging paradigm. *Biodiversity and Conservation* **6:** 109-129.

52. **Wood, D., and J. M. Lenné, eds.** 1999. *Agrobiodiversity: Characterization, Utilization and Management.* CABI, Wallingford, UK.

53

Sorghum On-Line Crop Information

Viswanathan Mahalakshmi, Belum V. S. Reddy, Ranajit Bandyopadhyay, Hari C. Sharma, Nanduri K. Rao, and Rodomiro Ortiz

The problem faced by many agricultural scientists and extension specialists is how to package and deliver technical information in a timely and orderly fashion. The sharing and dissemination of information has undergone a profound change as a part of the ongoing digital revolution. The current revolution in information technology has opened up new ways to disseminate information both quickly and in a timely manner. The International Crops Research Institute for the Semi-Arid Tropics (ICRISAT) conserves and enhances the productivity of sorghum, pearl millet, chickpea, pigeonpea, and groundnut, all of which are grown in the semi-arid tropics.

ICRISAT holds in trust 113,000 accessions of these crops under an agreement between the Food and Agricultural Organization (FAO) of the United Nations and the Consultative Group on International Agricultural Research (CGIAR). These germplasm accessions have been systematically characterized to maintain and distinguish their identity and to enhance their utilization. ICRISAT also has collected information on various aspects of these crops including screening methods and the host plant responses to biotic and abiotic constraints. The on-line crop information system was designed as a new means of transferring and sharing up-to-date information with farmers, researchers, extension workers, and students worldwide. This system consists of educational information, and diagnosis and control measures for management of common diseases, pests, parasitic weeds, and nutritional disorders of sorghum. Digital diagnostic systems for the diagnosis and management of diseases, insect pests, plant nutritional disorders, and parasitic weeds for these crops were developed to provide new and innovative methods of transferring timely and up-to-date information to farmers and researchers via the World Wide Web (WWW). We developed WWW-based learning modules for crop production. These digital learning systems overcome the barriers of fixed timing of courses and differing rates of progression through the material.

Sorghum [*Sorghum bicolor* (L.) Moench], one of ICRISAT's five mandate crops, is the fifth most important cereal crop grown in the semi-arid regions of the world after rice, wheat, maize, and barley. Sorghum grain yields on farmers' fields in Asia and Africa are generally low (500-800 kg ha^{-1}), primarily due to stresses from insects, diseases, weeds, and drought. Nearly 150 insect species have been reported as pests on sorghum (8, 12). Sorghum is a host for more than 100 plant pathogens, including fungi, bacteria, viruses, and nematodes (4). Sorghum also is attacked by the parasitic weed *Striga*. Though precise crop loss estimates are difficult to make, complete crop failures caused by *Striga* have been recorded and could occur again in many regions of Africa and India. *Striga* control not only would increase food production in *Striga*-infested areas, but also would allow production of sorghum again in areas that were abandoned due to *Striga* infestation (11). Sorghum is grown on very diverse soils such as Vertisols, Alfisols, Oxisols, Ultisols, Entisols, and Inceptisols. The range of soils is so wide that this crop suffers from both low availability of certain nutrients in some environments and excessive quantity of some of the same nutrients in other locations. Most of the symptoms of nutritional disorders in sorghum have been well documented and described by Grundon *et al.* (5). Hence, improvements in sorghum production in these harsh environments often are more readily achieved through better production practices than through additional inputs. Thus, knowledge of the factors limiting sorghum production and measures to overcome these constraints is essential for both researchers and extension workers.

Visible symptoms are the first clues to the identity of the cause of disorders due to pests, disease, nutritional disorders, and parasitic plants. A critical problem is how

FIGURE 53-1. Screen-capture image of ICRISAT sorghum on-line information.

to package and deliver the relatively large amounts of technical information on these systems to the user in an organized and timely manner. Digital diagnostic systems are one method for transferring timely, up-to-date information to farmers and researchers. In the past, this information sharing occurred through printed bulletins and reference manuals, with new information and updates provided through meetings, newsletters, and demonstrations. Recently, there have been efforts to disseminate such information on digital media. The digital media, like their printed predecessors, require distribution of the new or updated material to the user as it becomes available.

The recent revolution in information technology, with the Internet and the WWW, provides a new means for information dissemination that can be updated quickly and inexpensively. A Web-based decision support system for crop management, Pl@nteInfo® (http://www.planteinfo.dk) is one such system currently available for farmers in Denmark (7). A properly developed digital diagnostic system can be a powerful tool for the farmers, extension agents, and researchers. Such systems can help them with the day-to-day decisions that must be made to grow a crop successfully. In the past, ICRISAT trained sorghum researchers and extension workers on production practices of this crop through its in-service training program, and it still uses special training programs for specialized skills. Instructor-based training courses, though very effective in the transfer of information and the development of skills, are both expensive and require constant updating and continued investment. Our objective was to develop a World Wide Web site, http://grep.icrisat.cgiar.org/project1/sorghumhomepage.htm, where information on sorghum crop production could be obtained by researchers and extension workers worldwide. In addition, an on-line, digital diagnostic system is available to help farmers identify problems based on observed visual symptoms and to suggest remedial measures.

System Design and Development

Data and information source. To develop this system, we collated information from various sources onto a single platform. Information on diseases, pests, and *Striga* were obtained from the relevant ICRISAT information bulletins (11, 13, 14) and Frederiksen (3). The information on plant nutritional disorders (5) was provided by the Australian Centre for International Agricultural Research (ACIAR). The information on taxonomy was compiled from de Wet (1), Dogget (2), Harlan and de Wet (6), Lazarides *et al.* (9), and Murthy *et al.* (10). Information on sorghum germplasm and germplasm enhanced for its response to various biotic and abiotic constraints that is available through ICRISAT also was used to develop this site.

Computer Platform. As the information was intended for retrieval through the WWW from a Web server, it was not necessary to consider the host platform as a potential limitation for every client. Our information dissemination system is compatible with the two most popular browsers (Netscape Communicator™ and Internet Explorer™). We used industry standards, which facilitate consistency, to make the interface intuitive, predictable, and comfortable, and to allow users to access the information quickly and learn quickly. Screen layouts, dialog boxes, and buttons in the graphical user interface are similar to those commonly used at other sites on the WWW.

Database design and structure. Information was incorporated into a relational database management system. As the information was to be publicly available on the WWW, both security and accessibility were important considerations. Our corporate platform is based on Microsoft™; therefore, data and information were deployed in either MS ACCESS™ or MICROSOFT SQL™ 7.0.

Programming languages. The system was designed to respond to client queries and to filter selected information for the user. Therefore, query-based languages rather than static programs were used. Microsoft Active Server Page (ASP™) with scripting languages like VB Script and Java Script were used for client-based querying. For displaying information in a graphic format, programming was done in Sunsystem JAVA™ 2.0 and embedded.

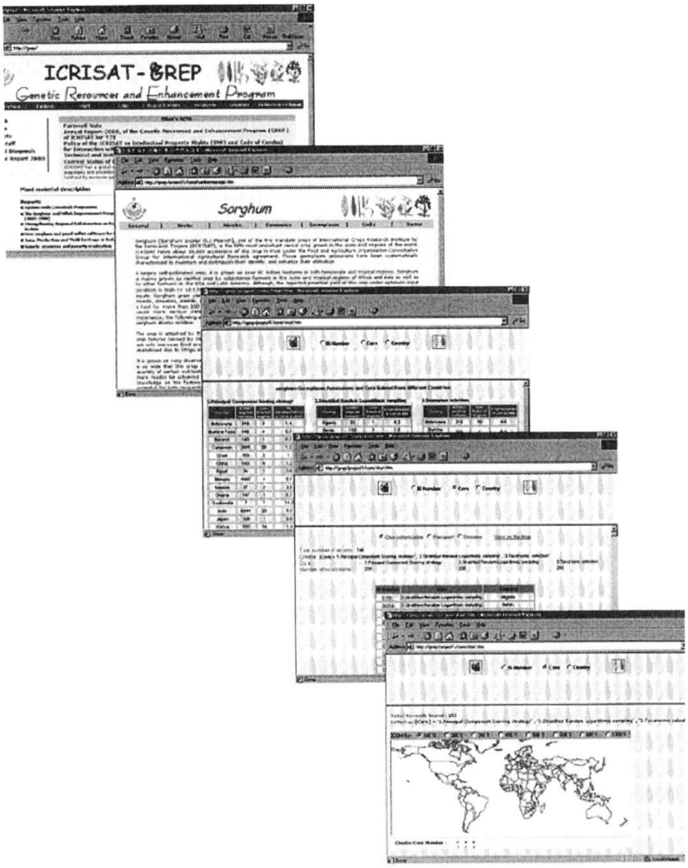

FIGURE 53-2. Screen-capture image output from a sample query on core collections.

System Outputs

Accessing the information. The online information is available on the Genetic Resources and Enhancement Program page, through the ICRISAT home page (http://www.cgiar.org/icrisat). Under the item "crops" on the menu bar on the layout of the opening page of GREP (http://grep.icrisat.org/text/research/grep/homepage.htm) are the choices of ICRISAT's mandate crops. Selecting sorghum transfers the user to a page that lists the available information on this crop (Fig. 53-1). From this page, the user can access information from the list by clicking on an icon and further querying on the item chosen (Fig 53-2). Informa-tion on other aspects of this crop can be accessed in a similar manner.

Information on germplasm, core collection, and breeding material. Selecting germplasm on the list results in a display of information in the left frame on the number of accessions by the country of origin. The available search options and the sources on which the search will be made are displayed. The passport table has information that was gathered at the time of collection, while the characterization table contains both botanical and agronomic descriptors. On-line help is generated dynamically and contains information on how to use the system and limits of the values for the different parameters. The statistical function and the frequency distribution modules allow in-depth analyses of data that can be viewed both graphically and that also can be displayed on a world map with a zoom option for the retrieval of information on the queried germplasm accessions. The core collection, which contains the maximum genetic diversity, is available on-line with complete information. Accessions in the core were chosen primarily to be representative of the whole collection, to be ecologically or genetically distinct from one another, and to maximize genetic diversity. This core collection serves as an entry point to the entire germplasm collection. Identification of 'core collections also simplifies the process of multiplication and distribution of seed to potential users by substantially reducing the number of accessions that need to be maintained in active collections.

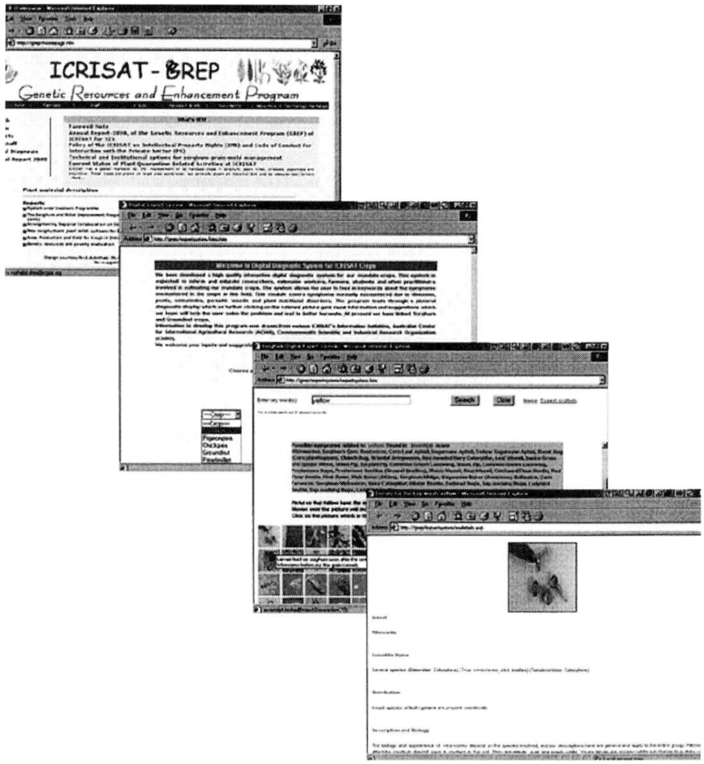

FIGURE 53-3. Screen-capture image output from a search on the digital diagnostic system.

Core collections were selected based on the available information and then subdivided further to cluster groups of similar accessions. The breeding module contains complete pedigree information on all of the enhanced material available from ICRISAT. Information on the performance of individual lines for traits of interest also can be displayed, and further information on parents used in the breeding program can be obtained by clicking on the icon for the parent, which is linked to the germplasm passport data. Seeds of both germplasm and breeding lines can be received by accepting the material transfer agreement online and placing the request.

Similarly, information on diseases, pests, parasitic weeds, and plant nutritional disorders is available through this on-line system from the main crop page, where querying is possible either with the common disease name or with the scientific name. There also is a two-module on-line learning system on sorghum that is based on sorghum production practices that were drawn from an earlier training manual. This manual also is available on a CD-ROM entitled "Computer Based Tutorials on Development of Cultivars and Seed Production Techniques in Sorghum." This module is a complete learning system including a self-assessment for some chapters. Another module identifies various resistance mechanisms and screening methods for pests and pathogens of sorghum. This module has the same features as the production practices module and allows the user to assess their mastery of the material.

Digital diagnostic system. This comprehensive system spans ICRISAT's mandate crops and can be accessed from http://www.icrisat.org/text/research/grep/homepage.htm. Clicking on the digital diagnostic system on this page transfers the user to the opening module of the digital diagnostic system (Fig. 53-3). The crop option is displayed both in picture and list box format (for quicker downloading when connection speeds are slow). Currently this system is available only for sorghum and peanut. As this information becomes available for other crops, appropriate links for them will be added here. This strategy was adopted to keep the modules similar for all of the crops and to make it easier for users to access the system. Selecting a crop either from the list box or clicking on the crop image generates a screen where the user can enter the symptoms observed.

Unlike the traditional "expert system" programs, in which a series of questions are asked to reduce the number of possibilities until the system identifies a suitable diagnosis, this system allows the user to feed in keywords associated with the symptoms. The reason for this choice was that a series of questions requires that a connection be established to the server to obtain the next set of options after each question is answered. This process can be both time- and resource-consuming, and, given the connection speeds of many clients, we thought the keyword approach would be more effective. The better the description of the symptoms, the more rapid the search, and the fewer the alternatives to choose between. The keywords are searched in the symptom section, and thumbnail images of the relevant diseases, pests, nutritional disorders, and parasitic weeds are displayed. When the mouse is moved over a thumbnail image, the related disorder is highlighted as a tool-tip. The user can view these images and select the one that is the closest to the observed symptoms. Clicking on a thumbnail image brings up the image and information on the problem and its control. The key words entered by the user also are highlighted. Clicking on the picture transfers the entire image to a new window in which the user can zoom in on a particular portion of the image by selecting the area to be enlarged. This option could be useful when comparing the observed specimen with an image. If the user is not satisfied with the output, then he/she can close this window and repeat the process by selecting an alternative thumbnail image until he/she is satisfied with the results.

Discussion

A comprehensive on-line information system that provides a wide array of information on sorghum crop production should improve the dissemination of skills and knowledge of this crop. This information system was developed as an interactive system through querying to provide specific information rather than flooding the user with a vast amount of information, which then must be sorted. The clients targeted are growers, researchers, students, and extension workers. Different users may access different information. To the extent that relevant information from other databases also might be of interest to the user, links have been established so that the user can determine the amount of detail displayed. Another advantage of the on-line system is that any module can be modified at any time, without needing to send the modifications to all of the known users. The overall goal for this online system was to facilitate the use of the knowledge and skills regarding sorghum crop production gathered through the years by various individuals and institutions by the widest range of users through the WWW.

Acknowledgements

ICRISAT thanks the Australian Centre for International Agriculture (ACIAR) for permission to share the information on the nutritional disorders of sorghum on the WWW. I thank S. A. Arshad, M. Raja Ganesh, Upendra Kota, Y. Leela Devi, V. Srikanth, and M. Ali for programming assistance.

References

1. **de Wet, J. M. J.** 1978. Systematics and evolution of *Sorghum* section *Sorghum* (Gramineae). *Amer. J. Bot.* **65:** 477-484.
2. **Dogget, H.** 1988. *Sorghum.* The Tropical Agricultural Series. John Wiley & Sons, New York City, New York.
3. **Frederiksen, R. A., ed.** 1986. *Compendium of Sorghum Dis-eases*, 1st ed. APS Press, St. Paul, Minnesota.
4. **Frederiksen, R. A., and G. N. Odvody, eds.** 2000. *Compendium of Sorghum Diseases*, 2nd ed. APS Press, St. Paul, Minnesota.
5. **Grundon, N. J., D. J. Edwards, P. N. Takkar, C. J. Asher, and R. B. Clark.** 1987. *Nutritional Disorder of Grain Sorghum.* Australian Centre for International Agricultural Research (ACIAR), Indian Council of Agricultural Research (ICAR) and University of Queensland, Brisbane. ACIAR monograph 2.
6. **Harlan, J. R., and J. M. J. de Wet.** 1972. A simplified classification of cultivated sorghum. *Crop. Sci.* **12:** 172-176.
7. **Jensen, A. L., P. S. Boll, I. Thysen, and B. K. Pathak.** 2000. Pl@nteInfo®–a web-based system for personalized decision support in crop management. *Comput. Electron. in Agric.* **25:** 271-292.
8. **Jotwani, M. G., W. R. Young, and G. L. Teetes.** 1980. *Elements of Integrated Control of Sorghum Pests.* FAO Plant Production and Protection Papers. FAO, Rome, Italy.
9. **Lazarides, M., J. B. Hacker, and M. H. Andrew.** 1991. Taxonomy, cytology and ecology of indigenous Australian sorghums. *Austr. Syst. Bot.* **4:** 591-635.
10. **Murty, B. R., V. Arunachalam, and K. B. L. Saxena.** 1967. Classification and catalogue of old collection of sorghum. *Ind. J. Genet.* **27:** 1-394.
11. **Ramaiah, K. V., C. Parker, M. J. V. Rao, and L. J. Musselman.** 1983. Striga *Identification and Control Handbook.* Information Bulletin no. 15, ICRISAT, Patancheru (A.P.), India.
12. **Sharma, H. C.** 1993. Host plant resistance to insects in sorghum and its role in integrated pest management. *Crop Protect.* **12:** 11-34.
13. **Teetes, G. L., K. V. S. Reddy, K. Leischner, and L. R. House.** 1983. *Sorghum Insect Identification Handbook.* Information Bulletin no. 12, ICRISAT, Patancheru (A.P.), India.
14. **Williams, R. J., R. A. Frederiksen, and J. C. Girard.** 1978. *Sorghum and Pearl Millet Disease Identification Handbook.* Information Bulletin no. 2, ICRISAT, Patancheru (A.P.), India.

54

Sorghum Grain Quality for Increased Utilization

Ralph D. Waniska and Lloyd W. Rooney

Abstract

To increase utilization, sorghum grain needs to be improved beyond the elimination of deterioration caused by diseases and insects. Higher test weight and improved appearance of the grain are needed. Selecting against dark secondary-plant and glume colors decreases damage-induced pigmentation in the pericarp and yields milled products with fewer off colors and pigments. Traits of food-type sorghums include white or clear pericarp, tan secondary-plant-color, straw-colored glumes, nonpigmented testa, intermediate to hard endosperm texture, and ability to be milled into products with a bland flavor, white color, and no off colors. Harder-endosperm and spherical-shaped sorghums result in higher decortication yield and lighter-colored grits and flours. The ability to be milled into uniform small-particle-size flour also is desirable. The bran contains most of the pericarp and germ, sources of nutraceuticals, *e.g.*, antioxidants and fiber. The nutritive value, *e.g.*, protein, oil, and starch, and functionality should be improved, but not at the price of increased grain deterioration. Most cultivated sorghums no longer contain condensed tannins in the (pigmented) testa layer of the pericarp. Sorghums with improved properties will need to be handled to preserve their identity.

Introduction

Sorghum is used mainly as a livestock feed in the Western Hemisphere, but significant quantities are used for human food in Africa and Asia. The grain properties required to produce high-quality food products have been determined by collaborative efforts among research groups throughout the world (29, 37-39). This information has been used to develop sorghums with acceptable food quality as well as improved agronomic properties, disease and insect resistance, and stable yields. This chapter identifies opportunities to improve sorghum grain beyond grain molding and weathering and specifically addresses issues of appearance, milling, nutritional value, and tannins.

Appearance of Grain

Feed sorghums usually are red seeded with purple or red plant color that causes significant off color and, many times, off flavors in sorghum products used for food. Thus, research to incorporate desirable genes for food into sorghums useful for food processing has been the goal of research groups at ICRISAT and at Texas A&M University. ICRISAT has promoted the use of food sorghums as well, especially in southern Africa. Texas A&M University has released numerous sorghum lines with inherently improved food quality attributes. These and several A, B, and R lines are in part the basis for food-type sorghums grown in Texas, Mexico, and other areas (35). Indian scientists also have released many food-type sorghum hybrids for use in Asia (29).

Food-type sorghums have white or clear pericarp, tan secondary-plant-color, straw-colored glumes, nonpigmented testa, intermediate to hard endosperm texture, and can be milled into products with bland flavor, white color, and no off colors. These new hybrids are later, taller, and yield quite well under irrigation. Significant quantities of the resulting grain are now available in identity-preserved storage, and can be used to produce flour and other products that have a light color and a bland flavor. Sorghum for food in India is grown in the rabbi, or dry, season, which produces white grain free of blemishes that is highly prized for food processing. Sorghum grown in the Kharif, or wet, season has staining and molding that reduces food quality. Deterioration of sorghum caused by molding and/or weathering, *e.g.*, by microbes, insects, sunshine, or rain, decreases the value and quality of the

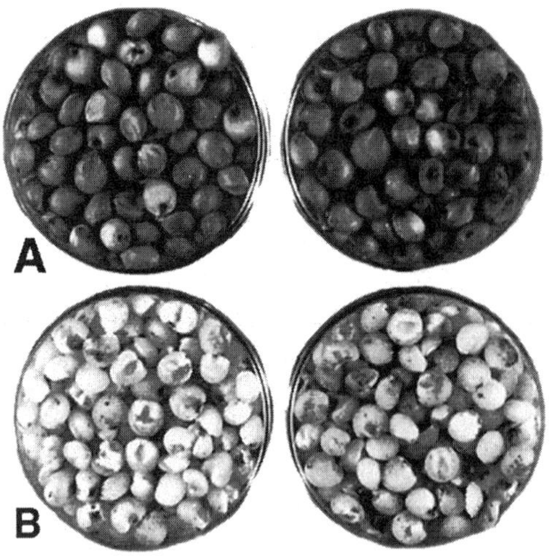

FIGURE 54-1. The effect of growing environment on the appearance of whole (top row) and decorticated (bottom row) caryopsis of a commercial hybrid with red pericarp and no pigmented testa. Phenolic compounds (nontannin) from the pericarp stain and darken the pericarp and endosperm when grown in a mold-conducive environment. The mold-free environment of Dumas, Texas yielded caryopses with fewer dark spots (left) than those from the mold-conducive environment of College Station, Texas (right).

grain (38, 43, 45, 47). Some food-type sorghums are moderately resistant to grain molding (7, 43).

The food-type sorghums also are preferred for livestock feeds and pet foods since they have light color, bland flavor, and lack the purple glumes that can cause black specks in eviscerated broilers. The specks in broiler carcasses cause greater rejection of broilers by USDA inspectors. The tan plant and straw-colored glume characters apparently solve this problem. In many cases, the light color also is advantageous for livestock feeds, although the trait is cosmetic and has no nutritional significance *per se*. Red-pericarp sorghums with tan secondary-plant color and light-colored glumes also have useful properties for feed and food processing (35).

Not all white-appearing sorghums are food type (35, 47). White kernels produced on plants with dark-pigmented secondary plant and glume colors often are stained in the field (47). Grain quality is acceptable when the grain matures during hot, dry weather, but high humidity and precipitation cause quality problems (29). These problems often occur in dry areas, *e.g.*, Mali, where local sorghums often have large purple glumes that can stain the grain during late-season rains. Staining decreases the yield of decorticated grain and flour when milled to obtain light-colored products (40). In Mali, changing from purple to tan plant color improved the grain quality and food processing potential of N'tenemissa, a white pericarp, tan-plant, partially photosensitive variety.

Sorghums with red pericarp color, including those with a pigmented testa, often are used to produce malt and in local alcoholic beverages and porridges. Special methods to inactivate the tannins have been devised. In some cases, consumers prefer the brown or red sorghum products. Sorghums in the United States have a red or white pericarp color, and the seed industry refers to sorghums as cream, red, bronze, and yellow. Cream sorghums have a white pericarp, purple plant color, no testa, and yellow or hetero-yellow endosperm. Hetero-yellow sorghum hybrids have a yellow endosperm pollinator with a nonyellow female as the parents. The yellow endosperm term means that kernels appear yellow because they have a thin, colorless pericarp, no testa, and a bright yellow endosperm. Bronze sorghum hybrids produce grain with hetero-yellow endosperm a thin, red pericarp, and no testa. Some new hybrids have bright red pericarp color due to the presence of an intensifier gene. Genetics of kernel characteristics are discussed in more detail elsewhere in this volume (48).

The development of new food-type hybrids along with their identity preservation in the marketing system provides significant amounts of high-quality sorghums for food processing. This availability could lead to increased use of sorghum in many products over the next few years. Sorghum flour sometimes is preferred to maize flour because of its light color and bland flavor (35). In Central America, tan-plant, white sorghums are preferred to local white sorghums with purple-plant color for making tortillas because the resulting products are lighter in color.

Dry Milling (Decortication and Milling)

During traditional dehulling or decortication of the pericarp from the endosperm, the sorghum grain is usually wetted with excess water, drained, and pounded in a mortar with additional water added as required (35, 40). The hulls, or bran, are removed by washing in excess water or by air drying and winnowing, and the decorticated material is ground by additional pounding into flour. Commercially, tempering is not usually done unless degermination is intended. Usually, the pericarp is removed without addition of water, which avoids additional staining of the endosperm and the need to dry milled stocks.

Dry milling operations for sorghum vary from cracking whole grain to produce a coarse milled product;

dehulling; degerming to produce pearled whole or broken kernels; and abrasive or attrition milling to produce a moderately fine flour (12, 28, 30, 34, 36). Sorghum usually is not tempered prior to abrasively removing the outer layers of the kernel with rice milling equipment or specially designed pearlers that contain stones or resinoid discs. Typically, 5-20% of the initial weight is removed as bran depending upon the degree of refinement desired. The decorticated grain may be used directly, may be cracked and sieved to produce various sizes of grits, or may be ground to flour.

Spherical sorghum kernels with a high proportion of hard endosperm are ideally suited for decortication by pearling procedures. The abraded material initially is mostly pericarp, and is high in fiber, ash, fat, and protein. As more peripheral endosperm is abraded, starch content of the bran increases and the other components decrease. Soft endosperm sorghums disintegrate more during abrasive milling, and more endosperm is lost in the bran fraction, which decreases the decortication yield. Grains damaged by insects and molds often are stained by the growth of the fungi. The grain colonized by fungi also is enzymatically softened, resulting in decreased decortication yields (Fig. 54-1). Sorghums with similar endosperm hardness vary in the ease of pericarp removal (Fig. 54-2) and the color of the decorticated grain (2). ATx635 imparts significantly improved milling properties to its hybrids.

Flour is produced in an attrition stone or plate mill with sifting to remove large pieces of bran. Flour quality is highest when stone or plate milling is used. Hammer milling is energy efficient, but produces grainy, sandy flours that are not preferred by consumers unless the flour is processed further to reduce the sandy particles, *e.g.*, porridge production or fermentation. The appearance of decorticated, ground grain also is reduced by black specks from the hilum.

A wheat break and roller milling system is used to produce sorghum flour. Sorghum is tempered to 15-16% moisture just prior to roller milling. Tempering improves the separation of bran from flour and sometimes gives acceptable products (12). Sorghums with soft, floury endosperm are more easily crushed and the flour separated from the pericarp. When refined products are required, the quality and yields of flour and grits produced by wheat flour mills are sometimes undesirable. Sorghums with soft, floury endosperm are easily crushed and separated from the pericarp. Sorghum grain is harder than wheat kernels, and hammer-milled sorghum flour consists mostly (> 97%) of particles ≥ 180 µm in diameter, while wheat flour only contains particles ≤ 125 µm in

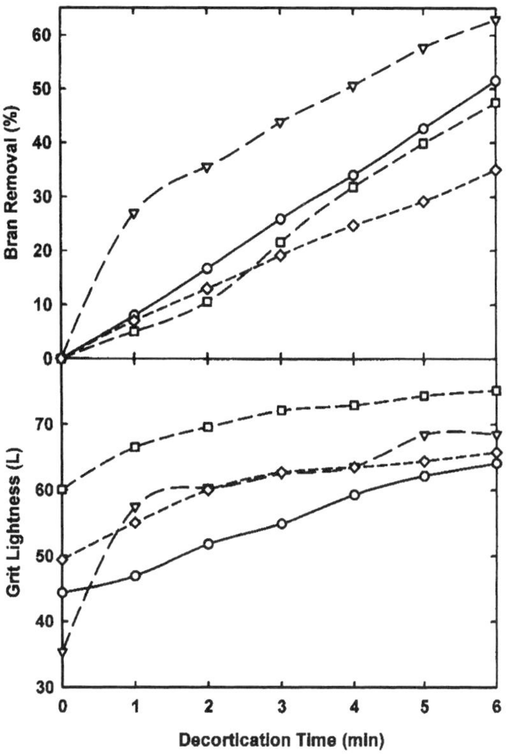

FIGURE 54-2. Effect of decortication time on bran removal and lightness of decorticated sorghum grain beginning with whole grain (time zero). Results presented are the average of three experiments. White pericarp (□), black pericarp (▽), red pericarp (◇), and tan pericarp (□).

diameter (2; Table 54-1). Roller milling generates fewer large flour particles, but ≥ 97% of the particles are ≥ 125 µm in diameter. Sequential milling using decortication, hammer milling, and roller milling, and results in fewer large flour particles, but 83% of the particles remain 125 µm in diameter. The food-type sorghums have slightly higher test weights (density of a volume), true density (grain density), and slightly higher yields of decorticated grain than the red-pericarp sorghums. The combination of decortication and roller milling of white food-type sorghums yields flour that appears white and from which most of the black specks from the hilum have been removed.

In India, sorghum flour (≥ 95% of grain) is obtained by using a stone mill with sifting after milling to remove large pericarp pieces. The coarse, particle-size flour (atta) is used for roti, dosa, and porridges. In many areas, sorghum is decorticated by traditional hand pounding in a

TABLE 54-1. Effects of type of dry mill on weight distribution of flour particles produced from a decorticated (30%), intermediate-textured sorghum. Least significant difference ($p \leq 0.05$) = 0.4.

Dry Mill Type	Amount (%) of Flour on U.S. Standard Sieves					
	# 40 < 425 μm	# 60 < 250 μm	# 80 < 180 μm	# 100 < 150 μm	# 120 < 125 μm	Pan > 125 μm
Hammer	83	9.1	5.9	1.3	0.5	0.2
Roller	8.8	44	34	7.0	3.0	2.5
Hammer+ Roller	0.2	0.3	51	20	11	16

mortar and pestle followed by attrition milling into flour. Botswana has a large number of small mills that produce sorghum flour based on the abrasive decorticators made in Botswana (34) followed by grinding. Mills in Nigeria, Mexico, the United States, Kenya, Sudan, and South Africa produce a wide variety of sorghum meals, flours, grits, and decorticated kernels for porridges, couscous, and baking. Production of processed sorghum products often is limited by lack of grain of sufficient quality. Sorghum often costs more than maize, *e.g.*, in South Africa sorghum products are taxed at a rate of 14%, while maize products are not taxed at all.

The advantage of food-type sorghums over red-pericarp sorghums is in the yield of decorticated grain and in the production of flour with a uniform color (30, 40; Fig. 54-3). High quality decorticated grits and white flour have lightness (L) values of 70 and 85, respectively. Lightness values range from 0 (black) through 100 (opaque, white) (2, 12). Red-pericarp sorghums had substantially reduced dry milling yields relative to white sorghums to attain similar levels of whiteness. Also, yields of flour and grits of the white food sorghums, *e.g.*, W-902W, are inversely related to the amount of weathering that occurred in the field prior to harvest. Dorado had bright white grain free of weathering. Thus, for milling comparisons, yields at constant lightness values should be used.

The particle sizes of flours produced from sorghum are coarser than those from wheat (36, 38). All the particles of wheat flour pass through a 120-mesh sieve, while less than 20% of sorghum flour particles are usually this small (Table 54-1). Elimination of the larger particles from sorghum flour would decrease the grittiness observed in many baked products. Hammer-milled sorghum flour contains larger flour particles (most will not pass through an 80-mesh sieve) than does flours produced by other milling procedures. Decortication followed by hammer milling followed by roller milling yielded smaller particles in the flour prepared from an intermediate texture sorghum. Roller milling produced flour with the smallest particle size from sorghums with hard endosperm (Table 54-1).

Nutritional Value

The nutritive value and functionality of the components (protein, oil, and starch) of sorghum grain should be improved, but not at the price of increased grain deterioration. A modest amount of research has occurred on the proteins and somewhat less on the starches present in sorghum, but little research has been conducted to improve the content or functionality of other components, *e.g.*, lipids, fiber, minerals, and antioxidants. Protein and starch content and composition vary due to genotype and agronomic conditions, *e.g.*, water availability, soil fertileity, temperature, and environmental conditions, during grain development (47). Sorghum proteins are located in the endosperm (80%), germ (16%), and pericarp (3%). Kafirins, or prolamins, then glutelins comprise the major protein fractions in sorghum and are located primarily within the protein bodies and protein matrix of the endosperm, respectively (47).

Sorghum protein is deficient in the essential amino acid lysine, with only 40% of the recommended daily intake (RDI) for young children consuming a diet composed solely of sorghum (11). Protein quality is critically important in developing countries where human diets consist mainly of cereals. Supplementation of a sorghum-based diet with relatively small amounts of legumes (15-20% of the cereal weight) can overcome most amino acid deficiencies (11, 47). The albumin, globulin, and glutelin fractions contain more lysine and other essential amino acids than do the kafirins. Cultivars with improved protein quality contain more of these proteins, with a corresponding lower proportion of kafirins (47). Cultivars that were induced, selected, and/or bred to contain a larger germ-to-endosperm ratio also yield more albumins, globulins, and glutelins (13).

Brown (tannin-containing) and regular (without tannin) sorghums have similar amino acid composition (47). The high-lysine cultivars, however, contain approximately 50% more lysine (47) and approximately 70% of the RDI for lysine. No high-lysine sorghums are in commercial production, even though their discovery occurred nearly two decades ago. Nutritionally, tannin sorghums have lower protein quality than do their regular counterparts due to reduced protein utilization. In monogastric animals, high-tannin sorghums consistently reduce protein retention and growth by 5-10% (6).

High-lysine sorghum from Ethiopia has a soft, floury endosperm texture, a shriveled kernel structure, and is susceptible to deterioration in environments where it rains during caryopsis development (37). Sorghums with improved protein digestibility (51) and chemically induced, high-lysine sorghum (13) have intermediate-soft to soft endosperm textures. Both types have reduced grain yields and increased grain deterioration due to molds and weathering in many areas.

The sorghum kafirins have been the subject of much research and appear to be related to endosperm hardness and decreased protein digestibility (10, 21, 22, 49). The amounts of α- and γ-kafirins and of protein bodies were greater in vitreous than in floury endosperm areas (21, 22, 49), and in hard relative to soft endosperm sorghums. The floury endosperm contains more albumins and globulins, but glutelins were evenly distributed throughout the endosperm. The extent and characteristics of cross-linking of kafirins in the mature caryopses are greater than acknowledged previously (10). Most (70%) kafirins, even those extracted in aqueous alcohol, are disulfide-linked dimers, trimers, or polymers. Subsequent extraction following sonication yielded > 90% disulfide-linked polymers of kafirins. β-kafirin appears to be involved in the formation of the largest polymers.

The digestibility of kafirins decreases during caryopsis development and after cooking, apparently due to intermolecular disulfide cross-linking. The α- and β-kafirins cross-link as the caryopsis desiccates (32). The location of β- and γ-kafirid near the surface of the protein bodies and their higher content of disulfide groups enables these kafirins to cross-link during cooking, as indicated by a fibrous-looking, slow-digesting, protein network that remains on the surface of pepsin-treated protein bodies. The α-kafirins are located in the center of the protein bodies and are the slowest to digest after cooking. Protein digestibility of sorghum is similar to that of other cereals when reducing agents such as metabisulfite or mercaptoethanol, which break disulfide bonds, are added and cooked with the sorghum (18, 19, 50).

Improved protein digestibility in sorghum, with easier-to-digest kafirins, is caused by more invaginations in the protein bodies, less kafirin in the endosperm, and less protein cross-linking during cooking (51). Since more amino acids are hydrolyzed and absorbed from kafirins, the lysine deficiency is greatest when less kafirin is hydrolyzed. Increased digestibility of kafirins increases the essential amino acid imbalance of sorghum because kafirin is deficient in essential amino acids (47). These sorghums have a ring of corneous (vitreous) endosperm surrounded (outside and inside) by opaque, soft endosperm areas (51). The softer endosperm near the periphery increases the chance for grain mold and insect damage.

Kafirin produced by the chemically mutated, high-lysine sorghum had increased digestibility; however, these traits appear to be recessive and the grain is susceptible to molds. Hence, the higher-lysine and increased digestibility traits must be transferred to another background to be commercially useful. Considerable research still is needed to produce sorghums with improved amino acid content, protein digestibility, harder endosperm texture, and grain mold resistance, since increased digestibility almost always increases the susceptibility of the grain to postmaturity molds and deterioration.

An alternative approach to improve digestibility is to properly process high-yielding, intermediate- to hard-endosperm sorghums. Many indigenous peoples effect-

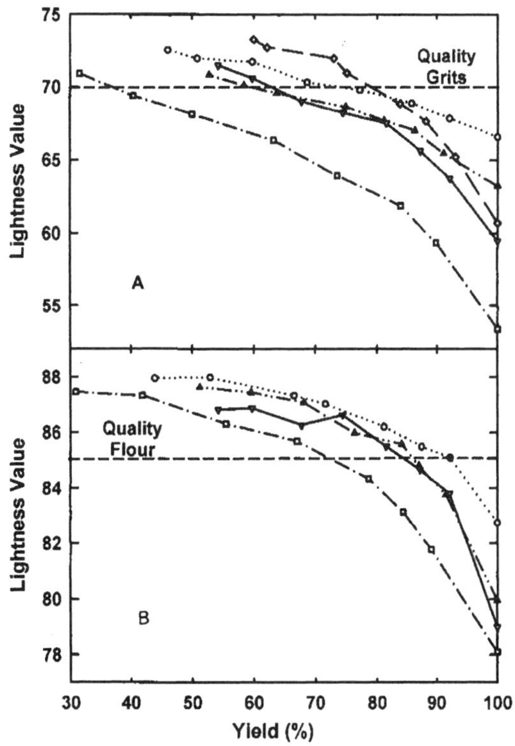

FIGURE 54-3. Effect of decortication rate on lightness (L-value) of decorticated sorghum grain and sorghum flour. A. Decorticated grain. B. Flour. Results presented are the average of three replicates. Dorado (□), ATx399 × RTx430 (▫), Warner 902W-1 (Δ), Warner 902W-3 (∇), and ATx631 × SC103-12 (◇).

tively process the grain and use it in foods (40). The excellent health of sorghum consumers around the world strongly indicates that sorghum is a highly nutritious cereal when it is part of a balanced diet. Improved food-type sorghums developed over the last several decades have high yield, hard endosperm, better mold resistance, and decreased pigments in the pericarp, testa, and associated tissues.

Tannins and Other Phenolic Compounds

Sorghums that contain tannins, proanthocyanidins, are called tannin or brown sorghum even though the pericarp color may be white, yellow, or red. Grain appearance need not be related to tannin presence or content. Most commercial sorghums do not contain tannins, but all sorghums contain phenols and flavonoids. Sorghums are classified as type I, no tannins; type II, tannins in pigmented testa; or type III, tannins in pigmented testa and pericarp. Type II and III sorghums have dominant *B1* and *B2* genes and a pigmented testa (seed coat). Type III sorghums have a dominant spreader (*S*) gene and more tannins than do type II sorghums (42). Phenolic compounds and tannins have been reported at much higher levels, 10-100×, in pericarp, glumes, and leaf sheaths than in the endosperm. Tannins, when present, apparently occur only in the pericarp and pigmented testa layers (17) and in the glumes (9; Fig. 54-4). The presence of a pigmented testa is easily seen after scraping and removing the pericarp.

Tannins protect sorghum grain against insects, birds, fungi, and weathering, *i.e.*, environmental deterioration (5, 45), and reduce the rate of preharvest germination or early sprouting. These beneficial effects ensure that brown sorghums will continue to be produced in certain pest-ridden areas (5). These agronomic advantages are accompanied by nutritional disadvantages and reduced food quality (3, 4, 6, 8, 15, 31, 41). The antinutritional effects of tannins include diminished growth rate, protein digestibility, and feed efficiency in rats, hamsters, swine, poultry, and ruminants (20, 23-26). Tannins in sorghum reduce digestibility and efficiency of utilization of absorbed nutrients from 3-15%. Most animals consume brown or tannin sorghums at the same or slightly higher rates but do not gain as much weight.

Extracted tannins bind proteins and inhibit many enzymes in *in vitro* assays (14, 16). Tannins from sorghum stimulate production and secretion of proline-rich, salivary proteins that bind tannins during mastication in many mammals (1, 23); and some tannin-protein complexes are recovered intact in the feces (6). Tannin labeled with ^{14}C was not absorbed by chicks; however, ^{14}C-labeled unidentified polyphenols were absorbed and recovered in serum and other tissues. Apparently, lower molecular weight, absorbable polyphenols cause the systemic effects traditionally attributed to tannins (6).

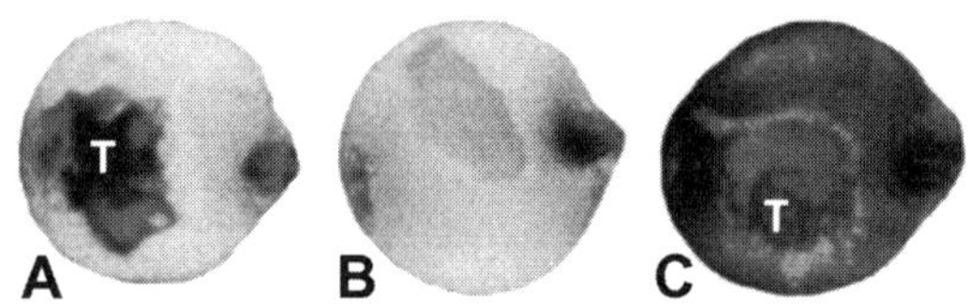

FIGURE 54-4. Caryopses of three sorghum kernels were scraped with a knife to expose the light-colored endosperm. The sorghum caryopses on the left (A) and right (C) had a dark layer between the pericarp and the endosperm. The dark layer, the pigmented testa (T), contains tannins. The pericarp color of the kernels on left (A) and center (B) is white, while the pericarp color of the sorghum on the right (C) is red.

Several approaches for detoxification of the tannins in brown sorghums have been proposed. The most common and practical ways are by decortication and malting. Decortication removes the pericarp and testa and, therefore, most tannins (34). However, decortication is inefficient because most brown sorghums have soft endosperms (30). Parboiling of brown sorghum prior to decortication is a better alternative for tannin removal (52). Chemical décortication with alkali solutions followed by neutralization with acid decreases the amount of assayable tannins (5, 44). Steeping alone or steeping and malting lowers measurable tannins by up to 43% (33). Sorghums treated with moisture and wood ashes have up to 97% fewer measurable tannins, improved protein digestibility, and result in increased weight gain in rats (27). Similar effects are observed when brown sorghums are treated with Magadi soda. Several experimental alkaline treatments to detoxify brown sorghums have been proposed (5, 44).

The colors of sorghum grain and flour play an important role in its acceptance. In general, white sorghums produce the most acceptable colored food products. The color of the food is the result of factors such as pericarp color, pigmented testa, endosperm color, presence of tannins, degree of milling, and the pH of the food system (40, 46, 47). Pigmentation in the pericarp and testa is due primarily to phenolic compounds, including tannins, with the color intensity depending on the pH. Anthocyanins are unstable in acid medium and are easily converted to the corresponding anthocyanidin under slight acidic conditions (17). Cooking in alkali, *e.g.*, alkaline tô or tortillas, usually increases color formation, especially in nonwhite sorghums. In these light-colored food products, dark-colored glume or secondary-plant colors also detract from the appearance of food products. Uniformly light-colored food products usually are prepared by using grain with straw-colored rather than darker colored glumes.

Tannins and other phenolic compounds in sorghum have antioxidant properties and are currently being considered as nutraceuticals (15). Whole sorghum grain and sorghum bran, which is rich in fiber and phenols, are currently being evaluated in several food systems at Texas A&M University.

Conclusion

Increased grain yield must occur along with increased grain quality attributes. Fewer stains and off-colored products will result if grain is produced on plants with tan-secondary plant color and straw-colored glumes. White- or clear-pericarp color is preferred, but red-pericarp grain grown on tan plants has better grain mold resistance. The variation in the ease of pericarp removal and the ability to mill the grain into a fine flour need to be determined and exploited. Grains with a hard endosperm generally have less grain molding and yield a finer flour. The bland, white-colored, milled products of sorghum can be utilized in many food systems with little change in flavor and texture. Increased lysine content and protein nutritional value in hard endosperm sorghums are needed to maitain parity with high-quality protein maize. The incorporation of the bioactivity of tannins and phenols (without their color and antinutritive properties) should not detract from food quality attributes while providing biochemical mechanisms for disease and insect resistance. Sorghums with improved utilization potential need to be handled in a manner that retains their identities from production through utilization.

Acknowledgments

We thank former students, colleagues, and scientists from all over the world, especially those from the Texas A&M University Sorghum Improvement Program, who have provided inspiration, reliable information, and help for more than 30 years. The long-term support of the Texas Agricultural Experiment Station and private industry is appreciated. This research has been partially funded since 1979 by the Title XII Sorghum and Millet Collaborative Research Support Program, INTSORMIL, under Grant AID/DSAN/XII/G-0149 from the Agency for International Development, Washington, DC.

References

1. **Asquith, T. N., J. Uhlig, H. Mehansho, L. Putman, D. M. Carlson, and L. G. Butler.** 1987. Binding of condensed tannins to salivary proline-rich glycoproteins: The role of carbohydrate. *J. Agric. Food Chem.* **35:** 331-334.
2. **Awika, J. M., E. L. Suhendro, and L. W. Rooney.** 2002. Milling value of sorghums compared by adjusting yields to a constant product color. *Cereal Chem.* **79:** 249-251.
3. **Butler, L. G.** 1989. Effects of condensed tannins on animal nutrition, p. 391-402. *In* R. W. Hemingway and J. J. Karchesy (eds.), *Chemistry and Significance of Condensed Tannins.* Plenum Press, New York City, New York.
4. **Butler, L. G.** 1989. New perspectives on the antinutritional effects of tannins, p. 402-409. *In* J. E. Kinsella and W. G. Soucie (eds.), *Food Proteins.* American Oil Chemists Society, Champaign, Illinois.
5. **Butler, L. G.** 1990. The nature and amelioration of the antinutritional effects of tannins in sorghum grain, p. 191-205. *In* E. T. Mertz, G. Ejeta, L. W. Rooney, R. Schaffert, and J. Yohe (eds.), *Sorghum Nutritional Quality.* INTSORMIL, Lincoln, Nebraska.
6. **Butler, L. G., and J. C. Rogler.** 1992. Biochemical mechanisms of the antinutritional effects of tannins, in phenolic compounds in food and their effects on health. *Amer. Chem. Soc. Symp.Ser.* **506:** 289-304.
7. **Chandrashekar, A., and H. Mazhar.** 1999. The biochemical basis and implications of grain strength in sorghum and maize. *J. Cereal Sci.* **30:** 193-207.
8. **Chung, K. T., T. Y. Wong, C. I. Wei, Y. W. Huang, and Y. Lin.** 1998. Tannins and human health: A review. *Crit. Rev. Food Sci. Nutr.* **38:** 421-464.
9. **Doherty, C. A., R. D. Waniska, L. W. Rooney, C. F. Earp, and J. H. Poe.** 1987. Free phenolic compounds and tannins in sorghum caryopsis and glumes during development. *Cer. Chem.* **64:** 42-46.
10. **El Nour, I. N. A., A. D. B. Peruffo, and A. Curioni.** 1998. Characterization of sorghum kafirins in relation to their cross-linking behavior. *J. Cer. Sci.* **28:** 197-207.
11. **FAO/WHO.** 1991. *Protein Quality: Evaluation*; FAO/WHO Nutrition Meetings, Report Series 51; Food Agriculture Organization/World Health Organization, Rome, Italy.
12. **Gomez, M. I.** 1993. Comparative evaluation and optimization of a milling system for small grains, p. 463-471. *In* J. R. N. Taylor, P. G. Randall, and J. H. Viljoen (eds.), *Proc. Cereal Science and Technology: Impact on a Changing Africa.* CSIR, Pretoria, South Africa.
13. **Guiragossian, V., B. A. K. Chibber, S. Vanscoyoc, R. Jambunathan, E. T. Mertz, and J. D. Axtell.** 1978. Characteristics of proteins from normal, high lysine, and high tannin sorghums. *J. Agric. Food Chem.* **26:** 219-223.
14. **Hagerman, A. E., and L. G. Butler.** 1994. Assay of condensed tannins or flavonoid oligomers and related flavonoids in plants. *Method Enzymol.* **234:** 429-437.
15. **Hagerman, A. E., K. M. Reidl, G. A. Jones, K. N. Sovik, N. T. Ritchard, P. W. Hartzfield, and T. L. Tiechel.** 1998. High molecular weight plant polyphenolics (tannins) as biological antioxidants. *J. Agric. Food Chem.* **46:** 1887-1892.

16. **Hagerman, A. E., Y. Zhao, and S. Johnson.** 1997. Methods for determination of condensed and hydrolyzable tannins. *Amer. Chem. Soc. Symp. Ser.* **662:** 209-222.
17. **Hahn, D. H., and L. W. Rooney.** 1986. Effect of genotype on tannins and phenols of sorghum. *Cer. Chem.* **63:** 4-8.
18. **Hamaker, B. R., A. W. Kirleis, L. G. Butler, J. D. Axtell, and E. T. Mertz.** 1987. Improving the *in vitro* protein digestibility of sorghum with reducing agents. *Proc. Natl. Acad. Sci. USA* **84:** 626-628.
19. **Kirleis, A. W.** 1990. The prolamins of sorghum: their role in protein digestibility, p. 168-176. *In* E. T. Mertz, G. Ejeta, L. W. Rooney, R. Schaffert, and J. Yohe (eds.), *Sorghum Nutritional Quality.* INTSORMIL, Lincoln, Nebraska.
20. **Knabe, D. A.** 1990. Sorghum as swine feed, p. 95-107. *In* E. T. Mertz, G. Ejeta, L. W. Rooney, R. Schaffert, and J. Yohe (eds.), *Sorghum Nutritional Quality.* INTSORMIL, Lincoln, Nebraska.
21. **Mazhar, H., and A. Chandrashekar.** 1993. Differences in kafirin composition during endosperm development and germination in sorghum cultivars varying in hardness. *Cer. Chem.* **70:** 667-671.
22. **Mazhar, H., and A. Chandrashekar.** 1995. Quantification and distribution of kafirins in the kernels of sorghum cultivars varying in endosperm hardness. *J. Cereal Sci.* **21:** 155-162.
23. **Mehansho, H., D. K. Ann, L. G. Butler, J. C. Rogler, and D. M. Carlson.** 1987. Induction of proline rich proteins in hamster salivary glands by isoproterenol treatment and an unusual growth inhibition by tannins. *J. Biol. Chem.* **262:** 12344-12350.
24. **Mehansho, H., L. G. Butler, and D. M. Carlson.** 1987. Dietary tannins and salivary praline-rich proteins: Interactions, induction, and defense mechanisms. *Annu. Rev. Nutr.* **7:** 423-440.
25. **Mole, S., J. C. Rogler, and L. G. Butler.** 1993. Growth reduction by dietary tannins: Different effects due to different tannins. *Biochem. Systemat. Ecol.* **21:** 667-677.
26. **Mole, S., J. C. Rogler, C. J. Morell, and L. G. Butler.** 1990. Herbivore growth reduction by tannins: Use of Waldbauer ratio techniques and manipulation of salivary protein production to elucidate mechanisms of action. *Biochem. Systemat. Ecol.* **18:** 183-197.
27. **Mukuru, S. Z., L. G. Butler, J. C. Rogler, A. W. Kirleis, G. Ejeta, J. D. Axtell, and E. T. Mertz.** 1992. Traditional processing of high-tannin sorghum grain in Uganda and its effect on tannin, protein digestibility, and rat growth. *J. Agric. Food Chem.* **40:** 1172-1175.
28. **Munck, L.** 1995. New milling technologies and products, p. 223-281. *In* D. A. V. Dendy (ed.), *Sorghum and Millets: Chemistry and Technology.* American Association of Cereal Chemists Press, St. Paul, Minnesota.
29. **Murty, D. S., and K. A. Kumar.** 1995. Traditional uses of sorghum and millets, p. 185-221. *In* D. A. V. Dendy (ed.), *Sorghum and Millets: Chemistry and Technology.* American Association of Cereal Chemists Press, St. Paul, Minnesota.
30. **Nwasaru, M. A., R. D. Reichert, and S. Z. Mukuru.** 1988. Factors affecting the abrasive dehulling efficiency of high tannin sorghum. *Cer. Chem.* **65:** 171-174.
31. **Nyachoti, C. M., J. L. Atkinson, and S. Leeson.** 1997. Sorghum tannins: A review. *World Poult. Sci. J.* **53:** 5-21.
32. **Oria, M. P., B. R. Hamaker, and J. M. Schull.** 1995. In vitro protein digestibility of developing and mature sorghum grain in relation to α-, β-, and γ-kafirin disulfide crosslinking. *J. Cer. Sci.* **22:** 85-93.
33. **Osuntogun, B. A., S. R. A. Adewusi, A. Adewusi, J. O. Ogundiwin, and C. C. Nwasike.** 1989. Effect of cultivar, steeping, and malting on tannin, total polyphenol, and cyanide content of Nigerian sorghum. *Cer. Chem.* **66:** 87-89.
34. **Reichert, R. D.** 1982. Sorghum dry milling, p. 547-564. *In* L. R. House, L. K. Mughogho, J. M. Peacock, and J. V. Mertin (eds.), *Sorghum in the Eighties*, Vol. II. ICRISAT, Patancheru (A. P.), India.
35. **Rooney, L. W.** 1996. Attributes of improved quality sorghums for value-added marketing. *Proc. ASTA Annu. Corn & Sorg. Res. Conf. (Chicago, IL)* **51:** 112-124.
36. **Rooney, L. W., M. N. Khan, and C. F. Earp.** 1980. The technology of sorghum products, p. 513-554. *In* G. E. Inglett and L. Munck (eds.), *Cereals for Food and Beverages: Recent Progress in Cereal Chemistry.* Academic Press, New York City, New York.
37. **Rooney, L. W., A. W. Kirleis, and D. S. Murty.** 1986. Traditional foods from sorghum: Their production, evaluation and nutritional value, p. 317-353. *In* Y. Pomeranz (ed.), *Advances in Cereal Science and Technology*, Vol. VIII. American Association of Cereal Chemists Press, St. Paul, Minnesota.
38. **Rooney, L. W., and D. S. Murty, eds.** 1982. *Proc. of an International Symposium on Sorghum Grain Quality.* ICRISAT, Patancheru (A. P.), India.
39. **Rooney, L. W., and D. S. Murty.** 1982. Evaluation of sorghum food quality, p. 571-588. *In* L. R. House, L. K. Mughogho, J. M. Peacock, and J. V. Mertin (eds.), *Sorghum in the Eighties*, Vol. II. ICRISAT, Patancheru (A.P.), India.
40. **Rooney, L. W., and R. D. Waniska.** 2000. Sorghum food and industrial utilization, p. 689-725. *In* C. W. Smith and R. A. Frederiksen (eds.), *Sorghum: Origin, History, Technology, and Production.* John Wiley, New York City, New York.
41. **Salunkhe, D. K., S. J. Jadhav, S. S. Kadam, and J. K. Chavan.** 1982. Chemical, biochemical and biological significance of polyphenols in cereals and legumes. *Crit. Rev. Food Sci. Nutr.* **17:** 277-305.
42. **Serna-Saldivar, S. O., and L. W. Rooney.** 1995. Structure and chemistry of sorghum and millets, p. 69-124. *In* D. A. V. Dendy (ed.), *Sorghum and Millets: Chemistry and Technology.* American Association of Cereal Chemists Press, St. Paul, Minnesota.
43. **Stenhouse, J. W., R. Bandyopadhyay, S. D. Singh, and V. S. Subramanian.** 1998. Breeding for grain mold resistance in sorghum. *Proc. Intl. Conf. on Genetic Improvement of Sorghum and Pearl Millet*: 326-336.
44. **Waichungo, W. W., and D. L. Holt.** 1995. Use of ammonium hydroxide to reduce the level of assayable tannin high-tannin sorghum grain. *J. Agric. Food Chem.* **43:** 728-732.
45. **Waniska, R. D., G. A. Forbes, R. Bandyopadhyay, R. A. Frederiksen, and L. W. Rooney.** 1988. Cereal chem.-istry and grain mold resistance, p. 265-272. *In* W. A. J. de

Milliano, R. A. Frederiksen, and G. D. Bengston (eds.), *Sorghum and Millets Diseases: A Second World Review*. ICRISAT, Patancheru (A.P.), India.

46. **Waniska, R. D., L. Hugo, and L. W. Rooney.** 1992. Practical methods to determine the presence of tannins in sorghum. *J. Appl. Poult. Res.* **1:** 122-128.
47. **Waniska, R.W., and L. W. Rooney.** 2001. The sorghum caryopsis: Structure and chemistry, p. 649-688. *In* C. W. Smith and R. A. Frederiksen (eds.), *Sorghum: Origin, History, Technology, and Production*. John Wiley, New York City, New York.
48. **Waniska, R. D., R. T. Venkatesha, A. Chandrashekar, F. P. Bejosano, S. Krishnaveni, J. Jeoung, J. Jayaraj, S. Muthukrishnan, and G. H. Liang.** 2002. Antifungal proteins and other mechanisms for the control of sorghum stalk rot and grain mold, p. 287-297. *In* J. F. Leslie (ed.), *Sorghum and Millets Diseases*. Iowa State Press, Ames, Iowa.
49. **Watterson, J. J., J. M. Shull, and A. W. Kirleis.** 1993. Quantitation of α-, β-, and γ-kafirins in vitreous and opaque endosperm of *Sorghum bicolor*. *Cer. Chem.* **70:** 452-457.
50. **Watterson, J., J. M. Shull, A. A. Mohamed, V. Reddy, and A. W. Kirleis.** 1990. Isolation of a high cysteine kafirin protein and its cross reactivity with gamma zein antiserum. *J. Cer. Sci.* **12:** 137-144.
51. **Weaver, C. A., B. R. Hamaker, and J. D. Axtell.** 1998. Discovery of grain sorghum germplasm with high uncooked and cooked *in vitro* protein digestibilities. *Cer. Chem.* **75:** 665-670.
52. **Young, R., M. Haidara, L. W. Rooney, and R. D. Waniska.** 1990. Parboiled sorghum: development of a novel decorticated product. *J. Cer. Sci.* **11:** 277-289.

Part IX

Collaboration and Its Implementation

55

Public-Private Partnerships in International Agricultural Research: A Case for Promoting Technology Transfer and Enhancing Global Trade in Sorghum and Millet

Mark S. Condon and Jeffrey A. Dahlberg

Agricultural Technology and Transfer Mechanisms–A Call for Consensus

The world has embarked on a new agricultural revolution. The rapid integration of the global economy and the recent commercialization of new agricultural technologies are transforming the structure and economics of world agriculture. Biotechnology, for example, is bringing about the vertical integration of crop protection, seed producers, and food industries. This progression has the potential to generate greater efficiency and productivity in the world's agricultural and food sectors, especially for national and regional agricultural sectors that can readily support these schemes.

At the same time, however, there are substantial differences in opinion on what technologies and transfer mechanisms are appropriate for the sustenance and development of global agricultural systems (1, 2). Critics of the Green Revolution maintain that the massive transfer of technology during the 1960s and 1970s from industrialized countries to the developing world has resulted in dependency on imported technologies, environmental damage by pesticides and fertilizers, disruption of viable traditional agricultural systems, and a loss of biodiversity (5). On the other hand, private sector advances in biotechnology have been rejected by a number of policy makers, regulators, and public and consumer groups on alleged safety, environmental, and antitrust grounds (1, 3). Advocates for the maintenance and development of traditional technologies and agricultural systems also have been criticized in that such strategies do not support globally competitive agricultural production, economic development, or global food security. In addition, it is feared that reliance on labor-intensive, low-productivity traditional systems will result in the expansion of agriculture into environmentally sensitive areas and with an accompanying decrease in biodiversity. There is even a debate on the viability and beneficial impact of public research institutions and the Consultative Group for International Agricultural Research (CGIAR) in developing and disseminating beneficial agricultural technologies to the developing world, *e.g.*, the G.R.E.A.N. Initiative (http://grean.ifas.ufl.edu/).

The lack of consensus on what technologies and transfer mechanisms developed and utilized by public and private institutions are appropriate and necessary for global agricultural systems has impeded a coordinated strategic response to global agricultural development. Clearly, we need frameworks and programs that facilitate public-private partnerships in international agricultural research to narrow the divergent points of view, leverage limited resources, and better coordinate a strategic response by relevant stakeholders.

Technology Transfer and Agricultural Development

Several nations, despite decades of donor assistance for agricultural development, lag well behind the rest of the world in agricultural sector development (see Food and Agricultural Organization home site at www.fao.org). In many cases, countries have not taken steps toward global integration and cooperation in areas of agricultural research, regulation, and commerce. Such inertia does not bode well for global agricultural and economic development. As history has proven, agricultural development is a necessary precursor to industrialization in agrarian societies (4). Moreover, a globally competitive, efficient and sustainable agricultural sector is a necessary precursor to economic development

Within the critique of agricultural development schemes, there is a growing call for review and improvement of national, regional, and global systems for agricultural technology development and transfer, *e.g.*, the G.R.E.A.N. Initiative. In many countries, agricultural development is impeded by inefficiencies in diverse agricultural sectors extending from agricultural research and technology transfer to production, marketing, and regulatory systems. The relative lack of public-private partnerships in agricultural research also has meant that there is no need to change the strict regulation of both multinational and indigenous agricultural industries or the accompanying disincentives for private investment in agricultural and related sectors.

Both private and public sectors, at national and international levels, have much to offer in terms of technical expetise, technology, or monetary resources that can benefit farmers worldwide. Clearly the objective of both the public and private research sectors is the development and transfer of technology that will benefit farmers worldwide. However, few public or private technology providers have realized their full potential, and a mutually beneficial synergy with their counterparts under many contemporary schemes for technology transfer would strengthen all of the participants.

If the global community is to successfully develop sustainable global agricultural systems and feed a growing world population, agricultural technologies must be continually developed, improved, and distributed worldwide without undue restrictions due to policy, regulatory, infrastructural, or philosophical obstacles. Unfortunately, many national and international systems for agricultural technology transfer still inadvertently prevent beneficial technologies from reaching a many farmers.

Public-Private Partnerships in Research for Global Development and the Transfer of Agricultural Technologies

Preliminary discussions among public and private stakeholders revealed that public-private partnerships in research that facilitate global development and agricultural technology transfer is a viable conceptual and operational "umbrella" that could result in coordinated activities involving both sectors to the benefit of the developing world. Obstacles to the free movement of agricultural technologies often stem from perceived resource and benefit inequities of new agricultural technologies between public and private sectors, as well as between developing and industrialized countries. Global public-private partnerships in the development and transfer of technology will aim to narrow the "North-South" technology gap.

The concept of public-private partnerships in agricultural research has broad support from international donor organizations, private industry, universities, and governments; see the Queensland Department of Primary Industries at www.dpi.qld.gov.au for some examples. Some entities, recognizing the potential benefits of such collaboration, have already engaged in this activity. However, very few public or private organizations have successfully developed collective partnership schemes at regional or international levels for agricultural technology development and transfer that simultaneously leverage public, private, and donor financial support.

Both public and private sectors need to design and implement an initiative that will lead to public-private research partnerships and to improved mechanisms for the development and transfer of agricultural technologies. At present, it is increasingly difficult to justify and maintain current research and development expenditures given the suboptimal adoption rates and levels of distribution of some private sector technologies. Public sector research funds are diminishing worldwide and may no longer be sufficient to ensure global food security (6). In addition, the appropriateness and benefit of some new agricultural technologies developed by the private sector are a subject of real or perceived public concern (1, 3; see also various Web sites, *e.g.*, www.worldwatch.org or www.greenpeace.org).

A Call for Public and Private Partnerships in Sorghum and Millet Research

Both public and private sectors should explore the development of an initiative or plan of action that will result in collective public-private partnerships in international agricultural research. This initiative should give priority to research activities that will improve or redefine mechanisms for the global development and transfer of agricultural technologies, including those relevant to sorghum and millet breeding, production, marketing, and utilization.

To increase the benefits that sorghum and millet provide for many farmers and consumers, we must identify and prioritize issue areas under this initiative that will form the basis for future collaborative research projects. These areas probably will include, but should not be limited to:

- Biotechnology.
- Design of new, or improvement of old, public and private mechanisms for global agricultural technology transfer.
- Intellectual property rights.
- Development of improved varieties, cultivars, and hybrids.
- Improved test and field inspection methodologies for the detection for economically important seed borne pathogens.
- Seed treatment(s).
- Regulatory system harmonization and reform that enhances the development and dissemination of sorghum and millet technologies, *e.g.*, phytosanitary certificates and quarantine issues.
- Increased epidemiological research on economically important host-pathogen combinations relative to sorghum and millet.
- Identification and support basic research of mutual benefit to public and private research priorities.
- Joint ventures in proprietary research.

To ensure the viability and sustainability of public-private collaboration in these research efforts, we must, in partnership, continually seek sustained funding from public and private sources for priority research activities.

References

1. **Avery, D. T.** 1995. *Saving the Planet with Pesticides and Plastics: The Environmental Triumph of High-Yield Farming.* Hudson Institute, Indianapolis, Indiana.
2. **Brown, L. R.** 1997. Can we raise enough grain yields fast enough? *World Watch* **10 (4):** 8-17.
3. **Charles, D.** 2001. *Lords of the Harvest: Biotech, Big Money, and the Future of Food.* Perseus Publishing, Cambridge, Massachusetts.
4. **Harlan, J. R.** 1975. *Crops and Man.* ASA and CSSA, Madison, Wisconsin.
5. **Lappe, F. M., J. Collins, and C. Fowler.** 1977. *Food First: Beyond the Myth of Scarcity.* Houghton-Mifflin, Boston, Massachusetts.
6. **National Academy Press.** 2000. *Transgenic Plants and World Agriculture.* Report prepared under the auspices of the Royal Society of London, the U.S. National Academy of Sciences, the Brazilian Academy of Sciences, the Chinese Academy of Sciences, the Indian National Science Academy, the Mexican Academy of Sciences, and the Third World Academy of Sciences. National Academy Press, Washington, D.C.

56

Changing Paradigms in the Design and Implementation of Collaborative Research

Richard A. Frederiksen and John M. Yohe

Abstract

In 1977, the US Agency for International Development (USAID) began new and innovative programs for work with agricultural scientists and sociologists in developing countries. These programs were the Collaborative Research Support Programs (CRSPs), which permitted US scientists to work with Developing Country (LDC) collaborators on long-term research issues. These programs were on crop improvement, *e.g.*, sorghum, millet, peanuts, beans and cowpeas; on animal programs, *e.g.*, fisheries, and small and large ruminant animals; and on soils, sustainable agriculture, and integrated pest management. All of the CRSPs have provided benefits to US agriculture as well as to the collaborating countries. These programs have been extremely valuable and need to be continued. Many of these programs have now finished some 20 years of work. We analyze the role that these programs have played, identify their strengths and weaknesses, and suggest some modifications that might be considered for future programs.

Introduction

The CRSPs were established by USAID in the late 1970s. Many of these programs will be considered for renewal by USAID between 2000 and 2002. When these CRSPs were designed, they were prepared, almost exclusively, by members of the US academic faculty who had strong interests in the areas of collaboration, matching support from their home institutions, and expertise with and commitment to the commodity or science involved. Over the years these programs have been in place, essentially 22 years for the oldest CRSPs, there has been concern that they have become institutionalized. Changes to strengthen programs that have been in operation over such a long period of time is challenging, especially since for the most part, the operational structure of the CRSPs has changed little from their inception.

CRSPs have encountered problems from the US commodity producers that led in part to the Bumpers Amendment (No. 1129). This amendment became law in May 1986, and it was to " ... prevent American tax dollars from being used to help foreign countries who are trying to take our export markets" (5). While this amendment was prepared to address concerns of the soybean growers, the peanut growers also were concerned about the extent of foreign aid assisting their competitors. At this writing, however, the peanut growers are highly supportive of the CRSPs. Sorghum researchers have worked successfully with the National Grain Sorghum Producers (NGSP) to maintain support for the INTSORMIL programs, and NGSP Directors have testified to Congress on behalf of the CRSPs.

Collaborative Research Model

In the 1970s, the US Congress proposed a new approach to reducing world hunger through the Title XII Amendment to the Foreign Assistance Act of 1961. This new approach was to be accomplished, in part, by involving key agricultural scientists from the United States, primarily from the Land Grant University system, to assist in the development of food production in LDCs. In this Title XII Amendment to the Foreign Assistance Act of 1961, Congress authorized the President to "... provide assistance on such terms and conditions as he shall determine ... to provide program support for long-term collaborative university research on food production and distribution, storage, marketing and consumption." The idea at the time was to encourage development of a technology transfer system for agricultural workers in LDCs in a manner similar with that of the successful American Land Grant University system. Between 1977 and 1993,

TABLE 56-1. A list of Collaborative Research Support Programs (CRSPs) and the year of their initiation by the US Agency for International Development (USAID).

CRSP Program	Years of Operation
Small Ruminants	1978-
International Sorghum & Millet (INTSORMIL)	1979-
Bean/Cowpea	1980-
Soils Management	1981-
Nutrition	1981-1982
Peanut	1982-
Pond Dynamics	1982-
Fisheries Stock Assessment	1985-1993
Sustainable Agriculture & Natural Resource Management (SANREM)	1992-
Integrated Pest Management (IPM-CRSP)	1993-
Basis	1996-

eleven CRSPs were established (Table 56-1). A general outline of the mission of these CRSPs is summarized in Yohe *et al.* (6).

Through the CRSPs, many faculty members at Land Grant universities have been involved in collaborative research with scientists in developing countries. Since Congress specified that the collaborative work was to benefit US agriculture as well, these programs have been valuable programs for all of the participating institutions, both domestically and internationally. As early as 1979, memoranda of understanding (MOUs) with various national agricultural research programs (NARs) or, in some cases, research universities were prepared. Yohe and Frederiksen (7) prepared a partial case history of the INTSORMIL program. They describe in some detail the history of the program in Sudan and how the collaboration enhanced the ability of the Sudanese agricultural scientists to pursue their research agenda while also benefiting US agriculture. Sorghum, for example, is the most important food commodity in Sudan, and much of Sudanese culture is based on this important crop. Sudan is extremely hot and dry during the summer growing season, and sorghum tolerates these hot, dry conditions better than any other food crop in Sudan. The benefits of collaboration in Sudan were obvious to sorghum growers in the Southwestern United States, where similar conditions prevail.

There is no single CRSP model (6), and each of these collaborative programs has developed different approaches. Some established most of their research in the LDCs, others less. Some based their work on regional programs, while others were more specific, scientist-to-scientist programs. Surprisingly, perhaps because of their continuity and longevity, all of these models appear to have worked. This success may have several causes, including:

- Programs were designed to continue for several years. In this respect, collaborators got to know each other and generally found ways to exchange ideas and develop programs based on mutual interest.
- In general, the top researchers from premier units in both the United States and the NARs participated in the collaboration. Rapport was generally excellent and, consequently, collaboration was profitable for all involved.
- Both programs benefited from the collaboration. For INTSORMIL, US scientists working in the Sudan found highly desirable genetic germplasm. These lines directly contributed to the success of US grain sorghum hybrids. Sudan benefited from the development of hybrids and from other technologies, *e.g.*, the control of *Striga*, a major parasitic plant.
- There was a 25% matching buy-in from US institutions participating in the program. Consequently, the US partner had a vested interest in the success of the project. Effective collaboration commits both to the success of the research.

We believe that there will be new opportunities for programs, particularly in Africa, based on the success of the CRSP programs. Many of the current CRSPs will need major revisions by 2002, or a similar date. At the time they were developed, the idea of making the work long-term and including key scientists with a vested interest in the program as full partners added to the credibility of the research. Soil scientists worked with soil scientists, breeders with breeders, and other scientists with their appropriate counterparts. These examples stretched across many different disciplines and, in time, resulted in the development of Inter-CRSP programs. In essence, the CRSPs succeeded in their mission to develop collaborative programs between US scientists and those in the developing countries. Reviews and evaluations of these programs have universally declared them effective and successful. These reviews documented many examples of advances in agricultural productivity for both the United States and the LDC partners.

One recent example is of particular note. In 1995, sorghum ergot appeared in the Americas for the first time. Prior to that time, this plant disease was known only in Africa and Asia. Beginning in Brazil the pathogen spread rapidly throughout Latin America and reached the United States in March 1997. By October 1997, it had spread throughout the US sorghum belt. INTSORMIL scientists, in particular Dr. Gary Odvody at Texas A&M University, addressed this plant disease almost as soon as it was found in Brazil. He, along with INTSORMIL colleagues in Africa, had worked on the disease for nearly a decade and knew much of the basic biology of the causal pathogen and some of the applied methods that could be used to lessen the impact of this disease in the US sorghum seed production areas. INTSORMIL plant pathologists organized an international conference on sorghum ergot that was held in Brazil and invited sorghum ergot workers from Europe, Asia, and Africa to share their experiences with researchers and sorghum seed producers in the Americas. A follow-up conference with the National Grain Sorghum Producers followed shortly in Amarillo, Texas to ensure that US sorghum producers and seed suppliers could be informed in detail by global experts as rapidly as possible. Ergot was not a welcome disease, particularly at a time when the cash value of feed commodities was low, but the response by Dr. Odvody and other members of the INTSORMIL research team was clearly a very valuable benefit to US agriculture.

Naturally there have been many variations in CRSP programs and activities. One of the strengths of the US universities, particularly in the Land Grant system, is the ability to provide advanced training for graduate students. In several countries, researchers were trained in graduate programs before they could effectively participate in collaborative research. For example, in Niger there were relatively few trained staff in the national research program (INRAN). INTSORMIL principal investtigators identified individuals and provided them with funding to complete their education in the US. Most of these individuals studied in the United States, many at Purdue University. The INTSORMIL team at Purdue is overtly proud of their role in this program. The INTSORMIL program in Sudan, in contrast, was quite different (7) in that there was relatively large number of trained scientists already in place and effective collaboration could begin almost immediately.

There is a need to continue collaborations between scientists in developed and developing countries. Our objective in this paper is to suggest approaches that might enhance or improve both the concept and meaningfulness of these collaborations. There is little question that the CRSP concept as formulated and modified over the past 22 years has led to as valuable a program as any funded by USAID. The programs have been cost effective, progresssive, and altruistic. External evaluation panels and USAID-sponsored reviews document these conclusions based on periodic reviews of each of the CRSPs. However, these successes do not mean that there are not serious challenges for the future that remain to be addressed.

The Equity Problem

In the CRSP model there are two partners, one with essentially all of the financial resources and the other with most of the needs. Few, if any, scientists from developing countries have been included in the initial planning of the CRSP programs. For example, the planning for the SANREM CRSP in 1991 included no participants from any of the prospective collaborators. It is difficult for the NAR/PL (National Agricultural Researcher/ Project Leader) to participate as a full partner unless they contribute to the planning process. Planning equity can be even more challenging to achieve when the commodities, disciplines, and to a very large extent, the research participants, by nature of the design of the program, are predestined to work together. In our initial program in Sudan, we did not consider the equity problem. We almost literally arrived in the country and began discussions with potential Sudanese PIs as to how we could best collaborate to move forward. Is it any wonder that we were well received! In Sudan, US researchers had much to gain from the Sudanese participants because of the long Sudanese history of sorghum research and because the Sudanese research program had developed excellent germplasm that could be effectively utilized by the sorghum industry in the United States. If the US and Sudanese roles had been reversed, the outcome almost certainly would have been very similar. It is a pity that the model of active planning participation by the developing country scientists was not tested experimentally. Today it probably could be attempted in several regions in which INTSORMIL has been active.

In Niger, it was not possible to design a program similar to the one in Sudan because there were not enough trained researchers in the various disciplines that are essential for a successful crop improvement program. In Niger, INTSORMIL PIs worked with INRAN to select candidates for baccalaureate and advanced degrees. Over a period of 15 years, collaborators grew up within the INTSORMIL program to become effective collaborators in programs such as those developed in Sudan in 1980. INTSORMIL's recent expansion into Ethiopia is balanced in terms of disciplines but continues primarily as a "donor"-based model of collaborative research. The INTSORMIL projects in Mali, Zimbabwe, Botswana, Eritrea, and Latin America also are more similar to the Nigerian program than to the Sudanese one.

Retention of Scientists in National Programs

It has never been a secret that retention of scientists in many national programs is a challenge. Many scientists, often the best and most highly motivated, are attracted to better-paying positions in other regions or by different employers. Based on our experience, it may require as many as five candidates with appropriate graduate degrees for one to be retained in the NAR program. When the goal of a commodity CRSP, *e.g.*, INTSORMIL, is to develop mini-crop improvement programs based on balanced interdisciplinary teams, the goal may not be achievable unless these highly trained individuals can be convinced to remain with the national program.

During training it is implicit that the individual(s) should return to the national program and continue collaboration. Once the individual has returned home, however, the challenges and frustrations in doing research in his/her own country may seem insurmountable, and it becomes much easier to accept employment elsewhere with an "international" salary that is paid on time and is accompanied by support staff and a budget for materials and travel. All too often, national scientists must work with a limited, or nonexistent, budget for travel within the country, encounter difficulties in obtaining supplies and spare parts, and at times, the lack of essential infrastructure, *e.g.*, continuous electrical power for the laboratory. To further compound problems, the national program scientist, when in training, often does research as a part of the NAR's program, and specific knowledge as well as discipline expertise is lost if the individual is not retained. The NAR programs must find a way to break this cycle if they expect to retain the best and most highly qualified scientists. Addressing this problem in future collaboration will increase the productivity of both the US and developing country research teams.

What are the Proper Research Questions?

The identification of significant, addressable research questions might be the biggest challenge for those designing future collaborative projects. In Western culture, economic considerations often are paramount, but are these always the correct questions? In Sudan, INTSORMIL conducted several anthropological studies of the agricultural operations of families that relied on subsistence farming. The anthropologists who worked in the fields with the farmers learned much of the needs and problems confronted by these individuals. They had the humility to understand that they needed to learn more about the culture and traditions of these producers (3).

Recently, Leslie and Frederiksen (2) asked whether the information and techniques from contemporary boilogy could be used in asking questions for developing countries. Many graduate students are now conducting research that requires molecular biological or other sophisticated techniques for success. Based on the information available at that time, the participants at the Bellagio conference on "Application of Genetics and Biotechnology to the Characterization of Fungal Pathogens of Sorghum and Millet" held in November 1993 concluded that much fundamental biology remained to be done before many of the more sophisticated tools in our arsenal could be employed effectively by the NAR programs. Norman Borlaug (1), in his foreword to that document, states, "I would be remiss if I did not express my doubts as to whether much of the information on the molecular level can be applied in the developing countries of Africa within the next two decades. Without exception, all countries of central Africa have serious limitations in budgeting for research, equipment and operations. All too often, over the past three decades, new theories, hypotheses and partially tested technologies from donor countries have been recommended and introduced into developing countries without first being adequately evaluated in the country where they originated. Consequently, failures discrediting the scientific organizations have occurred, and budgeting for more appropriate research has suffered." Yet while acknowledging these concerns, neither do we want our colleagues and collaborators to be "scientific cripples" who neither understand nor can apply these technologies. Identifying the balance, and when it has shifted, thus becomes critical, and assuring NAR scientists of our support as they work to modernize their systems from within becomes a must.

Much of the work in the crop improvement CRSPs has met with mixed success. For some commodities, direct transfer of germplasm is possible, but for most, local adaptation is needed. Heterosis is a form of plant technology, albeit nearly half a century old, that may have a 20-40% production benefit. Hybrids are extremely important in Asia and some regions of Africa, and the benefits derived from hybrids could be applied in other parts of Africa as well. Currently, hybrids are being considered for poor countries, such as Niger, but is this the only critical need? Certainly higher production is both needed and important, but there also is real need for better storage, for improved processing and marketing technologies of many of the more traditional commodities, and for the identification of economic bottlenecks that need to be broken (4).

Responses to the Problems

The collaborative research programs have been remarkably beneficial, but there are concerns. Funding for USAID programs is being re-evaluated. Normally, individual programs will do everything within their power to continue their programs, and, all too frequently, as similar as possible to the existing program. This is only natural. INTSORMIL will soon experience an important period of renewal with potential for significant change as many of the senior PIs who have been with the program since near its inception depart through retirement.

Recently, we have discussed the problems of equity, personnel retention, and research question formulation with scientists from developing countries. There is no reason why scientists from these countries can not or should not participate in the design of collaborative research projects. Planning sessions can be held in either developing or developed countries. If actual face-to-face collaboration in planning is not possible, then e-mail and other forms of electronic communication usually are available even the in poorest of countries. More collaboration needs to be done at the initial planning stages, because it is through these activities that ownership of the research on the part of the developing country scientists is initiated. If at all possible, more of this type of work needs to be done in the developing country and with developing country tools and resources. Consequently, developed country scientists need to be more aware of the needs of the developing country and have the patience to work through the planning process with their host country counterpart(s).

Better planning will not solve the problem of attrition of recently trained scientists. Often, the difference between remaining with a National Program and taking a position with an entity such as FAO or the African Development Bank is the need for a regular salary. A perk that would be extremely beneficial is an endowment for the education of the children of a staff member after a certain number of years of service.

One-on-one collaboration has excellent benefits and needs to continue; yet regional needs are increasing in importance, and US PIs are going to be stretched over more geographic distance and host country counterparts. There is little reason to duplicate research in adjacent countries such as Niger and Mali, Chad and even Senegal, and a great need to use limited resources in a complementary and synergistic manner. One way to expand and tighten regional cooperation is through regional committees/networks such as ROCFREMI and ROCARS in west Africa. Regional coordination also forces CRSPs to coordinate with the International Agricultural Research Centers. Coordination of efforts such as these is no easy task, but is especially important since these regional projects often are the first to lose funding.

Finally, there are new obligations on the part of the NARS scientists. They must aggressively pursue collaboration in areas of the research planning, and should expect to develop research plans, write project reports, and publish their findings in appropriate journals as do their US counterparts. Many scientists will not welcome these responsibilities, but if they are to be peers in agricultural research they need to acquire both the stature and the expertise that comes from these kinds of activities. For those scientists in developing countries who want to make a difference, we in the CRSPs must provide continued support and treasure their accomplishments.

References

1. **Borlaug, N.** 1995. Foreword, p. vii-viii. *In* J. F. Leslie and R. A. Frederiksen (eds.), *Disease Analysis through Genetics and Biotechnology: Interdisciplinary Bridges to Improved Sorghum and Millet Crops.* Iowa State University Press, Ames, Iowa.
2. **Leslie, J. F., and R. A. Frederiksen, eds.** 1995. *Disease Analysis through Genetics and Biotechnology: Interdisciplinary Bridges to Improved Sorghum and Millet Crops.* Iowa State University Press, Ames, Iowa.
3. **Reeves, E. B., and T. R. Frankenburger.** 1981. Socioeconomic constraints to the production, distribution, and consumption of millet, sorghum and cash crops in North Kordofan, Sudan. Report 1, Department of Sociology, University of Kentucky, Lexington, KY.
4. **Sanders, J. H., B. I. Shapiro, and S. Ramaswamy.** 1996. *The Economics of Agricultural Technology in Semiarid Sub-Saharan Africa.* Johns Hopkins University Press, Baltimore, Maryland.
5. **Thompson, P. B.** 1992. *The Ethics of Trade and Aid: U.S. Food Policy, Foreign Competition, and the Social Contract.* Cambridge University Press, New York City, New York.
6. **Yohe, J. M., P. Barnes-McConnell, H. S. Egna, J. Rowntree, J. Oxley, R. G. Hanson, D. Commins, and A. Kirksey.** 1995. The CRSPs: International Collaborative Research Support Programs, p. 321-338. *In* J. F. Leslie and R. A. Frederiksen (eds.), *Disease Analysis through Genetics and Biotechnology: Interdisciplinary Bridges to Improved Sorghum and Millet Crops.* Iowa State University Press, Ames, Iowa.
7. **Yohe, J. M., and R. A. Frederiksen.** 1995. Developing interdisciplinary research in a collaborative context, p. 307-319. *In* J. F. Leslie and R. A. Frederiksen (eds.), *Disease Analysis through Genetics and Biotechnology: Interdisciplinary Bridges to Improved Sorghum and Millet Crops.* Iowa State University Press, Ames, Iowa.

57

Private Sector and Public Institution Interactions on Sorghum and Pearl Millet Disease Management

Marc A. Johnson

Voluntary interaction requires mutual attraction. If two entities are the same, there is no gain from interaction and, therefore, no basis for attraction. Private sector and public institution interactions will occur for disease management because private and public entities differ. The bases for public-private differences lie in missions and means of communication. A corollary to these postulates is that people who choose to operate in either the public or private sector have less than full appreciation for their counterparts in the other sector, but this, too, can be explained by differences in institutional missions and means of communication.

Missions of public institutions entail working for the public good, including enhancing the welfare of all citizens. Broad missions include economic development to raise levels of resources per capita and to increase the availability of amenities that improve living conditions. Providing food security, for example, fulfills a basic need by making calories and nutrients nonlimiting factors for humans in their personal and professional lives. Human development is to make nutrition, education, and access to knowledge available to everyone at a level such that each individual can choose how they wish to express their potential. Opportunity distribution, including income distribution plus some nonmonetized items like rights of choice and access, guarantees everyone a place in society. Environmental quality maintains air, water, forest, and land resource quality for the long term.

The means of communication in the public sector is called politics: the process of communicating one's desires for resources that are not exchanged in the marketplace. In public institutions, decisions are influenced by interest groups, and decisions are made by large committees. Therefore, balance, consistency, and follow through are not strengths.

The mission of private firms is to produce and distribute products and services that yield profit in the marketplace. Producers know what it takes to assemble and pay for inputs to produce and deliver a product. Consumers know what they are willing to pay for the product. In the marketplace, producers and consumers come together to work out a price that is satisfactory to consummate an exchange.

The means of communication in the private sector is the monetary price. Competition among producers stimulates activity to creatively produce products at the lowest practical cost. Competition among consumers holds prices offered to levels reflecting their maximal willingness to pay. Since many individuals participate in this price discovery process, there is balance, consistency, and follow-through.

In relation to food systems, public entities plan on longer-term time scales and can invest in areas that may not show forseeable profit, as long as the activity is directed at fulfillment of the broad missions that increase the public good. Private companies, in contrast, take available knowledge and implement it immediately. Private firms will invest in risky research only if a large market potential exists, *e.g.*, pharmaceutical companies. I cannot name a publicly controlled food distribution system that does not spawn either legal side private markets or illegal black markets to assure consumers that they can obtain what they want.

The private sector is attracted to the public sector to obtain research knowledge generated through long-term efforts that have reached maturity and that are ready for implementation. The public sector is attracted to the private sector to obtain their capacity to commercialize technology and to deliver goods to consumers.

Within this context, private sector and public institution interactions in sorghum and millet disease management can now be examined. Sorghum and millet are important cereal grains grown extensively in semi-arid zones at lower volumes than the big three cereals, wheat, maize, and rice. Sorghum and millet often are replaced in production with maize and oilseeds when sufficient moisture is available. These conditions describe commodities of unstable supply and demand and low volume relative to the three major cereals. In other words, seed companies and specialized merchandising service firms will find sorghum and millet to be higher cost and riskier to serve than these other three cereals. Because sorghum and millet are important food and feed sources and result in economic development in semi-arid regions, public institutions are justified in continued investments in research for these grains, but sorghum and millet seed companies are less likely to invest individually.

Public and private entities will participate differentially in various areas of work. Private producers, seed companies, and processing firms normally discuss industry issues with public institutions. The public institutions investigate the underlying causes of these issues and identify potential solutions, including genetic resistance work. Private entities will use the knowledge and germplasm developed in the public sector to develop and distribute seeds.

Basic research on biological and chemical disease controls will be done publicly, and applied research, testing, and demonstration will be done by both the private and public sectors. Public institutions will continue to play a role in objective, third-party testing of disease control methods. The agronomy of sorghum and millet production also will be studied publicly because there are no products to sell from this work.

Developments in grain processing and utilization are more likely to involve private sector research investment. Processors tend to have a substantial market and the development of a proprietary product may provide a strong rate of return. The public sector will continue to play an important role in developing special genetic characteristics in grain that contribute to special attributes in products. The public sector also will provide research in cereal chemistry and process microbiology and engineering to aid in the evolution of cereal manufacturing methods.

Finally, the public sector will continue to conduct research on the role of sorghum and millets in human nutrition and the effects of cereal elements on human health.

In conclusion, I expect the public sector to be more comprehensively involved in research on sorghum and millet than in research on corn and oilseeds. The private sector will be responsible for seed development, grain production, and processing, but the public sector will supply research information and technology at every step along the way. Should public research programs shift away from comprehensive support for sorghum and millet, collectives of private sector firms should seriously consider contributing to a directed, general research fund for genetic resistance and disease control. All of the members of such a collective would have access to the most basic knowledge for application to specific products.

The rate of return to public agricultural research always has been shown to be high.

Acknowledgments

I thank Larry Claflin, Ken Kofoid, John Leslie, John Reese, Mitch Tuinstra, Gerald Wilde, and Robert Zeigler for their assistance.

58

Examples of ICRISAT's Research and Development Partnerships in Sorghum and Millet Improvement[1]

Rodomiro Ortiz

ICRISAT, established in 1972, has an eco-regional mandate for the semi-arid tropics, particularly in Africa, Asia, and Latin America, and a global crop research focus on sorghum, millets, and legumes such as groundnuts (or peanuts), chickpea, and pigeonpea. This eco-geographical domain provides a home to about one sixth of the world's population, of which 300 million are extremely poor (daily income < US$ 1).

International Agricultural Research and ICRISAT

The CGIAR recognizes two important pillars in international agricultural research, namely, integrated gene management and integrated natural resources management (21), which together with socioeconomics and policy analysis constitute the focus of the core of the research agenda of ICRISAT. By working in these domains, ICRISAT and its research partners in the NARS, advanced research institutes, civil society, and sister centers of the CGIAR strive together to apply science to improve agriculture in areas of the world where "marginal crops are grown on marginal lands with marginal resources." Enhanced, sustained crop productivity will improve human health and rural household food security.

ICRISAT's New Vision and Mission. ICRISAT's research agenda is to help the people of the semi-arid tropics to understand the potential of semi-arid tropic resources, to generate and assess knowledge and technology that promotes better use of semi-arid tropic resources, and to share this technology with partners. ICRISAT's priorities for the next 10 years are to build on the quality of science, strengthen strategic alliances, focus on Africa, increase financial stability, and achieve organizational renewal. ICRISAT scientists implement this research agenda through a rolling 3-year Medium-Term Plan, which expands the traditional focus on genetic resources and environmental management to include both human and ecosystem welfare. This new research-for-

[1]Abbreviations: AFLP–Amplified Fragment Length Polymorphism; ASARECA–Association for Strengthening Agricultural Research in Eastern and Central Africa; CGIAR–Consultative Group on International Agriculture; CIRAD–Centre de Coopération Internationale en Recherche Agronomique pour le Developement (Montpellier, France); CORAF/WECARD–West and Central Africa Committee for Agriculture Research Directors (Dakar, Senegal); DFID–Department for International Development (London, United Kingdom); EARSAM–East Africa Sorghum and Millet Network; ECARSAM–Eastern and Central Africa Regional Sorghum and Millet Network; ICAR–Indian Council for Agricultural Research; ICRISAT–International Crops Research Institute for the Semi-Arid Tropics (Patancheru, India); IER–Institut d'Economie Rurale (Mali); IFAD–International Fund for Agricultural Development; IITA–International Institute of Tropical Agriculture (Ibadan, Nigeria); INTSORMIL–International Sorghum and Millet Collaborative Research Support Program (Lincoln, Nebraska); LCRI–Lake Chad Research Institute (Nigeria); MAHYCO–Maharashtra Hybrid Seeds Company (India); MOU–Memorandum of Understanding; MStV–Maize Stripe Tenuivirus; NARO–National Agricultural Research Organization (Uganda); NARS–National Agricultural Research Systems; NGOs–Non-Governmental Organizations; NGSP–National Grain Sorghum Producers (Lubbock, Texas); QDPI–Queensland Department of Primary Industries (Toowoomba, Australia); QTL–Quantitative Trait Loci; R&D–Research and Development; RAPD–Random Amplified Polymorphic DNA; RFLP–Restriction Fragment Length Polymorphism; ROCARS–West and Central African Sorghum Improvement Network (Bamako, Mali); ROCAFREMI–West and Central African Millet Improvement Network (Niamey, Niger); SACCAR–Southern African Center for Cooperation in Agriculture and Natural Resources Research and Training; SADC–Southern Africa Development Community; SCRI–Scottish Crop Research Institute (Dundee, United Kingdom); SG 2000–Sasakawa Global 2000; SICNA–Sorghum Improvement Conference of North America; SMINET–Sorghum and Millet Improvement Network; SMIP–Sorghum and Millets Improvement Program; SSR–Simple Sequence Repeats; TAMU–Texas A & M University (College Station, Texas); USAID–United States Agency for International Development; USDA–United States Department of Agriculture.

development paradigm should permit the conservation and enhancement of biological diversity in a rapidly changing world with a rapidly growing human population (15).

The New Role of ICRISAT in Improving Agriculture in the Semi-Arid Tropics. ICRISAT has moved into the new millennium by addressing the challenges of sustaining food security, alleviating poverty, protecting the environment, and being proactive in its new triple role as bridge, broker, and catalyst in rural development. ICRISAT serves as a research bridge when it fosters the international sharing of knowledge, skills, and technology, *e.g.*, by making available on-line crop and resource information databases for the semi-arid tropics or through technology exchange. ICRISAT acts as an international broker when it helps its partners negotiate acceptable terms for intellectual, genetic, and other proprietary assets, *e.g.*, through material transfer agreements for international public goods. Finally, ICRISAT scientists help test creative approaches, *e.g.*, participatory research methods for enhanced management of semi-arid tropic resources, or new breeding techniques or diagnostic tools for pests and diseases resulting from advances in applied genomics of semi-arid tropics crops and their pathogens.

ICRISAT's "Old" and "New" Partners

ICRISAT's R&D since the mid-1990s has been driven by the need to help the developing world in the semi-arid tropics increase farm productivity and food security, reduce poverty, and protect the environment through optimal use of natural resources through partnership-based international agricultural research, particularly with NARS and advanced research institutes. ICRISAT shares cultivars, integrated pest management, and other products of its research with farmers directly or through a broad range of partnerships, particularly with NARS. More recently, the private sector and NGOs have become new ICRISAT partners. A good example of the range of partnerships in sorghum and millet research is the diverse affiliations of the members of the organizing committee of the "Global 2000: Sorghum and Millet Diseases III" conference on which this volume is based.

In this era of globalization and mass flow of information, we aim to share freely our knowledge with our research partners and stakeholders. In the short-term, ICRISAT would like to be known as an R&D organization that provides timely information about agriculture in the semi-arid tropics. ICRISAT is not the only institution working in this area, however, and includes printed and electronic materials produced by many advanced research institutes on its Web site for crops for the semi-arid tropics. ICRISAT also has taken a proactive role in "learning-by-doing" as a new means for providing training to its partners in R&D. The "good old days" of group training are being replaced with professional internships and "hands-on" research in topics relevant to the country from which the trainee comes.

Farmers as Research Clients and Partners in R&D. Farmers, driven by the need to ensure a sustainable food supply, have maintained dynamic and productive agro-biodiversity in their fields for hundreds of years. During domestication and evolution of plant species within farming systems, farmers selected against the original agro-biodiversity if it did not meet their needs. Farmers change agro-biodiversity in their fields by introducing new species or new forms of existing crops or by selecting either mutants that arise *de novo* or hybrid genotypes that are derived from natural outcrosses. Farmers also have accumulated native or traditional knowledge of crop management, including methods for selecting promising material and managing pests and diseases. For these reasons, farmers both influence and help implement ICRISAT's R&D agenda, and are important partners in ICRISAT's endeavors. ICRISAT has involved farmers in different regions of the world, *e.g.*, Rajasthan in India, and Mali and Namibia in Africa, in the assessment and development of a wide range of new, improved, locally adapted sorghum and millet technologies for the semi-arid tropics.

The Private Sector in R&D and the Development of Public Goods. ICRISAT has attracted private sector investment to support its research agenda (especially for hybrid breeding) and to provide additional support for purchases of equipment or the development of professional skills and tool, *e.g.*, on-line crop information systems, while simultaneously ensuring that any technology or knowledge ensuing from such investments remain in the public domain. As a plant pathology example, representatives of the Indian private sector with a stake in sorghum grain as a raw material are interested in research on sorghum grain mold in conjunction with ICRISAT, NARS and NGOs. Such funds are supplements to, not replacements for, funds provided by the public sector, foundations or even NGOs. The evolution of public/private sector relationships will require careful management (13) for mutual productivity, especially with respect to intellectual property rights.

Regional Research Networks. Results of global agricultural research may require applied or adaptive adjustments for them to have local relevance. As communication systems have improved, research results have been disseminated widely through technical publications and through informal linkages, *e.g.*, research collaboration between institutions, and at regional, national and international levels. These exchanges are essential for research and technology development, but do not always result in either rapid or uniform distribution of new information and technology to farmers. In the 1980s, ICRISAT developed formal networks to establish critical masses of scientists to solve specific problems and to reduce or prevent duplicative research efforts. These networks strengthened agricultural research in both developed and developing countries, and enhanced interactions and the exchange of knowledge and materials among members. The agricultural research networks formed by ICRISAT are composed of individuals or institutions who are committed to collaborate in addressing one, or more, common agricultural problems (http://www.icrisat.org/text/partnerships/partnerships.htm).

For sorghum and millet, ICRISAT provides direct technical oversight and backstopping to ROCARS for sorghum in west and central Africa (http://grep.icrisat.cgiar.org/archives/wcasrn.htm), ROCAFREMI for pearl millet in west and central Africa (http://grep.icrisat.cgiar.org/archives/wcamrn.htm), and SMINET for sorghum and pearl millet in southern Africa (http://grep.icrisat.cgiar.org/archives/sminet.htm). Other projects, *e.g.*, the SADC/ICRISAT Sorghum and Millet Improvement Program for Southern Africa (http://grep.icrisat.cgiar.org/archives/smip.html), also have significant networking components. A primary objective of these networks is to increase collaboration between NARS and ICRISAT, with special attention given to capacity building. Effective partnerships, shared responsibility, and joint ownership are the pillars of network operations. The success and impact of these networks depends on joint responsibility and action among members and sponsors.

INTSORMIL and ICRISAT: A Long-lasting R&D Partnership for the Benefit of Farmers in the Semi-Arid Tropics

ICRISAT works collaboratively with other R&D organizations to improve the livelihood of the poor residents of the semi-arid tropics. ICRISAT's goal is to help provide "a food secure world for all." The partnership between INTSORMIL and ICRISAT is an important, evolving, and on-going collaboration with the goal of improving the quality of life for those who rely on sorghum and millets to sustain life.

History of the collaboration. Since the inception of INTSORMIL in 1979, INTSORMIL and ICRISAT have had extensive scientific and organizational contacts. Initially, collaboration was informal, between individual ICRISAT scientists based in Sub-Saharan Africa and at CIMMYT, and INTSORMIL scientists based in the United States who had research activities in Africa and Latin America. The two organizations signed a Memorandum of Understanding in 1983, to increase the effectiveness of mutual efforts, and to capitalize on INTSORMIL's strengths in training and basic research and ICRISAT's strength in applied research and its presence in the semi-arid tropics. The MOU has evolved and expanded with time. For example, in 1985 INTSORMIL began providing degree training, funded by USAID, for SADC Sorghum and Millet Improvement Program scientists in southern and eastern Africa.

Since the early 1990s, collaboration between ICRISAT and INTSORMIL has focused on regional sorghum and millet research networks in southern and western Africa and in Central America, through which both organizations provide various forms of planning, research and training support to the NARS. Both organizations recently assisted ASARECA in establishing a sorghum and millet network for eastern and central Africa, and the Horn of Africa.

ICRISAT and INTSORMIL also have collaborated in planning and review meetings, and in the sponsorship of a number of major, global sorghum and pearl millet research meetings. For example, the proceedings of the previous joint meetings on sorghum and millets pathology (6, 25, 27) usually were edited jointly by ICRISAT and INTSORMIL scientists and often published by either ICRISAT or INTSORMIL. Through such efforts both organizations have strengthened the global planning, execution, and reporting of research related to sorghum and millets, while retaining their individual integrity, funding, and operational procedures.

Outputs of collaboration. Collaborative research has resulted in technical advances in male-sterile cytoplasm(s), striga resistance, management of insect pests and diseases, sorghum germplasm evaluation protocols, and grain quality in sorghum. Greater human resources for sorghum and millet research have resulted from degree training programs executed through SMIP and the ICRISAT-Mali bilateral program. Staff sabbatical opportunities for both ICRISAT and NARS scientists at INTSORMIL institutions have benefited the staff member involved and strengthened existing collaborations. Greater dissemination and exchange of research information on sorghum and millet, and the identification of priority areas for collaborative research among NARS, INTSORMIL, and ICRISAT have resulted from the joint organization of workshops and review meetings.

Priorities for the future. Both ICRISAT and INTSORMIL will continue to contribute to the alleviation of constraints to production and utilization of both sorghum and millets. However, priority areas for collaboration will change in the coming decade, in response to changing requests from regional research networks and NARS. Some areas of change and opportunity include:

- Conservation of biodiversity and natural resources.
  - Develop regional gene banks and regional germplasm collections.
  - Increase research on production agronomy.
- Technology generation
  - Use molecular biology approaches to relieve biotic and abiotic constraints.
  - Hybrid parent development and testing to support NARS hybrid breeding programs.
  - Integrated pest management technology for use by resource-poor farmers.
  - Grain and food processing technologies to improve markets and increase opportunities for commercialization.
- Socio-economics and policy research.
  - Information dissemination on food processing and food product diversification.
  - Increased opportunities for sorghum and millet in developing feed markets.
  - Increased adoption of improved technologies.
- Information and networking.
  - Joint international workshop on "Progress in Marker-Assisted Selection and New Opportunities."
  - Regional workshops on "Traditional Sorghum and Millet Seed Systems and Potential for Privatization" and "Constraints and Potential to Increasing Adoption of Improved Technologies."
- Human resource development
  - Provision of training in plant breeding and utilization of genetic resources.
  - Assistance in obtaining bilateral or regional funding for research on priority problems.

Examples of Collaborations in Progress

In the following section are 17 examples of recent, ongoing, and soon to start R&D partnerships involving ICRISAT, NARS, advanced research institutes, NGOs, farmers, and the private sector that involve sorghum and millets in some manner. These topics are relevant to ICRISAT's mission(s) and agenda(s) and attracted funding from external sponsors, who are supporting the improvement of agriculture in the developing world.

1. NARS, NGOs, Advanced Research Institutes, the Private Sector and ICRISAT: Grain Mold in Sorghum

Grain mold disease, caused by a complex of fungi, is one of the most important biotic constraints to sorghum improvement, production, and utilization worldwide (2). As the utilization of sorghum for human food declines, nonfood uses for the crop become more important. Several approaches to control grain mold have been attempted, but the problem continues to be a priority (5). The results of previous research suggest that pre- and postharvest issues and technical and institutional factors are interdependent. In part, this interdependency reflects the new consumers in a greatly enlarged postharvest chain, and the potential of these new participants to alter quality management measures. Yet grain mold also must be viewed in the context of new market demands. There are both technological and institutional opportunities for grain mold management, since the development of new sorghum markets means that quality measures, including grain mold resistance, become more important in crop improvement strategies.

During 1999–2000, Department for International Development, UK., funded ICRISAT, the Central Food Technological Research Institute (Mysore, India), and the Natural Resources Institute and Long Ashton Research Station (Bristol, UK) to undertake a project: "Review of Technical and Institutional Options for Sorghum Grain Mold Management and the Potential Impact on the Poor." Researchers from private and public organizations, including INTSORMIL, the poultry industry, poultry and cattle feed manufacturers, alcohol and brewing industries, NGOs, and seed companies were consulted as part of the review. Project conclusions include: (i) That there is a large range of technical and nontechnical, basic and applied research problems that must be addressed to solve the grain mold problem. (ii) That in some cases the poor will clearly be the direct beneficiaries, but in other cases the research will "enable" intermediary organizations to increase both direct and indirect livelihood opportunities for the poor. These intermediary organizations may be NGOs or in the private-for-profit sector. (iii) That institutional partnerships will be essential to address the technical issues. The poverty relief resulting from the new research is strongly linked to the long-term market prospects for sorghum as an industrial raw material and the ability of the private sector to develop adequate markets.

The challenge is to attract a sufficiently large commitment from a consortium of donors to enable the necessary research and its follow-up activities. This program will:

- ❖ Cut across conventional research sectors–crop protection, postharvest crop management, livestock, and plant sciences.
- ❖ Have upstream, strategic elements and downstream adaptive and developmental research elements.
- ❖ Blur the conventional distinction between public and private goods research.
- ❖ At times, be more concerned with establishing working relationships between key players and in devising institutional management and interaction protocols than in producing conventional outputs.

The next step for this collaboration is to seek seed funds to develop this approach into a concept note with stakeholder input.

2. ICRISAT, INTSORMIL (TAMU), and USDA: Ergot Control in Sorghum

Ergot is a highly infectious disease of sorghum, which adversely affects seed quality and is potentially toxic to humans and livestock (1, 19). Sorghum ergot can spread rapidly and in an uncontrollable manner. The disease, once found only in Asia and Africa, spread through the Americas and Australia from 1995-1997. Ergot is most damaging in hybrid seed production fields, and puts at risk the availability and profitability of F_1 hybrid seeds.

ICRISAT has invested considerable resources and effort on sorghum ergot. In the 1980s, ICRISAT conducted significant research on ergot biology, sorghum flowering biology, host-pathogen-environment interactions, development of resistance screening techniques, and identification of resistance sources. ICRISAT also conducted collaborative research on ergot with the Ethiopian and Rwandan national programs. After the appearance of sorghum ergot in the United States, TAMU, INTSORMIL, and the Agricultural Research Service of the USDA took leadership roles in studying the disease. Linkages between ICRISAT, TAMU, INTSORMIL, and USDA were established in 1997 and used to develop a program for management of sorghum ergot. As part of this linkage, ICRISAT sent an ICRISAT pathologist to work with TAMU and USDA scientists. This collaboration resulted in the identification of repetitive microcyclic sporulation by *Claviceps africana*, which partially explains the explosive nature of sorghum ergot epidemics, and in the identification of the more virulent ergot species, *C. africana*, in India, where it may be replacing the endemic pathogen *Claviceps sorghi*.

When ergot arrived in the Americas, EMBRAPA, INTSORMIL, and ICRISAT organized a global conference on ergot of sorghum in Brazil in 1997, with the proceedings published by INTSORMIL (4). A colored flier with pertinent information on the disease was prepared by ICRISAT with funding from the NGSP. About 20,000 copies of the flier–in English, Spanish, and Portuguese–have been distributed (plus electronically through the Internet) to sorghum farmers, the seed industry, researchers, and policy makers throughout the Americas.

For their contributions to sorghum ergot research, the ICRISAT pathologist and his collaborators from South Africa and Zimbabwe were presented the "Outstanding Achievement Award" by the NGSP Board and SICNA during the Twenty-first Biennial Grain Sorghum Research and Utilization Conference in 1999 for their collaborative research with INTSORMIL scientists on sorghum ergot.

3. CIRAD and ICRISAT: Improving Sorghum for Adaptation and Pest Resistance, Particularly in West Africa

From 1989-1999, the government of France provided major support to the West African Sorghum Improvement Program by posting in Mali four scientists from CIRAD, who worked in close collaboration with ICRISAT. The French government provided a plant breeder, an entomologist, an agronomist, and a striga specialist. Responsibilities of ICRISAT and CIRAD staff were complementary, especially with respect to breeding programs in Nigeria (hybrid development) and Mali (population improvement) and in terms of genetic backgrounds (caudatum in Nigeria and guinea in Mali).

The CIRAD breeder focused on the local guinea population. This work is being continued by ICRISAT with mass selection for grain maturity and panicle characteristics to form base material for cultivar development. The CIRAD entomologist identified and helped combine resistances to head bug and midge and determined that resistance to these pests was additive. In agronomy, the value of legume-sorghum rotations and intercrops, especially when sown before cotton was established. On-farm research focused on the development of fertility maintenance and restoration techniques. The *Striga* specialist inventoried parasitic plants and proposed cultural and chemical control methods against *Striga*.

ICRISAT continues to improve the guinea populations and to increase genetic diversity within them through the introgression of caudatum material into the guinea populations. Recently, an eco-physiologist from CIRAD joined ICRISAT-Mali to work on sorghum and focused on sorghum breeding based on plant ideotype, yield potential, and stability. Initially, this project will simulate the effect of photoperiodism on the agro-ecological adaptation of sorghum, and its significance from both plant breeding and crop management perspectives.

4. SCRI, ILRI, MAHYCO, Indian Agricultural Universities, and ICRISAT: Characterization and Management of MStV in Sorghum

Maize stripe tenuivirus is a serious pathogen of sorghum in the Deccan Plateau of India, where the stover is used for animal feed and the grain for both human food and animal feed. MStV causes significant losses in grain and fodder yield and quality in post-rainy season sorghum. The price of diseased sorghum fodder is reduced 20-25% relative to healthy fodder. Annual losses due to MStV are ≥ US$ 74 million in grain yield and ≥ US$ 78 million in fodder.

ICRISAT works with SCRI, ILRI, MAHYCO, the University of Agricultural Sciences (Karnataka, India), and Marathawada Agricultural University (Maharashtra, India) to develop cost-effective diagnostic tools for MStV, to assess biodiversity among MStV isolates, and to identify sources of resistance to MStV for use in breeding programs. Genotypes with resistance to more than one biotype of MStV will be supplied to NARS and NGOs to develop MStV-resistant cultivars suitable for rural and peri-urban areas. The technology of producing diagnostic reagents and their use to identify sources of MStV resistance also is to be transferred to scientists in the NARS and NGOs.

5. Advanced Research Institutes, INTSORMIL, and ICRISAT: DNA Markers for Terminal Drought Tolerance (Stay Green Trait) and Charcoal Rot Resistance in Sorghum

The post-rainy season at ICRISAT is very conducive for terminal drought screening. ICRISAT collaborates with QDPI and Purdue University to phenotype recombinant inbred lines that have line B35 as the source of the stay green trait, and with Texas Tech University for the screening of populations with the stay green marker that originated in E-36-1. In both cases, the objective is to identify markers for use in a marker-assisted backcross breeding program for the stay green trait. Two populations developed for striga resistance (10) also were evaluated for terminal drought tolerance (and stay green trait) and charcoal rot resistance.

6. IER, INTSORMIL and ICRISAT: Sorghum Hybrids for West Africa

Sorghum hybrids can increase productivity, thereby enhancing food security, and serve as a foundation for intensified production systems in western Africa. At present, sorghum hybrids are not available in western Africa because there are no hybrid parents with the adaptive traits of the guinea race. These traits include resistance to the major diseases, *e.g.*, anthracnose (caused by *Colletotrichum graminicola*), resistance to the grain mold-head bug complex, photoperiod sensitivity, and the grain characteristics required for storage, processing, and cooking. IER and ICRISAT are developing hybrid parents based on the guinea race of sorghum. With funding from the Rockefeller Foundation, the first guinea male-sterile lines and data from an assessment of heterosis levels will be available by 2003.

7. India and ICRISAT Partnership: Development of *Striga asiatica*-Resistant Sorghum Cultivars

Striga asiatica, a parasitic weed of sorghum, is a yield-limiting factor in many parts of Asia and eastern Africa, especially in marginal soils where fertility is low. Twenty-one million ha of land in Africa are estimated to be infested by striga, resulting in an annual grain yield loss of 4.1 million tons of grain and an estimated overall loss of US$ 7 billion in revenues. Reduction or elimination of strigol production and faster root growth by the host contribute to field resistance (7).

ICRISAT and scientists at Punjab Rao Deshmukh Krishi Vidyapeeth University, (Akola, India) have identified the striga-resistant sources N 13, 555, and Framida, and are breeding high-yielding striga-resistant restorer lines, cultivars, and seed parents. Some of the restorer lines have been utilized by African NARS, *e.g.*, Sudan, for developing striga-resistant cultivars. In India, ICSV 145 was released for general cultivation in striga endemic areas. Striga-resistant seed parents are being used by the national programs, and high-yielding striga-resistant hybrids have been identified.

8. University of Hohenheim, University of Tübingen, and ICRISAT: DNA Markers for Improving Sorghum Tolerance to Striga

Striga resistance is a complex quantitative trait, whose transfer into adapted sorghum cultivars has been limited due to inadequate information on its genetic basis, and the difficulty of evaluating resistance in the field (9, 10, 12). Genes for qualitative and quantitative resistance to *Striga hermonthica* in sorghum have been identified and mapped. Two mapping populations (RIPs), each consisting of 226 F_3-derived F_5 lines (F_3 families), were developed from IS 9830 (a low producer of stimulant for striga germination) × E 36-1 (striga susceptible, but terminal drought tolerant) (RIP 1), and from N 13 (impedes parasite attachment to its roots by "mechanical" barriers) × E 36-1 (RIP 2). Marker analyses were performed with bulked DNA from 20 plants per F_3 family. RIP 1 was genotyped for 177 marker loci (151 AFLP, 16 RFLP, 4 SSR, 6 RAPD), which were distributed in a genomic map of 872 cM length, and contained 147 markers distributed over 10 meaningful linkage groups (at LOD 5.1). For RIP 2, 243 marker loci (197 AFLP, 20 RFLP, 9 SSR, 17 RAPD) were distributed in a genomic map of 939 cM, and contained 191 markers distributed over 11 meaningful linkage groups (at LOD 4.5). The total lengths of the genomic maps were lower than expected.

Germination distance, a measure of the proportion of striga seeds germinating near a host genotype in a laboratory agar-gel assay, was determined for RIP 1. Two to eight QTLs were identified for germination distance and sufficed to explain most of the observed genetic variation. Both the RIP 1 and the RIP 2 populations were evaluated for field reactions to striga. The area under the striga number progress curve (a measure of infestation severity) depended on location, with 0-6 QTLs segregateing in RIP 1 and 3-8 QTLs segregating in RIP 2. None of the QTLs segregated at all locations in either RIP. Work continues to add more markers to the genetic maps, to refine the QTL analyses, and to more accurately measure the effects of the QTLs.

9. University of Georgia/USDA and ICRISAT: Characterization of Novel Mechanisms of Resistance to Striga in Pearl Millet

Cereals and grain legumes are the principal food crops grown in west Africa. In Benin, Cameroon, Gambia, Ghana, Nigeria, and Togo, 40-77% of the grain fields are infested with striga. Yield losses average 10-31% in infested fields, but in areas of heavy infestation, losses of 90-100% may occur in some years; loss in total grain yield averages 3-26%.

Host plant resistance is the most economic control measure, but little information is available on genetic resistance to striga in pearl millet. The wild grassy subspecies *Pennisetum glaucum* ssp. *monodii* and *P. glaucum* ssp. *stenostachyum* (the primary gene pool of commercial pearl millet) possess resistance to a number of major diseases, but this germplasm has not been evaluated previously for resistance to *Striga* spp.

USAID funded ICRISAT and the University of Georgia/USDA to evaluate wild germplasm in the primary, secondary, and tertiary gene pools of pearl millet (*P. glaucum*) for genes conferring resistance to *S. hermonthica* (26). Selection for striga resistance is confounded by its relationship with variables not directly related to resistance *per se*, *e.g.*, date of flowering and downy mildew incidence. Early maturity is a useful criterion for reduced striga, but is not desirable in all cropping systems. A selection index was developed that identifies accessions with lower than expected striga counts, based on regression equations, and gives additional weight to accessions with a low overall striga count. Variation in reaction among wild *Pennisetums* at different locations or in different years is large. Field trials are essential, as results from laboratory-based agar-gel assays that measure differences in stimulant production are not correlated with differences in striga infestation in the field.

Eighty-four *P. glaucum* ssp. *monodii* lines, wild primary gene pool accessions with the greatest resistance in prior evaluations, were moved from the Coastal Plain Experiment Station at Tifton, Georgia to sick plots in Maiduguri (Nigeria), Cinzana (Mali), and Sadore (Niger). Few lines were resistant to both striga and downy mildew. One hundred pearl millet entries, including hybrids, new seed parents and downy mildew resistant pollinators developed in Nigeria, and improved cultivars/local landraces from West Africa are being evaluated at Maiduguri and Babura, Nigeria. From these, parental lines will be selected for the study of the heritability of resistance. These crosses will be advanced by a single seed descent method for advanced progeny analysis, and progeny populations used to develop genetic maps identify QTLs. During this process, some resistance genes will be transferred to cultivated pearl millet types.

10. INTSORMIL (TAMU) and ICRISAT: Molecular Characterization of Downy Mildew in Pearl Millet

Scientists at TAMU and ICRISAT are using DNA markers to characterize genetic variability in pathogen populations as a means of understanding virulence diversity (11). At ICRISAT, DNA fingerprinting with SSR is used to detect genetic diversity in *Sclerospora graminicola* (20, 23, 24), with $(GATA)_4$ generating the most polymorphism. This SSR also was used to detect a new virulent pathotype specific to a recently introduced commercial hybrid cultivar MLBH 104. AFLP markers also have been used to detect genetic variability among *S. graminicola* isolates from India, and result in 50-70 bands per primer-pair combination, with 35-60% of the bands polymorphic.

AFLP analysis is being adapted to work with small numbers of spores and protocols that permit "whole genome amplification," originally developed for use in human diagnostics and forensics, are being tested for suitability for use with *S. graminicola*. Emphasis is being placed on the degenerate oligonucleotide primer (DOP)-PCR protocol, primarily because this protocol gives fairly uniform amplification of sequences throughout the genome. Analysis of DOP-PCR products amplified from *S. graminicola* DNA with RAPD or microsatellite primers has identified a high level of polymorphism in Indian and Malian populations.

11. British Advanced Research Institutes, NARS, and ICRISAT: Molecular Mapping of the Pearl Millet Genome and Marker-Assisted Selection for Pyramiding Resistance Genes for Powdery Mildew

In 1990 M. D. Gale (John Innes Institute) and J. R. Witcombe (ICRISAT) decided to develop a molecular marker based genetic linkage map of pearl millet (16). This joint venture provided John Innes Institute scientists with another grass species to use in their comparative mapping studies, and provided ICRISAT scientists with markers that could be used to effectively manipulate recessive traits, *e.g.*, disease resistance, in an otherwise conventional pedigree and backcross breeding program to generate hybrid parental lines. The collaboration now includes a companion project with the University of Wales (Bangor) to map QTLs associated with resistance to pearl millet downy mildew, and another with the Institute of Grassland and Environmental Research (Aberystwyth) that is identifying QTLs for drought and seedling heat tolerance (8).

As a result of this group of projects, there is now a usable pearl millet linkage map based on diverse molecular markers (RFLP, AFLP, SSR, and some isozymes), and the linkage group relationships between the pearl millet genome and those of rice and foxtail millet are known. This linkage map has been used to detect QTLs contributing to resistance to downy mildew populations in both Africa and Asia (14), as well as flowering time, plant height, seedling heat tolerance, grain and stover yield components, drought tolerance of these grain and stover yield components, and components of the ruminant nutritional quality of pearl millet straw.

The mapped QTLs are being used in marker-assisted selection programs that have incorporated new downy mildew resistance (3), and a QTL for increased plant height and biomass that is not associated with delayed maturity, into the elite pollinator H 77/833-2 and into elite seed parent maintainer line 843B. With these new versions of the parental lines of the widely grown single-cross hybrid HHB 67 (843A × H 77/833-2), the potential for a downy mildew epidemic should be greatly reduced. Germplasm with other important traits, *e.g.*, more drought-tolerant versions of H 77/833-2 and 841B, and more digestible versions of 841B, also is being developed.

12. NARS, Extension Services, Farmers Organizations, IITA and ICRISAT: SOSAT C88–A New Pearl Millet Cultivar for Nigeria

SOSAT-C88, a pearl millet synthetic cultivar developed by ICRISAT and LCRI, was released in Nigeria in January 2000 and is based on a collaboration begun in 1995. SOSAT-C88 was developed by recombining 19 S_1-selected progenies at Cinzana (Mali) in 1988 from a cross between the Sauna and Sanio land races and was first tested in Nigeria in 1995.

Based on data from three years and 15 environments, SOSAT-C88 produced 20% more grain than Farmers' Local and 15% more than ex-Borno, an improved check in the trial, while reducing downy mildew incidence (by more than half) and flowering 7-10 days earlier than the controls. In farmer-managed farmer-implemented on-farm trials for two years in 1997 and 1998 in six states of northern Nigeria, SOSAT-C88 produced 35% more grain than Farmers' Local, flowered at least six days earlier, and had a lower incidence of downy mildew. Farmers preferred SOSAT-C88 because of its well-filled panicle size, thick stem, downy mildew resistance, high grain yield, and food quality. SOSAT-C88 also produced the most grain and dry fodder in collaborative study with IITA of intercropping of cowpea and pearl millet.

A checklist survey of 230 farmers was conducted by LCRI in the millet-growing zones of Nigeria. Farmers listed the traits they used to select a cultivar: 99% preferred SOSAT-C88 because of its superior food quality and food taste, 72% preferred its earliness, 91% preferred larger grain, and 95% preferred shorter maturity. SOSAT-C88 was rated as more tolerant/resistant to pests/diseases than their local land race by 84% of poll respondents. However, SOSAT-C88 was perceived to tiller poorly (69% of respondents) and to have reduced stalk production (56% of respondents).

SOSAT-C88 was included in Small Plot Adaptation Trials conducted by Agricultural Development Programs in ten Nigerian states. Over 5000 small plot adaptation trials of SOSAT-C88 were conducted in northern Nigeria, where SOSAT-C88 is the cultivar preferred by most of the farmers. Two NGOs (SG 2000 and the Sokoto Agricultural Community Development Project) also are demonstrating SOSAT-C88 in Nigeria. This cultivar has become very popular in some states because of the large number of demonstrations by extension and SG 2000.

Three seed companies (Premier Seeds, UAC Seeds, and Nagoma Seeds) have been multiplying the seed of SOSAT-C88 and usually sell this seed for a premium price. Although the exact quantity of seed produced and sold is not known, more than 10,000 farmers were expected to sow SOSAT-C88 in Nigeria during its first year of release.

13. NARS, Advanced Research Institutes and ICRISAT: Understanding Finger Millet Blast, and Improving its Management and Host Plant Resistance

NARO, ICRISAT, Horticulture Research International (UK) and Queen's University of Belfast (Belfast, UK) are characterizing genetic and pathogenic diversity in *Pyricularia grisea*, the causal agent of finger millet blast, populations in Uganda and Kenya. The information will be used to develop sustainable disease management strategies to relieve the constraints imposed by blast. So far 194 *P. grisea* samples have been taken from finger millet panicles, necks, stems, or leaves at 34 sites from eight districts and one research station, in three provinces of western Kenya.

14. Indian Private Seed Sector and ICRISAT: Improving Pearl Millet and Sorghum Hybrids

ICRISAT sorghum programs in Asia used to develop finished products, *e.g.*, cultivars and hybrids, which, along with their parental lines were supplied to the public sector and to private sector seed companies who moved these materials to the farmers after testing for local adaptation. As private seed sector research capacity in India has increased, ICRISAT now aims to develop intermediate products, *e.g.*, seed parents and restorers instead of finished products. At present, 60-80% of the hybrids marketed by the private sector in India have ICRISAT-bred seed parents and/or restorers either as the parents of the hybrids or in their pedigrees. ICRISAT-private sector partnership sorghum hybrids are now grown on 1-1.5 million ha in India. For example, JKSH 22, a grain sorghum hybrid developed by J K Agri-genetics (Secunderabad), is based on modified ICRISAT breeding lines and is now grown on approximately 200,000 ha in the states of peninsular India, which constitutes about 10% of the total kharif hybrid sorghum area of these states. Eight Indian seed companies now provide US$ 33,000 of annual support to ICRISAT to promote sorghum research to produce high-yielding, stable, and temperature insensitive male-sterile lines (2003), high-yielding shoot-fly-resistant male-sterile lines (2003), high-yielding dual-purpose restorer lines (2003), information on the roles of cytoplasms (2002), information on GCA of restorers and CMS lines (2004), information on genetics of fertility restoration (2000), a database on the pedigrees of the lines (2004), and to improve the skills of the staff of the private sector (2004). Reddy *et al.* (17) describe the pathway to partnership between private sector seed companies and ICRISAT on sorghum and millet research in India.

Downy mildew, caused by *S. graminicola*, is the most devastating disease of pearl millet. ICRISAT now evaluates hybrids for two or three private companies (not the same company every year) for downy mildew resistance (seedling inoculation) at full cost-recovery. This activity enables private seed companies to have information on the resistance levels of their hybrids against pathotypes for which they do not have field/laboratory access. Single-cross hybrids are especially vulnerable to downy mildew, yet more than half of the 10 million ha of pearl millet grown in India is planted with single-cross hybrids. Most of the hybrid seed is produced and marketed by private seed companies who, therefore, have an intense interest in diversifying the genetic base of their hybrid cultivars and in developing downy mildew resistant parental lines.

Nine Indian seed companies are contributing approximately US$ 5000 each per year for a five-year term to support an umbrella of pearl millet hybrid parent research activities focused on genetic and cytoplasmic diversification, reduction of potential vulnerability of hybrids to downy mildew epidemics, and more extensive testing of breeding lines for resistance to different downy mildew pathotypes. Some of the companies also provide logistical support for field evaluation of hybrid parents at their research facilities. Research partnerships of this kind increase the amount of relevant data available and speed the dissemination of seed parents and breeding lines to NARS through both public and private sectors.

15. CORAF and ICRISAT: ROCARS and ROCAFREMI Research Networks for Sorghum and Millet in West Africa

ICRISAT, ROCAFREMI (the west and central African millet network), and ROCARS (the west and central African sorghum network) have contributed extensively to the development of a regional strategy for cereal improvement. Priority areas are: (i) establishment of a regional gene bank; (ii) improvement of germplasm, including the use of marker-assisted selection; (iii) development of commercial seed production; (iv) developing cost-effective methods to control insect pests, striga, and diseases; (v) identifying technologies to increase and maintain soil fertility; (vi) developing genetic and management options to alleviate drought; (vii) introducing new cropping systems; and (viii) improving grain and food processing. ICRISAT and INTSORMIL have contributed to the planning and execution of several research projects of both networks.

ROCARS. Sorghum production in West and Central Africa is beset by biotic, abiotic, socioeconomic, and institutional constraints. Striga, grain mold, anthracnose, charcoal rot, and leaf blight are among the major biological constraints. ROCARS, in partnership with ICRISAT, INTSORMIL, CIRAD, and SG 2000, is conducting regional trials. The most important achievements so far are: (i) establishing the relationship between head-bug attack and grain mold infections, (ii) identifying sorghum cultivars tolerant to grain molds and to striga, (iii) conducting on-farm tests of crop residue management to control anthracnose, and (iv) developing an integrated pest management package to control charcoal rot, striga, and anthracnose.

ROCAFREMI. ROCAFREMI was begun to boost millet production, to increase food security in the region, and to establish a sustainable mechanism for collaborative research and information exchange between those involved in millet research and development in the region. Ongoing projects include: (i) improvement of millet-based cropping systems, (ii) millet promotion through improved processing, (iii) integrated management of millet pests, and (iv) cultivar selection and seed multiplication with farmers' participation.

16. SADC/SACCAR and ICRISAT: Sorghum and Millet Improvement Program for Southern Africa

The SADC/ICRISAT Sorghum and Millet Improvement Program began in 1983 under the auspices of SACCAR. Its main purpose was to improve food security in drought-prone areas of SADC by increasing the productivity of sorghum and pearl millet, and to foster efficient utilization and conservation of natural resources. SMIP was conceived as a four-phase program. Phase I (1983-1988) focused on germplasm collection and exchange, cultivar development, training for national scientists and technicians, and infrastructure development. In Phase II (1988-1993), activities were expanded to include economics and policy studies, and food technology research. This research agenda was modified in Phase III (1993-1998) with greater emphasis placed on technology exchange, and collaboration with a wider range of partners including NGOs and the private sector. This change was made to increase the delivery of research products, after 10 years of work. These efforts continued in Phase IV (1999-2003) with the establishment of SMINET, the regional sorghum and millet network. SMINET provides a mechanism for sharing technologies across the region, broadens the range of partners, improves the access of regional researchers to the global knowledge base and germplasm resources for sorghum and pearl millet, and facilitates technology exchange between countries. Phase IV goes beyond crop improvement in that it addresses crop management and links farmers with improved marketing opportunities. Current SMIP activities involve as many as 110 organizations in 11 countries.

TABLE 58-1. Adoption of NARS-SMIP cultivars in the SADC region.

Country	% of National Sorghum/Millet Area	
	Sorghum	Pearl Millet
Angola	17	10
Botswana	20	30
Lesotho	3	Not grown
Malawi	9	7
Mozambique	7	6
Namibia	Not grown	49
Swaziland	50	Not grown
Tanzania	2	< 2
Zambia	38	19
Zimbabwe	30	27

SMIP has contributed substantially to agricultural R&D in the SADC region, including strengthened national research programs, higher crop productivity, pest and disease resistance, better understanding of environmental and socioeconomic factors in agriculture, and improved food security and farm income.

SMIP has distributed over 130,000 sorghum and pearl millet genotypes to SADC national programs, helping them develop a range of new cultivars for specific environments. To date, 46 improved NARS-SMIP cultivars have been released in eight SADC countries, where they are being widely adopted (Table 58-1). The improved cultivars have 30% higher yields than traditional land-races, with the potential for further gains if improved crop management methods also are adopted. Equally important, the improved cultivars mature significantly earlier, which reduces the risk of drought-induced crop failure and, therefore, improves food security.

Impact assessments by NARS/SMIP teams have documented that the benefits derived from these cultivars

TABLE 58-2. Countries in southern Africa in which sorghum productivity has improved as a result of ICRISAT partnerships with NARS and INTSORMIL.

Country	Release	Year Released	Approximate adoption area (ha)	Area (as % of total)
Botswana	Phofu[a]	1994	37,500	25
	BSH 1	1994	3,000	2
	Mahube	1994	1,500	1
Malawi	Pilira 1	1993	3,000	5
	Pilira 2	1993	3,750	7
Mozambique	Macia[a]	1987	22,000	5
	Chokwe	1999	22,000	5
Tanzania	Pato	1995	68,000	10
Zambia	Kuyuma[b]	1989	15,750	35
	MMSH 413[b]	1992	2,250	5
	MMSH 375	1992	2,250	5
	Sima	1989	2,250	5
Zimbabwe	SV 2	1987	63,000	36
	Macia[a]	1998	26,250	15
	SV 1	1987	7,000	4

[a]Same cultivar released in five different countries under different names. First released in Mozambique.
[b]Little impact in the region outside Zambia.

far exceed the donors' investments in the project. In Namibia, investments of approximately US$ 3 million by SMIP and the national program in developing and releasing Okashana-1 have yielded benefits with a value > US$ 11 million by 1998. The Internal Rate of Return to these investments was 50% (18). In Zambia, investments in sorghum breeding have yielded an initial rate of return of 10%, but more widespread adoption is expected to increase this value. An adoption rate of 50% would yield an IRR of 15% with benefits > US$ 6.4 million. In Zimbabwe, investments made by SMIP and the national program in developing and disseminating two cultivars (sorghum SV2 and pearl millet PMV3) have an initial rate of return of 27% at 1995 adoption levels, a value that rises to 34% at a more realistic adoption level of 50%. The net benefits from these two cultivars ranges from US$ 7.8-29 million, depending on the adoption rates. Impacts due to sorghum (Table 58-2) have generally been much higher than those for pearl millet, which are limited primarily to Okashana-1 in Namibia and PMV 2 in Zimbabwe, which is cultivated on 30% of the area planted to pearl millet.

This project pioneered the use of farmer-participatory research methods to develop crop management technology. For example, a common problem was extension recommendations that were inappropriate for smallholder farming conditions and, consequently, rarely adopted. By using new tools, *e.g.* simulation modeling, the project has helped to develop and test a range of practices that are practical and relevant in terms of labor/input requirements and cropping system priorities.

Researchers have identified the key pests and diseases in the region. Genetic resistance or tolerance to major pests and diseases has been bred into a wide range of genetically improved materials that have been widely distributed and are being used to further develop cultivars and hybrids.

Integrated pest management components for the control of the armored bush cricket (*Acanthoplus speiseri* Brancsik and *A. discoidalis* J. Iris) have been developed and tested on-farm. Although cricket populations were 25-30% lower than normal, due to poor rainfall, yield losses were still 10-40% in Zambia and 5-15% in Namibia, but ≤ 5% in IPM on-farm trials. Technology exchange on IPM strategies was promoted through farmer participation in Zambia at a SMIP-sponsored farmers' field day.

Other major SMIP outputs include: (i) Development of infrastructure suitable for crop research at its centers and at a number of NARS research stations. (ii) Provision of support for the training of 96 NARS scientists (34 Ph.D., 53 M.Sc., and 9 B.Sc.) from nine SADC countries. (iii) Collection, evaluation, and characterization of local germplasm–1,769 sorghum samples and 2,345 pearl millet samples collected between 1983 and 1992. (iv) Evaluation of over 4,075 sorghum and pearl millet cultivars for 15 grain quality traits related to grain use in food, malting, feed, and flour milling.

SMIP's Phase IV agenda builds on past successes in cultivar development. The main emphasis now is on seed delivery, crop management, grain marketing and commercialization. At present a greater share of SMIP's resources are targeted to Botswana, Mozambique, Tanzania, and Zimbabwe, the countries in the region that are the most dependent on sorghum and pearl millet.

17. ASARECA and ICRISAT: The Eastern and Central African Research Network for Sorghum and Millet

ICRISAT and the NARS of the region have worked to improve sorghum and millet production in eastern Africa. The EARSAM network ended in 1993, and a new network, ECARSAM was developed by ASARECA. ICRISAT will be the implementing agency for ECARSAM, and both ICRISAT and INTSORMIL will provide technical backstopping. ECARSAM's overall goal is to improve income and food security for small-scale farmers through increased productivity and commercialization of sorghum and millets. The network will promote adoption of improved technologies, by increasing knowledge of striga, insect pests, diseases, and drought, as they affect sorghum and millet, and develop integrated management packages for each constraint. International organizations, advanced research institutes and development investors (probably the European Union) will be accorded observer status on the Steering Committee. ECARSAM will work with other ASARECA networks to maintain research efficiency and to implement a systems approach.

Conclusion

As noted by Swindale (22), research cooperation for improving farming systems in the semi-arid tropics has been effective when driven by science opportunities rather than by supply-driven partnerships for their own sake. An initial careful analysis of a collaborative project generates confidences in both partners and sponsors. Successful project planning and execution require the participation of NARS, civil society, and the private sector, *e.g.*, local seed suppliers and village store owners, as well as organizations such as INTSORMIL and ICRISAT to ensure that projects remain relevant, timely, and on track.

The goal of ICRISAT's research is to help provide sustainable food security and reduce poverty in the developing world by adopting a regional people-centered systems approach, and by close collaboration with its research partners to apply modern science to the problems of the rural poor. ICRISAT puts great importance on technology exchange, information sharing, knowledge management, and partnerships as it implements its R&D agenda. Hence, ICRISAT investigations and collaborations are not solely for the sake of science, but also provide new technology to improve the livelihoods of those living in the semi-arid tropics. In this regard, human welfare, the embodiment of "Science with a Human Face," is the top priority in ICRISAT agenda for improving agriculture in the semi-arid tropics and the livelihood of the rural poor.

Acknowledgements

ICRISAT thanks partners, stakeholders and development investors for their support and active participation in our research partnerships. Persons providing information used in compiling this report include: I. Akintayo, R. Bandyopadhyay, F. R. Bidinger, P. J. Bramel, C. Clerget, S. C. Gupta, C. Tom Hash, G. M. Heinrich, Dale E. Hess, R. B. Jones, K. Anand Kumar, V. Mahalakshmi, A. Mayeux, M. A. Mgonja, E. N. Minja, E. S. Monyo, J. Ndjeunga, A. B. Obilana, B. Ouendeba, K. N. Rai, N. Kameswara Rao, H. F. W. Rattunde, B. S. V. Reddy, D. V. R. Reddy, David D. Rohrbach, N. Seetharama, H. C. Sharma, Ram P. Thakur, E. Weltzien-Rattunde, and O. Youm.

References

1. **Bandyopadhyay, R., D. E. Frederickson, N. W. McLaren, G. N. Odvody, and M. J. Ryley.** 1998. Ergot: a new disease threat to sorghum in the Americas and Australia. *Plant Dis.* **82:** 356–367.
2. **Bandyopadhyay, R., C. R. Little, R. D. Waniska, and D. R. Butler.** 2002. Sorghum grain mold: through the 1990s into the new millennium, p. 173-183. *In* J. F. Leslie (ed.), *Sorghum and Millets Diseases.* Iowa State Press, Ames, Iowa.
3. **Breese, W. A., C. T. Hash, K. M. Devos, and C. J. Howarth.** 2002. Pearl millet genomics and breeding for resistance to downy mildew, p. 243-246. *In* J. F. Leslie (ed.), *Sorghum and Millets Diseases.* Iowa State Press, Ames, Iowa.
4. **Casela, C. R., and J. A. Dahlberg, eds.** 1999. *Proc. Global Conf. on Ergot of Sorghum (1-8 June 1997, Sete Lagoas, Brazil).* INTSORMIL, University of Nebraska, Lincoln, Nebraska.
5. **Chrandrashekar, A., R. Bandyopadhyay, and A. Hall, eds.** 2000. *Technical and Institutional Options for Sorghum Grain Mold Management: Proc. Intl. Consultation (18-19 May 2000, ICRISAT, Patancheru, India).* ICRISAT, Patancheru (A.P.), India.
6. **de Milliano, W. A. J., R. A. Frederiksen, and G. D. Bengston, eds.** 1992. *Sorghum and Millets Diseases: A Second World Review.* ICRISAT, Patancheru, (A.P.), India.
7. **Ejeta, G., L. G. Butler, D. E. Hess, T. Obilana, and B. S. V. Reddy.** 1997. Breeding for Striga resistance in sorghum, p. 504-516. *Proc. Intl. Conf. on Genetic Improvement of Sorghum and Pearl Millet (22-27 September 1996, Lubbock, Texas)*: 504-516.
8. **Hash, C. T., and J. R. Witcombe.** 2002. Gene management and breeding for downy mildew resistance, p. 27-36. *In* J. F. Leslie (ed.), *Sorghum and Millets Diseases.* Iowa State Press, Ames, Iowa.
9. **Haussmann, B. I. G., D. E. Hess, B. S. V. Reddy, S. Z. Mukuru, N. Seetharama, M. Kayentao, G. O. Omanya, H. G. Welz, and H. H. Geiger.** 2000. QTL for striga resistance in sorghum populations derived from IS 9830 and N 13, p. 159-171. *In* B. I. G. Haussmann, D. E. Hess, M. L. Koyama, L. Grivet, H. F. W. Rattunde, and H. H. Geiger (eds.), *Breeding for Striga Resistance in Cereals.* Margraf Verlag, Weikersheim, Germany.
10. **Haussmann, B. I. G., D. E. Hess, B. S. V. Reddy, H. G. Welz, and H. H. Geiger.** 2000. Analysis of resistance to *Striga hermonthica* in diallel crosses of sorghum. *Euphytica* **116:** 33-40.
11. **Hess, D. E., R. P. Thakur, C. T. Hash, P. Sérémé and C. W. Magill.** 2002. Pearl millet downy mildew: problems and control strategies for a new millennium, p. 37-42. *In* J. F. Leslie (ed.), *Sorghum and MilletsDiseases.* Iowa State Press, Ames, Iowa.
12. **Hess, D. E., J. Kroschel, D. Traoré, A. E. M. Elzein, P. S. Marley, A. A. Abbasher and C. Diarra.** 2002. Striga: biological control strategies for a new millennium, p. 165-

170. *In* J. F. Leslie (ed.), *Sorghum and Millets Diseases*. Iowa State Press, Ames, Iowa.

13. **Johnson, M. A.** 2002. Private sector and public institution interactions on sorghum and pearl millet disease management, p. 349-350. *In* J. F. Leslie (ed.), *Sorghum and Millets Diseases*. Iowa State Press, Ames, Iowa.
14. **Jones, E. S., C. J. Liu, M. D. Gale, C. T. Hash, and J. R. Witcombe.** 1995. Mapping quantitative trait loci for downy mildew resistance in pearl millet. *Theor. Appl. Genet.* **91:** 448-456.
15. **Lenné, J. M., and R. Ortiz**. 2002. Agrobiodiversity in pest management, p. 309-319. *In* J. F. Leslie (ed.), *Sorghum and Millets Diseases*. Iowa State Press, Ames, Iowa.
16. **Liu, C. J., J. R. Witcombe, T. S. Pittaway, M. Nash, C. T. Hash, C. S. Busso, and M. D. Gale.** 1994. An RFLP-based genetic map of pearl millet (*Pennisetum glaucum*). *Theor. Appl. Genet.* **89:** 481-487.
17. **Reddy, B. V. S., A. J. Hall, and K. N. Rai.** 2001. The long road to partnership: Private support of public research on sorghum and pearl millet. p. 27-34. *In* A. J. Hall, B. Yoganand, V. Rasheed Sulaiman, and N. G. Clark (eds.), *Sharing Perspectives on Public–Private Sector Interaction: Proc. of a Workshop (10 April 2001, Patancheru, India).* ICRISAT, Patancheru (A.P.), India.
18. **Rohrbach, D. D., W. R. Lechner, S. A. Ipinge, and E. S. Monyo.** 1999. *Impact from Investments in Crop Breeding: The Case of Okashana-1 in Namibia*. ICRISAT Impact Series no. 4. ICRISAT, Patancheru (A.P.), India.
19. **Ryley, M. J., D. J. Herde, S. A. Bhuiyan, R. G. Henzell, and D. R. Jordan.** 2002. An overview of the biology of sorghum ergot, p. 141-150. *In* J. F. Leslie (ed.) *Sorghum and Millets Diseases*. Iowa State Press, Ames, Iowa.
20. **Sastry, J. G., W. Ramakrishna, S. Sivaramakrishnan, R. P. Thakur, V. S. Gupta, and P. K. Ranjekar.** 1995. DNA fingerprinting detects genetic variability in the pearl millet downy mildew pathogen (*Sclerespora graminicola*). *Theor. Appl. Genet.* **91:** 856-861.
21. **Shah, M., and M. Strong.** 1999. *Food in the 21st Century: From Science to Sustainable Agriculture*. CGIAR System Review Secretariat, The World Bank, Washington, DC.
22. **Swindale, L.** 2000. Crop and livestock synergies, p. 16-18. In *Synergies in Science: Intercenter Collaboration to Eradicate Hunger and Poverty*. CGIAR, Washington, DC.
23. **Thakur, R. P., C. W. Magill, S. Sivaramakrishnan, C. T. Hash, H. S. Shetty, and D. E. Hess.** 2002. Variability in *Sclerospora graminicola*, the pearl millet downy mildew pathogen, 51-56. *In* J. F. Leslie (ed.), *Sorghum and MilletsDiseases*. Iowa State Press, Ames, Iowa.
24. **Thakur, R. P., V. P. Rao, J. G. Sastry, S. Sivaramakrishnan, K. N. Amruthesh, and L. D. Barbind.** 1999. Evidence for a new virulent pathotype of *Sclerospora graminicola* on pearl millet. *J. Mycol. Plant Pathol.* **29:** 61-69.
25. **Williams, R. J., R. A. Frederiksen, L. K. Mughogho, and G. B. Bengston, eds.** 1980. *Proc. Intl. Workshop on Sorghum Diseases (11-15 December 1978, Hyderabad, India).* ICRISAT, Patancheru (A.P.), India.
26. **Wilson, J. P., D. E. Hess and W. W. Hanna.** 2000. Resistance to *Striga hermonthica* in the primary gene pool of *Pennisetum glaucum*, p. 147-156. *In* B. I. G. Haussmann, D. E. Hess, M. L. Koyama, L. Grivet, H. F. W. Rattunde, and H. H. Geiger (eds.), *Breeding for Striga Resistance in Cereals.* Margraf Verlag, Weikersheim, Germany.
27. **Witcombe, J. R., and S. R. Beckerman, eds.** 1987. *Proc. Intl. Pearl Millet Workshop (7-11 April 1986, Patancheru, India*). ICRISAT, Patancheru (A.P.), India.

59

PROFIT–Productive Rotations On Farms In Texas

Gary C. Peterson, Bonnie B. Pendleton, and George L. Teetes

Abstract

Economic viability in twenty-first century agriculture requires Texas farmers to use a diverse, sustainable, environmentally friendly, and economically profitable production system, "PROFIT." Farmers require information to make decisions that affect profitability and sustainability while adapting to rapidly changing needs and opportunities. The realities of the twenty-first century require a new research and education paradigm, "PROFIT," with sorghum as the keystone to directly address future agricultural problems and opportunities. PROFIT is funded by the Texas Legislature to improve sorghum profit-ability (by 20%) and to increase sorghum yields (by 20-25%). PROFIT research and information delivery is an integrated, coordinated activity involving personnel from multiple agencies–Texas Agricultural Experiment Station (TAES), Texas Cooperative Extension (TCE), Texas Tech University (TTU), and the United States Department of Agriculture–Agricultural Research Service (USDA-ARS)–at locations across Texas. PROFIT has two kinds of projects–regional and research support. Regional projects join research and extension in on-farm or on-station activities to demonstrate different production systems and the role of sorghum in the system. Technology in regional programs is available for immediate adoption by farmers. Research support projects conduct longer-term research and technology development for the regional programs. An economic analysis of PROFIT projects will be conducted to validate potential impact. An Internet site (http://sorghum.tamu.edu) is available for dissemination of PROFIT or other information on sorghum. Farmers' representatives were involved in developing PROFIT and participate in project selection and evaluation.

The PROFIT Initiative

Texas farmers need an integrated production system comprising specific crops that are profitable and well adapted to different regions of the state. The research and education mission of the Texas A&M University System (TAMUS) Agriculture Program is to develop and deliver information conducive to the production of a safe, high-quality food, feed, and fiber supply with minimal environmental disruption. Economic viability in twenty-first century agriculture requires Texas farmers to use a diverse, sustainable, environmentally friendly, and economically profitable production system, "PROFIT." Establishment of PROFIT by Texas farmers and the TAMUS Agriculture Program is both consistent with and an important part of Texas A&M University's mission and purpose as a land grant institution.

The PROFIT (Productive Rotations On Farms In Texas) initiative resulted from a Grain Sorghum Research Task Force appointed in 1998 by the director of the TAES. The Task Force was charged to evaluate the future of grain sorghum research conducted by TAES, to determine TAES options, and to develop a strategic plan for sorghum research. The impetus for the Task Force was the anticipated retirement of key TAES sorghum improvement scientists and the need to justify continuation of their research activities while optimizing the utilization of TAES resources. Specifically, the Task Force was to: (i) determine the need to continue TAES sorghum research, (ii) identify current and future priority research issues, (iii) identify the disciplines needed to fulfill mission objectives, (iv) recommend how the program should be coordinated and the resources required for implementation, and (v) anticipate the expected impact. Sorghum farmers and the Texas Grain Sorghum Association (TGSA) participated in the evaluation of TAES sorghum research and in developing the rationale for continuing and expanding sorghum activity by the TAMUS Agriculture Program.

The Task Force assessed the value of sorghum to the Texas farmer and economy. Initially many of us perceived that sorghum was declining in importance and that there was little justification to continue the current level

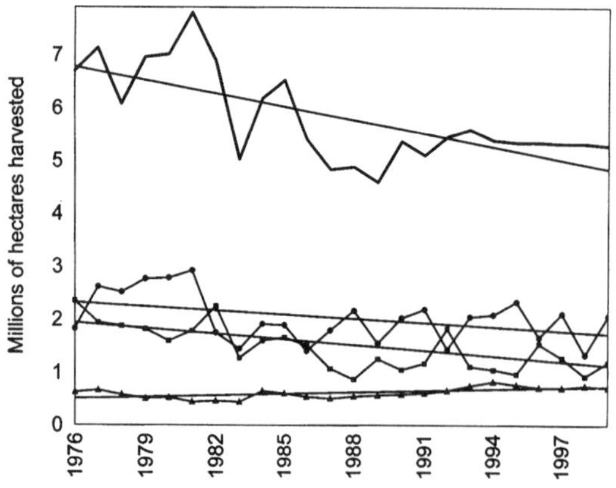

FIGURE 59-1. Hectarage of cotton (•), sorghum (■), maize (▲), and total field crops (———) in Texas, 1976–1999 (1).

of TAES research on sorghum. However, some task force members insisted that sorghum was too important to Texas agriculture for TAMUS not to have a sorghum research and information delivery program. They argued the numerous well-known advantages of sorghum and examined crop production statistics for Texas (1). Conlusions drawn from examining long-term production trends for sorghum and cotton were crucial in developing the Task Force's recommendations.

The Task Force concluded that sorghum is: (i) the keystone to production of Texas row crops, (ii) the most important crop for dryland production, and (iii) the only crop that can be rotated with all crops anywhere in the state. Sorghum is the highest hectarage summer grain crop in Texas (1), and hectarage has remained relatively constant at approximately 1.2 million hectares a year (Fig. 59-1). Twenty-year (1976-1996) averages indicated that both sorghum and cotton production have declined (1). The area planted to maize increased slightly during the same period of time. Thus, in spite of the perception that sorghum hectarage and importance have decreased, the percentage of decline is approximately equal to that of the overall decrease in field crop hectarage in the state. The Task Force concluded that to maintain profitable row crop production in Texas, the TAMUS Agriculture Program should increase sorghum research and technology delivery to improve the role of sorghum in the production system and thus increase the profitability and sustainability of the production system. No other crop can be substituted for sorghum in the production system of Texas.

Sorghum's Advantages

There are significant advantages to including sorghum in the Texas crop production system including: (i) profit in integrated production systems, (ii) compatibility in integrated production systems, (iii) wide geographic adaptation, (iv) water-use efficiency, (v) benefit to the environment and agroecosystems, (vi) excellent risk aversion, (vii) low inputs, (viii) soil and water conservation, and (ix) genetic diversity. Sorghum commonly generates at least 25% ($1 billion/year) of the total row crop revenue in Texas (1). In Texas crop production systems, sorghum, primarily because of the advantage of including it in crop rotations, also contributes to the profitability of other crops, an impact that has never been analyzed economically.

Additionally, most farming in Texas is dry land, and about 75% of sorghum in the state is produced dry land (1). There is only an 11% difference between planted and harvested sorghum hectarage, which is the lowest percentage of difference when compared with other row crops (Table 59-1). There is no other crop that can be substituted for sorghum in Texas cropping systems. Cropping systems are determined by environment, soil type, and price or marketing conditions. Sorghum is grown in Texas because it is more water-use efficient and more likely than maize, or other crops, to produce a harvestable crop under adverse conditions. Sorghum is the only cereal crop that can be rotated with broadleaf crops during the summer in most parts of Texas. As both the Texas environment and marketing factors favor sorghum, sorghum will continue to be a keystone crop and the only one suitable for production in all regions of Texas.

Constraints to Sorghum Profitability

Constraints affecting sorghum must be addressed by the TAES through research to define and integrate sorghum into the PROFIT program. TCE will then deliver this research-based technology to farmers. The goal of enhancing sorghum's role in PROFIT can be achieved by: (i) developing production/rotation strategies to optimize inputs and the genetic potential of hybrids, (ii) genetic improvement of grain quality, (iii) increased end-use demand through better quality and improved processsing, (iv) plant protection through resistant hybrids and integrated pest management, (v) plant stress adaptation, (vi) increasing harvestable grain yield through better environmental adaptation, (vii) improved plant health through more efficient nutrient utilization, (viii) establishment of integrated multidisciplinary research, and (ix) effective delivery of information by TCE.

To achieve PROFIT's goals, research and technology transfer are required for: (i) crop-management strategies that include agronomy, plant physiology, and soil chemistry; (ii) improvements in grain quality, end uses, and desirable utilization characteristics that include cereal chemistry and livestock and poultry nutrition; (iii) genetic improvements that include plant breeding and molecular biology (TAES should become the national/international center for sorghum biotechnology research); and (iv) improvements in plant health (plant protection/integrated pest management) that include entomology, plant pathology, and weed science. Personnel from the following units could be involved: (i) departments of Agricultural Economics, Animal Science, Biochemistry and Biophysics, Entomology, Plant Pathology and Microbiology, Poultry Science, and Soil and Crop Sciences; (ii) the Crop Biotechnology Center; and (iii) Texas A&M University Agricultural Research and Extension centers at Amarillo, Chillicothe/Vernon, Corpus Christi, Lubbock, Uvalde, and Weslaco.

Cropping system needs, and the role of sorghum in these systems, vary by geographic region. Integrated expertise in research and extension is required to address the role of sorghum in these various cropping systems. Moreover, greater extension involvement is essential to maximize the role of sorghum in PROFIT. Through greater extension involvement, farmers have access to up-to-date information on improved production practices, *e.g.*, adjustments to planting population, row spacing, weed control, and fertilizer application, and may more easily gain information on new hybrid genetic options. Regional meetings involving TAES, TCE, and farmers were held with the goals of: (i) identifying research and extension goals for PROFIT, (ii) creating regional sorghum production system teams/PROFIT program teams, and (iii) redirecting personnel to sorghum research and education. At these meetings, farmers discussed the role of sorghum in their farming operations, problems in sorghum production that required research and/or extension involvement, and future directions for sorghum research and extension. Research/extension production teams and programs were developed to address the stated problems. If the farmers did not believe that particular research or extension activities would be beneficial, then those activities were not considered further.

Resources for PROFIT

To conduct a broad-based sorghum research and technology transfer program and to ensure its success, resources (personnel and program funds) must be identified and dedicated to the PROFIT program. Current TAES full-time equivalents (FTEs) assigned to sorghum research are estimated at 6.95 in 10 disciplinary areas (Table 59-2). Anticipated needs for a successful role of sorghum in the PROFIT program require an additional 6.15 FTEs in research. Current TCE FTEs assigned to sorghum technology delivery were not considered adequate to meet the needs of a keystone crop, and an additional 4 FTEs in extension with primary sorghum responsibility are needed to enable TCE to adequately perform its technology transfer mission.

TABLE 59-1. Comparison of planted and harvested hectares of the four most important Texas row crops grown under dryland conditions from 1976-1999 (1).

Crop	Million Hectares		%
	Planted	Harvested	Difference
Sorghum	1.13	1.01	11
Corn	0.40	0.32	13
Cotton	1.46	1.17	19
Wheat	2.15	1.05	50

Some resources required for PROFIT could come from existing personnel and programs and from the redirection/reorientation of existing resources. PROFIT would provide the impetus for sorghum research and educational efforts to obtain additional funds. An estimated additional $1.2 million of program funds (excluding salaries and overhead) are needed to cover research and extension activities for the sorghum component of the PROFIT program. A legislative initiative for funding integrated sorghum cropping systems research and educational efforts was recommended by the Grain Sorghum Research Task Force as justified based on the potential impact on farmers and agriculture in Texas. To validate the effect of the PROFIT program, the economic impact of these additional funds for research and extension is to be measured and correlated to total impact on the Texas economy.

Legislative Action

A legislative initiative was prepared and presented to the 1999 Texas legislature. The request for $1.2 million for PROFIT was to: (i) increase sorghum profitability by 20%, (ii) increase yield by 20-25%, (iii) improve grain quality, and (iv) reduce risk. No additional faculty were to be hired and all funds were to be used to support PROFIT activities. Involvement of farmers, the TGSA, and other commodity groups was critical to persuade the legislature to fund PROFIT. PROFIT would not exist without the dedication and willingness of farmers and the TGSA to explain to legislators the need for a sustainable, economically profitable, environmentally friendly production system. In so doing, farmers committed themselves to the success of PROFIT and to work with TAES

TABLE 59-2. Current and additional full-time faculty equivalents required for the PROFIT program.

Expertise	Full-Time Equivalents (FTEs)	
	Current	Additional
Agronomy	0.3	0.7
Biotechnology	1.0	1.0
Economics/marketing	0.1	0.9
Entomology	0.7	0.3
Physiology	0.2	0.8
Plant breeding	2.5	-0-
Plant pathology	0.8	0.2
Soil fertility	0.3	0.7
Utilization	0.8	0.8
Weed science	0.25	0.75
TAEX program specialists	-0-	4.0
Total	6.95	10.15

and TCE to ensure that appropriate research and technology activities were conceived, organized, and evaluated, and that the impact of these activities was properly documented.

Through General Appropriations Act: House Bill 1, 76th Legislative Regular Session, $750,000 in base funds were appropriated to initiate PROFIT. With base funds, PROFIT became a recurring item and does not need to be reauthorized each session. So long as progress is made towards PROFIT goals and funds are adequately accountted for, the TAMUS Agriculture Program will have $750,000 per year dedicated to sorghum research and technology transfer.

PROFIT Participation and Projects

PROFIT was conceived as an inclusive program. PROFIT can be successful only through participation across disciplines, agencies, and locations. To give all TAMUS Agriculture Program faculty interested in sorghum an opportunity to participate in PROFIT, a survey was sent to all TAMUS agriculture units. The objective of the survey was to identify sorghum research and information delivery activity and expertise within the TAMUS Agriculture Program. Information requested included: (i) the unit's past and current activities in sorghum (and personnel involved), and (ii) what the unit proposed to do in the future if additional funds were available. Survey responses were used to develop a database of TAMUS sorghum-related activity and personnel. PROFIT funds were appropriated by the legislature for discretionary use by the TAES Director. There were no prior arrangements or understandings as to how the money would be distributed among units or faculty other than that all funds would be used to strengthen activity of the TAMUS Agriculture Program.

A request for proposals (RFP) was used to solicit projects for funding consideration. The RFP requested that interdisciplinary and interorganizational teams develop innovative and effective research and information delivery programs to directly benefit producers, but did not suggest specific research or information delivery activities. Single investigator or agency proposals were discouraged (but not prohibited). Thus, personnel were encouraged to present their best suggestions for programs to improve the role of sorghum in the Texas cropping system. Proposals were expected to have a clear relationship to the improvement of the value of sorghum in a cropping system. Proposals also were expected to show clear linkages across locations and agencies and to describe the activity in such a manner that the contribution of each component (and/or individual) could be measured and the impact documented.

Two kinds of projects were requested: regional projects for the two primary sorghum-growing regions (Coastal Bend/Central Texas and High Plains) and research support projects. The regional projects were expected to propose activity best suited to that region. Utility of results from regional projects in other areas was to be determined after the goals of the regional program were achieved. Regional projects were expected to have technology ready for adoption by farmers, and to include the expertise necessary to meet the goals. Longer-term research support projects were expected to provide new knowledge and technology to improve the role and value of sorghum in one or more of the regional cropping systems, and to describe how and why the research could improve farmer's profits.

An Internal/External Review Committee, composed of representatives of TAES, TCE, TTU, farmers, and the TGSA, was appointed in September 1999 by the director of the TAES. Members of the review committee were selected to represent the agencies or groups involved in PROFIT. This committee was charged with evaluating the proposals, recommending projects for funding, and soliciting proposals for activities considered critical for PROFIT's success. Farmer representatives were adamant as to the kinds of activities that would improve their profits, and their comments led to significant revisions in some proposals prior to funding. Of 37 proposals received, 22 were recommended for funding. No proposals were received in three areas–plant pathology directed primarily at stalk rots, weed control in South Texas, and an economic project directed at policy/lending–and proposals to meet these needs were solicited from appropriate persons.

The funded proposals constitute a well-rounded project portfolio, and involve 44 scientists from TAES, TCE, TTU and USDA-ARS, and county extension agents at various locations throughout the state. Two regional projects were selected–one for Central/South Texas and the other for the High Plains. Eleven TAES or TCE faculty members or county extension agents from five locations are involved in the Central/South Texas project. The High Plains regional project includes 13 TAES, TCE, or TTU faculty members at Lubbock and 10 TAES, TCE, or USDA-ARS personnel from Amarillo. Research support projects include 14 faculty members at three locations. The information delivery project is led by two faculty members at Lubbock. The policy and lending economics project is led by one faculty member.

Program Management

During the implementation process it became apparent that an individual(s) should be designated as "PROFIT Coordinator." Duties of the position include developing presentations or documentation of PROFIT activities for presentation to clientele, serving as an initial contact for and source of information about the program, and ensuring routine reporting, interaction with the administration, and scheduling of meetings and tours occurred in a timely manner. At present there are three co-coordinators–Gary Peterson from TAMUS, Bonnie Pendleton from West Texas A&M University, and Bruce Maunder from the National Grain Sorghum Producers (NGSP). The co-coordinators are volunteers, and no compensation is given to the individuals for the effort required for the coordination activity.

An internal/external Advisory Board was appointed in the summer of 2000 by the Director of the TAES to evaluate PROFIT projects and provide recommendations on projects, recommended changes, and general oversight to ensure PROFIT evolves as needed and meets the goals and objectives agreed upon. The Advisory Board is composed of representatives from TAES, TCE, TTU, TGSA, Plains Cotton Growers, South Texas Cotton and Grain, and private seed industry. The intent in structuring the Advisory Board was to receive input from the state agencies involved in PROFIT, sorghum farmers, sorghum producers, cotton producers, and private seed industry. In addition to representing clientele, the Advisory Board composition provides for an unbiased evaluation of PROFIT projects, new proposals, and progress. The Advisory Board works with the PROFIT co-coordinators to provide overall guidance for PROFIT.

PROFIT was conceived as a dynamic program that would evolve with the supported activities and that could capitalize on new opportunities as they arose. This evolution will require additional funds to enable the new activities to contribute to PROFIT while maintaining existing activities. A new initiative, designated PROFIT II, was presented to the 2001 Texas Legislature. PROFIT II was to fund: (i) crop management specialists to work directly with producers, (ii) internships for students to work with TCE specialists, researchers, and crop management specialists to receive on-the-job training in cropping systems, (iii) a statewide sorghum TCE specialist, and (iv) additional funds for research and coordination. The 2001 Texas Legislature did not appropriate funds to support PROFIT II. To enable PROFIT to fulfill its mission of improving the role and productivity of sorghum in Texas cropping systems, the PROFIT Advisory Board recommended in June 2001, that some funds from PROFIT be used to hire a state sorghum/cropping systems specialist and an integrated crop management specialist. PROFIT funds used to support the two new TCE specialists were re-directed from research support projects. Discussions are underway to develop subsequent initiatives to provide funds for other program components, *e.g.* expansion of research and technology transfer activity to include aspects of sorghum not included in the current program and the incorporation of other commodities.

Acknowledgments

Grain Sorghum Research Task Force (1998): TAES–Gary Peterson (Chair), Richard Frederiksen, Wayne Greene, John Matocha, Gary Odvody, Andrew Paterson, Bill Rooney, Lloyd Rooney, Darrell Rosenow, George Teetes; TTU–Eduardo Segarra; TGSA–Travis Taylor (Executive Director); Administrative Liaison–Frank Gilstrap (Associate Director, TAES).

Internal/External Review Committee (1999): TAES–Gary Peterson, George Teetes, Bonnie Pendleton, Bobby Eddleman; TCE–James Supak; TTU–Robert Albin; TGSA–Travis Taylor (Executive Director), Bill Kubecka, Chris Cogburn; Administrative Liaison–Frank Gilstrap (Associate Director, TAES).

PROFIT Advisory Board (2000-present): Gary Peterson (Chair, TAES), Bonnie Pendleton (West Texas A&M University), Eduardo Segarra (TTU), Travis Miller (TCE), Jerry O'Rear (Garrison and Townsend Seed Company), Steve Veritt (Executive Director, Plains Cotton), Jeff Nunley (Executive Director, South Texas Cotton and Grain), Ray Watson (Executive Director, TGSA), Bill Kubecka and Chris Cogburn (farmers, TGSA members), Bruce Maunder (Research Advisor, NGSP); Administrative Liaison–Frank Gilstrap (Associate Director, TAES).

Reference

1. **Texas Agricultural Statistics.** 1976-1999. Texas Agricultural Statistics Service, Texas Department of Agriculture and United Stated States Department of Agriculture, Austin, Texas.

Part X

Country and Regional Disease Reports

60

Recommendations from the Working Group for the Americas at the Third Global Conference on Sorghum and Millets Diseases

Reported by *Laura M. Giorda*

Considerable progress has been made in addressing the disease research recommendations made during the previous global meeting despite declining numbers of research scientists and diminishing research funds. Among these accomplishments are identification and global utilization of the stay green trait to increase drought tolerance and to reduce the incidence of stalk rots. Adoption and implementation of molecular detection and identification protocols have advanced our research capabilities in ways and to an extent that was not thought possible at the last meeting in 1988. These techniques have allowed discrimination of distinct species within *Fusarium* section *Liseola*, which cause head blight, grain mold and weathering deterioration of sorghum grain, and some stalk rots. Host plant resistance developed for sorghum downy mildew, viruses, anthracnose, and head smut has minimized the threat posed by these diseases both in the Americas and in other parts of the world. Similar advances have been made in resistance to grain mold and foliar diseases.

In addition, linkages established between scientists at the prior congress enabled a rapid response to the global threat of sorghum ergot when it appeared in Brazil. However, the primary constraint to progress in meeting stated recommendations continues to be a lack of adequate research funds and a relatively small number of researchers, many of whom have responsibilities other than for sorghum. Comprehensive disease management strategies also are needed to control multiple disease problems and to conserve identified and deployed host plant resistance and other components of existing disease management systems. Despite our successes, much remains to be done to address the disease problems outlined in this volume and in these recommendations. Molecular approaches have altered our views of pathogens and have begun to impact breeding practices. If we continue to build on our successes we should be pleasantly surprised at our progress when we gather again a decade from now.

Sorghum Diseases to Which Research Priority Should be Given

- **Foliar Diseases:** Anthracnose, leaf blight, rust, sooty stripe, and bacterial stripe.
- **Root and Stalk Rots:** Fusarium stalk rot and charcoal rot.
- **Grain and Panicle Diseases:** Grain mold (caused by *Fusarium* and *Curvularia*), grain weathering, ergot, head blight, and head smut.
- **Systemic Diseases:** Sorghum downy mildew and viruses, especially MDMV.

Overall Priority Ratings (ranked):

1. **Grain mold, head blight, and weathering.** Determine the role of endophytic *Fusarium* spp. in the infections responsible for grain mold and head blight. Epidemiological studies, especially as related to environmental factors. Identify and describe causal agents, and determine their distribution and population structure. Identify phenotypic traits and molecular markers that can be used in resistance screening programs.
2. **Stalk rots.** Further studies of pathogen and population dynamics associated with environmental interactions and cropping systems. Identify and describe causal agents, and determine their distribution and population structure. Identify suitable markers for use in resistance screening programs.
3. **Ergot.** Identify origin of primary inoculum for ergot epidemics, including determining how the pathogen

survives between seasons and the role of sclerotia and sphacelia in the pathogen's life cycle in the Americas. Regional calibration of epidemiology and disease prediction models. Develop an integrated disease control program that optimizes fungicide use, and employs nonchemical control methods to the greatest extent possible.

4. **Anthracnose.** Continue collaborative screening for host plant resistance at strategic locations that represent diverse climates and pathogen populations both within the Americas and elsewhere.

Pearl Millet

As pearl millet cultivation expands into non-traditional environments, diseases are becoming a more important constraint to the production of high quality grain and forage. Since the area planted to pearl millet in South America is increasing rapidly, especially in Brazil, and because South and North America are not epidemiologically isolated regions, we recommend that:

- The threat of the introduction of an exotic plant pathogen, *e.g.*, pearl millet downy mildew, into the Americas be reduced by restricting the introduction of pearl millet germplasm into South America from the eastern hemisphere. At present, such introductions are not regulated.
- Collaborative research linkages be developed between North and South American agencies, and that hemisphere-wide strategies be established for the management of rust and leaf blight.
- That the potential for fungal and mycotoxin contamination be reduced through the development of resistant germplasm and appropriate cultural and storage practices.

61

Recommendations from the African Working Group at the Third Global Conference on Sorghum and Millets Diseases

Reported by *Dale E. Hess*

This group met during the Third Global Conference on Sorghum and Millet Diseases at Guanajuato, Mexico, 23-29 September 2000, to update the research recommendations made at the Second Global Conference on Sorghum and Millets held in Harare, Zimbabwe, 1-11 March 1988. The meeting was open to all interested conference participants, and no list of participants peculiar to this meeting is available. Participation of African scientists in this meeting was limited by the number who could obtain the necessary visas and travel documents to enter Mexico. More details on many of these recommendations can be found in the country and regional reports included in this volume.

Progress Since the Second Global Conference

Foliar Diseases

- Screening methodologies were developed for, and resistance sources identified and utilized in breeding for resistance to:
  - Sorghum foliar diseases such as anthracnose, downy mildew, sooty stripe.
  - Pearl millet downy mildew.
- Cultural management approach investigated:
  - Stover management to control sorghum anthracnose.
- Chemical control treatment developed:
  - Seed treatment (Apron Plus®) to control downy mildew and reduce growth variability in pearl millet.

Panicle Diseases

- Resistance sources identified and available for use:
  - Sorghum grain molds and long smut.
  - Pearl millet ergot and smut.

New Diseases

- Viral diseases in west and central Africa.
  - Maize mosaic virus, maize stripe virus, and an unidentified poty virus.
- Ladder leaf spot (Causal agent *Cercospora fusimaculans* Atk.) in west and central Africa.
- Nematode diseases.
  - *Pratylenchus zeae* (root lesion nematode), *Meloidogyne javanica* (root knot nematode), and *Meloidogyne incognita* (root knot nematode).

Continuing Needs

- New sources of host plant resistance for all diseases, but especially for sorghum ergot and grain mold.
- Cultural management methodologies for all sorghum foliar diseases.
- Pathogen variability should be evaluated in sorghum ergot and anthracnose. Some work is currently underway on pearl millet downy mildew.
- Evaluation of the significance of viral diseases of sorghum and Dactuliophora leaf spot (causal agent *Dactuliophora elongata* Leakey) of pearl millet.
- Screening for resistance to root and stalk rots of pearl millet and sorghum.
- Identification of the causal agents of the prevalent seedling diseases; evaluate control options (host plant resistance, cultural, and chemical).
- Identify the primary inoculum source(s) of sorghum ergot.
- Control sorghum ergot through pollen management (pollen/pathogen studies).
- Modeling of environment-sorghum ergot interactions.
- *Striga* spp. continue to constitute an important constraint to pearl millet and sorghum production in sub-Saharan Africa.

- Surveys to define the importance and distribution of:
  - Viral diseases.
  - Head blight.
  - Nematodes, especially the interaction between nematodes and stalk-rot causing fungi.
- Some problems are of restricted importance and should be addressed locally.
  - Sooty stripe.
  - Leaf blight.
  - Yellow leaf blotch.
  - Rust.
  - Leaf spots.
  - Smuts
- Provide training for LDC scientific and extension personnel in all areas of plant pathology.
- Develop/strengthen collaboration among researchers both within Africa and between Africa and other continents.

62

Recommendations from the Working Group for Asia/Australia at the Third Global Conference on Sorghum and Millets Diseases

Reported by *Sunkara Indira*

This group of Asian and Australian scientists met during the Third Global Conference on Sorghum and Millet Diseases at Guanajuato, Mexico, 23-29 September 2000, to update the research recommendations made at the Second Global Conference on Sorghum and Millets held in Harare, Zimbabwe, 1-11 March 1988. The meeting was open to all interested conference participants, and no list of participants peculiar to this meeting is available. More details on many of the diseases and recommendations can be found in the country and regional reports included in this volume. The committee identified the following disease priorities:

- Grain mold is a serious problem and economically important in India and Thailand.
- Ergot is a major problem in India and Australia.
- Stalk rots are a serious problem in Australia, Indonesia, Malaysia, the Philippines, and India.
- Viruses are gaining in importance in India, Japan, and Australia.
- Smuts are economically important in China and Russia.
- Anthracnose, leaf blight and rust are important in Asia, especially India, on fodder sorghum.
- Bacterial leaf diseases are a serious problem in Russia.

Recommendations for Future Work

- Better communication between researchers in the area in terms of identification of disease priorities and research schemes, thereby enabling better utilization of unique and complementary advantages of various regions.
- Continued research on pathogen diversity.
- Further incorporation of molecular approaches and molecular markers into conventional programs that are breeding for disease resistance in sorghum and pearl millet.
- Changing socioeconomic scenarios and potential new markets and end users for sorghum and pearl millet in some countries may alter disease management strategies. In particular, grain quality standards need to be set, and the overall quality of grain available in the market improved if sorghum and millet are to be effectively utilized in new products.

63

Sorghum Diseases in Brazil

Carlos R. Casela, Alexandre S. Ferreira, Fredolino G. Santos, and Fernando B. Guimarães

Introduction

The area planted to sorghum in Brazil more than doubled from 212,000 ha in 1994/95 to 558,000 ha in the 1998/99 season. In the 1998/99 season, grain sorghum production reached one million tons and the area planted to forage sorghum was estimated at 200,000 ha. Estimates indicate that Brazilian sorghum production will reach 10% of the total maize production by 2005. The central west region of Brazil is responsible for 65% of the total grain sorghum production in the country. In these states sorghum is sown in February-March and is cultivated as a second crop after soybean in the "safrinha" season, sometimes without any additional fertilization. Southeastern and southern Brazil represent 30% and 8% of the total cultivated sorghum area, while the semi-arid northeast represents only 1-2% of the total area planted to sorghum in Brazil. The state of Goiás, in central Brazil, is the major sorghum producer, with 37% of the total sorghum production in the country. In Rio Grande do Sul, sorghum is planted as a summer crop beginning in November.

Due to wide range of climate and soil conditions in which sorghum is cultivated in Brazil, the crop is vulnerable to various pathogens, some of which can be very damaging and cause severe losses. Annual surveys conducted by EMBRAPA indicate that anthracnose (caused by *Colletotrichum graminicola*), ergot (caused by *Claviceps africana*), rust (caused by *Puccinia purpurea*), downy mildew (caused by *Peronosclerospora sorghi*), leaf blight (caused by *Exserohilum turcicum*), and charcoal rot (caused by *Macrophomina phaseolina*) are the major sorghum diseases in Brazil, with several other diseases of minor importance also occurring. Annual yield losses due to these diseases are estimated at 15% in Brazil, averaged over the summer and winter seasons.

Ergot or Sugary Disease (*Claviceps africana*)

Sorghum ergot was first reported in Brazil in 1995 simultaneously in the states of São Paulo and Minas Gerais, in areas approximately 400 km apart. Within 30 days, additional reports of the disease were made from different locations in Minas Gerais, São Paulo, Mato Grosso do Sul, Goiás, and Paraná. The pathogen was identified as *Claviceps africana* (8). Due to the impact of the disease on sorghum seed production, an integrated effort, which includes the seed industry, EMBRAPA, and the University of Paso Fundo in the state of Rio Grande do Sul, began to develop strategies for management of the disease. The fungicides tebuconazole, triadimenol, and a combination of propiconazole and difeconazole effectively controlled *C. africana* (7). Control required three applications at 5-7-day intervals, beginning at anthesis and continuing until the completion of flowering (9). Recent results have indicated no significant differences in the effectiveness of the control of *C. africana* when the intervals between applications of triadimenol (46.6 a.i. g/ha) are of 5, 7, or 9 days. Two applications of the fungicide separated by 5-9 days are now known to suffice for good control of the disease (5).

Anthracnose (*Colletotrichum graminicola*)

Sorghum anthracnose continues to be a major disease in Brazil. The disease is present in all sorghum-growing areas of the country, even in the safrinha season. Understanding pathogen variability has been a key factor in the successful management of anthracnose in Brazil. Dilatory resistance suffices for successful control of the disease in some portions of the country, but a specific host-pathogen interaction is known for some sorghum genotypes and pathogen isolates (6). Important sorghum elite lines of the sorghum breeding program of EMBRAPA have a differential interaction for dilatory resistance, with some of the races of *C. graminicola* that are normally present in the population of this pathogen in Brazil (Table 63-1). These results suggest that at least a portion of the dilatory resistance of sorghum to anthracnose is race specific,

TABLE 63-1. Area under disease progress curve (AUDPC) for seven sorghum genotypes inoculated with three races of *Colletotrichum graminicola* (6).

Sorghum Line	*Colletotrichum* Race		
	12A	14C	21E
CMS×S203B	970 a[1]	730 a	840 ab
CMS×S207R	710 ab	170 b	370 bc
CMS×S112B	390 bc	120 b	430 bc
CMS×S101B	230 bc	180 b	230 c
CMS×S202B	210 c	1060 a	930 a
CMS×S221B	160 c	210 b	130 c
CMS×S173R	120 c	83 b	130 c

[1]Numbers within a column followed by a common letter are not significantly different at the $p = 0.05$ level as determined by Tukey's Multiple Range Test.

and, consequently, could be of low durability. Some sorghum genotypes, *e.g.*, BR005 (SC326-6) and its derivatives, however, have exhibited stable dilatory resistance for an extended period of time throughout the country.

The hypersensitive type of resistance has been used in breeding programs through combinations of A and R lines for which there is no associated virulence in the pathogen population. Disease nurseries with elite lines and experimental hybrids are planted in hot spot areas of anthracnose and evaluated for resistance. At the same time, single-spore isolates obtained from each field are inoculated on elite lines in the greenhouse to identify combinations of A and R lines with nonassociated virulence. New combinations are selected based on pathogenicity and virulence association coefficients (PAC and VAC), as defined by Browder and Eversmeyer (1). Based on these coefficients, desirable combinations for disease control are characterized by a high PAC and a low VAC; in such a case most of the population would avirulent. High values of PAC and VAC indicate that most of the population already is virulent to both lines and that a combination of those lines would be of little value in the control of anthracnose. A low PAC and a low VAC indicate that most of the protection is afforded by only one of the two lines, rather than by both lines. This method has been very useful for the identification of sorghum hybrids with potentially durable, vertical resistance to *C. graminicola*. Several combinations, investigated in 1996 and 1997, were identified as potentially useful based on their high PAC and low VAC values.

We are now developing additional hybrids (Table 63-2). For example, combinations involving CMS×S197A and CMS×S219A are potentially useful for the control of the disease (4). Experimental hybrids involving these A and R lines have been effectively evaluated in highly infected areas and their predicted high level of resistance confirmed. Additional confirmation of the effectiveness of this analysis in predicting durable hypersensitive resistance to anthracnose comes from the commercial hybrid BR305. High PAC and low VAC values were obtained for its A and R lines (2). Released in 1995, its resistance has remained stable, despite being cultivated in areas of very severe anthracnose, especially in southern Brazil.

Another potential strategy to develop durable resistance to *C. graminicola* is to plant cultivar mixtures. Susceptible cultivars have lower levels of infection when mixed with resistant genotypes (6). We inoculated mixtures of *C. graminicola* races differing in virulence for five (experiment 1) or six generations (experiment 2) on the susceptible sorghum cultivar BR009. Significant changes in race composition were observed ($\chi^2 = 94$, $p = 0.001$ in experiment 1, and $\chi^2 = 67$, $p = 0.001$ in experiment 2). In both experiments isolates of the race with the

TABLE 63-2. Pathogenicity association coefficients (PAC) and virulence association coefficients (VAC) (1) of *Colletotrichum graminicola* to 10 combinations of sorghum genotypes estimated from samples from three locations in Brazil in 1996 and 1997.

Sorghum Line	S. Lagoas–1996		S. Lagoas–1997		Capinópolis		Cravinhos		Pelotas	
	PAC	VAC	PAC	VAC	PAC	VAC	PAC	VAC	PAC	VAC
CMSXS197A × BR005R	0.674	0.021	0.612	0.021	0.800	-0-	0.47	0.042	0.583	0.002
CMSXS219A × BR005R	0.761	0.021	0.612	0.021	0.767	0.033	0.489	0.067	0.542	0.002
CMSXS197A × CMSX×S178R	0.696	0.021	0.633	0.021	0.800	-0-	0.822	-0-	0.625	0.02
CMSX×S197A × CMSX×S184R	0.565	-0-	0.531	0.041	0.700	-0-	0.533	0.067	0.500	0.042
CMSX×S197A × CMSX×S185R	0.761	0.043	0.694	-0-	0.933	-0-	0.711	0.044	0.542	-0-
CMSX×S197A × CMSX×S186R	0.565	0.021	0.766	-0-	0.833	-0-	0.778	0.067	0.625	0.021
CMSX×S219A × CMSX×S178R	0.783	0.021	0.632	0.021	0.767	0.033	0.8	-0-	0.583	0.021
CMSX×S219A × CMSX×S184R	0.696	0.021	0.49	0.021	0.733	0.067	0.556	0.089	0.458	0.042
CMS×XS219A × CMS×XS185R	0.840	0.021	0.694	-0-	0.833	-0-	0.644	0.022	0.625	0.042
CMS×S219A × CMS×S186R	0.609	-0-	0.776	-0-	0.867	-0-	0.667	0.022	0.688	0.063

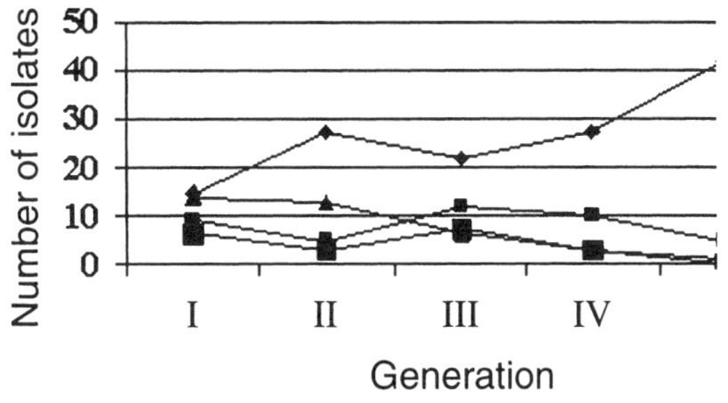

FIGURE 63-1. Changes in the composition of a mixture of four races of *Colletotrichum graminicola* inoculated on the susceptible cultivar BR009 over five generations (Experiment 1). --□-- = race 30A, --■-- = race 31B, --▲-- = race 31C, and --◆-- race 31E.

narrowest virulence range, 31A, dominated the more broadly virulent races 31B, 31C, and 31E, indicating that these races may differ in survival/competitive ability (Figs. 63-1 and 63-2). These results reflect differences in fitness among races of *C. graminicola* and are consistent with field observations on race distribution in areas of predominance of anthracnose (3). The predominance of race 30A, the least virulent, and the lowest frequency of race 31E, with the highest virulence, after five and six generations on the susceptible genotypes BR009, suggests that virulence and survival ability of races of *C. graminicola* are not independent variables. More extensive data from field and greenhouse experiments involveing additional races of the pathogen, with different virulence ranges, and sorghum genotypes are necessary to confirm this relationship. Information on the survival/competitive ability of this pathogen races will be important in the development of gene deployment strategies, so that the opportunity for the evolution of highly virulent races with improved survival/competitive abilities can be minimized.

Rust (*Puccinia purpurea*)

Rust has become an important disease in the last five years, primarily in central Brazil. The disease has been economically important in sorghum areas of Goiás and northern São Paulo in the safrinha season. Several resistant lines and hybrids have been identified by screening resistant materials under natural infection through multi-location tests of the National Uniform Sorghum Disease Nursery (NUSDN).

Downy Mildew (*Peronosclerospora sorghi*)

Sorghum downy mildew is most severe in Rio Grande do Sul, São Paulo, and Minas Gerais. Most Brazilian sorghum hybrids are resistant to downy mildew. The pathogen population has apparently remained stable, as no major changes in the reactions of sorghum elite lines have been observed in the last 10 years. There is need, however, to intensify the evaluation of male sterile lines with good resistance, especially for the forage sorghum breeding program. Tests for downy mildew resistance have been performed both in the greenhouse and under field conditions at EMBRAPA facilities in Rio Grande do Sul and Minas Gerais. Lines with less than 5% incidence have been identified and incorporated into the breeding program for resistance to this disease.

Leaf Blight (*Exserohilum turcicum*)

Leaf blight has increased in both incidence and severity in central Brazil, primarily during the safrinha season when it is favored by moderate temperatures and heavy dews. Resistant lines and hybrids have been identified by EMBRAPA through the evaluation and selection of genotypes in the NSDUN, in the main areas of occurrence of the disease.

Charcoal Rot (*Macrophomina phaseolina*)

Charcoal rot is potentially destructive in areas of central Brazil during the safrinha season, due to warm, dry weather that coincides with the postflowering period of the crop. Due to the sporadic occurrence of this disease, we have developed no special screening technique for this disease in Brazil, beyond the evaluation of the NUSDN in areas and years of occurrence of the disease.

Research Needs

The already known high variability of the sorghum anthracnose fungus and the ability of the pathogen to overcome host resistance was recently confirmed by the observed change in the reaction of the resistant elite line

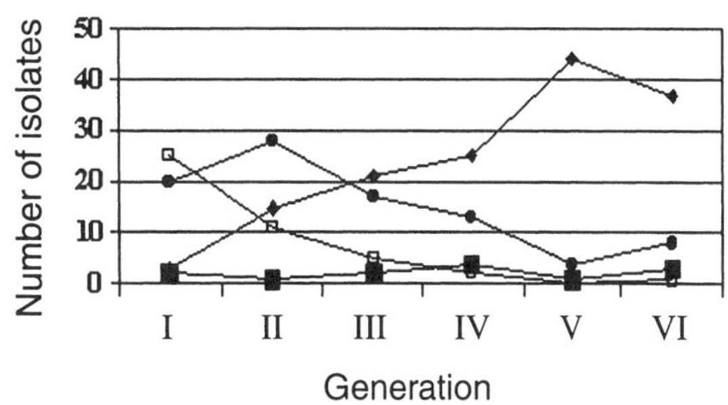

FIGURE 63-2. Changes in the composition of a mix-ture of four races of *Colletotrichum graminicola* inoculated on the susceptible cultivar BR009 over six generations (Experiment 2). --□-- = race 30A, --■-- = race 31B, --▲-- = race 31C, and --◆-- race 31E.

CMSXS210 from resistant to susceptible in central Brazil. Thus, monitoring the pathogen population in both the normal and the safrinha seasons is essential for the development of stable resistance to this pathogen under Brazilian conditions. Evaluation and development of gene management strategies, such as multigene cultivars, genetic mixtures, and gene rotation, to ensure more stable and durable resistance also is important due to the high selection pressure imposed by sorghum pathogens, especially since dilatory resistance to *C. graminicola* may not always be an effective primary strategy for the control of this disease. Multigene cultivars and genetic mixtures in double or triple hybrids are possible alternatives to increase the durability of resistance. The prediction of changes in races of plant pathogens through a program of rotation of resistance genes also might provide more stable resistance in the Cerrado region. As the sorghum-growing area in Brazil is relatively small, and since sorghum is grown in very well defined habitats, the development of such a program as a collaborative effort with the private sector is an intriguing possibility, and such preliminary studies of cultivar mixtures and gene rotation have already begun.

With respect to ergot, it is important to develop more efficient chemical controls. The dissemination of downy mildew to sorghum-growing areas other than Rio Grande do Sul and Paraná indicates that more effort is needed to understand the variability of virulence of *P. sorghi* and that further development of resistant hybrids will be needed. The increasing incidence and severity of rust, leaf blight, and charcoal rot in central Brazil indicates that a continuous screening for resistance to these pathogens remains necessary.

References

1. **Browder, L. E., and M. G. Eversmeyer.** 1977. Pathogenicity associations in *Puccinia recondita tritici*. *Phytopathology* **67:** 766-771.
2. **Casela, C. R., A. S. Ferreira, and F. G. Santos.** 1998. Associação de virulência de *Colletotrichum graminicola* à resistência genética em sorgo. *Fitopatol. Brasi.* **23:**143-146.
3. **Casela, C. R., A. S. Ferreira, and F. G. Santos.** 2001. Differences in competitive ability among races of *Colletotrichum graminicola* in mixtures. *Fitopatol. Brasi.* **26:** 217-219.
4. **Casela, C. R., F. G. Santos, and A. S. Ferreira.** 2000. Associações de patogenicidade e diversidade fenotípica de *Colletotrichum graminicola*, agente causal da antracnose em sorgo. *Fitopatol. Brasil.* **25:** 517-521.
5. **Ferreira, A. S., and C. R. Casela.** 1998. Efeito de fungicidas e dose no controle da doença açucarada do sorgo (ergot). *Fitopatologia. Brasileira.* **31:** 243 (Suppl.).
6. **Guimarães, F. B., C. R. Casela, X. R. Vale, L. Zambolim, and F. G. Santos.** 1998. Resistência dilatória de genótipos de sorgo a diferentes raças de *Colletotrichum graminicola*. *Summa Phytopathol.* **24:** 136-140.
7. **Pinto, N. F. J. A., A. S. Ferreira, and C. R. Casela.** 1999. Chemical control of sugary disease of sorghum (*Claviceps africana*). *Proc. Global Conf. on Ergot of Sorghum (1-8 June 1997, Sete Lagoas, Brazil)*: 158-166.
8. **Reis, E. M., P. G. Mantle, and H. A. Hassan.** 1996. First report in the Americas of sorghum ergot disease, caused by a pathogen diagnosed as *Claviceps africana*. *Plant Dis.* **80:** 463.
9. **Ribas, P. M.** 1999. Strategies for the control of ergot in sorghum seed production in Brazil, p. 26-32

64

Sorghum and Pearl Millet Diseases in the Horn of Africa

J. Peter Esele

Abstract

Sorghum and pearl millet are basic staple food crops in the Horn of Africa. Their production is largely by smallholders and limited by numerous constraints. One of the most common constraints is disease. On sorghum, the most common and important diseases are smuts, grain mold, ergot, and the foliar infections of anthracnose, downy mildew, leaf blight, and gray leaf spot. In pearl millet, the most important diseases in the region are downy mildew, ergot, and smuts. For both crops, *Striga* is an important parasite. Some advances have been made during the last decade in developing control strategies for these diseases, primarily through collaborations with INTSORMIL and ICRISAT.

Sorghum Diseases

Introduction. Sorghum is the basic staple food in the Horn of Africa and is mainly cultivated by small-scale farmers. The major constraints to increased production and utilization of this crop are generally common to all the countries in the region (Table 64-1).

Food security problems have prompted many African national agricultural programs to actively encourage farmers to grow sorghum, with the national programs in turn encouraged to follow this policy by international research programs such as ICRISAT and INTSORMIL. ICRISAT, INTSORMIL, and the national programs currently conduct joint research and disseminate technologies including the exchange and improvement of germplasm. However, limited resources in most of the countries in the region have slowed the pace at which technology development and dissemination is occurring. Smallholder sorghum is very susceptible to a complex of pests and diseases, the exact components of which depend on location, cultivar, and farming practice. Most of the traditional land races are tolerant of many of these problems, often through escape associated with late maturity or photoperiodicity, or other plant characteristics. The improved materials, unfortunately, often succumb to these problems due to their earliness and soft stems or grain.

In the East and Central Africa Regional Sorghum and Millets Network proposal, submitted for funding to the European Union, of the fourteen identified priority research areas, crop protection ranks fifth after technology transfer, *Striga*, variety development, and germplasm exchange (1; Table 64-1). At a traveling workshop from 21-28 September 1997 in Ethiopia organized by INTSORMIL, the national programs identified the most important diseases across the region as smuts (both covered and loose but with a bias towards covered), grain mold, anthracnose, and leaf blight. The workshop participants also recommended that to realize significant control, research should be multidisciplinary, involving pathologists, breeders, agronomists, seed technologists, and socioeconomists. The other major problem identified was *Striga*.

Smuts. Sorghum smuts in the Horn of Africa have been a problem for a long time. Of the four sorghum smuts, three are common–covered kernel smut (caused by *Sporisorium sorghi*), loose smut (caused by *Sphacelotheca cruenta*), and head smut (caused by *Sporisorium relianum*). All the smuts are economically important, causing 5-80% yield losses, and collectively, they are the most damaging of all the diseases of sorghum in the region. Covered kernel smut and loose smut, which are both seedborne, are the most common. Most farmers keep their own seed for planting during the next season, and very few farmers treat seed before planting. Loose smut also is propagated through ratooned sorghum. Head smut often occurs on late-maturing varieties, which are commonly grown by peasant farmers. Thus, the traditional

TABLE 64-1. Horn of Africa regional priority areas for networking; order based on summation of priority assigned by each country with equal weight given to each country, adapted from ASARECA (1).

Networking Area	Country					
	Rwanda	Uganda	Kenya	Ethiopia	Tanzania	Eritrea
Technology transfer	1	1	1	1	1	2
Variety development	1	1	2	1	1	1
Striga	1	1	3	1	1	2
Germplasm exchange	1	1	2	3	1	1
Short-term training	1	2	2	2	3	1
Long-term training	1	2	2	2	3	1
Crop protection	3	2	2	1	2	1
Drought	4	2	1	1	2	1
Processing and utilization	2	2	2	1	1	3
Seed production and marketing	2	1	1	3	1	3
Information exchange	1	1	3	4	1	2
Soil/Water conservation	2	3	2	2	1	2
Soil fertility	2	3	4	2	1	2
Socioeconomics	2	3	4	2	2	3

practices employed by the farmers are largely responsible for the continued occurrence of sorghum smuts in the region.

In a recent survey in Uganda, very few farmers had any technical knowledge of these diseases or their control. *S. sorghi* and *S. cruenta* can be reduced significantly through fungicidal seed treatment, since they are both seedborne. In Uganda, research on smuts has been limited, until recently, to diagnosing diseases, recording outbreaks, and advising farmers on cultural control methods (use of disease-free seed and field sanitation). Little is known of the exact losses due to the diseases, of the variation within the pathogen populations, or of the number, distribution, or effectiveness of resistance genes in the host germplasm. In a survey conducted in 2000 (Hayden *et al., unpublished*) in Teso, Uganda, the smuts were the most important disease. The efficacy of chemical seed treatment on the locally available seeds is not known. However, work has now begun on covered and loose smuts throughout the country. In 2000, collaborative work on this problem began between Serere Agricultural and Animal Production Research Institute (Soroti, Uganda) and the Natural Resources Institute (Chatam, UK) (NRI). The project seeks to identify and develop sustainable control strategies for smuts and to identify pathogenic variation and resistant varieties in Uganda.

In Kenya, NRI has just concluded work on the control of covered kernel smut through seed treatment with the fungicide Thiram®. The strategy is being disseminated to farmers through demonstrations and the education of schoolchildren.

In Ethiopia, research is being conducted on the control of covered smut through seed treatment with cow urine, goat urine, and some plant products, although Thiram® and Apron® Plus seed dressings also have been tested. The most promising of these control methods are the use of goat urine and Apron® Plus (G. Tegegne, personal communication).

Grain Mold. Very few traditional sorghum cultivars succumb to grain mold. This resistance probably is attributable to their long maturity periods, which results in grain formation coinciding with dry weather. In addition, most traditional sorghums have high levels of phenolic compounds and/or have seed coats that provide resistance to grain molds. However, the general trend for all of the national programs in the region is to introduce farmers to the high-yielding, good-quality, early-maturing cultivars. Unfortunately these sorghums, by their very natures, are susceptible to grain mold.

Although many genera of fungi have been reported as causal agents for grain mold, the most frequently occurring are *Fusarium, Curvularia*, *Phoma*, *Alternaria*, *Colletotrichum*, *Aspergillus*, and *Penicillium*. Little research has been conducted in this region on grain mold. Most of the cultivars recommended to farmers have reasonable levels of resistance, except for Tegemeo (Tanzania and Kenya) or Epuripur (Uganda). In Sudan, areas receiving rainfall in excess of 800 mm often report grain mold damage. There is a program in Uganda to breed for grain mold resistance with the local sorghums and those provided by the Texas A&M University program based on collaborative efforts of the National Agricultural Research Organization (NARO) and INTSORMIL (4).

Ergot. The ergot disease, caused by *Claviceps africana*, occurs in the region but is not of major economic importance. It is most commonly reported in Sudan on parental lines of the hybrid Hageen Durra 1. In 1999, ergot devastated foundation seed of this hybrid. Sudan has implemented a strict control strategy that includes seed dressing and spraying to combat future outbreaks of the disease and to protect the hybrid seed industry. The disease was not observed in nearby fields in which open pollinated varieties were cultivated. The limited use of hybrids in most of the other countries in the region probably explains why the disease is of minor importance in those countries.

Foliar Diseases. The most frequently occurring leaf diseases of sorghum in the region are anthracnose (caused by *Colletotrichum graminicola*), leaf blight (caused by *Exserohilum turcicum*), gray leaf spot (caused by *Cercospora sorghi*), and downy mildew (caused by *Peronosclerospora sorghi*). Those of lesser importance include sooty stripe (caused by *Ramulispora sorghi*), zonate leaf spot (caused by *Gloeocercospora sorghi*), and rust (caused by *Puccinia purpurea*). These also are the major foliar diseases reported in many other sorghum-producing regions in the world. Disease incidence and severity vary across the region, depending largely on season, *i.e.*, the amount of wet humid weather, the sorghum variety planted, and the farming system. Leaf blight is important in the Southern Blue Nile region in Sudan on hybrids and the newly released cultivar, Tobat. In Uganda, Kenya, and Ethiopia, anthracnose is the major foliar disease. Downy mildew does not occur every season, but can be destructive.

There is apparently no significant national or regional research being conducted on these diseases at this time. Screening for resistance is done in Uganda, Ethiopia, and Kenya, especially for anthracnose. However, in collaboration with INTSORMIL and ICRISAT, a number of international nurseries, *e.g.*, the International Sorghum Anthracnose Virulence Nursery (ISAVN), the International Sorghum Downy Mildew Nursery (ISDMN), and the Extended Anthracnose Virulence Nursery (EAVN), have been planted across the region.

Striga. *Striga* is, perhaps, the greatest biological constraint to food production in Africa, being a more serious problem than insects, birds, or plant diseases (3). Yield losses from damage by *Striga* often are significant. Lagoke *et al.* (7) estimated annual cereal grain losses attributable to *Striga* to average 40% across Africa. In Ethiopia and Sudan, losses of 65-100% are common in heavily infested fields (3). In Uganda, *Striga* is probably the most important biological constraint to sorghum production, as it is present in all the sorghum-growing areas (2). In two districts surveyed in 1997 by Serere Station researchers, *Striga* was found in 83% of the fields surveyed, causing yield losses estimated of 60-85% (Esele and Ebiyau, *unpublished*). A similar situation exists in western Kenya. The weed can survive the low moisture and low soil-fertility conditions that exist in most of the sorghum-growing areas. In Ethiopia and Sudan, the continuous cropping of sorghum makes the losses due to *Striga* even worse.

INTSORMIL has a collaborative program with national programs in the region aimed at identifying practical approaches to *Striga* eradication. Approaches being evaluated include a combination of host plant resistance and improved agronomic practices, *e.g.*, rotation with trap crops, mechanical and chemical weed control, and nitrogen fertilization, and the preliminary results are promising. Sudan has the most experience in *Striga* research in the region.

Pearl Millet Diseases

Introduction. Pearl millet is a staple cereal crop in most marginal areas of the region where the climates are harsh with hot (> 30°C) temperatures and < 500 mm annual rainfall. Grain yields in the farmers' fields average 500-550 kg/ha; although these vary by country and year, depending primarily on weather and variety. In the Horn of Africa, pearl millet is grown exclusively for grain, with Sudan being the largest producer followed by Ethiopia and Eritrea (1).

Downy mildew, ergot, and smut are the most widespread diseases of pearl millet (5). *Striga* also is an important problem in pearl millet production. Generally, the diseases of pearl millet in the region have received lit-tle research attention, except in Sudan. ICRISAT has been the major collaborator in this area, mainly through the distribution and evaluation of international nurseries, of which the Pearl Millet Downy Mildew Nursery has been the most common.

Downy Mildew. Downy mildew (caused by *Sclerospora graminicola*) is the most important and widespread disease of pearl millet in the Horn of Africa. The disease has been known in the region since the early 1950s but has recently become a serious problem in all of the pearl millet-growing areas. In Sudan, the disease occurs at incidences of 11-82% (E. Omer, *unpublished*). Screening trials in Sudan have established that all of the popular cultivars are susceptible to the disease. In Eritrea, downy mildew devastated the pearl millet crop in 1997, when conditions were quite favorable with warm, humid weather, and may have been responsible for the famine in the country that year. The epidemic occurred at the seedling stage, from which recovery is particularly difficult.

Considerable information is available on the biology, epidemiology, and control of *S. graminicola* (8, 9, 11). The amount of variation within populations of this pathogen in the Horn of Africa is unknown and is an important area that needs immediate attention.

Strategies for the control of downy mildew in pearl millet have been developed in collaboration with ICRISAT and INTSORMIL. In the Horn of Africa, the efficacy of seed treatment with metalaxyl has been demonstrated only at research stations. The most widely recommended practice is the use of host plant resistance, with numerous sources of resistance identified among accessions of germplasm by ICRISAT (5). These accessions are being tested further in downy mildew nurseries. Field-screening techniques, involving sick plots, infector rows, or a combination of the two, have been adopted, especially in Sudan, Uganda, and Kenya. In Sudan, exotic materials from ICRISAT, India, and Zimbabwe outperformed local material. Cultivar Okashana Z (SDMV 93032), developed by the Sorghum/Millet Improvement Program (SMIP) through a number of trials, had both excellent disease resistance and superior agronomic performance. This cultivar matures early, has large grain size, and, even under intense disease pressure, outyields all the other cultivars. This cultivar has been released under the name Ashana ("Our Dinner"). In Uganda, cultivars ICMV 221, ICPT 8203, and SDMV 92040 have been identified as downy mildew resistant. They also are early, drought tolerant, have very high yields, and are in the process of being released.

Ergot. Ergot (*Claviceps fusiformis*) on pearl millet is common in the Horn of Africa, but epidemics such as those that occurred in India in the 1950s have not been recorded. The disease reduces yield, as florets do not form seed, and it also reduces grain quality due to contamination with fungal sclerotia that contain high levels of alkaloids that are toxic to both humans and domesticated animals.

The disease is most important in late-planted crops. In Uganda, the disease can spread quickly across a field, especially on rainy days. This observation suggests that the spread of ergot is facilitated by wind and rain splash. Unfortunately no research or management practice studies have been conducted in the region, but the use of host plant resistance and appropriate cultural practices, *e.g.*, sclerotia-free seed, early planting, and field sanitation, hold some promise.

Smut. Smut (*Tolyposporium penicillariae*) is the second most important disease of pearl millet in the Horn of Africa. This disease occurs in Sudan, Ethiopia, and Eritrea. It has been reported to be most damaging on tillers, but the main head usually escapes the infection. Infected florets first appear as top-shaped green sori, later turning brown and rupturing easily at harvest. These sori are similar to the long smut structures seen in sorghum.

Fungicidal control of smut in pearl millet does not seem to be practical for small-scale farmers in Africa. Instead the many lines with stable resistance appear to be the best control strategy (10).

Other diseases of minor importance include rust (caused by *Puccinia penniseti*) and virus-like diseases. Rust normally occurs late in plant development, but usually has little effect on grain yield. We have observed virus-like diseases in some of our fields in Uganda, but have paid little attention to them. Symptoms include mosaic and streak.

Striga. *Striga* is an important problem in pearl millet production, especially in Sudan. The most common form is *Striga hermontheca*, although reports of *Striga asiatica* have been made from Kenya and Uganda. Independent research on *Striga* on pearl millet has not been done, but what has been done with sorghum often has been repeated in pearl millet. Resistance to *Striga* has not been confirmed in pearl millet (5). King and Zummo (6) reported that in Africa there are varieties of *S. hermontheca* that appear to be specific for pearl millet. These two areas, confirmation of existence of resistance of pearl millet to *Striga* and the existence of varieties specific for pearl millet, require further investigation and will be useful sources of information for breeding for *Striga* resistance.

In general the most important diseases of sorghum in the Horn of Africa are the grain diseases, especially the smuts, followed by foliar diseases, of which anthracnose is the most important. Downy mildew, although it occurs only occasionally, can be very damaging. In pearl millet, both downy mildew and ergot are important. For both crops, *Striga* is the biggest constraint to production. Considerable research effort is needed on the management of all of the major production constraints that occur in the region.

References

1. **ASARECA (Association for Strengthening Agricultural Research in Eastern and Central Africa).** 1998. A proposal for an Eastern and Central Africa Regional Sorghum and Millet (ECARSAM) Network. Submitted to the European Union. 49 pp.
2. **Baguma, S. D.** 1996. *Striga* research activities in Uganda, 1994-1995. *Striga Newsl.* **6**: 5-6.
3. **Ejeta, G., L. Butler, and A. G. T. Babiker.** 1992. *New Approaches to the Control of* Striga. Striga *Research at Purdue University.* Purdue Univ., West Lafayette, Indiana.

4. **Esele, J. P.** 1991. *The Genetics of Grain Mold Resistance in Sorghum [*Sorghum bicolor *(L) Moench]*. Ph.D. Dissertation, Texas A & M Univ., College Station, Texas.
5. **King, S. B.** 1992. World review of pearl millet diseases: Knowledge and future research needs, p. 95-108. *In* W. A. J. de Milliano, R. A. Frederiksen, and G. D. Bengston (eds.), *Sorghum and Millets Diseases: A Second World Review.* ICRISAT, Patancheru (A.P.), India.
6. **King, S. B., and N. Zummo.** 1977. Physiologic specialization in *Striga hermontheca* in western Africa. *Plant Dis. Reptr.* **61**:770-773.
7. **Lagoke, S. T., V. Parkinson, and R. M. Agumbiade.** 1991. Parasitic weeds and their control methods in Africa, p. 3-15. *In* S. K. Kin (ed.), *Combating* Striga *in Africa. Proc. Intl.Workshop organized by IITA, ICRISAT & IDRC (22-28 August 1998).* IITA, Ibadan, Nigeria.
8. **Shetty, H. S.** 1987. Biology and epidemiology of downy mildew of pearl millet, p. 147-160. *In* J. R. Witcombe and S. R. Beckerman (eds.), *Proc. Intl. Pearl Millet Workshop (7-11 April 1986, Patancheru, India).* ICRISAT, Patancheru (A.P.), India.
9. **Singh, S. D., S. Ball, and D. P. Thakur**. 1987. Problems and strategies in the control of downy mildew, p. 161-172. *In* J. R. Witcombe and S. R. Beckerman (eds.), *Proc. Intl. Pearl Millet Workshop (7-11 April 1986, Patancheru, India).* ICRISAT, Patancheru (A.P.), India.
10. **Thakur, R. P., K. V. Subba Rao, R. J. Williams, S. C. Gupta, D. P. Thakur, N. V. Sundaram, J. A. Frowd, and E. J. Guthrie.** 1986. Identification of stable resistance to smut in pearl millet. *Plant Dis.* **70**:38-41.
11. **Williams, R. J.** 1984. Downy mildew of tropical cereals. *Adv. Plant Pathol.* **2**:1-103.

65

Sorghum Diseases in Argentina

Laura M. Giorda

Sorghum hybrids were introduced in Argentina in 1957, which was followed by a rapid increase in the planted area until 1984, when the planted area began dropping, as sorghum was replaced by oil crops, *e.g.*, soybean and sunflower, in some rotations. Yields, however, have increased from an average of 3.1 t/ha (1984/85) to about 4.6 t/ha (1999/2000). In the 1999/2000 growing season, 827,000 hectares of sorghum were planted, with total sorghum production reaching 3.35 million t (4). Grain sorghum is grown primarily as a feed crop in dry land areas and is usually cultivated in a rotation system, primarily with soybean. With increased no-tillage cultivation, management of the sorghum crop has changed, and consequently, the pathogen and pest populations are expected to shift as well.

Forage and grain sorghum in Argentina are cultivated throughout the subtropical and temperate regions between latitudes 22-40° S and longitudes 56-66° W in three agroecological growing regions (Fig. 65-1): the subtropical north (between latitudes 22° and 30°) and the temperate central and southern regions (between latitudes 30-35° S and 35-40° S, respectively). More than 62% of the total area planted is in the central region of the country, with production concentrated in the provinces of Córdoba, Santa Fe, La Pampa, Buenos Aires, Santiago del Estero, and Entre Ríos (91% of total production).

Thirty-four diseases caused by more than 45 pathogens, including fungi, bacteria, and viruses, have been identified from Argentina (Table 65-1). However, the major sorghum diseases in Argentina are grain mold, stalk and root rot (induced primarily by *F. moniliforme sensu lato*), bacterial leaf stripe, sorghum downy mildew, maize dwarf mosaic, and ergot (2, 3). The major diseases can be found throughout the entire sorghum-growing region, although their prevalence and severity depend upon the environmental conditions, the sorghum cultivar planted, and rainfall distribution. Diseases such as head blight, zonate leaf spot, and rust have occasionally been important in the northern and central regions, and some of the Argentine commercial hybrids are resistant to these diseases. In the 1999/2000 growing season, rust was severe in some portions of Córdoba.

Foliar diseases commonly occur late in the growing

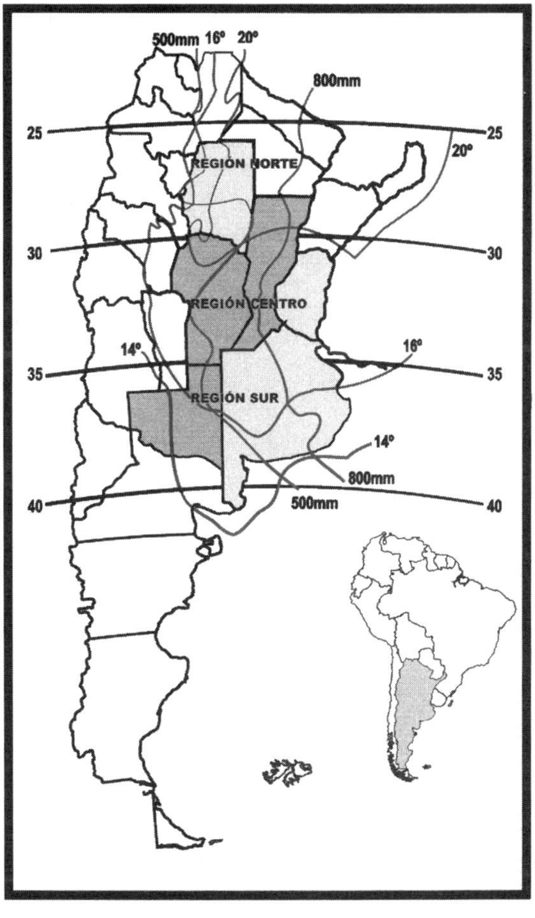

FIGURE 65-1. Map of Argentina identifying the three main sorghum-growing regions and the areas of highest sorghum production.

TABLE 65-1. Sorghum diseases in Argentina.

Disease	Importance[a]	Pathogen(s)
Panicle and Seed Diseases		
Grain mold	1	*Curvularia lunata* (Wakker) Boedijn, *Fusarium* spp., *Fusarium moniliforme sensu lato, Fusarium proliferatum* (Matsushima) Nirenberg, *Fusarium subglutinans* (Wollenweber & Reinking) Nelson, Toussoun & Marasas
Grain weathering	2	*Alternaria* spp., *Cladosporium* spp., *Colletotrichum graminicola* (Cesati) Wilson, *C. lunata, F. moniliforme sensu lato, Fusarium semitectum* Berk & Ravenel, *Fusarium* spp., *Phoma sorghina* (Saccardo) Boerema *et al*
Head blight	2	*F. moniliforme sensu lato, Fusarium* spp.
Ergot	2	*Claviceps africana* Frederickson, Mantle & de Milliano
Panicle and grain anthracnose	4	*C. graminicola*
Storage molds	3	*Penicillum* spp., *Aspergillus* spp.
Downy Mildew		
Sorghum downy mildew[b]	2	*Peronosclerospora sorghi* (Weston & Uppal) Shaw
Crazy top	4	*Sclerophthora macrospora* (Saccardo) Thirum *et al*
Root and Stalk Diseases		
Fusarium stalk rot	1	*Fusarium graminearum* Schwabe, *F. moniliforme sensu lato, F. proliferatum, F. subglutinans*
Charcoal rot	3	*Macrophomina phaseolina* (Tassi) Gold
Stalk rot	2	*C. graminicola, Fusarium* spp., *Helminthosporium sativum* Pamm, King & Bakke, *Nigrospora sphaerica* (Saccardo) Mason, *Pythium* spp., *Rhizoctonia solani* Kuhn, *Tetraploa ellisii* Cooke
Acremonium wilt	5	*Acremonium strictum* Gams
Pokkah boeng (twisted top)	4	*F. moniliforme sensu lato, F. subglutinans*
Virus Diseases		
Maize dwarf mosaic virus (MDMV)[b]	2	Potyvirus
Sugarcane mosaic virus (SCMV)	4	Potyvirus
Mal de Río Cuarto virus (MRCV)	5	Fijivirus
Foliar Diseases and Leaf Sheath Diseases		
Bacterial leaf stripe	1	*Pseudomonas andropogoni* (Smith) Stapp
Bacterial leaf streak	3	*Xanthomonas campestris* pv. *holcicola* (Elliot) Dye
Bacterial spot	4	*Pseudomonas syringae* pv. *syringae* (Van Hall) Dye
Rust	2	*Puccinia purpurea* Cooke
Leaf blight	3	*Exserohilum turcicum* (Pass.) Leonard & Suggs
Grey leaf spot	4	*Cercospora sorghi* Ellis & Everh
Zonate leaf spot	3	*Gloeocercospora sorghi* Bain & Edgerton
Anthracnose	4	*C. graminicola*
Sooty stripe	5	*Ramulispora sorghi* (Ellis & Everh) Olive & Lefebvre
Oval leaf spot	5	*R. sorghicola* Harris
Dry leaf spot	5	*Phoma insidiosa* Tassi
Dotted leaf spot	5	*Phyllosticta* sp., *Physoderma* sp.
Target leaf spot	5	*Bipolaris sorghicola* (Lefebvre & Sherwin) Shoem.
Sheath blight	5	*R. solani*
Sheath rot	5	*Sclerotium rolfsii* Saccardo
Smuts		
Head smut	5	*Sporisorium reilianum* (Kuhn) Langdon & Fullerton
Covered kernel smut	5	*Sporisorium sorghi* (Ehrenberg) Link
Loose kernel smut	5	*Sporisorium cruenta* (Kuhn) Potter
Seedling Diseases		
Poor stand establishment (pre- and post-emergence damping off)	3	*Pythium* spp., *Fusarium* spp., *R. solani*

[a]Rating scale (after Frederiksen *et al.*, 1) refers to the presence of the disease in any Argentine sorghum growing region: 1, important; 2, sporadically important; 3, occasional problem; 4, minor problem; and 5, rarely observed.

[b]Occurs annually, but not of economic significance because of resistant cultivars.

season. Although their effect on grain yield may be significant, depending upon the cultivar, location, and year, they always reduce the quality of the forage. Most Argentine sorghum (76% of the crop) is consumed on farm through grazing or as silage; only about 17% is exported, and the remaining 7% is dry milled.

Grain mold and grain weathering are most prevalent when low-tannin sorghums are associated with high relative humidity during flowering and maturation of the crop (some portions of the northern and central sorghum growing regions). Tolerance (2, 5) in some commercial sorghum hybrids and chemical control during anthesis are the major ways in which these diseases are managed in Argentina. Chemical control can be achieved by using tebuconazole (25%) in doses of 0.0625 to 0.125 kg a.i./ha, applied at 50% of anthesis. Although the percentage of brown sorghums has decreased during the last five years in favor of red hybrids, brown sorghums are still planted in areas with high bird pressure and with environmental conditions that are conducive to grain deterioration, primarily the northern and east central regions.

Sorghum downy mildew and maize dwarf mosaic virus (MDMV) are widespread in Argentina, but losses to these pathogens have not been economically significant because of the development and spread of resistant and/or tolerant commercial hybrids.

The prevalence and incidence of bacterial diseases, mainly leaf stripe, has increased since the second world review throughout the sorghum-growing region in Argentina, and there are a few commercial sorghum cultivars that are resistant to these diseases.

With respect to ergot, this disease first appeared in Argentina in 1996. Since then its incidence has increased in farmers' fields, where it is associated with late planting (end of November through the beginning January). The recommended management for this disease in Argentina is early planting of commercial hybrids and chemical control in plots for seed production with either tria-dimenol (25%) at 0.125 kg a.i./ha or tebuconazole (25%) 0.25 kg a.i./ha.

Stalk and root rots in Argentina are usually associated with drought and/or photosynthetic stress induced by greenbug damage during grain filling. Losses from stalk rot have been reduced over the past few years because of changes to no-till cultivation, which reduces the soil water deficit, the development of tolerant/resistant cultivars, and the development of machinery to mechanically harvest lodged sorghum.

In summary, some sorghum diseases occur annually in Argentina but cause no economic damage because resistance to them is widely available in commercial hybrids. Others, *e.g.*, rust and head blight, are of intermittent economic significance. The increased use of grain and forage sorghum for on-farm consumption, as well as the inclusion of sorghum in a rotation system to improve soil characteristics, should lead to increased yield, improved forage quality, and better adaptation of the crop for use in a no-tillage system.

References

1. **Frederiksen, R. A., D. T. Rosenow, G. L. Teetes, and J. W. Johnson.** 1976. Disease and insect rating schemes for sorghum. *Sorg. Newsl.* **19:** 117-123.
2. **Giorda, L. M.** 1997. Enfermedades, p. 40-44. *In* L. M. Giorda (ed.), *Sorgo granífero*, Cuaderno de Actualización Técnica No. 7. INTA Centro Regional Córdoba, EEA, Man-fredi, Córdoba, Argentina.
3. **Giorda, L. M., M. J. Martinez, M. Nassetta, and S. Palacios.** 1999. Sorghum ergot in Argentina. *Proc. Global Conf. on Ergot of Sorghum (1-8 June 1997, Sete Lagoas, Brazil)*: 79-83.
4. **Granda, J.** 2000. Fue buen negocio el sorgo? Campaña 99/00. Forrajes y granos. Agribusiness J. **5(53):** 98-99.
5. **Martinez, M. J., L. M. Giorda, M. Nassetta, K. Frutero, and J. Zygadlo.** 2002. Phytoalexin synthesis by sorghum grain in response to grain mold, p. 474-475. *In* J. F. Leslie (ed.), *Sorghum and Millets Diseases*. Iowa State Press, Ames, Iowa.

66

Diseases of Sorghum and Pearl Millet in Asia

S. Indira, Xiude Xu, Nipon Iamsupasit, H. S. Shetty, N. S. Vasanthi, S. D. Singh, and Ranajit Bandyopadhyay

Abstract

Sorghum and pearl millet are essential crops in arid and semi-arid regions of Asia. Sorghum is threatened mainly by grain molds, stalk rots, and ergot, while pearl millet is threatened by downy mildew, ergot, smut, and rust, among others. New biotechnology methods, *e.g.*, induced systemic resistance, are attracting more attention than are conventional technologies, *e.g.*, host plant resistance, plant breeding and cultural practices. The introduction of new technologies based on shorter duration and higher-yielding genotypes requires new strategies to avoid disease stress. In this presentation, we review progress made in Asia during the past ten years, evaluate the effectiveness of available strategies for disease management, and identify areas in which further research is needed.

Sorghum

Diseases take a heavy toll of sorghum production in most of the countries in Asia. In India, both grain mold (caused by a complex of fungi), a major disease associated with late rains, and ergot/sugary disease (caused by *Claviceps sorghi* and *C. africana*) reduce yield in seed production plots. Leaf spots (caused by *Cercospora sorghi*, *Ramularia sorghicola*, and *Bipolaris sorghicola*) and smuts (caused by *Sphacelotheca sorghi, Sporisorium cruenta, Sporisorium reilianum*, and *Sporisorium ehrenbergii*) are important on local cultivars throughout the country, and downy mildew (caused by *Peronosclerospora sorghi*) is prevalent in the states of Karnataka and Tamil Nadu. During the post-rainy season, charcoal rot (caused by *Macrophomina phaseolina*) is severe under dry land conditions and rust (caused by *Puccinia purpurea*) under irrigated conditions. In China, 31 fungal, five bacterial, and one viral disease have been reported (6). Head smut, caused by *S. reilianum*, can result in 80% yield loss (120); red stripe, caused by maize dwarf mosaic virus (MDMV), can result in 20-90% yield loss (119); and sheath blight, caused by *Rhizoctonia solani*, can result in 3-5% yield losses (6). In Japan, sugarcane mosaic virus (SCMV) and bacterial leaf stripe (caused by *Pseudomonas andropogonis*) have been reported (51). Two fungal diseases, Hayake-byou (caused by *Kabatiella sorghi*) and yellow spot (caused by *Cercospora koepkei* var. *sorghi*), were first reported in Japan in 1940 and may be unique to these sorghum-growing areas (51). In Thailand, grain mold, leaf blight, anthracnose, rust, and zonate, target, tar, and gray leaf spots are common (14).

In Indonesia, Malaysia, and the Philippines, the most common diseases are anthracnose, sheath blight, tar spot, gray leaf spot, kernel smut, rust, grain molds, stalk rot, bacterial stalk rot, seedling rot, downy mildew, zonate leaf spot, and viruses (84). In Syria, wilt caused by *Acremonium strictum* and *Fusarium moniliforme sensu lato* are common (3).

In general, the relative importance of individual diseases depends upon geographic location, with anthracnose and stalk rots a problem in most countries and grain molds the only other "very important" disease in more than one country (Table 66-1). There also have been a significant number of new diseases reported from Asia during the last 10 years (Table 66-2).

Grain molds. A complex of several fungal pathogens causes grain mold (2, 79), and a pictorial guide to the fungi commonly associated with sorghum grain is available (70). In Thailand, even though sorghum is grown as a second crop or as an intercrop, grain mold is a major disease (16). The combined effect of *F. moniliforme sensu lato* and ear head bug, *Calocoris angustatus*, is to increase the incidence of grain mold (34, 87). Variability has been reported in cultural characteristics and pathogenicity among the isolates of one of the components of grain mold, *Curvularia lunata* (101). Airborne spores of *Fusarium* spp., *Curvularia* spp., and *Alternaria* spp. can be recovered during grain development and increase in number following grain maturity (8). Genotype × environment interactions are highly significant for grain mold incidence and grain weight (34, 98).

TABLE 66-1. Perceived disease significance rankings[a] in various Asian countries.

Disease	China	India	Iran	Myanmar	Pakistan	Thailand
Head smut	1	-	-	-	-	-
Loose kernel smut	-	-	-	-	-	5
Anthracnose	3	3	5	2	3	5
Charcoal/stalk rot	3	1	5	4	5	5
Grain molds	-	1	5	1	4	1
Downy mildew	-	4	4	4	5	-
Leaf blight	-	4	5	2	2	5
Ergot	-	3	5	-	3	3
Sorghum stripe	-	5	5	-	-	-
Stalk/root rot	-	5	5	-	-	5
Striga	-	3	5	2	-	-
Bacterial leaf stripe	-	-	-	-	3	-

[a]Rated on a scale of 1-5 with 1 = very important and 5 = less important.

In screening of sorghum genotypes against toxigenic species of *Fusarium*, zearalenone levels were highest (260 ppb) in CSH1 and lowest (20 ppb) in CSH 9, CSV 1, CSH 5, and CSV 13 (78). No known toxins, including fumonisins, were produced by the *F. moniliforme sensu lato* strains examined (1). The amino acid spectrum in diseased sorghum grain is altered by hydrolysis of the weakened protein matrix (102). Simultaneous saccharification and fermentation processes with *Aspergillus niger* and *Saccharomyces cerevisiae* produced more ethanol from damaged sorghum grain than from rice (103).

Resistance to grain mold is associated with fewer floating kernels, lower water imbibition, reduced grain hardness (99), higher wax content on the seed surface (100), higher seed phenols, darker glumes (5), increased flavan-4-ols (37, 38), lower electrical conductivity in seed leachates (97), and lower ergosterol (39, 40). With respect to chemical control, propiconazole reduced grain mold, and garlic extract reduced infection by *C. lunata* (1). Harvesting at physiological maturity reduces grain mold, enhances seed germination, and improves grain quality (23). Sorghum parental lines with resistance to grain mold include: GMRP (Grain mold resistance lines from Parbhani, Maharashtra)-4, B 75219, IS 24995, and B 58586 as good combiners (25), and white grain zera zera photoperiod insensitive accessions (92). Grain mold-resistant lines include GMRP numbers 4, 8, 9, 13, 27, and 44; and SR (lines from Surat, Gujarat) numbers 155-12, 202-16, 382-1-2, 532-1-5-3, and GJ 35-15-15 (1). IS 13958, IS 25038, IS 23585, and IS 23599 from ICRISAT are being used as parental sources with grain mold resistance in Thailand (14). Crosses IS 13599 × R 5015, IOS 15715 × 6065, IS 23599 × KU 439 (15), U Thong 1 × ISC 23585, and late Hegari × IS 21454 (63) showed resistance to grain mold.

Ergot. Ergot is a constraint in seed production where it causes 10-80% yield loss. Three *Claviceps* species may cause ergot in Asia–*C. sorghi* in India (54), *C. sorghicola* in Japan (77), and *C. africana* in all countries that have reported sorghum ergot (77). Etiology, epidemiology, and management strategies for this disease were recently reviewed by Sangitrao *et al.* (85), who concluded that delayed sowing and nonsynchronization of seed parents are the major factors contributing to the incidence, development, and spread of disease.

Isolates of *C. sorghi* from India and Japan vary in their morphology and pathogenicity, and were similar to *C. africana*, which now appears to be indigenous to India (13, 77), southern Japan, and Thailand (66). Intraspecific variation in *C. africana*, as measured by differences in DNA sequence, supports the separation of the three species of *Claviceps* from Asia at the species level (77). In addition, random amplified polymorphic DNA (RAPD) patterns and rDNA sequences of some Indian isolates are identical to those of *C. africana*, thereby confirming that *C. africana* is present in India (77). Sorghum parental lines SRT 18 B, NSV 13, AKMS 14B, AKR 150, RS 29, SRT 26B, SB 1085, and MR 750 are sources of resistance to ergot. An organosulphur compound, ajoene, from *Allium sativum*, which has antifungal properties, also can control ergot (89).

Foliar diseases. Research conducted during the last decade has emphasized host plant resistance and variability in the anthracnose pathogen (72). A collaborative International Sorghum Anthracnose Virulence Nursery established in 1992 for the evaluation of pathogenic variability was planted at six All India Coordinated Sorghum Improvement Program (AICSIP) locations. There were differences in virulence and aggressiveness that suggest that *C. graminicola* can be subdivided into distinct pathotypes (110). A field screening technique has been developed and used to identify stable resistance sources (75). Genotype IS 18758 was resistant to 16 isolates of *C. graminicola* (112). *C. gloeosporioides* has been reported from yellow sorghum (67). The reduction in photosynthetic rate is higher in plants affected by zonate leaf spot than in plants affected by either sooty stripe or anthracnose (44).

Multiple resistance to foliar diseases often is attributed to increased cuticular and laminar thickness (42), and higher levels of sugars, phenols, soluble proteins, and amino acids (43). Varieties No. 6 C, UPCA-S1, UPCA-S2, KD 5, Keris, Padeli, Hegari-Genjah, Sangkheer, and Mandau, and a local variety, Sangkhur, are resistant to rust (84). Sources of rust resistance identified in purple-colored sorghums include IS 2300, IS 3443, IS 31446, IS 18758, and IS 7023 (93). New resistance sources for foliar diseases and downy mildew are DMRS × QL 3, 296B × IS 3547, (DMRS × QL 3) × 36772B, 296 B × QL 3, and IS 3443 × QL 3, and new sources for resistance to charcoal rot and leaf rust are E36-1 × GRS-1 and E36-1 × M35-1 (1). Seed treatment with metalaxyl controls downy mildew but may reduce seed germination (61). Aureomycin and aureofungin also were effective but did not inhibit germination (62). Few accessions have been identified as resistant to target leaf spot in China. Developing varietal resistance to banded leaf and sheath blight are major programs in Japan (51).

TABLE 66-2. New disease reports in the past 10 years on sorghum in various Asian countries.

Disease	Species Responsible	Country(ies)
Leaf blight	*Drechslera longirostrata*	India
Target leaf spot	*Bipolaris sorghicola*	China, Thailand, Korea
Leaf lesions	*Neottiosporina sorghicola*	China
Leaf blotch	*Bacillus subtilis*	Gulbarga and Karnataka, India
Leaf blight	*Bipolaris australiensis*	India
Bacterial leaf streak	*Xanthomonas campestris* pv. *holcicola*	Pakistan
Stalk rot	*Rhizoctonia zeae*	Turkey
Stalk rot and top rot	*Erwinia chrysanthemi* var. *zeae*	India
Columnar smut	*Sporisorium sorghicola*	China
Loose kernel smut	*Sporisorium cruenta*	Thailand
Glume blight	*Alternaria alternata* *Pencillium oxalicum*	China
Acremonium wilt	*Acremonium strictum*	China
Top rot	*F. subglutinans sensu lato*	China

Stalk rots. Stalk rots continue to be of importance, with most of the recent research focusing on disease management. Isolates of *Macrophomina phaseolina*, the causal agent of charcoal rot, from Sholapur grew faster and were more virulent than those from Bijapur and Dharwad (1). Early sowing, high plant populations (74), moisture stress (73), and colonization of stalks by *M. phaseolina* at maturity (75) all aggravate lodging and are correlated with soft stalks, the number of nodes crossed, root damage, plant senescence, and grain yield losses. Sorghum/pigeon pea intercropping, conventional incorporation of crop residue, deep sowing with or without fertilizer, and mulching with wheat straw all reduced charcoal rot and lodging while increasing yields. Seed treatment with Ridomil, Benlate, Brassicol, and Dithane M-45 decreased disease incidence and increased yields (21).

Resistance to the stalk rot complex, composed of *C. graminicola*, *F. moniliforme sensu lato*, *M. phaseolina*, and *Erwinia chrysanthemi*, was identified in sorghum accessions BJ 111, CSH 6, E36-1, PQ 16-2, PQ 54, PQ 60-2, PQ 76, Q 101, Q102, Q 103, and Q 104. Varieties SPV 655 and 1055 were resistant to charcoal rot (19). Resistant nonsenescent sorghum has more green area at physiological maturity than does sensitive sorghum, confirming the relationship between resistance and the "stay-green" phenotype (46).

Local germplasm resistant to charcoal rot and rust are Lakkadi, Bidar local, Kemp jowar, Jewargi local, Basavanapada, pop sorghum, Kadabinajola, and IS no. 2209, 4703, 5409, 19697, 23598. Some charcoal rot-resistant sorghum lines are CRP (charcoal rot resistant lines from Parbhani, Maharashtra) numbers 3, 7, 11, 12, 15, 17, 23, 24, 27, 29, 31, 42, 43, SPV 459, SPV 1159, RSCR 7, RSCR 14, and RSP 50-1-2 (1). New resistance sources to charcoal rot have been developed in crosses such as SB 401 A × AGP 289, SB 401 A × AGP 353, SB 401 A × SPV 489, MS 31-2A × Bilegund, and E36-1 × SPV 86 (1). Molecular markers and the use of culture filtrate extract are in progress to help to identify resistance to stalk rots in India.

Smuts. Smuts are a major problem in sorghum production in China. Head smut (caused by *S. reilianum*) is widespread, economically important, and may result in 80% yield losses (121). Head smut resistance could not be maintained indefinitely because of the presence of three different races (121). About 20 lines, A 421, B421, A7050, AT × 378, A232E, B232 E, BICS 49, Jin 5/Hui 7, Jin 5-28, Ying-Hui no. 2, Lian-Tang-Ai, 516, 302, LR 2939, 168, SA 281, GW 0552, and GW 2978, were immune to race 3 and are being used in breeding programs. Three resistant hybrids, *i.e.*, Liaoza 10, Jinza no. 93, and Liaoza 6, also have been released. Loose kernel smut (caused by *S. cruentum*), with yield losses up to 20-30%, is primarily seed borne (6). Seed dressing with fungicide(s) and the use of resistant varieties are both effective means of control. Floral smut (caused by *Ustilago kenjiana*), a nonsystemic disease, could cause losses of up to 40% in susceptible varieties (33), with temperature and moisture being the major factors determining pathogen infection. Columnar smut (caused by *Sporisorium sorghicola*) is a newly described species (64) and is economically important in China where it is both soil and seedborne. Jawar-86, BR 116, and BR 123 were resistant to covered kernel smut under semi-arid conditions (50). Seed dressing with Myclobutanil, Raxil, Triadimenol, and Triadimefon is effective for the control of smuts. Reports of the control of long smut (caused by *S. ehrenbergii*) by seed treatment with Thiram (58) require further confirmation (22).

Viruses. The importance of viruses on sorghum in India has increased dramatically in the last decade. The three viruses present in India are maize mosaic virus, sorghum stripe virus, and maize stripe mosaic virus, which are described in more detail elsewhere in this volume (69).

Multiple disease resistance. Significant efforts have been made to identify multiple-disease resistant genotypes (Table 66-3). Genotypes ICSV 705, which is resistant to shoot fly, and ICSV 745, which is resistant to midge, also were resistant to sugary disease, rust, grain molds, and rough leaf spot. Genotypes IS 21444 and IS 17610 were resistant to both head bugs and grain molds, thus indicating the availability of multiple disease and insect resistance sources. Genotypes IS 8283 and IS 8607 were both resistant to six diseases: downy mildew, grain molds, rust, gray leaf spot, anthracnose, and zonate leaf spot.

TABLE 66-3. Multiple disease resistance in available sorghum accessions

Disease–Causal Organism	Sorghum Accession
Cercospora sorghi, *Colletotrichum graminicola*, *Gloeocercospora sorghi*, and *Ramulispora sorghi*	MPC 123 & 2219A × 3274-1-2
C. graminicola, *Macrophomina phaseolina*, *Peronosclerospora sorghi*, and *Puccinia purpurea*	SPV 462
Ascochyta sorghi, *Bipolaris sorghicola*, *C. sorghi*, *C. graminicola*, *Phyllachora* spp., *P. purpurea*, and *Ramulispora sorghina*	IS 8283, IS 8607, M 817 & MR 825
P. purpurea and *Setosphaeria turcica*	IS 2300, IS 3443, IS 7023, IS 31446, C40, C242 & C603
P. sorghi, *Sphacelia sorghi*, *P. purpurea*, and grain molds (*Fusarium moniliforme sensu lato*, *Curvularia lunata*, etc.)	Co25, TNS 23 & TNS 28
C. graminicola, *Erwinia chrysanthemi*, *F. moniliforme sensu lato*, and *M. phaseolina*	Q101, Q102, Q103, E36-1, BJ111 & PQ60-2
Fusarium oxysporum, *Fusarium semitectum*, *Fusarium subglutinans sensu lato*, and *M. phaseolina*	Q101, Q102, Q103, Q104 & PQ54

Pearl Millet

Pearl millet (*Pennisetum glaucum*) is an important cereal crop in semi-arid and arid areas of the world. In India, it is the fourth most important food crop. The crop is commonly subjected to severe epidemics of downy mildew, ergot, rust, and smut (7). The pearl millet improvement program at the Indian Agricultural Research Institute includes genetic studies of downy mildew and smut resistance, variety identification based on protein profiles generated by polyacrylamide gel electrophoresis, analysis of genetic variability, and *in vitro* selection for disease tolerance and nutritional improvement. Restriction fragment length polymorphisms (RFLPs) are being used in marker-assisted selection breeding programs for resistance to downy mildew, for drought tolerance, and for broadening the genetic base of male sterility (28). At present, the program emphasizes the management of economically important diseases such as downy mildew (caused by *Sclerospora graminicola*), ergot (caused by *Claviceps fusiformis*), smut (caused by *Tolyposporium penicillariae*), and rusts (caused by *Puccinia penniseti*, and *Puccinia substriata* var. *penicillariae*).

Downy mildew. Downy mildew, caused by the variable pathogen *S. graminicola* (109), is a major biotic constraint to pearl millet production that is subject to a diverse series of controls (32). Based on a downy mildew incidence level of 62%, the estimated yield loss compared to a crop with no visible disease was 34%. Relative to a chemically protected crop, the yield reduction for an unprotected crop was 57% (29). Losses result from complete or partial conversion of florets to leafy structures, termed "green ear," instead of grain (91). Downy mildew infection also reduces photosynthesis (45). The pathogen varies in virulence (60), with root, coleoptile, and leaves all possible sites for infection by zoospores, but pathogen growth is limited in seedlings of resistant genotypes (86). The hypersensitive response in seedlings is a component of an early defense response (24). The defense-related enzymes β-1,3-glucanase, chitinase, peroxidase, phenylalanine ammonia lyase, superoxide dismutase, lipoxygenase, and nucleases are expressed during the hypersensitive reaction, and their activity in resistant cultivars suggests their potential utility as biochemical markers (52, 53). Immunological studies implicate pathogen-specific, surface proteins in host recognition phenomena (60). An enzyme-linked immunosorbant assay (ELISA) has been developed for *in vivo* quantification of *S. graminicola* (57).

Sugars do not induce germination of sporangia, and glucose has the least inhibitory effect. The amino acids indole acetic acid and NAA inhibit sporangial germination, but maleic hydrazide, 2,4-D, and gibberellic acid do not. L-glutamic acid is most inhibitory and L-leucine the least (4). During downy mildew infection, the ascorbic acid content of leaves and ear heads significantly increases (10); the pool of available amino acids decreases; phenol accumulation increases (11); and the total sugar, starch, and protein of leaves and stems of infected tissues increases (12), as does respiration, which reduces photosynthetic activity (9). Green ear disease extracts contain high levels of toxic metabolites and have high levels of cellulase activity that lead to increased loss of electrolytes from leaf tissues (26).

Downy mildew resistance is conferred by alleles at at least two loci (48), although some modifier genes also appear to be important (20). In the cultivar Pennisetum 292, the inheritance of resistance is controlled by *Rsg1*, a single dominant gene (96). The resistant hybrids may become susceptible to downy mildew, as the hybrids have a narrow genetic base. The restorer parent of pearl millet (*P. glaucum*) single-cross grain hybrid ICMH 451 has been released and has higher yield and resistance to downy mildew (59). The best sowing technique is transplantation or direct sowing into ridges. The recovery trait in a pearl millet line with a high level of recovery resistance, ICMAI (89), also designated RECRES 81A, was effective against the major pathotypes of downy mildew (95). Based on estimates of genetic resistance parameters, additive and dominance effects are occurring. Epistasis was of the dominant duplicate type. Inheritance of resistance is complex, as indicated by heritability estimates and the degree of genetic advance, which ranged from 12-59% and 61-280%, respectively (47). Breeding for resistance to pearl millet downy mildew remains a major challenge that is now being met with unusual breeding schemes and marker-assisted selection (17, 31, 118).

There is a direct correlation between a seedborne oosporic index and disease incidence in the field (107). Oospores can remain viable for up to 13 years, but the highest percent germination occurred with one-year-old spores (71). Variation in pathogenicity among single-oospore isolates of *S. graminicola* has been reported (115).

Seed treatment or foliar sprays with metalaxyl reduced disease intensity and enhanced yields (94) but were less effective than a combination of seed treatment and foliar spray (36, 83). However, applications of metalaxyl often are not cost effective, as there are reports of development of resistant strains of *S. graminicola* (88). Crops sown in a NW-SE direction had 40% less disease incidence at 70 days after sowing (29). The effect(s) of antagonistic microorganisms *e.g.*, *Pseudomonas fluorescens*, *Bacillus subtilis*, *Trichoderma harzianum*, and *Chaetomium globosum*, on the causal agent of downy mildew are being examined in India. Agents such as sublethal doses of *S. graminicola* and *Pernosclerospora sorghi*, and chemicals such as β-aminobutyric acid, benzothiadiazole, acetyl salcylic acid, salcylic acid, and $CaCl_2$, can induce systemic resistance at the seedling stage (55) that remains durable through both the vegetative and the reproductive stages. The protection offered by this induced resistance is not dependent on the cultivar used and is not related to their constitutive level of resistance (55, 56). Top cross hybrids have been evaluated for increased stable yield and durable resistance to downy mildew (105), although only one source of cytoplasm was evaluated for downy mildew susceptibility (123). Quantitative trait loci (QTLs) that contribute to resistance to downy mildew have been identified from mapping populations in Asia (30). There were no QTLs that were simultaneously effective against pathogen populations from India, Nigeria, Niger, and Senegal, demonstrating that resistance is pathotype specific. For all but one of the QTLs, resistance was inherited from the resistant parent as a dominant trait (41).

Ergot. Pearl millet ergot, *C. fusiformis* causes losses of up to 70% (111) and is widespread in Asia and Africa (80). The infected ovary is transformed into a sclerotium, reducing the quality and quantity of the grain. There is a significant negative correlation between number and weight of sclerotia and of the grain (114). Pollen management and exploitation of post-pollination stigmatic constriction can control ergot (65). The time between full protogyny and anthesis initiation, shorter stigma, and higher seed set were characteristic of the resistant cultivars (108). Breeding for ergot resistance and work on induced mutagenesis (116), a factor independent of flowering events governing resistance in A lines and their hybrids (113), has been conducted at Pantnagar.

Smuts. Pearl millet smut, *T. penicillariae*, was at epidemic levels in 1989 (104, 122), resulting in losses of up to 20%. Solarization with white polyethylene and grass mulch reduced smut disease severity (82). Resistance to smut is inherited in both a dominant and an additive manner. Resistance in most ergot-susceptible lines was independent of the timing of flowering events, but the reverse was true in ergot-resistant lines (106). Six lines with stable resistance to smut were identified for use in resistance breeding (108). High-yielding tall and early hybrids with resistance include 81A × SR/ICMPS-101-1 and S10A × 77/181-4 (49).

Rust. Pearl millet rust, *P. penniseti*, is widespread in Asia, Africa, and the United States, and is of major importance on multicut forage hybrids, where it can cause heavy losses (117). The germplasm accessions ICML 17, 18, 19, 20, and 21 (P1537582 to 537586) from Nigeria, the United States, Mali, India, and Cameroon, respectively, all had high levels of resistance (90). The resistance is a single gene dominant trait (81). Variability in *P. substriata* var. *indica* isolates also has been observed (76).

Fungal associations. Fungi (83% pathogenic) causing seed deterioration and soft rot include *A. alternata, C. lunata, D. sacchari, Exerohilum halodes, F. moniliforme sensu lato, F. semitectum*, and *Phoma spp*. The number of these fungi tended to increase from grain filling to physiological maturity, and more fungi were found on seeds than on glumes (35).

Mycotoxins. Pearl millet seeds carried *F. semitectum, F. moniliforme sensu lato*, and *F. fusarioides*, of which strains of *F. moniliforme sensu lato* and *F. fusarioides* produced moniliformin (18). A toxin from a 14-day-old culture filtrate of *C. lunata* was most active at pH 5.5. The active principle was heat stable, water-soluble, and affected the cell permeability of susceptible host tissues (27). Of the 22 fungal species isolated from two stored and cooked grain samples of pearl millet, *Aspergillus flavus* and *A. parasiticus* dominated. The levels of aflatoxins B_1 ranged from 14-2100 μg/kg and 18-550 μg/kg in stored and cooked samples, respectively (68).

Future Needs

- ❖ Host-pathogen interaction studies of the mechanism of resistance, disease dynamics, and the development of forecasting models.
- ❖ Studies of the biology and epidemiology of the pathogen, host matrix influence, and soil factors in disease development.
- ❖ Understanding the diversity within the pathogen populations, and the factors underlying changes in virulence.
- ❖ Development of screening techniques to identify the sources of durable resistance, breed improved resistance sources, and deploy these sources as part of cost-effective integrated pest management strategies.
- ❖ Development of near isogenic lines for disease resistance genes, and to understand how these genes are related to host specificity and races.
- ❖ Development of effective alternatives to resistance breeding, *e.g.*, seed treatment with bioagents or induction of systemic resistance.
- ❖ Adoption of disease control measures into the socioeconomic, marketing, organizational, and institutional framework within which farmers operate.

References

1. **AICSIP (All India Coordinated Sorghum Improvement Project).** Plant pathology progress reports 1994-1999, National Research Center for Sorghum, Hyderabad, India.
2. **Adiver, S. S., and K. H. Anahosur.** 1994. Association of *Curvularia* and *Exserohilum* species, sorghum molds and their effect on seed germination and seedling growth. *Curr. Res.–Univ. Agric. Sci. (Banglore)* **23:** 38-40.
3. **Al-Ahmad, M.** 1993. Vascular and vascular-like diseases of grain sorghum in Syria. *Arab J. Plant Protect.* **11:** 39-44.
4. **Appaji, S., and D. P. Thakur.** 1994. Effect of sugars, growth regulators and amino acids on *in vitro* sporangial germination of pearl millet downy mildew. *Ind. Phytopathol.* **47:** 386-390.
5. **Audilakshmi, S., J. W. Stenhouse, T. P. Reddy, and M. V. R. Prasad.** 1999. Grain mold resistance and associated characters of sorghum genotypes. *Euphytica* **107:** 91-103.
6. **Bai, J.-K.** 1997. *Cereal Crop Diseases.* Agricultural Publishing House, Beijing, China.
7. **Balasubramanian, K. A.** 1992. Pearl millet diseases in India: Knowledge and future research needs, p. 123-127. *In* W. A. J. de Milliano, R. A. Frederiksen, and G. D. Bengston (eds.), *Sorghum and Millets Diseases: A Second World Review.* ICRISAT, Patancheru (A.P.), India.
8. **Bandyopadhyay, R., L. K. Mughogho, and M. V. Satyanarayana.** 1991. Occurrence of airborne spores of fungi causing grain mold over a sorghum crop. *Mycol. Res.* **95:** 1315-1320.
9. **Bhatia, J. N., and D. P. Thakur.** 1990. Physiological pattern of pearl millet plant in relation to green ear disease, p. 34-38. *In* M. P. Srivastava and G. S. Saharan (eds.), *Plant Pathological Research: Problems and Progress.* Indian Phytopathological Society (North Zone), Hisar, Haryana, India.
10. **Bhatia, J. N., and D. P. Thakur.** 1992. Ascorbic acid control of healthy and diseased parts of pearl millet in relation to downy mildew disease. *Plant Dis. Res.* **7:** 240-241.
11. **Bhatia, J. N., and D. P. Thakur.** 1992. Biochemical components of pearl millet (*Pennisetum glaucum* L. R. Br.) tissues in healthy and diseased state induced by *Sclerospora graminicola* (Sacc.) Schroet. *Biochem. Physiol. der Pflanz.* **188:** 349-353.
12. **Bhatia, J. N., and D. P. Thakur.** 1994. Biochemical components of pearl millet in relation to downy mildew disease. *Ind. J. Mycol. Plant Pathol.* **24:** 216-219.
13. **Bogo, A., and P. G. Mantle.** 1999. *Claviceps africana* discovered in India. *Plant Dis.* **83:** 79.

14. **Boon-Long, T.** 1992. Sorghum diseases in Thailand, p. 41-43. *In* W. A. J. de Milliano, R. A. Frederiksen, and G. D. Bengston (eds.), *Sorghum and Millets Diseases: A Second World Review*. ICRISAT, Patancheru (A.P.), India.
15. **Boon-Long, T., K. Wirojwatanakul, P. Tubthong, and K. Lertprasertrat.** 1994. Sorghum grain mold disease resistance screening, p. 46-56. In *Annual Report 1994: Field Crops Disease Working Group*. Pathology and Microbiology Division, Min. Agric., Bangkok, Thailand.
16. **Boon-Long, T., K. Wirojwatanakul, P. Tubthong, K, Lertprasertrat, and S. Titatarn**. 1992. Sorghum grain mold resistance screening, p. 72. In *The 23rd Thai National Corn and Sorghum Reporting Session*. Min. Agric., Bangkok, Thailand.
17. **Breese, W. A., C. T. Hash, K. M. Devos, and C. J. Howarth.** 2002. Pearl millet genomics and breeding for resistance to downy mildew, p. 243-246. *In* J. F. Leslie (ed.), *Sorghum and Millets Diseases*. Iowa State Press, Ames, Iowa.
18. **Chandran, R. N., O. P. Mall, and S. K. Chauhan.** 1993. Moniliformin production by *Fusaria* infecting bajra. *Natl. Acad. Sci. Lett.* **16:** 285-286.
19. **Desai, S. A.** 1998. Evaluation of advanced sorghum varieties against charcoal rot disease in Karnataka. *Karnataka J. Agric. Sci.* **11:** 104-107.
20. **Deswal, D.P., and O. P. Govila.** 1994. Genetics of resistance to downy mildew (*Sclerospora graminicola*) in pearl millet (*Pennisetum glaucum*). *Ind. J. Agric. Sci.* **64:** 661-663.
21. **Evangelista, C. C., and N. G. Tangonon.** 1991. Root and stalk rot incidence in non-senescent sorghum genotypes and their chemical control. *R&D Philippines*, p. 27-28.
22. **Frederiksen, R. A., and G. N. Odvody, eds.** 2000. *Compendium of Sorghum Diseases*, 2nd ed. APS Press, St. Paul, Minnesota.
23. **Garud, T. B., B. M. Shinde, S. Ismail, and B. N. Aglave.** 1998. Effect of delayed harvesting on grain mold development in sorghum. *Intl. Sorg. Millets Newsl.* **39:** 106-107.
24. **Geetha, S., S. A. Shetty, H. S. Shetty, and H. S. Prakash.** 1996. Arachidonic acid-induced hypersensitive cell death as an assay of downy mildew resistance in pearl millet. *Ann. Appl. Biol.* **129:** 91-96.
25. **Ghorade, R. B., V. B. Shekar, B. D. Gite, and B. A. Sakhare.** 1997. Some general combiners for grain mould resistance in sorghum. *J. Soils Crops* **7:** 8-11.
26. **Gour, H. N., and A. Kumar.** 1992. Downy mildew of pearl millet: II. Extraction of toxins and alterations in cell permeability. *Biochem. Physiol. Pflanz.* **188:** 45-50.
27. **Gour, H. N., P. D. Nitharwal, and S. Agarwal.** 1992. Partial purification of toxin from *Curvularia lunata* (Wakker) Boedijn. *Biochem. Physiol. Pflanz.* **188:** 128-135.
28. **Govila, O. P.** 1991. Pearl millet improvement program at the Indian Agricultural Research Institute (IARI), p. 25-26. In *Report of Meeting. Rockefeller Foundation Conference: the Establishment of a Sorghum and Millet RFLP Network to Support Breeding in Developing Countries, 6-10 May 1991*. Rockefeller Foundation, New York City, New York.
29. **Gupta, G. K., and D. Singh.** 1996. Effectiveness of water absorbents, FYM application, and sowing direction in downy mildew reduction in pearl millet under arid environment. *Ind. Phytopathol.* **49:** 41-47.
30. **Hash, C. T., and J. R. Witcombe.** 1994. Pearl millet mapping populations at ICRISAT. *Use of Molecular Markers in Sorghum and Pearl Millet Breeding for Developing Countries. Proc. ODA Plant Sciences Res. Program Conf. (29 March–1 April 1993, Norwich, UK)*: 69-75.
31. **Hash, C. T. and J. R. Witcombe.** 2002. Gene management and breeding for downy mildew resistance, p. 27-36. *In* J. F. Leslie (ed.), *Sorghum and Millets Diseases*. Iowa State Press, Ames, Iowa.
32. **Hess, D. E., R. P. Thakur, C. T. Hash, P. Sérémé, and C. W. Magill.** 2002. Pearl millet downy mildew: Problems and control strategies for a new millennium, p. 37-42. *In* J. F. Leslie (ed.), *Sorghum and Millets Diseases*. Iowa State Press, Ames, Iowa.
33. **Hua, Z., and J. Bai.** 1989. On the floral smut (*Ustilago kenjiana*), of sorghum. *Jilin Agric. Sci.* **55:** 37-41.
34. **Indira, S., M. Jitendra, and B. S. Rana.** 1991. Genotype × environmental interaction for grain mold resistance and seed weight in sorghum. *Ind. Phytopathol.* **44:** 523-525.
35. **Ingle, R. W., and J. G. Raut.** 1993. Fungi associated with glume and seed in pearl millet at different seed development stages. *Seed Res.* **21:** 131-132.
36. **Jain, A. K., S. K. Tripathi, and J. C. Gupta.** 1993. Metalaxyl on downy mildew of pearl millet. *Curr. Res.–Univ. Agric. Sci. (Bangalore)* **22:** 133-134.
37. **Jambunathan, R., and M. S. Kherdekar.** 1991. Flavan-4-ols concentration in leaf tissues of grain mold susceptible and resistant sorghum plants at different stages of leaf development. *J. Agric. Food Chem.* **39:** 1163-1165.
38. **Jambunathan, R., M. S. Kherdekar, and R. Bandyopadhyay.** 1990. Flavan-4-ols concentration in mold-susceptible and mold-resistant sorghum at different stages of grain development. *J. Agric. Food Chem.* **38:** 545-548.
39. **Jambunathan, R., M. S. Kherdekar, and J. W. Stenhouse.** 1992. Sorghum grain hardness and its relationship to mold susceptibility and mold resistance. *J. Agric. Food Chem.* **40:** 1403-1408.
40. **Jambunathan, R., M. S. Kherdekar, and P. Vaidya.** 1991. Ergosterol concentration in mold-susceptible and mold resistant sorghum at different stages of grain development and its relationship to flavan-4-ols. *J. Agric. Food Chem.* **39:** 1866-1870.
41. **Jones, E. S., C. J. Liu, M. D. Gale, C. T. Hash, and J. R. Witcombe.** 1995. Mapping quantitative trait loci for downy mildew resistance in pearl millet. *Theor. Appl. Genet.* **91:** 448-456.
42. **Kalappanavar, I. K., and R. V. Hiremath.** 1998. Variation in histological parameters in resistant and susceptible sorghum genotypes. *Karnataka J. Agric. Sci.* **11:** 90-95.
43. **Kalappanavar, I. K., and R. V. Hiremath.** 1998. Biochemical factors for resistance to foliar diseases of sorghum. *Karnataka J. Agric. Sci.* **11:** 240-241.
44. **Kalappanavar, I. K., R. V. Hiremath, and B. B. C. Gowdar.** 1998. Changes in physiological activities influenced by disease situation in sorghum genotypes. *Karnataka J. Agric. Sci.* **11:** 237-239.
45. **Karande, S. M.** 1994. Photosynthesis in downy mildew diseased pearl millet, p. 381-384. In *Plant Productivity Under Environmental Stress: National Seminar on Approaches to Plant Production under Environmental Stress (Jaipur, Rajasthan, India, Jan 1992)*. Agro Botanical Publishers, Bikaner, Rajasthan, India.

46. **Karunakar, R. I., S. Pande, K. Satyaprasad, and P. Ramarao.** 1993. Varietal reaction of non-senescent sorghums to the pathogens causing root and stalk rot of sorghum in India. *Inl. J. Pest Manag.* **39:** 343-346.
47. **Kataria, R. P., H. P. Yadav, C. R. Beniwal, and M. S. Narwal.** 1994. Genetics of incidence of downy mildew (*Sclerospora graminicola*) in pearl millet (*Pennisetum glaucum*). *Ind. J. Agric. Sci.* **64:** 664-666.
48. **Kataria, R., H. P. Yadav, and S. Dass.** 1993. Inheritance of downy mildew disease resistance in pearl millet [*Pennisetum typhoides* (B) S & H]. *Agric. Sci. Digest* **13:** 128-130.
49. **Khairwal, I. S., S. Singh, and O. Parkash.** 1990. Path analysis in pearl millet. *Haryana Agric. Univ. J. Res.* **20:** 76-77.
50. **Khaleeque, M. I., A. Alam, M. Inam-ul-Haq, and M. Ahmad.** 1995. Reaction of ten sorghum cultivars against grain smut under semi-arid conditions of Bahawalpur. *Pakistan J. Phytopathol.* **7:** 90-91.
51. **Kimigafukuro, T.** 1992. Sorghum and sorghum diseases in Japan, p. 31-34. *In* W. A. J. de Milliano, R. A. Frederiksen, and G. D. Bengston (eds.), *Sorghum and Millets Diseases: A Second World Review.* ICRISAT, Patancheru (A.P.), India.
52. **Kini, R. K., N. S. Vasanthi, S. U. Kumar, and H. S. Shetty.** 2000. Purification and properties of a major isoform of β-1,3-glucanase from pearl millet seedlings. *Eur. J. Plant Sci.* **150:** 139-145.
53. **Kini, R. K., N. S. Vasanthi, and H. S. Shetty.** 2000. Induction of β-1,3-glucanase in seedlings of pearl millet in response to infection by *Sclerospora graminicola. Eur. J. Plant Pathol.* **106:** 267-274.
54. **Kulkarni, B. G. P., V. S. Seshdri, and R. K. Hegde.** 1976. The perfect stage of *Sphacelia sorghi* McRae. *Mysore J. Agric. Sci.* **10:** 286-289.
55. **Kumar, V. U., C. K. Hindumathi, K. R. Kini, and H. S. Shetty.** 1998. Prior inoculation inducing resistance to downy mildew (*Sclerospora graminicola*) enhance growth, peroxidase, B-1,3-glucanases activity in pearl millet (*Pennisetum glaucum*). *J. Plant Pathol.* **80:** 203-209.
56. **Kumar, V. U, M. S. Meera, C. K. Hindumathi, and H. S. Shetty.** 1993. Induced systemic resistance protects pearl millet plants against downy mildew disease due to *Sclerospora graminicola. Crop Protect.* **12:** 458-462.
57. **Kumar, V. U., M. S. Meera, S. Shishupala, and H. S. Shetty.** 1993. Quantification of *Sclerospora graminicola* tissues of *Pennisetum glaucum* using an enzyme linked immunosorbent assay. *Plant Pathol.* **42:** 251-255.
58. **Kumar, A., and V. Nath.** 1991. Fungicidal control and disease rating scale of long smut of sorghum. *Ind. J. Agric. Sci.* **61:** 225-227.
59. **Kumar, K. A., K. N. Rai, D. J. Andrews, B. S. Talkdar, S. D. Singh, A. S. Rao, P. P. Babu, and B. P. Reddy.** 1995. Registration of ICMP 451 parental line of pearl millet. *Crop Sci.* **35:** 605.
60. **Kumar, V. U., S. Shishupala, H. S. Shetty, and S. U. Kumar.** 1993. Serological evidence for occurrence of races in *Sclerospora graminicola* and identification of race specific surface protein in host-pathogen recognition. *Can. J. Bot.* **71:** 1467-1471.
61. **Lakshmanan, P., and S. Mohan.** 1989. Effect of fungicides against sorghum downy mildew in Tamil Nadu. *Ind. J. Mycol. Plant Pathol.* **19:** 111-113.
62. **Lakshmanan, P., and S. Mohan.** 1989. Evaluation of tan sorghum cultures for multiple disease resistance. *Madras Agric. J.* **76:** 588-589.
63. **Lertprasertrat, K., U. Pleinsinchai, and C. Lairungreong.** 1996. Evaluation of sorghum lines for grain mold resistance, p. 1-9. In *Annual Report 1996. Sorghum and Local Field Crops.* Suphan Buri Field Crops Res. Center, Field Crops Res. Inst., Min. of Agric., Thailand.
64. **Liang, J., S. Wang, and J. Bai.** 1997. A new smut (*Sporisorium sorghicola*) occurring on sorghum. *J. Shenyang Agric. Univ.* **28:** 152-155.
65. **Mantle, P. G.** 1992. Ergot disease of pearl millet: Toxicological significance and a mechanism for disease escape, p. 129-132. *In* W. A. J. de Milliano, R. A. Frederiksen, and G. D. Bengston (eds.), *Sorghum and Millets Diseases: A Second World Review.* ICRISAT, Patancheru (A.P.), India.
66. **Mantle, P. G., and H. A. G. Hassan.** 1994. Widening geographical distribution of *Claviceps africana,* an important ovary pathogen of grain sorghum. *Intl. Sorg. Millets Newsl.* **35:** 97-98.
67. **Mathur, K., R. P. Thakur, and V. P. Rao.** 1997. A strain of *Colletotrichum gloeosporioides* causing anthracnose on yellow sorghum. *Intl. Sorg. Millets Newsl.* **38:** 69-70.
68. **Mishra, N. K., and S. K. Daradhiyar.** 1991. Mold flora and aflatoxin contamination of stored and cooked samples of pearl millet in the Paharia tribal belt of Santhal Pargana, India. *Appl. Environ. Microbiol.* **57:** 1223-1226.
69. **Narayana, Y. D., R. Bandyopadhyay, S. S. Navi, and V. Muniyappa.** 2002. Sorghum viruses in Asia and Africa, p. 431-440. *In* J. F. Leslie (ed.), *Sorghum and Millets Diseases.* Iowa State Press, Ames, Iowa.
70. **Navi, S. S., R. Badyopadhyay, A. J. Hall, and P. J. Bramel-Cox.** 1999. *Pictorial Guide for the Identification of Mold Fungi on Sorghum Grain.* Information Bulletin No. 59. ICRISAT, Patancheru (A.P.), India.
71. **Panchbhai, S. D., M. S. Reddy, and S. D. Singh.** 1991. A repeatable method of germination of oospores of *Sclerospora graminicola* and its significance in downy mildew disease. *Ind. J. Plant Protect.* **19:** 101-103.
72. **Pande, S., L. K. Mughogho, R. Bandyopadhyay, and R. I. Karunakar.** 1991. Variation in pathogenicity and cultural characteristics of sorghum isolates of *Colletotrichium graminicola* in India. *Plant Dis.* **75:** 778-783.
73. **Pande, S., L. K. Mughogho, and R. I. Karunakar.** 1990. Effect of moisture stress, plant population density and pathogen inoculation on charcoal stalk rot of sorghum. *Ann. Appl. Biol.* **116:** 221-232.
74. **Pande, S., L. K. Mughogho, and R. I. Karunakar.** 1992. Incidence of charcoal rot in sorghum cultivars as affected by sowing date and plant density. *Ind. J. Plant Protect.* **20:** 162-170.
75. **Pande, S., R. P. Thakur, R. I. Karunakar, R. Bandyopadhyay, and B. V. S. Reddy.** 1994. Development of screening methods and identification of stable resistance to anthracnose in sorghum. *Field Crops Res.* **38:** 157-166.

76. **Pannu, P. P. S., S. S. Sokhi, and K. S. Aulakh.** 1994. Variability in *Puccinia substriata* var. *indica* causing rust of pearl millet. *Intl. Sorg. Millets Newsl.* **35:** 68.
77. **Pažoutová, S., R. Bandyopadhyay, D. E. Frederickson, P. G. Mantle, and R. A. Frederiksen.** 2000. Relations among sorghum ergot strains from the United States, Mexico, Puerto Rico, Bolivia, India and Australia. *Plant Dis.* **84:** 437-442.
78. **Rajni, G.** 1995. Screening of sorghum cultivars for the elaboration of various toxigenic species of *Fusarium*. *J. Recent Adv. Appl. Sci.* **10:** 37-39.
79. **Rajni, G.** 1996. Seedborne fungi in sorghum. *Plant Gen. Resour. Newsl.* **106:** 39-40.
80. **Rakesh, K. P., M. S. Dang, and J. K. Kalra.** 1997. Estimation of yield losses due to ergot disease in pearl millet. *Ann. Agric. Res.* **18:** 143-145.
81. **Ramamoorthi, N., G. Thulaidas, G. Arjunan, and K. S. Jehangir.** 1995. Inheritance of resistance to rust in pearl millet. *Ind. J. Genet. Plant Breed.* **55:** 362-364.
82. **Rathi, A. S., M. S. Panwar, and M. P. Srivasatava.** 1995. Effect of soil solarization on smut of pearl millet. *Intl. Sorg. Millets Newsl.* **36:** 54-55.
83. **Reddy, G. S., K. C. S. Rao, M. S. Reddy, and S. D. Singh.** 1991. Assessment of yield losses due to downy mildew of pearl millet. *Ind. J. Plant Protect.* **19:** 81-83.
84. **Salleh, B., S. Nurdijati, M. S. Sudjadi, N. G. Tangonan, and S. Chuenchit.** 1995. Current status of sorghum cultivation and diseases in Southeast Asia, p. 105-115. *In* J. F. Leslie and R. A. Frederiksen (eds.), *Disease Analysis through Genetics and Biotechnology: Interdisciplinary Bridges to Improved Sorghum and Millet Crops.* Iowa State University Press, Ames, Iowa.
85. **Sangitrao, C., S. Indira, and R. Bandyopadhyay.** 1997. Sorghum ergot in India. *Proc. Global Conf. on Ergot of Sorghum (1-7 June 1997, Sete Lagoas, Brazil)*: 41-54.
86. **Sharada, M. S., S. A. Shetty, and H. S. Shetty.** 1995. Infection processes of *Sclerospora graminicola* in *Pennisetum glaucum* lines resistant and susceptible to downy mildew. *Mycol. Res.* **99:** 317-322.
87. **Sharma, H. C., M. V. Satyanarayana, S. D. Singh, and J. W. Stenhouse.** 2000. Inheritance to head bugs and its interactions with grain molds in *Sorghum bicolor*. *Euphytica* **112:** 167-173.
88. **Shetty, H. S., and R. Ahmed.** 1981. Physiological specialization in *Sclerospora graminicola*. *Ind. Phytopathol.* **34:** 307-309.
89. **Singh, S. D.** 1994. Biological control of sorghum ergot. Paper presented at the XXIVth Annual Group Meeting of the All India Coordinated Sorghum Improvement Project (18-20 April 1994, Pantnagar, India).
90. **Singh, S. D., S. B. King, and P. M. Reddy.** 1990. Registration of five pearl millet germplasm sources with stable resistance to rust. *Crop Sci.* **30:** 1165.
91. **Singh, S. D., S. B. King, and J. Werder.** 1993. *Downy Mildew Disease of Pearl Millet.* Information Bulletin No. 37. ICRISAT, Patancheru (A.P.), India.
92. **Singh, S. D., S. S. Navi, J. W. Stenhouse, and K. E. P. Rao.** 1995. Grain mold resistance in white grain sorghum. *Intl. Sorg. Millets Newsl.* **36:** 95-96.
93. **Singh, S. D., P. Sathiah, and K. E. P. Rao.** 1994. Sources of rust resistance in purple-colored sorghum. *Intl. Sorg. Millets Newsl.* **35:** 100-101.
94. **Singh, S. D., and H. S. Shetty.** 1990. Efficacy of systemic fungicide metalaxyl for the control of downy mildew (*Sclerospora graminicola*) of pearl millet (*Pennisetum glaucum*). *Ind. J. Agric. Sci.* **60:** 575-581.
95. **Singh, S. D., and B. S. Talukdar.** 1996. Recovery resistance to downy mildew in pearl millet parental lines ICMA 1 and ICMB 1. *Crop Sci.* **36:** 201-203.
96. **Singh. S. D., and B. S. Talukdar.** 1998. Inheritance of complete resistance to pearl millet downy mildew. *Plant Dis.* **82:** 791-793.
97. **Somani, R. B., and S. Indira.** 1999. Electrical conductivity of seed leachate in sorghum. *Intl. Sorg. Millet Newsl.* **39:** 107-109.
98. **Somani, R. B., and S. Indira.** 1999. Effect of grain molds on grain weight in sorghum. *Ind. J. Mycol. Plant Pathol.* **29:** 22-24.
99. **Somani, R. B., and S. Indira.** 2000. Relationship between water imbibition and grain hardness to grain mould resistance in sorghum. *Ind. Phytopathol.* **53:** 68-70.
100. **Somani, R. B., S. Indira, and B. S. Rana.** 1997. Detection of surface wax and its role in mold resistance in grain sorghum. *Proc. Ind. Phytopathol. Soc. Golden Jubilee Internatl. Conf. (10-15 November 1997, New Delhi)*: 187.
101. **Somani, R. B., R. W. Ingle, and S. S. Wanjari.** 1994. Variability in *Curvularia lunata* Boedijn causing grain mold in sorghum. *Intl. Sorg. Millets Newsl.* **35:** 106-107.
102. **Somani, R. B., R. B. Pandrangi, S. G. Wankhade, and D. B. Patil.** 1993. Amino acid spectra of healthy and moldy grains of sorghum hybrid SPH 388. *Ind. Phytopathol.* **46:** 249-250.
103. **Suresh, K., N. K. Sree, N., and L. V. Rao.** 1999. Utilization of damaged sorghum and rice grains for ethanol production by simultaneous saccharification and fermentation. *BioResource Technol.* **68:** 301-304.
104. **Sushil, S., O. P. Yadav, and S. Singh.** 1996. Pearl millet disease scenario in southwestern Haryana. *Intl. Sorg. Millets Newsl.* **37:** 86-88.
105. **Talukdar, B. S., S. D. Singh, and P. P. P. Babu.** 1996. Prospects of topcross hybrids in increasing and stabilizing grain yield in pearl millet. *Crop Improve.* **23:** 147-150.
106. **Thakur, R. P.** 1989. Flowering events in relation to smut susceptibility in pearl millet. *Plant Pathol.* **38:** 557-563.
107. **Thakur, R. P.** 1993. Host-plant resistance to diseases: Problems, progress and future needs, p. 244-251. *In* H. C. Sharma and M. V. Rao (eds.), *Pests and Pest Management in India: The Changing Scenario–Lead Papers of the National Seminar on Changing Scenario in Pests and Pest Management in India, (31 January–1 February 1992, Hyderabad, A. P., India).* Plant Protection Association of India. Hyderabad, A. P., India.
108. **Thakur, R. P., S. B. King, K. N. Rai, and V. P. Rao.** 1992. *Identification and Utilization of Smut Resistance in Pearl Millet.* Information Bulletin No. 16. ICRISAT, Patancheru, A.P., India.
109. **R. P. Thakur, C. W. Magill, S. Sivaramakrishnan, C. T. Hash, H. S. Shetty, and D. E. Hess.** 2002. Variability in *Sclerospora graminicola*, the pearl millet downy mildew pathogen, p. 51-56. *In* J. F. Leslie (ed.), *Sorghum and Millets Diseases*. Iowa State Press, Ames, Iowa.
110. **Thakur, R. P., K. Mathur, V. P. Rao, S. Chandra, S. Sivaramakrishnan, S. Kannan, R. V. Hiremath, H. C.**

Tailor, U. S. Kushwaha, R. R. Dwivedi, and S. Indira. 1998. Pathogenic and genetic characterization of six Indian populations of *Colletotrichum sublineolum*, the causal agent of sorghum anthracnose. *Ind. Phytopathol.* **51:** 338-348.

111. **Thakur, R. P., and K. N. Rai.** 2002. Pearl millet ergot research: Advances and implications, p. 57-63. *In* J. F. Leslie (ed.), *Sorghum and Millets Diseases*. Iowa State Press, Ames, Iowa.
112. **Thakur, R. P., and V. P. Rao.** 1999. *International Sorghum Anthracnose Virulence Nursery (ISAVN), A Summary Report (1992-1998).* Genetic Resources and Enhancement Program, ICRISAT, Patancheru (A.P.), India.
113. **Thakur, R. P., V. P. Rao, and S. B. King.** 1991. Influence of flowering event factors in cytoplasmic male sterile lines and F_1 hybrids on infection by *Claviceps fusiformis* in pearl millet. *Plant Dis.* **75:** 1217-1222.
114. **Thakur, R. P., V. P. Rao, and S. B. King.** 1992. Factors associated with ergot resistance in pearl millet. *Ind. Phytopathol.* **45:** 326-330.
115. **Thakur, R. P., and K. G. Shetty.** 1993. Variation in pathogenicity among single-oospore isolates of *Sclerospora graminicola*, the causal organism of downy mildew in pearl millet. *Plant Pathol.* **42:** 715-721.
116. **Tyagi, D. V. S., and R. S. Rawat.** 1992. Millets breeding, p. 137-144. In *25 Years of Plant Breeding Research at Pantanagar, 1967-1992.* Dept. of Plant Breeding, G. B. Plant Univ. of Agric., Pantnagar, Uttar Pradesh, India.
117. **Wilson, J. P.** 2002. Diseases of pearl millet in the Americas, p. 465-469. *In* J. F. Leslie (ed.), *Sorghum and Millets Diseases 2000.* Iowa State Press, Ames, Iowa.
118. **Wilson, J. P., and R. N. Gates.** 2002. The dynamic multiline population: An alternative approach to durable resistance, p. 65-69. *In* J. F. Leslie (ed.), *Sorghum and Millets Diseases*. Iowa State Press, Ames, Iowa.
119. **Xu, X., H. Dong, and Z. Liu.** 1996. The identification of sorghum red stripe. *Liaoning Agric. Sci.* **5:** 47-48.
120. **Xu, X., and Q.-S. Lu.** 1994. Pathogenicity measurement to American strain differential hosts using Chinese sorghum head smut. *Liaoning Agric. Sci.* **1:** 8-10.
121. **Xu, X., Q.-S. Lu, and T.-C. Zhao.** 1994. On physiological specialization of head smut [*Sporisorium reilianum* (Kuhn) Langdon & Fullerton] in sorghum. *Acta Phytopathol. Sin.* **24:** 58-61.
122. **Yadav, M. S., and J. C. Duhan.** 1993. Prevalence and severity of smut disease in pearl millet growing belt of Haryana. *Crop Res.* **6:** 517-519.
123. **Yadav, O. P., V. K. Manga, and G. K. Gupta.** 1993. Influence of A_1 cytoplasmic substitution on the downy mildew incidence of pearl millet. *Theor. Appl. Genet.* **87:** 558-560.

67

Virus Diseases of Sorghum and Millet in the Americas and Australia

Stanley G. Jensen and Laura M. Giorda

Abstract

Sorghum may be infected experimentally or naturally by 13 different groups of viruses. Some of these viruses have many strains. With the diseases caused by viruses, the disease results from the direct interaction of the host and viral genomes. Vectors that transmit the virus are diverse, and their biology complicates the understanding and management of virus diseases. Proper diagnosis of the disease-causing agent is critical to its control. Once the pathogen is properly identified, its vector can be identified more easily and the time and location for disease management and control determined.

Introduction

The Americas and Australia contain very diverse environments ranging from wet tropics to arid temperate regions, in almost all of which sorghum is grown, and which contain a variety of other plants and insects. These plants and insects may serve as reservoirs and vectors for various viruses that can infect sorghum. A virus-infected plant may be severely or modestly affected depending on the host genotype, the virus genotype, the time of infection, and the environmental conditions. Viruses are not free living and require a vector to transport them and introduce them into the sorghum plant. The biology of the vector can greatly complicate the understanding and development of virus diseases, and must be understood before control and reduction of virus disease losses can be effected. Of the four factors that impact viral diseases–the host, the vector, the environment and, the virus–the most difficult to understand is the genetics of the virus. During the 1990s there has been explosive development of techniques that that improve our understanding of the intrinsic properties of viruses, *e.g.*, their nucleic acid sequence, their gene function, their movement within the plant, and their replication cycle. Among other things, these new technologies allow viruses with similar properties to be grouped and then divided into taxonomic groups (*e.g.*, 20, 25, 26, 28, 32, 33, 42, 43, 46, 49, 50). These groups are based on the intrinsic properties of the virus rather than on properties such as host range or symptom expression that also depend on the virus interaction with the host genome. The first step in working with a new or threatening virus is to identify it so that control measures can be designed and implemented.

Generally, members of a group of viruses will share the same type of vector or vectors, and it is often the biology of the vector that determines the epidemiology of the virus in question (2, 6, 10, 36, 38). Unfortunately, the biology of the vector often is the most poorly understood part of the epidemiological equation. Many of the insect vectors of sorghum viruses, for example, do not, by themselves, cause significant economic damage, so there is little reason for detailed study. We do not understand as well as we need to the conditions of climate, weather, or cropping that may increase or decrease the vector's population, and thereby impact the epidemiology of a virus disease from year to year.

The host is the factor that can be altered the most easily to reduce the severity of viral diseases, and there are host plant resistance genes for many viral diseases. With increased characterization of the sorghum genome, the ability to map and manipulate genes for virus resistance is steadily increasing.

Viruses Causing Diseases in Sorghum

Although numerous viruses can infect sorghum (Table 67-1), only a subset are found naturally attacking the crop. For diagnostic and control purposes, we consider only those viruses known to naturally infect sorghum. In general, old or recognized virus problems that can cause

TABLE 67-1. Viruses naturally and experimentally transmitted to sorghum.

Virus	Vector(s)	Location	References
Bromovirus			
Brome Mosaic Virus (BMV)	Nematode, beetle[a]	USA[b]	23, 44
Cucumovirus			
Cucumber Mosaic Virus (CMV)	Aphid[c]	USA[b]	44
Fijivirus			
Mal de Rio Cuarto Virus (MRCV)	Planthopper[a]	Argentina	16, 29
Fiji Disease Virus (FDV)	Planthopper[a]	Australia	23
Furovirus			
Sorghum Chlorotic Spot Virus (SgCSV)	Fungus	USA	6, 24
Indian Peanut Clump Virus (IPCV)	Fungus	India	30
Mastrevirus			
Maize Streak Virus (MSV)	Leafhopper[a]	USA[b]	8, 23, 44
Luteovirus			
Barley Yellow Dwarf Virus (BYDV)	Aphid[a]	USA[b]	23, 44
Machlovirus			
Maize Chlorotic Mottle Virus (MCMV)	Beetles[a]	Peru	23, 31, 44
Nucleorhabdovirus			
Sorghum Stunt Mosaic Virus (SSMV)	Planthopper[a]	USA	7, 23, 44
Maize Mosaic Virus (MMV)	Planthopper[a]	Central America, USA	23, 44
		India	35
Potexvirus			
Foxtail Mosaic Virus (FoMV)		USA	39
Potyvirus			
Johnsongrass Mosaic Virus (JGMV)	Aphid[c]	Australia	19, 32, 42, 43
		USA	33, 42, 49
Maize Dwarf Mosaic Virus (MDMV)	Aphid[c]	Argentina	14, 15
		Central America, Venezuela	4, 48
		USA	42, 43
Sugarcane Mosaic Virus (SCMV)	Aphid[c]	Argentina	14, 15
		Australia	19, 32, 42
		Central America	48
		Colombia, Venezuela	3, 4, 11
		USA	32, 42, 43, 50
Sorghum Mosaic Virus (SrMV)	Aphid[c]	USA	42, 43, 49, 50
Sugarcane Streak Mosaic Virus (SCSMV)	Mite(?)	India[b]	20, 21
Zea Mosaic Virus (ZeMV)	Aphid	Israel[b]	41
Sobemovirus			
Panicum Mosaic Virus (PMV)		USA[b]	23, 44
Tenuivirus			
Iranian Wheat Stripe Virus (IWhSV)	Leafhopper[a]	Iran[b]	22
Maize Stripe Virus (MStpV)	Planthopper[a]	Australia	5, 23, 37, 44
		India	36, 37
		Venezuela	2, 23
Maize Yellow Stripe Virus (MYSV)	Leafhopper[a]	USA[b]	1
Rice Stripe Virus (RSV)	Leafhopper[a]	Japan[b]	44
Tritimovirus			
Wheat Streak Mosaic Virus (WSMV)	Mite[a]	USA	38
Waikavirus			
Maize Chlorotic Dwarf Virus (MCDV)	Leafhopper[d]	USA	23, 44
Unclassified Viruses			
MDMV	Aphid[c]	Cuba, Honduras	15, 48
MDMV-V	Aphid[c]	Venezuela	11
Sorghum Red Stripe Virus (SRSV)		India	47
Sorghum Yellow Banding Virus (SYBV)		USA	18, 27
SCMV	Aphid[c]	Cuba, Honduras	15, 48
Sugarcane Chlorotic Streak Virus (SCChSV)	Leafhopper	USA[b]	44

Legend for Table 67-1.

[a]Persistently transmitted by vector.
[b]Experimental transmission only.
[c]Non-persistently transmitted by vector.
[d]Semi-persistently transmitted by vector.

severe loss in yield in North America have declined in the last decade. Some new viruses have been discovered and some old viruses have been found in new places. The challenge of virus diseases of sorghum will not go away and may increase.

The potyviruses, with their diversity and broad distribution, still constitute the major virus threat to sorghum production. There are several genetically different species of potyviruses that cause these problems (Table 67-2). Sugar cane mosaic virus (SCMV), maize dwarf mosaic virus (MDMV), johnsongrass mosaic virus (JGMV), and sorghum mosaic virus (SrMV) are all sorghum-infecting species of aphid-transmitted potyviruses (32). Wheat streak mosaic virus (WSMV), a potyvirus transmitted by Eriophyid mites and widespread in wheat and maize, is a continuing threat to sorghum in the Great Plains of the United States (40).

In addition, two new potyviruses have been described recently that can infect sorghum and that may become a threat to sorghum grain production. Sugarcane streak mosaic virus (SCSMV) has been recovered from sugarcane in Pakistan (20) and India (21). This virus can infect sorghum that carries the Krish gene (27) for immunity to the other potyviruses. Based on similarities in nucleotide sequence, this virus should be grouped with WSMV. The vector of SCSMV is not known, but the similarity to WSMV suggests that the vector may be an Eriophyid mite. The virus has not yet been found in North America, but sugarcane cuttings circulate globally and the virus probably will spread unless stopped by quarantine officials. The type and distribution of the vector is the critical question.

Another recently described potyvirus is Zea mosaic virus (ZeMV), first described from Israel (41). Although this virus was isolated from maize, it infected every cultivar of sorghum inoculated and produced necrosis on more sorghum genotypes than do most of the other potyviruses. This virus is more severe on sorghum than it is on maize.

Sorghum yellow banding virus was first observed in south-central Texas (18), where it appeared to be soil associated, since it reappeared in the same fields year after year and natural transmission from field soil was demonstrated. There is some evidence that this virus also may have an aerial vector. Since the first report (17), this virus also has been found in California (24), the Texas panhandle (S. G. Jensen, *unpublished*), and possibly in Venezuela (12). The distribution of this virus probably is wider than reported, but because the disease develops symptoms so slowly it is not perceived as a major threat.

The maize stripe tenuivirus (MSTV) is transmitted to maize and sorghum by the delphacid planthopper *Peregrinis maidis*, a semitropical insect (2, 8). The virus occurs in semitropical regions of North, Central and South America. Infected sorghum plants may show severe symptoms of stunting and striping of the leaves. The virus also has been reported from Puerto Rico, Venezuela, and the Gulf Coast of the United States (23, 36).

Maize chlorotic dwarf virus (MCDV) is another virus that is primarily a threat to maize but that also can infect sorghum (13). The overwintering reservoir for the virus is the widespread perennial weed johnsongrass (9, 10). In the spring, MCDV is transmitted by leafhoppers to maize or sorghum. MCDV has been spreading and may now be found widely throughout the southern United States. However, unlike MSTV, MCDV is more severe on maize, with the symptoms on sorghum typically mild and economically insignificant (23).

Foxtail mosaic virus is a member of the potexvirus group and has recently been found naturally infecting sorghum in central Kansas (39). The vector is probably an aphid, but has not yet been identified. Although this virus has been found at the same location for several years, its economic importance is not known.

Mal de Rio Cuarto virus is a phytoreovirus that infects several grassy hosts including sorghum. In Argentina, where it is endemic, it is primarily a disease of maize and only incidentally a disease of wheat and sorghum (14, 16). The virus seems to have a wide host range and may develop into a more significant threat to sorghum if there is a change in pathogenicity or in the biology of the planthopper vector, *Delphacodes kuscheli*. Fortunately there already appears to be good resistance in some sorghum lines (16).

Sorghum yellow stunt is the only reported disease of sorghum caused by a phytoplasma. The symptoms resemble those caused by virus diseases, and the two could be confused easily. The disease has been reported from Gulf Coast of the United States, but has not been extensively studied (51).

Diagnosis of Viruses Causing Diseases of Sorghum

We begin again with the potyviruses, which are the largest taxonomic group of viruses infecting sorghum in North America (32, 33, 40). The distribution of the four species of potyviruses that pose the most serious threat to sorghum production in the Americas has been but poorly surveyed in most areas, probably because the most discriminating diagnostic tests are laborious, expensive, and time consuming. Listing of a potyvirus species as originating from an area is typically based on only a few samples. All potyviruses are mechanically transmitted rather easily to sorghum. In an electron micrograph, these viruses appear as long flexible rods. The purified coat protein is usually 32-39 kDa (28).

To assign a potyvirus to species, host range and symptom expression usually provide useful information (45). However, the use of sorghum genotypes and symptom responses to virus challenge may not be sufficient, because the sorghum genotypes may not be pure and the virus genotypes are variable. This combination may result in inconsistent reactions with the same virus species. However, the only member of the potyvirus group that will infect wheat is WSMV, the only potyvirus that will infect oats is JGMV, and the only virus that will infect sorghum carrying the Krish gene for potyvirus immunity is SCSMV (20). JGMV and MDMV both will infect johnsongrass, but not sugarcane. None of the SRMV or SCMV strains will infect johnsongrass. SRMV and SCMV, except for the SCMV-MD-B strains, infect sugarcane (23). Such cross-infection information usually suffices to tentatively identify a virus to species.

Serological tests can be used to differentiate virus species, but since virus species can share epitopes, cross reactions may occur, and tests must be conducted carefully and include the proper controls for the identifications to be reliable (42). The enzyme linked immunosorbent assay (ELISA) test is an amplification system that relies on an enzyme attached to an antibody to convert a colorless substrate into a colored product. When the virus binds to the antibody-enzyme complex, it triggers a reaction that results in a rapid change in color.

Western blots also can be performed if the virus is concentrated from plant sap, usually by centrifugation, and then enzymatically digested to release the coat protein. The protein mixture is resolved by electrophoresis in a polyacrylamide gel. The protein is then transferred to a solid surface, typically nitrocellulose paper, and the paper is soaked in an enzyme-linked antiserum to allow the antiserum to bind to the virus coat protein. In this test the enzyme converts the substrate molecules to a form that binds to the paper and creates a dark spot. This test gives two complementary results. A dark spot confirms that the antibody reacted with a protein, and the size of the protein, determined by its migration through the gel, should be the same as that of the diagnosed virus. Healthy plant sap, or plants infected with an unrelated virus, will not give a positive spot. False positives, which may be confusing in ELISA tests, are easily recognized on a western blot.

The most effective diagnostic tools currently available are the molecular biological tools that allow the comparison of the viral genome base sequence with that of known viral sequences (20, 41, 50). Viruses with sequence similarity ≥ 85% usually are in the same species, while viruses with ≤ 75% sequence similarity are usually in different species. By using sequences available through GenBank and other on-line databases, it is possible to make identifications based on comparisons with sequences on file rather than having to sequence both the known and unknown virus. In the case of viruses with RNA instead of DNA as the nucleic acid, the viral RNA is copied to DNA with reverse transcriptase, and the resulting DNA copy is amplified by the polymerase chain reaction (PCR) before sequencing (20, 41, 50).

If sequence data are not available, then sequence-specific probes can be used in hybridization experiments

TABLE 67-2. Species of potyviruses and strains of each infecting sorghum.

Species	Property	Strain (Location)
Johnsongrass Mosaic Virus (JGMV)	Will not infect sugarcane	MDMV-Ks1 (USA) MDMV-O (USA) SCMV-Jg (Australia)
Maize Dwarf Mosaic Virus (MDMV)	Will not infect sugarcane	MDMV-A (USA) MDMV-D (USA) MDMV-E (USA) MDMV-F (USA) MMV-Et (Yugoslavia) SCMV-Jg (USA)
Sugarcane Mosaic Virus (SCMV)	Will not infect johnsongrass	SCMV-A (USA) SCMV-B (USA) SCMV-D (USA) SCMV-BC (Australia) SCMV-E (USA) SCMV-Isis (India) SCMV-Sabi (Australia) SCMV-SC (Australia)
SCMV-MD-B	Will not infect sugarcane or johnsongrass	MDMV-B (USA)
Sugarcane Streak Mosaic Virus (SCSMV)	Will infect Krish sorghum	SCSMV (India)
Sorghum Mosaic Virus (SrMV)	Will not infect johnsongrass	SCMV-H (USA) SCMV-I (USA) SCMV-M (USA)
Wheat Streak Mosaic Virus (WSMV)		WSMV-sr (USA)
Zea Mosaic Virus (ZeMV)		ZeMV (Israel)

with the nucleic acid of the unidentified virus. If hybridization occurs, indicating that the nucleic acid sequences of the probe and the unknown virus are complementary, then the two viruses are identical or very closely related. A test that is becoming more common requires two short DNA fragments, termed primers, that can be used to initiate replication of a virus-specific nucleic acid via PCR (50). If a product of the correct size results from the PCR, then the virus usually can be quickly identified. Unfortunately, if the virus is a new virus there will be no published sequences and neither probes nor primers will be available.

Most plant viruses have RNA genomes, and at some stage in the replicative cycle a complementary copy, the minus strand, must be synthesized. Thus, at some time in the cycle a double-stranded RNA (dsRNA) must be synthesized. Isolation and analysis of this dsRNA (34) also can be used to identify the virus and to determine its capabilities.

Another very effective diagnostic technique is minipurification (28). In this case the virus is purified by a protocol that is similar to western blotting, but without the antiserum. The proteins are separated by polyacrylamide gel electrophoresis, stained, and compared to those isolated from a healthy sorghum plant. The presence of a large viral coat protein in the infected plant, but not in the healthy plant, is evidence for a virus as the causal agent of the disease, and is indicative of the family to which the virus belongs.

Distribution and Epidemiology of Sorghum Viruses

The distribution of sorghum viruses is continually being reevaluated as new problems are encountered and new diagnostic methods applied. For example, consider the distribution of the four species of potyviruses that have been well documented from sorghum in the United States (Table 67-2). SCMV, in various forms, is well-established in the sugarcane growing areas along the Gulf Coast. Some of these strains recently have been split from SCMV and designated as a new species, SrMV (32, 45). SrMV occurs most commonly in sorghum fields along the Gulf Coast of Louisiana and Texas when the sorghum is planted near sugarcane, the apparent source of the virus.

MDMV was identified in the eastern corn belt of the United States in 1964 and is now established in the southern half of the country, where it also infects johnsongrass, but not sugarcane (23). Shortly after the discovery of MDMV, a second virus was found that was similar to MDMV, except that it could infect neither johnsongrass nor sugarcane (23). This virus was originally named MDMV-B, although we now know that MDMV-B is a strain of SCMV that cannot infect sugarcane and is more correctly designated as SCMV-MD-B (32). This example illustrates the hazards of using symptomology or host range alone for virus identification.

MDMV strains are common wherever johnsongrass occurs, *i.e.*, south of a latitude (38-39°N) equivalent to central Kansas (9, 10). Johnsongrass is assumed to be an alternate perennial host for MDMV, and other winter annuals and perennial grasses, *e.g.*, cheat grass (*Bromus secalinus)*, downy brome (*Bromus tectorum*), and eastern gamagrass (*Tripsacum dactyloides*), that survive further north can be alternate hosts for SCMV-MD-B, which does not infect johnsongrass. However, these other grasses are more susceptible to MDMV than to SCMV-MD-B, which leaves the hypothesis for the predominance of SCMV-MD-B strains in temperate latitudes and elevations without clear support.

JGMV was first identified in the United States in southern Texas and named MDMV, strain "O" (32, 33). This virus has been established in central Kansas for many years and is now recognized as a separate species. This virus also is thought to overwinter on johnsongrass. Although JGMV is being diagnosed with greater frequency in the Texas high plains, little is known about the epidemiology of this virus or its potential economical impact.

Johnsongrass also can host MCDV as well as two species of potyviruses. Aphids transmit potyviruses and MCDV is transmitted by leafhoppers, but it is not uncommon for maize or sorghum growing near Johnsongrass to be doubly infected (9, 10). As MCDV is not typically severe on sorghum, the double infection often goes unnoticed and only the potyvirus is reported.

For some sorghum viruses, the key to the distribution may be the availability of a vector. For example, maize stripe virus is limited by the semitropical distribution of its vector, the planthopper *Peregrinis maidis.* Wheat steak mosaic virus (WSMV) also can infect sorghum (40). This virus is transmitted by the Eriophyid mite *Aceria tosichella*, also known as the wheat curl mite, which is common in the winter wheat production areas of the high plains of the United States. Although the virus may infect sorghum in other areas, sorghum and wheat production overlap on the high plains, and this proximity combined with an ideal climate for mite reproduction make it likely that this virus will have its greatest impact on sorghum production in this area.

Management of Sorghum Virus Diseases

Control of sorghum viruses may take many forms, of which host plant resistance is the most commonly employed. Genes for resistance to potyviruses are limited, but the opportunities are abundant. For example, there are at least eight different types of reactions that can occur when a potyvirus infects sorghum, including at least four different types of necrotic reaction: (i) temperature-sensitive red leaf, (ii) temperature-insensitive red stripe, (iii) local lesion, and (iv) symptomless plants. There also are genes affecting the pattern and severity of mosaic symptoms, which are forms of partial resistance, and the Krish gene for immunity to most potyviruses. There also are genes that affect virus titer in the tissue, and others that can reduce the incidence of viral infection by reducing the ease with which a vector can transmit the virus to the sorghum plant. Mapping these genes will greatly facilitate their use in breeding programs to control these diseases. Genes for resistance to other viruses in sorghum are not as well described, but, based on laboratory and/or field reactions, good resistance or tolerance to many of them is probably available if these viruses become economically important.

The primary alternative to host plant resistance is the removal of the alternate or overwintering virus source. In the case of MDMV, JGMV, and MCDV, such control means the destruction of johnsongrass in road sides, ditch banks, and waterways where it grows as a perennial, but this cultural control is very limited since there are regions where johnsongrass eradication is all but impossible. For WSMV, the typical cultural control in wheat is to remove any green grassy vegetation that serves as an alternate host for the virus and allows the mite to survive the hot, dry summer. In the case of sorghum, the best cultural practice is to avoid planting sorghum downwind and close to wheat before it matures. The first response to a virus disease problem is to alter cultural practices to reduce the amount of infection. Although cultural practices may not be as effective as host plant resistance, the development and deployment of genetic control may require years of effort.

Research Needs for Management of Virus Diseases of Sorghum

The greatest need for managing virus problems in sorghum is to identify the virus that is causing the problem. Current diagnostic techniques are accurate, but they are sometimes expensive, time consuming, and demanding of training and special equipment. Rapid, simple, and cheap diagnostic tests are needed to conduct the widespread surveys that are needed, as commercial kits, at present, are too expensive for this purpose. Well-conducted surveys are required to estimate the amount of virus disease present and to distinguish virus-infected plants from those that are growing poorly due to abiotic conditions. More efforts to develop diagnostics for sorghum viruses are needed to ensure that survey results are correctly interpreted and that breeding programs are focused on individual viruses of economic import.

Genetics of resistance to several important sorghum virus diseases is lacking or incomplete because these viruses are difficult to work with under field conditions. Natural infection by different viruses also can result in similar symptoms, but the basis for resistance to these viruses can be quite different. Better screening trials and improved techniques to identify and map genes for resistance or tolerance to economically important virus diseases would facilitate their utilization by commercial seed companies. At present, seed companies are incorporating genes for virus resistance into their material through conventional breeding strategies. At present, seed companies are unlikely to incorporate genes for virus resistance into their material unless there is overwhelming demand or the genes are easy to incorporate and deploy.

References

1. **Ammar, E. D., R. E. Gingery, D. T. Gordon, and A. E. About.** 1990. Tubular helical structures and fine filaments associated with the leaf hopper-borne maize yellow stripe virus. *Phytopathology* **80:** 303-309.
2. **Ammar, E. D., R. E. Gingery, and L. V. Madden.** 1995. Transmission efficiency of three isolates of maize stripe tenuivirus in relation to virus titre in the planthopper vector. *Plant Pathol.* **44:** 239-243.
3. **Barragan-Quijano, E.** 1991. Logros y avances en la investtigación del sorgo en Colombia, p. 123-128. *In* F. Gaitán (ed.), *Sorgo Para el Futuro. Seminario Internacional Sobre el Cultivo del Sorgo.* INTSORMIL, ICA, FENALCE, GRANJA EL ALCARAVAN, Cali, Colombia.
4. **Blanco, R. A., and M. J. Garrido. 1996.** Identificación de algunos virus que infectan sorgo forrajero en Maracay. *Rev. Fac. Agron. (Luz.)* **13:** 273-284.
5. **Bradfute, O. E., and J. H. Tsai.** 1990. Rapid identification of maize stripe virus. *Phytopathology* **80:** 715-719.
6. **Cooper, J. I., and M. J. C. Asher, eds.** 1987. *Viruses with Fungal Vectors.* Lavenham Press, Ltd., Sudbury, Suffolk, UK.
7. **Creamer, R.** 1992. Purification and protein characterization of sorghum stunt mosaic rhabdovirus. *Phytopathology* **82:** 1473-1476.
8. **Damsteegt, V. D.** 1983. Maize streak virus: I. host range and vulnerability of maize germ plasm. *Plant Dis.* **67:** 734-737.

9. **Eberwine, J. W., and E. S. Hagood.** 1995. Effect of johnsongrass (*Sorghum halepense*) control on the severity of virus diseases of corn (*Zea mays*). *Weed Technol.* **9:** 73-79.
10. **Eberwine, J. W., E. S. Hagood, and S. A. Tolin.** 1998. Quantification of viral disease incidence in corn (*Zea mays*) as affected by johnsongrass (*Sorghum halepense*) control. *Weed Technol.* **12:** 121-127.
11. **Garrido, M. J., A. Ordosgoitti, and G. E. Trujillo.** 1993. Cultivares de sorgo diferenciales para el virus del mosaico enanizante del maíz-raza venezolana y el virus del mosaico de la caña deazucar. *Rev. Fac. Agron. (Maracay)* **19:** 65-74.
12. **Garrido, M. J. and G. E. Trujillo.** 2001. Detection of sorghum yellow banding virus infecting sorghum in Venezuela. *Plant Dis.* **85:** 230.
13. **Gingery, R. E., and L. R. Nault.** 1990. Severe maize chlorotic dwarf disease caused by double infection of mild virus strains. *Phytopathology* **80:** 687-691.
14. **Giorda, L. M.** 1997. Enfermedades, p. 40-46. *In* L. M. Giorda (ed.), *Sorgo Granífero*. Centro Regional Cordoba, EEA, Manfredi, Cordoba, Argentina.
15. **Giorda, L. M., and M. J. Martinez.** 1995. Opportunities for biotechnology research on sorghum stalk rot and virus diseases. *Afric. Crop J.* **3:** 191-201.
16. **Giorda, L. M., and J. O. Muñoz.** 1993. A new Reovirus disease naturally infecting sorghum. *Sorghum Newsl.* **34:** 24.
17. **Giorda, L. M., R. W. Toler, and J. D. Alexander.** 1988. Host range studies of sorghum yellow banding, a newly recognized disease of grain sorghum and sorghum × sudangrass hybrids in Texas. *Phytopathology* **78:** 627A.
18. **Giorda, L. M., R. W. Toler, and G. N. Odvody.** 1992. Sorghum yellow banding virus (SYBV), a new polyhedral virus on *Sorghum spp.*, p. 292-293. *In* W. A. J. de Milliano, R. A. Frederiksen, and G. D. Bengston (eds.), *Sorghum and Millet Diseases: A Second World Review*. ICRISAT, Patancheru (A.P.), India.
19. **Gough, K. H., and D. D. Shukla.** 1981. Coat protein of potyvirus I. Comparison of four Australian strains of sugarcane mosaic virus. *Virology* **111:** 455-462.
20. **Hall, J. S., B. J. Adams, T. J. Parsons, R. C. French, L. C. Lane, and S. G. Jensen.** 1997. Molecular cloning, sequencing, and phylogenetic relationships of a new potyvirus, sugar cane streak mosaic virus (SCSMV) and a re-evaluation of the classification of the *Potyviridae*. *Molec. Phylogenet. Evol.* **10:** 323-332.
21. **Hema, M., M. Venkatramana, H. S. Savithri, and P. Sreenivasulu.** 1999. Biological, antigenic and genomic relationships among the virus isolates causing mosaic disease of sugarcane in south India. *Curr. Sci.* **77:** 698-702.
22. **Heydarnejad, J., and K. Izadpanah**. 1992. Isolation and partial characterization of a tenuivirus from wheat in Iran. *J. Phytopathol.* **136:** 279-287.
23. **Jensen, S. G.** 1999. Diseases caused by viruses and virus-like organisms, p. 43-50. *In* R. A. Frederiksen and G. N. Odvody (eds.), *Compendium of Sorghum Diseases*, 2nd ed. APS Press, St. Paul, Minnesota.
24. **Kendall, T. L., W. G. Langenberg, and S. A. Lommel.** 1988. Molecular characterization of sorghum chlorotic spot virus, a proposed furovirus. *J. Gen. Virol.* **69:** 1335-1345.
25. **Kitajima, E. W., A. C. DeAvila, and R. O. Resende.** 1997. Taxonomía de virus de plantas. *Fitopatol. Brasil.* **22:** 5-24.
26. **Klassen, V. A., and B. W. Falk.** 1989. Characterization of a California isolate of sorghum yellow banding virus. *Phytopathology* **79:** 646-650.
27. **Krishnaswamy, N, V. S. Raman, and P. Chandrasekharan.** 1956. An interspecific hybrid of grain sorghum and johnsongrass *S. halepense* (2n = 20) × *S. roxburghii* (2n = 20). *Curr. Sci.* **25:** 195-197.
28. **Lane, L. C.** 1986. Propagation and purification of RNA plant viruses. *Meth. Enzymol.* **118:** 687-696.
29. **Lenardón, S. L., G. J. March, and J. A. Ornaghi.** 1999. Virus del mal de Rio Cuarto en Maíz, p. Maiz 2-1-10. *In* S. L. Lenardon (ed.), *Proyecto de Investigaciones en Fitovirología. Enfermedades Causadas por Virus y Fitoplasmas en Cultivos Extensivos-Intensivos*. IFFIVE, INTA-JICA, Córdoba, Argentina.
30. **Littlefield, L. J., P. Delfosse, J. H. Whallon, Z. M. Hassan, J. L. Sherwood, and D. V. R. Reddy.** 1997. Anatomy of sporosori of *Polymyxa graminis*, the vector of Indian peanut clump virus in roots of *Sorghum bicolor*. *Can. J. Plant Pathol.* **19:** 281-288.
31. **Lommel, S. A., T. L. Kendall, N. F. Siu, and R. C. Nutter.** 1991. Characterization of maize chlorotic mottle virus. *Phytopathology* **81:** 819-823.
32. **McKern, N. M., D. D. Shukla, R. W. Toler, S. G. Jensen, M. Tosic, R. E. Ford, O. Leon, and C. W. Ward.** 1991. Confirmation that the sugarcane mosaic virus subgroup consists of four distinct potyvirus by using peptide profile of coat protein. *Phytopathology* **81:** 1025-1029.
33. **McKern, N. M., P. M. Whitaker, R. E. Ford, S. G. Jensen, and D. D. Shukla.** 1990. Coat protein properties indicate that MDMV-O and Ks1 are strains of johnsongrass mosaic virus. *Phytopathology* **80:** 907-912.
34. **Morris, T. J., and J. A. Dodds.** 1979. Isolation and analysis of double stranded RNA and fungal tissue. *Phytopathology* **69:** 854-858.
35. **Naidu, R. A., R. Harikrishnan, S. K. Manohar, D. V. R. Reddy, A. S. Ratna, S. B. King, and R. Bandyopadhyay.** 1989. The occurrence of maize mosaic virus on sorghum in India. *Ann. Appl. Biol.* **114:** 301-310.
36. **Narayana, Y. D., and V. Muniyappa.** 1996. Virus-vector relationships of a planthopper (*Peregrinus maidis*)-borne sorghum stripe tenuivirus. *Intl. J. Pest Manag.* **42:** 165-170.
37. **Peterschmitt, M., A. S. Ratna, W. R. Sachs, D. V. R. Reddy, and L. K. Mughogho.** 1991. Occurrence of an isolate of maize stripe virus on sorghum in India. *Ann. Appl. Biol.* **118:** 57-70.
38. **Rodriguez, C. M., L. V. Madden, L. R. Nault, and R. Louie.** 1993. Spread of maize chlorotic dwarf virus from infected corn and johnsongrass by *Graminella nigrifrons*. *Plant Dis.* **77:** 55-60.
39. **Seifers, D. L., T. L. Harvey, S. Haber, Y.-M. She, I. Chernvshevich, W. Ens, and K. G. Standing.** 1999. Natural infection of sorghum by foxtail mosaic virus in Kansas. *Plant Dis.* **83:** 905-912.
40. **Seifers, D. L., T. L. Harvey, K. D. Kofoid, and W. D. Stegtmeier.** 1996. Natural infection of pearl millet and sorghum by wheat streak mosaic virus in Kansas. *Plant Dis.* **80:** 179-185.

41. **Seifers, D. L., R. Salmon, V. Marie-Jean, B. Alliot, P. Signoret, S. Haber, A. Loboda, W. Ens, Y.-M. She, and K. G. Standing.** 2000. Characterization of a novel potyvirus isolated from maize in Israel. *Phytopathology* **90:** 505-513.
42. **Shukla, D. D., M. Tosic, J. Jilka, R. E. Ford, R. W. Toler, and M. A. C. Langham.** 1989. Taxonomy of potyviruses infecting maize, sorghum and sugarcane in Australia and the United States as determined by reactivities of polyclonal antibodies directed towards virus-specific N-termini of coat proteins. *Phytopathology* **79:** 223-229.
43. **Shukla, D. D., and C. W. Ward.** 1989. Structure of potyvirus coat protein and its application in the taxonomy of the potyvirus group. *Adv. Virus Res.* **36:** 273-314.
44. **Toler, R. W.** 1986. Diseases caused by viruses and virus-like organisms, p. 42-50. *In* R. A. Frederiksen (ed.), *Compendium of Sorghum Diseases*, 1st ed. APS Press, St. Paul, Minnesota.
45. **Tosic, M., R. E. Ford, D. D. Shukla, and J. Jilka.** 1990. Differentiation of sugarcane, maize dwarf, johnsongrass, and sorghum mosaic viruses on reactions of oats and some sorghum cultivars. *Plant Dis.* **74:** 549-552.
46. **van Regenmortel, M. H. V., C. M. Fauquet, D. H. L. Bishop, E. B. Carstens, M. K. Estes, S. M. Lemmon, J. Maniloff, M. A. Mayo, D. J. McGeoch, C. R. Pringle, and R. B. Wickner, eds.** 2000. *Virus Taxonomy. The Seventh Report of the International Committee on Taxonomy of Viruses.* Academic Press, San Diego, California.
47. **Vyavabhare,V. M., T. B. Garud, and V. R. Mali.** 1989. Effect of sorghum red stripe virus on sorghum genotype. *Ind. Phytopathol.* **42:** 16-20.
48. **Wall, G. C., and D. H. Meckenstock.** 1992. Sorghum diseases in Central America and the Caribbean basin, p. 67-73. *In* W. A. J. de Milliano, R. A. Frederiksen, and G. D. Bengston (eds.), *Sorghum and Millet Diseases: A Second World Review.* ICRISAT, Patancheru (A.P.), India.
49. **Ward, C. W., and D. D. Shukla.** 1991. Taxonomy of potyviruses: Current problems and some solutions. *Intervirology* **32:** 269-296.
50. **Yang, Z. N., and T. E. Mirkov.** 1997. Sequence and relationships of sugarcane mosaic and sorghum mosaic virus strains and development of RT-PCR-based RFLPs for strain discrimination. *Phytopathology* **87:** 932-939.
51. **Zummo, N., O. E. Bradfute, D. Robinson, and K. C. Freeman.** 1975. Yellow sorghum stunt: A disease symptom of sweet sorghum associated with a mycoplasma-like body in the U.S. *Plant Dis. Rept.* **59:** 114-116.

68
The Status of Sorghum Diseases in Russia

Aliya S. Kasakova and Alexander A. Oleynick

Abstract

Sorghum is one of the most important fodder crops in the arid regions of southern Russia. Diseases are both common and economically important production constraints. The most frequent of these diseases (in order of importance) include smuts, foliar diseases, seed rot and seedling diseases, stalk rot, grain mold, head blight, and rust. During the last decade, some advances have been made in breeding for resistance to these diseases and in developing effective disease control strategies.

Introduction

Sorghum is one of the most important crops for arid regions of southern Russia. Sorghum commonly is used as animal feed, and, more recently, white grain sorghum has been used for human food. Sorghum is grown in dry lands, both steppe and semidesert. Precipitation during the growing season ranges from 250-350 mm, with 3200-3600 degree-days. Russia currently plants 1.5-2.0 million hectares of sorghum, with the area planted to sorghum gradually increasing. About 50% of the area planted to sorghum is planted to sweet sorghums and is used for silage.

Sorghum diseases are economically important and can result in significant yield losses, with the most important being smuts, foliar diseases, seed rot and seedling diseases, stalk rot, grain mold, head blight, and rust. Smuts, foliar diseases, and seed rot and seedling diseases are widespread. The economic significance of individual diseases varies by year and is heavily dependent upon weather. Smuts can cause losses of 50-90%, foliar diseases losses of up to 50%, and seed rot and seedling diseases losses of 20-70%. Losses attributed to other diseases either have not been rigorously quantified, or the diseases are relatively rare.

Smuts

Sorghum smuts have been a problem in Russia for a long time. Of the four sorghum smuts, two are common–head smut (caused by *Sporisorium relianum*) and covered kernel smut (caused by *Sporisorium sorgi*). The smuts occur commonly in all sorghum-growing regions of Russia and the Ukraine (2, 3, 6, 9, 10). Both smuts are economically important, causing sorghum grain yield losses of 17-86% and green mass losses of up to 19% in the Volga region of Russia. These diseases remain important because farmers do not use available chemical disease control. Covered kernel smut, which is seedborne, is the most common. Seed treatment with Premis, Vitavacs, or Vitatiuram before planting is needed to control this disease. Head smut, which is soilborne, requires both seed treatment with the preparations that are used to control covered kernel smut and adequate field sanitation. Proper seed treatment and crop rotation by farmers probably would reduce the losses attributed to these diseases.

One means to control smuts is through the development of resistant host varieties and hybrids. In ARRISC, smut research is a high priority, and germplasm with high levels of resistance has been identified (Table 68-1). Smut-resistant lines are available for grain, sweet, and forage sorghums. The most resistant lines are for grain sorghum and sudangrass. All broom corn sorghum varieties and hybrids are susceptible to smut. Some lines of grain and sweet sorghum are resistant to both head smut and covered kernel smut.

Significant research work is needed on smuts in Russia. Little information is available on losses due to the various smuts, on the variation within the pathogen populations, or on the effectiveness of chemical seed treatment for seeds of local varieties and hybrids. More information on the distribution and effectiveness of smut-resistance genes in the host germplasm also is needed.

TABLE 68-1. Smut resistance of the ARRISC sorghum collection (based on results of 250 lines evaluated for at least three years).

Resistance	% Lines Examined	
Level	Covered Kernel Smut	Head Smut
High	29	25
Moderate	60	63
Susceptible	12	12

Foliar Diseases

In Russia, sorghum foliar diseases occur in all sorghum-growing areas (1, 4, 8, 10-12). Causal organisms include both bacteria and fungi. The most frequently occurring bacterial leaf diseases are leaf spot (caused by *Pseudomonas syringae*), leaf streak (caused by *Xanthomonas holcicola*), and leaf stripe (caused by *Pseudomonas andropogonis*) (7). Different sorghum species in Russia vary in their resistance to bacterial leaf diseases (Table 68-2), with *S. bicoloria*, *S. durra*, and *S. nervosa* being the most resistant. All sweet sorghum lines are susceptible to bacterial leaf diseases.

Leaf diseases caused by fungi include leaf blight (caused by *Exserohilum turcicum*), leaf anthracnose (caused by *Colletotrichum graminicola*), and gray leaf spot (caused by *Cercospora sorghi*), which are common in many other sorghum-producing regions of the world. Disease incidence varies by region across the country and depends on the weather (number of wet humid days) during vegetative growth, the sorghum variety planted, and the farming system. In some years, foliar diseases cause extensive damage on susceptible cultivars. Economic losses occur due to decreases in grain yield of 20-50% and quality degradation of the green mass. Screening for foliar disease resistance is a routine part of the breeding program, and ARRISC has a special nursery for the evaluation of elite materials following artificial infection.

Seed Rot and Seedling Diseases

Russia is at the northern edge of the world's sorghum growing regions, and problems with plant stand establishment are common when sorghum is planted in cold, wet soil. In some cases, cool or cold weather returns after planting. High soil moisture permits the attack of seeds and seedlings by many soilborne and seedborne fungi, *e.g.*, *Aspergillus* spp., *Fusarium* spp., *Phoma* spp., and *Pythium* spp. Seed rot and seedling diseases can be major factors in stand establishment. Damaged seed germinate poorly, if at all. Infection during or shortly after germination also may cause pre-emergence damping-off, killing seedlings before they emerge. In some years, seed rot and seedling diseases are so severe that large areas must be replanted.

TABLE 68-2. Average resistance of more than 1800 lines from eight sorghum species to bacterial leaf diseases (after 12).

Sorghum Type	Hot/Wet Season[a]	Cool/Wet Season[b]
Guinea	0	44
Kafir	5.4	21
Bicolor	6.1	62
Durra	8.1	37
Nervosa	9.0	19
Sweet sorghum	0	0
Grass sorghum	17	13
Broom sorghum	8.5	13

[a]Precipitation during the growing season (May-September) was 335 mm, and the daily mean temperature was 21.8°C.
[b]Precipitation during the growing season (May-September) was 362 mm, and daily mean temperature was 18.5°C.

TABLE 68-3. Disease damage (%) to potential sources of sorghum resistance for smut and bacterial leaf diseases (average of 5 years).

Sorghum Type	Source	Covered Kernel Smut	Head Smut	Bacterial Leaf Diseases
Grain	Pishevoe	0	0	0
	Sorgogradskoe	0	10	2
	Zernogradskoe 54	6.7	5.1	2
	ZSK-150	0	5.3	6
	ZSK-155	0	0	4
	ZSK-164	0	0	3
	ZSK-212	0	0	2
	ZSK-221	0	0	4
Sweet	A-83 x D2276	0	6.4	2
	3212	7.5	5.5	2
	3215	6.8	7.5	2
	3262	7.1	6.8	3

Sorghum Disease Resistance Breeding Program

Breeding for host resistance to diseases is an integral part of all sorghum breeding programs in ARRISC, with plant pathologists and breeders working together in teams. All breeding material, collections, and new commercial varieties and hybrids are evaluated for disease resistance (5). Smut and bacterial leaf resistance usually is evaluated after infection in a special nursery. Some breeding lines of grain sorghum have no apparent smut damage after infection, and some sweet sorghums have smut damage only of 5.0-7.5% (Table 68-3).

Conclusions

At present, most newly released sorghum varieties and hybrids are moderately to highly resistant to the existing major diseases. A number of sources of resistance to these diseases have yet to be utilized commercially.

References

1. **Chumaevskaya, M. Y.** 1972. Leaf spot of sorghum and sudan grass. *Zatsita Rastenii* **8:** 29-32.

2. **Kulpinova, M. N.** 1995. Head smut of sorghum. *Zatsita rastenii* **8:** 25-27.
3. **Oleynick, A. A.** 1990. Sorghum samples resistant to head smut. *Selektsia i Semenovodstvo* **5:** 36-37.
4. **Oleynick, A. A.** 1991. Diseases resistance of sorghum plants. *Zatsita rastenii* **10:** 10-11.
5. **Oleynick, A. A., N. Y. Kolomiets, P. A. Mangush, and Y. N. Klepko.** 2001. Sorghum breeding for diseases resistance. *Zatsita i Karantin Rasteniy* **9:** 17-18.
6. **Oleynick, A. A., V. V. Metlin.** 1995. Genetic sources of sweet sorghum resistance for diseases, p. 118-121. In *Proc. DZNIISH (Don Region Research Institute of Agriculture) "Crop Breeding, Seed Production and Technology of Growing."* VNIPTIMESH, Zernograd, Russia.
7. **Pastushenko, L. T.** 1962. Bacterial leaf stripe of sorghum and sudan grass in Ukraine. *J. Ukraine Inst. Microbiol.* **24:** 34-45.
8. **Silaev, A. I.** 1980. Sorghum smut resistance in the Volga region, p. 85-92. In *Bulletene VNII zatsiti rastenii.* Agropromizdat, Leningrad, Russia.
9. **Shepel, N. A.** 1989. *Sorghum–Intensive Crop.* Tavriya, Simpheropol, Ukraine.
10. **Vlasova, V. I., N. P. Nikolaeva.** 1972. Sorghum bacterial diseases in the Stavropol region. *Zatsita Rastenii ot Vrediteley i Bolezney* **35:** 21-27.
11. **Voronkevich, I. V., E. S. Phohrutdinova.** 1961. Bacterial leaf spot of sorghum, sudan grass and millet. Nauchnie dokladi vishey shkoli. *Biologicheskie nauki* **2:** 18-22.
12. **Yakushevskii, Y. S., L. K. Ivanukovich, N. P. Suhotskaya.** 1974. Evaluating sorghum species and varieties for bacterial diseases. *Bull. Appl. Bot., Genet and Plant Breed. [Proc. of the All-Russia Res. Inst. of Plant Growth in the name of N. I. Vavilov (Leningrad, Russia)].* **53:** 124-125.

69

Current Status of Sorghum Diseases in Venezuela

Gino Malaguti

Even though sorghum is a tough plant that is adapted to variable climate and soil conditions, in Venezuela, as in many other countries, it is susceptible to various diseases, with variable symptomology, incidence, and levels of loss. This diversity is partially related to different environmental conditions and cultivation practices in the different parts of the country. In Venezuela, there are two main sorghum-growing regions–one is a dry region, with sandy soil, located in the north central part of the country (Guárico and Anzoategui states), where planting occurs at the beginning of the rainy season. In the second region, with higher humidity and more fertile soils, sorghum is planted at the end of the rainy season, commonly following maize (Portuguesa and Barinas States). Recently, the previously clear differences between the dry- and rainy-season planting areas has begun to blur.

Most of the diseases known on sorghum are found in Venezuela, with variable severity and economic impact, and can be placed in one of three categories: (i) destructtive diseases capable of preventing grain production and killing the plant; (ii) weakening diseases that generally do not kill the plant, but can reduce production substantially; and (iii) sporadic diseases that occur irregularly, but that can cause severe losses when they do occur.

Destructive Diseases

Anthracnose, caused by *Colletotrichum graminicola* is the most important disease, especially as a stalk and head rot. This disease occurs in all regions of the country and often results in lodging of entire fields. Research on the reaction of different cultivars to the pathogen is in progress, but there is not a consistent resistance breeding effort.

Charcoal stalk rot, caused by *Macrophomina phaseolina*, usually appears as large patches in a field and is related to poor or deficient soil conditions, drought, or other stress factors. Microsclerotia are very abundant in the infected tissues, but no pycnidia have been found.

Downy mildew, caused by *Peronosclerospora sorghi*, has been decreasing due to the use of resistant cultivars. In areas where inoculum is abundant, from previous diseased crops or collateral wild hosts, seed dressing with metalaxil usually suffices to avoid plant infection by germinating oospores (4).

Fusarium stem rot occurs in very humid areas. *Fusarium moniliforme sensu lato* can invade the stem and the axis of the panicle, reducing grain formation and causing the plant to collapse.

Weakening Diseases

Both foliar and viral diseases are in this category. The most common foliar diseases are gray leaf spot (caused by *Cercospora sorghi*), leaf blight (caused by *Exserohilum turcicum*), zonate leaf spot (caused by *Gloeocercospora sorghi*), target leaf spot (caused by *Bipolaris sorghicola*), and leaf anthracnose (caused by *C. graminicola*). *G. sorghi* can produce abundant sclerotia in leaf tissue.

C. sorghi is of particular mycological interest due to its variable symptoms, including the "scaliform lesions" commonly observed in the wild sorghum *Sorghum verticilliflorum*. As suggested by Frederiksen (1), the special "chain leaf spot" or "scaliform lesions" are caused by a fungus that is related to, but distinct from, *C. sorghi* and that was recently described as *Phaeoramularia ciccaronei* based on morphological and physiological differences. *P. ciccaronei* does not infect maize, although *C. sorghi* can (5). The rust fungus, *Puccinia purpurea*, also is common on leaves and leaf sheaths, especially near crop maturity, and leaves may be covered by the uredinia and telia of the fungus. *P. purpurea* may be controlled by the hyperparasite *Eudarluca caricis* (anamorph, *Sphaerellopsis filum*) (1).

Among viral diseases, sugarcane mosaic virus (SCMV) is the best known and infects many gramineous plants, including sorghum. In very susceptible cultivars, SCMV causes severe dwarfing, leaf necrosis, and stem necrosis. Several strains of the virus have been identified in Venezuela. At present, the most severe viral disease of sorghum is caused maize dwarf mosaic virus (MDMV) of which there are two strains, A and V. The more virulent V strain was first isolated in Venezuela and kills all plants of susceptible cultivars. Many important sorghum breeding lines in the United States, *e.g.*, Tx430, are susceptible to MDMV-V. A large number of resistant sorghum lines have been identified in Venezuela (2).

MDMV-V infects Venezuelan johnsongrass, but not johnsongrass from the United States. This virus disease commonly reduces yield 70-80%, and occasionally 100%, in experimental trials, if the plants were inoculated early (two weeks after planting). Other viruses in Venezuela are sorghum yellow banding virus (SYBV), sugarcane mosaic virus D (SCMV-D), and maize stripe virus (MSTP V). The use of tolerant or resistant hybrids has reduced the incidence and severity of these diseases in Venezuela (2).

Sporadic Diseases

Acremonium wilt is observed occasionally and may cause severe damage. Many bacteria have been isolated from sorghum in Venezuela, but they do not represent a serious economic problem. *Pseudomonas andropogonis* is the most frequently isolated sorghum-associated bacterium.

Smuts. The only smut present in Venezuela is the covered kernel smut, induced by the fungus *Sporisorium* (*Sphacelotheca*) *sorghi*, which is sometimes found in forage sorghum. Covered kernel smut was the first sorghum disease to be widely investigated in 1948, when sorghum was first introduced into Venezuela. Interest in this disease waned when control through seed dressing was discovered.

Ergot. Ergot, caused by *Claviceps africana*, is very important in the fields in which hybrid sorghum seed is produced, but is only occasionally important in commercial grain sorghum fields where the flowers are rapidly self fertilized. Macroconidia, secondary conidia, and microconidia are all abundant in the honeydew exudate. The infected ovary turns into a stromatic mass, producing conidiophores and conidia. Typical, well-developed, elongated sclerotia, which can give rise to the ascogenous forms, have not been observed. The disease appeared suddenly in a severe epiphytotic form in 1996 and 1997 (3), but it now is present only occasionally. A clear environmental factor to explain the observed disease pattern has not been identified. However, in 1997, known chemical controls were effective in some hybrid seed production fields (6).

Grain mold. Grain mold can be severe when plant maturity and harvesting occur during periods of high rainfall. Disease severity can increase during transportation and storage. The fungi commonly found on grain usually infect during plant growth and invade the spike, spikelet, peduncle, and rachis. Grains with mold may contain dangerous mycotoxins. *Fusarium moniliforme sensu lato* is the fungus most commonly recovered from grain molded heads in Venezuela. In some cases, the incidence of this species was ≥ 80% in the harvested grain. *Aspergillus flavus* and *Aspergillus terreus* also are common, as are the apparently atoxigenic *Curvularia* spp., *Cladosporium* spp., and *Fusarium semitectum*.

Final Considerations

In Venezuela, grain sorghum is used only for animal (poultry and swine) feed. The country produces 35-40% of its needs (about 1,000,000 metric tons). About 600,000 metric tons is imported annually from the United States. The imported material usually is red grain (yellow standard) that contains less tannin than the brown grain type that dominates in Venezuela.

The grain sorghum growing area in Venezuela increased very rapidly in the 1970s and 1980s; in 1978, more than 300,000 ha were planted to sorghum, but this area has now decreased due primarily to local circumstances. In 1990, only approximately 150,000 ha were planted to sorghum, but the area planted to sorghum is now increasing again, and in 1998, approximately 230,000 ha were planted to sorghum. Venezuela has developed outstanding tropical sorghum hybrids and is producing most of the hybrid seed needed locally, although some seed is still imported.

In general, diseases are not limiting sorghum production in Venezuela, even if some of these diseases can be difficult to control due to high levels of inoculum in collateral hosts such as wild sorghums, *e.g.*, *Sorghum halepense*, *S. verticilliflorum*, and *Sorghum arundinaceum*, and the many natural hybrids that occur between the wild species and the cultivated one. *Panicum maximum*, both tall and dwarf, also is a common host for sorghum pathogens. These weeds, in addition to being very effecttive sorghum competitors, are perennial hosts of most pathogens, including viruses. In some years they can cover entire fields and adjoining areas. In some cases, yield losses attributed to a specific disease, *e.g.*, anthracnose, downy mildew, charcoal rot, or viruses, have reached 80-90% of the total production. However, a general estimate of losses due to diseases across the crop in Venezuela is 12-15% of the total production.

References

1. **Frederiksen, R. A., ed.** 1986. *Compendium of Sorghum Diseases*, 1st ed. APS Press, St. Paul, Minnesota.
2. **Garrido, M., G. E. Y. Trujillo, and R. Cuello de Uzcateguí.** 1994. Identificación de aislamientos virales procedentes de zonas productoras de sorgo. *Agron. Trop. (Venezuela)* **44:** 263-278.
3. **Malaguti, G., and N. Pons.** 1997. Ergot, honeydew or sugary disease of sorghum in Venezuela. *Proc. Global Conf. on Ergot of Sorghum (1-8 June 1997, Sete Lagoas, Brazil)*: 88-96.
4. **Malaguti, G.** 1978. Sorghum downy mildew in the Americas, p. 184-194. *In* R. J. Williams, R. A. Frederiksen, and L. K. Mughogho (eds.), *Sorghum Diseases, A World Review*. ICRISAT, Patancheru (A.P.), India.
5. **Pons, N.** 1993. *Phaeoramularia ciccaronei* sp. nov., fungus causing the chain leaf spot in *Sorghum* spp. *Fitopatol. Venez.* **6:** 2-7.
6. **Reis E. M., and M. M. Casa Blum.** 1997. Integrated management of sugary disease of sorghum: Practical aspects. *Proc. Global Conf. on Ergot of Sorghum (1-8 June 1997, Sete Lagoas, Brazil)*: 161-166.

70
Sorghum and Pearl Millet Diseases in West and Central Africa

Paul S. Marley, Mamourou Diourté, Adama Neya, Stephen K. Nutsugah, P. Sérémé, Seriba O. Katilé, Dale E. Hess, Demba F. Mbaye, and Zachee Ngoko

Abstract

In west and central Africa, sorghum (*Sorghum bicolor*) and pearl millet (*Pennisetum glaucum*) are subject to attack by numerous diseases and the witch weed, *Striga hermonthica*. The most damaging diseases on sorghum are anthracnose (caused by *Colletotrichum graminicola*) and grain mold caused by a complex of fungi, while on pearl millet, downy mildew (caused by *Sclerospora graminicola*) is the most important. *Striga hermonthica* continues to be a major constraint to sorghum and millet production. Research efforts in the last decade have led to a better understanding of variability in *C. graminicola* and *S. graminicola* and to the identification of resistance sources and development of control strategies for the major diseases and *Striga*. A grain mold/head bug relationship has been established, and ladder leaf spot, viruses, and new problems attributed to nematodes in farmers' fields are reported, along with a description of the practices used in west and central Africa for the management of these diseases.

Introduction

Sorghum [*Sorghum bicolor* (L.) Moench] and pearl millet [*Pennisetum glaucum* (L.) R. Br.] are very important food crops in the western and central Africa, *i.e.*, Benin Republic, Burkina Faso, Cameroon, Central African Republic, Cote d'Ivore, Gabon, Gambia, Ghana, Guinea, Liberia, Mali, Mauritania, Niger, Nigeria, Senegal, Sierra Leone, and Togo. These crops are well adapted to the semi-arid tropics, with production concentrated in three agroecological zones–Sahelian (400-600 mm annual rainfall), Sudanese (600-1000 mm annual rainfall), and Guinea-Savanna (1000-1300 mm annual rainfall)–that cut across the region.

The total area sown to sorghum and pearl millet in west and central Africa is estimated at 12.5 and 14.9 million hectares, with average yields of 0.78 and 0.56 tons/ha, respectively. The total production of these crops is reported to be 10.2 and 9.5 million tons, respectively, of which Nigeria is the major producer (10). These crops are primarily produced for human consumption, but small quantities are utilized for animal feeds, the production of beer and other malted beverages, and the production of sweeteners, especially in Nigeria.

The production of sorghum and pearl millet in west and central Africa is limited by many constraints including biotic factors. Some of the more common diseases are caused by fungi, bacteria, viruses, and nematodes (Table 70-1). A few of these diseases are currently of major importance, some have potential importance as emerging diseases, and many are limited in distribution or are sporadic in occurrence.

Sorghum Diseases of Major Importance

Foliar diseases. *Anthracnose.* Anthracnose, caused by *Colletotrichum graminicola* (syn. *Colletotrichum sublineolum)*, is the most important foliar disease on both local and improved sorghum varieties in west and central Africa. It is most prevalent and destructive in the northern Guinean and Sudanese zones (26). Although three symptom types (panicle, foliar, and stalk) are known in west and central Africa, recent reports suggest that panicle anthracnose also is present in farmers' fields (2, 16), where it may cause yield losses following foliar infection of 47-67% on susceptible varieties (12, 36). There is genetic variation within the foliar population of *C. graminicola* in west and central Africa (25, 35). Although variation in Burkina Faso is limited (22), work at Samaru, Nigeria identified five physiological races from 50

TABLE 70-1. Ranked list of the prevalence and importance of sorghum and millet diseases in west and central Africa.

Disease	Country									
	Benin	Burkina Faso	Cameroon	Central African Republic	Cote d'Ivoire	Guinea	Mali	Niger	Nigeria	Togo
Sorghum										
Anthracnose	1[a]	2	1	2	2	1	1	2	1	1
Grain Mold	3	3	1	2	1	2	3	2	1	1
Covered Kernel Smut	3	3	2	1	2	2	1	3	1	3
Long Smut	2	3	1	2	3	1	1	2	3	3
Loose Smut	3	3	2	1	3	1	1	3	1	3
Zonate Leaf Spot	2	1	2	1	3	3	2	3	3	3
Downy Mildew	2	3	3	2	3	1	3	3	3	1
Stalk Rot	3	2	2	3	3	1	2.5	3	1	3
Head Blight	3	2	2	2	3	2	3	3	2	3
Head Smut	3	3	1		3	1	1.5	3	3	3
Sooty Stripe	3	2	3	2	3	3	1	3	2	3
Banded Leaf and Sheath Blight	2	3	2	3	3	2	3	3	3	2
Ergot	2	3	3	2	3	2	2	3	3	3
Gray Leaf Spot	3	2	1	3	3	2	3	3	3	3
Rust	2	3	3	2	3	3	3	3	3	1
Bacterial Streak	2	3	2	2	3	2	3	3	3	3
Bacterial Stripe	3	3	2	2	3	2	3	3	3	3
Charcoal Rot	3	2	2	3	3	3	2	3	3	3
Leaf Blight	3	3	2	3	3	1	3	3	3	3
Pokkah Boeng	2	2	3	3	3	2	3	3	3	3
MDMV	2	2	3	3	3	2	3	3	3	3
MStV	2	2	3	3	3	2	3	3	3	3
MMV	2	3	3	3	3	2	3	3	3	3
Ladder Leaf Spot	2	2	3	3	3	2	2.5	3	3	3
Oval Leaf Spot	3	2	2	2	3	3	3	3	3	3
Rough Leaf Spot	3	3	3	3	3	2	3	3	3	3
Tar Spot	3	3	3	3	3	2	3	3	3	3
Millet										
Downy Mildew	2	1	1	3	1	1	1	1	1	1
Smut	2	2	2	1	3	1	1	2	2	1
Ergot	2	2	2	1	3	2	2	2	2	1
Blast	1	2	3	3	3	3	3	3	3	2
Rust	2	2	3	3	3	3	3	3	3	3

[a]Rating: 1, important; 2, potentially important; and 3, not now important.

foliar isolates collected from several sorghum-growing areas (21).

Management of anthracnose has depended on fungicides and the planting of resistant varieties. Many resistant lines have been identified within the region and include 98 FA-EARS-T-101, CSM 417, Gaya Early, Gnofing, ICSV 400 BF, IS 5359, IS 5360, IS 20302, IS 24733, IS 24721, ICSV 901 NG, Malisor 84-5, Naga white, Sariasso 10, Siripe 1, and Yarruruka (17, 18, 22). Lines with resistance to foliar, panicle, and grain anthracnose also have been identified (8).

Gray Leaf Spot. This disease is caused by *Cercospora sorghi* and is most common on local varieties in the Guinea savanna zone (26). This disease can cause yield losses that reach 36% in highly susceptible genotypes, *e.g.*, SK 5912, although Thomas (34) noted that the disease may not be of any economic importance. Mixed infections of this disease and zonate leaf spot (caused by *Gloeocercospora sorghi*) have been observed in Mali and Nigeria.

Disease management is based on host plant resistance (where available). Genotypes such as Gaya Early,

ICSV 901 NG, ICSV 902 NG, Naga white, Sariasso 3, Sariasso 10, and Yarruruka have been identified as having resistance to gray leaf spot (18; Table 70-2).

Panicle diseases. *Grain Mold.* Grain mold is one of the most important diseases limiting sorghum production in western and central Africa, although earlier reports (34) had indicated that it was rare in farmers' fields. Grain mold is common on early-maturing local and improved varieties grown in the Sudanese-Sahelian Zone, and also occurs on medium-maturing local and improved varieties in the southern Guinea Savanna and the southern fringes of the northern Guinea Savanna (15). Grain mold is caused by a complex of fungi including *Aspergillus flavus*, *Curvularia lunata*, *Fusarium moniliforme sensu lato*, *F. pallido-roseum*, and *Phoma sorghina* (20, 31), and species of *Acremonium*, *Bipolaris*, *Cercospora*, *Exserohilum*, *Gloeocercospora*, *Nigrospora*, *Phomopsis*, and *Verticillium* (23). Grain molds degrade crop yield and quality by discoloration, reduced seed size, and poorer feed processing quality (15). Further, moldy grain may contain mycotoxins such as aflatoxins, zearalenone, and patulin (31).

Predisposition to grain mold through head bug activity has been established in the region (15, 19, 30). Head bugs, especially *Eurystylus oldi*, puncture the grain during feeding and oviposition, thus predisposing it to infection and mold development. In a regional research effort, two cultivars (IS 14384 and CCGM 39/17-2-2) were identified as having high levels of resistance to both grain mold and head bugs (29). Further, dominant grain mold fungi were identified as *Phoma* spp. (predominantly *P. sorghina*) and *Fusarium* spp. (predominantly *F. moniliforme sensu lato*). Most importantly, this study suggests hypotheses to explain the observed relationship between grain mold, weather, and headbugs. In particular, high relative humidity can affect grain mold by providing suitable conditions for infection and by enhancing fungal sporulation (and thus the inoculum load).

Management of grain mold in western and central Africa (15) has included the use of host plant resistance, appropriate cultural practices, *e.g.*, date of planting, and integrated management options when high-yielding local resistant varieties are planted.

Smuts. Smuts are the most widespread group of diseases in sorghum-growing areas of west and central Africa (14) and cause damage on both traditional and improved cultivars. There are four smuts on rain fed sorghum: grain, or covered kernel, smut (caused by *Sporisorium sorghi*); loose smut (caused by *Sporisorium cruentum*); head smut (caused by *Sporisorium reilianum*); and long smut (caused by *Sporisorium ehrenbergii*). Covered kernel smut is the only post-rainy season smut of sorghum in the Lake Chad area (24). The prevalence of smuts across the agroecological zones of west and central Africa varies. Covered kernel smut is prevalent in the Sahelian, Sudanese, and northern and southern Guinea Savanna zones. Although covered kernel smut was once endemic, it can be successfully controlled with seed dressing chemicals, even though its incidence has recently been increasing (14, 27). Similarly, loose smut, whose occurrence had been light and sporadic, is now widespread with increaseing incidence in the Sudanese and Sahelian zones (14).

The increased incidence of covered kernel and loose smuts may be attributed to many factors including increased cost and erratic supply of agricultural inputs (28), the lack of suitable seed treatment chemicals (12), and the presence of an alternate host, *Cynodon dactylon*, for the covered kernel smut fungus (11).

Smut management is based on seed dressing fungicides, host plant resistance (13), and the use of local, plant-based pesticides as seed dressing, *e.g.*, *Canavalia ensiformis*, *Securidaca longipedonculata*, *Parkia biglo-*

TABLE 70-2. Mean disease severity of early-maturing and medium-maturing sorghum genotypes to anthracnose and gray leaf spot grown at Ajiwa and Bagauda, Nigeria, respectively, from 1995 to 1997 (after 18).

Genotype	Disease[a]	
	Anthracnose	Gray Leaf Spot
Early maturing		
CSM 415	1.0	5.7
IRAT 204	6.1	1.0
ICSV 400	2.4	2.3
CSM 417	1.0	4.1
Malisor 84-5	1.0	3.6
Naga White	1.0	1.8
Yarrurkua	1.0	2.3
90 SN 7	3.1	2.9
ICSV 901 NG	1.0	1.0
Malisor 92-1	2.7	2.8
Medium maturing		
CGM 19/9-11	1.0	1.8
ICSV 902 NG	4.0	1.0
Kassouroka	1.0	3.9
CEM 328/2-1-1-3	2.7	1.0
CEM 326/11-5-1-1	3.1	1.0
Sariaso 10	2.7	1.0
Sabakoun	1.0	4.3
Sariaso 9	1.0	4.1
Gaya Early	1.0	2.3
Sariaso 3	3.5	1.0

[a]Mean disease severity calculated from disease severity scores from 1995, 1996, and 1997. Reaction class of host: resistant, ≤ 1.0; moderately resistant, 1.1-3.0; and susceptible, 3.1-9.0.

bosa, *Lonchocarpus laxiflorus*, and *Adenum obosum*. In preliminary studies, these materials both reduced covered kernel smut incidence and increased grain yield 15-60% in Mali (Diourté, *unpublished*). Resistant varieties identified include ICSV 2, ICSV 1002, ICSV 1049, and NSV 74 (13).

Long smut is predominantly a disease of the dry Sudanese and Sahelian zones (14, 26). In the last two decades, the disease has occurred in most western and central African countries, but most often in Mali, Niger, and Burkina Faso. Management of the disease is by use of resistant varieties, where available, and by late planting (14).

Head smut is most prevalent in the southern Guinea Savanna zone of west and central Africa with sporadic occurrence in the northern Guinea Savanna and the Sudanese and Sahelian zones (14). Head smut usually occurs on late-maturing cultivars, and disease management is through the planting of resistant varieties (14).

Parasitic Weeds

Striga hermonthica is a major biotic constraint to sorghum production in western and central Africa, where many local and improved varieties are susceptible to the parasite. Yield losses, particularly under drought conditions, may be very high (90-100%) (6). Host plant resistance can be a major component of integrated striga management when it occurs in adapted, productive cultivars, and resistant varieties are available in the region. A workshop was held in 1999 at IITA in Nigeria (5) to assess the current status of cereal breeding for striga resistance and to develop new strategies for dealing with this weed. Management approaches in the region are based on cultural control strategies (7). Biological control of *Striga* spp. is a research area that holds much promise (9).

Sorghum Diseases of Minor Importance

Sooty stripe, ergot, and head blight are potentially important diseases in western and central Africa. Sooty stripe, caused by *Ramulispora sorghi*, is common in Burkina Faso, where it may cause yield losses of more than 20%. Ergot, caused by *Claviceps africana*, occurs on seed production plots in Nigeria. Head blight, caused by *F. moniliforme sensu lato*, is prevalent in the northern Guinea and Sudanese zones of Nigeria, where it occasionally causes significant yield loss (12).

Oval leaf spot (caused by *Ramulispora sorghicola*), zonate leaf spot, and leaf blight (caused by *Exserohilum turcicum*) occur, but at low levels, in many farmers' fields in the northern Guinea and Sudanese Zones (34), while rough leaf spot, caused by *Ascochyta sorghina*, is more common in the southern Guinea Savanna. A high incidence of ladder leaf spot, caused by *Cercospora fuscimaculans*, occurs in the southern Guinea Savanna of Nigeria.

A relationship between sorghum yellow leaf blotch caused by *Xanthomonas* sp. and spittle bugs (*Locris rubens* and *Poophilus costalis*) has been reported in western and central Africa (1), where the feeding bugs transmit the bacterium *Xanthomonas* sp. Maize Streak Virus (MStV) and Maize Mosaic Virus (MMV) are known to occur on sorghum in Nigeria (26).

Eighteen genera of plant parasitic nematodes are associated with sorghum and other cereals in the Nigerian savanna (4, 37). Species belonging to five genera–*Aphelenchoides*, *Ditylenchus*, *Helicotylenchus*, *Pratylenchus*, and *Tylenchus*–occurred in more than 50% of the surveyed fields. *Pratylenchus* and *Helicotylenchus* species had the highest incidence, occurring in 90% of the sorghum fields surveyed. This study also established a correlation between the incidence of Pokkah boeng (induced by *Fusarium* sp.) and the incidence of nematode symptoms across all fields surveyed (3).

Pearl Millet Diseases of Major Importance

Foliar diseases. *Downy mildew.* Downy mildew, caused by *Sclerospora graminicola*, is the major disease of millet in western and central Africa, and may affect all of the above-ground parts of the plant. Yield losses due to this disease range from 21-60% in Burkina Faso (40), 6-72% in Nigeria, and 3.5-21% in Mali (41).

Pathogenic variability in populations of *S. graminicola* has been reported in the region (32, 33). Resistant varieties include: SOSAT-C 88 and Synth 16 C1 across the region; IKMP 1, IKMP 2, IKMP 5, IKMV 8201, ICMV IS 88102 in Burkina Faso; India Na 05 and Toroniou C1 in Mali; and MDN 90, MDN 92, and B Composite in Nigeria (41). Management of downy mildew in western and central Africa is principally through the use and integration of seed dressing fungicides and resistant varieties. Plant-derived pesticides (seed dressing) and roguing and burial of diseased plants are techniques of local importance (41).

Panicle diseases. *Smut*: Smut, caused by *Moesziomyces penicillariae*, syn. *Tolyposporium penicillariae*, is the second most important disease of millet after downy mildew. Yield loss varies within western and central Africa, but does not exceed 10% even during severe epidemics. Management is through the planting of resistant types and varieties. These lines include ICMV 89201, SE 360, SID 87, SE 13, SID 377, and MDN 90 (41).

Parasitic weeds. *Striga hermonthica* is a frequently occurring parasitic weed on pearl millet, but most land races are very tolerant to attack; however, in times of drought, striga infestation can be devastating. Management of striga on pearl millet relies upon rotation and/or intercropping with legumes, weeding, use of fertilizer, and date of planting (7). Recently, wild pearl millet relatives have been examined for resistance to *S. hermonthica* (38).

Pearl Millet Diseases of Minor Importance

Several minor diseases of pearl millet occur, including ergot, caused by *Claviceps fusiformis*, which is sporadic in occurrence and rarely of economic importance; leaf blast, caused by *Pyricularia grisea*, which generally occurs on exotic short pearl millet types and is most apparent in years of abundant rainfall; Dactuliophora leaf spot, caused by *Dactuliophora elongata*, in Nigeria, Niger, and Mali (39); rust, caused by *Puccinia penniseti*, which is sporadic in areas with rainfall of approximately 800 mm; bacterial streak, caused by *Pseudomonas rubrillineans*, in southern Mali and northern parts of Nigeria (43); and zonate leaf spot, caused by *Gloeocercospora* sp., in Mali and Niger. Causal organisms of a midrib infection on millet in Nigeria include *Curvularia penniseti*, *Curvularia eragrostidis*, *Curvularia intermedia*, and *Curvularia verrucosa*. All *Curvularia* spp. induced typical midrib spot symptoms on maize, sorghum, *Pennisetum* sp., *Eragrostis* sp., and dauro, gero and maiwa types of millet in host range studies, but not on *Digitaria* sp. (42).

Recommended Future Research Activities

The research activities of the National Agricultural Research Systems in western and central Africa are focused on the development of adapted, high-yielding sorghum and pearl millet varieties that are resistant or tolerant to the major diseases and striga. These goals are being pursued through regional networks, *e.g.*, West and Central Africa Sorghum Research Network (WCASRN) and West and Central Africa Millet Regional Network (WCAMRN), and international research groups, *e.g.*, INTSORMIL and ICRISAT. Future activities in sorghum will focus on:

- Evaluation of variation in *Colletotrichum graminicola*, the anthracnose pathogen, and the identification of factors associated with host plant resistance.
- Development of crop residue management systems to reduce disease incidence and intensity.
- Epidemiology of grain molds.
- Identification of genetic factors conferring resistance to grain molds.
- Expansion of farmer-conducted integrated *Striga* management trials.
- Development of enhanced integrated disease management strategies.
- Increasing scientific capacity through the training of scientists, technicians, extension staff, and producers.

Future activities in pearl millet will focus on:

- Evaluation of genetic variation in *Sclerospora graminicola* downy mildew populations.
- Breeding new, disease-resistant, high-yielding, adapted pearl millet varieties.
- Development of integrated disease management strategies for pearl millet pests and diseases.
- Production of information bulletins and pamphlets on the important diseases of pearl millet.
- Increasing scientific capacity through the training of scientists, technicians, extension staff, and producers.

References

1. **Ajayi, O., and A. D. Akpa.** 1996. Interactions between spittle bugs and bacteria on sorghum in West Africa, CD-ROM. *In* Ratnadass, A. O. Ajayi, P. S. Marley, and I. Akintayo (eds.), *Insect Pests of Sorghum in West and Central Africa*. ICRISAT, Bamako, Mali.
2. **Bandyopadhyay, R., D. E. Hess, and I. Sissoko.** 1996. *Summary Report of Sorghum Pathology Activities in Mali.* ICRISAT, Bamako, Mali.
3. **Chindo, P. S.** 1998. *Studies on Plant Parasitic Nematodes Associated with Maize, Sorghum and Millet in the Nigerian Savannas.* Ph.D. Thesis, Ahmadu Bello Univ., Samaru, Zaria, Nigeria.
4. **Chindo, P. S., G. K. Weber, and K. A. Elemo.** 1997. Sustainability of intensified cereal-based systems in the savannas: Nematodes as a constraint, p. 343-353. *In* B. Badu-Apraku, M. O. Akoroda, M. Ouédraogo, and F. M. Quinn (eds.), *Contributing to Food Self-Sufficiency: Maize Research and Development in West and Central Africa.* International Institute for Tropical Agriculture, (IITA), Ibadan, Nigeria.
5. **Haussmann, B. I. G., D. E. Hess, M. L. Koyama, L. Grivet, H. F. W. Rattunde, and H. H. Geiger, eds.** 2000. *Breeding for Striga Resistance in Cereals.* Margraf Verlag, Weikersheim, Germany.
6. **Haussmann, B. I. G., D. E. Hess, H. G. Welz, and H. H. Geiger.** 2000. Improved methodologies for breeding striga-resistant sorghums. *Field Crops Res.* **66:** 195-211.
7. **Hess, D. E., and B. Dembélé.** 1999. Cultural management of *Striga* on cereals, p. 23-32. *In* D. E. Hess and J. M. Lenné (eds.), *Report on the ICRISAT Sector Review for Striga control in Sorghum and Millet.* ICRISAT, Bamako, Mali.
8. **Hess, D. E., R. Bandyopadhyay, and I. Sissoko.** 2000. Reactions of sorghum genotypes to leaf, panicle and grain anthracnose (*Colletotrichum graminicola*) under field

conditions in Mali, p. 163-177. *In* I. Akintayo, and J. Sedgo (eds.), *Towards Sustainable Sorghum Production, Utilization and Commercialization in West and Central Africa*. ICRISAT, Bamako, Mali.

9. **Hess, D. E., J. Kroschel, D. Traoré, A. E. M. Elzein, P. S. Marley, A. A. Abbasher, and C. Diarra.** 2002. Striga: Biological control strategies for a new millennium, p. 165-170. *In* J. F. Leslie (ed.), *Sorghum and Millets Pathology 2000*. Iowa State Press, Ames, Iowa.
10. **ICRISAT and FAO.** 1996. *The World Sorghum and Millet Economies, Facts, Trends and Outlook*. ICRISAT, Bamako, Mali, and Food and Agriculture Organization of the United Nations, Rome, Italy.
11. **Marley, P. S.** 1995. *Cynodon dactylon*: Alternative host to *Sporisorium sorghi* Link causal organism of sorghum covered smut. *Crop Protect.* **14:** 491-493.
12. **Marley, P. S.** 1997. *Sorghum Pathology. Cereals Research Program Report to the 1997 Annual Cropping Scheme Meeting*. Inst. for Agric. Res., Ahmadu Bello Univ., Samaru, Zaria.
13. **Marley, P. S., and D. A. Aba.** 1996. Varietal resistance of *Sorghum bicolor* (L.) Moench germplasm to *Sporisorium sorghi* Link in Nigeria. *Afr. J. Plant Protect.* **6:** 83-89.
14. **Marley, P. S., and D. A. Aba.** 1999. Current status of sorghum smuts in Nigeria. *J. Sustain. Agric.* **15:** 199-132.
15. **Marley, P. S., and O. Ajayi.** 1999. Sorghum grain mold and the influence of headbug *Eurystulus oldi* in West and Central Africa. *J. Sustain. Agric.* **13:** 35-44.
16. **Marley, P. S., and R. Bandyopadhyay.** 1996. *Prevalence of Panicle Anthracnose on Farmers' Fields in Nigeria. Report of Survey on Panicle Anthracnose*. ICRISAT, Patancheru (A.P.), India.
17. **Marley, P. S., R. Bandyopadhyay, R. Tabo, and O. Ajayi.** 2000. Availability of sorghum germplasm with multiple resistance to two major diseases in the Nigerian savanna. *Trop. Sci.* **41:** 44-49.
18. **Marley, P. S., K. A. Elemo, D. A. Aba, I. Onu, and I. Akintayo.** 2001. Reactions of sorghum genotypes to anthracnose and gray leaf spot diseases under Sudan and Sahel savanna field conditions of Nigeria. *J. Sustain. Agric.* **18:** 105-116.
19. **Marley, P. S., and A. M. Malgwi.** 1999. Influence of headbugs (*Eurystylus* sp.) on sorghum grain mould in the Nigerian savanna. *J. Agric. Sci.* **132:** 71-75.
20. **Marley, P. S., S. K. Manzo, and A. C. Ebenebe.** 1997. Comparative storability of three sorghum cultivars in Nigeria. *Trop. Sci.* **37:** 65-70.
21. **Marley, P. S., R. P. Thakur, and O. Ajayi.** 2001. Variation among foliar isolates of *Colletotrichum sublineolum* of sorghum in Nigeria. *Field Crops Res.* **69:** 133-142.
22. **Neya, A., and L. E. Normand.** 1998. Responses of sorghum genotypes to leaf anthracnose (*Colletotrichum graminicola*) under field conditions in Burkina Faso. *Crop Protect.* **17:** 47-53.
23. **Nutsugah, S. K., K. F. Ayensu, and K. F. Tsigbey.** 1999. Survey of seed-borne diseases in Ghana. *Proc. 1999 World Seed Conf. (6-8 Sept. 1999, Cambridge, UK.)*: 50.
24. **Olabanji, O. G., R. Tabo, D. J. Flower, O. Ajayi, F. Ushie, B. K. Kaigama, and M. C. Ikwelle**. 1996. Survey of Masakwa sorghum growing areas in northeastern Nigeria. *Intl. Sorg. Millet Newsl.* **37:** 61-63.
25. **Ozolua, K. O. O., P. D. Tyagi, and A. M. Emechebe.** 1986. Pathogenic variation in *Colletotrichum graminicola*, the causal agent of anthracnose of sorghum in northern Nigeria. *Samaru J. Agric. Res.* **4:** 79-84.
26. **Pande, S., R. Harikrishnan, M. D. Alegbejo, L. K. Mughogho, R. I. Karunakar, and O. Ajayi.** 1993. Prevalence of sorghum diseases in Nigeria. *Intl. J. Pest Manag.* **39:** 297-303.
27. **Pande, S., P. S. Marley, and O. Ajayi.** 1997. Increasing incidence of covered kernel (grain) smut disease of sorghum in northern Nigeria. *Intl. Sorg. Millet Newsl.* **38:** 59-61.
28. **Poswal, M. A. T., and A. D. Akpa.** 1991. Current trends in the use of traditional and organic methods for the control of crop pests and diseases in Nigeria. *Trop. Pest Manage.* **37:** 329-333.
29. **Ratnadass, A., D. R. Butler, P. S. Marley, O. Ajayi, R. Bandyopadhyay, M. A. Hamada, D. E. Hess, F. Assamoi, I. D. K. Atokple, J. Beyo, O. Cisse, D. Dakouo, M. Diakité, S. Dossou-Yovo, B. Le Diambo, I. Sissoko, M. B. Vopeyande, and I. Akintayo.** 1999. Interacting effect of headbugs, molds and climate on sorghum grains in West and Central Africa, p. 120-140. *In* I. Akintayo and J. Sedgo (eds.), *Towards Sustainable Sorghum Production, Utilization, and Commercialization in West Africa*. ICRISAT, Patancheru (A.P.), India.
30. **Ratnadass, A., Y. O. Doumbia, and O. Ajayi.** 1995. Bioecology of sorghum headbug *Eurystylus immaculatus* and crop losses in West Africa, p. 91-102. *In* K. F. Nwanze and O. Youm (eds), *Panicle Insect Pests of Sorghum and Pearl Millet*. ICRISAT, Patancheru (A.P.), India.
31. **Salifu, A.** 1981. Mycotoxins in short season sorghums in northern Nigeria. *Samaru J. Agric. Res.* **1:** 83-88.
32. **Selvaraj, J. C.** 1979. *Research on Pearl Millet Diseases in Nigeria. Samaru Conference Paper 28*. Inst. for Agric. Res., Ahmadu Bello Univ., Samaru, Zaria, Nigeria.
33. **Thakur, R. P., C. W. Magill, S. Sivaramakrishnan, C. T. Hash, H. S. Shetty, and D. E. Hess.** 2002. Variability in *Sclerospora graminicola*, the pearl millet downy mildew pathogen, p. 51-56. *In* J. F. Leslie (ed.), *Sorghum and Millets Diseases*. Iowa State Press, Ames, Iowa.
34. **Thomas, M. D.** 1992. Sorghum diseases in Western Africa, p. 25-29. *In* W. A. J. de Milliano, R. A. Frederiksen, and G. D. Bengston (eds.), *Sorghum and Millets Diseases: A Second World Review*. ICRISAT, Patancheru (A.P.), India.
35. **Thomas, M. D., and R. A. Frederiksen.** 1995. Fertility tests on sorghum isolates of *Colletotrichum graminicola* from West Africa. *Intl. Sorg. Millets Newsl.* **36:** 79.
36. **Thomas, M. D., I. Sissoko, and M. Sacko.** 1996. Development of leaf anthracnose and its effect on yield and grain weight of sorghum in West Africa. *Plant Dis.* **80:** 151-153.
37. **Weber, G., P. S. Chindo, K. A. Elemo, and S. Oikeh.** 1995. Nematodes as production constraints in intensifying cereal-based cropping systems of the northern Guinea

savanna. Resource and Crop Management Research Monograph No.17. IITA, Ibadan, Nigeria.

38. **Wilson, J. P., D. E. Hess, and W. W. Hanna.** 2000. Resistance to *Striga hermonthica* in wild accessions of *Pennisetum glaucum*. *Phytopathology* **90:** 1169-1172.
39. **Wilson, J. P., D. E. Hess, and K. A. Kumar.** 2000. Dactuliophora leaf spot of pearl millet in Niger and Mali. *Plant Dis.* **84:** 201.
40. **Zangré, R., M. C. Sédogo, P. Sérémé, C. Dabiré, and Z. Cissé.** 1993. Principales contraintes à la production du mil et priorités de recherche au Burkina Faso, p. 85-86. *In* O. Youm (ed.), *Proc. Regional Pearl Millet Workshop (19-23 September 1990, Niamey, Niger).* ICRISAT, Patancheru (A.P.), India.
41. **Zarafi, A. B.** 1997. *Final Report on WCAMRN Project P3 Activities Conducted in Nigeria During the 1992-1996 Cropping Seasons.* ROCAFREMI/IAR, Inst. for Agric. Res., Ahmadu Bello Univ., Samaru, Zaria, Nigeria.
42. **Zarafi, A. B., A. M. Emechebe, and A. D. Akpa.** 1995. Identification and host range of causal agent of millet midrib spot disease. *Samaru J. of Agric. Res.* **12:** 25-29.
43. **Zummo, N.** 1976. Yellow leaf blotch, a new bacterial disease of sorghum maize and millet in West Africa. *Plant Dis. Rept.* **60:** 798-799.

71

Diseases of Sorghum and Pearl Millet in Some Southern African Countries

Esther Mtisi and Neal W. McLaren

Abstract

Sorghum *(Sorghum bicolor* L.) is an important cereal crop grown mostly by small-scale farmers in semi-arid areas of Southern Africa. Over 30 diseases have been recorded on sorghum in the region. Of these, ergot (caused by *Claviceps africana*), anthracnose (caused by *Colletotrichum graminicola*), leaf blight (caused by *Exserohilum turcicum*), sooty stripe (caused by *Ramulispora sorghi*), downy mildew (caused by *Peronosclerospora sorghi*), grain molds (caused by *Fusarium* spp., *Curvularia lunata*, *Phoma sorghina*, *Alternaria* spp., *Bipolaris* spp., and *Dreschslera* spp.), rust (caused by *Puccinia purpurea*), covered kernel smut (caused by *Sporisorium sorghi*), seedling diseases (caused by *Fusarium* spp., *Pythium* spp., and *Drechslera* spp.), and *Striga* weed are the major constraints to sorghum production. The importance of individual diseases varies by country. New disease records include *Pantoea agglomerans* and *Pseudomonas syringae* in pearl millet and sugarcane mosaic virus (SCMV) in sorghum.

Introduction

Sorghum (*Sorghum bicolor* L.) is an important cereal crop in most countries in southern Africa. The crop is mostly cultivated by small-scale resource-poor farmers located in semi-arid areas. However, in countries such as South Africa and Zimbabwe, both small-scale and commercial farmers grow the crop. Most small-scale farmers in the region usually grow local land races, but, with the establishment of SADC-ICRISAT Center at Matopos, Zimbabwe in the 1980s, there has been a shift towards growing improved varieties and hybrids of sorghum even by small-scale farmers (28). This change has brought with it disease problems that were not previously of economic importance, *e.g.*, ergot (caused by *C. africana*) in hybrid seed production fields (23). Our objective in this chapter is to briefly review the current status of sorghum and pearl millet diseases in Southern Africa, work in progress, and future trends.

Current Disease Status and Activities in the Region

Diseases of sorghum and millets in the region have been extensively reviewed (6, 7), and relatively little has changed during the past decade with respect to the range of diseases observed (16, 19). Most of the work in progress follows from recommendations made during the Second Global Conference on Sorghum and Millet Diseases held in Harare in 1988 (7). New problems include economically important seedling diseases caused by *Curvularia* spp., *Dreschslera* spp., *Fusarium* spp., and *Pythium* spp. in South Africa (20); the appearance of bacterial leaf spot on pearl millet (15); SCMV and maize dwarf mosaic virus (MDMV) on sorghum in Zambia (29); SCMV on sorghum in Botswana; and sorghum dwarf mosaic virus (SDMV) in South Africa, where disease incidence is geographically limited but of local importance.

Some countries have reported a significant change in the incidence and severity of some diseases of sorghum and millet. In Tanzania, downy mildew (caused by *P. sorghi*), ergot (caused by *C. africana*), and sooty stripe (caused by *R. sorghi*) seem to be increasing (19). In Zimbabwe, ladder spot (caused by *Cercospora fusimaculans*) has increased on susceptible varieties at the Henderson and Panmure research stations. The incidence of false mildew (caused by *Beniowskia sphaeroidea*) has decreased. In general, ergot, anthracnose, striga, leaf blight, covered kernel smut (caused by *Sporisorium sorghi*), downy mildew, sooty stripe, rust and grain molds remain the most important diseases of sorghum in southern Africa. In addition, Zimbabwe has made an effort to develop a list of diseases currently occurring in the country (18).

Much of the research in the region focuses on the identification of disease-resistant germplasm. During the latter part of the 1990s, screening techniques were developed and disease hot spot locations were identified for ergot (Masumba, Zambia), anthracnose (Mansa, Zambia and Henderson, Zimbabwe), grain molds (Ifakara, Tanzania), sorghum downy mildew (Golden Valley, Zambia; Matopos and Panmure, Zimbabwe), leaf blight (Golden Valley, Zambia, and Henderson, Zimbabwe), *Striga asiatica* (Hombolo, Tanzania and Henderson, Zimbabwe), and *Striga hermotheca* (Ukiriguru, Tanzania) (6). Superior germplasm was evaluated at these hot spot locations for tolerance to these diseases and *Striga*. By using this strategy, sources of resistance have been identified and resistant/tolerant lines incorporated into breeding programs (6). The facility still exists for those who would like to cooperate with scientists in the region on similar problems.

Ergot (*Claviceps africana*). Ergot is an important constraint to hybrid sorghum production in Botswana, Malawi, Tanzania, Zambia, Zimbabwe, and South Africa. During 1996, yield losses were estimated at 10-15% nationally, with specific losses of 32% (Heilbron), 43% (Klerksdorp), and 45% (Standerton) in South Africa. In the 1999/2000 season, the disease was common in sorghum hybrid and seed production fields. It cost seed companies over R2,000,000 (8.8R = US$1.00) just to clean seed. In contrast, there was very little ergot at the Henderson Research Station (Mazowe) in Zimbabwe, even though favorable conditions prevailed during the 1999/2000 season. Extensive work has been carried out by scientists in the region on ergot description (8, 12, 22), epidemiology (8-11, 13, 14, 24, 25), disease control and germplasm resistance (21, 23, 24, 27), and the identification of hosts and alternate hosts (10).

Anthracnose (*Colletotrichum graminicola*). This disease is economically important in farmers' fields in Zambia and at research stations in Botswana, Tanzania, Zambia (Mansa), and, to a lesser extent, at the Henderson Research Station in Zimbabwe. In South Africa, this disease is limited to the eastern production areas (Standerton), but when conditions are favorable for its development, this disease can be of economic importance. Breeders and pathologists in most countries screen germplasm for tolerance to anthracnose (6).

Striga. Three species of Striga have been identified. *S. asiatica* is important in farmers' fields in some parts of Zimbabwe. In Tanzania, *Striga forbesii*, and *S. hermotheca* are restricted to Ilonga and Ukiriguru, while *S. asiatica* is important at Hombolo, Ilonga, Manhala, Mwele, Naliendele, and Ikiriguru. Visual losses of up to 100% have been observed in sorghum fields (16). In Tanzania, SAR29 and its hybrid, SPL38A × SAR29, and Weijita were found to be resistant to *S. asiatica* and *S. hermotheca* (16).

Leaf blight (*Exserohilum turcicum*). Leaf blight is a common disease in southern African countries, but is of economic importance only in the high potential yield areas of Zimbabwe and South Africa. In South Africa severe losses due to leaf blight occurred at Mpumalanga during 1999/2000 cropping season. Breeders and pathologists in most countries screen germplasm for tolerance to leaf blight (6).

Smuts. Covered kernel smut (caused by *Sporisorium sorghi*) is important in farmers' fields in Botswana and Tanzania (16). In communal fields in Zimbabwe, this disease has become important enough to warrant control measures (26), and seed treatments for this purpose have been recently evaluated. Other smuts, *Sphacelotheca reiliana*, *Sphacelotheca sorghi*, *Sphacelotheca cruenta*, and *Tolyposporium ehrenbergii*, occur in most countries, but are not presently of economic importance.

Downy Mildew (*Perenosclerospora sorghi*). Downy mildew is common at the Sebele research station in Botswana; the Ilonnga, Singida, and Ukiriguru research stations in Tanzania; most research stations in Zambia; and the Henderson, Matopos, and Panmure stations in Zimbabwe. In South Africa, the disease is limited to the northern productions areas where it can be economically important. Research has been conducted on epidemiology of, genetic variability in, and control of this pathogen (1, 4, 5). The geographic distribution of downy mildew (3) and its ability to attack a number of grasses also were examined (2).

Sooty stripe (*Ramulispora sorghi*). Sooty stripe is found in Zimbabwe (Save Valley), Zambia (Golden Valley), and at the research stations in Tanzania. In 1995, unlike other years, a high incidence of sooty stripe was reported at Mansa (Zambia), a location better known for its high incidence of anthracnose. Breeders and pathologists in most countries screen germplasm for tolerance to sooty stripe (6).

Rust (*Puccunia purpurea*). Rust is common at research stations in Tanzania and Zimbabwe. In South Africa, disease severity has been increasing the past few years and was particularly severe during the 1998/99 and 1999/2000 cropping seasons. Breeders and pathologists in most countries screen germplasm for tolerance to rust (6).

Grain molds (*Fusarium* spp., *Curvularia lunata*, *Phoma sorghina*, *Alternaria* spp., *Bipolaris* spp., and *Drechslera* spp.). Grain molds are very important in

Tanzania and South Africa. In South Africa during the 1999/2000 cropping season, seed companies and producers experienced seed losses worth several million Rand due to poor seed germination. Characterization of *Fusarium* section *Liseola* was conducted in Tanzania (17), and grain-mold resistant germplasm has been identified (16).

Bacteria. Frederickson reported leaf spots of pearl millet caused by *Pontoea agglomerans* and *Pseudomonas syringae* (15), but these diseases are not of economic importance.

Viruses. Research in viruses has primarily been externally based. Identification of SCMV and MDMV from the region was done in the United States (29).

Other Diseases. In total, over 30 diseases have been reported in the region (18, 19), but many of these are of minor importance.

Most southern African countries participate in international collaborative research with ICRISAT, NRI (UK), DANIDA (Denmark), and/or INTSORMIL (USA). Through these cooperative research efforts, some countries have been able to train personnel and to execute projects focused on sorghum and millet pathology. For example, through INTSORMIL Project TAM-228 sorghum lines with desirable characters, both agronomic and disease tolerance, have been identified and incorporated by some breeders into their crossing blocks.

Future Needs

- Information on yield and economic losses attributable to most sorghum diseases is lacking in most countries.
- Sorghum virus research is lacking in most southern African countries. For example, in Zimbabwe there is no virologist in the Department of Research and Specialist Services.
- Quarantine services and seed health testing need to be strengthened. The appearance of *Pantoea agglomerans* in an International *Pseudomonas syringae* nursery in Zimbabwe may have been due to poor seed health testing and quarantine services.
- A shortage of human and material resources is a limiting factor at most research stations. In some cases, the research and the priorities are donor driven.
- The impact of major weather events, *e.g.*, cyclone Elaine, on sorghum disease incidence and distribution needs to be analyzed.

References

1. **Bock, C. H.** 1995. *Studies of the Epidemiology, Variability and Control of Sorghum Downy Mildew* Peronosclerospora sorghi *(Weston & Uppal C.G. Shaw) on Sorghum and Maize in Africa.* Ph.D. dissertation. Univ. of Reading, Reading, UK.
2. **Bock, C. H., M. J. Jeger, and N. Bosque-Perez.** 1996. Host range of sorghum downy mildew in Africa. *Intl. Sorg. Millets Newsl.* **37:** 56-58.
3. **Bock, C. H., M. J. Jeger, L. K. Mughogho, K. F. Cardwell, V. Adenele, E. Mtisi, A. D. Akpa, G. Kaula, D. Mukasambina, and C. B. Meyers.** 1998. Occurrence and distribution of *Peronosclerospora sorghi* [Weston & Uppal (Shaw)] in selected countries of west and southern Africa. *Crop Protect.* **17:** 427-439.
4. **Bock, C. H., M. J. Jeger, L. K. Mughogho, K. F. Cardwell, and E. Mtisi.** 1999. Effect of dew point, temperature and conidium age on germination, germ tube growth and infection of maize and sorghum by *Peronosclerospora sorghi. Mycol. Res.* **103:** 854-859.
5. **Bock, C. H., M. J. Jeger, L. K. Mughogho, E. Mtisi, and K. F. Cardwell.** 1998. Production of conidia by *Peronosclerospora sorghi* in Zimbabwe. *Plant Pathol.* **47:** 243-251.
6. **de Milliano, W. A. J.** 1990. Sorghum pathology in the SADCC region. In *Proc. 6th Regional Workshop on Sorghum and Millets for Southern Africa (Sept 18-22, 1989, Bulawayo, Zimbabwe)*: 91-104.
7. **de Milliano, W. A. J., R. A. Frederiksen, and G. D. Bengston, eds.** 1992. *Sorghum and Millets Diseases: A Second World Review.* ICRISAT, Pantancheru (A.P.), India.
8. **Frederickson, D. E.** 1990. *Ergot Disease of Sorghum.* Ph.D. Dissertation. Univ. of London, London, UK.
9. **Frederickson, D. E.** 1999. The life cycle, spread and survival of the sorghum pathogen. *Proc. Global Conf. on Ergot of Sorghum (1-8 June 1997, Sete Lagoas, Brazil)*: 135-139.
10. **Frederickson, D. E., and P. G. Mantle.** 1996. Pearl millet as an alternate host of sorghum ergot pathogen *Claviceps africana. Intl. Sorg. Millets Newsl.* **37:** 83-85.
11. **Frederickson, D. E., P. G. Mantle, and W. A. J. de Milliano.** 1989. Secondary conidiation of *Spacelia sorghi* in sorghum, a novel factor in the epidemiology of ergot disease. *Mycol. Res.* **93:** 497-502.
12. **Frederickson, D. E., P. G. Mantle, and W. A. J. de Milliano.** 1991. *Claviceps africana* sp. nov., the distinctive ergot pathogen of sorghum in Africa. *Mycol. Res.* **95:** 1101-1107.
13. **Frederickson, D. E., P. G. Mantle, W. A. J. de Milliano.** 1993. Wind-borne spread of ergot disease (*Claviceps africana*) in sorghum A-lines in Zimbabwe. *Plant Pathol.* **42:** 238-377.
14. **Frederickson, D. E., P. G. Mantle, and W. A. J. de Milliano.** 1994. Susceptibility to ergot in Zimbabwe of sorghums that remained uninfected in their native climate in Ethiopia and Rwanda. *Plant Pathol.* **43:** 27-32.

15. **Frederickson, D. E., E. S. Monyo, S. B. King, and G. N. Odvody.** 1997. A disease of pearl millet in Zimbabwe caused by *Pantoea agglomerans*. *Plant Dis.* **81:** 959.
16. **Mansuetus, S. B. A.** 1995. Sorghum diseases in Tanzania. *Afr. Crop Sci.* J. **3:** 203-207.
17. **Mansuetus, A. S. B., G. N. Odvody, R. A. Frederiksen, and J. F. Leslie.** 1997. Biological species of *Gibberella fujikuroi* (*Fusarium* section *Liseola*) recovered from sorghum in Tanzania. *Mycol. Res.* **101:** 815-820.
18. **Masuka, A. J., D. L. Cole, and C. Mguni, eds.** 1998. *List of Plant Diseases in Zimbabwe*. Dept. of Research and Specialist Services, Causeway, Harare, Zimbabwe.
19. **Mbwaga, A. M., S. Pande, W. A. J. de Milliano, and R. J. Karunakar.** 1993. Diseases and parasitic weeds of sorghum in Tanzania: Occurrence and incidence, 1986-1990. *Crop Protect.* **12:** 183-188.
20. **McLaren, N. W.** 1987. *Studies on Pre- and Post-emergence Damping-off and Seedling Blight.* M.Sc. (Agric.) Thesis. Univ. of Natal, Pietemaritzburg, South Africa.
21. **McLaren, N. W.** 1992. Quantifying resistance of sorghum genotypes to the sugary disease pathogen (*Claviceps africana*). *Plant Dis.* **76:** 986-988.
22. **McLaren, N. W.** 1993. Effect of sugary disease exudates on germination, seedling development and predisposition to seedling diseases of sorghum (*Sorghum bicolor*). *So. Afr. J. Plant Soil* **10:** 12-16.
23. **McLaren, N. W.** 1994. Efficacy of systemic fungicides and timing of preventive sprays in control of sugary disease of sorghum (*Sorghum bicolor*). *So. Afr. J. Plant Soil* **11:** 30-33.
24. **McLaren, N. W.** 1999. Host × pathogen × environment interaction in sorghum ergot disease. *Proc. Global Conf. on Ergot of Sorghum (June 1-8, 1997, Sete Lagoas, Brazil)*: 114-134.
25. **McLaren, N. W., and F. C. Wehner.** 1990. Relationship between climatic variables during early flowering of sorghum and incidence of sugary disease caused by *Sphacelia sorghi*. *J. Phytopathol.* **130:** 82-88.
26. **Mtisi, E.** 1997. Evaluation of seed treatments for the control of covered kernel smut on sorghum. *Afr. Crop Sci. J.* **3:** 943-946.
27. **Nowell, D. C., and N. W. McLaren.** 1999. Experiences with ergot of sorghum in southern Africa. *Proc. Global Conf. on Ergot of Sorghum (1-8 June 1997, Sete Lagoas, Brazil)*: 36-40.
28. **SACCAR (Southern African Centre for Cooperation in Agriculture Research).** 1995. Report on the impact assessment of the SADC/ICRISAT Sorghum and Millet Improvement Program. SACCAR, Gaborone, Botswana.
29. **Toler, R. W., B. Nath, D. Roose, and G. M. Kaula.** 1989. Occurrence of Maize Dwarf Mosaic Virus on sorghum in Zambia. *Plant Dis.* **73:** 273.

72

Sorghum Viruses in Asia and Africa

Y. D. Narayana, Ranajit Bandyopadhyay, S. S. Navi, and V. Muniyappa

Abstract

Viruses infecting sorghum are widely distributed in Asia and Africa and are of economic importance. In recent years, apparent virus disease incidence has increased, probably due to changes in agricultural practices associated with the introduction of hybrids and new varieties. A total of 23 viruses are known pathogens either in the field or following experimental transmission to sorghum. Of these 23, five are economically important: sugarcane mosaic virus (SCMV), maize dwarf mosaic virus (MDMV), maize stripe virus (MStV), maize mosaic virus (MMV), and sorghum mosaic virus (SrMV). These viruses cause various symptoms including mosaic, chlorosis, streaks, stripes, reddening of leaves, stunting, delayed flowering, smaller heads, and quantitative and qualitative losses in grain and stover. Serological assays and biotechnological approaches using polyclonal or monoclonal antisera have been widely applied and can be used in developing countries for early diagnosis of viral diseases. Graminaceous native and weed grasses serve as virus refuges, and their vectors play a key role in virus disease epidemics. In some cases, virus spread is reduced by cultural practices that reduce vector populations or by applying insecticides mixed with oil early in the cropping season. The most economic and practical way to control sorghum viral diseases is with tolerant/resistant cultivars, but these are not always available in adapted hybrids or varieties. The genetic basis of resistance to virus diseases depends on host-virus-vector relationships.

Introduction

Viruses and virus-like diseases infecting sorghum are distributed worldwide and cause varying levels of economic losses. In recent years, the incidence of virus diseases in Asia and Africa has increased. The first virus disease described on sorghum was sugarcane mosaic virus (SCMV) (10). In 1962, viral diseases of sorghum were reviewed and described as causing little damage worldwide (79), but later reviews (32, 84) identified 16 viruses as pathogens of sorghum. Since then, 23 viruses have been identified within 12 virus groups, and other

TABLE 72-1. Distribution sorghum viruses by country in Asia and Africa.

Country	Virus Diseases	Reference
Asia		
China	MDMV	94
India	MDMV(Strains-A, H), MMV, MStV, SCMV (Strains - A, B, D, E, F and H), SoMV-Ind & SrMV	7, 26, 49, 51, 62, 65, 66
Indonesia	MDMV	4
Iran	Iranian Maize Mosaic, IranianWheat Stripe tenuivirus, Johnsongrass Chlorotic, MMV, SCMV, JGMV-JG & Stripe Mosaic	29-31
Japan	SCMV & MDMV-A	41
Korea	MDMV	12
Philippines	SCMV (Redleaf) and FDV	6, 14, 28, 58
Thailand	MStV & MDMV	71
Africa		
Botswana	MStV	24
Burkino Faso	MDMV, MMV & MStV	89
Egypt	MYSV & BYDV	1
Ethiopia	MSV, MDMV & JGMV	43
Ivory Coast	MDMV, MMV & MStV	89
Kenya	MStV & SCMV	24, 44
Mali	MDMV, MMV, MStV & SCMV	89
Malawi	MSV	16
Nigeria	MMV, MSV, MStV, Potyvirus, SCMV & Sugarcane Fiji disease	17, 59
Swaziland	MDMV	16
Zambia	MDMV & Brome mosaic virus	42, 87

Abbreviations: BYDV, Barley Yellow Dwarf Virus; FDV, Sugarcane Fiji Virus; JGMV, Johnsongrass Mosaic Virus; MDMV, Maize Dwarf Mosaic Virus; MMV, Maize Mosaic Virus; MSV, Maize Streak Virus; MStV, Maize Stripe Virus; MYSV, Maize Yellow Stripe Virus; and SCMV, Sugarcane Mosaic Virus.

viruses are known that have not been classified (22). However, most viruses infecting sorghum belong to one of six virus groups. The potyviruses, maize dwarf mosaic virus (MDMV), SCMV, and their derivatives, are the most widespread viral pathogens of sorghum (21). Some of these viruses also are important in developing countries (65, 92).

Geographic Distribution

Sorghum viruses are widely distributed in Asia and Africa (Table 72-1), and causal agents of many other symptoms remain to be identified and confirmed. MDMV and SCMV are distributed worldwide. The sorghum strains of maize stripe virus (MStV), maize mosaic virus (MMV), and sorghum mosaic potyvirus (SrMV) are distributed widely in India (7, 19, 51, 62, 66). The incidence and distribution of MStV is increasing in India (52). MStV and MDMV are common in Thailand (71), MDMV in both China (94) and Korea (12), and SCMV in Japan (41). SCMV has been reported on sorghum in many Indian states (37, 39, 45), and many derivatives of SCMV, as well as MDMV and its derivatives, have been reported from several regions in India (49, 65, 66), where they infect sugarcane, maize, sorghum, and johnsongrass (*Sorghum halepense*) (Table 72-1). Most researchers (*e.g.*, 19, 26, 66) use the base name "maize dwarf mosaic virus" and then add strain designations, *e.g.*, MDMV-A, MDMV-B, MDMV-C, MDMV-D, MDMV-E, MDMV-F, and MDMV-O. Only MDMV strains A and B are naturally transmitted to sorghum. In India, SCMV on sorghum causes red stripe symptoms as strains of a derivative, johnsongrass mosaic virus (JGMV) (45), are common, widely dispersed, and can result in considerable economic loss (91).

In South Africa, brome mosaic virus (BMV), MDMV, and SCMV have been reported and appear to be increasing in importance (16). Sugarcane Fiji disease *fijivirus* (FDV) occurs on sorghum in the Philippines and Nigeria but does not cause losses (28, 58). Virus and virus-like diseases on sorghum that differ serologically from the viruses on sorghum in India also have been observed in Zambia (87). Low incidence of MStV, MMV, and MDMV has been reported in western Africa (Burkina Faso, Cote d'Ivoire, and Mali) and Cameroon, in central Africa, on maize, sorghum, and nondomesticated grasses (Poaceae) (89). MDMV, johnsongrass mosaic virus (JGMV), maize mosaic chlorotic stunt virus, and maize streak virus (MSV) also have been identified by using serological tests and electron microscopy from samples of maize and several grasses from western, central, and southern parts of Ethiopia (43). Similarly, ELISA tests have been used to confirm the presence of MStV, MMV, and a potyvirus in four major sorghum growing climatic zones (Sahel, Sudan, northern Guinea, and southern Guinea) of Nigeria (59), and the presence of MStV in Botswana and Mauritius (9).

The serological relationship of MDMV to SCMV initially suggested that MDMV should be named the Johnsongrass strain of SCMV or SCMV-Jg (73). Subsequently, analyses of amino-terminal serology (46), divided the 17 potyviruses related to SCMV that infect maize, sorghum, and sugarcane into four groups: JGMV, SrMV, MDMV, and SCMV (75, 81). SCMV and MDMV are the definitive members of family potyviridae (76) and can infect sugarcane, maize, sorghum and other Poaceous plant species (11, 33).

Economic Importance

Sorghum viruses in Asia and Africa cause considerable losses in grain yield and fodder quality. The loss in sorghum grain yield to virus diseases usually is reflected in lower total grain weight and test weight (21, 85). In general, the extent of loss depends on the stage of the crop at infection, with earlier infections resulting in greater losses (53). For example, MDMV-infected sorghum plants suffered a 16% reduction in plant height and a 13% reduction in head size (82). Similarly, sorghum red stripe virus (SRSV), a derivative of JGMV found in Maharashtra and Karnataka, reduces both grain (26-52%) and fodder (40-60%) yields depending upon the cultivator and the time of infection (3, 92). MStV and MMV also cause considerable losses in vegetative growth and grain yield of sorghum grown in India (53), with plant height, head length, and grain yield per plant all reduced (34). MMV also has been reported to reduce plant biomass by 10-53%, while increasing the total soluble sugars and N in infected plants (93). Reduction in stover yield is par-ticularly important and can cause significant losses in in-come from crop-livestock production system in rural and peri-urban India since sorghum crop residue is an impor-tant feed for cattle in this system (64).

Symptomology

Plant viruses are identified by the symptoms they induce in host plants and their morphology, mode of transmission, and biochemical and serological properties. In some cases, the reaction on an indicator host or host differenttials aids in identification (60). The earliest symptoms of viral diseases under field conditions often resemble those associated with herbicide, insecticide, or fungicide damage, or with genetic abnormalities (47, 68). These symptoms may include chlorosis, yellowing, chlorotic streaks, stripes or rings, reddening of leaves, necrotic spots, dwarfing/stunting, rosetting, excessive tillering, sterility and delayed flowering, and smaller heads with few or no seeds. Mycoplasma/rikettsia and genetic and nutritional abnormalities can be distinguished from viral pathogens by some researchers based on growth in axenic culture, electron microscopy, and/or immunofluorescent microscopy (67).

Symptoms of MDMV infection depend on genotype, virus strain, temperature, and time of infection. The classic symptoms are red leaf, red stripe, necrosis, stunting, delay in flowering, and reductions in head length, seed size, number of seeds/head, and grain yield. The mosaic pattern develops into narrow, light green, or chlorotic streaks of leaves; red discoloration of leaf sheaths; and peduncle stunting in later stages of plant growth. The degree of dwarfing or stunting due to MDMV-A can be determined by comparing the height of diseased and healthy plants (21).

Sorghum plants infected with SCMV or its derivatives have light green mottling or mosaic, often with red lesions. The infected plants are stunted and the leaf area, number of heads, number of seeds per head, and head length may be reduced. The red stripe virus disease on sorghum is attributed to JGMV, which is characterized by chlorotic and necrotic red stripes, either broken or continuous, that parallel the veins of younger leaves and result in mosaic mottling of older leaves and premature withering of the entire plant (45).

A sorghum plant infected by MSTV is characterized by continuous chlorotic stripes and bands on leaves, severe stunting, and panicles that abort or contain little grain (62). Symptoms of MMV on sorghum in the Indian subcontinent include fine continuous chlorotic or broken streaks between the veins that may become necrotic as the disease progresses. When the infection becomes severe, the whole leaf is chlorotic, the plant is severely stunted, and the panicles produced contain few seeds (51). MSV, when it infects sorghum, results in pale circular spots on the lowest portion of young leaves, followed by pronounced chlorosis, narrow broken stripes along the veins, and stunted plant growth (8, 83). The symptoms of SFDV observed on sorghum in the Philippines and Nigeria include elongated galls or enations on stems and the undersurface of leaves, shortened stalks with stiff malformed leaves, and plants that turn prematurely brown without producing panicles (28, 58).

Diagnosis of Viruses

Identification of sorghum viruses on a routine basis is required to develop effective control measures. Traditional approaches for detecting and identifying plant viruses include symptoms, transmissibility, host range, reaction on indicator hosts, vector relationships, and immunological, physical, and chemical properties. Early diagnosis techniques include microscopic studies of cellular inclusions and electron microscopy combined with serological assays (86). Serological assays, *e.g.*, ELISA, that use either polyclonal or monoclonal virus-specific antibodies have been widely used for early detection of many plant viruses and their derivatives (25, 75, 78). Potyvirus identification and classification often is hampered by variable proportions of cross-reacting antibodies in the antiserum. This limitation can be overcome by developing monoclonal and virus-specific anti-bodies (75). The serological assays, when combined with new biotechnological approaches, can quickly and effectively identify viruses and their derivatives. Such assays have

TABLE 72-2. Characteristics of viruses infecting sorghum in Asia and Africa.

Group	Virus	Shape	Size (nm)	Vector	Reference
Potyvirus	Maize dwarf mosaic virus	Filamentous	770	Aphids	11
	Sugarcane mosaic virus	Filamentous	755-770	Aphids	11
	Johnsongrass mosaic virus	Filamentous	773-778	Aphids	11
	Sorghum mosaic virus	Filamentous	702-714	Aphids	11
Tenuivirus	Maize stripe virus (Sorghum)	Flexuous branched filaments	< 10	*Peregrinus maidis*	62
Rhabdovirus	Maize mosaic virus (Sorghum)	Bacilliform	230 × 90	*Peregrinus maidis*	51
Geminivirus	Maize streak virus	Isometric hexagonal	18 × 20, 20 nm diam.	*Cicadulina mbila*	8
Reovirus	Sugarcane Fiji virus	Isometric	70 nm width	*Perkinsiella* spp.	28
Bromovirus	Brome mosaic virus	Icosahedral	25 nm width	Sap	42

been used in disease surveys to determine the incidence and distribution of sorghum viruses, to detect viruses in vectors, and to study mixed infection of plants by different virus strains (9, 18, 23, 54, 59, 63, 66, 90).

Causal Organisms

Twenty-three viruses are sorghum pathogens under either field or experimental conditions, of which nine (Table 72-2) have been particularly well characterized. There are several viruses on sorghum in Asia and Africa that remain to be identified and characterized.

Transmission

Plant viruses are obligate pathogens. Identification usually follows proof of the pathogenicity of the causal agent. Testing potential vectors is critical if the virus is not sap transmitted. Transmission methods that should be tested include mechanical sap inoculation through arthropod, nematode, fungal and insect vectors, and seeds. Sorghum viruses are transmitted by nonpersistent, semipersistent and persistent vectors that differ in the length of time needed by the vectors to acquire the virus from the source and to inoculate the virus. MDMV and SCMV infecting sorghum may be transmitted by sap and by various aphid species. Low incidence of seed transmission of MDMV has been reported for maize, but not for sorghum (74). These viruses are vectored by 23 species of aphids, including *Rhopalosiphum maidis*, *Rhopalosiphum padi*, *Myzus persicae*, *Aphis craccivora*, *Aphis nerri*, *Aphis gossypii*, *Hysterneura stariae*, *Melanaphis sacchari*, *Dactynotus ambrosiae*, and *Schizaphis graminum* (38). The number of vector species in Africa is not known. SCMV is transmitted in a nonpersistent manner by several aphid species with a probing period as short as 30-60 seconds. A single aphid can infect a plant with a single probe, but cannot transmit the virus to other plants without again probing an infected plant. Air currents can disperse viruliferous aphids over long distances. These vectors interact with ecological factors to produce viral epidemics in various sorghum growing regions (32).

Planthopper-borne viruses are major constraints for or potential threats to cereal production in Asia and Africa. MStV and MMV graminaceous hosts, including sorghum (24), and are persistently transmitted by planthoppers, but are neither mechanically nor seed transmitted. The recently reported maize yellow stripe virus has several tenuivirus characters, but is persistently transmitted by the leafhopper *Cicadulina chinai*. The latent period, *i.e.*, the period of time after the acquisition of a virus by a vector before the vector becomes infectious, of planthopper-borne viruses in the vector range from 7-21 days, and retention period, *i.e.*, the period in which the vector remains infectious, is up to 84 days (2).

The mode of transmission of MMV and MStV by *P. maidis* has been studied in detail (51, 62, 90). The vector acquires the virus after a minimum acquisition access feeding period of four hours. After the virus is acquired, its latent period in the vector is 8-22 days. The vector requires a minimum inoculation access period of 1 hour to transmit the virus. The transmission efficiency of the vector was 42% with a single viruliferous vector and was 100% when a group of five viruliferous vectors were used to inoculate seedlings (54). Multiplication of MStV in the vector was demonstrated with ELISA, and such reproduction leads to higher estimates of transmission (57). The vector transmits the virus in a persistent manner and retains the virus until its death. The nymphs and macropterous females of *P. maidis* were more efficient transmitters of MStV and MMV than were the males (35, 54). Transovarial transmission may be sufficiently frequent for the virus to persist through many generations of the vector without further access to infected plants (54, 90). Thus, a comprehensive understanding of the biology, persistence, and distribution of insects, and of virus-vector-host relationships is important in the development of disease management practices.

Epidemiology

The effectiveness of plant virus transmission and spread depends on several environmental factors, including the availability of alternate hosts, and the ability of a virus to multiply in the vector(s) and to be passed to the vector's progeny (56). MDMV and SCMV are nonpersistently transmitted by aphids, can survive on perennial hosts, and may intensify epidemics in early cropping seasons. The green bug, *Schizaphis graminum*, overwinters on small grain weed hosts and grasses and appears to be the primary agent responsible for spread of MDMV-A, although the virus is mechanically transmissible (32).

Alternate hosts also play an important role in the perpetuation of viruses infecting sorghum, and include maize, elephant grass, pearl millet, and sorghum (7). Johnsongrass is the most notorious of the alternate hosts. It serves as a perennial reservoir for and a primary inoculum source of MDMV-A (85), MMV (34), and MStV (52). These three viruses also naturally infect maize and noncultivated poaceous grasses in four western and central African countries (89). In Iran, MStV infects wheat, barley, oats, rice, rye, sorghum, and various noncultivated graminaceous hosts, *e.g.*, Johnsongrass (29). Four species, five subspecies, and one commercial sorghum accession were susceptible to MSV in experimental transmissions with the vector *Cicadulina mbila* (15).

The age of sorghum and maize plants affects susceptibility, symptom expression, severity of damage, and yield loss. For example, the highest percentage (62%) of MStV transmission occurred when seedlings at the two-leaf stage were exposed to viruliferous vectors (53). Early infection results in more infected leaves than a late infection, due to the systemic nature of the virus, and early infections cause greater damage to plants than do late infections (53).

Hopper-borne viruses are considered a major constraint of or threat to the production of sorghum in Asia and maize in Africa (56). The persistent and transovarial nature of virus transmission across insect generations contributes greatly to the perpetuation of the disease. Previously unknown viruses from indigenous plants, on which a virus produces little or no injury, can become destructive if a new, susceptible variety is introduced. For example, *P. maidis* is a vector of both MSV and MMV and a major pest on sorghum (77) that could become a significant threat to sorghum production.

Management of Sorghum Viruses

The management of plant viruses involves the integration of several methods. Effective control of sorghum viruses usually involves the destruction of johnsongrass, a widespread weed. Perennial grasses cultivated as pastures, if irrigated, often support large planthopper populations and increase the risk of virus transmission when adjacent crops and grasses are dry or drying. Avoiding growing sorghum under irrigation where other cereals were previously grown, rotation with crops that are not viral hosts, the use of selective insecticides to control the vector, and using virus nonhost, resistant, or tolerant cultivars all can reduce virus spread under field conditions (85).

Mixed infection of the same plant by two or more viruses may result in cross-protection, wherein one virus reduces infection by another. Related strains of the same virus are capable of cross protection, whereas distinct groups are not (40). Cross protection was tested among strains of viruses from the sugarcane mosaic virus subgroup, *i.e.*, SCMV, MDMV, JGMV, and SrMV, in 53 pairwise combinations. Cross protection was found only between SCMV-MDB and SCMV-BC, and only when SCMV-MDB was inoculated first and SCMV-BC used as a challenge strain.

Vector specificity, vector numbers and behavior, and the potential for mechanical and seed transmission all impact virus spread. Altering vector behavior with light, sticky traps, pheromone traps, and the planting of disease-tolerant or -resistant varieties that reduce virus multiplication all can improve virus disease control (95). Insecticides often fail to limit virus transmission because virus spread is complete before the insecticide kills the insect vector. In some cases insecticide/herbicide applications are associated with increased virus spread (72). Foliar application of a mixture of mineral oil and pyrethroids that interfere with virus transmission, however, has been successful (20).

Host plant resistance is the most effective way to manage viral diseases of sorghum, since resistant varieties and hybrids can easily be adopted by farmers. Tolerance is the primary mechanism for control of MDMV in grain sorghum and is available commercially (70). Field resistance may be used when virus inoculum pressure is low. In a test of field resistance, six sorghum lines–IS-2549C, IS-2816C, IS-12612C, IS-12666C, Rio, and TAM-2566–were infected by mechanical inoculation but remained healthy even with heavy infestations of viruliferous aphids in the field. Immunity to MDMV was first observed in *krish* sorghum in a cross between *S. halepense* and *S. roxburghi*. A sorghum genotype resistant to JGMV and all four Australian strains of SCMV, as well as immune to the isolate of MDMV-A from the United States, has been identified (32, 88).

Genetics of resistance to the variable sorghum potyviruses (47, 88) has been described (61). The *krish* gene conditions immunity, which depends upon a single gene, *K*, with resistance, *i.e.*, symptomless plants, dominant to susceptibility (13). Resistance based on *krish* has been used extensively since 1971 in breeding programs worldwide to develop cultivars resistant to JGMV, SCMV, SrMV, MDMV, and their derivatives. MDMV susceptibility in sorghum is governed by at least three dominant/recessive factors and four with epistatic effects (48). Sorghum lines Cempaka, CS110, and CS 116, based on germplasm from Nigeria, all had high levels of resistance to MDMV and might serve as a potential source of resistance to this virus. In another study, 11 Kenyan local sorghum lines suffered little damage from MDMV alone, and ICSH 193, a high-yielding hybrid from ICRISAT, was free of MDMV symptoms (82).

Coat-protein-mediated resistance occurs for potyviruses (5, 27, 36). Transgenic maize plants that express the MDMV-B coat protein (50) were resistant to MDMV-A, MDMV-B, and a mixed infection of MDMV and MCMV. Thus, coat protein-mediated resistance protects the host from related viruses and their derivatives. Similar strategies could provide alternatives to or supplements for existing genetic resistance. For example, introducing MDMV-A coat protein into a high-yielding female sor-

TABLE 72-3. Reactions of sorghum to sorghum stripe virus following greenhouse inoculation via *Peregrinus maidis*[a].

Source	Number of Genotypes in Each % Susceptibility Class[b]									
	10[c]	20	30	40	50	60	70	80	90	100
Germplasm[d]	-	3	3	3	4	2	4	5	1	3
Breeding Populations[e]	-	3	12	14	7	4	-	-	-	-
Hybrids & Varieties	-	3	7	14	11	14	6	-	1	-
Other Lines[d]	2	3	5	9	4	3	4	-	-	-
Advanced Varieties	1	7	13	7	7	3	2	-	-	-

[a]Second and third instar nymphs had a 4-day acquisition access period and a 10 day incubation period.
[b]24 seedlings of each cultivar were inoculated at the two-leaf stage.
[c]Percentage of infected plants.
[d]Source of lines–ICRISAT, Patancheru (A.P.), India.
[e]Source of lines–National Sorghum Research Center, Rajendranagar (A.P.), India.

ghum line (A/B), without introducing undesirable traits, could provide resistance to other sorghum potyviruses.

Attempts have been made to identify sources of resistance to MStV and MMV either by exposing the plants to natural infection in the field (34) or by inoculating seedlings in the greenhouse (34, 55). None of the sorghum genotypes tested was resistant to both viruses in the greenhouse evaluations (Table 72-3). There are several combinations of factors that favor reductions in disease including: (i) nonpreference of the vectors for the resistant plants, (ii) resistance to infection, and (iii) tolerance to disease development. Evaluation of elite sorghum germplasm accessions with appropriate and reliable techniques to identify resistance to important virus diseases remains part of sorghum improvement and breeding programs (69).

Conclusions

The incidence of viruses infecting sorghum is increasing, especially in Asia and Africa, due to changes in agricultural management practices and the widespread cultivation of a few high-yielding cultivars throughout the year. Virus and virus-like symptoms on sorghum have been repeatedly reported, but a significant number remain to be identified and characterized. The main constraints to research on virus diseases of sorghum are a lack of appropriately trained pathologists, and access to the materials and equipment that enable rapid detection and identification of viruses. The application of serological assays and other biotechnological approaches have several advantages relative to traditional methods for the early diagnosis and identification of viruses and their derivatives, and describing the ecology and epidemiology of sorghum virus diseases in developing countries. For example, the generation of reliable information on the epidemiology of MStV was possible only because serological assays for this virus were available (52, 54).

Determining the distribution of gramineaceous wild and weed grasses that could serve as a reservoir for both viruses and their vectors is needed to avoid disease epidemics and to develop effective management practices. The survival and multiplication of viruses in their vectors, their transovarial passage ability, and the availability of alternate and collateral hosts for both the virus and the vector need to be considered in the management of virus diseases of sorghum. Farmers should be encouraged to adopt the cultural practices that minimize viral diseases, *e.g.*, the modification of planting dates and the removal of alternate and collateral hosts on which the virus and the vector can multiply. Although combinations of insecticides with oils may effectively reduce disease spread, insecticides should be used judiciously since they may affect the applicator, the consumer, and other beneficial fauna.

When available, the most economical and practical approach for controlling sorghum virus diseases is through the planting of resistant or tolerant cultivars. The genetic basis for resistance to viruses infecting sorghum is incompletely understood, and needs further research. Biotechnological approaches should be useful in virus diagnosis and identification, and in breeding for disease resistance through marker-assisted selection. For some potyviruses, coat-protein-mediated resistance is known, and a better understanding of coat protein functions may lead to the design of a gene that blocks assembly of the virus, reduces its systemic spread within plants, or alters its transmission by insect vectors. Coat-protein-mediated resistance presently is the most attractive alternative for supplementing existing genetic resistance. It is likely that biotechnology, coupled with conventional methods, will result in new management practices for sorghum viruses in Africa and Asia.

Acknowledgments

Research described in this chapter was partially funded by the United Kingdom Department for International Development (DFID, R 7346, the Crop Protection Programme) for the benefit of developing countries. The views expressed are not necessarily those of DFID.

References

1. **Abul-Ata, A. E., J. C. Thouvenel, K. M. Makkouk, and M. M. Satour.** 1992. Barley yellow dwarf virus in Egypt: Natural incidence and transmission and wild hosts. *Arab Pl. Prot. J.* **10:** 226-231.
2. **Ammar, E. D., R. E. Gingery, and L. R. Nault.** 1987. Interactions between maize mosaic and maize stripe viruses in their vector, *Peregrinus maidis* and in maize. *Phytopathology* **77:** 1051-1056.
3. **Anahosur, K. H., K. A. Kulkarni, and R. Parameswarappa.** 1980. Red stripe disease of sorghum in Karnataka. *Sorg. Newsl.* **23:** 122.
4. **Baliadi, A., R. Saleh, and M. Hadi.** 1995. Screening of resistance to maize and sorghum varieties/lines to maize dwarf mosaic virus, p. 36-39. *1991 Report of Research Results*. Malang Res. Inst. of Food Crops, Malang, Indonesia.
5. **Beachy, R. N., S. Loesch-Fries, and N. E. Tunmer.** 1990. Coat protein mediated resistance against virus infection. *Annu. Rev. Phytopathol.* **28:** 451-474.
6. **Benegno, D. R., and D. C. Vergara.** 1977. Red stripe of sorghum in the Philippines. *Philip. Agric.* **61:** 157-165.
7. **Bhargava, K. S.** 1975. Sugarcane mosaic-retrospect and prospect. *Ind. Phytopathol.* **28:** 1-11.
8. **Bock, K. R.** 1974. Maize streak virus. *CMI/AAB Descriptions of Plant Viruses*. No. 133.
9. **Bradfute, O. E., and J. H. Tsai.** 1990. Rapid identification of maize stripe virus. *Phytopathology* **80:** 715-719.
10. **Brandes, E. W., and P. J. Klapaak.** 1923. Cultivated and wild hosts of sugarcane or grass mosaic. *J. Agric. Res.* **24:** 257-262.
11. **Brunt, A. A., K. Crabtree, M. J. Dallwitz, A. J. Gibbs, and L. Watson, eds.** 1996. *Viruses of Plants. Descriptions and Lists from VIDE Database*. CAB International, Wallingford, Oxon, UK.
12. **Choi, C. W., D. J. Gallenberg, and W. S. Gardner.** 1986. Response of selected sorghum hybrids to inoculation with MDMV A and B strains. *Phytopathology* **76:** 1126-1127.
13. **Conde, B. D., R. F. Moore, D. S. Fletches, and D. S. Teakle.** 1976. Inheritance of the resistance of Krish sorghum to SCMV. *Austr. J. Agric. Res.* **27:** 45-52.
14. **Dalmacio, S. C.** 1979. Sorghum diseases in the Philippines, p. 70-72. *In* R. J. Williams, R. A. Frederiksen, L. K. Mughogho, and G. D. Bengston (eds.), *Sorghum Diseases: A World Review*. ICRISAT, Patancheru (A.P.), India.
15. **Damsteegt, V. D.** 1978. Maize streak virus: Additional hosts and vectors. *Phytopathol. News* **12:** 460.
16. **de Milliano, W. A. J.** 1992. Sorghum diseases in southern Africa, p. 9-19. *In* W. A. J. de Milliano, R. A. Frederiksen, and G. D. Bengston (eds.), *Sorghum and Millets Diseases: A Second World Review*. ICRISAT, Patancheru (A.P.), India.
17. **Fajemisin, J. M., G. E. Cook, F. Okusanya, and S. A. Shoyina.** 1976. Maize streak epiphytotic in Nigeria. *Plant Dis. Rept.* **60:** 443-447.
18. **Falk, B. W., and J. H. Tsai.** 1985. Serological detection and evidence for multiplication of maize mosaic virus in planthopper *Peregrinus maidis*. *Phytopathology* **75:** 852-855.
19. **Garud, T. B., and V. R. Mali.** 1983. A red stripe virus disease of sorghum in India. *Ind. Phytopathol.* **36:** 545-546.
20. **Gibson, R.W., and A. D. Rice.** 1986. The combined use of mineral oils and pyrethroids to control plant virus transmitted by non- and semi-persistently by *Myzus persicae*. *Ann. Appl. Biol.* **109:** 465-472.
21. **Giorda, L. M.** 1983. *Identification and Evaluation of an Isolate of Sugarcane Mosaic Virus*. M.S. thesis, Texas A&M Univ., College Station, Texas.
22. **Giorda, L. M., and M. J. Martinez.** 1995. Opportunities for biotechnology research on research on sorghum stalk rot of virus diseases. *Afr. Crop Sci. J.* **3:** 191-201.
23. **Giorda, L. M, and R. W. Toler.** 1985. Effect of single and mixed infection of MDMV strain-A, SCMV-H and an isolate of SCMV-H on 27 sorghum accessions of grain sorghum. *Sorg. Newsl.* **38:** 103-105.
24. **Gingery, R. E., L. R. Nault, and O. E. Bradfute.** 1981. Maize stripe virus characteristics of a member of a new virus class. *Virology* **112:** 99-108.
25. **Hamilton, R. I., J. R. Edwardson, R. I. B. Francki, H. T. Hsu, R. Hull, R. Koenig, and R. G. Milne.** 1981. Guidelines for the identification and characterization of plant viruses. *J. Gen. Virol.* **54:** 223-241.
26. **Hema, M., P. Sreenivasalu, K. Gopinath, G. Kiramai, and T. Satyanarayana.** 1997. Partial characterization of a potyvirus causing mosaic disease of sugarcane in Andhra Pradesh. *Ind. J. Virol.* **13:** 125-129.
27. **Hull, R.** 1994. Resistance to plant viruses: Obtaining genes by non-conventional approaches. *Euphytica* **75:** 195-205.
28. **Hutchinson, P. B. and R. I. B. Francki.** 1973. Sugarcane Fiji Virus Disease. *CMI/AAB Descriptions of Plant Viruses*. No. 119.
29. **Hyderneyed, J., and K. Izadpanah.** 1992. Isolation and partial characterization of a tenuivirus from wheat in Iran. *J. Phytopathol.* **136:** 279-287.
30. **Izadpanah, K., A. A. Ahmadi, S. Parvin, and S. A. Jafari.** 1993. Transmission, particle size, and additional hosts of the rhabdovirus causing maize mosaic in Shiraz, Iran. *Phytopathol. Z.* **107:** 283-288.
31. **Izadpanah, K., W. Huth, D. E. Lesemann, and H. J. Vetten.** 1993. Properties of johnsongrass chlorotic stripe mosaic virus. *J. Phytopathol.* **137:** 105-117.
32. **Jensen, S.** 2000. Diseases caused by viruses and virus-like organisms, p. 43-50. *In* R. A. Frederiksen and G. N. Odvody (eds.), *Compendium of Sorghum Diseases*, 2nd ed. APS Press, St. Paul, Minnesota.
33. **S. G. Jensen, S. G., and L. M. Giorda.** 2002. Virus diseases of sorghum and millet in the Americas and Australia, p. 403-410. *In* J. F. Leslie (ed.), *Sorghum and Millets Diseases*. Iowa State Press, Ames, Iowa.
34. **Jyothi, R.** 1994. Detection, host reactions and epidemiology of maize mosaic virus on sorghum. M.Sc. Thesis, Univ. of Agric. Sci., Bangalore, Karnataka, India.

35. **Jyothi, R., V. Muniyappa, and Y. D. Narayana.** 1995. Virus-vector relationships of maize mosaic virus and planthopper (*Peregrinus maidis*) in sorghum. *Ind. J. Virol.* **11:** 61-66.
36. **Kavanagh, T. A., and C. Spillane.** 1995. Strategies for engineering virus resistance in transgenic plants. *Euphytica* **85:** 149-158.
37. **Khurana, S. M. P., and K. Singh.** 1972. Sugarcane mosaic virus strain F & C in India and new sorghum differentials. *Sugarcane Pathol. Newsl.* **9:** 6-8.
38. **Knoke, J. K., R. J. Anderson, R. Louie, L. V. Madden, and W. R. Findley.** 1983. Insect vectors of maize virus and maize chlorotic dwarf virus, p. 130-138. *In* D. T. Gordon, J. K. Knoke, L. R. Nault, and R. M. Ritter (eds.), *Proc. Intl. Maize Virus Disease Colloq. and Work-shop (2-6 August 1982, Wooster, Ohio).* OARDC, Ohio State Univ., Wooster, Ohio.
39. **Kondaiah, E., and M. V. Naidu.** 1985. SCMV strain N: A new strain of sugarcane mosaic virus. *Sugarcane* **4:** 11-14.
40. **Krstic, B., R. E. Ford, D. D. Shukla, and M. Tosic.** 1995. Cross protection studies between strains of sugarcane mosaic, maize dwarf mosaic, johnsongrass mosaic and sorghum mosaic potyviruses. *Plant Dis.* **79:** 135-138.
41. **Kumigofukuro, T.** 1992. Sorghum and sorghum diseases of Japan, p. 31-34. *In* W. A. J. de Milliano, R. A. Frederiksen, and G. D. Bengston (eds.), *Sorghum and Millets Diseases: A Second World Review.* ICRISAT, Patancheru (A.P.), India.
42. **Lane, L.** 1977. Brome mosaic virus. *CMI/AAB Descriptions of Plant Viruses.* No. 180.
43. **Lencho, A., A. Abraham, B. Bekele, and M. Tesera.** 1997. Identification of viruses from maize and grass hosts. *Pest Manage. J. Ethiopia* **1:** 73-76.
44. **Louie, R.** 1980. Sugarcane mosaic virus in Kenya. *Plant Dis.* **64:** 944-947.
45. **Mali, V. R., and V. Y. Bhagwat.** 1975. Virus and virus-like diseases of sorghum in India. *Sorg. Newsl.* **18:** 41-42.
46. **McKern, N. M., D. D. Shukla, R. W. Toler, R.W, Jenson, M. Tosic, R. E. Ford, O. Leon, and C. W. Ward.** 1991. Confirmation that SCMV subgroup consists of four distinct potyviruses by using peptides profiles of coat proteins. *Phytopathology* **81:** 1025-1029.
47. **Merkle, M. G.** 1986. Pesticide injury, p. 60-61. *In* R. A. Frederiksen (ed.), *Compendium of Sorghum Diseases*, 1st ed. APS Press, St. Paul, Minnesota.
48. **Mijavec, C. A.** 1991. Studies of resistance of sorghum (*Sorghum bicolor*) genotypes to maize dwarf mosaic virus. Zbornik Radova Poljoprivrednog Faculteta, Univ. u Beogradu (Yugoslavia) **36:** 9-37.
49. **Mishra, M., G. P. Rao, P. P. Upadhyaya, and M. Mishra.** 1998. Identification and characterization of maize dwarf mosaic in India. *Ind. J. Virol.* **14:** 75-79.
50. **Murry, L. E., L. G. Elliot, S. A. Capitant, J. A. West, K. K. Hanson, L. Scarafia, S. Johnston, C. Deluca-Flahety, S. Nichola, D. Cunanan, P. S. Dietrich, I. J. Mettler, S. Dewards, D. A. Warwick, C. Rhodes, R. M. Sinibaldi, and K. J. Brunke.** 1993. Transgenic corn plants expressing MDMV-strain B coat protein is resistant to mixed infection of MDMV and maize chlorotic mottle virus. *Bio/Technology* **11:** 1559-1564.
51. **Naidu, R. A., R. Harikrishnan, S. K. Manohar, D. V. R. Reddy, A. S. Ratna, S. B. King, and R. Bandyopadhyay.** 1989. Maize mosaic virus on sorghum. *Ann. Appl. Biol.* **114:** 301-310.
52. **Narayana, Y. D., and V. Muniyappa.** 1995. Survey, symptomotology and detection of an isolate of maize stripe virus on sorghum in Karnataka. *Ind. Phytopathol.* **48:** 171-176.
53. **Narayana, Y. D., and V. Muniyappa.** 1995. Effects of sorghum stripe virus on plant growth and grain yield of sorghum. *Ind. J. Virol.* **11:** 53-58.
54. **Narayana, Y. D., and V. Muniyappa.** 1996. Virus-vector relationships of a planthopper (*Peregrinus maidis*)-borne sorghum stripe tenuivirus. *Intl. J. Pest Manag.* **42:** 165-170.
55. **Narayana, Y. D., and V. Muniyappa.** 1996. Evaluation techniques for the efficient transmission of sorghum stripe virus by vector (*Peregrinus maidis*) and screening for disease resistance. *Trop. Agric.* **73:** 119-123.
56. **Nault, L. R., and E. D. Ammar.** 1989. Leafhopper and planthopper transmission of plant viruses. *Annu. Rev. Entomol.* **34:** 503-529.
57. **Nault, L. R., and D. T. Gordon.** 1988. Multiplication of maize stripe virus in *Peregrinus maidis*. *Phytopathology* **78:** 991-995.
58. **Ocefemia, G. C., and M. S. Celino.** 1955. Note on the presence of symptoms on sorghum like those of Fiji disease of sugarcane. *Philip. Agric.* **38:** 558-559.
59. **Pande, S., R. Harikrishnan, M. D. Alegbejo, L. K. Mughogho, R. I. Karunakar, and O. Ajay.** 1993. Prevalence of sorghum diseases in Nigeria. *Intl. J. Pest Manag.* **39:** 297-303.
60. **Persley, D. M., R. G. Henzell, R. S. Greter, D. S. Teakle, and R. W. Toler.** 1985. Use of a set of differential sorghum inbred lines to compare isolates of SCMV from sorghum and maize in nine countries. *Plant Dis.* **69:** 1046-1049.
61. **Persely, D. M., R. F. Moore, and D. S. Fletcher.** 1977. The inheritance of red leaf reaction of grain sorghum to SCMV infection. *Austr. J. Agric. Res.* **28:** 853-858.
62. **Peterschmitt, M., A. S. Ratna, W. R. Sacks, D. V. R. Reddy, and L. K. Mughogho.** 1991. Occurrence of an isolate of maize stripe on sorghum in India. *Ann. Appl. Biol.* **118:** 57-70.
63. **Peterschmitt, M., B. Reynaud, G. Sommermeyer, and P. Baudin.** 1991. Characterization of maize streak virus isolates using monoclonal and polyclonal antibodies and by transmission to few hosts. *Plant Dis.* **75:** 27-32.
64. **Rama Devi, K., R. Bandyopadhyay, A. J. Hall, S. Indira, S. Pande, and P. Jaiswal.** 2000. *Farmers' Perceptions of the Effect of Plant Diseases on the Yield and Nutritive Value of Crop Residues Used for Periurban Dairy Production on the Deccan Plateau: Findings from Participatory Rural Appraisals.* Information Bulletin no. 60, ICRISAT, Patancheru (A.P.), India.
65. **Rao, G. P.** 1998. Identification of sugarcane mosaic potyvirus on sorghum and bajra in India. *Ind. J. Virol.* **14:** 81-84.

66. **Rao, G. P., R. K. Jain, and A. Varma.** 1998. Identification of sugarcane mosaic and maize mosaic potyviruses infecting poaceous crops in India. *Ind. Phytopathol.* **51:** 10-16.
67. **Rocha, A. S. T., and C. Hiruki.** 1986. Detection of mycoplasma-like organisms in situ by indirect immunofluorescence microscopy. *Phytopathology* **76:** 864-868.
68. **Rosenow, D. T.** 1986. Genetic disorders, p. 60. *In* R. A. Frederiksen (ed.), *Compendium of Sorghum Diseases*, 1st ed. APS Press, St. Paul, Minnesota.
69. **Rosenow, D. T.** 1992. Using germplasm from the world collection in breeding for disease resistance, p. 319-324. *In* W. A. J. de Milliano, R. A. Frederiksen, and G. D. Bengston (eds.), *Sorghum and Millets Diseases: A Second World Review.* ICRISAT, Patancheru (A.P.), India.
70. **Schafer, J. F.** 1971. Tolerance to plant disease. *Annu. Rev. Phytopathol.* **9:** 235-252.
71. **Sdoodee, R., D. S. Teakle, and R. Louie.** 1997. Preliminary identification of maize stripe tenuivirus in Thailand. *Plant Dis.* **81:** 226.
72. **Seifers, D. L., and T. L. Harvey.** 1989. Effect of carbofuran on transmission of maize dwarf mosaic virus in sorghum mechanically and by the aphid *Schizaphis graminum. Plant Dis.* **73:** 61-63.
73. **Shephard, R. J.** 1965. Properties of mosaic virus of corn and johnsongrass and its relation to sugarcane mosaic virus. *Phytopathology* **55:** 1250-1256.
74. **Shephard, R. J. and Q. J. Holderman.** 1965. Seed transmission of the johnsongrass strain of sugarcane mosaic virus in corn. *Plant Dis. Rept.* **49:** 468-469.
75. **Shukla, D. D., J. Jilka, M. Tosic, and R. E. Ford.** 1989. A novel approach to the serology of potyviruses involving affinity purified polyclonal antibodies directed towards virus-specific N-termini of coat proteins. *J. Gen. Virol.* **70:** 13-23.
76. **Shukla, D. D., and C. W. Ward.** 1994. Recent developments in identification and classifications of viruses comprising of sugarcane mosaic virus subgroups of potyviruses, p. 185-198. *In* G. P. Rao, A. G. Gillaspie, Jr., P. P. Upadhyay, A. Bergamin-Folho, V. P. Agnihotri, and C. T. Chen (eds.), *Current Trends in Sugarcane Pathology.* Intl. Books and Periodicals Supply Service, Delhi, India.
77. **Singh, B. U., and B. S. Rana.** 1992. Stability of resistance to corn planthopper, *Peregrinus maidis* (Ashmead) in sorghum germplasm. *Insect Sci. Applic.* **13:** 251-263.
78. **Sudarshana, M. R., and D. V. R. Reddy.** 1989. Penicillinase-based enzyme-linked-immunosorbent assay for the detection of plant viruses. *J. Virol. Meth.* **26:** 45-52.
79. **Tarr, S. A. J.** 1962. *Diseases of Sorghum, Sudangrass, and Broom Corn.* Oxford University Press, Oxford, UK.
80. **Teakle, D. S., and A. J. Pritchard.** 1971. Resistance of krish sorghum to four strains of sugarcane mosaic virus in Queensland. *Plant Dis. Rept.* **55:** 596-598.
81. **Teakle, D. S., D. D. Shukla, and R. E. Ford.** 1989. Sugarcane mosaic virus. *CMI/AAB Descriptions of Plant Viruses.* No. 342.
82. **Theuri, J. M., J. G. M. Njuguina, N. W. Ochanda, and V. Guiragossian.** 1992. Maize dwarf mosaic on sorghum in Kenya: Some preliminary observations, p. 291. *In* W. A. J. de Milliano, R. A. Frederiksen, and G. D. Bengston (eds.), *Sorghum and Millets Diseases: A Second World Review.* ICRISAT, Patancheru (A.P.), India.
83. **Thotapilly, G., N. A. Bosque-Perz, and H. W. Rossel.** 1993. Viruses and virus diseases of maize in tropical Africa. *Plant Pathol.* **42:** 494-509.
84. **Toler, R. W.** 1980. Virus and viral diseases of sorghum, p. 395-408. *In* R. J. Williams, R. A. Frederiksen, L. K. Mughogho, and G. D. Bengston (eds.), *Sorghum Diseases: A World Review.* ICRISAT, Patancheru (A.P.), India.
85. **Toler, R. W.** 1985. Maize dwarf mosaic the most important disease of sorghum. *Plant Dis.* **69:** 1011-1015.
86. **Toler, R. W., and L. M. Giorda.** 1992. Detection and identification of viruses and virus diseases of sorghum, p. 153-159. *In* W. A. J. de Milliano, R. A. Frederiksen, and G. D. Bengston (eds.), *Sorghum and Millets Diseases: A Second World Review.* ICRISAT, Patancheru (A.P.), India.
87. **Toler, R. W., B. Nath, D. Roose, and G. M. Kaula.** 1989. Occurrence of maize dwarf mosaic on sorghum in Zambia. *Plant Dis.* **73:** 273.
88. **Tosic, M., R. E. Ford, D. D. Shukla, and J. Jilka.** 1990. Differentiation of sugarcane, maize dwarf, johnsongrass and sorghum mosaic viruses based on reactions of oat and some sorghum cultivars. *Plant Dis.* **74:** 540-552.
89. **Traore, O., and G. Konate.** 1997. Maize viruses and their reservoir hosts in western and central Africa. *Cahiers Agric.* **6:** 257-262.
90. **Tsai, J. H., and T. A. Zitter.** 1982. Characteristics of maize stripe virus transmission by the corn delphacid. *J. Econ. Entomol.* **75:** 397-400.
91. **Varma, A.** 1989. Top ten filamentous viruses of the Indian subcontinent, p. 371-378. *In* R. G. Milne (ed.), *The Plant Viruses*, Vol. 4. Plenum Press, New York City, New York.
92. **Vyavahare, V. M., T. B. Garud, and V. R. Mali.** 1989. Effect of sorghum red stripe virus on sorghum genotypes. *Ind. Phytopathol.* **42:** 16-20.
93. **Wani, S. P., S. Sivaramakrishnan, R. A. Naidu, M. A. Zambre, K. K. Lee, and S. Pande.** 1991. Biochemical changes in sorghum (*Sorghum bicolor* L. Monech) plants infected with maize mosaic virus. *Ind. J. Microbiol.* **31:** 387-395.
94. **Zhu, F. C., Y. T. Chen, H. F. Zhang, and J. H. Tsai.** 1983. Identification, transmission, host range and epidemiology of maize dwarf mosaic virus in Northwestern China, p. 194-202. *In* D. T. Gordon, J. K. Knoke, L. R. Nault, and R. M. Ritter (eds.), *Proc. Intl. Maize Virus Disease Colloq. and Workshop (2-6 August 1982, Wooster, Ohio).* OARDC, Ohio State Univ., Wooster, Ohio.
95. **Zitter, T. A., and J. N. Simons.** 1980. Management of viruses by alteration of vector efficiency and by cultural practices. *Annu. Rev. Phytopathol.* **18:** 289-310.

73

Status of Sorghum and Pearl Millet Diseases in Australia

Malcolm J. Ryley, Denis M. Persley, David R. Jordan, and Robert G. Henzell

Abstract

Ergot, caused by *Claviceps africana*, is the most important disease of sorghum in Australia and is impacting all sectors of the industry. Rust, caused by *Puccinia purpurea*, and leaf blight, caused by *Exserohilum turcicum*, are the most serious leaf diseases, causing direct yield losses and contributing to lodging in some years. Lodging, in association with stalk rot caused by *Fusarium moniliforme sensu lato* and *Macrophomina phaseolina*, is a significant problem. Selection for the "stay green" trait is improving the lodging resistance of commercial sorghum cultivars. Ergot, caused by *Claviceps fusiformis*, and long smut, caused by *Moesiomyces bullatus*, are both known to occur on native grasses in Australia and are considered to be the greatest threat to a potential Australian pearl millet grain industry.

Sorghum

Introduction. Grain sorghum is the main summer grain crop in the cropping systems of central and southern Queensland and northern New South Wales (20-30°S). From 1993-2000, grain sorghum production has ranged from 1.1-1.8 million t on 240,000-522,000 ha annually. Sorghum is grown under dry land conditions in areas with an unreliable annual rainfall of 500-700 mm and is planted between October and mid-February, depending on available soil moisture and temperature. Planting outside this period is considered risky, due to potential damage from cold weather during emergence leading to seedling damping-off in early plantings, and from midge [*Stenodiplosis sorghicola* (Coquillett)], frost, and ergot (caused by *Claviceps africana* Frederickson, Mantle & de Milliano) in late-flowering crops. Sorghum is used almost exclusively as feed for livestock, particularly poultry, beef cattle, pigs, and dairy cattle, with a very small quantity being used in breakfast cereals and snack food. The use of sorghum for stock feed in Australia increased from 340,000 t in 1985-86 to 1.0 million t in 1999-2000, with the remainder (750,000 t) being exported.

Most of the 5,000-6,000 t of hybrid seed used in Australia is produced during the summer months under irrigation in central and southern New South Wales (30°S-35°S). Production of some short-day-length cultivars occurs during the winter months in irrigated areas on the Ord River, Western Australia, the Burdekin River on the northern Queensland coast, and at Lakeland Downs in northern Queensland. Breeders' nurseries and parent seed production blocks often are included at the winter nursery localities.

In Australia, over 50 fungal, bacterial and viral pathogens have been recorded on grain sorghum [11, 39, 41; records of Herbaria BRIP (Plant Pathology Herbarium, Queensland Department of Primary Industries, Indooroopilly, Queensland, Australia), DAR (New South Wales Agriculture, Orange, New South Wales, Australia), PERTH (Agriculture Western Australia, South Perth, Western Australia, Australia)]. Most of the diseases caused by these pathogens do not contribute to significant yield loss, but other common diseases such as ergot, rust, leaf spots, and stalk rots cause serious yield losses in different regions in some years.

One of the most important diseases of sorghum, sorghum downy mildew caused by *Peronosclerospora sorghi* (Weston & Uppal) C.G. Shaw (4), has not been recorded in Australia. According to the distribution map published in Bock and Jeger (4), *P. sorghi* has been recorded in Queensland. However, this record is incorrect because specimens of a downy mildew on maize, which were tentatively identified as *P. sorghi* at the Commonwealth Mycological Institute, were re-examined and now are considered to be *Peronosclerospora maydis* (Racib.) C.G. Shaw (Stamps, *personal communication*). Ramsey and Jones (31) confirmed the identity of this downy mildew as *P. maydis*. Wang *et al.* (46) developed a set of rules based on published literature on the influence of environmental factors on sporulation and infection of the pathogen to determine the potential risk from sorghum downy mildew in Australia. The analysis indicated that between November and February there is a low risk of downy mildew on sorghum in Australia, with this risk increasing during the summer and peaking in May for most of Queensland. Coastal environments in Queensland had a higher risk for downy mildew than did the drier inland areas. Another downy mildew disease, crazy top caused by *Sclerophthora macrospora* (Sacc.) Thirum., Shaw & Naras., is occasionally seen in sorghum crops that have been flooded.

Leaf diseases. The two most important fungal leaf diseases of sorghum in Australia are rust, caused by *Puccinia purpurea* Cooke, and leaf blight, caused by *Exserohilum turcicum* (Pass.) K. J. Leonard & E. G. Suggs. Target leaf spot, caused by *Bipolaris sorghicola* (Lefebrve & Sherwin) Alcorn; zonate leaf spot, caused by *Gloeocercospora sorghi* Bain. & Edgerton ex Deighton; and gray leaf spot, caused by *Cercospora sorghi* Ellis & Everh. are of relatively minor importance (Table 73-1). Anthracnose, caused by *Colletotrichum graminicola* (Ces.) G. W. Wilson, can cause severe infections on above-ground plant parts in wet years. Three bacterial leaf diseases have been recorded on sorghum in Australia, namely bacterial leaf stripe, caused by *Burkholderia andropogonis* (Smith) Gillis *et al.*, bacterial leaf spot, caused by *Pseudomonas syringae* pv. *syringae* van Hall, and leaf streak, caused by *Xanthomonas campestris* pv. *holcicola* (Elliott) Dye. These bacterial diseases are an occasional minor problem, particularly in the more humid coastal environments. Of the other bacteria recorded as causing leaf diseases on sorghum (5), *Erwinia chrysanthemi* Burkholder, McFadden & Dimock has been recorded in Australia on sorghum, but not as a leaf disease, and *Acidovorax avenae* subsp. *avenae* (Manns) Willems *et al.* and *Herbaspirillum rubisubalbicans* (Christopher & Edgerton) Baldani *et al.* have been recorded on grass hosts other than sorghum (11, 41; records of Herbarium BRIP).

Of the fungal leaf diseases, rust usually has the most significant impact on grain production. The disease is seen every year on most sorghum crops, usually after flowering, but only occasionally causes losses since most commercial cultivars have adequate levels of resistance. Frequent periods of cool, wet weather during the early vegetative growth phases can lead to epiphytotics of rust, resulting in very high infection levels during flowering and early grain fill, with a corresponding impact on yield. The 1998/1999 growing season in southern Queensland was characterized by such conditions during January and February, and cultivars that had been considered to be highly resistant to rust were severely infected. It is possible that a change in the race spectrum might have been partially responsible for the severe epidemic (Boucher, *personal communication*). In addition to direct yield losses, high infection levels during early flowering and grain fill also can predispose plants to lodging by decreasing green leaf area. Recent genetic mapping studies by Tao *et al.* (43) have identified four quantitative trait loci (QTLs) for rust resistance segregating in a recombinant inbred line population. The two most important QTLs explained 43% and 25% of the variation for rust resistance within the population, respectively.

TABLE 73-1. Leaf diseases recorded on *Sorghum bicolor* in Australia.[a]

Pathogen	Disease	Importance[b]
Bipolaris bicolor	Leaf spot	3
Bipolaris hawaiiensis	Leaf spot	3
Bipolaris maydis	Leaf spot	3
Bipolaris papendorfii	Leaf spot	3
Bipolaris sorghicola	Target leaf spot	2
Botryodiplodia theobromae	Leaf spot	3
Burkholderia andropogonis	Bacterial leaf stripe	3
Cercospora sorghi	Gray leaf spot	2
Cercospora sp.	Leaf spot	3
Colletotrichum graminicola	Anthracnose	2
Curvularia eragrostidis	Leaf spot	3
Curvularia pallescens	Leaf spot	3
Curvularia penniseti	Leaf spot	3
Curvularia sorghina	Leaf fleck	3
Curvularia verruculosa	Leaf spot	3
Curvularia sp. aff. *lunata*	Leaf spot	3
Exserohilum rostratum	Leaf spot	3
Exserohilum turcicum	Leaf blight	1
Gloeocercospora sorghi	Zonate leaf spot	2
Leptosphaeria spp.	On leaves	3
Phoma sorghina	Leaf spot	3
Pleospora infectoria	Leaf spot	3
Pseudomonas syringae pv. *syringae*	Bacterial leaf spot	3
Puccinia purpurea	Rust	1
Ramulispora sorghi	Sooty stripe	3
Sphaeronaema macrorostratum	Leaf spot	3
Xanthomonas campestris pv. *holcicola*	Leaf streak	3

[a]Compiled from from Fahy and Persley (10), Shivas (38), Simmonds (40), and records of Herbaria BRIP, DAR and PERTH.

[b]1 = common and occasionally severe, 2 = common and rarely severe, 3 = uncommon.

Resistance offers the best long-term solution for the management of leaf diseases, and there is adequate resistance to the major leaf diseases in the gene pools of Australian breeding programs (Henzell, *unpublished*). Breeding lines are screened in field trials for resistance to rust and the other important leaf diseases whenever there is sufficient natural disease pressure under field conditions.

Head smut. Of the four important smuts that attack sorghum, only head smut, caused by *Sporisorium reilianum* (J. Kühn) Langdon & Fullerton, and covered kernel smut, caused by *Sporisorium sorghi* Ehren. ex Link in Willd., have been recorded in Australia. Races 1 and 3 of *S. reilianum* have been identified in Australia (10), but the current race situation is unknown. In the 1980s and early 1990s, infection levels of 5-10% were recorded on early-planted crops in southern Queensland (10), especially when successive crops of sorghum were grown in the same affected field in successive seasons. Head smut has been of minor importance in recent years due to adequate resistance in commercial hybrids.

Stalk rots. The unreliability of rainfall in most sorghum production areas in Australia leads to frequent water deficits, with the consequence that lodging and associated stalk rot is a significant factor in sorghum production in some seasons, particularly in central Queensland (16). In southern Queensland in the 1998/1999 season, the incidence and severity of stalk rots, particularly those associated with *Fusarium moniliforme sensu lato*, was the worst in 20 years (Boucher, *personal communication*). Physiological stress in the plant, which results in an imbalance in the source-sink relationship, is hypothesized to be a predisposing factor for lodging (9, 16). The stress may be caused by a soil moisture deficit and then be exacerbated by other factors, *e.g.*, root rots or severe infection by leaf diseases such as rust.

Henzell *et al.* (16) provided strong evidence that stalk rotting pathogens are often, but not always, associated with lodged stalks, and that postflowering water stress was a prerequisite for stalk rot development. However, Dodman *et al.* (9) hypothesized that moisture stress alone was insufficient for lodging, and that invasion of the stalk by fungi was necessary for the complete development of the syndrome. *Fusarium moniliforme sensu lato* and *Macrophomina phaseolina* (Tassi) Goid. have been identified as the major pathogens involved in stalk rot in Australia (16), but very little direct screening for resistance to these pathogens has been undertaken. Although researchers in Australia have attempted to conduct such screenings (9), environmental conditions (particularly soil moisture) must be manipulated to stress the test plants. Rainout shelters and plastic mulch have been used to create these stresses, but the costs and limitations in the number of entries which can be screened make these techniques impractical for most breeding programs (9).

Lodging resistance is a priority of the public and breeding programs in Australia. There is a high correlation between the "stay green" characteristic and lodging resistance (16, 32), and resistance to stalk rot pathogens (32). In the Queensland Department of Primary Industries (QDPI) breeding program, selection for stay green is being used to indirectly identify breeding lines with lodging resistance. Pioneer HiBred utilizes field sites in southern Queensland with high levels of *F. moniliforme sensu lato* to screen hybrids, and is experimenting with inoculating breeding lines with *F. moniliforme sensu lato* and *M. phaseolina* under induced stress conditions. Leslie and Marasas (22) consider that most "*F. moniliforme*" isolates from sorghum belong in *Fusarium thapsinum* Klittich *et al.*, *F. andiyazi* Marasas *et al.*, *F. nygamai* Burgess & Trimboli, *F. proliferatum* (Matsushima) Nirenberg, or *F. verticillioides* (Sacc.) Nirenberg, with the first species being the most common. Little is known about the causal agent(s) of Fusarium stalk rot and Fusarium head blight in Australia, a situation which needs to be rectified.

Grain mold. Grain mold occurs when sorghum grain deteriorates as a result of a parasitic or saprophytic fungal association (12). This association may begin in the early stages of grain development, when fungi such as *F. moniliforme sensu lato*, *Curvularia lunata* (Wakker) Boedjin, and *Phoma sorghina* (Sacc.) Boerema, Doren. & van Kest invade spikelet tissues (12). A range of fungi, including species of *Alternaria*, *Aspergillus*, *Bipolaris*, *Curvularia*, *Fusarium* and *Phoma* have been isolated from "moldy grain" in Australia (records of Herbaria BRIP, DAR). Grain mold is considered to be a relatively minor problem in Australia because most crops mature during hot, dry weather, but significant outbreaks have occurred in central Queensland when crops matured under wet, humid conditions.

Head blight. Head blight caused by *F. moniliforme sensu lato* is an occasional serious problem in Australia, particularly in the more coastal production areas and in central Queensland. The disease has been linked with problems of small seed size in many crops in recent years (Boucher, *personal communication*). The biology of the disease is poorly understood, but prolonged periods of wet weather prior to anthesis apparently favor head blight development (15). Although selection for the stay green trait has indirectly led to an improvement in resistance to head blight (31; Henzell, *unpublished*), some commercial hybrids with strong stay green characters are not completely resistant to head blight. Pioneer HiBred may, in the future, screen for resistance to the pathogen by artificial inoculation (Boucher, *personal communication*).

Johnsongrass mosaic virus. Johnsongrass mosaic virus (JGMV), an aphid-transmitted potyvirus, occurs throughout the Australian grain sorghum production area (13). It is particularly common whenever the johnsongrass [*Sorghum halepense* (L.) Pers.] alternative host also is common. Yield losses in hybrids that develop milder mosaic symptoms is minimal, but significant losses also can occur in hybrids that develop severe mosaic, necrotic red stripe, or red leaf reactions (17). High resistance to infection, derived from Krish sorghum, has been incorporated into parental lines in Australia (18).

A variant of JGMV [JGMV(K)] that could overcome the single gene Krish resistance was detected in hybrids shortly after their release (30), and is now the dominant JGMV strain in grain sorghum in Australia (13). JGMV(K) is very similar to standard JGMV in biological and molecular properties (13, 42), except for its virulence towards hosts with Krish resistance genotypes. Many lines carrying the Krish resistance gene develop severe symptoms following JGMV(K) infection. Resistance and tolerance (mild symptoms) to JGMV(K) have been sought as part of the QDPI sorghum breeding program. Several advanced F_4 lines based on parental line 40932 have both low levels of infection and mild symptoms. Line QL19, derived from the resistance source Q7539 (18), has a high level of resistance under field conditions and lines incorporating this resistance have had infection levels that are less than or equal to those of QL19 when exposed to high levels of JGMV(K) inoculum in the field. Use of this germplasm in the development of parental lines will reduce the risk of economic damage to hybrids by JGMV. In addition, the University of Queensland is currently attempting to develop JGMV resistant genotypes via transformation using viral coat protein mediated resistance (21).

Ergot. Since its discovery in Australia in 1996 (34), ergot has had a significant impact on all sectors of the Australian sorghum industry. Although the sclerotia and conidia of many ergot isolates in Australia have morphological characters similar to those described for *C. africana* (14), others differ significantly, which suggests that other species also may be present (35). A recent analysis (20) of > 50 Australian isolates of *Claviceps* from *Sorghum* spp. revealed greater genetic diversity than that reported by Pazoutova *et al.* (29), who evaluated only four isolates from Australia. The Australian isolates of *C. africana* are similar to those from India and Asia (20, 29), raising the possibility that the pathogen may have entered Australia on several occasions. Some genotypes also may have evolved locally (20). Sorghum ergot is now endemic in Australia, and is found wherever grain and forage sorghums are grown, in seed production blocks in central and southern New South Wales and near the Ord River in Western Australia, on the common weeds *S. halepense* and *Sorghum almum* Parodi, and on forage sorghum throughout the year in frost-free areas (36). We think that these alternative hosts are the main inoculum sources for outbreaks of ergot early in the growing season, as has been suggested for *C. africana* in Africa (1). Several perennial *Sorghum* species native to northern Australia also are susceptible to *C. africana* after artificial inoculation (Ryley and Herde, *unpublished*), and also may be important alternative hosts in this region.

The greatest impact of sorghum ergot in Australia, as in other countries, has been on seed production and breeder's nurseries (1, 35, 36). All seed production blocks and nurseries are at risk from ergot, with those in the more humid areas at greater risk than those in drier areas. The triazole fungicides triadimenol and tebuconazole are used under permit for sorghum ergot control in these situations. They are applied regularly in the high-risk situations and when necessary in drier locations. Seed companies also have increased the ratio of B/R lines to A lines in seed production blocks to improve the effectiveness of pollination. Considerable research has been conducted in Australia on fungicide efficacy (33, 36) and on spray application technology. Acibenzolar-S-methyl, which activates systemic induced resistance, has potential, but a few other fungicides, including azoxystrobin, are effective (33). The efficacy of fungicides delivered by tractor-mounted sprayers is improved if flat fan nozzles are used instead of hollow cone nozzles (Ryley, *unpublished*). Work is planned to identify optimum dose × water rate combinations and the efficacy of aerial application, which is the only practical method of applying fungicides to large seed production blocks in Australia. Montes *et al.* (27) report that acibenzolar-S-methyl is effective in controlling *C. africana* but reduces seed set at 100 ppm, and that aerial application of triazoles is ineffective.

Sorghum ergot outbreaks in Australia have, so far, been localized, both geographically and temporally. In Australia, sorghum is sown when soil moisture levels are adequate, and plantings tend to be staggered by region because of the unreliable rainfall. According to an ergot risk model (44, 45) based on long-term weather data, crops that flower late in the growing season (April onwards) are at greatest risk for ergot because of the cooler temperatures and higher relative humidity levels. However, ergot outbreaks can occur at any time during the growing season. In southern Queensland and northern New South Wales, ergot outbreaks occur during the summer if there is a period of at least three days of cool, wet weather during flowering. In the 1999/2000 season ergot was recorded on only 10-15% of the crop on the central Darling Downs, but in every crop on the eastern Darling Downs, with the difference being explained by more frequent periods of ergot-conducive weather. Field experiments conducted during the last 3 years have confirmed that ergot severity is highest at low temperatures (mean hourly temperature < 21°C) and high relative humidity during flowering, and that there is a good correlation between rainfall events during warmer weather and ergot outbreaks (37, 45). The amount of viable pollen present during flowering also is a critical variable in ergot infection, and genotypes may vary in their response to temperature for this character.

Ergot levels in the main sorghum panicles range from a trace (< 0.1% infected florets/panicle) to ~ 30%, with the most severely affected crops being those that flowered unevenly due to planting problems, inadequate soil moisture, within field soil differences, or low quality seed. Ergot is more common on tillers, which flower during cooler weather, than on the primary panicles, and particularly when harvesting of the main crop is delayed by rainfall. Farmers experienced considerable difficulty during the first few growing seasons with honeydew on tillers that clogged and damaged harvesters. Many sorghum farmers in southern Queensland are now killing tillers before they flower to avoid these harvesting problems.

Early reports indicated that the dihydroergosine found in sclerotia of *C. africana* had low toxicity (24), but in 1997 cases of agalactica in sows and cows fed a diet that included ergot-contaminated sorghum were reported in central Queensland (3). Blaney *et al.* (3) established a link between these symptoms and high levels of dihydroergosine in the ergot-contaminated sorghum grain. Consequently, a regulatory limit of 0.3% sclerotia by weight (including immature sclerotia) was set as the stock feed standard. At concentrations higher than 0.3%, *C. africana* sclerotia caused mortality in chickens, agalactica in sows and cows (3), and reduced growth rates in feedlot cattle (26). Some end-users, particularly cattle feed lots, are reluctant to use any ergot-contaminated grain.

The quantification of ergot levels in grain samples remains a problem because of: (i) sampling errors due to uneven distribution of sclerotia in loads, (ii) inaccurate identification of sclerotial material, and (iii) variation in the maturity, and therefore alkaloid content, of sclerotia between loads and samples. A simple, accurate and rapid test and protocol is being developed to quantify the alkaloid concentration in samples (Blaney, *personal communication*). Mechanical harvesting can reduce sclerotial concentrations by up to 75%, which in many cases would reduce the contamination level to less than the 0.3% limit (Ryley, *unpublished*). Despite the high ergot levels observed in some crops in the 1999/2000 cropping season, the major sorghum grain buyer in Australia (Grainco Australia Ltd.) did not reject any loads because of high ergot levels. However, it is possible that growers with highly contaminated grain found alternate buyers, or mixed contaminated material with ergot-free grain.

Before ergot was identified in Australia, forage sorghum was usually allowed to flower before being grazed or cut for silage. Most forage sorghum genotypes are highly susceptible to ergot, and the disease can be found in virtually every flowering forage crop irrespective of locality and time of year. There has been considerable concern among producers about the toxicity of ergot-infected forage sorghum to grazing animals, particularly dairy cattle, and after cutting and baling. This uncertainty, and the lack of research on this problem, has led to the practice of cutting or grazing before flowering.

At present, options for the management of ergot in commercial grain sorghum are limited to agronomic practices, *e.g.*, adjusting planting times and techniques to avoid cool weather during flowering and to ensure even flowering, sowing high quality seed, and killing tillers before they flower (36). Genotypes with some resistance to ergot have been identified (6, 7, 25, 36), but the report of very high resistance levels in the line IS8525 (6, 7) has led to a resurgence in the search for new and better sources of resistance. We have confirmed the high resistance of IS8525, even under conditions that reduce pollen production by high-pollen-producing lines (36), and should lead to an understanding of the physiological mechanism of resistance. We also are constructing genetic maps to identify QTLs controlling resistance in two recombinant inbred line populations derived from crosses between IS8525 and susceptible lines, with the goal of developing molecular markers to assist in the selection and breeding of ergot resistant hybrids.

Considerable research has been conducted in Australia since 1996 on the biology, toxicity and management of sorghum ergot, in addition to the breeding efforts. Although sorghum ergot has not been as devastating in Australia as was first feared, it is imperative that a strong research effort in these areas be maintained.

Pearl Millet

Introduction. Pearl millet [*Pennisetum glaucum* (L.) R. Br.] currently is not grown as a grain crop in Australia. Several cultivars have been used as a summer forage crop for cattle in southern and central Queensland. Low com-

TABLE 73-2. Diseases of pearl millet recorded in Australia.[a]

Pathogen	Disease
Alternaria sp.	Leaf spot
Bipolaris bicolor	Leaf spot
Bipolaris nodulosa	Leaf spot
Bipolaris sp.	Leaf spot
Cercospora sp.	Leaf spot
Claviceps africana	Ergot
Claviceps fusiformis	Ergot
Curvularia protruberata	Leaf spot
Exserohilum rostratum	Leaf spot
Moesziomyces bullatus	Long smut
Pseudomonas syringae pv. *syringae*	Bacterial leaf spot
Puccinia substriata	Rust
Sugarcane mosaic virus (SCMV)	Sugarcane mosaic

[a]Compiled from from Fahy and Persley (10), Shivas (38), Simmonds (41), and records of Herbaria BRIP, DAR and PERTH.

modity prices for sorghum, lodging problems associated with many commercial sorghum cultivars, and concern by many farmers about the impact of sorghum ergot on the sorghum crop have led to research on alternative grains. A project was funded by the Australian Grains Research and Development Corporation and QDPI to develop a pearl millet grain industry in Australia. Pearl millet is being evaluated as a replacement for sorghum in semi-arid environments, due to its high drought and heat tolerance, and high yield under low rainfall conditions. It also has the advantage of being a high quality feed for poultry. Pearl millet could be grown on approximately 20,000 ha in central and southwestern Queensland that currently produces dryland sorghum. The New Crops program at QDPI, (Biloela, Queensland) has imported open-pollinated and hybrid dwarf grain lines for assessment in a variety of Queensland environments.

Recorded diseases. Relatively few diseases have been recorded on pearl millet in Australia (Table 73-2). Several fungal and bacterial leaf pathogens, including *Bipolaris* species, *Cercospora* sp., *Exserohilum rostratum* (Drechs.) K. J. Leonard & E. G. Suggs, and *Pseudomonas syringae* pv. *syringae* have been reported, but none have caused serious concern. An aecial rust found on eggplant (*Solanum melongena* L.) at Bundaberg, Queensland in 2001 was suspected to be pearl millet rust, *Puccinia substriata* Ell. & Barth., but its pathogenicity on pearl millet has not been confirmed. Specimens of the rust were deposited at Herbarium BRIP. Long smut, caused by *Moesiomyces bullatus* (J. Schrot.) K. Vanky (syn. *Tolyposporium penicillariae* Bref.), has been recorded once on pearl millet in experimental plots at Berrimah, Northern Territory, over 1,500 km from the potential grain millet production areas of southern and central Queensland. This smut also has been recorded on two other grasses in Queensland, namely *Paspalum distichum* L. and *Leersia hexandra* Swartz (records of Herbarium BRIP).

There are two records of ergot on pearl millet in Australia. The first, in August 1996, was on a self-sown plant in a field of grain sorghum that was heavily infected by *C. africana*. Drops of honeydew were exuding from two florets on a single head, and the morphological characteristics of the macroconidia and secondary conidia were very similar to those of the *C. africana* in the surrounding ergot-infected sorghum. Attempts to artificially infect pearl millet with a suspension of *C. africana* macroconidia were unsuccessful (Ryley, *unpublished*). The other record of ergot is from a pearl millet A line growing in trial plots at Biloela, Queensland. The macroconidial characters of this pathogen were identical to those described for *Claviceps fusiformis* Loveless (23). This species has been recorded on other common grasses in Queensland, including *Cenchrus ciliaris* L. (buffel grass), which is used in improved pastures in the 200-1000 mm annual rainfall belt in Queensland. Ergot has been an occasional serious problem in buffel grass seed production in inland drier areas of Queensland (Ryley, *unpublished*).

Potentially important diseases. *Sclerospora graminicola* (Sacc.) Schroet., the causal agent of downy mildew, which is perhaps the most serious disease of pearl millet (19), has not been recorded on any host in Australia. Although *S. graminicola* may be seedborne, as oospores on the seed surfaces and as internal hyphae (38), it is unlikely that the pathogen will enter Australia on imported pearl millet seed, because of strict quarantine restrictions. If the pathogen were to become established in Australia, then downy mildew could become an important disease, particularly on late-planted pearl millet, which is grown under cooler conditions that favor sporulation and infection (2). Charcoal rot caused by *Macrophomina phaseolina* is a common and often serious disease of summer crops, such as, sorghum, maize, and pulses in southern and central Queensland. This pathogen may survive on the roots of symptomless weed plants (Fuhlbohm, *unpublished*), as well as in infected crop stubble and as microsclerotia in the soil (28). There is a high probability that in low rainfall years charcoal rot will be an important problem in the drier areas targeted for pearl millet grain production.

As discussed above, *M. bullatus* has been recorded only once on pearl millet in Australia, but has been found on other grasses in Queensland. The disease is common in Africa and India (2, 8) and is likely to be an occasional problem in Queensland considering its recorded occurrence on two other grasses in southern Queensland. Diseases, such as those caused by endemic pathogens, *e.g. Bipolaris* spp., *Cercospora* sp., and *Puccinia substriata* Ell. & Barth., and as yet unrecorded pathogens, *e.g.* species of *Beniowskia*, *Ramulispora*, *Pyricularia* and *Phoma* (8, 19), may be important in wetter than normal years.

Ergot probably is the greatest threat to a grain pearl millet grain industry in Australia. *C. fusiformis* is endemic on grasses in the targeted production areas and has caused serious problems in seed production of some of these grasses. Sorghum ergot is a continuing problem in sorghum planted late in summer (February-March) in central Queensland, and a similar scenario may hold for pearl millet ergot. The moderate temperatures and high humidity that favor ergot infection of pearl millet (40) prevail during these late summer-early autumn months.

Host plant resistance offers the best long-term strategy to manage the major diseases of pearl millet. Germplasm that has resistance to some of the major pathogens of pearl millet has been identified (19). Should any of these pathogens cause serious losses in grain pearl millet this material could be incorporated into the Queensland breeding program.

Acknowledgments

We thank the Australian Grains Research & Development Corporation for financial support of various sorghum pathology projects.

References

1. **Bandyopadhyay, R., D. E. Frederickson, N. W. McLaren, G. N. Odvody, and M. J. Ryley.** 1998. Ergot: a new disease threat to sorghum in the Americas and Australia. *Plant Dis.* **82:** 356-367.
2. **Balasubramanian, K. A.** 1992. Pearl millet diseases in India: Knowledge and future research needs, p. 123-127. *In* W. A. J. de Milliano, R. A. Frederiksen, and G. D. Bengston (eds.), *Sorghum and Millets Diseases: A Second World Review.* ICRISAT, Patancheru (A.P.), India.
3. **Blaney, B. J., R. A. McKenzie, J. R. Walters, W. S. Bewg, M. J. Ryley, and R. Maryam.** 2000. Sorghum ergot (*Claviceps africana*) associated with agalactica and feed refusal in pigs and dairy cattle. *Austr. Vet. J.* **78:** 102-107.
4. **Bock, C. H., and M. J. Jeger.** 1996. Downy mildew of sorghum. *Intl. Sorg. Millets Newsl.* **37:** 33-51.
5. **Claflin, L. E., B. A. Ramundo, J. E. Leach, and M. Qhobela.** 1992. Bacterial diseases of sorghum, p. 135-151. *In* W. A. J. de Milliano, R. A. Frederiksen, and G. D. Bengston (eds.), *Sorghum and Millets Diseases: A Second World Review.* ICRISAT, Patancheru (A.P.), India.
6. **Dahlberg, J., R. Bandyopadhyay, W. Rooney, G. N. Odvody, and D. E. Frederickson.** 1998. Host-plant strategies within the United States. In: *Proc. of the Conf. on the Status of Sorghum Ergot in North America, (24-26 June, 1998, Corpus Christi, Texas).* http:/www.sorghumgrowers.com/Research/confernc/dahlberg.htm (last checked May 2002)
7. **Dahlberg, J., R. Bandyopadhyay, W. Rooney, G. N. Odvody, and P. Madera-Torres.** 2001. Evaluation of sorghum germplasm used in US breeding programmes for sources of sugary disease resistance. *Plant Pathol.* **50:** 681-689.
8. **de Milliano, W. A. J.** 1992. Pearl millet diseases in southern Africa, p. 115-122. *In* W. A. J. de Milliano, R. A. Frederiksen, and G. D. Bengston (eds.), *Sorghum and Millets Diseases: A Second World Review.* ICRISAT, Patancheru (A.P.), India.
9. **Dodman, R. L., R. L. Brengman, and N. R. Obst.** 1992. The detection of resistance to stalk rot in sorghum, p. 260-267. *In* M. A. Foale, R. G. Henzell and P. N. Vance (eds.), *Proc. of the 2nd Australian Sorghum Conference (4-6 February 1992, Gatton, Queensland).* Australian Inst. of Agric. Sci., Occas. Pub. 68, Melbourne, Victoria, Australia
10. **Dodman, R. L., N. R. Obst, and R. G. Henzell.** 1985. Races of sorghum head smut in south-east Queensland. *Australas. Plant Pathol.* **14:** 45.
11. **Fahy, P. C., and G. J. Persley.** 1983. *Plant Bacterial Diseases–A Diagnostic Guide.* Academic Press, Sydney, Australia.
12. **Forbes, G. A., R. Bandyopadhyay, and G. Garcia.** 1992. A review of sorghum grain mold, p. 253-264. *In* W. A. J. de Milliano, R. A. Frederiksen, and G. D. Bengston (eds.), *Sorghum and Millets Diseases: A Second World Review.* ICRISAT, Patancheru (A.P.), India.
13. **Franzmann, B. A., D. M. Persley, and D. A. H. Murray.** 1996. Protecting sorghum against pests and diseases, p. 103-112. *In* M. A. Foale, R. G. Henzell, and J. F. Kneipp (eds), *Proc. 3rd Australian Sorghum Conf. (20-22 February 1996, Tamworth, New South Wales).* Australian Inst. of Agric. Sci., Occas. Pub. 93, Melbourne, Victoria, Australia.
14. **Frederickson, D. E., P. G. Mantle, and W. A. J. de Milliano.** 1991. *Claviceps africana* sp. nov.; the distinctive ergot pathogen of sorghum in Africa. *Mycol Res.* **95:** 1101-1107.
15. **Frederiksen, R. A.** 2000. Head blight, p. 40. *In* R. A. Frederiksen and G. N. Odvody (eds.), *Compendium of Sorghum Diseases*, 2nd ed. APS Press, St. Paul, Minnesota.
16. **Henzell, R. G., R. L. Dodman, A. A. Done, R. L. Brengman, and P. E. Mayers.** 1984. Lodging, stalk rot, and root rot in sorghum in Australia, p. 225-236. *In* L. K. Mughogho (ed.), *Sorghum Root and Stalk Rots–a Critical Review.* ICRISAT, Patancheru (A.P.), India.
17. **Henzell, R. G., D. M. Persley, D. S. Fletcher, R. S. Greber, and L. van Slobbe.** 1979. The effects of sugarcane mosaic virus on the yield of eleven grain sorghum cultivars. *Austr. J. Exp. Agric. Animal Husb.* **19:** 225-232.
18. **Henzell, R. G., D. M. Persley, R. S. Greber, D. S. Fletcher, and L. van Slobbe.** 1982. Development of grain sorghum lines with resistance to sugarcane mosaic and other sorghum diseases. *Plant Dis.* **66:** 900-901.
19. **King, S. B.** 1992. World review of pearl millet diseases: Knowledge and future research needs, p. 95-108. *In* W. A. J. de Milliano, R. A. Frederiksen, and G. D. Bengston (eds.), *Sorghum and Millets Diseases: A Second World Review.* ICRISAT, Patancheru (A.P.), India
20. **Komolong, B., S. Chakraborty, M. J. Ryley, and D. Yates.** 2002. Four genotypes of *Claviceps* spp. cause sorghum ergot in Australia, p. 87-89. *In* J. F. Leslie (ed.), *Sorghum and Millets Diseases.* Iowa State Press, Ames, Iowa.

21. **Laidlaw, H. K. C., D. M. Persley, C. K. Pallaghy, and I. D. Godwin.** 2002. Development of durable Johnsongrass Mosaic Virus (JGMV) resistance in sorghum via a transgenic approach, p. 473-474. *In* J. F. Leslie (ed.), *Sorghum and Millets Diseases*. Iowa State University Press, Ames, Iowa.
22. **Leslie, J. F., and W. F. O. Marasas.** 2002. Will the real *Fusarium moniliforme* please stand up!, p. 201-209. *In* J. F. Leslie (ed.), *Sorghum and Millets Diseases*. Iowa State Press, Ames, Iowa.
23. **Loveless, A. R.** 1967. *Claviceps fusiformis* sp. nov. the causal agent of agalactica in sows. *Trans. Brit. Mycol. Soc.* **50:** 15-18.
24. **Mantle, P. G., and Waight, E. S.** 1968. Dihydroergosine, A new naturally occurring alkaloid from the sclerotia of *Sphacelia sorghi* (McRae). *Nature* **218:** 986-988.
25. **McLaren, N. W.** 2002. Genotype × environment effects on the response of sorghum to ergot and repercussions for disease screening. p. 95-101. *In* J. F. Leslie (ed.), *Sorghum and Millets Diseases*. Iowa State Press, Ames, Iowa.
26. **McLennan, S. R., B. J. Blaney, J. F. Kidd, J. A. Connell, and V. Doogan.** 2001. Effects of sorghum ergot on the performance of cattle in feedlots. *In* A. K. Borrell and R. G. Henzell (eds.), *Proc. 4th Australian Sorghum Conf. (5-8 February 2001, The Kooralbyn Hotel Resort,)*. CD Rom, ISBN: 0724221638. Range Media Pty Ltd., Toowoomba, Queensland, Australia.
27. **Montes, N., G. N. Odvody, and H. Williams-Alanis.** 2002. Advances in *Claviceps africana* chemical control, p. 105-110. *In* J. F. Leslie (ed.), *Sorghum and Millets Diseases*. Iowa State Press, Ames, Iowa.
28. **Pande, S.** 2002. Charcoal rot, pp. 30-32. *In* R. A. Frederiksen and G. N. Odvody (eds.), *Compendium of Sorghum Diseases*, 2nd ed. APS Press, St. Paul, Minnesota.
29. **Pazoutova, S., R. Bandyopadhyay, D. E. Frederickson, P. G. Mantle, and R. A. Frederiksen.** 2000. Relations among sorghum ergot isolates from the Americas, Africa, India and Australia. *Plant Dis.* **84:** 437-442.
30. **Persley, D. M., R. S. Greber, and R. G. Henzell.** 1987. Isolates of sugarcane mosaic virus – Johnsongrass strain infecting Krish resistant grain sorghum genotypes in Australia. *Sorg. Newsl.* **30:** 72-73.
31. **Ramsey, M. D., and D. R. Jones.** 1988. *Peronosclerospora maydis* found on maize, sweet corn and plume sorghum in Far North Queensland. *Plant Pathol.* **37:** 581-587.
32. **Rosenow, D. T.** 1984. Breeding for resistance to root and stalk rots in Texas, p. 209-218. *In* L. K. Mughogho (ed.), *Sorghum Root and Stalk Rots–A Critical Review*. ICRISAT, Patancheru (A.P.), India.
33. **Ryley, M.** 1998. Efficacy of fungicides for control of *Claviceps africana* in Australia. In: *Proc. of the Conf. on the Status of Sorghum Ergot in North America, (24-26 June, 1998, Corpus Christi, Texas)*. http:/www.sorghumgrowers.com/Research/confernc/ryley.htm (last checked May 2002).
34. **Ryley, M. J., J. L. Alcorn, J. K. Kochman, J. K., G. A. Kong, and S. M. Thompson.** 1996. Ergot on *Sorghum* spp. in Australia. *Australas. Plant Pathol.* **25:** 214.
35. **Ryley, M. J., and R. G. Henzell.** 1999. Sorghum ergot in Australia, p. 182-190. *Proc. Global Conf. on Ergot of Sorghum (1-8 June 1997, Sete Lagoas, Brazil)*: 182-190.
36. **Ryley M. J., R. G. Henzell, S. Chakraborty, S. Bhuiyan, and D. Herde.** 2001. The biology and management of sorghum ergot in Australia. *In* A. K. Borrell and R .G. Henzell, (eds.), *Proc. 4th Australian Sorghum Conf. (5-8 February 2001, The Kooralbyn Hotel Resort)*. CD Rom ISBN: 0724221638. Range Media Pty Ltd., Toowoomba, Queensland, Australia.
37. **Ryley, M. J., D. J. Herde, S. A. Bhuiyan, R. G. Henzell, and D. R. Jordan.** 2002. Overview of the biology of sorghum ergot, p. 141-150. *In* J. F. Leslie (ed.), *Sorghum and Millets Diseases*. Iowa State Press, Ames, Iowa.
38. **Shetty, H. S., P. Neergaard, and S. B. Mathur.** 1977. Demonstration of seed transmission of downy mildew on green ear disease, *Sclerospora graminicola* in pearl millet, *Pennisetum typhoides*. *Proc. Indian Nat. Acad.* **43:** 201-205.
39. **Shivas, R. G.** 1989. Fungal and bacterial diseases of plants in Western Australia. *J. Roy. Soc. West. Austral.* **72:** 1-62.
40. **Siddiqui, M. R., and I. D. Khan.** 1973. Dynamics of inoculum and environment in relation to ergot incidence on *Pennisetum typhoides* (Burm.) Stapf and Hubbard. *Trans. Mycol. Soc. Jpn.* **14:** 280-288.
41. **Simmonds, J. H.** 1966. *Host Index of Plant Diseases in Queensland*. Queensland Dept. of Primary Industries, Brisbane, Queensland, Australia.
42. **Suranto, K. H. Gough, D. D. Shukla, and C. K. Pallaghy.** 1998. Coat protein sequence of Krish-infecting strain of johnsongrass mosaic potyvirus. *Arch. Virol.* **143:** 1015-1020.
43. **Tao, Y. Z., D. R. Jordan, R. G. Henzell, and C. L. McIntyre.** 1998. Identification of genomic regions for rust resistance in sorghum. *Euphytica* **103:** 287-292.
44. **Wang, E., H. Meinke, and M. J. Ryley.** 2000. Event frequency and severity of sorghum ergot in Australia. *Austral. J. Agric. Res.* **51:** 457-466.
45. **Wang, E., H. Meinke, M. J. Ryley, D. Herde, and R. G. Henzell.** 2000. On the relation between weather variables and sorghum ergot infection. *Austral. J. Agric. Res.* **51:** 313-324.
46. **Wang, E., M. J. Ryley, and H. Meinke.** 2000. Prediction of sorghum downy mildew risk in Australia using daily weather data. *Australas. Plant Pathol.* **29:** 108-119.

74
Recurring and Emerging Sorghum Diseases in North America

James P. Stack

Abstract

Diseases impact the yield and profitability of sorghum production in North America. Some diseases are widespread affecting all sectors of the sorghum industry, *i.e.*, grain sorghum, forage sorghum, and seed production, while other diseases are limited to a specific geographic area or one sector of the industry. Among the most important established and recurring diseases impacting grain production in Mexico are leaf blight, Fusarium stalk rot, charcoal rot, anthracnose and head smut (caused by *Sporisorium reilianum*) over most of the sorghum producing areas and head blight (caused by *Fusarium moniliforme sensu lato*) in the eastern and southern production areas. The three most important diseases in Mexico with respect to incidence and severity are leaf blight, Fusarium head blight, and Fusarium stalk rot. Among the most important established and recurring diseases impacting grain production and quality in the United States are: sooty stripe (caused by *Ramulispora sorghi*), Fusarium stalk rot, and grain molds in the Northern Great Plains; grain molds, charcoal rot (caused by *Macrophomina phaseolina*), maize dwarf mosaic virus, and downy mildew (caused by *Peronosclerospora sorghi*) in the Southern Great Plains; and sorghum rust and seed decay, seedling disease, and root rot caused by *Pythium* spp. across much of the sorghum producing area of North America. The most important emerging disease of sorghum in North America is sorghum ergot (caused by *Claviceps africana*), which has the potential to significantly impact seed production in Texas, the germplasm improvement areas of Texas, Mexico, and Puerto Rico, and commercial grain production in Mexico. The economic impact, yield loss, and disease management practices vary with disease, crop, *i.e.*, grain, forage, or seed production, cropping system, *i.e.*, irrigated or dry land, and rotation or continuous, and with geographic region.

Introduction

The introduction of new plant pathogens can have very serious consequences, *e.g.*, chestnut blight in North America (24), Dutch elm disease in Europe (24), and sorghum ergot worldwide (1). In some cases, pathogen introduction can be traced to movement of infested agricultural products, *e.g.*, chestnut blight, but in other cases, passive dispersal of pathogens over long distances, *e.g.*, coffee rust across oceans, by extreme weather events also has occurred (4, 6, 17). Plant pathogenic fungi can be dispersed over great distances in single dispersal events, *e.g.*, poplar rust–3000 km in 2-3 days, or in multiple dispersal events; *e.g.*, wheat rust–2400 km in 2-3 months (22). Agriculture in the twenty-first century is global in nature, as can be seen easily with the seed production industry in general and for sorghum in particular. One consequence of the international distribution of seed is the unintentional dissemination of pathogens (23), both those that are seedborne as well as those that are carried in debris mixed with the seed. Hence, the potential exists for the continued redistribution of sorghum pathogens around the world.

Periodic surveys of recurring and emerging diseases in any region are important for developing effective and sustainable disease management strategies. Information on disease distribution and pathogen population attributes, *e.g.*, predominant virulence genes, level of resistance to fungicides, etc., can be critical to the effective deployment of host resistance genes and the appropriate use of fungicides (5). Disease survey information also is important for establishing research priorities. Unfortunately, in recent years funding for such survey work has been reduced and decades-old disease distribution data is used to establish research priorities and could lead to inappropriate recommendations. An international effort for the periodic, systematic collection of disease distribution and pathogen population data is needed. These data would facilitate the development and implementation of

TABLE 74-1. Relative prevalence, importance, and distribution of sorghum diseases in Mexico in 2000.

Disease	Causal Agent	Economic Impact[a]	Prevalence[b]	Locations
Fusarium stalk rot	*Fusarium moniliforme sensu lato*	+++	+++	Colima, Guanajuato, Jalisco, Michoacan, Nayarit, Nuevo Leon
Head blight	*Fusarium moniliforme sensu lato*	+++	+++	Guanajuato, Michoacan, Nuevo Leon, Tamaulipas
Leaf blight	*Exserohilum turcicum*	+++	+++	Guanajuato, Jalisco, Michoacan, Morelos, Nayarit, Nuevo Leon, Sinaloa, Tamaulipas, Veracruz
Charcoal rot	*Macrophomina phaseolina*	+++	++	Colima, Guanajuato, Jalisco, Michoacan, Nuevo Leon, Tamaulipas, Veracruz
Head smut	*Sporisorium reilianum*	+++	++	Colima, Guanajuato, Michoacan, Nayarit, Nuevo Leon, Sinaloa, State of Mexico, Tamaulipas
Rust	*Puccinia purpurea*	++	+++	Guanajuato, Jalisco, Michoacan, Nuevo Leon, Tamaulipas
Anthracnose leaf blight	*Colletotrichum graminicola*	++	++	Guanajuato, Jalisco, Nayarit, Nuevo Leon, Tamaulipas, Veracruz
Bacterial leaf streak	*Xanthomonas campestris* pv. *holcicola*	++	++	Guanajuato, Jalisco, Morelos, Sinaloa, Sonora, Yucatan
Ergot	*Claviceps sorghi*	++	++	Campeche, Chiapas, Guanajuato, Jalisco, Michoacan, Morelos, Sinaloa, Sonora, Tamaulipas
Grain mold	*Curvularia* spp, *Fusarium* spp., *Alternaria* spp.	++	++	Colima, Guanajuato, Michoacan, Nayarit, Tamaulipas, Veracruz
Gray leaf spot	*Cercospora sorghi*	++	++	Campeche, Guanajuato, Jalisco, Michoacan, Nayarit, Nuevo Leon, Tamaulipas, Veracruz, Yucatan
	Maize Dwarf Mosaic Virus	++	++	Chiapas, Colima, Guanajuato, Jalisco, Morelos, Tamaulipas, Veracruz
Zonate leaf spot	*Gloeocercospora sorghi*	++	++	Chiapas, Jalisco, Michoacan, Morelos, Nuevo Leon, Tamaulipas, Veracruz, Yucatan
Anthracnose stalk rot	*Colletotrichum graminicola*	++	+	Guanajuato, Michoacan, Nayarit, Nuevo Leon, Tamaulipas, Veracruz
Downy mildew	*Peronosclerospora sorghi*	++	+	Chiapas, Colima, Jalisco, Michoacan, Morelos, Nayarit, Nuevo Leon, Tamaulipas, Veracruz
Acremonium wilt	*Acremonium strictum*	+	+	Guanajuato, Morelos, Tamaulipas
Bacterial leaf stripe	*Pseudomonas andropogonis*	+	+	Campeche, Chiapas, Guanajuato, Jalisco, Michoacan, Tamaulipas, Yucatan
Crazy top	*Sclerophthora macrospora*	+	+	Chiapas, Guanajuato, Jalisco, Michoacan, Morelos, Nuevo Leon, Tamaulipas, Veracruz
Mycoplasm	Yellow sorghum stunt	+	+	Guanajuato, Morelos, Veracruz
Pokkah boeng	*Fusarium moniliforme sensu lato*	+	+	Guanajuato, Jalisco, Michoacan, Tamaulipas
Seedling blight	*Fusarium* spp., *Pythium* spp., *Rhizoctonia* spp., *Aspergillus* spp.	+	+	Guanajuato, Jalisco, Tamaulipas
Storage rots	*Alternaria* spp., *Aspergillus* spp., *Botrytis* spp., *Curvularia* spp., *Phoma* spp., *Penicillium* spp.,	+	+	Guanajuato, Jalisco, Michoacan, Tamaulipas
Bacterial sun spot	*Pseudomonas* spp.	±	+	Guanajuato, Morelos
Covered kernel smut	*Sporisorium sorghi*	±	+	Coahuila, Guanajuato, Jalisco, Morelos, Sinaloa, Tamaulipas
Loose kernel smut	*Sphacelotheca cruenta*	±	+	Guanajuato, Jalisco, Michoacan, Morelos, Sinaloa, Tamaulipas
Sooty stripe	*Ramulispora sorghi*	±	+	Campeche, Jalisco, Tamaulipas, Veracruz

[a]±, no significant loss; +, minor importance; ++, moderate importance; and +++, at times a major deterrent to crop production.
[b]+, occasionally present; ++, commonly present; and +++, found on most plants in most fields.

effective disease management strategies, including quarantine, and the establishment of appropriate research priorities.

The Sorghum Industry in North America

The sorghum industry in North America includes grain sorghum, food-grade sorghum, forage sorghum, commercial seed production, and noncommercial germplasm improvement. Sorghum is grown over a wide area in Mexico including states along both coasts and in the central highlands (27). The highest grain yields in the world have been produced in the Bajío region of the state of Guanajuato. Although sorghum is grown over a wide area in the United States, stretching from California to Georgia, most (> 90%) is produced in the central Great Plains region of Kansas, Texas, Nebraska, Oklahoma, and Missouri, with Kansas and Texas together accounting for over two-thirds of the total. Yields vary greatly within and among production areas as well as among different cropping systems. Much of the grain is either exported or used as animal feed, with Japan and Mexico being the major importers. There also are approximately 30,000 hectares of seed production in the United States, mostly in the high plains of Texas, with average yields of ninety 8.8 kg units/ha. Approximately 20% of this seed is exported.

Sorghum production in North America occurs over a region covering almost forty degrees of latitude. Weather, both temperature and precipitation, and the length of the growing season varies by production area. This variation results in sequential planting and harvesting of sorghum from central Mexico to the northern Great Plains of the United States, and there is sorghum growing somewhere in North America virtually year round. These overlapping sorghum growing seasons significantly impact the development of epidemics of some diseases, *e.g.* sorghum ergot. Variation in cropping systems also is important. For example, the increase in reduced tillage in the northern Great Plains has contributed to increased diseases caused by pathogens that survive in crop residue, *e.g.*, sooty stripe in Nebraska.

Recurring Sorghum Diseases in Mexico

Mexico is both a major sorghum producer and importer (27). Sorghum diseases in Mexico were reviewed and reported at the last International Sorghum and Millet Diseases meeting in Harare, Zimbabwe in 1988 (18), and several of these remain important production constraints (Table 74-1). The most important established sorghum diseases in Mexico remain leaf blight, Fusarium stalk rot, charcoal rot, anthracnose, and head smut (caused by *Sporisorium reilianum*) in most of the sorghum-producing areas and head blight (*Fusarium thapsinum*) in the eastern and southern production areas.

Among the significant changes is a reduction in downy mildew incidence, which was considered common in nine states in 1988 but is now observed only occasionally. Bacterial leaf streak (*Xanthomonas campestris* pv. *holcicola*) is common in several states. Sorghum rust (caused by *Puccinia purpurea*) was common in 1988 and is now very prevalent, *i.e.*, present on most plants in most fields on most plants, in several states and considered of moderate importance. Acremonium wilt (caused by *Acremonium strictum*) has remained constant in distribution, but its severity has increased in the three states where it occurs. The most significant change is the appearance of sorghum ergot. Ergot (caused by *Claviceps africana*) was not present in Mexico in 1988; however, it is now considered common and of moderate importance in at least six states.

The three most important diseases in Mexico with respect to incidence and severity are leaf blight, Fusarium head blight, and Fusarium stalk rot. These diseases occur in most of the sorghum growing areas of Mexico and commonly cause yield loss. The primary pathogen causing head blight and stalk rot is commonly termed "*Fusarium moniliforme*," but the species involved and their relative importance remains to be determined (14).

Recurring Sorghum Diseases in the United States

Sorghum diseases in the United States were reviewed by Frederiksen and Duncan at the last International Sorghum and Millets Diseases meeting in 1988 (10). A similar approach was adopted for collecting the disease distribution and prevalence data for this report. Additional consideration was given to the relative importance of a disease to a particular segment of the sorghum industry, *i.e.*, seed production, grain production, and forage production. Surveys were sent to academic and industry personnel in the dominant sorghum-producing states requesting information on the prevalent diseases and their relative importance, emerging diseases and their potential threat, and sorghum production statistics including industry structure. Not all surveys were returned, and even in returned surveys, not all of the information requested was provided. Some respondents indicated that sorghum disease surveys were never conducted and recorded for that location or it had been many years since such a disease survey was conducted. Consequently, a complete picture for all the sorghum-growing areas of North America is not available. Where possible, disease distribution and prevalence data are presented by state. For some diseases, extrapolations can be made from these data for other areas.

Georgia. Sorghum production in Georgia (32,000 ha) is approximately 50% grain sorghum and 50% forage sorghum, which is processed as silage. The incidence, severity, and distribution of sorghum diseases in Georgia has not changed dramatically since the previous report (10). The most significant change has been the introduction and apparent establishment of sorghum ergot. Ergot occurred to a limited extent late in the growing seasons (August and September) of 1997-2000.

Kansas. Kansas is the leading producer of grain (1.4 million ha) and forage (ca. 32,000 ha) sorghum in the United States. Most of the sorghum grain produced in Kansas is for export or animal feed. Commercial seed production (400 ha) also is important. The environment is diverse particularly with respect to precipitation, which decreases from approximately 900 mm/year in the eastern part of the state to less than 230 mm/year in the west. Most sorghum is grown under dry land conditions with various cropping systems, *e.g.*, no tillage, and rotation with corn, soybeans, or wheat. Although many diseases have been observed in Kansas over the past ten years, only a few are considered to be yield limiting (Table 74-2). The most significant changes from the previous report are the inclusion of seed and seedling diseases (caused by *Pythium* and *Fusarium* spp.), sooty stripe (caused by *Ramulispora sorghi*), and the emergence of ergot (*Claviceps africana*) as a potential problem. Research at Kansas State University indicates that yield loss to sooty stripe is variable (10-26%), but significant, and dependent upon the hybrid (12, 28). The Kansas Department of Agriculture conducts systematic disease surveys of all major crops grown in the state, including sorghum (see www.ink.org/public/kda/phealth/phprot/dispts/html (last checked May 2002).

Missouri. Most sorghum grown in Missouri is for grain for export or animal feed. Commercial seed production is not important in this state. Forage sorghum is grown as a minor crop and is used locally. Sorghum is not an intensively managed crop in Missouri and detailed information on the incidence and severity of diseases or the pathogens causing those diseases is not available. Root and stalk diseases are the most significant disease problems in Missouri. Seedling blights rank second in importance. Downy mildews, viruses, grain molds, and foliar diseases all occur, but are not considered yield limiting in most years.

Nebraska. Most of the sorghum produced in Nebraska (ca. 220,000 ha) is grain for export or animal feed. Although significant germplasm improvement research and development occurs, commercial seed production is not important in this state. Forage sorghum (ca. 50,000 ha) is a minor crop and is used locally. The rainfall gradient from east to west is similar to that seen in Kansas, and cropping systems are similar as well. Many diseases occur annually but accurate statistics on incidence, severity, and yield loss are not available. Some diseases are considered yield limiting including, seed and seedling diseases (caused by *Pythium* spp. and *Fusarium* spp.), stalk rots (caused primarily by *Fusarium* spp.), sooty stripe (caused by *Ramulispora sorghi*), and in some years zonate leaf spot (caused by *Gloeocercospora sorghi*). Bacterial stripe (caused by *Burkholderia andropogonis*), bacterial leaf spot (caused by *Pseudomonas syringae*), and grain mold (caused by many fungal species) occur every year on sorghum in Nebraska, but their impact on yield has not been determined recently. The most significant changes from the previous report are the single occurrence of ergot (caused by *Claviceps africana*) in one county in 1997 and the increasing prevalence of sooty stripe.

Oklahoma. Most sorghum production in Oklahoma (140, 000 ha) is grain for export (25%) or domestic animal feed (75%). Commercial seed production (600 ha) and forage production (28,000 ha) are of secondary importance in this

TABLE 74-2. Disease distribution, frequency of occurrence, area affected, and impact of prevalent diseases in Kansas. All diseases occur ever year except for ergot.

Disease	Causal Agent	Geographic Distribution	Affected Area[a]	Severity[b]
Seed rot and seedling blight	*Pythium* spp.	Central and Eastern	–	30 kg/ha
Fusarium seedling blight	*Fusarium* spp.	Central and Eastern	–	–
Sooty stripe	*Ramulispora sorghi*	All	80	30
Rust	*Puccinia purpurea*	Eastern and Southern	Varies	15
Crazy top	*Sclerophthora macrospora*	Eastern and Southern	1	100
Fusarium root rot	*Fusarium* spp.	All	–	–
Fusarium stalk rot	*Fusarium* spp.	All	75	5-15
Charcoal rot	*Macrophomina phaseolina*	Eastern and Southern	25	30
Ergot	*Claviceps africana*	Eastern and Southern	1	–

[a]Percentage of fields in the region affected by the disease.
[b]Estimated percentage loss in an affected field, unless otherwise noted.

state. Most sorghum is grown in rotation with wheat and soybeans, and a small proportion, approximately 16%, is irrigated. Several diseases are constraints to sorghum production in Oklahoma (Table 74-3), however, no systematic disease survey has been conducted recently. The diseases perceived as the most serious at present are maize dwarf mosaic virus (MDMV), stalk rots, and seed decay and seedling blights. Diseases of greatest concern for the future are MDMV, leaf blight, and ergot.

TABLE 74-3. Disease distribution, frequency of occurrence, area affected, and impact of prevalent diseases in Oklahoma. All diseases occur every year except for stalk rots, whose frequency of occurrence is unknown.

Disease	Causal Agent	Geographic Distribution	Affected Fields[a]	Affected Area[b]	Severity[c]
Seed rot and seedling decay	*Pythium* spp.	Northern	–	–	–
Leaf blight	*Exserohilum turcicum*	Northeastern	25	75	25
Stalk rots	–	All	5	10-30	10
Maize dwarf mosaic virus		All	10	40	5-10

[a]Percentage of fields in the region affected by the disease.
[b]Percentage of plants in the infected fields affected by the disease.
[c]Estimated percentage yield loss in an affected field.

Texas. Texas is the leading commercial sorghum seed producer (ca. 25,600 ha) and second in grain production (1.2 million ha). Forage sorghum and a small amount of food-grade sorghum for human consumption also are produced. Sorghum production in Texas spans 1300 km from the Rio Grande valley on the Mexican border to the border of the panhandle with Oklahoma, and consists of significantly different production environments and cropping systems. No systematic disease survey has been conducted since the last report, but many diseases have been observed. Some diseases, *e.g.*, zonate leaf spot and rust, are widespread and occur every year but are not perceived to cause significant yield losses. Other diseases reduce yield and or quality to varying degrees and at varying frequencies. Charcoal rot (*Macrophomina phaseolina*) remains the most important disease across all production areas with yield losses of 10-100%, depending on production area and year. The most significant change in the central blacklands production area from the previous report is the recognition of *Pythium* root and seedling rot as a significant constraint to production (Table 74-4). The most significant changes in the coastal bend production area from the previous report are the decrease in incidence and severity of sorghum downy mildew (caused by *Peronosclerospora sorghi*) and anthracnose (caused by *Colletotrichum graminicola*) and the emergence of ergot (caused by *Claviceps africana*) as a commonly occurring disease (Table 74-5).

Puerto Rico/St. Croix. Puerto Rico is an important location for germplasm improvement and seed increase. Breeders from the United States Department of Agriculture, universities, and commercial seed companies all have winter nurseries in Puerto Rico. Forage sorghum is a very minor crop. Quarantined sorghum is grown on St. Croix. Although many diseases are observed in Puerto Rico, only ergot is perceived as a production constraint. Among the commonly observed diseases in Puerto Rico and St. Croix are: downy mildew, anthracnose, rust, zonate leaf spot, and various grain molds. Bacterial whorl and stalk rot (*Erwinia chrysanthemi*) also occur. Johnsongrass is prevalent on the island and serves as an important alternative host in the epidemiology of several diseases. Charcoal rot is infrequent on both islands.

Emerging Sorghum Diseases in North America

In both the United States and Mexico, the emerging disease of most concern is sorghum ergot (1). The disease has been reported throughout much of Mexico including the states of Tamaulipas, Colima, Gaunajuato, Jalisco, and Sinaloa (26). It also has been reported in most sorghum growing areas of the United States from Texas to Nebraska and from New Mexico to Georgia. *C. africana* is well established in Mexico and Texas and sorghum ergot has developed every year since 1997. Sorghum ergot also has occurred regularly since 1997 in Kansas, Georgia, and Puerto Rico but at lower incidence and severity. Whether *C. africana* is endemic in these regions or is brought or blown in each growing season is not known.

Although, widespread in occurrence, ergot is perceived as a production constraint in only a few specific areas. This disease poses its largest threat to commercial seed production and to germplasm improvement programs in North America (20). Whether this potential threat is realized will depend upon many aspects of the disease cycle that remain unknown, *e.g.*, how far north the pathogen can survive winter conditions, the mechanism(s) and dynamic(s) of pathogen dispersal, and the significance of alternate hosts to epidemic initiation and development.

TABLE 74-4. Disease distribution, frequency of occurrence, and impact of prevalent diseases in the High Plains, Upper Coast, and Central Blacklands regions of Texas.

Disease	Causal Agent	Geographic Distribution	Frequency	Affected Fields[a]	Affected Area[b]	Severity[c]
Root rot and seed decay	*Pythium* spp.	Central Blacklands	Infrequent	5	20	0-10
Rough leaf spot	*Ascochyta sorghina*	Central Blacklands	Infrequent	5	10	0
Banded leaf and sheath blight	*Rhizoctonia solani* AG1	Central Blacklands	Infrequent	5	10	0
Charcoal rot	*Macrophomina phaseolina*	Central Blacklands	Frequent	1-25	10-100	1-100
		High Plains	Annual	20	10-100	10-100
		Upper Coast	Frequent	1-25	10-100	1-100
Target leaf spot	*Bipolaris sorghicola*	Upper Coast	Infrequent	5	10	0
Zonate leaf spot	*Gloeocercospora sorghi*	Upper Coast	Annual	15	30	?
Rust	*Puccinia purpurea*	Upper Coast	Annual	85	90	?

[a]Percentage of fields in the region affected by the disease.
[b]Percentage of plants in the infected fields affected by the disease.
[c]Estimated percentage yield loss in an affected field.

Of the three fungal species associated with sorghum ergot worldwide, only *C. africana* has been identified in North America. This species differs in many ways from *C. sorghi*, which previously caused sorghum ergot in India (2, 3). Minor genetic variation in *C. africana* from the population in the United States has been detected with molecular techniques (21, 25), but no significant variation in virulence has been observed.

The epidemiology and basic pathogen biology of *C. africana* in South Africa has been determined (7, 8, 9), and much of this information appears to be applicable to North America. Disease prediction models developed for South Africa (15, 16) have accurately predicted sorghum ergot incidence in parts of the United States. Thus, it is reasonable to assume that ergot epidemics in North America will be similar to epidemics in South Africa and that effective disease management strategies in South Africa also may be effective in North America.

Conclusions

Sorghum is an important cereal crop in North America and diseases remain a serious constraint to production. The variation in climate over the forty degrees of latitude from Mexico to the northern Great Plains of the United States allows for the development of a wide array of diseases that vary in impact and frequency of occurrence. In general, diseases caused by fungi are the most important. In Mexico, leaf blight, Fusarium head blight, and Fusarium stalk rot are considered the most important yield-limiting diseases of the last decade. In the southern Great Plains of the United States, charcoal rot, grain molds, and *Pythium* seed and seedling decay are the most important diseases. In the northern Great Plains, sooty stripe, stalk rots, and grain molds are the most significant. Grain molds are considered a major disease worldwide with many pathogens involved (19). Several diseases, *e.g.*, rust, zonate leaf spot, and bacterial leaf stripe, occur regularly across a wide area, but are not perceived to significantly reduce yield.

TABLE 74-5. Incidence, severity, and frequency of occurrence of the prevalent diseases impacting sorghum production in the Coastal Bend region of Texas. All diseases occur ever year except for anthracnose.

Disease	Causal Agent	Affected Area[a]	Severity[b]
Grain mold	*Curvularia lunata* *Fusarium* spp.	80	10 – QL
Charcoal rot	*Macrophomina phaseolina*	25	10 – YL
Head smut	*Sporisorium reilianum*	10	2 – YL
Downy mildew	*Peronosclerospora sorghi*	5	1 – YL
Ergot	*Claviceps africana*	20	1 – YL
Anthracnose	*Colletotrichum graminicola*	< 1	< 1 – YL
Maize dwarf mosaic virus		20	1 – YL

[a]Percentage of plants in the infected fields affected by the disease.
[b]Estimated percentage yield loss in an affected field; QL, quality loss; YL, yield loss.

Throughout North America, the emerging disease with the greatest potential to influence the sorghum industry is sorghum ergot, and its greatest impact probably will occur in seed production fields. More detailed information on the life history of *C. africana* in North America is needed before an accurate estimate can be made of its potential impact on grain sorghum production in the northern Great Plains. The rapid response and close cooperation amongst scientists from Mexico and the United States has minimized the potentially devastating impact of this disease on sorghum production and trade in North America.

In most areas regular disease surveys are not conducted, and in some areas a detailed survey of sorghum diseases has not been conducted for many years. Thus, disease distribution and impact data as well as the pathogen attribute data may be decades old. The application of molecular techniques to the population analysis of several pathogen species has revised our understanding of the geographic distribution of some important sorghum pathogens (11, 13, 21, 25). In several states, sorghum is a neglected crop receiving little attention. Some scientists providing data for this chapter were unable to provide yield loss data for many diseases, which means that we do not know the economic impact of several common sorghum diseases. Consequently, the establishment of interregional disease management strategies and the setting of research priorities may be based on incomplete or inaccurate information.

Acknowledgments

The following individuals provided region-specific data on sorghum diseases and production statistics: Jesus Naro, INIFAP; Jeff Wilson, USDA; Doug Jardine and Larry Claflin, Kansas State University; Laura Sweets, University of Missouri; Bruce Maunder and Jeff Dahlberg, National Grain Sorghum Producers; Phillip Pratt, Oklahoma State University; Larry Dunkle, Purdue University; and Gary Odvody, Tom Isakeit, and Richard Frederiksen, Texas A&M University.

References

1. **Bandyopadhyay, R., D. E. Frederickson, N. W. McLaren, and G. N. Odvody.** 1996. Ergot–A global threat to sorghum. *Intl. Sorg. Millets Newsl.* **37:** 1-32.
2. **Bandyopadhyay, R.** 1992. Sorghum ergot, p. 235-244. *In* W. A. J. de Milliano, R. A. Frederiksen, and G. D. Bengston (eds.), *Sorghum and Millets Diseases: A Second World Review.* ICRISAT, Patancheru (A.P.), India.
3. **Bandyopadhyay, R., L. K. Mughogho, and G. Tegegne.** 1991. Source of primary inoculum and spread of ergot, p. 24-25. In 1990 *ICRISAT Cereals Program Annual Report.* ICRISAT, Patancheru (A.P.), India.
4. **Bowden, J., P. H. Gregory, and C. G. Johnson.** 1971. Possible wind transport of coffee leaf rust across the Atlantic Ocean. *Nature* **229:** 500-501.
5. **Brown, J. K. M.** 1998. Surveys of variation in pathogen populations and their application to disease control, p. 73-102. *In* D. G. Jones (eds.), *The Epidemiology of Plant Diseases.* Kluwer Academic Publishers, Dordrecht, The Netherlands.
6. **Davis, J. M.** 1987. Modeling the long-range transport of plant pathogens in the atmosphere. *Annu. Rev. Phytopathol.* **25:** 169-188.
7. **Frederickson, D. E., P. G. Mantle, and W. A. J. de Milliano.** 1989. Secondary conidiation of *Sphacelia sorghi* on sorghum, a novel factor in the epidemiology of ergot disease. *Mycol. Res.* **93:** 497-502.
8. **Frederickson, D. E., P. G. Mantle, and W. A. J. de Milliano.** 1991. *Claviceps africana* sp. nov., the distinctive ergot pathogen of sorghum in Africa. *Mycol. Res.* **95:** 1101-1107.
9. **Frederickson, D. E., P. G. Mantle, and W. A. J. de Milliano.** 1991. Windborne spread of ergot disease (*Claviceps africana)* in sorghum. *Plant Pathol.* **42:** 368-377.
10. **Frederiksen, R. A., and R. R. Duncan.** 1992. Sorghum diseases in North America, p. 85-88. *In* W. A. J. de Milliano, R. A. Frederiksen, and G. D. Bengston (eds.), *Sorghum and Millets Diseases: A Second World Review.* ICRISAT, Patancheru (A.P.), India.
11. **Guthrie, P. A., C. W. Magill, R. A. Frederiksen, and G. N. Odvody.** 2000. Random amplified polymorphic DNA markers: A system for identifying and differentiating isolates of *Colletotrichum graminicola. Phytopathology* **82:** 832-835.
12. **Jardine, D. J., and W. B. Gordon.** 1997. Assessment of grain sorghum yield loss by sooty stripe (*Ramulsipora sorghi*). *Phytopatholology* **87:** S48.
13. **Klittich, C. J. R., J. F. Leslie, P. E. Nelson, and W. F. O. Marasas.** 1997. *Fusarium thapsinum* (*Gibberella thapsina*): A new species in section *Liseola* from sorghum. *Mycologia* **89:** 643-652.
14. **Leslie, J. F., and W. F. O. Marasas.** 2002. Will the real *Fusarium moniliforme* please stand up!, p. 201-209. *In* J. F. Leslie (ed.), *Sorghum and Millets Diseases.* Iowa State Press, Ames, Iowa.
15. **McLaren, N. W., and F. C. Wehner.** 1990. Relationship between climatic variables during early flowering of sorghum and the incidence of sugary disease caused by *Sphacelia sorghi. J. Phytopathol.* **130:** 82-88.
16. **McLaren, N. W., and F. C. Wehner.** 1992. Pre-flowering low temperature predisposition of sorghum to sugary disease (*Claviceps africana*). *J. Phytopathol.* **135:** 328-334.
17. **Nagarajan, S., and D. V. Singh.** 1990. Long-distance dispersion of rust pathogens. *Annu. Rev. Phytopathol.* **28:** 139-153.
18. **Narro, J., V. A. Betancourt, and J. I. Aguirre.** 1992. Sorghum diseases in Mexico, p. 75-84. *In* W. A. J. de Milliano, R. A. Frederiksen, and G. D. Bengston (eds.), *Sorghum and Millets Diseases: A Second World Review.* ICRISAT, Patancheru (A.P.), India.
19. **Navi, S. S., R. Bandyopadhyay, A. J. Hall, and P. J. Bramel-Cox.** 1999. *A Pictorial Guide for the Identifica-*

tion of Mold Fungi on Sorghum Grain. Information Bulletin no. 59. ICRISAT, Patancheru (A.P.), India.

20. **Odvody, G. N., and T. Isakeit.** 1999. Sorghum ergot in the United States. *Proc. Global Conf. on Ergot of Sorghum (1-8 June 1997, Sete Lagoas, Brazil)*: 68-75.
21. **Pazoutova, S., R. Bandyopadhyay, D. E. Frederickson, P. G. Mantle, and R. A. Frederiksen.** 2000. Relations among sorghum ergot isolates from the Americas, Africa, India, and Australia. *Plant Dis.* **84:** 437-442.
22. **Pedgley, D. E.** 1986. Long distance dispersal of spores, p. 346-365. *In* K. J. Leonard and W. E. Fry (eds.), *Plant Disease Epidemiology, Vol. 1, Population Dynamics and Management*. Macmillan, New York City, New York.
23. **Rennie, W. J.** 1998. Seedborne diseases, pp. 295-307. *In* D. G. Jones (ed.), *The Epidemiology of Plant Diseases*. Kluwer Academic Publishers, Dordrecht, The Netherlands.
24. **Schumann, G. L.** 1991. *Plant Diseases: Their Biology and Social Impact*. APS Press, St. Paul, Minnesota.
25. **Tooley, P. W., N. R. O'Neil, E. D. Goley, and M. M. Carras.** 2000. Assessment of diversity in *Claviceps africana* and other *Claviceps* species by RAM and AFLP analyses. *Phytopathology* **90:** 1126-1130.
26. **Torres-Montalvo, J. H., and N. Montes-Garcia.** 1999. Sorghum ergot in Mexico. *Proc. Global Conf. on Ergot of Sorghum (1-8 June 1997, Sete Lagoas, Brazil)*: 101-108.
27. **Williams-Alanis, H.** 2002. Sorghum and millet diseases in Mexico, p.457-463. In J. F. Leslie (ed.), *Sorghum and Millets Diseases 2000*. Iowa State Press, Ames, Iowa.
28. **Xu, X., L. E. Claflin, B. A. Ramundo, and D. J. Jardine.** 1995. Optimizing inoculum production for (*Ramulispora sorghi*). *Phytopathology* **85:** 1169.

75

Sorghum and Millet in Mexico

Héctor Williams-Alanis

Abstract

Sorghum is one the most important crops in Mexico and is cultivated on the third largest area, after maize and dry beans. The area planted to sorghum and the amount of grain produced have generally increased over the last 40 years. Sorghum grain is used mainly for animal feed, but 44% of the national production is imported to satisfy annual demand. In contrast, millet is an almost unknown crop in Mexico. However, there are excellent prospects for millet production, primarily in arid and semi-arid regions. Any millet grain produced probably would be used to supplement sorghum in animal feeding, as millet has a higher nutritional value. Both sorghum and millet could be, but have not been, used for human food.

Sorghum

Sorghum was originally introduced to dry land regions of Mexico because of its drought tolerance. However, it has been adapted to all irrigated and dry land areas in the country. Sorghum has been cultivated in Mexico since the nineteenth century, with the first publication by the Secretary of Economic Development (Secretaría de Fomento) published in 1882 on the uses for "Kafir maize" detailing sorghum's uses and drought resistance (29). Between 1896 and 1898, two more articles about sorghum were published in "Mexican Progress" (Progreso de México) (29), and in 1914 Rómulo Escobar predicted that sorghum would be used for human food in Mexico (29).

The first reported sorghum research in Mexico was conducted in 1944 by the Special Studies Office (Oficina de Estudios Especiales) established as a part of the Secretary of Agriculture (Secretaría de Agricultura) as a result of an agreement between the Rockefeller Foundation and the Mexican government (23). Research on sorghum production was conducted to try to establish this crop in marginal areas with limited precipitation. The early trials were not enthusiastically received by farmers (21). In 1950, the first production practices and recommendations bulletin was published (22). The Instituto Nacional de Investigaciones Agrícolas (INIA), now Instituto Nacional de Investigaciones Forestales, Agrícolas y Pecuarias (INIFAP), was founded on 6 December 1960, by integrating the Institute for Agricultural Research (Instituto de Investigaciones Agrícolas) and the Special Studies Office, and research on sorghum production was continued under its direction, mainly in the "El Bajío" region, state of Guanajuato (23).

From a technological point of view, sorghum production in Mexico began to be successful following the first use of hybrids in Texas in the late 1950s. After discovering the potential of these hybrids, farmers in both northern and central Mexico adopted them quickly. At the same time,

TABLE 75-1. Sorghum imports by major sorghum-importing countries (ton × 1000). After SAGAR (26).

Country	Cropping Season										
	89/90	90/91	91/92	92/93	93/94	94/95	95/96	96/97	97/98	98/99	Mean
Mexico	3.02	3.03	5.00	4.02	3.09	2.54	1.76	2.09	3.34	3.15	2.86
Japan	3.92	3.59	3.25	3.22	2.85	2.41	2.54	2.77	2.77	2.45	2.72
European Union	0.00	0.24	0.20	0.38	0.21	0.59	0.98	0.16	0.23	0.30	0.41
Israel	0.04	0.07	0.05	0.22	0.07	0.21	0.36	0.46	0.08	0.10	0.21
Taiwan	0.36	0.17	0.18	0.01	0.03	0.03	0.09	0.08	0.07	0.04	0.05
Others	0.16	0.65	0.68	003	0.82	0.63	0.73	0.59	0.34	0.44	0.65
Total	8.98	7.75	9.36	8.88	7.07	6.41	6.46	6.15	6.83	6.48	6.90

TABLE 75-2. Area of cultivation (ha × 1000) for the main grain crops in Mexico, after SAGAR (26).

Crop	Year									
	1990	1991	1992	1993	1994	1995	1996	1997	1998	Mean
Field corn	7,900	7,700	8,000	8,200	9,200	9,100	8,600	9,100	8,500	8,500
Dry beans	2,300	2,200	1,900	2,200	2,400	2,400	2,200	2,300	2,400	2,200
Sorghum	1,900	1,500	1,500	970	1,400	1,600	2,300	2,100	2,200	1,700
Total	14,100	13,400	13,200	13,000	14,700	14,700	14,700	15,000	14,500	14,100

there was great demand for grain sorghum by the livestock feed industry (12).

Sorghum research in Mexico has been partially responsible for one of the most successful episodes in Mexican agricultural history and is sometimes termed the "Second Green Revolution." Measured in terms of adoption and yield increments, sorghum has been more successful in Mexico than wheat, the major crop of the Green Revolution. The collection of sorghum production statistics began in Mexico in 1958, and show spectacular growth. From 1965-1980, the overall cultivated area expanded at a rate of 1.5% per year, but the area planted to grain sorghum increased at a rate of 13% per year, reaching 1.5 millions hectares in 1984, almost 25% of the area planted to maize and twice the area planted to wheat. In terms of grain production, the growth rate was even faster at 18% per year (13).

INIFAP and Mexican universities conduct sorghum production research in Mexico in areas such as plant breeding, integrated pest management, sorghum grain and forage nutritive quality, and other applied research areas. Collectively these groups recommend sorghum management practices to farmers in different areas of Mexico. INIFAP has released grain sorghum hybrids, recommended for the main sorghum producing areas in Mexico: INIA-Purépecha, BJ-83, RB-3006, RB-3030, RB-4000, and RB-4040. Universities in the United States and the ICRISAT Latin American Sorghum Improvement Project (LASIP) also have made critical contributions of genetic germplasm and other innovations to the Mexican sorghum research program.

Sorghum production problems in Mexico. Sorghum is subject to several production constraints in Mexico:

- Drought, due to scarce and irregular precipitation (300-600 mm annually). Estimates of arid and semi-arid areas in Mexico range from 80-105 million ha (27). Consequently, 74% of the grain sorghum grown in Mexico is cultivated either in dryland regions or under limited irrigation conditions, primarily in the northern part of the country.
- Inadequate soil and crop management for soil moisture conservation and weed control. This problem is due to limited water storage and supply, or the lack or inadequate use of suitable technology.
- Low sorghum profitability, as annual increases in the cost of production inputs may be higher than the increase in grain price.
- Improper commercialization, primarily in Northeasttern Mexico, which produces 2 million tons of sorghum annually but that has a harvest season of only two months, is distant from the consumption centers, and has poor transport systems.
- Soils with low fertility, high salinity and poor drainage, often due to decades of monoculture and improper crop management.
- Diseases such as head smut, caused by *Sporisorium reilianum* (Kuhn); charcoal rot, caused by *Macrophomina phaseolina* (Tassi) Goid; grain mold, caused by *Curvularia* spp. and *Fusarium* spp.; leaf rust, caused by *Puccinia purpurea* Cooke; head blight, caused by *Fusarium moniliforme sensu lato*; leaf blight, caused by *Exserohilum turcicum* Pass; and ergot, caused by *Claviceps africana* Frederikson, Mantle & de Milliano.
- Insect pests, predominantly sorghum midge *Contarinia sorghicola* Coquillet and the brown stink bug *Oebalus pugnax* Fabricius.

Mexico in world sorghum production. Mexico, the United States, China, India, and Nigeria are the five lead-

TABLE 75-3. Sorghum production in Mexico, after SAGAR (26).

	1961	1965	1970	1975	1980	1985	1990	1991	1992	1993	1994	1995	1996	1997	1998	1999	Mean
Area (ha × 1000)	120	310	970	1,500	1,500	1,900	1,800	1,400	1,400	880	1,300	1,400	2,200	1,900	2,000	2,000	1,600
Production (ton × 1000)	290	750	2,700	4,100	4,700	6,600	6,000	4,300	5,400	2,600	3,700	4,200	6,800	5,700	6,500	6,000	5,000
Avg. yield (ton/ha)	2.5	2.4	2.8	3.0	3.0	3.5	3.3	3.1	3.9	2.9	3.0	3.0	3.1	3.0	3.3	3.0	3.2

TABLE 75-4. Sorghum production in Mexico in different crop seasons (ha × 1000), after SAGAR (26).

Season	1990	1991	1992	1993	1994	1995	1996	1997	1998	Average
Fall-Winter	1,900	1,600	2,500	1,200	1,600	910	2,300	2,400	2,300	1,800
Spring-Summer	4,100	2,800	2,900	1,400	2,100	3,300	4,500	3,400	4,200	3,200
Total	6,000	4,300	5,400	2,600	3,700	4,200	6,800	5,700	6,500	5,000

ing sorghum producing countries in the world. Collectively, these five countries produce about 70% of the world's sorghum. Mexico alone is responsible for about 10% of the world's total (26). However, this production is not enough to satisfy national demand, and Mexico imports > 2.8 million tons of sorghum per year, making Mexico the largest importer of sorghum in the world (Table 75-1).

Sorghum in Mexico. Sorghum is one the most important crops in Mexico (Tables 75-2 and 75-3), being the third largest in cultivated area (1.8 million hectares) and the second largest in grain production (5.8 million tons per year). About two-thirds of the sorghum is produced in the spring-summer season (April-September), and the remainder in the fall-winter season (October-March) (Table 75-4). Three quarters of the total sorghum area in Mexico is in dry land areas (26). The most important sorghum producing states in Mexico, as measured by both cultivated area and by total production, are Tamaulipas, Guanajuato, Sinaloa, Michoacán, and Jalisco (Table 75-5). Average grain sorghum yield in Mexico is 3.2 t/ha, almost 2 t/ha higher than world average (1.3 t/ha). Forage sorghum is cultivated in 137,000 ha averaging 24.3 tons of forage per ha (26).

Sorghum uses in Mexico. Grain sorghum is used directly for animal feed and for feed manufacture. Sorghum grain comprises 60-80% of the total material in poultry and pork diets (26). Maize is more expensive than sorghum and is used primarily for human consumption and not for animal feed. The basic food commodities consumed by the Mexican people–eggs, milk, and beef–are all produced primarily with sorghum. An undetermined amount of grain sorghum is utilized as malt for beer production. Sorghum also may be used directly as human food in combination with maize to make tortillas.

Agroecological Regions. Mexico has a mosaic of climatic conditions, including desert, semi-tropical and tropical areas (18). Sorghum production in Mexico occurs in four broad agroecological regions: El Bajío, Northern and Northeastern Mexico, North Pacific, and Southern Mexico.

The El Bajío region has an intermediate altitude, 1200-1850 meters above sea level (masl), and a temperate climate. This region includes the states of Guanajuato, Jalisco, Michoacán, and Querétaro in which approximately 400,000 ha are planted to grain sorghum with an annual production of approximately 2.2 million tons. Half of the cultivated area is irrigated. The crop is planted during the spring-summer season. The climate in this region is well-suited for sorghum, and all genotypes have maximum yields in irrigated areas and satisfactory yields at dry land sites. The average sorghum yield is 6.2 and 8.2 t/ha in Guanajuato and Querétaro, respectively (Table 75-5). Foliar diseases, such as leaf rust, head blight, leaf blight, and ergot, are important in this region.

The Northern and Northeastern regions are at altitudes > 1000 masl with primarily arid and semi-arid climates that are hot (35-40°C) in summer. Winter (January-March) is the most important planting season for sorghum. This region includes approximately 950,000 ha planted to sorghum, primarily in Tamaulipas, but also including small areas in the states of Nuevo León, Coahuila, and Chihuahua. Annual grain sorghum production in this region is approximately 2 million tons. Drought, poor quality soils, and high summer temperatures all reduce grain sorghum yield, which is lower here than in other regions. Diseases, *e.g.*, charcoal rot and head smut, reduce sorghum production in arid and semi-arid areas, and leaf rust, grain mold, and head blight are important constraints in the small more humid areas of southern Tamaulipas (28).

The North Pacific region includes low altitude areas from Sinaloa and Sonora states with arid and semi-arid climates. This region contains about 200,000 ha planted to sorghum and yields about 500,000 tons of grain sorghum annually. Sinaloa is the most important sorghum-producing state in this region. Planting usually occurs in the summer, July-August, on dry land with limited rainfall (300-600 mm annually) and relatively low average yields (1.9 t/ha). However, both grain and the stover are used as livestock feed. In general, sorghum diseases are not important in this region.

The Southern region includes dry and humid tropical areas in southern and southeastern Mexico with altitudes from 0-800 masl. Sorghum cultivation in this area is increasing and reached 130,000 ha and 425,000 tons of grain for each of the last five years. Climate and soil conditions range from poor soils and scarce rainfall to rich soils and excellent precipitation. Grain mold, leaf rust, head blight, and leaf blight are common problems in this region. In some locations, ergot is important during the winter planting season.

The future of sorghum in Mexico. In recent years, Mexico has been increasing sorghum production. Drought, high temperature tolerance, and adaptation to diverse soil types make sorghum a good option for farmers. Sorghum's future in Mexico is in rain fed regions where both the grain and the forage can be used in livestock production. The current high demand for grain sorghum by feed manufacturers requires importation of sorghum grain equivalent to 44% of national production. Research in sorghum breeding for drought tolerance and agronomic practices to conserve soil moisture are important for sorghum's future in Mexico.

Millet

Millet is an almost unknown crop in Mexico; however, it has great potential for establishment in the country because of the available climate(s). In the past, small areas of millet have been cultivated for grain, forage and bird seed. There are no statistics available on production area and yield for millets in Mexico.

A little research has been conducted on millet since 1967 in different regions by INIFAP, some private and state universities, and ICRISAT as part of the Latin-American Sorghum Improvement Project (LASIP). Previously developed millet hybrids and varieties have been introduced from ICRISAT, INTSORMIL, and universities in the United States. Drought and high temperature tolerance, early maturity, and adaptability to different soil types, are some of millet's characters that suggest a potentially wide planting of this crop in Mexico. The present goal is to integrate millet into crop rotations in arid and semi-arid areas of Mexico and to relieve some of the demand for grain sorghum from the feed industry, by substituting pearl millet, a grain with higher nutritive value, for sorghum.

Research on millet production in Mexico is dispersed because of the lack of an organized research program similar to those on maize, sorghum, and dry beans. Most results of millet research have been published as meeting abstracts and in notes, but relatively little as journal articles. Several different millet species have been evaluated, including pearl millet (*Pennisetum glaucum*) and the minor millets–teff *(Eragrostis tef)*, foxtail millet (*Setaria italica*), and proso millet (*Panicum miliaceum*). In general, the objective of these studies was to determine grain and forage yield and quality, planting dates and methods, population density, and fertilization rates.

Grain and Forage Production. Millets are well adapted for and have adequate yields of grain and forage in several regions in Mexico.

The Valles Altos region is located in central and northern Mexico at an altitude of 1800-2300 masl. This region has a frost-free period of just 110 to 140 days (April to September). Millet's advantage over maize and sorghum in this region is its early maturity, which avoids frost. Drought tolerance also is an important characteristic for this region.

Millet genotypes were evaluated in trials conducted by ICRISAT-LASIP at four sites located at 2240 masl in El Valle de Mexico, under dry land, rain fed conditions. Millets tested included teff, foxtail millet lines PI 391634 and NESE-2A, and proso millet line PI 367683. Grain yield ranged from 1.2 ton/ha for teff to 5.5 ton/ha for foxtail millet. In semi-commercial tests grain yield ranged from 2.9 to 4.6 ton/ha (20).

In 1982, four representative millets from the *Panicum* and *Setaria* genera were evaluated under dry land conditions at three planting dates at Colegio de Postgraduados at Montecillo, México. The first and second planting dates (July 1 and 16) had 178 mm of rain during the cropping season with the *Panicum* varieties outyielding the *Setaria* species 2.8:1.2 and 1.4:0.8 t/ha, respectively. At the August planting date, the *Panicum* lines yielded only 0.2 ton/ha while the *Setaria* lines yielded 0.7 ton/ha. The best sorghum varieties produced 1.2-3.2 ton/ha of grain in control trials (15).

At INIA-CIAMEC (Centro de Investigaciones Agrícolas de la Mesa Central) in the state of Tlaxcala, pearl and foxtail millet produced 35% more grain than native barley or oat Tulancingo, and 50% more than the sor-

TABLE 75-5. 1994-1998 average grain sorghum production by the main grain sorghum-producing states in Mexico, after SAGAR (26).

State	Area (ha)	Production (ton)	Average yield (kg/ha)
Tamaulipas	852,000	1,300,000	2200
Guanajuato	194,000	1,200,000	6200
Sinaloa	189,000	370,000	1900
Michoacan	134,000	570,000	4200
Jalisco	76,000	375,000	4900
Nuevo León	40,000	81,000	2100
Nayarit	35,000	161,000	4500
Morelos	34,000	116,000	3400
Sonora	29,000	101,000	3400
Veracruz	22,000	64,000	2900
Querétaro	6,000	49,000	8300

ghum line VA-120 (7). Four crops were evaluated at Sierra de Chihuahua Experiment Station-INIFAP, in Chihuahua, from 1984 to 1986 under dry land conditions. Grain yields were: sorghum VA-110, 2.2 ton/ha; proso millet, 1.5 ton/ha; sunflower Peredovik, 1.0 ton/ha, and drybean "Ojo de cabra," 0.8 ton/ha (4).

Northern Mexico. Drought is a very important issue in this region, because most of it is located in arid and semi-arid areas. The middle zone in San Luis Potosí is being considered for millet cultivation. Traditional crops, *e.g.*, maize and dry beans, yield poorly and have problems associated with drought in some areas and with poor soils in others. Proso millet and foxtail millet lines from ICRISAT were evaluated under dry land conditions at Palma de la Cruz Experiment Station (INIFAP). The best millet genotype, NESE 45, produced 1.4 ton/ha of grain and required 77 days to reach maturity, compared to barley, variety Cerro Prieto, which yielded 0.9 ton/ha of grain (17). In other trials, the best millet varieties were foxtail millet NESE 45 and NESE 42, yielding up to 3.5 ton/ha, with only 300 mm of rainfall during the cropping season (16). Foxtail millet genotypes required only 80 days to reach maturity.

Dry beans and maize are conventional crops in approximately 500,000 dry land ha in Durango. Dry beans are affected by diseases, and crop rotation is now required in this region. At Valle de Guadiana Experiment Station-INIFAP, the best millets evaluated were foxtail millet, averaging 2.5 ton/ha of grain, and proso and pearl millets, which averaged 1.5 ton/ha of grain. However, millet yields were lower than maize because drought conditions were not severe (6).

Breeding for forage millets has been conducted at Instituto Tecnológico y de Estudios Superiores de Monterrey (ITESM) in Nuevo León, since 1967. Forage millet variety ITESM-83, obtained from this program, can produce 17 t/ha of fresh forage, (24). Millet also has higher water, soil, and time efficiency than either maize or sorghum. Millet can produce almost a ton of protein per hectare in only 51 days, while maize and sorghum require twice this time (11).

Southern Mexico. Maize is very susceptible to drought and often fails to make a crop in Oaxaca. In 1987, 37 varieties from nine crops were evaluated at three locations. In Tlacolutla, a limited rainfall location, similar yields were obtained from minor millets, sunflower, wheat, and sorghum. The maize yield was statistically lower. In Ocotlán, the favorable rainfall location, the yield of the alternative crops yield was not statistically higher than that of maize. Thus, for limited rainfall locations, wheat, minor millets, sorghum, and sunflower, are good alternatives to maize.

Research to evaluate amaranth, millet, flax, lentil, sunflower, colza, safflower, chickpea, cowpea, maize, and a maize-dry bean intercrop association was conducted at Mixteca Alta, Oaxaca, in the 1984, 1986 and 1987 spring-summer seasons. Water consumption during the crop season was 565, 413, 356, 387, 352, 424, and 595 mm for flax, chickpea, sunflower, colza, lentil, millet, and maize, respectively. In 1986, millet and chickpeas produced 1.93 and 1.98 tons of grain/ha, respectively, and were superior to the maize-dry bean intercrop, which produced 1.67 ton/ha (9). It is likely that millet can result in higher, more reliable income than maize at locations with limited rainfall (10). In another study, wheat, dry beans, minor millets, and chickpeas were evaluated at two dry land locations in Oaxaca. Crop yields were– wheat 0.8 t/ha, minor millets 0.5 t/ha, and dry beans and chickpeas 0.4 t/ha (3).

In central Oaxaca valleys, maize fails to make a crop in one of every five years and in some areas, *e.g.*, Tlacolutla, maize fails in two out of three years. Pearl millet genotypes from India, selected for their drought tolerance, were evaluated in 1985 and 1986. NELC-C4 millet produced 5.8 t/ha of dry forage and required 100 days to reach maturity (25). The best plant density was 280,000 plants/ha, and average grain yield was 1.6 ton/ha.

Low precipitation (< 400 mm annually) limits dry land agriculture in Michoacán, where pearl millet genotypes from ICRISAT were evaluated. ICMV 84400 and ICTP 8203 were the best genotypes producing 1.8 and 1.5 ton of grain per ha, respectively. Average dry matter forage production for both genotypes was 10 t/ha (14).

Human consumption. *Millet bread.* Bread and cookies can be baked when flour from foxtail millet is mixed 15:85 millet:wheat. The composite flour yielded a bread with a volume and texture similar to that of whole wheat bread (5). Good quality cookies with a 40:60 millet: wheat composite flour. Both the bread and the cookies made with millet had good acceptance (19).

Millet tortillas. Maize tortillas are a staple food in Mexico. In 1983, the food technology laboratory at CIFAP Mexico (Centro de Investigaciones Forestales, Agrícolas y Pecuarias del Valle de México), evaluated millet grain as tortilla ingredient. Millet has excellent organoleptic characteristics when a 10:90 millet:corn composite flour is used to make tortillas (8). Tortillas made solely from minor millet grain by using a maize recipe had good appearance and taste. Millet grain is acceptable for human consumption due to its low tannin and phenol content (20).

Millet in the United States. The area planted to pearl millet in the United States, primarily for forage production, is approximately 0.6 million ha. Use of pearl millet grain on a commercial basis began in the early 1990s, with the largest areas planted in Georgia and Florida. Millet breeding research programs have been established at Kansas State University (Hays), USDA/ARS (Tifton, Georgia), and the University of Nebraska-Lincoln (1). These research programs also are important for Mexican agriculture, since some areas in Mexico have similar soils and climates. Recent research advances in millet production should be transferable to Mexico, in much the same way as those for sorghum production have been made in the past.

Future of Millet. Millet has excellent prospects in Mexico, primarily in arid and semi-arid regions, and in areas with poor soils. The most important features for millet in Mexico are drought tolerance, early maturity, and adaptation to a variety of soil conditions. Millet grain can complement or replace sorghum in animal feed, and producing the grain locally would help to decrease the grain sorghum deficit in Mexico that requires the annual importation of ≥ 2.8 million tons of grain sorghum. Demand for millet also might increase when industry begins to exploit millet's increased nutritive value relative to sorghum.

References

1. **Andrews, D. J., W. W. Hanna, J. F. Rajewski, and V. P. Collins.** 1996. Advances in grain pearl millet: Utilization and production research, p. 170-177. *In* J. Janick (ed.), *Progress in New Crops*. ASHS Press, Alexandria, Virginia.
2. **Arredondo, V. C.** 1988. Módulos de evaluación de cultivos de alternativa a maíz bajo condiciones de temporal. *Proc. of Primera Reunión Científica Forestal y Agropecuaria (11 October 1988, Oaxaca, Oax., Mexico)*: 15.
3. **Arredondo, V. C., and V. J. Ruíz.** 1988. Parcelas de observación de diversos cultivos bajo condiciones de humedad residual. *Proc. of Primera Reunión Científica Forestal y Agropecuaria (11 October 1988, Oaxaca, Oax., Mexico)*: 16.
4. **Avila, M. M. R.** 1994. Evaluación de cuatro especies cultivables, en un área agrícola de alta siniestralidad por sequía en el norte de México. *Memorias XV Congreso de Fitogenética (25-30 September 1994, Monterrey, N.L., Mexico)*: 483.
5. **Calderon, D. G.** 1987. Mezclas de trigo y mijo en panificación y galletas. *Agric. Téc. Méx.* **13:** 189-196.
6. **Castillo, R. A., R. V. M. Castro, and H. J. Lopez.** 1988. Mijo de grano, una posible alternativa para las zonas semiáridas de altura de Durango. *Proc. of Primera Reunión Científica Forestal y Agropecuaria (11 October 1988, Oaxaca, Oax., Mexico)*: 48.
7. **CIAMEC.** 1981. Logros y aportaciones de 1981 a la investigación agrícola en el estado de Tlaxcala. Page 51. SARH, INIA. Chapingo, Edo de México, Mexico.
8. **CIAMEC.** 1983. Informe de nixtamalización y elaboración de tortillas en mezclas maíz-mijo. SARH, INIA. Laboratorio de Tecnología de Alimentos. Chapingo, Mex. Mexico.
9. **Cuellar, D. G.** 1988. Introducción y observación de cultivos de alternativa en la Mixteca. *Proc. of Primera Reunión Científica Forestal y Agropecuaria (11 October 1988, Oaxaca, Oax., Mexico)*: 42.
10. **Cuellar, D. G.** 1989. Evaluación de cultivos de alternativa a maíz en la Mixteca Oaxaqueña. *Memorias Segunda Reunión Científica Forestal y Agropecuaria (9-10 November 1989, Xoxocotlán, Oax., Mexico)*: 69.
11. **De Alba, G. and F. Leal.** 1972. Mijo perla (*Pennisetum glaucum* L.), un cultivo nuevo en México, p. 42-43. *XIII Informe de Investigación. 1971-1972.* Instituto Tecnológico y de Estudios Superiores de Monterrey, Escuela de Agricultura y Ganadería, Monterrey, N.L., México.
12. **De Walt, B. R., and D. Barkin.** 1984. La crisis alimentaria mexicana e investigaciones en sorgo. *Memorias Primera Reunión Nacional Sobre Sorgo. Potencial y Uso del Sorgo Granífero (22-26 October 1984, Marín, N.L., México)*: 98-122.
13. **De Walt, B. R., and D. Barkin.** 1985. El sorgo y la crisis alimentaria mexicana, p. 153-167. In C. Paul and B. R. De Walt (ed.), *El Sorgo en Sistemas de Producción en América Latina*. San Andrés, la Libertad, El Salvador and INTSORMIL-CIMMYT. México, D. F. México.
14. **Espinoza, A. J.** 1992. Evaluación de 13 líneas de mijo perla bajo temporal en el Valle de Apatzingán, Michoacán. *Memorias 5ª Reunión Científica Forestal y Agropecuaria (10 November 1992, Morelia, Mich., Mexico)*: 61.
15. **Estrada, G. J. A., O. A. Muñoz, C. A. Carballo, C. A., and S. J. H. Hernández.** 1982. Resistencia a sequía en sorgo y mijo, p. 239. In *Avances en La Investigación*. Colegio de Postgraduados, Montecillo, Edo. de Mex., México.
16. **Hernández, A. J. A.** 1992. Introducción y generación de técnicas de producción en mijos para grano. *Tercera Reunión Científica Forestal y Agropecuaria de San Luis Potosí (5 June 1992, San Luis Potosí, S.L.P., Mexico)*: 193.
17. **Hernández, A. J. A., and M. A. Perales de la C.** 1988. Adaptación de mijo para grano (*Setaria italica* L. and *Panicum miliaceum*) en zonas de baja precipitación de San Luís Potosí. *Resúmenes del X11 Congreso de Fitogenética (18-22 July 1988, Chapingo, Edo. de México, México)*: 207.
18. **Medina, G. G., C. J. A. Ruiz, and P. R. A. Martínez.** 1998. *Los Climas de México. Una Estratificación Ambiental Basada en el Componente Climático. Libro técnico No. 1.* INIFAP, Guadalajara, Jal., México.
19. **Morales, G. M., and S. A. C. Moaut.** 1989. Uso de los cultivos de alternativa para la alimentación humana en la Mixteca Oaxaqueña. *Memorias Segunda Reunión Científica Forestal y Agropecuaria (9-10 November 1989, Xoxocotlán, Oax., Mexico)*: 100.
20. **Paul, C. L., and Y. D. Ojeda.** 1987. La productividad de mijos menores en el altiplano de México. *XXXIII Reunión Anual de PCCMCA (30 March–4 April 1987, Cd. de Guatemala, Guatemala)*: 102-114.
21. **Pitner, J. B., J. L. Lazo de la Vega, and D. N. Sánchez.** 1954. *El Cultivo del Sorgo. Folleto Técnico No. 15.* Oficina de Estudios Especiales, SAG, México, D. F., México.
22. **Pitner, J. B., D. N. Sánchez, and F. J. L. Puertas.** 1950. *Sorgo Para Grano. Folleto de Divulgación No. 11.* Oficina de Estudios Especiales, SAG, México, D. F., México.
23. **Reyes, C. P.** 1985. *Diseño de Experimentos Aplicados.* Editorial Trillas, México, D. F., México.
24. **Robles, S. R., and G. A. Romo.** 1988. Mejoramiento y uniformización de caracteres en mijo perla (*Pennisetum glaucum* L.) para producción de forraje en Apodaca, N.L, p.

101-102. *XX Informe de Investigación. 1985-1986.* Instituto Tecnológico y de Estudios Superiores de Monterrey, Escuela de Agricultura y Ganadería, Monterrey, N.L., México.

25. **Ruiz, V. J.** 1989. Evaluación de tres variedades de mijo perla y tres densidades de siembra en los valles centrales de Oaxaca. *Memorias Segunda Reunión Científica Forestal y Agropecuaria (9-10 November 1989, Xoxocotlán, Oax., Mexico)*: 65.
26. **SAGAR (Secretaría de Agricultura, Ganadería y Recursos Hidráulicos).** 2000. *Situación Actual y Perspectiva de la Producción de Sorgo en México–1990-1999.* Centro de Estadística Agropecuaria, Mexico, D.F., Mexico.
27. **Universidad Autónoma de Chapingo, ed.** 1999. *La Desertificación en el Altiplano Mexicano.* Comisión Nacional de las Zonas Aridas, México, D.F., México.
28. **Williams, A. H., H. R. Rodríguez, and G. N. Montes.** 1995. 20 años de investigación en sorgo en el Campo Experimental Río Bravo. *Germen* **11:** 1-35.
29. **Zavala, G. F.** 1984. *Estudios Sobre el Crecimiento y Desarrollo del Sorgo en México. Folleto de Divulgación No. 7.* Universidad Autónoma de Nuevo León, Facultad de Agronomía, Marin, N.L., Mexico.

76

Diseases of Pearl Millet in the Americas

Jeffrey P. Wilson

Introduction

The extent of pearl millet cultivation in the western hemisphere is greatly under reported since the crop presently is grown primarily for pasture. Approximately 3 million ha of pearl millet are grown as a high-quality forage crop in the United States and Brazil, and considerable interest in grain hybrids exists. Many pathogens, including bacteria, fungi, viruses, and nematodes, attack pearl millet in the Americas. The diseases of greatest economic impact in the Americas are rust, caused by *Puccinia substriata* var. *indica*, and the leaf blight complex, caused primarily by *Pyricularia grisea*. These diseases are found primarily in central Brazil (6) and the southeastern United States. Preharvest and storage grain molds can be significant in the United States. Information on the distribution of diseases throughout Central and South America is primarily anecdotal. As cultivation of the crop expands, it is likely that production constraints from diseases will assume greater importance.

Pearl millet diseases in the Americas have not been reviewed in either of the previous Global Conferences on Sorghum and Pearl Millet Diseases. Although most of the research cited in this review is from work reported within the past 10 years, selected earlier works are cited to provide a more complete historical perspective for the Western Hemisphere.

Bacterial Diseases

Bacterial diseases occur in the Americas, but appear to be of limited consequence. Bacterial spot, caused by *Pseudomonas syringae*, was reported to be pathogenic to pearl millet in Iowa (13). A pathogenic *Acidovorax avenae* (=*Pseudomonas avenae*) was isolated in Georgia from pearl millet with bacterial stripe symptoms (Gitaitis and Wilson, *unpublished*), and the bacterium also has been reported in Parana, Brazil (1). Elsewhere in the Americas, symptoms resembling those of a bacterial disease have been observed in the desert highlands of Aguascalientes, Mexico (Hash, *personal communication*).

Fungal Diseases

Fungi are the most important pathogens of pearl millet. In the Americas, these diseases can be incidental or severe, depending upon the pathogen, environment, and host susceptibility.

Leaf blight. Several fungi cause necrotic leaf spots and foliar blights of pearl millet and often occur as a leaf blight complex, which involves several primary pathogenic fungi and probably some secondary invaders. Fungal components of the leaf blight complex in Georgia include *P. grisea, Cercospora penniseti, Exserohilum rostratum, Drechslera dematioidea, Phyllosticta penicillariae, Bipolaris setariae*, and possibly pathogenic *Curvularia* spp. (37, 41). Of these fungi, *P. penicillariae* was identified as a foliar pathogen of pearl millet in the United States for the first time in 1989 (34).

The primary component of the leaf blight complex is *P. grisea*. Pyricularia leaf spot was first identified in the United States in 1968 (26). The disease has been observed in Guatemala on a set of introductions received from ICRISAT in the late 1980s (Hash, *personal communication*), and in the Cerrado region of Brazil (3, 4). The pathogens in the leaf blight complex in Brazil include *P. grisea, Bipolaris* spp., and *Phyllosticta* spp. The latter two pathogens frequently are seedborne (3).

The leaf blight complex can reduce digestible dry matter yields of forage pearl millet by 18% in an average year (37). Resistance has been incorporated into elite forage varieties (10), is simply inherited (15), and has been stable and effective to date.

Rust. Rust, caused by *P. substriata* var. *indica*, is the most severe disease of pearl millet in the southeastern United States and has been the subject of intensive study. The first rust epidemic in the United States occurred in 1972 (24). No resistant varieties were grown until the release of the forage hybrid Tifleaf 2 in 1988. This hybrid had the Rr_1 resistance gene (10). In 1993, the grain hybrid HGM 100, which also carried the Rr_1 gene, was released (8) thereby increasing the acreage protected by this single resistance gene. Pathotypes virulent to *Rr*$_1$ were detected in 1991 (29), and the gene was commercially ineffective by 1994. At least 11 different pathogenic races exist within the *P. substriata* var. *indica* population in the southeastern United States (19). Since only the *Rr*$_1$ resistance gene has been used commercially, this diversity is both remarkable and unexplained.

Eggplant is an alternate host of *P. substriata* var. *indica* (23), but the most common solanaceous weeds in the United States are not (23, 47). Natural infection of eggplant was observed in 1997 (48). The importance of eggplant as an alternate host for rust epidemics in pearl millet and its impact on pathogen diversity in the Americas are not known.

Elsewhere in the Americas, aecia of *P. substriata* var. *indica* have been observed on eggplant in Brazil (7). Rust also occurs on pearl millet in the Cerrado region of Brazil (3), and appears to be both widespread and serious (C. T. Hash, personal communication). Rust also has been observed on well-established perennial pearl millet × napiergrass hybrids in St. Croix, Virgin Islands (Hanna, *personal communication*), although no rust was observed on nearby pearl millet. The importance of the Caribbean as an overwintering site is unknown, but it is likely to be important.

Rust infection reduces both dry matter yield and digestibility of forage, resulting in a logarithmic decrease in digestible dry matter yield in response to increasing rust severity (38, 40). Losses of 25% of digestible dry matter yield can result from as little as 10% rust severity at harvest. In an average year, the digestible dry matter yield of a rust-infected hybrid is approximately 43% less than that of a resistant hybrid. Losses in grain yield are not detected until rust severity at harvest exceeds about 50% of the foliage infected (44). Apparently, carbohydrate reserves in the stem suffice to support grain fill and a moderate rust infection.

In light of the ephemeral nature of single, major genes for hypersensitive resistance, gene deployment strategies are being re-examined. The current strategy is to incorporate different resistance genes in forage and grain varieties. Thus, if a shift in virulence gene frequencies occurs in the pathogen population, only one class of hybrids will be rendered susceptible. The pearl millet grain hybrid breeding program in Georgia currently emphasizes resistance from Tift 89D$_2$ (9), which is independent of the *Rr*$_1$ locus (15), however, alternative sources of resistance are being incorporated into elite germplasm.

Resistance breeding strategies in forage hybrids currently are based on a broad genetic base, and include combinations of genes for slow-rusting coupled with race-specific genes for resistance deployed in a dynamic multiline population. Slow-rusting will be useful as a complement to enhance the effectiveness and longevity of race-specific genes. Slow-rusting resistance in ICMP 501 is conferred by 2-3 genes with additive, dominant and epistatic expression (31). This resistance has been incorporated into an elite forage pollinator inbred line (38).

Race-specific genes from diverse sources are being combined in a dynamic multiline population (39). Pearl millet populations heterogeneous for disease resistance reduce epidemic development (20, 45). When compared to the mean of their component inbreds, intermated mixtures give more consistent resistance and higher dry matter yields than do physical mixtures (20). Because forage hybrids do not have the same requirements for phenotypic uniformity as do grain hybrids, the dynamic multiline population concept was first developed for use in forage hybrids (43).

Smut. Smut, caused by *Moesziomyces penicillariae*, was first identified in the United States in 1962 (25). Research on this disease has been limited since most pearl millet in the United States is grown for forage, and neither grain production nor smut resistance have been significant selection criteria. The pathogen is genetically diverse, and can exist as individual pathogenic isolates or as sporidial clones, which presumably require a strain of a compatible mating type for the production of pathogenic dikaryons (33). At least six mating or compatibility groups have been identified, and differences in combining ability for virulence exist among these isolates.

The *tr* allele, which confers a trichomeless phenotype when homozygous, also results in reduced smut infection (27). A longer moisture period is required to obtain a specific level of disease severity on the trichomeless inbreds. More importantly, flag leaf sheaths are wrapped more tightly around the nonemerged panicles in trichomeless varieties than in varieties with trichomes (30). The tight flag leaf sheath makes it more difficult for sporidia suspended in dew and rainwater from washing from the flag leaf into the boot, at which time the plant is highly susceptible to infection. Excluding the pathogen from the infection court allows the plant to escape infection and is potentially a durable resistance mechanism that would be unaffected by variation for pathogenicity in the fungus. In the absence of disease, the trichomeless character has no effect on grain yield (42). Elsewhere in the Americas, smut has been observed at Sete Lagoas, in Minas Gerais, Brazil (Hash, *personal communication*).

Grain molds. The dominant preharvest grain mold fungi in Georgia include *Fusarium semitectum*, *Alternaria* spp., and *Curvularia* spp. (46). Other less-frequently isolated fungi include *Aspergillus flavus*, *B. setariae*, *Cladosporium* spp., *D. dematioidea*, *Epicoccum* spp., *E. rostrata*, *Fusarium chlamydosporum*, *F. equiseti*, *F. moniliforme sensu lato*, *Gloeocercospora sorghi*, *Penicillium* spp., and *P. penicillariae*. Frequencies of individual species are dependent on the environmental conditions 2-4 weeks prior to harvest. Timely harvest of grain reduces preharvest grain molds (35). Isolation of *F. semitectum* and *F. chlamydosporum* from grain increased linearly as harvest was delayed. Low levels of aflatoxins, trichothecenes and zearalenone have been detected in pearl millet grain, but mycotoxin contamination is not significantly affected by the harvest date. In these experiments, levels of Fusarium mycotoxins were associated with contamination by *F. chlamydosporum*.

Fungi causing grain molds in storage include *F. semitectum*, *F. chlamydosporum*, *Alternaria* spp., *A. flavus*, and *Helminthosporium* spp. Although not detected at high levels at harvest, aflatoxins can become a significant problem in stored pearl millet (12).

Seedling blights and stalk rots. Seed-borne fungi *F. moniliforme sensu lato*, *E. rostratum*, and *B. setariae* cause pre-emergence damping off in cool soils (< 15°C) (28). *Phyllosticta penicillariae* was first found as a pathogen on pearl millet in the United States due to its ability to cause a severe seedling blight (34).

In Georgia, stalk rot and lodging commonly accompany severe rust infections (38, 44). Many fungi have been isolated from rotted stalks that had not yet lodged (36). Of the dominant stalk-invading fungi, *F. moniliforme sensu lato* was nearly uniformly distributed in the nodes throughout the stalks. Infection by *F. graminearum* and *Alternaria* spp. were more common in the upper nodes, while infection by *F. semitectum* and *Nigrospora* spp. was more common in the lower nodes.

Other stalk rot disorders have been observed in the Americas. *F. moniliforme sensu lato* causes top-rot symptoms in Georgia (22). Rhizoctonia blight, caused by *Rhizoctonia solani*, and southern blight or white mold, caused by *Sclerotium rolfsii*, have been reported in Florida (21). *Macrophomina phaseolina* has been reported in Pernambuco, Brazil (13).

Viral Diseases

Panicum mosaic virus (PMV) has been isolated from switchgrass in Kansas, but inoculated pearl millet was immune to PMV (18), and the extent of natural infection of pearl millet is not known. Buzen *et al.* (2) reported that PMV and its satellite virus increased on pearl millet [*Setaria italica* (L.) Beauv.] (sic.), but there is a discrepancy in their host identification since *S. italica* is foxtail millet. Day *et al.* (5), who also referred to the host plant as pearl millet sent a voucher specimen to me that I confirmed to be pearl millet. Masuta *et al.* (14) and Qui *et al.* (16) use pearl millet to increase PMV and satellite PMV.

Wheat streak mosaic virus also has been recovered from pearl millet with virus-like symptoms in Kansas (17). The virus was transmitted mechanically and by the wheat curl mite. The extent of infection under field conditions is unknown.

Nematode Diseases

Nematode diseases are probably an under-appreciated yield constraint to pearl millet production. Nematodes pathogenic to pearl millet in Georgia include dagger nematode (*Xiphinema americanum*), ring nematode (*Criconemella ornata*), root knot nematodes (*Melodogyne incognita, M. javonica,* and *M. arenaria*), root lesion nematodes (*Pratylenchus mulchandi*, *P. brachyurus*, and *P. zeae*), sting nematode (*Belonolaimus longicaudatus*), and stubby-root nematode (*Paratrichodorus minor*). Although it is difficult to evaluate the relative severity of damage and yield reduction caused by individual nematode species in the field, the effect of the nematode complex on yield of pearl millet can be estimated by the use of nematicides. Treatments with the nematicides phenamiphos and aldicarb result in a 40-44% yield increase in forage dry matter (11). The effect of nematicides on yield of grain hybrids is presently unknown.

Conclusions

With very few exceptions, information on the geographic distribution and the extent of losses due to pearl millet diseases in the Americas is limited. Knowledge of pathogen and disease distribution, regardless of their perceived importance, is essential for the establishment of rational quarantine and regulatory policies. The severity of specific diseases depends on the cultivar and environment, and diseases of marginal importance in one region may be signifycant in others. Quarantine efforts may be less effective because information in the literature concerning the distribution and identification of many diseases of pearl millet is incomplete and not easily summarized. Recent efforts by the USDA-ARS to regenerate pearl millet accessions in the National Plant Germplasm System in post-entry quarantine has required more accurate information regarding pathogens of pearl millet and their geographic distribution. This information has recently been summarized (32). As pearl millet cultivation expands into nontraditional areas, it is inevitable that pathogens will become more widespread, and losses due to diseases will increase.

References

1. **Almeida, I. M. G., M. A. R. de Oliveira, V. A. Malavolta, Jr., and J. R. Neto.** 1996. Milheto, novo hospedeiro de

Acidovorax avenae subsp. *avenae* no Brazil. *Summa Phytopathol.* **22:** 66.

2. **Buzen, F.G., Jr., C. L. Niblett, G. R. Hooper, J. Hubbard, and M. A. Newman.** 1984. Further characterization of Panicum mosaic virus and its associated satellite virus. *Phytopathology* **74:** 313-318.
3. **d'A Charchar, M. J., and J. R. N. dos Anjos.** 1999. Common diseases in pearl millet (*Pennisetum americanum*) in the Cerrado. *Proc. Intl. Pearl Millet Workshop (9-10 June 1999, Planaltina, Brazil)*: 206-207.
4. **d'A Charchar, J. J., J. R. N. dos Anjos, and A. M. de Carvalho.** 1998. *Pyricularia grisea* infecto milheto (*Pennisetum americanum*) nos Cerrados. *Fitopatol. Brasil.* **23 (Suppl.):** 233.
5. **Day, J., H. Ban, S. Patel, S. B. Larson, and A. McPherson.** 1994. Characterization of crystals of satellite panicum mosaic virus. *J. Molec. Biol.* **238:** 849-851.
6. **de'A Lira, M., G. A. Maciel, J. N. Tabosa, M. R. Alves, J. P. de'O Santos, E. V. de Freitas, and A. S. Arcoverde.** 1983. *Cultivo do Milheto [*Pennisetum americanum *(L.) Leeke]. Instruções Técnicas do IPA No. 8.* Secretaria de Agricultura, Empresa Pernambucana de Pesquisa, Agropecuária, IPA, Recife, Pernambuco, Brazil.
7. **Figueiredo, M. B., B. P. Bastos Cruz, A. P. da Silveira, and N. Makishima.** 1971. Uma ferrugem cuja forma ecidial ocorre sobre berinjela (*Solanum melongena* L.). *Arq. Inst. Biol. Sao Paulo* **38:** 173-175.
8. **Hanna, W. W.** 1993. Registration of pearl millet parental lines Tift 8677 and A_1/B_1 Tift $90D_2E_1$. *Crop Sci.* **33:** 1119.
9. **Hanna, W. W., and H. D. Wells.** 1993. Registration of parental line Tift 89D2, rust-resistant pearl millet. *Crop Sci.* **33:** 361-362.
10. **Hanna, W. W., H. D. Wells, G. W. Burton, and W. G. Monson.** 1988. Registration of 'Tifleaf 2' pearl millet. *Crop Sci.* **28:** 1023.
11. **Johnson, A. W., and G. W. Burton.** 1977. Influence of nematicides on nematodes and yield of sorghum-sudangrass hybrids and millets. *Plant Dis. Rept.* **61:** 1013-1017.
12. **Jurjevic, Z., D. M. Wilson, J. P. Wilson, and H. Casper.** 1999. Fungal successions and mycotoxins in stored pearl millet. *Proc. of the Aflatoxin Elimination Workshop, (20-22 October 1999, Atlanta, Georgia)*: 34.
13. **Kendrick, J. B.** 1926. Holcus bacterial spot of species of *Holcus* and *Zea mays*. *Phytopathology* **16:** 236-237.
14. **Masuta, C., D. Zuidema, B. G. Hunter, L. A. Heaton, D. S. Sopher, and A. O. Jackson.** 1987. Analysis of the genome of satellite panicum mosaic virus. *Virology* **159:** 329-338.
15. **Morgan, R. N., J. P. Wilson, W. W. Hanna, and P. Ozias-Akins.** 1998. Molecular markers for rust and pyricularia leaf spot disease resistance in pearl millet. *Theor. Appl. Genet.* **96:** 413-420.
16. **Qui, W. P., M. Turina, and K.-B. G. Scholthof.** 1998. The capsid of satellite panicum mosaic virus (SPMV) is dispensable for infectivity but affects the symptom phenotype on pearl millet plants. *Phytopathology* **88:** S74.
17. **Seifers, D. L., T. L. Harvey, K. D. Kofoid, and W. D. Stegmeier.** 1996. Natural infection of pearl millet and sorghum by wheat streak mosaic virus in Kansas. *Plant Dis.* **80:** 179-185.
18. **Sill, W. H., Jr., and R. C. Pickett.** 1957. A new virus disease of switchgrass, *Panicum virgatum* L. *Plant Dis. Rept.* **41:** 241-249.
19. **Tapsoba, H., and J. P. Wilson.** 1996. Pathogenic variation in *Puccinia substriata* var. *indica* in the southeastern United States and screening for resistance in pearl millet germplasm. *Plant Dis.* **80:** 395-397.
20. **Tapsoba, H., and J. P. Wilson.** 1999. Increasing complexity of resistance in host mixtures through intermating to manage rust of pearl millet. *Phytopathology* **89:** 450-455.
21. **Weber, G. F.** 1963. *Pellicularia* spp. pathogenic on sorghum and pearl millet in Florida. *Plant Dis. Rept.* **47:** 654-656.
22. **Wells, H. D.** 1956. Top rot of pearl millet caused by *Fusarium moniliforme*. *Plant Dis. Rept.* **40:** 387.
23. **Wells, H. D.** 1978. Eggplant may provide primary inoculum for rust of pearl millet caused by *Puccinia substriata* var. *indica*. *Plant Dis. Rept.* **62:** 469-470.
24. **Wells, H. D., G. W. Burton, and J. F. Hennen.** 1973. *Puccinia substriata* var. *indica* in the southeast. *Plant Dis. Rep.* **57:** 262.
25. **Wells, H. D., G. W. Burton, and D. K. Ourecky.** 1963. Tolyposporium smut, a new disease on pearl millet, *Pennisetum glaucum*, in the United States. *Plant Dis. Rept.* **47:** 16-19.
26. **Wells, H. D., G. W. Burton, and J. B. Powell.** 1969. Piricularia [sic] leaf spot of pearl millet. *Phytopathology* **59:** 1057.
27. **Wells, H. D., W. W. Hanna, and G. W. Burton.** 1987. Effects of inoculation and pollination on smut development in near-isogenic lines of pearl millet. *Phytopathology* **77:** 293-296.
28. **Wells, H. D., and E. E. Winstead.** 1965. Seed-borne fungi in Georgia-grown and western-grown pearl millet seed on sale in Georgia during 1960. *Plant Dis. Rept.* **49:** 487-489.
29. **Wilson, J. P.** 1993. Identification of virulence in *Puccinia substriata* var. *indica* to Rr_1 in pearl millet. *Plant Dis.* **77:** 100.
30. **Wilson, J. P.** 1995. Mechanisms associated with the *tr* allele contributing to reduced smut susceptibility of pearl millet. *Phytopathology* **85:** 966-969.
31. **Wilson, J. P.** 1997. Inheritance of partial rust resistance in pearl millet. *Plant Breed.* 116: 239-243.
32. **Wilson, J. P.** 1999. *Pearl Millet Diseases: A Compilation of Information on the Known Pathogens of Pearl Millet,* Pennisetum glaucum *(L.) R. Br.* U.S. Department of Agriculture, Agric. Res. Service, Agricultural Handbook No. 716.
33. **Wilson, J. P., and K. Bondari.** 1990. Quantitative genetic analysis of pathogenicity and virulence of *Moesziomyces penicillariae*. *Phytopathology* **80:** 1173-1177.
34. **Wilson, J. P., and G. W. Burton.** 1990. *Phyllosticta penicillariae* on pearl millet in the United States. *Plant Dis.* **74:** 331.
35. **Wilson, J. P., H. H. Casper, and D. M. Wilson.** 1995. Effect of delayed harvest on contamination of pearl millet grain with mycotoxin-producing fungi and mycotoxins. *Mycopathologia* **132:** 27-30.
36. **Wilson, J. P., B. M. Cunfer, and D. V. Phillips.** 1999. Double-cropping and crop rotation effects on diseases and grain yield of pearl millet. *J. Prod. Agric.* **12:** 198-202.

37. **Wilson, J. P., and R. N. Gates.** 1993. Forage yield losses in hybrid pearl millet due to leaf blight caused primarily by *Pyricularia grisea. Phytopathology* **83:** 739-743.
38. **Wilson, J. P., and R. N. Gates.** 1999. Disease resistance and biomass stability of forage pearl millet hybrids with partial rust resistance. *Plant Dis.* **83:** 733-738.
39. **Wilson, J. P., and R. N. Gates.** 2002. The dynamic multi-line population: an alternative approach to durable resistance?, p. 65-69. *In* J. F. Leslie (ed.), *Sorghum and MilletsDiseases*. Iowa State Press, Ames, Iowa.
40. **Wilson, J. P., R. N. Gates, and W. W. Hanna.** 1991. Effect of rust on yield and digestibility of pearl millet forage. *Phytopathology* **81:** 233-236.
41. **Wilson, J. P., and W. W. Hanna.** 1992. Effects of gene and cytoplasm substitutions in pearl millet on leaf blight epidemics and infection by *Pyricularia grisea. Phytopathology* **82:** 839-842.
42. **Wilson, J. P., and W. W. Hanna.** 1998. Smut resistance and grain yield of pearl millet hybrids near-isogenic at the *Tr* locus. *Crop Sci.* **38:** 649-651.
43. **Wilson, J. P., W. W. Hanna, and K. Bondari.** 1993. Directed use of germplasm resources in breeding rust resistant pearl millet. *Trends in Agric. Sci. (Plant Pathol.)* **1:** 67-74.
44. **Wilson, J. P., W. W. Hanna, and G. Gascho.** 1996. Pearl millet grain yield loss from rust infection. *J. Prod. Agric.* **9:** 543-545.
45. **Wilson, J. P., W. W. Hanna, and R. N. Gates.** 1994. Stability of forage yield and quality in pearl millet hybrids heterogeneous for rust resistance. *Euphytica* **72:** 163-170.
46. **Wilson, J. P., W. W. Hanna, D. M. Wilson, R. W. Beaver, and H. H. Casper.** 1993. Fungal and mycotoxin contamination of pearl millet grain in response to environmental conditions in Georgia. *Plant Dis.* **77:** 121-124.
47. **Wilson, J. P., S. C. Phatak, and G. Lovell.** 1996. Aecial host range of *Puccinia substriata* var. *indica. Plant Dis.* **80:** 806-808.
48. **Wilson, J. P., and W. Williamson.** 1997. Natural infection of eggplant by *Puccinia substriata* var. *indica* in the United States. *Plant Dis.* **81:** 1093.

Part XI

Abstracts

77

Fungal Contaminants and Mycotoxins on Stored Pearl Millet Grain

Z. Jurjevic, David M. Wilson, Howard H. Casper, and Jeffrey P. Wilson

Storage conditions affecting fungal and mycotoxin contamination in pearl millet grain were evaluated. Grain harvested in 1996, 1997, and 1998 was stored for 3-9 weeks at 20 or 25°C, at 86, 91, or 100% relative humidity, an air or N_2 atmosphere, and at one of three grain moisture levels (9-11%, 17-20%, or 20-22%). The most common contaminants included *Fusarium chlamydosporum* (19%), *Curvularia* spp. (14%), *F. semitectum* (16%), *Alternaria* spp. (9%), "*Helminthosporium*" complex spp. (6%), *Aspergillus flavus* (8%), and *F. moniliforme sensu lato* (3%). The year of grain production was the primary source of variation affectting fungal contamination. The frequency of *A. flavus* isolation increased at 100% relative humidity and in the 17-20% and 20-22% grain moisture treatments, particularly at 25°C. Aflatoxin contamination averaged 170 ng/g over all treatments, and increased up to 800 ng/g in grain in the high moisture treatments at 25°C. Frequency of *F. chlamydosporum* isolation increased at 86 and 91% relative humidity. Low levels of nivalenol, deoxynivalenol, and zearalenone were detected.

78

A Physiological Approach to Resistance Breeding for Control of Seed Rot and Seedling Diseases of Grain Sorghum

Aliya S. Kasakova, E. V. Ionova, V. I. Pakhomov, E. I. Lipkovich, M. M. Yanina, M. A. Taranov, and M. G. Fedorishchenko

In Russia, cool wet springs result in seed rot and seeding diseases that affect sorghum stand establishment and yield. Our objectives were to develop a simple, indirect, physiological method for screening sorghum lines for resistance/tolerance to seedling rots and to identify nonchemical control methods. We examined 20 sorghum varieties and lines under laboratory and field conditions. The rate of seed swelling, seed emergence, α-amylase activity, and dry matter utilization were studied in plants grown under normal and stress conditions. The accessions most resistant to seed rot and seedling diseases also were the most drought resistant during emergence and seedling growth ($r^2 = 0.71$). We think these resistances are correlated because seed emergence in cool soil, whether wet or dry, depends upon the same mechanisms: rapid seed swelling, high α-amylase activity, and a high coefficient of dry matter util-zation. Thus, methods such as screening for the rate of seed emergence and seedling growth under artificial osmotic stress should be useful in identifying drought-resistant lines of sorghum. Some presowing seed treatments reduced seed rot and seedling diseases. Seeds treated with a physiologically active preparation (Amistim) prepared from mycorrhizal fungi associated with ginseng (*Panax ginseng*) plant roots reduced seed and seedling diseases and increased field emergence by 5-10% at low soil temperatures after early planting. Patented treatments of seed with an electromagnetic field of super-high frequency or with an alternating electromagnetic field (50 Hz) increased laboratory germination by 6.6-8.0% and field emergence by 9.1-10.6% at optimal, early, and late sorghum planting dates.

79

Development of Durable Johnsongrass Mosaic Virus (JGMV) Resistance in Sorghum via a Transgenic Approach

Hunter K. C. Laidlaw, Denis M. Persley, Charles K. Pallaghy, and Ian D. Godwin

Johnsongrass Mosaic Virus (JGMV) is a significant constraint to sorghum (*Sorghum bicolor*) production in Australia, particularly in northern regions when environmental conditions favor the build-up of the aphid vector. The QDPI breeding program introduced resistance from the Indian introduction cv. Krish, but this gene was overcome by JGMV immediately after commercial release. Hence, there are two strains of JGMV in Australia–JGMV-JG and the "Krish-breaker" strain JGMV-K. The primary aim of this research is to produce durable resistance or tolerance to these JGMV strains by using a genetic engineering approach. One approach to pathogen-derived resistance has been to transform the host plant with segments of the viral genome, such as the coat protein or the large nuclear inclusion protein coding sequence. Transgenic sorghum with these viral sequences will be assessed following JGMV infection and the responses compared to those observed by a La Trobe University research group working with maize. The sequence diversity of these viral genome regions will be determined by sequencing isolates from different sorghum growing regions and from a number of different seasons. This comparison should give some indication as to the potential durability of this form of resistance. A new selectable marker, the phosphate-6-mannose isomerase (PMI) system developed by Novartis, for use in the regeneration of transgenic sorghum also will be investigated. If successful, this system would enable the discontinuation of the *bar* (herbicide resistance) system, thus avoiding problems resulting from outcrossing with Johnsongrass (*Sorghum halepense*) and the potential transfer of herbicide resistance to this weedy relative.

80
Farmer Participatory Studies on Finger Millet in Western Kenya

Felister W. Makini and Nicholas J. Hayden

Finger millet (*Eleusine coracana*) is an important subsistence staple food where an estimated 65,000 ha are cultivated. Farmer participatory studies were undertaken on the production, uses and constraints to this crop. Farmers recognize blast (caused by *Pyricularia grisea*) as a major disease, which they associate with high humidity and variety susceptibility. A large number of varieties (22) were grown in the farms surveyed. The range of varieties reflects the numerous end uses of this crop as well as farmers' recognition that different varieties perform better in different seasons. Farmers often grew more than one variety (either in separate plots or as variety mixtures) but one variety, Enaikuru, dominated. In a survey of 200 farms in January/February 1999, 59% of the farmers grew Enaikuru as a single cultivar and more than 90% of finger millet mixtures included this cultivar. The need for blast resistant varieties was identified by the farmers during participatory research activities. The yields of released varieties P224 (blast tolerant) and Gulu E (moderately resistant) were similar to those of Enaikuru during on-farm evaluation under low disease pressure (1997 short rains). In the long rains (1998), yields were higher for Enaikuru than for either of the released varieties, but Gulu E was the most resistant to blast and yields were not correlated to blast resistance scores. Crop debris may contain inoculum that can cause blast in subsequent finger millet crops. The fungus may survive on crop debris in the soil for up to five months, although there is considerable reduction in inoculum potential two months after incorporation. Seed also is a potential source of inoculum since the blast fungus was found in 20/21 samples (mean incidence 9.6%).

81
Differentiation of *Fusarium verticillioides* and *Fusarium proliferatum* Isolates Causing Sorghum Grain Mold by Random Amplified Polymorphic DNA (RAPD) Analysis

Maria J. Martinez, Luis R. Conci and Laura M. Giorda

Pathogenicity and RAPD markers were used to assess genetic diversity among six isolates of *Fusarium verticillioides* (Sacc.) Nirenberg and two of *Fusarium proliferatum* (Matsushima) Nirenberg collected from different locations in Argentina, and to identify molecular markers correlated with pathogenicity. Two cultivars, resistant and susceptible to grain mold, were used to evaluate the pathogenicity of these isolates in incidence and germination tests. The amount of genetic variation was determined by banding patterns resulting from PCR with 15 RAPD primers. Genetic distances between isolates were used to generate a dendrogram showing their relatedness. Results from the pathogenicity tests indicated significant differences between the cultivars and among the isolates. Isolates did not cluster in groups based on their pathogenic reactions, so further research is needed to identify a marker associated with the ability to cause grain mold.

82
Phytoalexin Synthesis by Sorghum Grain in Response to Grain Mold

Maria J. Martinez, Laura M. Giorda, Mirtha Nassetta, Karino Frutero, and Julio Zygadlo

Infection of sorghum panicles during anthesis by *Fusarium verticillioides* (Sacc.) Nirenberg resulted in the accumulation of a pigment complex by two sorghum cultivars. Compounds in the complex included the flavonoids apigenindin, apigenin, luteolin, naringenin, and kaepferol. The concentration of the five compounds depended upon grain maturity, cultivar, and whether the grain was inoculated with the pathogen. The most commonly identified compound in both resistant and susceptible cultivars was the 3-deoxyanthocyanidin apigenindin. The concentration of this flavanoid 30 days after flowering in the inoculated resistant cultivar was up to 30 times higher than in the noninoculated control. Apigenin, luteolin, and kaempferol were found only in the resistant cultivar and their levels were always higher in the inoculated plants. In *in vitro* bioassays for fungitoxicity, mixtures of these flavonoids at the 0.01-1 mM level, were more effective in inhibiting mycelial and conidial growth than were any of the compounds alone. Thus, these flavonoids appear to have a synergistic effect in terms of resistance to grain mold, and may function as phytoalexins that are induced in response to fungal colonization.

83
Effect of Sorghum Sowing Dates on Grain Mold Development in Morelos State, Mexico

Roberto Montes-Belmont, Lectina Bravo-Luna, Hilda Elizabet Flores-Moctezuma, Ignacio Méndez-Ramírez and Raúl Nava-Juárez

In 1998 and 1999 field sorghum experiments were established at Tlayca, Morelos, Mexico to monitor grain mold resulting from two planting dates in June and two in July. In 1998, the first three sowing dates had similar levels (94-98%) of grain mold incidence, while the fourth had a lower level (73% incidence). In 1999, the disease had a higher incidence on the June sowing dates than it did for those in July. We found a highly significant correlation between grain mold and climatic factors.

84
Present Status of Sorghum Pathology in Cameroon

Zachee Ngoko

Sorghum [*Sorghum bicolor* (L.) Moench] disease surveys conducted by various researchers since the 1970s revealed that gray leaf spot (caused by *Cercospora sorghi*), sorghum downy mildew (caused by *Peronosclerospora sorghi*), sorghum leaf blight (caused by *Exserohilum turcicum*), foliar anthracnose (caused by *Colletotrichum graminicola*), oval leaf spot (caused by *Ramulispora sorghicola*), zonate leaf spot (caused by *Gloeocercospora sorghi*), long smut (caused by *Sporisorium erhenbergii*), loose smut (caused by *Sporisorium cruenta*), head smut (caused by *Sporisorium reilianum*), and grain mold (caused by *Fusarium* spp., *Curvularia* spp. and *Alternaria* spp.) are the major diseases observed on sorghum in northern Cameroon. Most research has been directed toward breeding for higher yielding genotypes. Less emphasis has been placed on breeding for disease resistance.

85
Identification of Host Plant Resistance to Ergot in Sorghum

Jacob Reed, Mitchell R. Tuinstra, and Larry E. Claflin

Ergot (caused by *Claviceps africana*) of sorghum (*Sorghum bicolor*) has become an international problem. Although genetic resistance may contribute to the control of this disease, only a few sources of resistance have been reported. Sorghum germplasm lines IS 8525, IS 14131, and IS 14257 have been reported to be resistant to ergot; however, the physiological basis for this resistance is not adequately understood. To evaluate the resistance mechanisms of these accessions, the entries were characterized in male-sterile (A3 cytoplasm) genetic backgrounds. IS8525 was crossed to both A3KS70 and A3Tx7000, and IS 14131 and IS 14257 were crossed with A3Tx430 to produce male-sterile test-cross hybrids. After spray inoculation, parent lines, male-sterile hybrids and susceptible checks were evaluated for ergot resistance in Guayanilla, Puerto Rico and under greenhouse conditions in Manhattan, Kansas during the spring of 2000. IS 8525 exhibited excellent resistance *per se*, but male-sterile hybrids were only moderately resistant to ergot. IS 14131 and IS 14257 both exhibited excellent resistance, but, more importantly, male-sterile hybrids of these accessions were significantly more resistant to ergot than the susceptible checks at both locations. The expression of ergot resistance in IS 14131 and IS 14257 *per se* and in corresponding male-sterile test-cross hybrids suggests that these lines may carry genes for resistance to ergot.

86
Relationship of Stay Green to Charcoal Rot and Lodging in Sorghum

Darrell T. Rosenow, Niaba Teme, Charles A. Woodfin, Gary N. Odvody, and Gary C. Peterson

Charcoal rot develops in stalks of sorghum that have been predisposed by moisture stress during the grain fill stage of growth. A post-flowering drought resistance trait, termed stay green, has been identified. It reduces or delays premature senescence when plants are in moisture stress during the grain fill stage, and reduces associated stalk rot and lodging. Ten grain sorghum hybrids with differing drought stress response (stay green) and greenbug resistance reaction (E and I) were evaluated under severe post-flowering soil moisture stress. Charcoal rot infested toothpicks were used to inoculate plants. Hybrids possessing stay green had lower leaf and plant death (LPD) ratings (2.9 vs. 3.9 for early rating, and 3.9 vs. 4.7 for late rating) and less charcoal rot (4.1 vs. 4.8) in inoculated plants than in non-stay green hybrids. The difference, however, was smaller than expected, probably due to the severity of the moisture stress and the effect of inoculation directly into the stem pith, possibly bypassing a useful form of resistance. Non-inoculated plants of stay-green hybrids tended to have green stems and peduncles, even with 100% dead leaf area, and showed good resistance to lodging. Stay green hybrids showed excellent lodging resistance (2% vs. 47%) compared to non-stay-green hybrids. Greenbug damage was difficult to rate due to the rather low and spotty infestation and the early onset and severity of moisture-stress induced leaf death. Therefore, conclusions on the effect of greenbug resistance on charcoal rot and on the combined effect of stay green and greenbug resistance probably should not be drawn. The stay green trait significantly reduced stress, including stalk lodging and naturally occurring charcoal rot. Toothpick inoculation and rating only on internodes infected by the organism may be too harsh a test under extremely severe moisture stress.

87
The Importance of Public-Private Sector Partnerships to Indian Sorghum Farmers

Virender Sheorain and Andy H. Hall

National and international public sector organizations in India have made enormous advances in sorghum research over the last three decades. The main focus of this research has been on production technology and the development of germplasm with an improved yield potential and with tolerance to various biotic and abiotic stresses. Despite considerable technological advances, sorghum has been reduced in crop plantings in many areas. Although the reasons for this reduction are complex, the development of stronger markets for the crop is clearly needed. These markets are important for farmers in mixed crop/livestock systems where sorghum has been a key source of fodder and is one of only a few viable crops. Stronger markets could be particularly important for marginal farmers producing rainy season sorghum. There is a growing recognition that the private sector has an important role in developing markets for sorghum, particularly as a raw material for industrial use. The public and private sectors have complementary skills, resources and incentives, and the synergy from collaborative efforts could be very high. It is critical to understand that the private sector brings with it not only a demand for a raw material–although this is important–but also the end user's perspective, and the skills to commercialize new technologies. The private sector also has the ability to identify preferred characteristics for a new technology, *e.g.*, sorghum varieties with high starch and low fiber, and to identify commercially important constraints in need of further research. These interests and abilities complement the infrastructure, human and genetic resources held by the public sector, and would further the public sector's mandate to develop sorghum for the benefit of society as a whole. This type of public/private partnership could alter the dynamics of and increase support for markets that are critically important to poor farmers.

88
Resistance to Fusarium Stalk Rot in Grain Sorghum

Tesfaye T. Teferra, Mitchell R. Tuinstra, and Larry E. Claflin

Fusarium (*Fusarium* spp.) stalk rot is one of the most important diseases of grain sorghum (*Sorghum bicolor*). Genetic resistance is the most effective means for controlling this disease. The objective of this experiment was to evaluate Fusarium stalk rot resistance in grain sorghum. Six genetically diverse sorghum R-lines (Tx2737, Tx435, SC33, SC35, SC599 and SC1158) and three standard seed parents (SA3042, Wheatland and Redland) were intercrossed by using a Design II mating scheme. The parent lines and hybrids were grown in a randomized complete block design in Manhattan and Hesston, Kansas in 1999. Plants in each plot were inoculated with a culture of *F. verticillioides* at 14 days after flowering and were rated for differences in resistance at physiological maturity. Significant differences in disease reaction were noted among parent lines and hybrids. There was significant general combining ability (GCA) and specific combining ability (SCA) effects for two resistance traits. SC599 expressed the highest GCA effect for Fusarium stalk rot resistance, while SC33, a charcoal rot resistant check, was highly susceptible to Fusarium stalk rot. SC599 should provide an excellent source of genetic resistance to Fusarium stalk rot.

89
Access to the Next Generation of Sustainable Control of Covered Kernel Smut of Sorghum

K. S. L. Wilson and Nicholas J. Hayden

Covered kernel smut (caused by *Sporisorium sorghi*) is a major constraint to yield of smallholder sorghum in semi-arid Kenya, Tanzania and Uganda. Sorghum yields in Africa are the lowest globally (800 kg/ha) even though 41% of worldwide production area is located there. Covered kernel smut is a major constraint to sorghum yield and, therefore, a threat to rural livelihoods in semi-arid areas throughout east Africa. *S. sorghi* is seedborne and may be controlled by reducing seedborne inoculum. Control of covered kernel smut can be easily attained by using fungicide seed treatments. This technology has not been widely adopted in east Africa due to a lack of availability/affordability of the product or to limited farmer knowledge of biology and control strategies. Participatory research activities showed that most smallholder farmers save their own sorghum seed. The intermingling of smutted and clean sorghum panicles provides an opportunity for disintegration and contamination of seed with teliospores. Sustainable control of covered kernel smut may be achieved through fungicide seed treatments and/or clean head selection strategies. The challenge is to provide farmers with knowledge to enable them to make informed decisions on disease management options. We attempt to reduce this knowledge gap by using children as a conduit to spread information on disease management strategies throughout the community. In primary schools, role-play activities and ballads are used, while in secondary schools poster competitions are used.

Index

B

C

D

E

F

G

J

S

T

U

V

W

X

Y

Z